Introduction to Classical Mechanics
Second Edition

Atam P. Arya
West Virginia University

Pearson
Education

PRENTICE HALL
Upper Saddle River, New Jersey 07458

Library of Congress Cataloging-in-Publication Data

Arya, Atam Parkash.
    Introduction to classical mechanics / Atam P. Arya. -- 2nd ed.
       p. cm.
    Includes index.
    ISBN 0-13-505223-8
    1. Mechanics.   I. Title.
QC125.2.A79   1998
531--dc21   97-16622
                                        CIP

Executive editor: Alison Reeves
Assistant editor: Wendy Rivers
Production: ETP Harrison
Copy editor: ETP Harrison
Cover designer: Bruce Kenselaar
Manufacturing manager: Trudy Pisciotti

© 1998, 1990 by Prentice-Hall, Inc.
Simon & Schuster / A Viacom Company
Upper Saddle River, New Jersey 07458

The author and publisher of this book have used their best efforts in preparing this book. These efforts include the development, research, and testing of the theories and programs to determine their effectiveness. The author and publisher make no warranty of any kind, expressed or implied, with regard to these programs or the documentation contained in this book. The author and publisher shall not be liable in any event for incidental or consequential damages in connection with, or arising out of, the furnishing, performance, or use of these programs.

Printed in the United States of America

10 9 8 7 6 5 4 3 2 1

ISBN 0-13-505223-8

Prentice-Hall International (UK) Limited, *London*
Prentice-Hall of Australia Pty. Limited, *Sydney*
Prentice-Hall Canada Inc., *Toronto*
Prentice-Hall Hispanoamericana, S.A., *Mexico*
Prentice-Hall of India Private Limited, *New Delhi*
Prentice-Hall of Japan, Inc., *Tokyo*
Simon & Schuster Asia Pte Ltd., *Singapore*
Editora Prentice-Hall do Brasil, Ltda., *Rio de Janeiro*

TRADEMARK INFORMATION
Mathcad is a registered trademark of MathSoft, Inc.

Contents

Contents

Preface

This edition is written to present a reasonably complete account of classical mechanics at an intermediate level, with an option of using state-of-the-art computer-based technology. The text affords maximum flexibility in the selection and arrangement of topics for a two-semester, three credit-hour course at a sophomore or junior level. However, with proper selection and omission of material, it may be used for a one-semester course. Students with adequate preparation in general physics and calculus are ready to start this course.

Mechanics is the foundation of pure and applied sciences. Its principles apply to the vast range and variety of physical systems. I intend this text to take students who have had introductory mechanics in general physics to an intermediate level of mechanics, which will give them a strong basis for their future work in applied and pure sciences, especially advanced physics. Attention has been paid to the following topics of interest: (a) nonlinear oscillators (Chapter 4); (b) central force motion (Chapter 7), which includes the capture of comets, satellite orbits and maneuvers, stability of circular orbits, and interplanetary transfer orbits; (c) collisions in CMCS, which are discussed in detail (Chapter 8); (d) horizontal wind circulation (weather systems) (Chapter 11); and the relationship between conservative laws and symmetry principles (Chapter 12).

The aim of classical mechanics is, and always will be, to understand physical phenomena and laws of mechanics and to apply them to different, everyday situations. In order to achieve this it is necessary to: (a) perform mathematical calculations and solve problems, (b) make graphs resulting from different calculations, and (c) make interpretations of these. These three procedures, especially the first two, are so time consuming that one forgets or ignores the real process of learning and understanding physics. The latest developments in computer software are quite helpful in overcoming these difficulties, and we must use these resources to their maximum.

Keeping this in mind, the main purpose of this second edition is to use Mathcad to solve most examples and to create many figures. Mathcad is a unique way to work with numbers, formulas, and graphs. To solve problems or graph mathematical relations, you type them as you would write them on the blackboard or see them in a reference or textbook. You can solve almost any math problem you can think of, symbolically or numerically.

The usefulness of Mathcad is demonstrated by incorporating it in more than 90% of the over 60 solved examples, as well as in many new figures.

- All graphs are created using numbers that reflect actual physical situations. You may change and experiment with numbers without losing time.
- Graphs that ordinarily take hours can be created in minutes, and if mistakes are made, corrections can be made in only a few seconds.
- Mathematical calculations such as differentiations and integrations can be done by using the symbolic program instead of losing time by looking up information in standard tables.
- The solving and graphing that normally would have taken hours takes only minutes.

  Finally, you may use this book in three different ways.

- Use the book the conventional way and it will still provide insight into the subject matter.
- Use Mathcad in a limited way up to the extent provided in the book, and it will still provide many benefits.
- Use Mathcad and fully explore its power and benefits by incorporating it in your learning process of mechanics or physics in general.

ACKNOWLEDGMENTS

Finally, I am very thankful to the reviewers who gave many useful suggestions for this edition. Professor John P. Toutnghi, Seattle University; Professor George Rainey, California State Poly. University; Professor Robert R. Marchini, University of Memphis.

There are many individuals who encouraged me in completing this edition. It would have been impossible to complete this project without the help and the encouragement of Professor Larry E. Halliburton, Chairman Physics Department, West Virginia University; Mr. Ray Henderson of Prentice Hall, and my wife Pauline. Finally, my thanks to Alison Reeves, Executive Editor, who took the project to its completion.

                                                                                          A. P. A.

> *To*
> *Professor William W. Pratt*

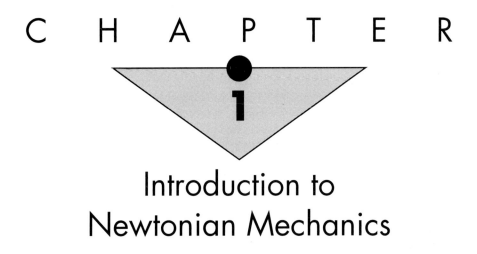

CHAPTER 1

Introduction to Newtonian Mechanics

1.1 INTRODUCTION

Mechanics is one of the oldest and most familiar branches of physics. It deals with bodies at rest and in motion and the conditions of rest and motion when bodies are under the influence of internal and external forces. The laws of mechanics apply to a whole range of objects, from microscopic to macroscopic, such as the motion of electrons in atoms and that of planets in space or even to the galaxies in distant parts of the universe.

Mechanics does not explain why bodies move; it simply shows how a body will move in a given situation and how to describe such motion. The study of mechanics may be divided into two parts: kinematics and dynamics. *Kinematics* is concerned with a purely geometrical description of the motion (or trajectories) of objects, disregarding the forces producing the motion. It deals with concepts and the interrelation between position, velocity, acceleration, and time. *Dynamics* is concerned with the forces that produce changes in motion or changes in other properties, such as the shape and size of objects. This leads us to the concepts of force and mass and the laws that govern the motion of objects. *Statics* deals with bodies at rest under the influence of external forces.

In antiquity significant gains were made in the theory of mechanics during Aristotle's time; however, it was not until the seventeenth century that the science of mechanics was truly founded by Galileo, Huygens, and Newton. They showed that objects move according to certain rules, and these rules were stated in the form of laws of motion. Essentially classical or Newtonian mechanics is the study of the consequences of the laws of motion as formulated by Newton in his *Philosophiae Naturalis Principia Mathematica* (the *Principia*) published in 1686.

Although Newton's laws provide a direct approach to the subject of classical mechanics, there are a number of other ways of formulating the principles of classical mechanics. Among these, the two most significant approaches are the formulations of Lagrange and Hamilton.

These two approaches take *energy* rather than force as the fundamental concept. In more than half of this text, we will use the classical approach of Newton, while in the later part of the text we will introduce Lagrange and Hamilton formulations.

Until the beginning of the twentieth century, Newton's laws were completely applicable to all well-known situations. The difficulties arose when these laws were applied to certain definite situations: (a) to very fast moving objects (objects moving with speeds approaching the speed of light) and (b) to objects of microscopic size such as electrons in atoms. These difficulties led to modifications in the laws of Newtonian mechanics: (a) to the formulation of the *special theory of relativity* for objects moving with high speeds, and (b) to the formulation of *quantum mechanics* for objects of microscopic size. The failure of classical mechanics in these situations is the result of inadequacies in classical concepts of space and time as discussed briefly in Chapter 16, Special Theory of Relativity.

Before we start an in-depth study of mechanics, we devote this chapter to summarizing briefly a few essential concepts of interest from introductory mechanics. We especially emphasize the importance of the role of Newton's laws of motion.

1.2 UNITS AND DIMENSIONS

Measurements in physics involve such quantities as velocity, force, energy, temperature, electric current, magnetic field, and many others. The most surprising aspect is that all these quantities can be expressed in terms of a few basic quantities, such as length L, mass M, and time T. These three quantities are called *fundamental* or *basic quantities* (*base units*); all others that are expressed in terms of these are called *derived quantities*.

Three Basic Standards: Length, Mass, and Time

Three different sets of units are in use. The most prevalent is that in which length is measured in *meters,* mass in *kilograms,* and time in *seconds,* hence the name *MKS system* (or *metric system*). As we will see, in practice there are five different quantities that are used as base units.

Standard of length: The meter.   The meter has been defined as the distance between the two marks on the ends of a platinum-iridium alloy metal bar kept in a temperature-controlled vault at the International Bureau of Weights and Measures in Sèvres, near Paris, France. In 1960, by international agreement, the General Conference on Weights and Measures changed the standard of length to an atomic constant by the following procedure. A glass tube is filled with krypton gas in which an electrical discharge is maintained. The standard *meter* is defined to be equal to exactly 1,650,763.73 wavelengths of orange-red light emitted in a vacuum from krypton-86 atoms. To improve the accuracy still further, a meter was redefined in 1983 as equal to a distance traveled by light in vacuum in a time interval of 1/299,792,458 of a second.

Standard of time: The second.   In the past, the spinning motion of Earth about its axis, as well as its orbital motion about the Sun, have been used to define a second. Thus, a second is defined to be 1/86,400 of a mean solar day. In October 1967, the time standard was redefined in terms of an atomic clock, which makes use of the periodic atomic vibrations of certain atoms. According to the cesium clock, a *second* is defined to be exactly equal to the time interval of 9,192,631,770 vibrations of radiation from cesium-133. This method has an accuracy of 1 part in 10^{11}. It is possible that two cesium clocks running over a period of 5000 years will differ by only 1 second.

Standard of mass: The kilogram.   A platinum-iridium cylinder is carefully stored in a repository at the International Bureau of Weights and Measures. The mass of the cylinder is defined to be exactly equal to a *kilogram.* This is the only base unit still defined by an artifact. The basic aim of scientists has been to define the three basic standards in such a way that they are accurately and easily reproducible in any laboratory.

Different Systems of Units

Besides MKS system, there are two others, all using five base units, which are briefly described below.

The CGS or Gaussian system.   In this system the unit of length is the *centimeter* $(=10^{-2}$ m), the unit of mass is the *gram* $(= 10^{-3}$ kg), and the unit of time is the *second.*

The British system.   This is used in the United States and may be referred to as U.S. engineering system. In this system the unit of length is the *foot* and the unit of time is the *second.* This system does not use mass as a basic unit; instead, *force* is used, the unit of which is the *pound* (lb). The unit of mass derived from the pound is called the *slug* $(= 32.17$ lb mass). The unit of temperature in the British system is the *degree Fahrenheit.*

The MKS or metric system.   In this system the unit of length is the *meter* (m), the unit of mass is the *kilogram* (kg), and the unit of time is the *second* (sec). These are the most commonly used units in the world. The other two base units are temperature in kelvins (K) and charge in coulombs (coul).

Five of the most commonly used base units in the different systems are listed here.

Units	MKS	CGS	USA
Length	$L: = 1 \cdot m$	$L: = 1 \cdot cm$	$L: = 1 \cdot ft$
Mass	$M: = 1 \cdot kg$	$M: = 1 \cdot gm$	$L: = 1 \cdot lb$
Time	$T: = 1 \cdot sec$	$T: = 1 \cdot sec$	$T: = 1 \cdot sec$
Temperature	$R: = 1 \cdot K$	$R: = 1 \cdot K$	$R: = 1 \cdot K$
Charge	$Q: = 1 \cdot coul$	$Q: = 1 \cdot coul$	$Q: = 1 \cdot coul$

International System of Units (SI).   The International System of Units, abbreviated SI after the French *Système international d'unités,* is the modern version of the metric system established by international agreement. For convenience it uses 7 base units: Five of these are the same as MKS already listed and the other two are:

Amount of substance	mole	$1 \cdot \text{mol}$
Luminous intensity	candela	$1 \cdot \text{cd}$

The SI also uses two supplementary units:

Plane angle	radian	$1 \cdot \text{rad}$
Solid angle	steradian	$1 \cdot \text{sr}$

Dimensions

Most physical quantities may be expressed in terms of length L, mass M, and time T, where L, M, and T are called dimensions. A quantity expressed as $L^a M^b T^c$ means that its length dimension is raised to the power a, its mass dimension is raised to the power b, and its time dimension is raised to the power c. Thus the dimensions of volume are L^3, that of acceleration are LT^{-2}, and that of force are MLT^{-2}.

    To add or substract two quantities in physics they must have the same dimensions. Similarly, no matter what system of units is used, all mathematical relations and equations must be dimensionally correct. That is, the quantities on both sides of the equations must have the same dimensions. For example, in the equation $x = v_0 + \frac{1}{2}at^2$, x has dimensions of L, $v_0 t$ has dimensions of $(L/T)T = L$, and $\frac{1}{2}at^2$ has dimensions of $\frac{1}{2}(L/T^2)(T^2) = L$. Thus dimensional analysis may be used to (1) check the correctness of the form of the equation, that is, every term in the equation must have the same dimensions, (2) to check an answer computed from an equation for plausibility in a given situation, and (3) to arrive at a formula if we know the dependence of a certain quantity on other physical quantities.

▷ Example 1.1 _____

The magnitude of the centripetal force Fc acting on an object is a function of mass M of the object, its velocity v, and the radius r of the circular path. By the method of dimensional analysis, find an expression for the centripetal force.

Solution
Since Fc is a function of M, v, and r, the values of a, b, and c are calculated in the expression for Fc.

$$Fc = M^a \cdot v^b \cdot r^c \qquad\qquad (i)$$

In terms of units, expression (i) takes the form (ii).

$$\text{kg} \cdot \frac{\text{m}}{\text{sec}^2} = \text{kg}^a \cdot \left(\frac{\text{m}}{\text{sec}}\right)^b \cdot \text{m}^c \qquad\qquad (ii)$$

We assume the values of a, b, and c
to be (guess values)

Guess

$$a := 1 \qquad b := 1 \qquad c := 1$$

Comparing the values of a, b, and c
on both sides of Eq. (ii), we get
Eqs. (iii).

Given

$$a = 1 \qquad b + c = 1 \qquad b - 2 = 0 \qquad \text{(iii)}$$

Let S represent the solution giving
the values of a, b, and c that satisfy
Eqs. (i) and (ii). The results are:
a =1, b = 2 and c = -1.

$$S := \text{Find}(a, b, c)$$

$$S = \begin{pmatrix} 1 \\ 2 \\ -1 \end{pmatrix}$$

Thus the proper equation for force
Fc and its units are

$$Fc = \frac{Mv^2}{r}$$

$$Fc = 1 \cdot kg \cdot m \cdot sec^{-2} \qquad \text{(iv)}$$

EXERCISE 1.1 The angular velocity ω of a simple pendulum is a function of its length L
and acceleration due to gravity g. Find an expression for the angular velocity ω and the time
period T of the pendulum by the method of dimensional analysis.

1.3 NEWTON'S LAWS AND INERTIAL SYSTEMS

Newton's laws may be stated in a brief and concise form as below:

Newton's First Law.   *Every object continues in its state of rest or uniform motion in a
straight line unless a net external force acts on it to change that state.*

Newton's Second Law.   *The rate of change of momentum of an object is directly proportional to the force applied and takes place in the direction of the force.*

Newton's Third Law.   *To every action there is always an equal and opposite reaction;
that is, whenever a body exerts a certain force on a second body, the second body exerts
an equal and opposite force on the first.*

These statements do look simple; but that is deceptive. Newton's laws are the results of a
combination of definitions, experimental observations from nature, and many intuitive concepts.
We cannot do justice to these concepts in a short space here, but we will try to expand our think-
ing horizon by discussing these statements in some detail.

The motion of objects in our immediate surroundings is complicated by ever present frictional and gravitational forces. Let us consider an isolated object that is moving with a constant
(or uniform) velocity in space. Describing it as an isolated object implies that it is far away from
any surrounding objects so that it does not interact with them; hence no net force (gravitational

or otherwise) acts on it. To describe the motion of the object we must draw a coordinate system with respect to which the object moves with uniform velocity. Such a coordinate system is called an *inertial system*. The essence of Newton's first law is that it is always possible to find a coordinate system with respect to which an isolated body moves with uniform velocity; that is, *Newton's first law asserts the existence of inertial systems*.

Newton's second law deals with such matters as: What happens when there is an interaction between objects? How do you represent interaction? And still further, what is inertia and how do we measure this property of an object? As we know, *inertia* is a property of a body that determines its resistance to acceleration or change in motion when that body interacts with another body. The quantitative measure of *inertia* is called *mass*.

Consider two bodies that are completely isolated from the surroundings but interact with one another. The interaction between these objects may result from being connected by means of a rubber band or a spring. The interaction results in acceleration of the bodies. Such accelerations may be measured by stretching the bodies apart by the same amount and then measuring the resultant accelerations. All possible measurements show that the accelerations of these two bodies are always in opposite directions and that the ratio of the accelerations are inversely proportional to the masses. That is,

$$\frac{a_A}{a_B} = -\frac{m_B}{m_A}$$

or
$$m_A a_A = -m_B a_B \tag{1.1}$$

Thus the effect of interaction is that the product of mass and acceleration is constant and denotes the *change in motion*. This product is called *force* and it represents interaction. Thus we may say that the force F_A acting on A due to interaction with B is

$$F_A = m_A a_A \tag{1.2}$$

while the force F_B acting on B due to interaction with A is

$$F_B = m_B a_B \tag{1.3}$$

Thus, in general, using vector notation we may write

$$\mathbf{F} = m\mathbf{a} \tag{1.4}$$

This equation is the definition of force when acting on a constant mass and holds good only in inertial systems. It is important to keep in mind that the force $\mathbf{F}$ arises because of an interaction or simply stands for an interaction. No acceleration could ever be produced without an interaction.

Let us now proceed to obtain the definition of force starting directly with the statement of Newton's second law given previously. Suppose an object of mass m is moving with velocity $\mathbf{v}$ so that the linear momentum $\mathbf{p}$ is defined as

$$\mathbf{p} = m\mathbf{v} \tag{1.5}$$

According to Newton's second law, the rate of change of momentum is defined as force **F**; that is,

$$\mathbf{F} = \frac{d\mathbf{p}}{dt} \tag{1.6}$$

This equation takes a much simpler form if mass m remains constant at all speeds. If **v** is very small as compared to the speed of light c ($= 3 \times 10^8$ m/s), the variation in mass m is negligible. Hence, we may write

$$\mathbf{F} = \frac{d}{dt}(m\mathbf{v}) = m\frac{d\mathbf{v}}{dt} = m\mathbf{a} \tag{1.7}$$

That is, force is equal to mass (or inertial mass) times acceleration provided m is constant. This is the same as Eq. (1.4). It should be clear that Newton's first law is a special case of the second law, when $\mathbf{F} = 0$.

Let us now say a few words about Newton's third law. According to Newton's third law, forces always exist in pairs. Thus, if two bodies A and B interact with one another, and if there is a force $\mathbf{F}_A$ acting on body A, then there must be a force $\mathbf{F}_B$ acting on body B, so

$$\mathbf{F}_A = -\mathbf{F}_B \tag{1.8}$$

Thus the law implies that forces always exist in pairs (a single force without its partner somewhere else is an impossibility) and that such forces are the result of interactions. We can never have an isolated object having acceleration. An object with acceleration must have a counterpart somewhere else with opposite acceleration that is inversely proportional to mass.

It should be clear that Eq. (1.8) implies that the forces are equal and opposite, but they do not *always* necessarily have the same line of action. These points are elaborated on in Chapter 8.

1.4   INERTIAL AND NONINERTIAL SYSTEMS: NONINERTIAL FORCES

As we mentioned earlier, the first law of motion defines a particular type of reference frame, called the inertial system; that is, the inertial system is one in which Newton's first law holds good. We would like to find a relation between the measurements made by an observer A in an inertial system S and another observer B in a noninertial system S', both observing a common object C that may be moving with acceleration. This situation is shown in Fig. 1.1. S being an inertial system means that observer A is moving with uniform velocity, while system S' being noninertial means that observer B has an acceleration.

Object C of mass M is accelerating. Observer A measures its acceleration to be a_A and observer B measures its acceleration to be a_B. Thus, according to observer A, the force acting on C is

$$F_A = Ma_A \tag{1.9}$$

while according to observer B the force acting on C is

$$F_B = Ma_B \tag{1.10}$$

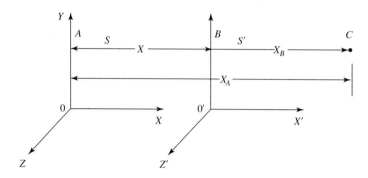

Figure 1.1   Moving object C being observed by an observer A in an inertial system S and another observer B in the noninertial system S'.

F_B would have been equal to F_A if the S' system was also a true inertial system. Let us find the relation between F_A and F_B. According to Fig. 1.1,

$$X_A(t) = X_B(t) + X(t) \tag{1.11}$$

Differentiating twice and rearranging,

$$\ddot{X}_B = \ddot{X}_A - \ddot{X} \tag{1.12}$$

Multiplying both sides by M,

$$M\ddot{X}_B = M\ddot{X}_A - M\ddot{X}$$

or

$$F_B = F_A - M\ddot{X} \tag{1.13}$$

Since observer A in system S is in a true inertial system, we may write

$$F_A = F_{\text{true}} = M\ddot{X}_A \tag{1.14}$$

while the force measured by observer B, who is in a noninertial system S', is not a true force but an apparent force given by

$$F_B = F_{\text{appt}} = M\ddot{X}_B \tag{1.15}$$

Thus we may write Eq. (1.13) as

$$F_{\text{appt}} = F_{\text{true}} - M\ddot{X} \tag{1.16}$$

Thus observer B will not measure a true force unless $\ddot{X} = 0$, in which case B will be moving with a uniform velocity with respect to A; hence S' itself will be a true inertial system. In general, for three-dimensional motion, we may write Eq. (1.16) as

$$\mathbf{F}_{\text{appt}} = \mathbf{F}_{\text{true}} - M\ddot{\mathbf{R}} \tag{1.17}$$

where $\ddot{\mathbf{R}}$ is the acceleration of the noninertial system S' with respect to the inertial system S or with respect to any other inertial system. If $\ddot{\mathbf{R}} = \mathbf{0}$, then $\mathbf{F}_{appt} = \mathbf{F}_{true}$, and hence both systems will be inertial. We may write Eq. (1.17) as

$$\mathbf{F}_{appt} = \mathbf{F}_{true} + \mathbf{F}_{fict} \qquad \textbf{(1.18a)}$$

where
$$\mathbf{F}_{fict} = -M\ddot{\mathbf{R}} \qquad \textbf{(1.18b)}$$

The last term is called a *noninertial force* or *fictitious force* because it is not a force in the true sense; no interactions are involved. It is simply a product of mass times acceleration.

1.5 SIMPLE APPLICATIONS OF NEWTON'S LAWS

A few simple applications of Newton's laws will be discussed in this section and the next.

Atwood Machine

A system of masses tied with a string and going over a pulley is called an Atwood machine, as shown in Fig. 1.2. We will assume that the pulley is frictionless and hence will not rotate. Mass m_2, being greater than mass m_1, will move downward and m_1 will move upward. The velocity $v = dx/dt$ is taken to be positive upward, while T (which is the same on both sides since the string is massless) is taken to be the tension in each string. Thus the motion of the two masses may be described by the following equations, $a = d^2x/dt^2$ being the acceleration for either mass. Acceleration a is the same on both sides since the string is "stretchless."

$$T - m_1 g = m_1 a \qquad \textbf{(1.19)}$$

$$m_2 g - T = m_2 a \qquad \textbf{(1.20)}$$

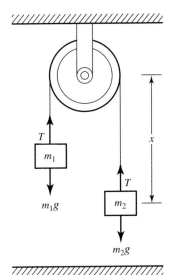

Figure 1.2   Atwood machine.

These equations may be solved to yield

$$a = \frac{m_2 - m_1}{m_1 + m_2} g \tag{1.21}$$

$$T = \frac{2m_1 m_2}{m_1 + m_2} g \tag{1.22}$$

If $m_1 = m_2$, we get $a = 0$, and $T = m_1 g = m_2 g$, which is the case for static equilibrium. On the other hand, if $m_2 \gg m_1$, we get $a \simeq g$ and $T \simeq 2m_1 g$.

Let us consider the case in which the pulley is not stationary but moves upward with an acceleration a, as shown in Fig. 1.3. In such a situation the total length of the string is

$$\ell = \pi R + (y - y_1) + (y - y_2) \tag{1.23}$$

Differentiating, we obtain

$$2\ddot{y} - \ddot{y}_1 - \ddot{y}_2 = 0 \tag{1.24}$$

But $\ddot{y} = a$ is the upward acceleration of the pulley. Hence

$$a = \tfrac{1}{2}(\ddot{y}_1 + \ddot{y}_2) \tag{1.25}$$

These principles can be extended to other situations involving many masses and pulleys.

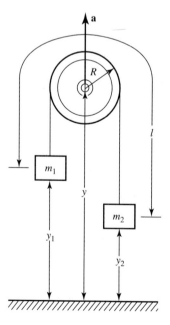

Figure 1.3 Motion of masses when the pulley has an upward acceleration a.

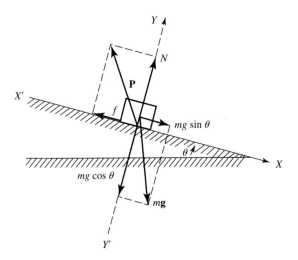

Figure 1.4   Forces acting on a mass m on an inclined plane.

Inclined Plane

Consider a mass m on an inclined plane that makes an angle θ with the horizontal, as shown in Fig. 1.4. The two forces acting on the mass m on the plane are the weight $m\mathbf{g}$ of mass m acting downward and the reaction $\mathbf{P}$ of the plane acting on mass m as shown. It is the resultant of these two forces that moves the mass up or down the plane, as illustrated for the two cases in Figs. 1.4 and 1.5. $\mathbf{P}$ can be resolved into two components; component N perpendicular to the surface of the plane is called the *normal reaction* (or normal force), and component f parallel to the surface of the plane is called the *frictional force*.

Let us consider the motion of mass m moving down the inclined plane, as shown in Fig. 1.4. Note that $m\mathbf{g}$ has been resolved into two components $mg \cos \theta$ and $mg \sin \theta$. Thus, for the motion of mass m, we may write

$$\sum F_x = mg \sin \theta - f = ma \tag{1.26}$$

$$\sum F_y = N - mg \cos \theta = 0 \tag{1.27}$$

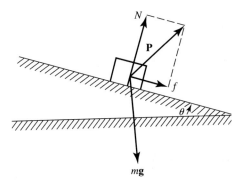

Figure 1.5   Motion of mass m moving upward on an inclined plane.

For a surface with a coefficient of friction μ, the frictional force f is found to be the product of μ and the normal reaction. That is,

$$f = \mu N = \mu mg \cos \theta \tag{1.28}$$

(If the body is at rest $\mu = \mu_s$, where μ_s, is the coefficient of static friction. If the body is in motion, $\mu = \mu_k$, where μ_k is the coefficient of kinetic friction. μ_k is always less than μ_s.) Substituting for f in Eq. (1.26) and solving for a,

$$a = g(\sin \theta - \mu \cos \theta) \tag{1.29}$$

If the mass m has an initial upward velocity along the plane, the direction of f will be opposite, as shown in Fig. 1.5, and the expression for the resulting acceleration (or deceleration) will be

$$a = g(\sin \theta + \mu \cos \theta) \tag{1.30}$$

Suppose mass m is sitting on the horizontal plane and the angle θ of the plane is increased steadily. When the angle reaches a certain value $\theta = \theta_f$, mass m just starts sliding. In this situation, when the motion starts $a = 0$, and we get

$$mg \sin \theta_f - f = 0$$

$$N - mg \cos \theta_f = 0$$

or
$$\frac{f}{N} = \tan \theta_f \tag{1.31}$$

But by definition $f = \mu_s N$, where μ_s is the coefficient of static friction; hence

$$\mu_s = \tan \theta_f \tag{1.32}$$

where θ_f is called the *angle of friction* or the *angle of repose*. If θ is greater than θ_f, the mass will not remain at rest. For the mass to remain at rest,

$$\tan \theta < \tan \theta_f = \mu_s \tag{1.33}$$

The same conclusion may be arrived at by considering Eq. (1.29), according to which the speed of the mass will increase if $a > 0$. This is possible only if

$$(\sin \theta - \mu \cos \theta) > 0$$

That is,
$$\theta > \tan^{-1} \mu \tag{1.34}$$

If $a = 0$, $\theta = \theta_f =$ the angle of friction. If $\theta < \theta_f$, a will be negative, and the particle will not move or will come to rest if already moving.

The Spinning Drum

In a spinning drum or well in an amusement park ride, the riders stand against the wall of the drum. When the drum starts spinning very fast, the bottom of the drum falls down but the riders stay pinned against the wall of the drum. We want to find the minimum angular velocity ω_{min} for which it is safe to remove the bottom.

The situation is as shown in Fig. 1.6, and with N being the unbalanced force the radial equation of motion is

$$N - Ma_r = 0 \qquad\qquad (1.35)$$

or we could say that the normal reaction N must provide the needed centripetal force F_c:

$$F_c = Ma_r = M\frac{v^2}{R} = MR\omega^2 \qquad\qquad (1.36)$$

Thus $$N = Ma_r = MR\omega^2 \qquad\qquad (1.37)$$

If f is the static frictional force, then

$$f \leq \mu_s N = \mu_s MR\omega^2 \qquad\qquad (1.38)$$

where μ_s is the coefficient of static friction between the rider and the surface of the drum. For the rider to stay pinned against the wall of the drum, f must be equal to Mg. Substituting this in Eq. (1.38) yields

$$Mg \leq \mu_s MR\omega^2 \qquad\qquad (1.39)$$

or $$\omega \geq \sqrt{\frac{g}{\mu_s R}} \qquad\qquad (1.40)$$

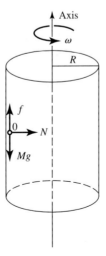

Figure 1.6 Spinning drum.

Thus the minimum safe value of ω is

$$\omega_{min} = \sqrt{\frac{g}{\mu_s R}} \tag{1.41}$$

Example 1.2

In Figure 1.6 the spinning drum has a radius of 2m and the coefficient of friction is 0.3.
Calculate the minimum speed at which it is safe to remove the floor of the spinning drum.

Solution

The different values given are

$$\mu := 0.3 \qquad R := 2 \cdot m \qquad g := 9.8 \cdot \frac{m}{sec^2}$$

From Eq. (1.41), the angular
frequency ω of the rotating drum is

$$\omega := \frac{\sqrt{g}}{\sqrt{\mu \cdot R}}$$

$$\omega = 4.041 \cdot sec^{-1} \cdot rad \qquad \omega = 231.558 \cdot sec^{-1} \cdot deg$$

Alternately, the revolution rate and
time period are

$$\frac{\omega}{2 \cdot \pi} = 0.643 \cdot sec^{-1} \qquad \frac{2 \cdot \pi}{\omega} = 1.555 \cdot sec$$

Thus if the drum makes 0.643 revolution per second, or it takes 1.555 seconds to
make one revolution, it will be safe to remove the drum floor from the bottom.

Should the drum speed be increased or decreased if (a) the coefficient of friction
increases or decreases and (b) the radius R is larger or smaller?

EXERCISE 1.2 In the example, suppose we want the drum to rotate at a speed of 2
revolutions per second and still be able to remove the floor safely by: **(a)** changing the
radius but keeping μ the same and **(b)** changing μ but keeping the radius the same.
What are the values of the radius and μ in the two cases?

Example 1.3

Two blocks of masses m and M are connected by a string and pass over a frictionless pulley. Mass m hangs
vertically and mass M moves on an inclined plane that makes an angle θ with the horizontal. If the coefficient of kinetic friction is μ_k, calculate the angle θ for which the blocks move with uniform velocity. Discuss the special case in which $m = M$.

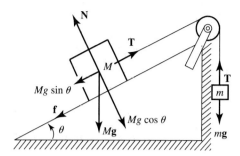

Figure Ex. 1.3

Solution

The situation is as shown in Fig. Ex. 1.3. From the force diagram, if the blocks are in equilibrium, that is, at rest or moving with a uniform velocity,

$$T - mg = 0 \tag{i}$$

$$N - Mg \cos \theta = 0 \tag{ii}$$

$$T - Mg \sin \theta - f = 0 \tag{iii}$$

and, from the definition of the coefficient of friction,

$$f = \mu_k N \tag{iv}$$

Combining Eqs. (ii) and (iv),

$$f = \mu_k Mg \cos \theta \tag{v}$$

Substituting for T from Eq. (i) and for f from Eq. (v) into Eq. (iii)

$$mg - Mg \sin \theta - \mu_k Mg \cos \theta = 0 \tag{vi}$$

First Method:
We can solve the given Eqs. (i), (ii), and (iii)
directly. From the force diagram, if the blocks are in
equilibrium, at rest, or moving with uniform velocity, the
equations after using the definition of the coefficient of friction
$f = \mu N$ are as shown.
Solving these equations for the unknowns T, N, and θ yields
the values as shown below.

Given

$T - m{\cdot}g = 0$

$N - M{\cdot}g{\cdot}\cos(\theta) = 0$

$T - M{\cdot}g{\cdot}\sin(\theta) - \mu{\cdot}N = 0$

$$\text{Find}(T,N,\theta) \rightarrow \begin{bmatrix} m\cdot g & m\cdot g \\ M\cdot g\cdot\cos\left[2\cdot\text{atan}\left[\frac{\left(M-\sqrt{M^2-m^2+\mu^2\cdot M^2}\right)}{(m+\mu\cdot M)}\right]\right] & M\cdot g\cdot\cos\left[2\cdot\text{atan}\left[\frac{\left(M+\sqrt{M^2-m^2+\mu^2\cdot M^2}\right)}{(m+\mu\cdot M)}\right]\right] \\ 2\cdot\text{atan}\left[\frac{\left(M-\sqrt{M^2-m^2+\mu^2\cdot M^2}\right)}{(m+\mu\cdot M)}\right] & 2\cdot\text{atan}\left[\frac{\left(M+\sqrt{M^2-m^2+\mu^2\cdot M^2}\right)}{(m+\mu\cdot M)}\right] \end{bmatrix}$$

For M = m, the two possible values of θ, given by Sθ, are calculated as shown.

$$M := m \qquad \mu := 0.3$$

$$S\theta := \begin{bmatrix} 2\cdot\text{atan}\left[\frac{1}{(2\cdot(-m-\mu\cdot M))}\cdot\left(-2\cdot M+2\cdot\sqrt{M^2-m^2+\mu^2\cdot M^2}\right)\right] \\ 2\cdot\text{atan}\left[\frac{1}{(2\cdot(-m-\mu\cdot M))}\cdot\left(-2\cdot M-2\cdot\sqrt{M^2-m^2+\mu^2\cdot M^2}\right)\right] \end{bmatrix}$$

$$S\theta = \begin{pmatrix} 0.988 \\ 1.571 \end{pmatrix}\cdot\text{rad} \qquad S\theta = \begin{pmatrix} 56.602 \\ 90 \end{pmatrix}\cdot\text{deg}$$

Second Method:

Let us now solve Eq. (vi) for the values

$$M := m \qquad \mu := 0.3$$

Eq. (vi) is written as

$$m\cdot g - M\cdot g\cdot\sin(\theta) - \mu\cdot M\cdot g\cdot\cos(\theta) = 0$$

The two possible values given by S are (solving above equation symbolically for θ)

$$S := \begin{bmatrix} 2\cdot\text{atan}\left[\frac{1}{(2\cdot(-m-\mu\cdot M))}\cdot\left(-2\cdot M+2\cdot\sqrt{M^2-m^2+\mu^2\cdot M^2}\right)\right] \\ 2\cdot\text{atan}\left[\frac{1}{(2\cdot(-m-\mu\cdot M))}\cdot\left(-2\cdot M-2\cdot\sqrt{M^2-m^2+\mu^2\cdot M^2}\right)\right] \end{bmatrix}$$

One value of θ is 56.6 degrees or 0.988 radians while the other value of θ = 90 degrees corresponds to the simple case of two masses hanging vertically from a pulley.

$$S = \begin{pmatrix} 0.988 \\ 1.571 \end{pmatrix}\cdot\text{rad} \qquad S = \begin{pmatrix} 56.602 \\ 90 \end{pmatrix}\cdot\text{deg}$$

EXERCISE 1.3 Find the ratio of the masses M/m so that the two blocks will move with uniform velocity for θ = 45 degrees and μ =0.3.

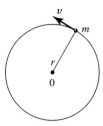

Figure 1.7 Mass m moving in a circle of radius r with a uniform speed v.

1.6 MOTION IN A CIRCLE AND GRAVITATION

Consider a mass m moving in a circle of radius r with a uniform speed v, as shown in Fig. 1.7. The acceleration of mass m is toward the center 0 and is given by

$$a_c = \frac{v^2}{r} \tag{1.42}$$

where a_c is called the *centripetal acceleration* and is produced by a constant force F_c, called the *centripetal force,* and given by

$$F_c = ma_c = m\frac{v^2}{r} \tag{1.43}$$

Note that F_c is not a force in the true sense because it is not produced as the result of interaction between objects. It simply happens to be the product of mass times acceleration.

According to Newton's universal law of gravitation, the gravitational force between mass m at a distance r from the center of Earth of mass M is

$$F_G = G\frac{Mm}{r^2} = mg \tag{1.44}$$

If this point mass is on the surface of Earth, which has a radius R, we may write

$$F = mg_0 = G\frac{Mm}{R^2} \tag{1.45}$$

That is,

$$g_0 = \frac{GM}{R^2} \quad \text{or} \quad G = \frac{g_0 R^2}{M} \tag{1.46}$$

Let us assume that m is a satellite or some other object moving with velocity v in a circle of radius r around Earth, as shown in Fig. 1.8. Gravitational force (toward the center of Earth) provides the necessary centripetal force to keep the mass moving in a circular orbit; that is,

$$F_c = F_G$$

$$\frac{mv^2}{r} = G\frac{Mm}{r^2} \tag{1.47}$$

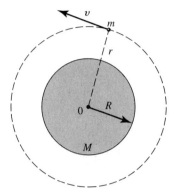

Figure 1.8   Mass m moving in a circle of radius r around Earth of mass M and radius R.

Substituting $v = 2\pi r/T$, where T is the time period of the circular orbit, in Eq. (1.47) and rearranging, we get

$$T^2 = \frac{4\pi^2}{GM}\, r^3 \tag{1.48}$$

or substituting $GM = gR^2$, we may write Eq. (1.48) as

$$T^2 = \frac{4\pi^2}{gR^2}\, r^3 \tag{1.49}$$

Equations (1.48) and (1.49) are statements of Kepler's third law.

Horizontal Circular Motion

A small mass m swings in a horizontal circle of radius r at the end of a string of length l_i, which makes an angle θ_i with the vertical as shown in Fig. 1.9. The string is slowly shortened by pulling it through a hole in its support until the final length is l_f and the string is making an angle θ_f with the vertical. Find an expression for l_f in terms of l_i, θ_i, and θ_f.

Suppose at some instant t the length of the string from the support to mass m is l_n, it makes an angle θ_n with the vertical, mass m is at a distance r_n from the axis of rotation, and v_n is the velocity of mass m. Let T_n be the tension in the string as shown. Since mass m is moving in a circle of radius r_n, the horizontal component of the tension in the string must provide the necessary centripetal acceleration; that is,

$$T_n \sin \theta_n = \frac{mv_n^2}{r_n} \tag{1.50}$$

while the vertical component of the tension balances the weight of mass m, that is,

$$T_n \cos \theta_n = mg \tag{1.51}$$

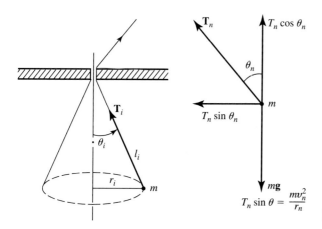

Figure 1.9

Dividing Eq. (1.50) by Eq. (1.51),

$$\tan \theta = \frac{v_n^2}{gr_n} \tag{1.52}$$

which may be written as

$$gr_n^3 \tan \theta_n = v_n^2 r_n^2 \tag{1.53}$$

Since no external torque acts on the system, the angular momentum of the system must be conserved; that is,

$$L_1 = L_2 \quad \text{or} \quad mr_1 v_1 = mr_2 v_2 \quad \text{or} \quad r_1 v_1 = r_2 v_2 \tag{1.54}$$

Thus, in general,

$$r_n v_n = \text{constant} \tag{1.55}$$

Combining Eqs. (1.53) and (1.55), we may conclude

$$r_i^3 \tan \theta_i = r_f^3 \tan \theta_f \tag{1.56}$$

But

$$r_n = l_n \sin \theta_n \tag{1.57}$$

Therefore, Eq. (1.56) takes the form

$$(l_i \sin \theta_i)^3 \tan \theta_i = (l_f \sin \theta_f)^3 \tan \theta_f \tag{1.58}$$

which is the required result and may be solved for l_f.

To illustrate the above concepts, we discuss a few examples involving the motion of the planets and the centripetal force.

Example 1.4 _____

(a) Calculate the variation in the value of g with distance from the center of Earth. **(b)** Apply Keplers law to the motion of the Moon around Earth and calculate the distance between them.

Solution

(a) From Eqs. (1.44) and (1.46), gr is the value of g at a distance r from the center of Earth, R0 is radius of Earth, and g0 is the value of g at Earth's surface.

$$gr = \frac{g0 \cdot R0^2}{r^2} = \frac{K}{r^2}$$

From the given values of g0 and R0, the value of K is calculated

$$g0 := 9.806 \qquad R0 := 6.38 \cdot 10^3$$

$$K := g0 \cdot R0^2 \qquad K = 3.991 \cdot 10^8$$

Thus the value of g in terms of r is

$$g = \frac{3.991 \cdot 10^8}{r^2}$$

We calculate g for 60 different values of r and then plot the results as shown in the graph.

$$N := 60 \qquad n := 0 .. N$$

Below are some values of g at different r, together with r_{60} and g_{60}, which correspond to the values near the Moon's surface.

$$r_n := R0 + R0 \cdot n \qquad g_n := \frac{K}{\left(r_n\right)^2}$$

$$r_0 = 6.38 \cdot 10^3 \qquad g_0 = 9.806$$

$$r_1 = 1.276 \cdot 10^4 \qquad g_1 = 2.451$$

$$r_{10} = 7.018 \cdot 10^4 \qquad g_{10} = 0.081$$

$$r_{60} = 3.892 \cdot 10^5 \qquad g_{60} = 0.003$$

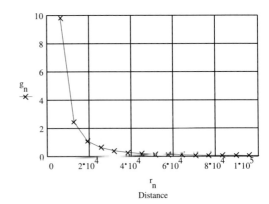

Acceleration g versus distance r

(b) Eq. (1.49) for T may be solved for r as shown.

$$T^2 = \left(\frac{4 \cdot \pi^2}{g \cdot R^2}\right) \cdot r^3$$

has solution(s)

$$
\begin{bmatrix}
\dfrac{1}{4} \cdot 4^{\left(\frac{2}{3}\right)} \cdot T^{\left(\frac{2}{3}\right)} \cdot g^{\left(\frac{1}{3}\right)} \cdot \dfrac{R^{\left(\frac{2}{3}\right)}}{\pi^{\left(\frac{2}{3}\right)}} \\[2em]
\dfrac{-1}{8} \cdot 4^{\left(\frac{2}{3}\right)} \cdot T^{\left(\frac{2}{3}\right)} \cdot g^{\left(\frac{1}{3}\right)} \cdot \dfrac{R^{\left(\frac{2}{3}\right)}}{\pi^{\left(\frac{2}{3}\right)}} + \dfrac{1}{8} \cdot i \cdot \sqrt{3} \cdot 4^{\left(\frac{2}{3}\right)} \cdot T^{\left(\frac{2}{3}\right)} \cdot g^{\left(\frac{1}{3}\right)} \cdot \dfrac{R^{\left(\frac{2}{3}\right)}}{\pi^{\left(\frac{2}{3}\right)}} \\[2em]
\dfrac{-1}{8} \cdot 4^{\left(\frac{2}{3}\right)} \cdot T^{\left(\frac{2}{3}\right)} \cdot g^{\left(\frac{1}{3}\right)} \cdot \dfrac{R^{\left(\frac{2}{3}\right)}}{\pi^{\left(\frac{2}{3}\right)}} - \dfrac{1}{8} \cdot i \cdot \sqrt{3} \cdot 4^{\left(\frac{2}{3}\right)} \cdot T^{\left(\frac{2}{3}\right)} \cdot g^{\left(\frac{1}{3}\right)} \cdot \dfrac{R^{\left(\frac{2}{3}\right)}}{\pi^{\left(\frac{2}{3}\right)}}
\end{bmatrix}
$$

T = 27.33 days is the period of revolution of the Moon around Earth and R is the radius of Earth.

$$T := (27.333) \cdot 24 \cdot 60 \cdot 60 \cdot sec \qquad R := 6.368 \cdot 10^6 \cdot m$$

Using the given values of T, g, and R, we obtain the value of r, the distance between Earth and the Moon.

$$T = 2.362 \cdot 10^6 \ \cdot sec \qquad g := 9.8 \cdot \frac{m}{sec^2}$$

$$r := \frac{1}{4} \cdot 4^{\left(\frac{2}{3}\right)} \cdot T^{\left(\frac{2}{3}\right)} \cdot g^{\left(\frac{1}{3}\right)} \cdot \frac{R^{\left(\frac{2}{3}\right)}}{\pi^{\left(\frac{2}{3}\right)}}$$

$$r = 3.829 \cdot 10^8 \ \cdot m \qquad\qquad r = 3.829 \cdot 10^5 \ \cdot km$$

EXERCISE 1.4 Repeat the example for another planet such as Mars or Venus.

Example 1.5 _____

A small mass swings in a horizontal circle of radius r at the end of a string
of length L_i and makes an angle θ_i with the vertical as shown in Figure 1.9. The string
is slowly shortened by pulling it through a hole in its support. Write expressions for r, v,
and time period of revolution T in terms of θ. Then graph these quantities.

Solution
We will use Eqs. (1.57) and (1.53), and $T = 2\pi/\omega$. Let us consider n = 25 values of θ
from a very small angle to 90 degree angle. Radius r_1 (for n = 1) is 0.063 meter, while the
vertical angle is 3.6 degrees. Three graphs are shown below.

(a) Which quantities become very small
near $\theta = 0$ degree angle and near $\theta =$
90 degree angle, and why?

$$n := 1 .. 25 \qquad \theta_n := \frac{\pi \cdot n}{50} \qquad g := 9.8$$

$$Li := 1.0 \qquad r0 := Li \cdot \sin(0) \qquad r0 = 0$$

(b) Which quantities become very large
at near $\theta = 0$ degree angle and near
$\theta = 90$ degree angle, and why?

$$r_n := Li \cdot \sin(\theta_n) \qquad v_n := \sqrt{\tan(\theta_n) \cdot g \cdot r_n}$$

$$T_n := \frac{2 \cdot \pi}{\left(\frac{v_n}{r_n}\right)} \qquad \theta_1 = 0.063 \text{ 'rad} \qquad \theta_1 = 3.6 \text{ 'deg}$$

$$\theta_{25} = 1.571 \text{ 'rad} \qquad \theta_{25} = 90 \text{ 'deg}$$

(c) What is the significance of the
points where two graphs intersect?

$r_1 = 0.063 \qquad v_1 = 0.197 \qquad T_1 = 2.005$

$r_{10} = 0.588 \quad v_{10} = 2.046 \qquad T_{10} = 1.805$

$r_{20} = 0.951 \quad v_{20} = 5.356 \qquad T_{20} = 1.116$

$r_{25} = 1 \qquad v_{25} = 4.001 \cdot 10^8 \quad T_{25} = 1.571 \cdot 10^{-8}$

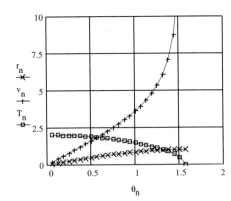

EXERCISE 1.5 Derive an expression for the tension Ft in the string as a function of
the angle of the string with the vertical. Graph the tension Ft and angular frequency ω
versus θ. Then answer all the questions in the example using this information.

PROBLEMS

1.1. The speed v of sound waves in air depends on the atmospheric pressure P and density ρ of the air. By using the method of dimensional analysis, find an expression for v in terms of P and ρ.

1.2. The velocity v of waves on a vibrating string depends on the tension T in the string and the mass per unit length λ of the string. Derive an expression for v by using the dimensional analysis method.

1.3. The time period T of a planet around the Sun of mass M is given by the following expression: $T^2 = 4\pi^2 r^3/MG$ where a is the radius of the circular orbit of the planet.
(a) What are the SI units of G?
(b) Derive the preceding expression by using the dimensional analysis method, that is, by assuming $T = T(r, M, G)$.

1.4. When a fluid flows in a pipe, the friction between the fluid and the surface of the pipe is given by the coefficient of viscosity η, defined by the equation $F/A = \eta(dv/ds)$, where F is the force of friction acting across an area A and dv/ds is the velocity gradient between layers of fluids.
(a) What are the units of η?
(b) If ΔP is the pressure difference and is directly proportional to Δl, by using the method of dimensional analysis, show that

$$\frac{\Delta P}{\Delta l} = \text{constant} \, \frac{\eta\phi}{r^4}$$

where ϕ is the volume flux of the fluid through the pipe and r is the radius of the pipe, as shown in Fig. P1.4.

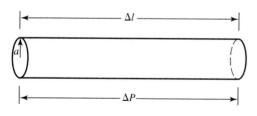

Figure P1.4

1.5. In Fig. 1.1, the mass of object C is 100 kg, the acceleration measured by the observer in the inertial system A is 100 m/s², while the observer in system B measures the acceleration as 90 m/s². What is the fictitious force? What is the acceleration of the noninertial system B? What could B do to achieve a true inertial system?

1.6. A mass m is given an initial velocity v_0 up an inclined plane of angle θ (θ is greater than the angle of friction). Find the distance the mass moves up the incline, the time it takes to reach this point, and the time it takes to return to its original position.

1.7. A box of mass m is connected by a rope that passes over a pulley to a box of mass M, as shown in Fig. P1.7. The coefficient of friction between m and the horizontal surface AB or the inclined surface BC is μ. Find the acceleration of the system and the tension in the rope for the portion when the mass is moving (a) between A and B, and (b) between B and C.

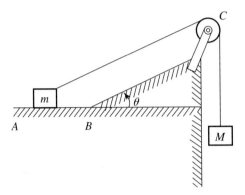

Figure P1.7

1.8. A man of mass M pushes horizontally a sled with a boy (sitting on it) of mass m. The coefficient of friction between the sled and the snow is μ, and the coefficient of friction between the man's feet and the snow is μ_s.
 (a) Draw a clear diagram showing all the forces acting on the sled and the man.
 (b) Calculate the horizontal and vertical components of the force when the man and the sled have an acceleration a.
 (c) What is the maximum acceleration the man can give to himself and the sled?

1.9. A man pushes a box of mass M with a force F using a stick AB of mass m and making an angle θ with the vertical, as shown in Fig. P1.9. The coefficient of friction between the box and the floor is μ.
 (a) Draw a clear diagram showing all the forces.
 (b) Calculate the value of F required to move the box with uniform velocity.
 (c) Show that, if θ is less than the angle of friction, the box cannot be started by just pushing.

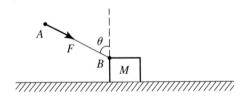

Figure P1.9

1.10. Consider a system of two masses and a pulley as shown in Fig. P1.10. Let $m_1 = 12$ kg, $m_2 = 8$ kg, the mass of the pulley $m = 10$ kg, and its radius $r = 10$ cm.
 (a) Show all the forces acting on the system.
 (b) Calculate T_1, T_2, and acceleration a. Assume the pulley to be a solid disk ($I_{\text{disk}} = mr^2/2$).

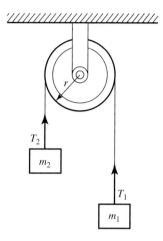

Figure P1.10

1.11. Repeat Problem 1.10 if the pulley were a hollow rim of the same mass and radius ($I_{\text{rim}} = mr^2$).

1.12. An automobile on a highway enters a curve of radius R and banking angle θ, as shown in Fig. P1.12. The coefficient of friction between the wheels and the road is μ. What are the maximum and minimum speeds with which a car can round the curve without skidding sideways?

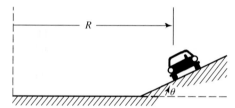

Figure P1.12

1.13. A mass M tied to a string of radius R is whirled in a vertical circle as shown in Fig. P1.13.
(a) Find the tension in the string at different points such as A, B, C, and D.
(b) What is the minimum velocity v_0 at the top point B so that the string won't slack?
(c) Graph T and v as a function of θ for given M and R.

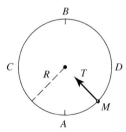

Figure P1.13

1.14. Two particles of masses m_1 and m_2 at a distance R from each other are under the influence of an attractive force F. If the two masses undergo uniform circular motion about each other with an angular velocity ω, show that

$$F = \left(\frac{m_1 m_2}{m_1 + m_2}\right)\omega^2 R = \frac{\omega^2 R}{(1/m_1) + (1/m_2)}$$

1.15. Calculate the height and velocity of a satellite that remains over the same point at all times as seen from Earth. Assume a circular orbit and express the height in terms of the radius of the Earth, R_e. Such a satellite, called a synchronous satellite, goes around Earth once every 24 h, so its position appears stationary with respect to a ground station. One such communication satellite was named Earlybird.

1.16. Consider a cone with an apex half-angle θ, as shown in Fig. P1.16. A particle of mass m slides without friction on the inside of the cone in a circular path in a horizontal plane with speed v. Draw a force diagram and calculate the radius of the circular path in terms of θ, v, and g.

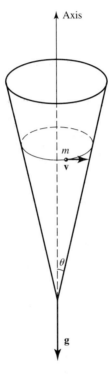

Figure P1.16

1.17. Repeat Problem 1.16 for the case when the surface is not frictionless and the coefficient of friction is μ.

1.18. Find the mass of the Sun, assuming that Earth moves in a circular orbit of radius 1.496×10^8 m and completes one revolution around the Sun in one year.

1.19. Find the distance between Earth and Mars by first calculating the distances of these planets from the Sun. The revolution period of Earth is 1.00 year and that of Mars is 1.88 years.

1.20. Repeat Problem 1.15 for a synchronous satellite going around Jupiter every 9 h, 50 min. Revolution period of Jupiter is 11.86 years, its mass is 317.80 times Earth's mass, and it's at a distance of 677.71×10^7 km from the Sun.

SUGGESTIONS FOR FURTHER READING

FORD, K. W., *Classical and Modern Physics,* Volume I. Lexington, Mass.: Xerox College Publishing, 1972.

FRENCH, A. P., *Newtonian Mechanics,* Chapters 7 and 12. New York: W. W. Norton and Co., Inc., 1971.

HALLIDAY, D., and RESNICK, R., *Physics for Students of Science and Engineering,* 2nd. ed. New York: John Wiley & Sons, Inc., 1981.

KITTEL, C., KNIGHT, W. D., and RUDERMAN, M. A., *Mechanics,* Berkeley Physics Course, Volume I. Chapter 3. New York: McGraw-Hill Book Co., 1965.

KLEPPNER, D., and KOLENKOW, R. J., *An Introduction to Mechanics,* Chapter 2. New York: McGraw-Hill Book Co., 1973.

SEARS, F. W., ZEMANSKY, M. W., and YOUNG, H. D., *University Physics,* 5th ed. Reading, Mass.: Addison-Wesley Publishing Co., 1980.

STEPHENSON, R. J., *Mechanics and Properties of Matter,* Chapter 2. New York: John Wiley & Sons, Inc., 1962.

SYMON, K. R., *Mechanics,* 3rd. ed., Chapter 1. Reading. Mass.: Addison-Wesley Publishing Co., 1971.

TAYLOR, E. F., *Introductory Mechanics,* Chapters 1, 2, and 3. New York: John Wiley & Sons, Inc., 1963.

TIPLER, P. A., *Physics,* 2nd. ed., Volume I. New York: Worth Publishers, Inc., 1982.

WEIDNER, R. T., in collaboration with BROWN, M. E., *Elementary Classical Physics,* Needham Heights, Mass.: Allyn and Bacon, 1985.

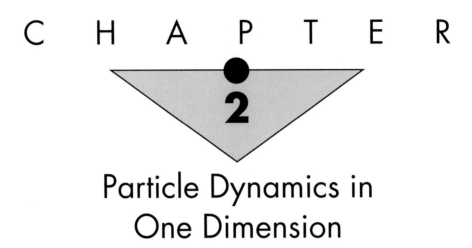

C H A P T E R

2

Particle Dynamics in One Dimension

2.1 INTRODUCTION

Suppose a particle of mass m is acted on by several forces $\mathbf{F}_1, \mathbf{F}_2, \ldots, \mathbf{F}_n$. The net force $\mathbf{F}$ acting on the particle is given by the superposition principle as

$$\mathbf{F} = \sum \mathbf{F}_n = \mathbf{F}_1 + \mathbf{F}_2 + \cdots + \mathbf{F}_n \tag{2.1}$$

and the motion of the particle is described by Newton's second law as

$$\mathbf{F} = \frac{d\mathbf{p}}{dt} \tag{2.2a}$$

where $\mathbf{p}$ is the linear momentum of the particle. Only when the mass m remains constant may we write

$$\mathbf{F} = m\frac{d^2\mathbf{r}}{dt^2} = m\mathbf{a} \tag{2.2b}$$

If we describe the motion in rectangular coordinates, Eq. (2.2b) may be written in the form of three components as

$$F_x = \sum F_{nx} = m\frac{d^2x}{dt^2} = m\ddot{x} = ma_x \tag{2.3}$$

with similar expressions for F_y and F_z. If the acceleration $\mathbf{a}$ or its components a_x, a_y, a_z are known, then Eq. (2.2b) may be used to solve for the force $\mathbf{F}$. In general, the situation in particle dynamics is just the reverse; that is, we know the net force $\mathbf{F}$ acting on a particle and we want to solve Eq. (2.2b) to find the position of the particle as a function of time t. In this chapter, since we are limiting our motion of the particle to one dimension, the only equation of interest is Eq. (2.3), which after dropping the subscript x may be written as

$$F = m \frac{d^2 x}{dt^2} \tag{2.4}$$

To be more explicit, we may write this equation as

$$F(x, \dot{x}, t) = m \frac{d^2 x}{dt^2} \tag{2.5}$$

where $\dot{x} = dx/dt = v$ is the velocity of the particle and Eq. (2.5) states that the force acting on the particle is a function of position, velocity, and time. Such a problem in which the applied force is a function of all three variables simultaneously is difficult to solve. On the other hand, if the applied force is a function of only one variable, the problem is much simplified. Hence we shall divide our discussion into the following four cases:

1. The applied force is constant; that is, F = constant, such as freely falling bodies and everyday motion.
2. The applied force is time dependent; that is, $F = (Ft)$, such as in the case of electromagnetic waves.
3. The applied force is velocity dependent; that is, $F = F(v)$, such as air resistance to falling or rising objects.
4. The applied force is position dependent; that is, $F = F(x)$, such as restoring force to vibrating springs.

Before we start solving Eq. (2.4) for these different cases, we may remind ourselves that since

$$\frac{d^2 x}{dt^2} = \frac{dv}{dt} = \frac{dv}{dx}\frac{dx}{dt} = v \frac{dv}{dx} \tag{2.6}$$

Eq. (2.4) may be written in the following different forms:

$$F = m \frac{d^2 x}{dt^2} \tag{2.7a}$$

$$F = m \frac{dv}{dt} \tag{2.7b}$$

$$F = mv \frac{dv}{dx} \tag{2.7c}$$

Also, since momentum p is defined as $p = mv = m(dx/dt)$, we may write Eq. (2.7a) as [or directly from Eq. (2.2a) as applied to the one-dimensional case]

$$F = \frac{dp}{dt} \tag{2.8}$$

That is, the applied force is equal to the rate of change of momentum. If the applied force acts between the time interval t_1 and t_2, then, by integrating Eq. (2.8), we get

$$p_2 - p_1 = \int_{t_1}^{t_2} F \, dt \tag{2.9}$$

which is the integral form of Newton's second law, while Eqs. (2.7) are the differential forms. The integral on the right side of Eq. (2.9) is the impulse delivered by a force F during a short time interval $(t_2 - t_1)$; that is, the change in the linear momentum is equal to the impulse delivered. Thus Eq. (2.9) is a statement of the *impulse–momentum theorem*.

2.2 CONSTANT APPLIED FORCE: *F* = CONSTANT

We are interested in studying the motion of a particle when the applied force acting on the particle is constant in time. Since F is constant, so will be the acceleration a, and we may write Newton's second law as

$$\frac{d^2x}{dt^2} = \frac{dv}{dt} = \frac{F}{m} = a = \text{constant} \tag{2.10}$$

The equation may be solved by direct integration provided we know the initial conditions. Solving Eq. (2.10) gives us the familiar results obtained in elementary mechanics, as we will show now. Let us assume that at $t = 0$, the initial velocity is v_0, and at time t the velocity is v. Thus, from Eq. (2.10),

$$\int_{v_0}^{v} dv = \int_{0}^{t} a \, dt$$

which on integration yields

$$v = v_0 + at \tag{2.11}$$

Substituting $v = dx/dt$ in Eq. (2.11) and again assuming the initial condition that $x = x_0$ at $t = 0$, we get by direct integration

$$x = x_0 + v_0 t + \tfrac{1}{2}at^2 \tag{2.12}$$

By eliminating t between Eqs. (2.11) and (2.12), we get

$$v^2 = v_0^2 + 2a(x - x_0) \tag{2.13}$$

Equations (2.11), (2.12), and (2.13) are the familiar equations that describe the translational motion of a particle in one dimension.

One of the most familiar examples of motion with constant force, hence constant acceleration, is the motion of freely falling bodies. In this case, a is replaced by g, the acceleration due to gravity, having the value $g = 9.8$ m/s^2 = 32.2 ft/s^2. The magnitude of the force of gravity acting downward is mg.

2.3 TIME-DEPENDENT FORCE: $F = F(t)$

In this case, the force being given by $F = F(t)$ implies that it is an explicit function of time; hence Newton's second law may be written as

$$m \frac{dv}{dt} = F(t) \tag{2.14}$$

which on integration gives, assuming that $v = v_0$ at $t = t_0$,

$$v = v_0 + \frac{1}{m} \int_{t_0}^{t} F(t) \, dt \tag{2.15}$$

Since $v = v(t) = dx(t)/dt$, Eq. (2.15) takes the form

$$\frac{dx(t)}{dt} = v_0 + \frac{1}{m} \int_{t_0}^{t} F(t) \, dt$$

or, integrating again,

$$x = x_0 + v_0(t - t_0) + \frac{1}{m} \int_{t_0}^{t} \left[\int_{t_0}^{t} F(t) \, dt \right] dt \tag{2.16}$$

Since there are two integrations, we may use two variables t' and t'' and write Eq. (2.16) as

$$x = x_0 + v_0(t - t_0) + \frac{1}{m} \int_{t_0}^{t} dt' \int_{t_0}^{t'} F(t'') \, dt'' \tag{2.17}$$

We will illustrate this discussion by applying it to the interaction of radio waves with electrons in the ionosphere, resulting in the reflection of radio waves from the ionosphere. The ionosphere is a region that surrounds Earth at a height of approximately 200 km (about 125 miles) from the surface of Earth. The ionosphere consists of positively charged ions and negatively charged electrons forming a neutral gas. When a radio wave, which is an electromagnetic wave, passes through the ionosphere, it interacts with the charged particles and accelerates them. We are interested in the motion of an electron of mass m and charge $-e$ initially at rest when it interacts with the incoming electromagnetic wave of electric field intensity E, given by

$$E = E_0 \sin(\omega t + \phi) \tag{2.18}$$

where ω is the oscillation frequency in radians per second of the incident electromagnetic wave and ϕ is the initial phase. The interaction results in a force F on the electron given by

$$F = -eE = -eE_0 \sin(\omega t + \phi) \tag{2.19}$$

while the acceleration of the electron is given by

$$a = \frac{F}{m} = -\frac{eE_0}{m}\sin(\omega t + \phi) \qquad (2.20)$$

Let $a_0 = eE_0/m$ be the maximum acceleration so that Eq. (2.20) becomes

$$a = -a_0\sin(\omega t + \phi) \qquad (2.21)$$

Since $a = dv/dt$, the equation of motion of the electron may be written as

$$\frac{dv}{dt} = -\frac{eE_0}{m}\sin(\omega t + \phi) \qquad (2.22)$$

Assuming initially the electron to be at rest, that is, $t = t_0 = 0$, $v_0 = 0$, the integration of Eq. (2.22) yields

$$v = -\frac{eE_0}{m\omega}\cos\phi + \frac{eE_0}{m\omega}\cos(\omega t + \phi) \qquad (2.23)$$

Since $v = dx/dt$, and assuming that $x = x_0$ at $t_0 = 0$, the integration of Eq. (2.23) yields

$$x = -\frac{eE_0}{m\omega^2}\sin\phi - \left(\frac{eE}{m\omega}\cos\phi\right)t + \frac{eE_0}{m\omega^2}\sin(\omega t + \phi) \qquad (2.24)$$

The first two terms indicate that the electron is drifting with a uniform velocity and this velocity is a function of the initial conditions only. Superimposed on this drifting motion of the electron is an oscillating motion represented by the last term. The oscillating frequency ω of the electron is independent of the initial conditions and is the same as the frequency of the incident electromagnetic waves. In the following, we want to investigate how such *coherent oscillations of free electrons can modify the propagation characteristics of incident electromagnetic waves.*

Comparing Eq. (2.19) for F with Eq. (2.24) for x, it becomes quite clear that the oscillating part of the displacement x is 180° out of phase with the applied force that results from the electric field of incident electromagnetic waves. Ordinarily, in a dielectric at low frequencies, the charges are displaced in the direction of the applied force, resulting in the polarization of the charges in phase with the applied force. In such situations, the resulting dielectric coefficient of the material is greater than 1. In the case of the ionosphere, it can be shown that the resulting polarization is 180° out of phase with the electric field; hence the *dielectric coefficient of the ionosphere is less than 1.* This result has two consequences.

1. The phase velocity v of electromagnetic waves in the ionosphere is greater than the speed of light c.
2. The refractive index of the ionosphere for incoming electromagnetic waves is less than the refractive index of the free space from where the waves are coming (the incident medium, which is a vacuum in this case).

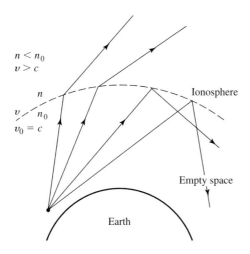

$n < n_0$
$v > c$

n Ionosphere

v n_0
$v_0 = c$

Empty space

Figure 2.1   Reflection of radiowaves
from the ionosphere. The total internal
reflection of electromagnetic waves
from the ionosphere.

Earth

This leads to the possibility of total internal reflection, that is, the reflection of incident electromagnetic waves from the ionosphere back to Earth, as illustrated in Fig. 2.1.

 Example 2.1 _____

A block of mass m is initially at rest on a frictionless surface at the origin. At time
t = 0, a decreasing force given by F = F0 exp(-λt), where λ = 0.5 is positive and less than
1, is applied. Calculate x(t) and v(t). Graph x, v, and F versus t.

Solution

From Newton's second law, after rearranging and integrating, we get the velocity v1 at time t1 to be

$$m \cdot \frac{d}{dt} v = F0 \cdot e^{-\lambda \cdot t}$$

$$\int_0^{v1} 1 \, dv = \int_0^{t1} \left(\frac{F0}{m} \right) \cdot e^{-\lambda \cdot t} \, dt$$

$$v1 = -F0 \cdot \frac{(\exp(-\lambda \cdot t1) - 1)}{(\lambda \cdot m)}$$

Once again rearranging and integrating, we get the displacement x1 to be

$$\int_0^{x1} 1 \, dx = \int_0^{t1} -F0 \cdot \frac{(\exp(-\lambda \cdot t) - 1)}{(\lambda \cdot m)} \, dt$$

$$x1 = F0 \cdot \frac{(\exp(-\lambda \cdot t1) + \lambda \cdot t1 - 1)}{(\lambda^2 \cdot m)}$$

Now we can write expressions for F_i, v_i, and x_i, keeping in mind that at $t = 0$, $F = Fo$. These calculations are made for $N = 100$ values even though only 15 values are shown in the graph. The values of F, x, and v at four different times are given below.

$$N := 100 \qquad i := 0..N \qquad t_i := i$$

$$m := 2 \qquad Fo := 1 \qquad \lambda := .5$$

$$F_i := Fo \cdot \left(e^{-\lambda \cdot t_i} \right) \qquad\qquad v_i := \frac{Fo}{m \cdot \lambda} \cdot \left(1 - e^{-\lambda \cdot t_i} \right)$$

$$x_i := \frac{Fo}{m \cdot \lambda^2} \cdot \left(-1 + e^{-\lambda \cdot t_i} \right) + \frac{Fo \cdot \lambda}{m \cdot \lambda^2} \cdot t_i$$

$$F_1 = 0.607 \qquad\qquad x_1 = 0.213 \qquad\qquad v_1 = 0.393$$

$$F_5 = 0.082 \qquad\qquad x_5 = 3.164 \qquad\qquad v_5 = 0.918$$

$$F_{25} = 3.727 \cdot 10^{-6} \qquad x_{25} = 23 \qquad\qquad v_{25} = 1$$

$$F_{50} = 1.389 \cdot 10^{-11} \qquad x_{50} = 48 \qquad\qquad v_{50} = 1$$

(a) Look at the variation in F, v, and x versus t and explain the conclusions you draw from such variations.

(b) What does the leveling of the values of F and v for high t mean?

(c) What changes in (b) will be observed if λ is 0.01, 1.0, and 5? (You can explain by regraphing for different values of λ.)

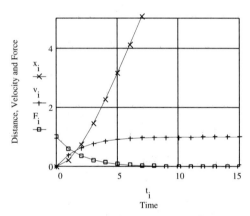

Motion due to a decreasing force

Exercise 2.1: A particle of mass m is at rest at the origin of the coordinate system. At $t = 0$, a force

$$F = F_0(1 - te^{-\lambda t})$$

is applied to the particle. Find the acceleration, velocity, and position of the particle as a function of time. Graph these values and answer (a), (b), and (c) in the example.

2.4 VELOCITY-DEPENDENT FORCE: $F = F(v)$

There are many situations of common everyday occurrence where, in addition to constant applied forces, forces are present that are a function of velocity. For example, when a body is in a
gravitational field, in addition to the gravitational force, there exists a force of air resistance on
the falling or rising body, and this resisting force is some complicated function of velocity. The
same is true of objects moving through fluids (gases and liquids). Such opposing forces to the
motion of objects through fluids are called *viscous forces* or *viscous resistance.* In these cases,
Newton's second law may be written in the following form:

$$F(v) = m\,\frac{dv}{dt} \tag{2.25}$$

or
$$F(v) = m\,\frac{dv}{dx}\frac{dx}{dt} = mv\,\frac{dv}{dx} \tag{2.26}$$

Knowing the form of the force $F(v)$, either of these two equations may be solved to analyze the
motion, that is, to calculate x as a function of t. Starting with Eq. (2.25), we may write

$$dt = m\,\frac{dv}{F(v)}$$

which on integration yields

$$t = t(v) = m \int \frac{dv}{F(v)} \tag{2.27}$$

Solving this gives v as a function of t; that is $v = v(t)$. Thus, knowing $v(t)$, we can solve for x.

$$v = \frac{dx}{dt} = v(t) \tag{2.28}$$

or
$$dx = v(t)\,dt$$

which on integration gives

$$x = x(t) = \int v(t)dt \tag{2.29}$$

Thus the problem is solved. Similarly, if we start with Eq. (2.26), we get

$$dx = m\,\frac{v\,dv}{F(v)} \tag{2.30}$$

which on integration yields

$$x = x(t) = m \int \frac{v\,dv}{F(v)} \tag{2.31}$$

Equations (2.29) and (2.31), which describe the displacement x as a function of t, may appear
to be quite different, but when evaluated, they yield the same relationships as can be demonstrated. We shall divide our discussion into two parts. First, we shall discuss those cases in which

there is no externally applied force besides the viscous resistance opposing the motion of the body. Later, we shall investigate more practical situations in which both types of forces, frictional as well as applied, are present.

Special Case

Suppose an automobile is moving with velocity v_0 on a smooth frictionless surface when its engine is suddenly shut off. Let us assume that the air resistance is proportional to velocity; that is,

$$F_r = F_r(v) = -kv \tag{2.32}$$

Assuming that at $t = 0$, $v = v_0$, calculate v and x as functions of t. We may write the differential equation of motion as

$$F_r(v) = -kv = m\frac{dv}{dt} \tag{2.33}$$

That is,
$$dt = -\frac{m}{k}\int_{v_0}^{v}\frac{dv}{v}$$

which on integration gives ($\int dv/v = \ln v$)

$$t = -\frac{m}{k}\ln\left(\frac{v}{v_0}\right) \tag{2.34}$$

After rearranging

$$v = v_0\, e^{-(k/m)t} \tag{2.35}$$

That is, the velocity decreases exponentially with time.

Substituting $v = dx/dt$ in Eq. (2.35) and rearranging, we get

$$dx = v_0\, e^{-(k/m)t}\, dt$$

which on integration, setting the limits as $t = 0$ when $x = 0$, and allowing the displacement to be x at time t, gives

$$x = \frac{mv_0}{k}[1 - e^{-(k/m)t}] \tag{2.36}$$

It is clear from Eqs. (2.35) and (2.36) that when $t = 0$, $v = v_0$, $x = 0$, as it should be. We note from Eq. (2.35) that $v = 0$ only when $t = \infty$, and then from Eq. (2.36), $x = mv_0/k = x_l$, where x_l is the limiting distance. The body never goes beyond this distance. (But it takes infinite time to reach there! We shall discuss this shortly.)

Reconsidering Eqs. (2.35) and (2.36), we know that the motion cannot continue forever and the automobile must come to rest long before the infinite time as calculated earlier. Let us assume that when the automobile reaches a certain minimum velocity v_ℓ it has almost reached the final distance x_ℓ. This will be true as long as v is less than the certain minimum value v_ℓ.

By substituting $v = v_\ell$ in Eq. (2.35), we can calculate the time t_ℓ it takes to reach velocity v_ℓ; that is,

$$v_\ell = v_0 e^{-(k/m)t_\ell} \tag{2.37}$$

or

$$t_\ell = \frac{m}{k} \ln\left(\frac{v_0}{v_\ell}\right) \tag{2.38}$$

Another interesting fact is revealed if we consider the motion in a short time interval when the retarding or the resistive force just begins acting on the moving body. To discuss this, let us expand the right sides of Eqs. (2.35) and (2.36) by using a Taylor series ($e^x = 1 + x + x^2/2! + x^3/3! + \cdots$). That is,

$$v = v_0 - \frac{kv_0}{m} t + \cdots = v_0 + \frac{F_{r0}}{m} t + \cdots$$

$$= v_0 + a_{r0}t \tag{2.39}$$

where $a_{r0} = F_{r0}/m$ is the acceleration at $t = 0$. Similarly,

$$x = v_0 t - \frac{1}{2}\frac{kv_0}{m} t^2 + \cdots = v_0 t + \frac{1}{2}\frac{F_{r0}}{m} t^2 + \cdots$$

$$= v_0 t + \tfrac{1}{2} a_{r0}t^2 + \cdots \tag{2.40}$$

If we ignore the higher terms, Eqs. (2.39) and (2.40) reveal that they describe the motion of a particle acted on by a constant force provided t is very small. Note that $-kv_0 = F_{r0} = ma_{r0}$, which is the force acting on the particle initially when $t = 0$; that is, these equations are simply the equations of motion of a particle under a constant force.

General Case

The preceding situation was limited to a simple case in which the retarding force $F_r(v)$ was proportional to velocity. In actual situations, F_r is a complicated function of velocity and the solutions cannot be obtained by simply using the integration tables. Instead it becomes necessary to do numerical integrations. But, in many cases over a wide range of velocities, it is possible, in practice, to use the following approximation in which the retarding force or frictional force is proportional to some power of velocity. That is,

$$F_r = (\mp)kv^n \tag{2.41}$$

where k is the positive constant of proportionality for the strength of the retarding force and n is a positive integer. If n is an odd integer, the negative sign in Eq. (2.41) must be used. If n is an even integer, v^n will be positive, and the sign $+$ or $-$ in Eq. (2.41) is chosen in such a way that it gives the direction of F_r to be opposite to that of v (F_r in the direction of v will be adding energy to the system instead of retarding it!). For small objects moving in air with velocities less than 25 m/s, it is found that, if we take $n \simeq 1$, we get good agreement with experimental results, while for velocities greater than 25 m/s but less than $\simeq 32$ m/s, the use of $n = 2$ gives good agreement with experimental values.

Let us apply these ideas to the case of a freely falling body, that is, to the vertical motion of an object in a resisting medium, the medium being air in this case. Let us assume that the air resistance is proportional to v, which can be written as $-kv$, independent of the sign of v. Thus the net force acting on the body is

$$F = F_g + F_r = -mg - kv \tag{2.42}$$

and the differential equation describing the motion of the freely falling body is

$$m\frac{dv}{dt} = -mg - kv \tag{2.43}$$

Taking $v = v_0$ at $t = 0$, we may write

$$\int dt = -\int_{v_0}^{v} \frac{m\,dv}{mg + kv} = -\frac{m}{k}\ln(mg + kv)\Big|_{v_0}^{v}$$

or

$$t = -\frac{m}{k}\ln\frac{mg + kv}{mg + kv_0} \tag{2.44}$$

Solving this equation for v, we get

$$v = -\frac{mg}{k} + \left(\frac{mg}{k} + v_0\right)e^{-(k/m)t} \tag{2.45}$$

Note that if the initial velocity is zero, that is, if at $t = 0$, $v_0 = 0$, we get

$$v = -\frac{mg}{k}[1 - e^{-(k/m)t}] \tag{2.46}$$

In Eq. (2.45), we may substitute $v = v(t) = dx/dt$ and the limits $x = x_0$ when $t = 0$ and $x = x$ when $t = t$. After integrating, we get the following result:

$$x = x_0 - \frac{mg}{k}t + \left(\frac{m^2g}{k^2} + \frac{mv_0}{k}\right)[1 - e^{-(k/m)t}] \tag{2.47}$$

Terminal Velocity

Let us consider Eq. (2.45) once again. As t increases, the exponential term decreases and drops to zero when t is very large as compared to m/k; that is, for $t \gg m/k$,

$$e^{-(k/m)t} = e^{-[t/(m/k)]} \rightarrow e^{-\infty} = 0$$

And for such large values of t, the velocity v reaches a limiting value, which from Eq. (2.45) is $v_t = -(mg/k)$. This limiting velocity v_t is called the terminal velocity of a falling body. The *terminal velocity* of a body is defined as the velocity of the body when the retarding force (the force of air resistance) is equal to the weight of the body, and the net force acting on the body is zero (hence there will be no acceleration of the body); that is,

$$F_r + F_g = 0 \quad \text{or} \quad -kv_t - mg = 0$$

or

$$v_t = -\frac{mg}{k} \tag{2.48}$$

The magnitude of the terminal velocity, which is equal to mg/k, is called the *terminal speed*. As an example, the terminal speed of raindrops varies anywhere from 3 to 7 m/s. Different bodies starting with different initial velocities will approach terminal velocities in different times. There are three different possibilities for initial velocities: $|\mathbf{v}_0| = 0$, $|\mathbf{v}_0| < |\mathbf{v}_t|$, and $|\mathbf{v}_0| > |\mathbf{v}_t|$. Figure 2.2 illustrates the time taken by the bodies to reach terminal velocity for the three cases.

Characteristic Time

It should be clear from the previous discussion that the dimensions of m/k are time. We define the quantity m/k as the *characteristic time* τ; that is, from Eq. (2.48) (ignoring the sign),

$$\tau = \frac{m}{k} = \frac{v_t}{g} \tag{2.49}$$

Using the definitions of v_t and τ, we rewrite Eqs. (2.45) and (2.47) as

$$v = -v_t + (v_t + v_0)e^{-t/\tau} \tag{2.50}$$

$$x = x_0 - v_t t + (g\tau^2 + v_0\tau)(1 - e^{-t/\tau}) \tag{2.51}$$

Let us consider a special case in which the body starts from rest. Substituting $v_0 = 0$ in Eq. (2.50), we obtain

$$v = -v_t(1 - e^{-t/\tau}) \tag{2.52}$$

and, if $t = \tau$, Eq. (2.52) takes the form

$$v = -v_t(1 - 1/e) = -0.63v_t$$

That is, in one characteristic time the body reaches 0.63 of its terminal speed. This allows us to define the *characteristic time* as that in which the body reaches 0.63 of its terminal speed. If $t = 2\tau$, then $v = -0.87v_t$, while for $t = 10\tau$, $v = -0.99995v_t$.

Let us go back to Eqs. (2.45) and (2.47) again. Using a series expansion, we can show that these equations reduce to the familiar equations of motion for the case of constant force. For $t \ll \tau (=m/k = v_t/g)$, we get

$$v = v_0 - gt \tag{2.53a}$$

and
$$x = x_0 + v_0 t - \tfrac{1}{2}gt^2 \tag{2.53b}$$

That is, for small t, the effect of air resistance is negligible. On the other hand, for $t \gg \tau (= m/k)$, Eqs. (2.45) and (2.47) reduce to

$$v = -v_t \tag{2.54a}$$

$$x = x_0 + v_0\tau + \left(\frac{m^2}{k^2}g - \frac{m}{k}gt\right)$$

$$= x_0 + v_0\tau + (\tau^2 g - g\tau t) \tag{2.54b}$$

Figure 2.2 _____

The graph illustrates the time it takes the bodies to reach terminal velocity for different initial velocities for given values of m, g, and k. vt is the terminal velocity, v01, v02, and v03 are three different initial velocities

$$n := 40 \qquad t := 0 .. n \qquad k := 5$$

v0 = 0

v02 < vt   (For absolute values) $m := 2.5 \qquad g := 9.8$

v03 > vt

$$vt := - \left(\frac{m \cdot g}{k} \right) \qquad vt = -4.9$$

Note that the terminal velocity is
vt = -4.9 m/sec $vo1 := 0 \qquad v1_t := - \left(\frac{m \cdot g}{k} \right) + \left(\frac{m \cdot g}{k} + vo1 \right) \cdot e^{- \left(\frac{k}{m} \right) \cdot \frac{t}{10}}$
and, depending on the initial
condition, the terminal velocity is
reached between 20 and 30 $vo2 := -2 \qquad v2_t := -\frac{m \cdot g}{k} + \left(\frac{m \cdot g}{k} + vo2 \right) \cdot e^{- \left(\frac{k}{m} \right) \cdot \frac{t}{10}}$
seconds. (In the graphs, t is divided
by 10.)

$$vo3 := -8 \qquad v3_t := -\frac{m \cdot g}{k} + \left(\frac{m \cdot g}{k} + vo3 \right) \cdot e^{- \left(\frac{k}{m} \right) \cdot \frac{t}{10}}$$

vt = -4.9

$v1_5 = -3.097 \qquad v2_5 = -3.833 \qquad v3_5 = -6.04$

$v1_{10} = -4.237 \quad v2_{10} = -4.508 \quad v3_{10} = -5.32$

$v1_{20} = -4.81 \quad v2_{20} = -4.847 \quad v3_{20} = -4.957$

$v1_{30} = -4.888 \quad v2_{30} = -4.893 \quad v3_{30} = -4.908$

$v1_{40} = -4.898 \quad v2_{40} = -4.899 \quad v3_{40} = -4.901$

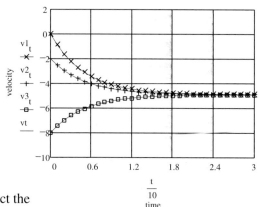

Time to reach terminal velocity

(a) What are the different factors that affect the terminal velocity?

(b) If the initial velocity is positive and upward, how will it affect the terminal velocity? How will this graph differ from the others before terminal velocity is reached? Graph this.

A Better Approximation

For small, compact, heavy bodies, a better approximation is that in which the retarding or viscous force is proportional to v^2. Thus the equation

$$F_r + F_g = m\frac{dv}{dt}$$

takes the form

$$\mp kv^2 - mg = m\frac{dv}{dt} \tag{2.55}$$

where $-kv^2$ is used for rising (or ascending) bodies, while $+kv^2$ is used for falling bodies. The terminal speed is given by

$$F_r + F_g = 0 \quad \text{or} \quad kv_t^2 = mg \tag{2.56}$$

or

$$v_t = \sqrt{\frac{mg}{k}} \tag{2.57}$$

and the characteristic time τ is given by

$$\tau = \frac{v_t}{g} = \sqrt{\frac{m}{kg}} \tag{2.58}$$

Following the procedure outlined before, Eq. (2.55) can be solved for v and x. The results obtained are

$$t = \int \frac{m\,dv}{-mg - kv^2}, \quad \text{for rising objects} \tag{2.59}$$

$$t = \int \frac{m\,dv}{-mg + kv^2}, \quad \text{for falling objects} \tag{2.60}$$

which give

$$t = -\tau \tan^{-1}\left(\frac{v}{v_t}\right) + C_1, \quad \text{for rising objects} \tag{2.61}$$

$$t = -\tau \tanh^{-1}\left(\frac{v}{v_t}\right) + C_2, \quad \text{for falling objects} \tag{2.62}$$

where C_1 and C_2 are constants to be determined from initial conditions. Solving these equations for v, we get

$$v = v_t \tan\frac{C_1 - t}{\tau}, \quad \text{for rising objects} \tag{2.63}$$

$$v = -v_t \tanh\frac{t - C_2}{\tau}, \quad \text{for falling objects} \tag{2.64}$$

As compared to a retarding force $\pm kv$, the terminal speed for the case of a retarding force $\pm kv^2$ is reached much faster, as illustrated in Fig. 2.3. In this case, when $t = 5\tau$, $v = -0.99991v_t$; that is, it reaches the same speed in half the time.

Figure 2.3

The graph shows the motion of a vertically falling object under a linear
force and a quadratic force with initial conditions $t_0 = 0$, $x_0 = 0$ and $v_0 = 0$.

  Linear retarding force: $F = -k \cdot x$ Quadratic retarding force: $F = -k \cdot v^2$ or $k \cdot v^2$

The terminal velocity $v2 = -0.808$ for a $N := 50$ $i := 0 .. N$ $t_i := \dfrac{i}{50}$
quadratic retarding force ($\pm kv^2$) is reached
much faster (in about 0.2 seconds) than $m := .1$ $g := 9.8$ $k := 1.5$
the terminal velocity $v1 = -0.653$ for a
linear retarding force ($-kv$) is reached in
about 0.3 seconds. To calculate v1 and $\tau 1 := \dfrac{m}{k}$ $\tau 2 := \sqrt{\dfrac{m}{k \cdot g}}$
v2 use Eqs. (2.46) and (2.64).

                                           $\tau 1 = 0.067$ $\tau 2 = 0.082$

(a) In a given time interval, which object $v1 := \dfrac{m \cdot g}{k}$ $v2 := \sqrt{\dfrac{m \cdot g}{k}}$
will travel a larger distance and why?

                                           $v1 = 0.653$ $v2 = 0.808$

(b) For the vertically falling objects
which of the two situations is more $v1_i := -(v1) \cdot \left[1 - e^{ -\left(\frac{k}{m}\right) \cdot t_i } \right]$ $v2_i := -v2 \cdot \tanh\left(\dfrac{t_i}{\tau 2} \right)$
desirable and why?

(c) What do the values of v1 and v2 at
time t = 20 and 30 indicate?

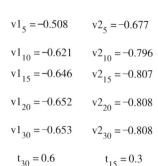
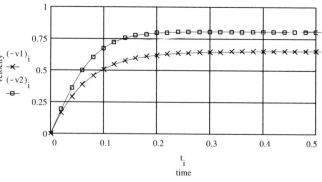

$v1_5 = -0.508$ $v2_5 = -0.677$

$v1_{10} = -0.621$ $v2_{10} = -0.796$

$v1_{15} = -0.646$ $v2_{15} = -0.807$

$v1_{20} = -0.652$ $v2_{20} = -0.808$

$v1_{30} = -0.653$ $v2_{30} = -0.808$

$t_{30} = 0.6$ $t_{15} = 0.3$

Speed for quadratic and linear force

Note that the terminal velocity v2 is reached much faster (in about 0.3 second) than the
terminal velocity v1 (in about 0.6 second). Explain why.

Example 2.2 _____

A ball of mass m is thrown with velocity vo on a horizontal surface where the retarding force is proportional to the square root of the instantaneous velocity. Calculate the velocity and the position of the ball as a function of time and graph the results.

Solution

The retarding force is given by
Rearrange this equation and integrate assuming that initial velocity at t = 0 is vo and v1 at time t1.

$$F = m \cdot \frac{dv}{dt} = -k \cdot v^{\frac{1}{2}}$$

$$\int_{v0}^{v1} \frac{1}{v^{\frac{1}{2}}}\, dv = \int_{0}^{t1} -\frac{k}{m}\, dt \quad \text{simplifies to}$$

Solving, we get the value of v1 at time t1 as

$$2 \cdot \sqrt{v1} - 2 \cdot \sqrt{v0} = -k \cdot \frac{t1}{m}$$

Integrating v1 we find the displacement x1 at time t1.

$$v1 = \left(\sqrt{v0} - \frac{k}{2} \cdot \frac{t1}{m}\right)^2 \qquad (i)$$

Now we may graph x and v as function of t using Eqs. (i) and (ii), rewritten as (iii) and (iv)

$$\int_{x0}^{x1} 1\, dx = \int_{0}^{t1} \left(\sqrt{v0} - \frac{k}{2} \cdot \frac{t}{m}\right)^2 dt$$

$$x1 - x0 = t1 \cdot v0 + \frac{1}{12} \cdot k^2 \cdot \frac{t1^3}{m^2} - \frac{1}{2} \cdot \sqrt{v0} \cdot k \cdot \frac{t1^2}{m} \qquad (ii)$$

$$i := 0..20 \qquad t_i := i \qquad xo := 0 \qquad vo := 20 \qquad k := 1.1 \qquad m := .5$$

$$v_i := vo - \frac{k}{m} \cdot t_i \cdot vo^{\frac{1}{2}} + \frac{k^2}{4 \cdot m^2} \cdot (t_i)^2 \qquad (iii)$$

$$x_i := xo + \left[vo \cdot t_i - \frac{1}{2} \cdot \frac{k}{m} \cdot (t_i)^2 \cdot \sqrt{vo} + \frac{1}{12} \cdot \frac{k^2}{m^2} \cdot (t_i)^3 \right] \qquad (iv)$$

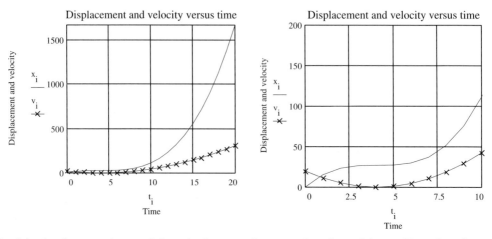

Explain the decrease in v and then the increase in v as a function of time t. How does it affect the value of x? (Refer to the zoomed graph on the right).

EXERCISE 2.2 Repeat the example for a retarding force that is proportional to the cube root of the instantaneous velocity.

2.5 POSITION-DEPENDENT FORCES: *F = F(x)*, CONSERVATIVE FORCES, AND POTENTIAL ENERGY

This is one of the most important cases considered so far. There are many situations in which motion depends on the position of the object. Examples of position-dependent forces are gravitational force, Coulomb force, and elastic (tension and compression) forces. The differential equation that describes the rectilinear motion of an object under the influence of a position-dependent force is

$$m \frac{d^2x}{dt^2} = F(x) \tag{2.65}$$

which may also be written in such a manner that v is a function of position; that is,

$$mv \frac{dv}{dx} = F(x) \tag{2.66}$$

or

$$\frac{d}{dx}\left(\frac{1}{2}mv^2\right) = F(x) \tag{2.67}$$

Since the kinetic energy of the particle is $K = \frac{1}{2}mv^2$, we may write Eq. (2.67) as

$$\frac{dK}{dx} = F(x) \tag{2.68}$$

which on integration gives

$$K - K_0 = \int_{x_0}^{x} F(x)\, dx \tag{2.69a}$$

or

$$\frac{1}{2}mv^2 - \frac{1}{2}mv_0^2 = \int_{x_0}^{x} F(x)\, dx \tag{2.69b}$$

The right side is equal to the work done when the particle is displaced from position x_0 to x.

It is convenient at this point to introduce *potential energy* or a *potential energy function* (or simply a *potential function*) $V(x)$ such that

$$-\frac{dV(x)}{dx} = F(x) \tag{2.70}$$

We define $V(x)$ as the work done by the force when the particle is displaced from x to some arbitrary chosen standard point x_S; that is,

$$V(x) = \int_{x}^{x_s} F(x)\, dx = -\int_{x_s}^{x} F(x)\, dx \tag{2.71}$$

which is consistent with Eq. (2.70). Thus the work done is going from x_0 to x is

$$\int_{x_0}^{x} F(x)\, dx = \int_{x_0}^{x} \left[-\frac{dV(x)}{dx} \right] dx = -\int_{x_0}^{x} dV(x)$$

$$= -\int_{x_0}^{x_s} dV(x) - \int_{x_s}^{x} dV(x)$$

$$= +V(x_0) - V(x) = -V(x) + V(x_0) \tag{2.72}$$

Combining Eqs. (2.69) and (2.72), we get

$$K + V(x) = K_0 + V(x_0) = \text{constant} = E \tag{2.73}$$

or

$$\frac{1}{2}m\left(\frac{dx}{dt}\right)^2 + V(x) = E \tag{2.74}$$

This equation states that *if a particle is moving under the action of a position-dependent force, then the sum of its kinetic energy and potential energy remains constant throughout its motion.* Such forces are called *conservative forces*. For nonconservative forces, $K + V \neq$ constant, and a potential energy function does not exist for such forces. An example of a nonconservative force is frictional force. [It may be pointed out that if $V(x)$ is replaced by $V(x)$ + constant, the preceding discussion still holds true. In other words, the sum of the kinetic and potential energy will still remain constant and will be equal to E.] E is the total energy, and Eq. (2.74) states the law of *conservation of energy,* which holds only if $F = F(x)$. A description of the motion of a particle may be obtained by solving the energy equation, Eq. (2.74); that is,

$$v\left(= \frac{dx}{dt} \right) = \pm \sqrt{\frac{2}{m} [E - V(x)]} \tag{2.75}$$

which on integration yields

$$t - t_0 = \int \frac{\pm\, dx}{\sqrt{(2/m)[E - V(x)]}} \tag{2.76}$$

and gives t as a function of x. [We shall not discuss the significance of the negative sign, which deals with time reversal.]

In considering the solution of Eq. (2.76), it is essential to note that only those values of x are possible for which the quantity $E - V(x)$ is positive. Negative values lead to imaginary solutions and hence are unacceptable. Also, the motion is limited to those values of x for which $E - V(x) \geq 0$; that is, the roots of this equation give the region or regions to which the motion is confined. This is demonstrated in Fig. 2.4. The function $\frac{1}{2}mx^2 + V(x)$ is called the *energy integral* of the equation of motion $m(dv/dt) = F(x)$, and such an integral is called a *constant of motion.* (This is the first integral of a second-order differential equation.)

Before we give specific examples of solving the equation of motion for $x(t)$, we shall show that much can be learned about the motion by simply plotting $V(x)$ versus x. Figure 2.5 shows the plot of a potential energy function for one-dimensional motion. As mentioned earlier, the motion of the particle is confined to those regions for which $E - V(x) > 0$ or $V(x) < E$. Let us keep Eq. (2.75) in mind and discuss different cases.

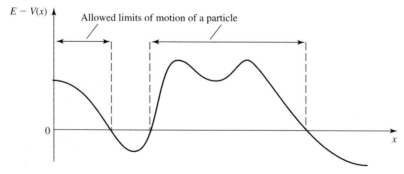

Figure 2.4   Allowed regions of motion for a particle in a position-dependent force field.

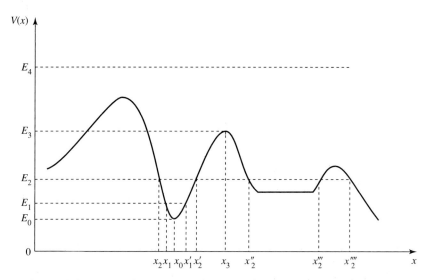

Figure 2.5   The solid curve corresponds to a potential function V(x), and E0, E1, . . . , are different energies of a particle moving in such a potential.

If $E = E_0$, as shown in Fig. 2.5, then $E_0 - V(x) = 0$ and, according to Eq. (2.75), $\dot{x} = 0$; that is, the particle stays at rest in equilibrium at $x = x_0$. Let us consider the case in which the particle energy is slightly greater than E_0, say E_1. For $x < x_1$ and $x > x_1'$, v will be imaginary; hence the particle cannot exist in these regions. Thus a particle with energy E_1 is constrained to move in the potential well (or valley) between x_1 and x_1'. A particle moving to the right is reflected back at x_1'; and when traveling to the left, it is reflected back at x_1. The points x_1 and x_1' are called the *turning points* and are obtained by solving $E_1 - V(x) = 0$. The velocity of the particle at these points is zero. Between x_1 and x_1', the velocity of the particle will change as $V(x)$ changes. We briefly explain the motion of a particle corresponding to different energies and moving in a potential $V(x)$, as shown in Fig. 2.5.

E_0: The particle is in stable equilibrium.

E_1: The particle moves between the turning points x_1 and x_1'.

E_2: The particle moves between the turning points x_2 and x_2' with changing velocity. While moving between the turning points x_2'' and x_2''', the particle has constant velocity and hence is in the region of neutral equilibrium. The particle can also exist in the region for $x > x_2'''$.

E_3: When a particle with this energy is at x_3, it is at a position of unstable equilibrium. It can also move in the valley on the left with a motion similar to that of a particle with energy E_2. Once it starts moving to the right, it keeps on moving, first with increasing velocity to x_2'' and then with constant velocity up to x_2'''.

E_4: A particle with this energy can move anywhere. When passing over the hills, it slows down, while over the valleys, it speeds up, as it should.

Continuing our discussion of position-dependent forces, we shall examine two special cases of interest in the next sections:

1. Motion under a linear restoring force
2. Variation of g in a gravitational field

Example 2.3 _____

A particle of mass m is subjected to a force $F = a - 2bx$, where a and b are constants. Find the potential energy $V = V(x)$. Then graph $F(x)$ and $V(x)$. Discuss the motion of the particle for different values of energy.

Solution
Substitute for F in the expression for V and integrate to get the value of V.

The motion is limited to the region $x = 0$ and $x = a/b$

For $E < 0$, the motion is limited to the left of $V(x)$ and cannot cross the barrier.

$$F = a - 2\cdot b\cdot x \qquad V = -\int F\,dx$$

$$V = -\int a - 2\cdot b\cdot x\,dx \quad \text{simplifies to} \quad V = -a\cdot x + b\cdot x^2 \quad (i)$$

For $V(x) = -a\cdot x + b\cdot x^2 = 0 \qquad x = \dfrac{a}{b}$

$i := 0..\,15 \qquad x_i := i \qquad a := 20 \qquad b := 2$

For $\dfrac{d}{dx}V(x) = -a + 2\cdot b\cdot x = 0 \qquad x = \dfrac{a}{2\cdot b}$

$$V_i := -a\cdot x_i + b\cdot(x_i)^2 \qquad F_i := a - 2\cdot b\cdot x_i \quad (ii)$$

At $x = \dfrac{a}{2\cdot b} \qquad Vmin = \dfrac{a^2}{4\cdot b}$

$F = 0$ at $x = 0$ and $\dfrac{a}{2\cdot b}$

Different values are calculated below

$\dfrac{a}{2\cdot b} = 5 \qquad \dfrac{a}{b} = 10 \qquad -\dfrac{a^2}{4\cdot b} = -50$

$V_0 = 0 \qquad V_5 = -50 \qquad V_{10} = 0$

$F_0 = 20 \qquad F_5 = 0 \qquad F_{10} = -20$

$(\min(V)) = -50 \qquad \max(V) = 150$

$\max(F) = 20 \qquad \min(F) = -40$

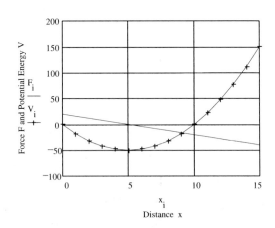

(a) Looking at the variation in the values of V and F versus x, what do you conclude from this variation?

(b) What are the values of F and V where these graphs intersect? What is the significance of this? Explain.

(c) What is the significance of x = 5 where F = 0 and V is minimum? Explain.

EXERCISE 2.3 Repeat the example for $F = a - 2bx^2$

2.6 MOTION UNDER A LINEAR RESTORING FORCE

Let the motion of a particle subject to a linear force be given by

$$F(x) = -kx \tag{2.77}$$

This equation is a statement of *Hooke's law*. A typical example of such a motion is that of a mass fastened to a spring. The resulting motion is simple harmonic, as we shall discuss in detail in Chapter 3. For the time being, we shall use the energy method discussed in the previous section to obtain the solution. Taking the standard point to be at the origin (also the equilibrium point), that is $x_s = 0$, we may write the potential energy to be

$$V(x) = -\int_{x_s}^{x} F(x)\, dx = -\int_{0}^{x} (-kx)\, dx$$

or $\qquad\qquad\qquad V(x) = \tfrac{1}{2}kx^2 \tag{2.78}$

Once again, the total energy is a constant of motion and may be arrived at in the same manner; that is,

$$mv \frac{dv}{dx} = F(x) = -kx \tag{2.79}$$

$$mv\, dv = -kx\, dx$$

which on integration gives

$$\tfrac{1}{2}mv^2 = -\tfrac{1}{2}kx^2 + \text{constant}$$

or $\qquad\qquad\qquad \tfrac{1}{2}mv^2 + \tfrac{1}{2}kx^2 = E = \text{total energy} \tag{2.80}$

We can now use this equation or Eq. (2.76) with $V(x)$ given by Eq. (2.78) to solve for the displacement x. For the conditions $t_0 = 0$ at $x = x_0$, Eq. (2.80) or (2.76) takes the form (keeping only the positive sign)

$$t = \int \frac{dx}{\sqrt{(2/m)(E - \frac{1}{2}kx^2)}} = \int \frac{dx}{\sqrt{\frac{2E}{m}\left(1 - \frac{k}{2E}x^2\right)}} \qquad (2.81)$$

Substituting

$$\sqrt{\frac{k}{2E}}\, x = \sin\theta \quad \text{and} \quad \sqrt{\frac{k}{2E}}\, dx = \cos\theta\, d\theta \qquad (2.82)$$

we get

$$t = +\sqrt{\frac{m}{k}} \int_{\theta_0}^{\theta} d\theta = \sqrt{\frac{m}{k}}\,(\theta - \theta_0) \qquad (2.83)$$

As usual, let the angular velocity be defined as

$$\omega = \sqrt{\frac{k}{m}}$$

Therefore, $$t = \frac{1}{\omega}(\theta - \theta_0)$$

or $$\theta = \omega t + \theta_0 \qquad (2.84)$$

Combining Eq. (2.84) and (2.82), we get

$$\sin^{-1}\left(\sqrt{\frac{k}{2E}}\, x\right) = \omega t + \theta_0$$

or $$x = A \sin(\omega t + \theta_0) \qquad (2.85)$$

where A is the amplitude given by

$$A = \sqrt{\frac{2E}{k}} \qquad (2.86)$$

Thus Eq. (2.85) states that the motion of the particle is simple harmonic with the coordinate x oscillating harmonically in time with amplitude A and frequency ω. A and θ can be determined from the initial conditions; that is, if E and x_0 are given, from Eqs. (2.86) and (2.85)

$$E = \frac{1}{2}kA^2 \qquad (2.87)$$

and $$x_0 = A \sin\theta_0 \qquad (2.88)$$

2.7 VARIATION OF *g* IN A GRAVITATIONAL FIELD

For small heights just above the surface of Earth, the value of *g* is almost constant and is equal to 9.8 m/s^2 = 32.2 ft/s^2. But at large distances above the surface of Earth, the value of *g* varies with distance and may be calculated in a simple manner. According to Newton's law of gravitation, the force between an object of mass *m* at a distance *x* from the center of Earth of mass *M* is

$$F(x) = -G\frac{Mm}{x^2} \tag{2.89}$$

If we neglect air resistance, the differential equation of motion of an object in a gravitional field may be written as

$$mv\frac{dv}{dx} = -G\frac{Mm}{x^2}$$

Since $v = \dot{x}$, we may write

$$m\int \dot{x}\,d\dot{x} = -GMm\int \frac{dx}{x^2}$$

which on integration gives

$$\frac{1}{2}m\dot{x}^2 - G\frac{Mm}{x} = \text{constant} = E \tag{2.90a}$$

or

$$\tfrac{1}{2}m\dot{x}^2 + V(x) = E \tag{2.90b}$$

where we have defined the *gravitational potential energy V(x)* to be

$$V(x) = -G\frac{Mm}{x} \tag{2.91}$$

[We can derive the same expression for the gravitational potential energy by the direct definition of the potential function. But it is convenient to take the initial conditions to be $x_s = \infty$, instead of $x_s = 0$, as we did previously. Thus

$$V(x) = -\int_{x_s}^{x} F(x)\,dx = -\int_{\infty}^{x} -\frac{GMm}{x^2}\,dx = -\frac{GMm}{x}$$

as defined previously.]

We may rewrite Eq. (2.90a) as

$$v = \dot{x} = \frac{dx}{dt} = \pm\sqrt{\frac{2}{m}\left(E + G\frac{Mm}{x}\right)} \tag{2.92}$$

where the positive sign corresponds to ascending motion, while the negative sign corresponds to descending motion. Equation (2.92) may be solved for *x(t)* by integrating; that is,

$$t = \sqrt{\frac{m}{2}}\int_{x_0}^{x} \frac{dx}{[E + (GMm/x)]^{1/2}} \tag{2.93}$$

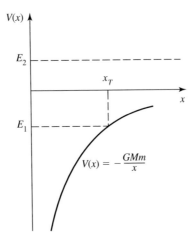

Figure 2.6   Particle of energy E in a gravitational potential $V(x)$ versus x.

The integration of this equation is not as simple as in the case of a linear restoring force. We shall not pursue this any further at the present time, but shall give a graphical interpretation and discuss some simple cases.

The plot of $V(x)$ versus x is shown in Fig. 2.6. It is quite clear that for negative values of E the motion is bound with a turning point at x_T. That is, when $E = E_1$ the body will go a height $x = x_T$, come to a stop, and turn around. For values of E greater than zero, there is no turning point and the body will never return to Earth. Thus, for a minimum energy $E = 0$, the velocity corresponds to $v = v_e$, the *escape velocity,* which we shall calculate below. We can calculate the turning point by substituting velocity $v = \dot{x} = 0$ and $x = x_T$ in Eq. (2.92) (by definition v is zero at the turning point), and we get

$$-G\frac{Mm}{x_T} = E \quad \text{or} \quad x_T = -G\frac{Mm}{E} \tag{2.94}$$

Since E is negative, x_T will be positive.

Let us consider Eq. (2.90a) once again:

$$\frac{1}{2}m\dot{x}^2 - G\frac{Mm}{x} = \text{constant} \tag{2.95}$$

Let a body be dropped from a height $x = x_0$ with zero initial velocity. That is, substituting $v = \dot{x} = 0$ at $x = x_0$ in Eq. (2.95),

$$-G\frac{Mm}{x_0} = \text{constant}$$

Using this value for the constant in Eq. (2.95) and rearranging, we get

$$\dot{x}^2 = 2GM\left(\frac{1}{x} - \frac{1}{x_0}\right) \tag{2.96}$$

Let g_0 be the value of the acceleration due to gravity on the surface of Earth, where $x = R$, so that

$$F_G = F_{g0} \quad \text{or} \quad -G\frac{Mm}{R^2} = -mg_0$$

That is,

$$g_0 = \frac{Gm}{R^2} \quad \text{or} \quad G = g_0\frac{R^2}{M} \tag{2.97}$$

Usually the distance is measured from the surface of Earth; hence we may write

$$x = R + r, \quad \dot{x} = \dot{r} \quad \text{and} \quad x_0 = R + h$$

where $h = x_0 - R$ equals the height (as measured from the surface of Earth) from which the body is dropped. Using this notation and Eqs. (2.96) and (2.97), we may write

$$v^2 = 2g_0 R^2\left(\frac{1}{R+r} - \frac{1}{R+h}\right) \tag{2.98}$$

Thus, for $r = 0$, that is, when the body reaches the surface, $v = v_0$; hence

$$v_0^2 = 2g_0 R^2\left(\frac{1}{R} - \frac{1}{R+h}\right) \tag{2.99}$$

and may be written as

$$v_0^2 = 2g_0 R\left[1 - \left(1 + \frac{h}{R}\right)^{-1}\right] \tag{2.100}$$

which reduces to $v_0 \simeq 2g_0 h$ if $h \ll R$, as it should. Equation (2.98) also applies to the case when a body is projected upward with a velocity v_0, and it will reach a height h when $v = 0$.

We can arrive at an expression for the *escape velocity* by substituting $h = \infty$ in Eq. (2.99), resulting in

$$v_e^2 = 2g_0 R = \frac{2GM}{R}$$

$$v_e = \sqrt{2g_0 R} = \sqrt{\frac{2GM}{R}} = 11 \text{ km/s} = 7 \text{ miles/s} \tag{2.101}$$

or we could say that at the surface of the Earth $E = 0$; hence potential energy must be equal to the kinetic energy. That is,

$$\frac{1}{2}m\dot{x}_e = \frac{GMm}{R}$$

$$v_e = \dot{x} = \sqrt{\frac{2GM}{R}}$$

PROBLEMS

2.1. Force F acting on a particle of mass m has the following dependencies:

    (a) $F(x, t) = f(x)g(t)$

    (b) $F(\dot{x}, t) = f(\dot{x})g(t)$

    (c) $F(\dot{x}, x) = f(\dot{x})g(x)$

    Write the differential equations describing these situations. Which of these differential equations can be solved to describe the motion of the particle? Explain.

2.2. A particle of mass m is acted on by the force **(a)**, **(b)**, **(c)**, **(d)**, or **(e)** as given below. Solve these equations to describe the motion of the particle.

    (a) $F(x, t) = k(x + t^2)$, for $t = 0$, $x = x_0$, and $v = v_0 = 0$

    (b) $F(\dot{x}, t) = kx^2\dot{x}$, for $t = 0$, $x = x_0$, and $v = v_0 = 0$

    (c) $F(\dot{x}, t) = k(a\dot{x} + t)$, for $t = 0$, $v = v_0$

    (d) $F(x, \dot{x}) = ax^2/\dot{x}$

    (e) $F(x, \dot{x}, t) = k(x + \dot{x}t)$

2.3. A block of mass m is initially at rest on a frictionless surface. At time $t = 0$, an increasing force given by by $F = kt^2$ is applied to the block. Find the velocity and the displacement of the block as a function of time and graph x and v versus t.

2.4. A block of mass m is initially at rest on a frictionless surface at the origin. At time $t = 0$, a force given by $F = F_0 t e^{-\lambda t}$ is applied. Calculate $x(t)$ and $v(t)$ and graph them. What are these values when **(a)** t is very small, and **(b)** t is very large?

2.5. A particle of mass m is at rest at $t = 0$ when it is subjected to a force $F = F_0 \sin(\omega t + \phi)$. **(a)** Calculate the values of $x(t)$ and $v(t)$ **(b)** Make plots of $x(t)$ and $v(t)$ versus t. What are the maximum and minimum values of x and v?

2.6. A particle of mass m is subjected to a force given by

$$F = F_0 e^{-\lambda t} \sin(\omega t + \phi)$$

Calculate the values of $v(t)$ and $x(t)$ and graph them. What is the magnitude of the terminal velocity in this case?

2.7. A particle of mass m is at rest at $t = 0$ when it is subjected to a force $F = F_0 \cos^2 \omega t$.

    (a) Calculate the values of $x(t)$ and $v(t)$

    (b) Make plots of $x(t)$ and $v(t)$ versus t.

    (c) Describe the outstanding characteristic of these graphs.

2.8. A ball of mass m is thrown with velocity v_0 on a horizontal surface where the retarding force is proportional to the square root of the instantaneous velocity. Calculate the velocity and the position of the ball as a function of time. Discuss any limitations.

2.9. An object of mass m is thrown up an inclined plane of an angle θ with an initial velocity v_0. If the motion is resisted by the retarding force $F_r = -kv$, how far will the mass travel before coming to rest? Assuming the same retarding force, how long will it take the object to travel back to the initial position?

2.10. Repeat Problem 2.9 for the retarding force $F_r = \pm kv^2$.

2.11. A boat is slowed down by a frictional force $F(v)$. Its velocity decreases according to the relation $v = k(t - t_s)^2$, where t_s is the time it takes to stop the boat and k is the constant. Calculate $F(v)$.

2.12. The motor of a speed boat is shut off when it has attained a speed of v_0. Now the boat is slowed down by a retarding force $F_r = Ce^{-kv}$. Calculate $v(t)$ and $x(t)$. How long will it take for the boat to stop, and how much distance will it travel before stopping?

2.13. A particle of mass m starting with an initial velocity v_0 is acted on by a force $F = m(kv + cv^2)$, where k and c are constants. Calculate the displacement as a function of time.

2.14. For the situation in Problem 2.12 graph F and x versus t.

2.15. For the situation in Problem 2.13 graph F and x versus t. Compare the results with those in Problem 2.14.

2.16. A body of mass m is dropped from a height h. Calculate the speed when it hits the ground if **(a)** there is no air resistance, **(b)** air resistance is proportional to the instantaneous velocity, and **(c)** air resistance is proportional to the square of the instantaneous velocity. Graph velocity versus time in each case and compare the results.

2.17. A projectile is thrown vertically with a velocity v_0. Calculate and compare times and maximum heights reached when air resistance is **(a)** zero, **(b)** proportional to the instantaneous velocity, and **(c)** proportional to the square of the instantaneous velocity. Graph distance versus time in all cases and compare the results.

2.18. A ball is thrown vertically upward with an initial velocity v_0. The air resistance is proportional to the square of the velocity. Show that the velocity with which the ball returns to the original position is

$$\frac{v_0 v_t}{\sqrt{v_0^2 + v_t^2}}$$

where v_t is the terminal velocity.

2.19. Derive Eq. (2.47); that is,

$$x = x_0 - \frac{mg}{k}t + \left(\frac{m^2 g}{k^2} + \frac{mv_0}{k}\right)(1 - e^{-(k/m)t})$$

2.20. Using Eqs. (2.45) and (2.47), show that for $t \ll \tau$, we get the familiar equations of motion, Eqs. (2.53a) and (2.53b).

2.21. Show that for the case $t \ll \tau$, Eqs. (2.45) and (2.47) reduce to Eqs. (2.54a) and (2.54b).

2.22. Starting with Eq. (2.55), derive Eqs. (2.61) and (2.62).

2.23. Starting with Eq. (2.60) and initial conditions that $v_0 = 0$, $x_0 = 0$ at $t = 0$, find the expressions for x and v.

2.24. Using the law of conservation of energy, derive the general equations of motion for freely falling bodies.

2.25. A particle of mass m is attracted toward the origin by a force that is inversely proportional to the square of the distance; that is, $F = -K/r^2$. If this mass is released from a distance L, show that it will take time t to reach the origin, where t is given by

$$t = \pi\left(\frac{mL^3}{8K}\right)^{1/2}$$

2.26. The velocity of a particle of mass m subjected to a certain force varies with the distance according to the relation $v = K/x^n$, where K is a constant. Assuming at $t = 0$, $x = x_0$, calculate the force acting on the particle as a function of **(a)** distance x, and **(b)** time t. **(c)** Calculate the position of the particle as a function of time t. **(d)** Graph in **(a)**, **(b)** and **(c)** and discuss any outstanding features.

2.27. A particle of mass m is subjected to a force given by $F = -ax + bx^2$, where a and b are constants.
   (a) Find the potential energy $V(x)$.
   (b) Make plots of $F(x)$ and $V(x)$.
   (c) Discuss the motion of the particle for different values of energy and also point out the regions in which the motion is forbidden.

2.28. A particle of mass m is subjected to a force represented by a potential function $V(x) = -ax^2 + bx^4$, where a and b constants.
   (a) Calculate $F(x)$.
   (b) Make plots of $F(x)$ and $V(x)$.
   (c) Discuss the motion of the particle for different values of energy. Also discuss the restrictions on the motion.

2.29. A particle at time $t = 0$ is at rest at a distance x_0 from the origin and is subjected to a force that is inversely proportional to the distance from the origin. Solve the equation of motion for this particle; that is, find $v(t)$ and $x(t)$.

2.30. A particle of mass m is subjected to a force

$$F(x) = -Cx + \frac{K}{x^3}$$

where C and K are constants.
   (a) Find the potential function $V(x)$.
   (b) Make plots of $F(x)$ and $V(x)$.
   (c) Discuss the nature of the motion for different values of E and also find the regions where the motion is not possible.

2.31. The force between two particles in a diatomic molecule is such that it may be represented by a potential function of the form

$$V(x) = -\frac{C_1}{x^6} + \frac{C_2}{x^{12}}$$

where C_1 and C_2 are the positive constants and x is the distance between the two atoms.
   (a) Find $F(x)$.
   (a) Make plots of $F(x)$ and $V(x)$.
   (c) Assume that one of the atoms in the molecules is very heavy and remains at rest at the origin. Discuss the possible motions of the other atom in the molecule.

2.32. An alpha particle when inside the nucleus is bound by the potential shown for $-R < x < R$, while outside the nucleus the interaction between the alpha particle and the nucleus is represented by the coulomb potential, as shown in Fig. P2.32.
   (a) Write a potential function for the different regions shown in the figure.
   (b) Calculate $F(x)$ and make a plot of $F(x)$ versus x.
   (c) Discuss the motion of the alpha particle for different values of energy and different regions.

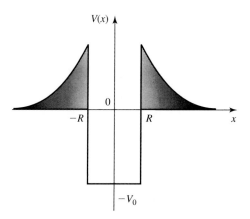

Figure P2.32

2.33. Show that for $h \ll R$, Eq. (2.100) reduces to $v_0 = 2g_0 h$.

2.34. Starting with Eq. (2.90a) and substituting

$$\cos \theta = \left(-\frac{E}{GMm} \right)^{1/2} x^{1/2}$$

show that $x = x_T \cos^2 \theta$, where

$$\theta + \frac{1}{2} \sin 2\theta = \sqrt{\frac{2GM}{x_T^2}} \, t$$

and $x = x_0 = x_T =$ turning point.

SUGGESTIONS FOR FURTHER READING

ARTHUR, W., and FENSTER, S. K., *Mechanics,* Chapter 3, New York: Holt, Rinehart and Winston, Inc., 1969.

BARGER, V., and OLSSON, M., *Classical Mechanics,* Chapter 1. New York: McGraw-Hill Book Co., 1973.

BECKER, R. A., *Introduction to Theoretical Mechanics,* Chapter 6, New York: McGraw-Hill Book Co., 1954.

DAVIS, A. DOUGLAS, *Classical Mechanics,* Chapter 2, New York: Academic Press, Inc., 1986.

FOWLES, G. R., *Analytical Mechanics,* Chapter 3, New York: Holt, Rinehart and Winston, Inc., 1962.

FRENCH, A. P., *Newtonian Mechanics,* Chapter 7, New York: W. W. Norton and Co., Inc., 1971.

HAUSER, W., *Introduction to the Principles of Mechanics,* Chapter 4. Reading, Mass.: Addison-Wesley Publishing Co., 1965.

KLEPPNER, D., and KOLENKOW, R. J., *An Introduction to Mechanics,* Chapter 4. New York: McGraw-Hill Book Co., 1973.

MARION, J. B., *Classical Dynamics,* 2nd ed., Chapter 2. New York: Academic Press, Inc., 1970.

ROSSBERG, K., *Analytical Mechanics,* Chapter 4. New York: John Wiley & Sons, Inc., 1983.

STEPHENSON, R. J., *Mechanics and Properties of Matter,* Chapter 2. New York: John Wiley & Sons, Inc., 1962.

SYMON, K. R., *Mechanics,* 3rd. ed., Chapter 2. Reading, Mass.: Addison-Wesley Publishing Co., 1971.

TAYLOR, E. F., *Introductory Mechanics,* Chapter 5. New York: John Wiley & Sons, Inc., 1963.

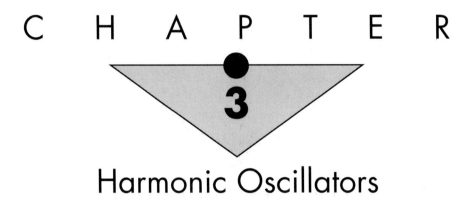

C H A P T E R

3

Harmonic Oscillators

3.1 INTRODUCTION

Consider a system in static or dynamic stable equilibrium. When such a system is displaced slightly from its equilibrium position, the resulting oscillatory motion is called *harmonic motion.* Such motions occur frequently in nature and are investigated, both from a practical as well as a theoretical point of view, in physics and engineering. A few examples of such motions are elastic springs, bending beams, pendula, vibrating strings, resonance of air cavities, and the motion of charges in certain electrical circuits and cavities.

To start, we shall study the motion of a *linear harmonic oscillator* (motion resulting from a small displacement of a system from its equilibrium) in one dimension. Unavoidable inclusion of friction in such motion leads to the investigation of a *damped harmonic oscillator.* To maitain oscillatory motion in the presence of friction, some external force must be applied. Such an oscillating system is called a *forced* or *driven oscillator.*

When the displacement of the system from equilibrium is *large,* the system is no longer linear. Such oscillating systems are called *nonlinear.* We divide our study into two parts. This chapter is mainly devoted to the study of a linear system including damped and forced harmonic oscillators. The study of nonlinear oscillations, the electrical equivalent of mechanical oscillators and multidimensional oscillators, will be investigated in Chapter 4. It may be pointed out that, in general, oscillations of systems occurring in nature are nonlinear. But their approximation to linear systems allows us to use strong analytical techniques developed for this purpose.

3.2 LINEAR AND NONLINEAR OSCILLATIONS

Consider a particle of mass m moving in an arbitrary conservative force field for which the potential energy $V(x)$ of the particle as a function of its displacement is represented by a heavy curve, as shown in Fig. 3.1. For a conservative force field, the total energy E of the particle is

$$E = K + V = \text{constant} \tag{3.1}$$

If $\dot{x}$ is the velocity of the particle,

$$E = \tfrac{1}{2} m\dot{x}^2 + V(x) \tag{3.2}$$

which when solved for $\dot{x}$ yields

$$\dot{x} = \frac{dx}{dt} = \pm \sqrt{\frac{2}{m}[E - V(x)]} \tag{3.3}$$

If $E = E_0$, as shown in Fig. 3.1, then $E_0 - V(x) = 0$ and $\dot{x} = 0$; that is, the particle stays at rest in a stable equilibrium at $x = x_0$. Let us consider the case in which the particle energy E_1 is slightly greater than E_0. For $x < x_1$ and $x > x_2$, $\dot{x}$ will be imaginary; hence the particle cannot exist in these regions. Thus a particle with energy E_1 is constrained to move in a potential well (or valley) between x_1 and x_2. The particle moving to the right is reflected back when it reaches x_2, and when traveling to the left it is reflected again at x_1. The points x_1 and x_2 are called *turning points,* and the velocity of the particle at these points is zero. These points are obtained by solving $E_1 - V(x) = 0$. In between these points, the velocity of m changes continuously depending on the value of $V(x)$. Hence a particle in a potential well moves back and forth and oscillates between x_1 and x_2 when its energy is greater than E_0.

The position $x(t)$ of a particle moving in potential well can be found by integrating Eq. (3.3); that is,

$$t_2 - t_1 = \sqrt{\frac{m}{2}} \int_{x_1}^{x_2} \frac{dx}{\sqrt{E - V(x)}} \tag{3.4}$$

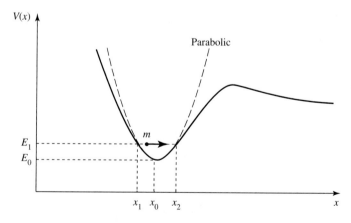

Figure 3.1   A particle of mass m and energy E is moving in an arbitrary potential energy function $V(x)$ shown by the solid heavy curve. The dotted curve is the parabolic potential approximation of the arbitrary potential.

while the time period T of one complete oscillation is given by

$$T = 2(t_2 - t_1) = \sqrt{2m} \int_{x_1}^{x_2} \frac{dx}{\sqrt{E - V(x)}} \qquad (3.5)$$

Equations (3.4) and (3.5) cannot be solved unless we know the form of the potential function $V(x)$. The motion of the particle can be limited to the region in the neighborhood of x_0, and for small displacements, such as these, it is possible to approximate the arbitrary potential function $V(x)$ by a parabolic potential shown by the dotted curve in Fig. 3.1. As we shall show later, this potential may be written as $V(x) = \frac{1}{2}k(x - x_0)^2$, where k is a constant, thereby enabling us to solve Eqs. (3.4) and (3.5).

Suppose a particle is oscillating about a point of stable equilibrium x_0, where the minimum potential is $V(x_0)$ at $x = x_0$. Let us expand the potential function $V(x)$ in a Taylor series about the point x_0.

$$V(x) = V(x_0) + \left(\frac{dV}{dx}\right)_{x=x_0}(x - x_0) + \frac{1}{2}\left(\frac{d^2V}{dx^2}\right)_{x=x_0}(x - x_0)^2$$

$$+ \frac{1}{6}\left(\frac{d^3V}{dx^3}\right)_{x=x_0}(x - x_0)^3 + \frac{1}{24}\left(\frac{d^4V}{dx^4}\right)_{x=x_0}(x - x_0)^4 + \cdots \qquad (3.6)$$

We limit our discussion to small displacements in symmetrical potentials. The term $X(x_0)$ is a constant term and can be dropped without affecting the results. Also, since x_0 is a point of minimum, for stable equilibrium in a symmetrical potential, the odd terms must be zero. [Note that if the expression resulting from the expansion of $F(x)$ were used the even terms would be zero.] Therefore,

$$\left(\frac{dV}{dx}\right)_{x=x_0} = 0, \qquad \left(\frac{d^3V}{dx^3}\right)_{x=x_0} = 0 \qquad (3.7a)$$

while

$$\left(\frac{d^2V}{dx^2}\right)_{x=x_0} > 0 \qquad (3.7b)$$

Define

$$(x - x_0) = x' \qquad (3.8)$$

$$\left(\frac{d^2V}{dx^2}\right)_{x=x_0} = k \qquad (3.9)$$

$$\frac{1}{6}\left(\frac{d^4V}{dx^4}\right)_{x=x_0} = +\epsilon \qquad (3.10)$$

Then the potential function may be written as

$$V(x') = \frac{1}{2}kx'^2 + \frac{1}{4}\epsilon x'^4 + \cdots \qquad (3.11)$$

Let us assume that the origin is located at the equilibrium point so that $x_0 = 0$ and $x' = x$, and by neglecting the higher-order terms in Eq. (3.11), we get

$$V(x) = \tfrac{1}{2}kx^2 + \tfrac{1}{4}\epsilon x^4 \tag{3.12}$$

Furthermore, since the motion of the particle is in a conservative force field, using the definition

$$F(x) = -\frac{dV}{dx}$$

and substituting for $V(x)$ from Eq. (3.12), we may write

$$F(x) = -kx - \epsilon x^3 \tag{3.13}$$

Linear Oscillations

In the first approximation, we can neglect all terms except the first in Eqs. (3.12) and (3.13) so that

$$V(x) = \tfrac{1}{2}kx^2 \tag{3.14}$$

$$F(x) = -kx \tag{3.15}$$

where

$$k = \left(\frac{d^2V}{dx^2}\right)_{x=x_0} = -\left(\frac{dF}{dx}\right)_{x=x_0} \tag{3.16}$$

Since $(d^2V/dx^2)_0$ is always positive, k will be positive also. Hence a force $F(x) = -kx$ is always directed toward the center and proportional to x. Such a force is called a *linear restoring force.* The potential corresponding to such a force is parabolic as given by Eq. (3.14) and shown by the dotted curves in Figs. 3.1 and 3.2 for different values of k. The corresponding linear forces are shown by the dotted lines.

Physical systems involving springs, pendula, and elastic deformation are described by Eqs. (3.14) and (3.15) and are said to obey *Hooke's law.* This is true only if the displacements are small and we remain within elastic limits, as shown in Fig. 3.2(a). Moreover, the results obtained are still approximate. We shall spend most of the time discussing linear oscillations resulting from approximate linear systems. k has been given several names, but is usually called the *spring constant* or *stiffness constant. k* is defined as the *force per unit length* with units of newtons per meter (N/m). $1/k$ is called the *compliance* of the spring.

Nonlinear Oscillations

If the displacement of the system from stable equilibrium is not small (or if we definitely want higher-order improvements in the linear approximation), we cannot drop the second term in Eqs. (3.12) and (3.13). Thus, according to Eq. (3.13), the force is no longer linear because of the presence of a x^3 term, while the potential is no longer parabolic because of the presence of

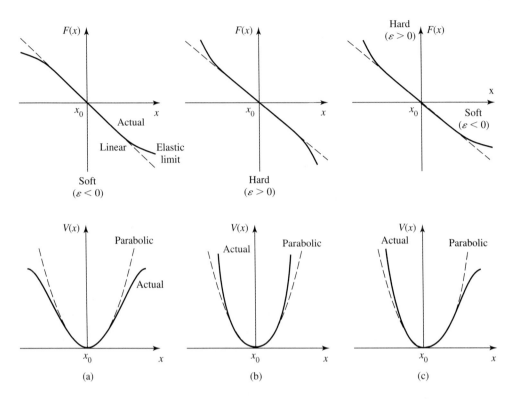

Figure 3.2   The plots of $F(x)$ versus x and $V(x)$ versus x for a variety of systems. The magnitude and sign of ϵ determines whether the system is hard or soft. For soft systems, $\epsilon < 0$, while for hard systems, $\epsilon > 0$.

a x^4 term. Different forms of forces and potentials are illustrated in Fig. 3.2 for systems with large displacements (hence no longer linear).

Let us further consider Eq. (3.13) for a nonlinear system; that is,

$$F(x) = -kx - \epsilon x^3 \tag{3.13}$$

We must remember that ϵ is a very small quantity as compared to k, but its magnitude and sign affect the linear term $-kx$, hence the resulting force $F(x)$. If $\epsilon < 0$, the magnitude of the force $F(x)$ will be less than the linear force kx alone and the system is said to be *soft*. On the other hand, if $\epsilon > 0$, the magnitude of the force $F(x)$ is greater than the linear force kx alone and the system is said to be *hard*. The forces and potentials of such systems are shown in Fig. 3.2.

3.3 LINEAR HARMONIC OSCILLATOR

Consider the prototype of a linear or simple harmonic oscillator shown in Fig. 3.3. It consists of a mass m tied to a spring having a force constant k. The spring-mass system oscillates in one dimension along the X-axis on a horizontal frictionless surface. The system obeys Hooke's law;

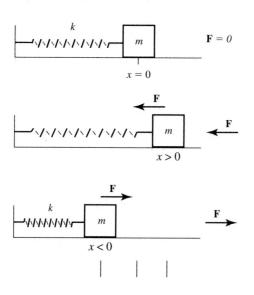

$$x = -A \qquad x = 0 \qquad x = +A$$
$$v = 0 \qquad v = v_{max} \qquad v = 0$$
$$a = +a_{max} \quad a = 0 \quad a = -a_{max}$$
$$F = +kA \qquad F = 0 \qquad F = -kA$$

Figure 3.3   Prototype of a linear harmonic oscillator showing the maximum and minimum values of x, v, a and F.

hence the system is linear. Measuring the displacement x from the equilibrium position, the potential energy $V(x)$ is

$$V(x) = \tfrac{1}{2}kx^2 \tag{3.17}$$

while the restoring force $F(x)$ is

$$F(x) = -kx \tag{3.18}$$

From Newton's second law, $F(x) = m(d^2x/dt^2)$; therefore,

$$m\,\frac{d^2x}{dt^2} = -kx \tag{3.19}$$

or

$$\frac{d^2x}{dt^2} + \omega_0^2 x = 0 \tag{3.20}$$

where

$$\omega_0 = \sqrt{\frac{k}{m}} \tag{3.21}$$

ω_0 is a constant and is called the *free natural angular frequency* (or *free oscillation frequency*) of the system.

Our aim is to solve Eq. (3.20) for $x(t)$. Before we do this, it must be pointed out that an equation of this form is frequently encountered both in physics and engineering; hence its solution must be thoroughly investigated. Equation (3.20) is a *second-order, linear, homogeneous differential equation.* The highest derivative that occurs in a differential equation is called its *order,* while a differential equation is *linear* if it does not contain terms higher than the first degree in the dependent variable [x is the dependent variable in Eq. (3.20)] and its derivative. Also, Eq. (3.20) is *homogeneous* because it does not contain terms other than the dependent variable and its derivatives. Thus the most general form of a differential equation of the *n*th order, linear and inhomogeneous, is

$$C_n \frac{d^n x}{dt^n} + C_{n-1} \frac{d^{n-1} x}{dt^{n-1}} + \cdots + C_1 \frac{dx}{dt} = b(t) \tag{3.22}$$

If $b(t) = 0$, the equation is homogeneous. The coefficients C_n, C_{n-1}, . . . , C_1 are constants that may or may not be independent of time, but we assume them to be independent of time.

We shall be dealing with second-order differential equations. We summarize next some properties of such equations, which will be helpful.

1. The general solution of any second-order differential equation depends on only two arbitrary constants. Suppose we choose C_1 and C_2 to be the arbitrary constants; then

$$x = x(t; C_1, C_2)$$

   C_1 and C_2 are arbitrary because any values of C_1 and C_2 will satisfy a second-order differential equation.
2. If $x_1(t)$ is any solution of a linear homogeneous differential equation, then $Cx_1(t)$ is also a solution, where C is an arbitrary constant.
3. If $x_1(t)$ and $x_2(t)$ are solutions of a linear homogeneous differential equation, then $x_1(t) + x_2(t)$ or any other linear combination $C_1 x_1(t) + C_2 x_2(t)$ is also a solution.

Let us now go back to Eq. (3.20) and try to find its solutions. To start, we may write it as

$$\ddot{x} + \omega_0^2 x = 0$$

Now, multiplying both sides by $2\dot{x}$,

$$2\dot{x}\ddot{x} = -2\omega_0^2 x\dot{x}$$

and integrating, we get

$$\dot{x}^2 = -\omega_0^2 x^2 + C$$

where C is a constant. When $x = A$, $\dot{x} = 0$; hence $C = \omega_0^2 A^2$. Thus

$$\dot{x}^2 = \omega_0^2 (A^2 - x^2) \tag{3.23}$$

which, after separating the variables, may be written as

$$\int \frac{dx}{\sqrt{A^2 - x^2}} = \omega_0 \int dt$$

Integrating this equation, we get

$$\sin^{-1}\left(\frac{x}{A}\right) = \omega_0 t + \phi$$

where ϕ is a constant, called the *initial phase* or *phase constant*. We may write this equation as

$$x = A \sin(\omega_0 t + \phi) \tag{3.24}$$

Thus the solution of Eq. (3.20), which is a second-order differential equation, is given by Eq. (3.24) and contains two arbitrary constants A and ϕ to be determined from the initial conditions. Equation (3.24) is a solution of a *linear oscillator* or *harmonic oscillator*. The graph of x versus t is shown in Fig. 3.4. x is called the *displacement;* the maximum displacement is called the *amplitude* of the oscillator and is equal to A. The quantity ω_0 is called the *angular frequency* and is given by Eq. (3.21). Also, $\omega_0 = 2\pi\nu_0$, where ν_0 is called the *frequency* of the oscillator. The *time period* T_0 of the oscillator is the time required to complete one oscillation. Thus, in time $t = T_0$ in Eq. (3.24), $\omega_0 t$ increases by 2π; that is,

$$\omega_0 T_0 = 2\pi$$

or
$$T_0 = \frac{2\pi}{\omega_0} = 2\pi\sqrt{\frac{m}{k}} \tag{3.25a}$$

or
$$\nu_0 = \frac{1}{T_0} = \frac{1}{2\pi}\sqrt{\frac{k}{m}} \tag{3.25b}$$

The expressions for velocity and acceleration may be obtained by differentiating Eq. (3.24); that is,

$$v = \dot{x} = \omega_0 A \cos(\omega_0 t + \phi) \tag{3.26a}$$

$$a = \ddot{x} = -\omega_0^2 \sin(\omega_0 t + \phi) = -\omega_0^2 x \tag{3.26b}$$

The plots of x, v, and a versus t are shown in Fig. 3.4.

The solution given by Eq. (3.24) may be written in a different form as follows:

$$x = A \sin(\omega_0 t + \phi)$$

$$= A \sin \omega_0 t \cos \phi + A \cos \omega_0 t \sin \phi$$

Substituting

$$A \cos \phi = B \tag{3.27a}$$

and
$$A \sin \phi = C \tag{3.27b}$$

we get

$$x = B \sin \omega_0 t + C \cos \omega_0 t \tag{3.28}$$

Figure 3.4

Below is the graph of the displacements x and x0 versus time t for two oscillators having the same frequency $\omega 0$. The phase difference between the two functions is $\phi = \pi/2$ or 90 degrees, as illustrated.

$$N := 100 \qquad n := 0..N \qquad t_n := \frac{n}{10}$$

$$A := .05 \qquad \omega 0 := 1.57 \qquad \phi := \frac{\pi}{2}$$

$$T := \frac{2 \cdot \pi}{\omega 0} \qquad T = 4.002 \quad v := \frac{1}{T} \qquad v = 0.25$$

$$x_n := A \cdot \sin\left(\omega 0 \cdot t_n + \phi\right) \qquad \max(x) = 0.05$$

The maximum values of the displacements (amplitudes) of x and x0, velocity v, and acceleration a are as shown. Also, the phase angles may be calculated as

$$v_n := \omega 0 \cdot A \cdot \cos\left(\omega 0 \cdot t_n + \phi\right) \qquad \max(v) = 0.078$$

$$a_n := -A \cdot \omega 0^2 \cdot \sin\left(\omega 0 \cdot t_n + \phi\right) \qquad \max(a) = 0.123$$

$$x0_n := A \cdot \sin\left(\omega 0 \cdot t_n\right) \qquad \max(x0) = 0.05$$

$$\left(x0_0 - x_0\right) \cdot 20 \cdot \omega 0 = -1.57 \cdot \text{rad}$$
$$\left(x_0 - x0_0\right) \cdot 20 \cdot \omega 0 = 89.954 \cdot \text{deg}$$

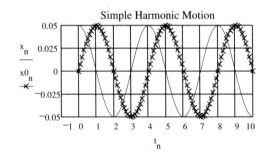

Below is the graph of displacement x, velocity v, and acceleration a versus time t.

(a) If we change the values of ϕ to 0, $\pi/4$, π, $3\pi/2$, and 2π, how do these graphs change?

(b) What are the phase relations between x and v and between x and a?

(c) If the phase of x changes, how will it affect the phase of v and a?

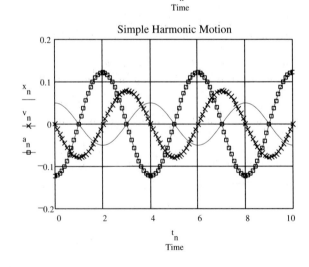

which is another form of the solution for a second-order differential equation. Squaring and adding Eqs. (3.27a) and (3.27b), we get

$$A = \sqrt{B^2 + C^2} \tag{3.29a}$$

and dividing Eq. (3.27b) by (3.27a), we get

$$\phi = \tan^{-1}\left(\frac{C}{B}\right) \tag{3.29b}$$

Equations (3.27) and (3.29) give the relations between the constants A, ϕ, and B, C. We may differentiate Eqs. (3.28) to get expressions for velocity and acceleration.

Equation (3.20) or any other second-order differential equation with constant coefficients may be solved by a trial solution of an exponential form, as explained next. Let the trial solution be

$$x = e^{\lambda t} \tag{3.30}$$

where λ is a constant to be determined. Substituting Eq. (3.30) into (3.20),

$$\lambda^2 e^{\lambda t} + \omega_0^2 e^{\lambda t} = 0$$

$$\lambda^2 + \omega_0^2 = 0 \tag{3.31}$$

This is called the *characteristic, initial,* or *auxiliary equation.* Thus

$$\lambda = \pm i\omega_0 = \pm i\sqrt{\frac{k}{m}} \tag{3.32}$$

where $i = \sqrt{-1}$. Thus, with two roots $\lambda_1 = +i\omega_0$ and $\lambda_2 = -i\omega_0$, the general solution is, using A_+ and A_- as two constants,

$$x = A_+ e^{+i\omega_0 t} + A_- e^{-i\omega_0 t} \tag{3.33}$$

All three solutions, Eqs. (3.24), (3.28), and (3.33), are equivalent, and any one can be derived from the other two. Each contains two constants. Equation (3.33) may be reduced to the other two by using the *Euler formulas*

$$e^{\pm i\theta} = \cos\theta \pm i\sin\theta \tag{3.34}$$

It may be well to remember that in a solution given by Eq. (3.33), the constants A_+ and A_- are complex quantities and sometime inconvenient to use. Still another way of writing this general solution is (with A and ϕ as two constants)

$$x = Ae^{i(\omega_0 t + \phi)}$$

$$= A\cos(\omega_0 t + \phi) + iA\sin(\omega_0 t + \phi) \tag{3.35}$$

where both the real and the imaginary parts of this equation are solutions of a general differential equation.

Energy of the Simple Harmonic Oscillator

For a simple harmonic oscillator, the displacement is

$$x = A \sin(\omega_0 t + \phi)$$

while the velocity is

$$\dot{x} = \frac{dx}{dt} = \omega_0 A \cos(\omega_0 t + \phi)$$

and the maximum value of the velocity v_0 is

$$v_0 = \omega_0 A = \sqrt{\frac{k}{m}} A$$

Hence the kinetic energy K of the oscillator is

$$K = \tfrac{1}{2}m\dot{x}^2 = \tfrac{1}{2}m\omega_0^2 A^2 \cos^2(\omega_0 t + \phi)$$

$$= K_0 \cos^2(\omega_0 t + \phi) \tag{3.36}$$

where K_0 is the maximum kinetic energy given by

$$K_0 = \tfrac{1}{2}m\omega_0^2 A^2 = \tfrac{1}{2}kA^2 \tag{3.37}$$

The potential energy of the system is equal to the work done by the applied force $F_a = -F = -(-kx) = kx$ in displacing the system from $x = 0$ to $x = x$. Thus

$$V(x) = W = \int_0^x F_a\, dx = \int_0^x kx\, dx = \tfrac{1}{2}kx^2 \tag{3.38}$$

Substituting for x:

$$V(x) = \tfrac{1}{2}kA^2 \sin^2(\omega_0 t + \phi)$$

$$V(x) = V_0 \sin^2(\omega_0 t + \phi) \tag{3.39}$$

where V_0 is the maximum potential energy when $x = A$; that is,

$$V_0 = \tfrac{1}{2}kA^2 \tag{3.40}$$

Thus the total energy E, which is always constant whenever there is a conservative force field, is

$$E = K + V = \tfrac{1}{2}m\dot{x}^2 + \tfrac{1}{2}kx^2 \tag{3.41}$$

This equation can be solved for $x(t)$ and provides more information about the problem under consideration, as we shall show next. From Eq. (3.41),

$$\dot{x} = \pm\left(\frac{2E}{m} - \frac{k}{m}x^2\right)^{1/2} \tag{3.42a}$$

or
$$\pm \int \frac{dx}{\sqrt{(2E/k) - x^2}} = \sqrt{\frac{k}{m}} \int dt \qquad \text{(3.42b)}$$

we get the solution for x to be

$$x = A \sin(\omega_0 t + \phi_1) \qquad \text{(3.43a)}$$

or
$$x = A \cos(\omega_0 t + \phi_2) \qquad \text{(3.43b)}$$

where ϕ_1 and ϕ_2 are constants, while the amplitude A is given by

$$A = \sqrt{\frac{2E}{k}} \qquad \text{(3.44)}$$

This relation tells us that x can vary between $+A$ and $-A$, that is, between

$$+\sqrt{\frac{2E}{k}} \qquad \text{and} \qquad -\sqrt{\frac{2E}{k}}$$

This has to be true because only then will x be real as given by Eqs. (3.42). The value of x then must lie between two limits that are determined by the energy E and the spring constant k.

To find the average values of V and K over one complete time period, we use the following general expression for the average value of quantity $f(t)$:

$$\langle f \rangle = \frac{1}{t_2 - t_1} \int_{t_1}^{t_2} f(t)\, dt \qquad \text{(3.45)}$$

That is,

$$\langle V \rangle = \frac{\int_0^T V\, dt}{\int_0^T dt} = \frac{\int_0^T V_0 \sin^2(\omega_0 t + \phi)\, dt}{T}$$

$$= \tfrac{1}{2} V_0 = \tfrac{1}{4} k A^2 \qquad \text{(3.46)}$$

and similarly $\qquad \langle K \rangle = \tfrac{1}{2} K_0 = \tfrac{1}{4} k A^2 \qquad \text{(3.47)}$

That is, $\qquad \langle V \rangle = \langle K \rangle = \tfrac{1}{2} E \qquad \text{(3.48)}$

If, instead of time averages, we calculate space averages over one complete time period, we get (see Problem 3.1)

$$\langle V \rangle_{\text{space}} = \tfrac{1}{6} k A^2, \qquad \langle K \rangle_{\text{space}} = \tfrac{1}{3} k A^2 \qquad \text{(3.49)}$$

and $\qquad \langle E \rangle_{\text{space}} = \langle V \rangle_{\text{space}} + \langle K \rangle_{\text{space}} = \langle E \rangle_{\text{time}} \qquad \text{(3.50)}$

3.4 DAMPED HARMONIC OSCILLATOR

Theoretically, a linear or a simple harmonic oscillator once set into motion will continue oscillating forever. Such oscillations are called *free oscillations*. In practice, however, in any physical situation there are dissipative or damping forces, and the oscillating system will lose energy with time. Thus the oscillating system is damped and eventually comes to rest. The differential equation for a linear oscillator given by Eq. (3.20) must be modified to include the effect of damping.

Once again we consider a mass m tied to a spring, as shown in Fig. 3.5, as a prototype and restrict its motion to one dimension. As the mass moves in a fluid, air or liquid, the frictional force is the viscous force that produces the damping. As long as the speed of the mass is small so as not to cause turbulence, the frictional force or damping force F_d may be assumed to be proportional to the velocity. That is,

$$F_d = -bv = -b\dot{x} \tag{3.51}$$

where b must be a positive constant. The net force F_{net} due to forces acting on mass m as shown in Fig. 3.5 is

$$F_{\text{net}} = F + F_d = -kx - b\dot{x} \tag{3.52}$$

Using Newton's second law and substituting $F_{\text{net}} = m\ddot{x}$ in Eq. (3.52), we get

$$m\ddot{x} + b\dot{x} + kx = 0 \tag{3.53}$$

which is a second-order differential equation for a damped harmonic oscillator. To solve this equation, we divide both sides by m and substitute

$$\gamma = \frac{b}{2m} \tag{3.54a}$$

and

$$\omega_0^2 = \frac{k}{m} \tag{3.54b}$$

to obtain

$$\ddot{x} + 2\gamma\dot{x} + \omega_0^2 = 0 \tag{3.55}$$

As before, let us try an exponential solution of the form

$$x = e^{\lambda t}, \quad \dot{x} = \lambda e^{\lambda t}, \quad \ddot{x} = \lambda^2 e^{\lambda t}$$

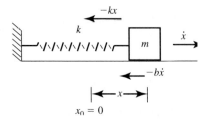

Figure 3.5   Forces acting on a prototype of a damped harmonic oscillator.

and substituting in Eq.(3.55), we get

$$e^{\lambda t}(\lambda^2 + 2\gamma\lambda + \omega_0^2) = 0$$

Since $e^{\lambda t} \neq 0$, we must have

$$(\lambda^2 + 2\gamma\lambda + \omega_0^2) = 0 \tag{3.56}$$

This auxiliary equation has the roots

$$\lambda_1 = -\gamma + \sqrt{\gamma^2 + \omega_0^2} \tag{3.57a}$$

and

$$\lambda_2 = -\gamma - \sqrt{\gamma^2 - \omega_0^2} \tag{3.57b}$$

Thus the general solution of Eq. (3.55) is, with A_1 and A_2 as arbitrary constants,

$$x(t) = A_1 e^{\lambda_1 t} + A_2 e^{\lambda_2 t}$$

or

$$x(t) = e^{-\gamma t}(A_1 e^{+\sqrt{\gamma^2 - \omega_0^2}\, t} + A_2 e^{-\sqrt{\gamma^2 - \omega_0^2}\, t}) \tag{3.58}$$

The following three cases of this solution are of special interest and will be discussed in some detail.

Case (a)	Underdamped (oscillatory)	$\omega_0^2 > \gamma^2$	λ_1 and λ_2 are imaginary roots
Case (b)	Critically damped (not oscillatory)	$\omega_0^2 = \gamma^2$	λ_1 and λ_2 are real and equal roots
Case (c)	Overdamped (not oscillatory)	$\omega_0^2 < \gamma^2$	λ_1 and λ_2 are real roots

Case (a)   Underdamped Oscillations, $\omega_0^2 > \gamma^2$:   For this case, it is convenient to make a substitution:

$$\omega_1 = \sqrt{\omega_0^2 - \gamma^2} = \sqrt{\frac{k}{m} - \frac{b^2}{4m^2}} \tag{3.59}$$

Thus the exponentials inside the parentheses in Eq. (3.58) are imaginary, and we may write this equation as

$$x(t) = e^{-\gamma t}(A_1 e^{+i\omega_1 t} + A_2 e^{-i\omega_1 t}) \tag{3.60}$$

which is a solution of an underdamped oscillator. Using the relation $e^{\pm i\theta} = \cos\theta \pm i\sin\theta$, we may write Eq. (3.60) as

$$x(t) = e^{-\gamma t}[i(A_1 - A_2)\sin\omega_1 t + (A_1 + A_2)\cos\omega_1 t]$$

Substituting $i(A_1 - A_2) = B$ and $(A_1 + A_2) = C$, we obtain an alternative solution:

$$x(t) = e^{-\gamma t}[B\sin\omega_1 t + C\cos\omega_1 t] \tag{3.61}$$

This may still be written in a slightly different form by making the following substitutions in Eq. (3.61).

$$A = \sqrt{B^2 + C^2} \quad \text{and} \quad \tan \phi = -\frac{C}{B}$$

Thus we obtain

$$x(t) = Ae^{-\gamma t} \cos(\omega_1 t + \phi) \tag{3.62}$$

Of the three solutions given by Eqs. (3.60), (3.61), and (3.62), we shall concentrate on Eq. (3.62). It may be pointed out that the constants A_1 and A_2 in Eq. (3.60) are complex quantities, while B and C in Eq. (3.61) and A and ϕ in Eq. (3.62) are all real quantities.

The solution given by Eq. (3.62) indicates that for a damped oscillator the motion is oscillatory, but the amplitude of the oscillations decays exponentially, as shown in Fig. 3.6. The *natural angular frequency,* ω_1, or the frequency of the damped oscillator is always less than the free oscillation frequency ω_0. The natural frequency ω_1 is not a frequency in the true sense of the word because the oscillator never passes through the same point twice with the same velocity; that is, the motion is not periodic. But if γ is very small, then $\omega_1 \simeq \omega_0$ (as shown later), and we can call ω_1 the "frequency." If γ is small, we can expand Eq. (3.59) (using the binomial expansion) as

$$\omega_1 = (\omega_0^2 - \gamma^2)^{1/2} = \omega_0 \left(1 - \frac{\gamma^2}{\omega_0^2}\right)^{1/2}$$

$$= \omega_0 \left(1 - \frac{\gamma^2}{2\omega_0^2} + \cdots\right) \tag{3.63a}$$

$$\simeq \omega_0 - \frac{\gamma^2}{2\omega_0^2} \tag{3.63b}$$

If $\gamma \ll \omega_0$,

$$\omega_1 \simeq \omega_0 \tag{3.63c}$$

According to Eq. (3.62) the case for $\phi = 0$ is shown in Fig. 3.6. Equation (3.62) states (and this is demonstrated in Fig. 3.6) that the maximum amplitude of the oscillations decreases exponentially with time because of the factor $e^{-\gamma t}$ and lies between the two curves given by

$$A_e(t) = \pm Ae^{-\gamma t} \tag{3.64}$$

where $A_e(t)$ is the envelope that limits the displacement of the oscillations. For comparison the x_0 graph represents the oscillations for a free oscillator, that is, for $\gamma = 0$. The graphs with different dampings are shown in Fig. 3.6. Plots x_4 and x_5 represent the envelope of the damped motion ($\gamma \neq 0$) that (due to the presence of the cosine term) touches the envelope at $\cos \omega_1 t = \pm 1$, that is, at times $\omega_1 t_n = n\pi$ or $t_n = n\pi/\omega_1$, where n is an integer. The period of the damped oscillation is $T_1 = 2\pi/\omega_1$. Since $\omega_1 < \omega_0$, that is, the damped frequency is smaller than the free

> Figure 3.6 _____

For undamped and underdamped oscillators, $\omega0^2 > \gamma^2$

$$N := 100 \qquad n := 0..N \qquad t_n := \frac{n}{5} \qquad A := 10 \qquad \omega0 := 1$$

Below is a graph of x versus t for the following degrees of freedom:

$$\gamma0 := 0 \qquad \omega01 := \sqrt{\omega0^2 - \gamma0^2} \qquad \omega01 = 1$$

x0   Undamped, $\gamma = 0$

$$\gamma1 := .05 \qquad \omega1 := \sqrt{\omega0^2 - \gamma1^2} \qquad \omega1 = 0.999$$

x1   Lightly damped, γ not 0

$$\gamma2 := .2 \qquad \omega2 := \sqrt{\omega0^2 - \gamma2^2} \qquad \omega2 = 0.98$$

x2   Moderately damped, γ not 0

$$\gamma3 := .7 \qquad \omega3 := \sqrt{\omega0^2 - \gamma3^2} \qquad \omega3 = 0.714$$

x3   Heavily damped, γ not 0

x4 and x5 are two envelopes for plot x1 showing that the amplitude of the oscillations decays exponentially.

$$x0_n := A \cdot \cos(\omega01 \cdot t_n) \qquad x1_n := A \cdot e^{-\gamma1 \cdot t_n} \cdot \cos(\omega1 \cdot t_n)$$

$$x2_n := A \cdot e^{-\gamma2 \cdot t_n} \cdot \cos(\omega2 \cdot t_n) \qquad x3_n := A \cdot e^{-\gamma3 \cdot t_n} \cdot \cos(\omega3 \cdot t_n)$$

$$x4_n := A \cdot e^{-\gamma1 \cdot t_n} \qquad\qquad x5_n := -A \cdot e^{-\gamma1 \cdot t_n}$$

(a) What equations (other than cosine and sine) can be used to obtain the same graph?

(b) How do the graphs of the four functions differ in their time periods, amplitude, and frequencies? Explain.

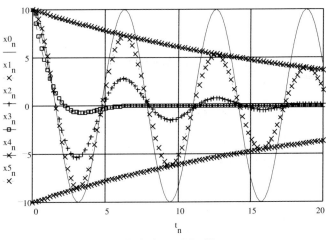

Underdamped Oscillators

frequency, the period T_1 of the damped oscillation is longer than the free period T_0. This is clear from the comparison of the ω values of the plots in Fig. 3.6. It is worth noting that the points of the curve that touch the envelope are $T_1/2 \, (=\pi/\omega_1)$ apart, but the maxima and the minima of the curve, even though separated by $T_1/2$, do not coincide with the points of maxima and minima of the undamped motion curve.

Figure 3.6 shows the plots of $x(t)$ versus t for different degrees of damping. The ratio γ/ω_1 determines the essential feature of these plots. If $\gamma/\omega_1 \ll 1$, the amplitude envelope $A_e(t)$ given by Eq. (3.64) changes very slowly with time, while the cosine term in $x(t)$ makes several zero crossings. Such a system is said to be *lightly damped*. On the other hand, if $\gamma/\omega_1 \gg 1$, the system is said to be *heavily damped* because $A_e(t)$ will decrease very rapidly and goes to zero, while the cosine term makes only a few zero crossing. In either case, the ratio of the two successive maxima is given by

$$\frac{Ae^{-\gamma t_1}}{Ae^{-\gamma t_2}} = \frac{Ae^{-\gamma t_m}}{Ae^{-\gamma(t_m + T_1)}} = e^{\gamma T_1} \tag{3.65}$$

where $t_1 = t_m$ is the time when the first maximum occurs and $t_2 = t_m + T_1$ is the time when the next maximum occurs, T_1 being the time period of the damped oscillation. The quantity exp (γT_1) is called the *decrement* of motion, while its logarithm, γT_1, is called the *logarithmic decrement*, δ; that is,

$$\delta = \ln e^{\gamma T_1} = \gamma T_1 = \left(\frac{b}{2m}\right)\left(\frac{2\pi}{\omega_1}\right) = \frac{b}{m}\frac{\pi}{\omega_1} \tag{3.66}$$

Case (b)   Critically Damped, $\omega_0^2 = \gamma^2$:   For this case, the two roots λ_1 and λ_2 given by Eqs. (3.57) are equal, that is,

$$\lambda_1 = \lambda_2 = -\gamma$$

and the general solution given by Eq. (3.58) takes the form

$$x(t) = (A_1 + A_2)e^{-\gamma t} = B_1 e^{-\gamma t}$$

where $(A_1 + A_2) = B_1 = $ constant. This is not a general solution because it contains only one constant. We can show that in such cases, if $e^{-\gamma t}$ is a solution,

$$x = te^{-\gamma t} \tag{3.67}$$

is also a solution. Substituting in the differential equation

$$\ddot{x} + 2\gamma\dot{x} + \omega_0^2 x = 0$$

we get

$$(\omega_0^2 - \gamma^2)e^{-\gamma t} = 0$$

Since $\omega_0 = \gamma$, the equation is satisfied, and $te^{-\gamma t}$ is also a solution. Thus, for a critically damped case, the general solution is a linear combination of $e^{-\gamma t}$ and $te^{-\gamma t}$; that is,

$$x(t) = (B_1 + B_2 t)e^{-\gamma t} \tag{3.68}$$

where B_1 and B_2 are constants to be determined by the initial conditions.

Figure 3.7 represents three cases of interest (for a critically damped oscillator) resulting from the solution given by Eq. (3.68). If we differentiate x with respect to t and equate to zero, we get the positions of maxima in the plot of x versus t. Thus $dx/dt = 0$ gives

$$t = -\frac{B_1}{B_2} + \frac{2m}{b} \tag{3.69}$$

If

$$\frac{B_1}{B_2} = \frac{2m}{b} \quad \text{and} \quad \frac{B_1}{B_2} < \frac{2m}{b}$$

the curves for x versus t have maxima at $t = 0$ and $t > 0$, as in Fig. 3.7.

If

$$\frac{B_1}{B_2} > \frac{2m}{b}$$

the curve does not have a maximum for $t > 0$, as demonstrated in Fig. 3.7.

Critical damping plays a very important role in the design of such instruments as galvanometers, hydraulic springs, and pointer reading meters. It is desired that the system attain an equilibrium position rapidly and smoothly in the presence of frictional damping.

Case (c)   Overdamped, $\omega_0^2 < \gamma^2$:   If the damping increases such that $\gamma^2 > \omega_0^2$, then the two roots λ_1 and λ_2 are real. If we represent

$$(\gamma^2 - \omega_0^2)^{1/2} = \omega_2$$

the general solution given by Eq. (3.58) takes the form

$$x(t) = e^{-\gamma t}[A_1 e^{\omega_2 t} + A_2 e^{-\omega_2 t}] \tag{3.70}$$

Note that ω_2 is no longer a frequency because the motion is no longer oscillatory. The exponents are real, and both terms on the right decay exponentially, one faster than the other. As shown in Fig. 3.8, for the case $x(0) \neq 0$, $\dot{x}(0) \neq 0$, the displacement goes to zero asymptotically, but not as rapidly as in the case of a critically damped system. For the case when $\dot{x}(0) > 0$ or < 0, Fig. 3.8, shows how $x(t)$ varies with time. For $\dot{x}(0) > 0$, $x(t)$ reaches maximum for $t > 0$. For $\dot{x}(0) < 0$, but small, $x(t)$ has no maximum for $t > 0$. For $\dot{x}(0) \ll 0$, but sufficiently large, $x(t)$ has a maximum for $t > 0$, as shown.

Figure 3.7 _____

For a critically damped oscillator (not oscillatory) $\omega 0^2 = \gamma^2$

Plots are for the same values of B1 and B2 but
different values of γ. For the present situation,
B2/B1 = 0.5, while γ = b/2m. The graphs for
three different values of γ and b are below.

$N := 50$ $n := 0 .. N$ $t_n := \dfrac{n}{2}$

$B1 := 6$ $B2 := 3$

(a) What is the effect of increasing the
value of γ. Explain

$\gamma 1 := .1$ $\gamma 2 := .34$ $\gamma 3 := .7$

$x1_n := \left(B1 + B2 \cdot t_n \right) \cdot e^{-\gamma 1 \cdot t_n}$

(b) What causes the changes in the values
of B1 and B2? Explain

$x2_n := \left(B1 + B2 \cdot t_n \right) \cdot e^{-\gamma 2 \cdot t_n}$

(c) How will a change in the value of b
and m change the values of B and γ?

$x3_n := \left(B1 + B2 \cdot t_n \right) \cdot e^{-\gamma 3 \cdot t_n}$

(d) What is the significance of the following
maximum values in the three cases?

min(x1) = 6   max(x1) = 13.48

min(x2) = 0.016   max(x2) = 6.406

min(x3) = $2.034 \cdot 10^{-6}$ max(x3) = 6

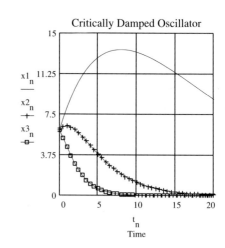

(e) How does the graph of a critically damped oscillator differ from that of an undamped
oscillator in terms of frequency of oscillations and amplitude?

Figure 3.8 _____

For an overdamped oscillator (not oscillatory)

Below is the graph of x versus t for three different initial conditions.

(a) How do different values of the coefficients B, C, and D affect different plots?

(b) What causes the change in the value of ω and how does it affect the graph?

(c) Explain why the two values of x1 and x3 have opposite signs.

$$\omega0^2 < \gamma^2 \qquad \text{At} \quad t = 0 \quad x \neq 0 \quad v \neq 0$$

$$N := 50 \qquad n := 0..N \qquad t_n := \frac{n}{10}$$

$$b := 6 \qquad k := 4 \qquad M := 1$$

$$\gamma := \frac{b}{2 \cdot M} \qquad \gamma = 3$$

$$\omega0 := \sqrt{\frac{k}{M}} \qquad \omega0 = 2$$

$$\omega := \left(\gamma^2 - \omega0^2\right)^{\frac{1}{2}} \qquad \omega = 2.236 \qquad (\omega < \gamma)$$

$$B1 := 10 \quad C1 := 1 \quad D1 := -3$$

$$B2 := -8 \quad C2 := 1 \quad D2 := 5$$

$$x1_n := \left(B1 \cdot e^{\omega \cdot t_n} + B2 \cdot e^{-\omega \cdot t_n}\right) \cdot e^{-\gamma t_n}$$

$$x2_n := \left(C1 \cdot e^{\omega \cdot t_n} + C2 \cdot e^{-\omega \cdot t_n}\right) \cdot e^{-\gamma t_n}$$

$$x3_n := \left(D1 \cdot e^{\omega \cdot t_n} + D2 \cdot e^{-\omega \cdot t_n}\right) \cdot e^{-\gamma t_n}$$

$$\max(x1) = 6.382 \qquad \max(x3) = 2$$

$$\min(x1) = 0.219 \qquad \min(x3) = -1.683$$

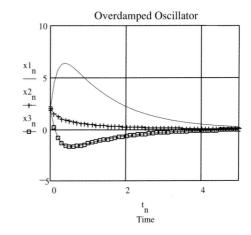

Overdamped Oscillator

Example 3.1 _____

Consider a damped harmonic oscillator and graph its motion using the following data and the initial conditions, where m = mass, k = spring constant, and b = damping constant

$$m := 0.05 \qquad k := 5.0 \qquad x0 := .01 \qquad v0 := 0 \qquad N := 20 \qquad t := 0..N$$

$$(a) \quad b1 := 0.1 \qquad (b) \; b2 := 1 \qquad (c) \quad b3 := 5$$

Solution

Approach using Eq. (3.63):

(a) Underdamped oscillator with b = 0.1

$$b1 := 0.1 \qquad \gamma1 := \frac{b1}{2 \cdot m} \qquad \gamma1 = 1 \qquad \omega0 := \sqrt{\frac{k}{m}} \qquad \omega0 = 10$$

$$\omega1 := \sqrt{\omega0^2 - \gamma1^2} \qquad \lambda11 := -\gamma1 + \sqrt{\gamma1^2 - \omega0^2} \qquad \lambda12 := -\gamma1 - \sqrt{\gamma1^2 - \omega0^2}$$

$$\omega1 = 9.95 \qquad \lambda11 = -1 + 9.95i \qquad \lambda12 = -1 - 9.95i$$

A and ϕ are determined from initial conditions. See the alternative approach below.

$$A := 0.1 \qquad \phi := -0.1 \qquad x11_t := A \cdot e^{-\gamma1 \cdot \frac{t}{3}} \cdot \cos\left(\omega1 \cdot \frac{t}{3} + \phi\right)$$

(b) Critically damped oscillator with b = 1.0

$$b2 := 1 \qquad \gamma2 := \frac{b2}{2 \cdot m} \qquad \gamma2 = 10 \qquad \omega0 := \sqrt{\frac{k}{m}} \qquad \omega0 = 10$$

$$\omega2 := \sqrt{\omega0^2 - \gamma2^2} \qquad \lambda21 := -\gamma2 + \sqrt{\gamma2^2 - \omega0^2} \qquad \lambda22 := -\gamma2 - \sqrt{\gamma2^2 - \omega0^2}$$

$$\omega2 = 0 \qquad \lambda21 = -10 \qquad \lambda22 = -10$$

$$A := 0.1 \qquad \phi := -0.1$$

$$x22_t := A \cdot e^{-\gamma2 \cdot \frac{t}{3}} \cdot \cos\left(\omega2 \cdot \frac{t}{3} + \phi\right)$$

(c) Overdamped oscillator with b = 5.0

$$b3 := 5 \qquad \gamma3 := \frac{b3}{2 \cdot m} \qquad \gamma3 = 50 \qquad \omega0 := \sqrt{\frac{k}{m}} \qquad \omega0 = 10$$

$$\omega3 := \sqrt{\omega0^2 - \gamma3^2} \qquad \lambda31 := -\gamma3 + \sqrt{\gamma3^2 - \omega0^2} \qquad \lambda32 := -\gamma3 - \sqrt{\gamma3^2 - \omega0^2}$$

$$\omega3 = 48.99i \qquad\qquad \lambda31 = -1.01 \qquad\qquad \lambda32 = -98.99$$

$$A := 0.1 \qquad\qquad \phi := -0.1 \cdot rad \qquad\qquad x33_t := A \cdot e^{-\gamma3 \cdot \frac{t}{3}} \cdot \cos\left(\omega3 \cdot \frac{t}{3} + \phi\right)$$

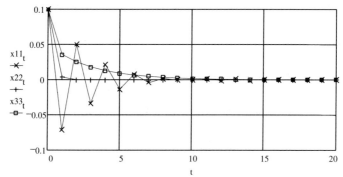

Under, critically, over damped

$$\omega1 = 9.95 \qquad\qquad \omega2 = 0 \qquad\qquad \omega3 = 48.99i$$

$$\max(x11) = 0.1 \qquad\qquad \max(x22) = 0.1 \qquad\qquad \max(x33) = 0.1 + 0.004i$$

$$\min(x11) = -0.071 \qquad\qquad \min(x22) = 0 \qquad\qquad \min(x33) = 5.915 \cdot 10^{-5}$$

Alternate approach

Using Eqs. (3.60), (3.61), and (3.62):

(a) Underdamped oscillator with $\gamma = 1.0$

Guess $A := .1 \cdot m$ $\phi := .1 \cdot rad$

Given $.1 = A \cdot \cos(\phi)$ $0 = -A \cdot \cos(\phi) - 10 \cdot A \cdot \sin(\phi)$ $\phi \leq 1 \cdot rad$

Evaluate the constants
by using the initial
conditions and Eq. (3.61). $Find(A, \phi) = \begin{pmatrix} 0.1 \\ -0.1 \end{pmatrix}$ $A := .1$ $A = 0.1$ $\phi := -.1$ $\phi = -0.1$

$$x1_t := A \cdot e^{-\gamma1 \cdot \frac{t}{3}} \cdot \cos\left(\omega1 \cdot \frac{t}{3} + \phi\right)$$

(b) Critically damped oscillator with $\gamma = 10$

Guess $\qquad\qquad$ A1 := .1 $\qquad\qquad$ A2 := .1

Given $\qquad\qquad$ A1 + A2 = .1 $\qquad\qquad$ - A1 - 99·A2 = 0

$\text{Find}(A1, A2) = \begin{pmatrix} 0.101 \\ -0.001 \end{pmatrix}$ $\qquad$ A1 := 0.10 $\qquad$ A2 := 1

$$x2_t := \left(A1 + A2 \cdot \frac{t}{3}\right) \cdot e^{-\gamma 2 \cdot \frac{t}{3}}$$

(c) Overdamped Oscillator with $\gamma = 50$

Guess $\qquad$ B1 := .1 $\qquad\qquad$ B2 := .1

Given $\qquad$ B1 = 0.10 $\qquad\qquad$ B2 - 10·B1 = 0

$\text{Find}(B1, B2) = \begin{pmatrix} 0.1 \\ 1 \end{pmatrix}$ $\quad$ B1 := 0.1 $\quad$ B2 := -0.001

$$x3_t := B1 \cdot e^{\lambda 31 \cdot \frac{t}{3}} + B2 \cdot e^{\lambda 32 \cdot \frac{t}{3}}$$

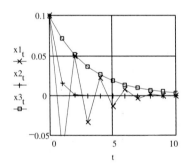 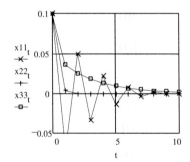

It is clear from these two graphs that both approaches give the same results. The third treatment mentioned in Section 3.4 also will yield the same results.

EXERCISE 3.1 Discuss the motion of a harmonic oscillator with the same initial condition as in the example, except that (a) the mass is changed to 1.0 kg while keeping the k the same and (b) the k is changed to 10 N/m while keeping the mass the same.

Energy Considerations

The total energy $E(t)$ of a damped harmonic system at any time t is given by

$$E(t) = E(0) + W_f \tag{3.71}$$

where $E(0)$ is the total energy at time $t = 0$ and W_f is the work done by friction in the time interval 0 to t. Assuming the dissipative frictional force $f = -b\dot{x} = -bv$, we can calculate W_f as follows:

$$W_f = \int f\,dx = \int f \frac{dx}{dt}\,dt = \int fv\,dt = \int_0^1 -bv^2\,dt \tag{3.72}$$

Thus the rate of energy loss by friction may be written as

$$\frac{dE}{dt} \left(= \frac{dW_f}{dt} \right) = -bv^2 \tag{3.72}$$

which is negative and represents the rate at which energy is being dissipated into heat. Since $W_f < 0$, E_t continuously decreases with time and may be calculated in the following manner:

$$E(t) = K(t) + U(t) = \tfrac{1}{2}m\dot{x}^2 + \tfrac{1}{2}kx^2 \tag{3.73}$$

From Eq. (3.62),

$$x(t) = Ae^{-\gamma t}\cos(\omega_1 t + \phi)$$

$$\dot{x}(t) = -\omega_1 A e^{-\gamma t}\left[\sin(\omega_1 t + \phi) + \frac{\gamma}{\omega_1}\cos(\omega_1 t + \phi) \right]$$

Let us assume that the system is lightly damped so that $\gamma/\omega_1 \ll 1$, and neglecting the second term on the right in the preceding expression for $\dot{x}$, we can substitute for x and $\dot{x}$ in Eq. (3.73):

$$E(t) = \tfrac{1}{2}A^2 e^{-2\gamma t}[m\omega_1^2 \sin^2(\omega_1 t + \phi) + k\cos^2(\omega_1 t + \phi)]$$

Since we assumed light damping, we may write $\omega_1^2 \simeq \omega_0^2 = k/m$; hence this equation takes the form

$$E(t) = \tfrac{1}{2}kA^2 e^{-2\gamma t} \tag{3.74}$$

while the initial energy of the system is obtained by substituting $t = 0$ in Eq. (3.74); that is,

$$E_0 = \tfrac{1}{2}kA^2 \tag{3.75}$$

Thus

$$E(t) = E_0 e^{-2\gamma t} \tag{3.76}$$

That is, the energy decreases (or decays) exponentially at a much faster rate ($e^{-2\gamma t}$) than the rate at which the amplitude decreases or decays ($e^{-\gamma t}$).

The time τ in which E decreases to $1/e$ ($= 0.368$) of its initial value is called the *characteristic time* or *decay constant* and may be evaluated by substituting $E(t) = E_0/e$ and $t = \tau$ in Eq. (3.76):

$$\frac{E_0}{e} = E_0 e^{-2\gamma\tau}$$

or $2\gamma\tau = 1$

That is,

$$\tau = \frac{1}{2\gamma} = \frac{2m}{2b} = \frac{m}{b} \tag{3.77}$$

If γ is very small , $\tau \to \infty$, and if γ is very large, $\tau \to 0$.
Also, using Eq. (3.76), we may write the *logarithmic derivative* of E as

$$\frac{d}{dt}(\ln E) = \frac{1}{E}\frac{dE}{dt} = -2\gamma \tag{3.78}$$

$(1/E)(dE/dt)$ represents the fractional rate of decrease in energy. Since the rate of energy loss is proportional to the square of the velocity [Eq. (3.72b)], the loss in energy is not uniform. dE/dt will be maximum when $\dot{x}$ is maximum (near the equilibrium), and it will drop to zero when $\dot{x}$ is zero near maximum amplitude. The plots of E and dE/dt are shown in Fig. 3.9.

3.5 QUALITY FACTOR

The quality factor Q, or simply Q value, is a frequently used term in mechanical oscillatory systems, as well as electrical oscillatory systems. Q is a dimensionless quantity and represents the degree of damping of an oscillator. The *quality factor* is defined as 2π times the ratio of the energy stored to the average energy loss per period. Thus

$$Q = 2\pi \frac{\text{energy stored in the oscillator}}{\text{average energy dissipated in one time period}} \tag{3.79}$$

If P is defined as the power loss or the rate at which the energy is dissipated, and the time period of oscillation $T_1 = 2\pi/\omega_1$, we can write the denominator as $PT_1 = P2\pi/\omega_1$, and hence Eq. (3.79) as

$$Q = 2\pi \frac{E}{P2\pi/\omega_1} = \frac{E}{P/\omega_1} \tag{3.80}$$

But $1/\omega_1$ is the time of motion for 1 radian. Thus

$$Q = \frac{\text{energy stored in the oscillator}}{\text{average energy dissipated per radian}} \tag{3.81}$$

As should be clear, for the lightly damped oscillator, Q will be very large, while for a heavily damped oscillator, it will be very small. We can calculate the Q value of the lightly damped oscillator as follows.
The energy of the oscillator and the rate at which it loses energy are given by [Eqs. (3.76) and (3.78)]

$$E(t) = E_0 e^{-2\gamma t} \tag{3.76}$$

Figure 3.9

Below are the graphs of energy E and the rate of energy loss ET = dE/dt versus time t for a damped oscillator $\omega0^2 > \gamma^2$

$$N := 100 \qquad n := 0..N \qquad t_n := \frac{n}{10} \qquad \phi := \frac{\pi}{2}$$

$$A := 1 \qquad k := 20 \qquad b := .5 \quad M := 1$$

For clarity, instead of graphing ET, we have graphed ET·5

$$\gamma := \frac{b}{2 \cdot M} \qquad \omega0 := \sqrt{\frac{k}{M}} \qquad \omega1 := \left(\sqrt{\omega0^2 - \gamma^2} \right)$$

$$\gamma = 0.25 \qquad \omega0 = 4.472 \qquad \omega1 = 4.465$$

(a) Why is E positive and ET negative?

$$E_n := \frac{1}{2} \cdot A^2 \cdot e^{-2 \cdot \gamma \cdot t_n} \cdot \left[M \cdot \omega1^2 \cdot \left(\sin\left(\omega1 \cdot t_n + \phi\right)\right) + k \cdot \cos\left(\omega1 \cdot t_n + \phi\right) \right]^2$$

(b) Is the rate of loss of energy ET in phase with or out of phase with energy E? What is the significance of this?

$$ET_n := -b \cdot \left[\left[\left(\omega1^2 \cdot A^2\right) \cdot e^{-\gamma \cdot t_n} \right] \cdot \left[\sin\left(\omega1 \cdot t_n + \phi\right) + \frac{\gamma}{\omega1} \cdot \left(\cos\left(\omega1 \cdot t_n + \phi\right)\right) \right]^2 \right]$$

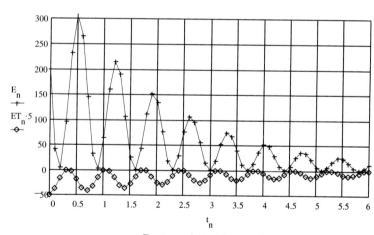

Energy and rate of energy loss

and
$$\frac{dE}{dt} = -2\gamma E \tag{3.78}$$

Thus the energy dissipated in time Δt will be

$$\Delta E = \left| \frac{dE}{dt} \right| \Delta t = 2\gamma E \, \Delta t \tag{3.82}$$

If Δt is the time for 1 radian of oscillation, $\Delta t = 1/\omega_1$; hence

$$Q = \frac{E}{\Delta E} = \frac{E}{2\gamma E/\omega_1} = \frac{\omega_1}{2\gamma} \tag{3.83}$$

For light damping, $\omega_1 \simeq \omega_0$; hence

$$Q = \frac{\omega_1}{2\gamma} \simeq \frac{\omega_0}{2\gamma} \tag{3.84}$$

If γ is small, Q will be large, and vice versa. Ordinary mechanical systems, such as loudspeakers and rubber bands, are heavily damped and may have Q values from 5 to 100. On the other hand, systems such as tuning forks and violin strings may have a Q value as high as 1000. A typical microwave cavity resonator has a Q value of about 10^4. Systems with extremely light damping are excited atoms ($Q \simeq 10^7$), excited nuclei ($Q \simeq 10^{12}$), and gas lasers ($Q \simeq 10^{14}$).

 Example 3.2 _____

Consider a critically damped oscillator of mass m, damping coefficient b, and initial displacement A. Calculate the rate of energy dissipation and the total energy dissipated during the time interval t = 0 and t = m/b.

Solution

According to Eq. (3.68), the solutions of the critically damped oscillator for x and v are (where $\gamma = b/2m$)

$$x = (B1 + B2 \cdot t) \cdot e^{-\gamma t}$$
$$v = (B2 - (B1 + B2 \cdot t) \cdot \gamma) \cdot \exp(-\gamma t)$$

Using the initial conditions at t = 0, x = A, and v = 0, the values of the two constants B1 and B2 are as shown

$$t = 0 \quad x = A \quad v = 0$$
$$B1 = A \qquad B2 = \gamma A$$

The resulting equations for x and v are (using $\gamma = b/2m$)

$$x = A\left(1 + \frac{b}{2 \cdot m} \cdot t\right) \cdot e^{-\left(\frac{b}{2 \cdot m}\right) \cdot t}$$

$$v = \frac{dx}{dt} = -\frac{b^2 \cdot A}{4 \cdot m^2} \cdot t \cdot e^{-\left(\frac{b}{2 \cdot m}\right) \cdot t}$$

Substituting the values of v above into Eq. (3.72), the time rate of change of energy dissipation through damping is

$$\frac{dW}{dt} = \left[-b \cdot v^2 = \frac{b^5 \cdot A^2 \cdot t^2}{16 \cdot m^4} \cdot e^{-\left(\frac{b}{m}\right) \cdot t} \right]$$

Integration with respect to t gives the energy dissipated in time t

$$Wi = \frac{-1}{8} \cdot \frac{\left(4 \cdot \gamma^2 \cdot t^2 \cdot \exp(-2 \cdot \gamma t) + 4 \cdot \gamma \cdot t \cdot \exp(-2 \cdot \gamma t) + 2 \cdot \exp(-2 \cdot \gamma t) \right)}{\gamma^3}$$

Using the values given in Example 3.1, we calculate the energy dissipated between t = 0 and t = m/b = 1/2γ.

$$t := 1 \qquad \gamma := \frac{1}{2 \cdot t}$$

$$Wi := \frac{-1}{8} \cdot \frac{\left(4 \cdot \gamma^2 \cdot t^2 \cdot \exp(-2 \cdot \gamma t) + 4 \cdot \gamma \cdot t \cdot \exp(-2 \cdot \gamma t) + 2 \cdot \exp(-2 \cdot \gamma t) \right)}{\gamma^3}$$

$$Wi = -1.839 \qquad \frac{Wi}{16} = -0.115$$

$$b := .1 \cdot \frac{\text{newton} \cdot \sec}{m} \qquad m := .05 \cdot kg \qquad \gamma := \frac{b}{2 \cdot m} \qquad \gamma = 1 \cdot \sec^{-1}$$

$$A := .01 \cdot m \qquad A = 5 \cdot 10^{-4} \cdot kg \qquad t := \frac{1}{2 \cdot \gamma} \qquad t = 0.5 \cdot \sec$$

P = rate of energy dissipation

$$P := \left(\frac{b^5}{16 \cdot m^4} \right) \cdot A^2 \cdot t^2 \cdot e^{-\left(\frac{b}{m}\right) \cdot t} \qquad P = 2.299 \cdot 10^{-9} \cdot kg^3 \cdot \sec^{-3}$$

$$W1 := \frac{P}{.025} \qquad W1 = 9.197 \cdot 10^{-8} \cdot kg^3 \cdot \sec^{-3}$$

EXERCISE 3.2 Make plots of W and P versus t for the situation given in Example 3.2.

3.6 FORCED HARMONIC OSCILLATOR (DRIVEN OSCILLATOR)

A free oscillator will oscillate forever. But, in reality, every system has some damping present (the energy is dissipated, say in the form of heat) and the system will eventually stop oscillating. To maintain the oscillations, energy from an external source must be supplied at a rate equal to the energy dissipated by the oscillator in the damping medium. Such motion in which energy is supplied externally is called *forced oscillations* or *driven oscillations,* while the system is called a *forced oscillator* or a *driven oscillator.* If the system is acted on by a driving force F_d then the net force, F_{net}, acting on the system is given by

$$F_{net} = F_s + F_f + F_d \qquad (3.85)$$

where
$$F_s = -kx, \quad F_f = -b\dot{x}$$

and from Newton's second law $F_{net} = m\ddot{x}$. Equation (3.85) cannot be solved unless we know the form of the applied force F_d. Since we have been limiting our discussion to linear oscillators, it is easier if we assume that the driving force has a sinusoidal form given by

$$F_d = F_0 \cos(\omega t + \theta_0) \tag{3.86}$$

We have good reasons to assume this form for the driving force. First, many actual situations involve just such a force, as, for example, the response of a bound electron when electromagnetic waves are incident on it, that is, in the scattering of light from bound electrons. Second, any periodic function of time can be represented as a sum of several harmonics (or sinusoidal) terms. Using the techniques of Fourier series, one can solve for the motion of the system under any periodic driving force (as discussed in Chapter 4.).

We may combine these equations and write the following equation that describes the motion of a driven harmonic oscillator:

$$m\ddot{x} + b\dot{x} + kx = F_0 \cos(\omega t + \theta_0) \tag{3.87}$$

This is an inhomogeneous, second-order, linear differential equation. The solution of Eq. (3.87) is given by the sum of two parts according to the following theorem:

> *If $x_i(t)$ is a particular solution of an inhomogeneous differential equation and the complementary function $x_h(t)$ is the solution of the corresponding homogeneous equation [that is, Eq. (3.87) with the right side equal to zero], then $x(t) = x_i(t) + x_h(t)$ is also a solution of the inhomogeneous differential equation.*

Thus the general solution of Eq. (3.87) is

$$x(t) = x_i(t) + x_h(t) \tag{3.88}$$

where x_h is the solution of the homogeneous equation

$$m\ddot{x} + b\dot{x} + kx = 0$$

From Section 3.4, the general solution of this homogeneous equation is given by any one of the three forms, Eq. (3.60), (3.61), or (3.62); hence

$$x_h(t) = e^{-\gamma t}[A_1 e^{+i\omega_1 t} + A_2 e^{-i\omega_1 t}] \tag{3.89a}$$

$$x_h(t) = e^{-\gamma t}(B \sin \omega_1 t + C \cos \omega_1 t) \tag{3.89b}$$

$$x_h(t) = A_h e^{-\gamma t} \cos(\omega_1 t + \phi_h) \tag{3.89c}$$

Since the oscillations of a damped oscillator eventually decay to zero, the x_h part of the solution is called the *transient term*. After a certain time, the x_h part of the solution is of no consequence; hence, for a steady-state solution we must concentrate on finding the particular solution $x_i(t)$.

According to Eq. (3.87), the applied force varies sinusoidally, so we expect the resulting steady-state solution $x_i(t)$ to vary sinusoidally. A solution of the form $x = A \cos \omega t$ would have been perfectly acceptable if the left side of the equation did not have an $\dot{x}$ term. To take care of this situation, we must have a solution of the form

$$x = A \cos(\omega t \pm \phi) \tag{3.90}$$

Let us assume a solution of the form

$$x_i = A \cos(\omega t - \phi) \tag{3.91}$$

To calculate A and ϕ, we substitute for x_i in Eq. (3.87), and after setting $\theta_0 = 0$, we get

$$-m\omega^2 A \cos(\omega t - \phi) - b\omega A \sin(\omega t - \phi) + kA \cos(\omega t - \phi) = F_0 \cos \omega t$$

Rearranging,

$$(kA \cos \phi - m\omega^2 A \cos \phi + b\omega A \sin \phi) \cos \omega t$$

$$- (kA \sin \phi - m\omega^2 A \sin \phi - b\omega A \cos \phi) \sin \omega t = F_0 \cos \omega t$$

For this to hold for all values of t, the coefficients of the $\cos \omega t$ and $\sin \omega t$ terms on each side must be separately equal. That is,

$$(k - m\omega^2) \cos \phi + b\omega \sin \phi = \frac{F_0}{A} \tag{3.92}$$

$$(k - m\omega^2) \sin \phi - b\omega \cos \phi = 0 \tag{3.93}$$

From Eq. (3.93), we obtain an expression for the phase angle to be

$$\tan \phi = \frac{b\omega}{k - m\omega^2} = \frac{b\omega/m}{k/m - \omega^2} \tag{3.94}$$

Using the usual notation $k/m = \omega_0^2$ and $\gamma = b/2m$, we get

$$\tan \phi = \frac{2\gamma\omega}{\omega_0^2 - \omega^2} \tag{3.95}$$

from which we obtain

$$\sin \phi = \frac{2\gamma\omega}{\sqrt{(\omega_0^2 - \omega^2)^2 + 4\gamma^2\omega^2}} \tag{3.96}$$

and

$$\cos \phi = \frac{\omega_0^2 - \omega^2}{\sqrt{(\omega_0^2 - \omega^2)^2 + 4\gamma^2\omega^2}} \tag{3.97}$$

If we substitute these in Eq. (3.92), we get

$$A = \frac{F_0/m}{\sqrt{(\omega_0^2 - \omega^2)^2 + 4\gamma^2\omega^2}} \tag{3.98}$$

Thus a particular solution of the inhomogeneous equation is

$$x_i(t) = \frac{F_0/m}{\sqrt{(\omega_0^2 - \omega^2)^2 + 4\gamma^2\omega^2}} \cos(\omega t - \phi) \tag{3.99}$$

where

$$\phi = \tan^{-1} \frac{2\gamma\omega}{\omega_0^2 - \omega^2} \tag{3.100}$$

[A slightly different procedure for obtaining the general solution is more convenient when the driving force F_d given by Eq. (3.86) is written in exponential form as

$$F_d = F_0 e^{i(\omega t + \theta_0)} \tag{3.101}$$

and we can obtain the same results.]

Using the solution given by Eq. (3.99) together with the homogeneous solution x_h, given by Eq. (3.89c), we get the general solution

$$x = x_h + x_i$$

$$= A_h e^{-\gamma t} \cos(\omega_1 t + \phi_h) + \frac{F_0/m}{\sqrt{(\omega_0^2 - \omega^2)^2 + 4\gamma^2\omega^2}} \cos(\omega t - \phi) \tag{3.102}$$

As required, this solution contains two arbitrary constants (of integration) A_h and ϕ_h, while ϕ is not a constant and is given by Eq. (3.100). The first part of the solution oscillates with a natural frequency ω_1. Because of the damping, the oscillations die out for large values of time, that is, for $t \gg 1/\gamma$. The homogeneous solution x_h is called the *transient solution,* while the particular solution x_i is the *steady-state solution.* The general solution x will be independent of the influence of the initial conditions except in the beginning when the transient term is still contributing. Figure 3.10 illustrates this for two special cases: (a) for $\omega < \omega_1$, that is, the driving frequency is less than the natural frequency; (b) for $\omega > \omega_1$, that is, the driving frequency is greater than the natural frequency. For both cases, the plots of the homogeneous solution x_h versus t as well as plots of the particular solution x_i versus t are shown. The resultant of these two, that is, the plots of $x_n = x_h + x_i$ versus t are also shown. As is clear from these plots, the transient solution x_h is effective only in the beginning and decays to zero as time passes, while the steady-state solution remains constant with time. Thus the transient solution effects the general solution only in the beginning. Furthermore, if $\omega < \omega_1$, the transient term x_h causes distortion of the resulting sinusoidal waveform as shown in Fig. 3.10(a). On the other hand, if $\omega > \omega_1$, the transient term x_h, instead of causing distortion, has the effect of modulating the oscillations due to the force function as shown in Fig. 3.10(b). Of course, in both cases, after the transient term has died out, the oscillations are governed by the force function. In addition to the relative values of ω and ω_1, initial conditions will also affect the detailed motion, but only in the beginning. It is important to note that the transient terms play an important role in electrical circuits. In designing such circuits, it is necessary to avoid peak voltages and currents when initially the circuits are closed.

Since for $t \gg 1/\gamma$, $x \simeq x_i$, we shall concentrate on the discussion of the steady-state solution, that is, the particular solution x_i given by Eqs. (3.99) and (3.100). This solution is independent of the initial conditions.

Figure 3.10 _____

Below are two special cases of the influence of the transient term on the steady-state solution.

$$N := 30 \qquad n := 0..N \qquad t_n := \frac{n}{5}$$

$$A := 1 \qquad \gamma := 1 \qquad \phi := 0 \qquad \omega 0 := 3$$

(a) When the driving frequency ω (=10) is greater than the natural frequency $\omega 1$ (2.828), this leads to *distortion*, as illustrated.

$$M := 1 \qquad \omega := 10 \qquad F0 := 50$$

$$\phi i := atan\left(\frac{2 \cdot \gamma \cdot \omega}{\omega 0^2 - \omega^2}\right) \qquad \omega 1 := \left(\left|\omega 0^2 - \gamma^2\right|\right)^{\frac{1}{2}}$$

$$\phi i = -0.216 \qquad \omega 1 = 2.828$$

$$xh_n := A \cdot e^{-\gamma t_n} \cdot cos\left(\omega 1 \cdot t_n + \phi\right) \qquad xi_n := \frac{\dfrac{F0}{M}}{\left[\left(\omega 0^2 - \omega^2\right)^2 + 4 \cdot \gamma^2 \cdot \omega^2\right]^{\frac{1}{2}}} \cdot cos\left(\omega \cdot t_n + \phi i\right)$$

$$x_n := xh_n + xi_n$$

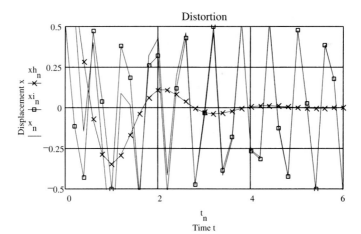

Distortion

Displacement x

xh_n
xi_n
x_n

t_n
Time t

(b) When the driving frequency ω (=3) is less than natural frequency $\omega 1$ (9.95), this leads to *modulation*, as illustrated.

$N := 30 \qquad n := 0..N \qquad t_n := \dfrac{n}{10}$

$A := 1 \qquad \gamma := 1$

$\phi := 0 \qquad \omega 0 := 10$

$M := 1 \qquad \omega := 3 \qquad F0 := 50$

$\phi i := \text{atan}\left(\dfrac{2 \cdot \gamma \cdot \omega}{\omega 0^2 - \omega^2}\right) \qquad\qquad \omega 1 := \left(\left|\omega 0^2 - \gamma^2\right|\right)^{\frac{1}{2}}$

$\phi i = 0.066 \qquad\qquad\qquad \omega 1 = 9.95$

$xh_n := A \cdot e^{-\gamma t_n} \cdot \cos\left(\omega 1 \cdot t_n + \phi\right) \qquad\qquad xi_n := \dfrac{\dfrac{F0}{M}}{\sqrt{\left(\omega 0^2 - \omega^2\right)^2 + 4 \cdot \gamma^2 \cdot \omega^2}} \cdot \cos\left(\omega \cdot t_n + \phi i\right)$

$x_n := xh_n + xi_n$

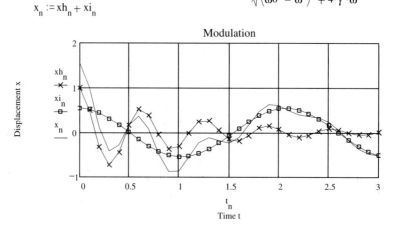

Modulation

If the driving frequency is equal to the natural frequency, what effect it will have on the amplitude?

 Example 3.3 _____

Consider a damped oscillator, for which $\gamma = \omega_0/4$, acted on by a driving force $F = F_0 \cos \omega t$. Find the general solution $x(t)$.

Solution

The second-order differential equation describing the driven oscillator is

$$m\ddot{x} + b\dot{x} + kx = F_0 \cos \omega t$$

or
$$\ddot{x} + 2\gamma\dot{x} + \omega_0^2 x = \frac{F_0}{m} \cos \omega t \tag{i}$$

where $\gamma = \omega_0/4$ and

$$\omega_1^2 = (\omega_0^2 - \gamma^2)^{1/2} \approx 0.97\omega_0 \tag{ii}$$

The transient (or the homogeneous) solution

$$x_h(t) = e^{-\gamma t}(A_1 \cos \omega_1 t + A_2 \sin \omega_1 t) \tag{iii}$$

takes the form

$$x_h(t) = e^{-\omega_0 t/4}(A_1 \cos 0.97t + A_2 \sin 0.97t) \tag{iv}$$

Let us assume the particular solution for the applied force $F = F_0 \cos \omega t$ to be

$$x_i(t) = B_1 \cos \omega t + B_2 \sin \omega t \tag{v}$$

$$\dot{x}_i(t) = -\omega B_1 \sin \omega t + \omega B_2 \cos \omega t \tag{vi}$$

$$\ddot{x}_i(t) = -\omega^2 B_1 \cos \omega t - \omega^2 B_2 \sin \omega t \tag{vii}$$

Substituting these three equations in Eq. (i) and rearranging gives

$$(-\omega^2 B_1 + 2\gamma\omega B_2 + \omega_0^2 B_1) \cos \omega t + (-\omega^2 B_2 - 2\gamma\omega B_1 + \omega_0^2 B_2) \sin \omega t = \frac{F_0}{m} \cos \omega t \tag{viii}$$

Equating the coefficients gives

$$(\omega_0^2 - \omega^2)B_1 + 2\gamma\omega B_2 = \frac{F_0}{m} \tag{ix}$$

$$-2\gamma\omega B_1 + (\omega_0^2 - \omega^2)B_2 = 0 \tag{x}$$

Solving these equations for B_1 and B_2 in terms of F_0 gives

$$B_1 = \frac{F_0(\omega_0^2 - \omega^2)}{m[(\omega_0^2 - \omega^2)^2 + 4\gamma^2\omega^2]} \tag{xi}$$

$$B_2 = \frac{2\gamma\omega F_0}{m[(\omega_0^2 - \omega^2)^2 + 4\gamma^2\omega^2]} \tag{xii}$$

Hence the general solution is given by

$$x(t) = x_h(t) + x_i(t)$$

$$= e^{-\omega_0 t/4}(A_1 \cos 0.97t + A_2 \sin 0.97t) + B_1 \cos \omega t + B_2 \sin \omega t \tag{xiii}$$

where B_1 and B_2 are given by Eqs. (xi) and (xii), while A_1 and A_2 are to be evaluated using the initial conditions in Eq. (xiii) by the usual procedure (see Example 3.1).

EXERCISE 3.3 Complete the example for the driving force $F = F_0 \sin \omega t$ with initial conditions $t = 0$, $x(0) = 0$, and $\dot{x}(0) = 0$, that is, calculate A_1 and A_2. Also graph x_h, x_i, and $x_n = x_h + x_i$.

3.7 AMPLITUDE RESONANCE

The amplitude A and the phase angle ϕ of steady-state motion according to Eqs. (3.98) and (3.100) are

$$A = \frac{F_0/m}{\sqrt{(\omega_0^2 - \omega^2)^2 + 4\gamma^2\omega^2}} \tag{3.103}$$

$$\phi = \tan^{-1}\frac{2\gamma\omega}{\omega_0^2 - \omega^2} \tag{3.104}$$

For a fixed value of ω_0, the variations in A and ϕ with the driving frequency ω for different values of γ are shown in Fig. 3.11. As illustrated, the behavior of these quantities strongly depends on the ratio ω/ω_0.

As stated earlier, ϕ represents the phase difference between the driving force F and the resulting motion x; that is, it represents a delay between the action and the response. As shown in Fig. 3.11, this phase lag, which is $\phi = 0$ when $\omega = 0$, increases to $\phi = \pi/2$ for $\omega = \omega_0$ and reaches $\phi = \pi$ as $\omega \to \infty$; that is, at very high frequencies the oscillations of the system are 180° out of phase with the driving force. It is interesting to note that as $\gamma \to 0$ the phase change occurs more and more rapidly, and in the extreme case when $\gamma = 0$, the phase changes suddenly from 0 to π at $\omega = \omega_0$.

From Fig. 3.11, it is clear that, depending on the values of γ, there is a certain driving frequency at which the amplitude A has a maximum value. The frequency at which the amplitude is maximum is called the *amplitude resonance frequency* ω_r. This frequency ω_r may be calculated from Eq. (3.103) by setting

$$\left.\frac{dA}{d\omega}\right|_{\omega=\omega_r} = 0 \tag{3.105}$$

Upon solving the resulting equation, we get

$$\omega = \omega_r = (\omega_0^2 - 2\gamma^2)^{1/2} \tag{3.106}$$

which states that as the damping coefficient decreases the resonance frequency increases, and in the limit as $\gamma \to 0$, $\omega_r \to \omega_0$, decreases the natural frequency of a free oscillator. In the case of extremely small damping, we can express the right side of Eq. (3.106) by using the binomial theorem; that is,

$$\omega_r = \omega_0\left(\frac{1}{2}\frac{2\gamma^2}{\omega_0^2} + \cdots\right)$$

or

$$\omega_r \simeq \omega_0 - \frac{\gamma^2}{\omega_0} \tag{3.107}$$

Equations (3.106) and (107) for a driven oscillator may be compared with the case of a damped oscillator discussed previously; that is,

$$\omega_1 = (\omega_0^2 - \gamma^2)^{1/2}$$

Figure 3.11 _____

(a) The graph below shows
amplitude resonance in the variation of
amplitude A versus frequency ratio $\omega/\omega 0$
for different values of γ.

$$N := 40 \qquad n := 0..N \qquad m := 0..4$$

$$\omega 0 := 2 \qquad \omega_n := \frac{n}{10} + .01$$

$$F0 := 6 \qquad M := 5$$

m
0
1
2
3
4

$\gamma_m :=$

0
.2
.4
.6
1

$$A_{n,m} := \frac{\dfrac{F0}{M}}{\sqrt{\left[\left[\omega 0^2 - \left(\omega_n\right)^2\right]^2 + 4\cdot\left(\gamma_m\right)^2\cdot\left(\omega_n\right)^2\right]}}$$

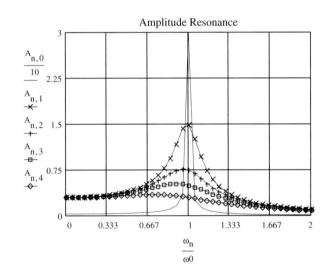

(b) The graph below shows the
phase angle variation
in the phase angle ϕ versus frequency
ratio $\omega/\omega 0$.

$$N := 39 \qquad n := 1..N \qquad i := 41..80 \qquad m := 1..3$$

$$\omega 0 := 4 \qquad \omega_n := \frac{n}{10} \qquad \omega_i := \frac{i}{10}$$

m
1
2
3

$\gamma_m :=$

1
2
3

$$\gamma 1 := 1 \qquad \gamma 2 := 2 \qquad \gamma 3 := 3$$

$$\phi_{n,m} := \operatorname{atan}\left[\dfrac{2 \cdot \gamma_m \cdot \omega_n}{\omega 0^2 - \left(\omega_n\right)^2}\right] \qquad \phi_{i,m} := \operatorname{atan}\left[\dfrac{2 \cdot \gamma_m \cdot \omega_i}{\omega 0^2 - \left(\omega_i\right)^2}\right] + \pi$$

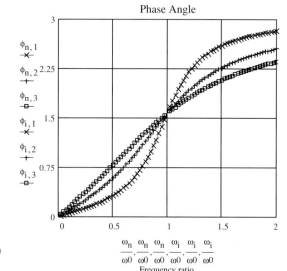

What is the effect of using
values of γ greater than 1 in (a)
and less than 1 in (b).

and, for small γ,

$$\omega_1 \simeq \omega_0 - \frac{\gamma^2}{2\omega_0} \tag{3.108}$$

while for a free oscillator $\omega_0^2 = k/m$. Thus ω_1 lies to the right of ω_r, while ω_0 is still farther away from ω_r, as shown in Fig. 3.12 where $A^2/A_0^2 = 1$ at $\omega = \omega_r$.

Thus the maximum amplitude $A = A_0$ that occurs at $\omega = \omega_r$ may be obtained from Eqs. (3.103) and (3.106) to be

$$A = \frac{F_0/m}{2\gamma \sqrt{\omega_0^2 - \gamma^2}} \tag{3.109}$$

In the case of small damping, we assume that $\gamma \to 0$; hence

$$A_0 \simeq \frac{F_0}{2m\gamma\omega_0} \simeq \frac{F_0}{b\omega_0} \tag{3.110}$$

It is clear that if b is small or $\gamma \to 0$, the amplitude A_0 becomes very large. For undamped systems, $b = 0$ and hence $A_0 = \infty$; but there are hardly any systems that are undamped.

Figure 3.12 _____

The graph below shows the relative positions of the resonance frequency ωr, the natural frequency $\omega 1$, and the natural free frequency $\omega 0$. $Q(\omega)$ are the positions of the resonance amplitude for frequencies ωr, $\omega 0$, and $\omega 1$. The graphs are the ratio of the square of the amplitudes versus $\omega n - \omega r$.

$$N := 50 \quad n := 0..N \qquad \omega_n := \frac{n}{5} \qquad F0 := 2 \qquad \omega 0 := 7 \qquad \gamma := 1 \qquad M := 1$$

$$\omega r1 := \omega 0 \cdot \left(1 - 2 \cdot \frac{\gamma^2}{\omega 0^2}\right)^{\frac{1}{2}} \qquad \omega 11 := \left(\omega 0^2 - \gamma^2\right)^{\frac{1}{2}} \qquad A_n := \frac{\dfrac{F0}{M}}{\sqrt{\left[\omega 0^2 - \left(\omega_n\right)^2\right]^2 + 4 \cdot \gamma^2 \cdot \left(\omega_n\right)^2}}$$

$$\omega r1 = 6.856 \qquad\qquad \omega 11 = 6.928$$

$$\omega r := \omega 0 - \frac{\gamma^2}{\omega 0} \qquad \omega 1 := \omega 0 - \frac{\gamma^2}{2 \cdot \omega 0} \qquad Q(\omega) := \frac{\left[\dfrac{\dfrac{F0}{M}}{\sqrt{\left[\omega 0^2 - (\omega)^2\right]^2 + 4 \cdot \gamma^2 \cdot (\omega)^2}}\right]^2}{\left(\dfrac{F0}{2 \cdot M \cdot \gamma \cdot \omega 0}\right)^2}$$

$$\omega r = 6.857 \qquad\qquad \omega 1 = 6.929$$

$$A0 := \frac{F0}{2 \cdot M \cdot \gamma \cdot \omega 0} \qquad A0 = 0.143 \qquad\qquad i := 0..2$$

y_i are different values of Q for different frequencies. x_i and y_i are plotted against $x_i - \omega r$. Note the small differences in the positions of the resonance frequencies.

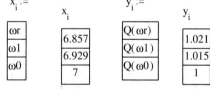

$$x_i := \begin{array}{|c|} \hline \omega r \\ \hline \omega 1 \\ \hline \omega 0 \\ \hline \end{array} \qquad x_i = \begin{array}{|c|} \hline 6.857 \\ \hline 6.929 \\ \hline 7 \\ \hline \end{array} \qquad y_i := \begin{array}{|c|} \hline Q(\omega r) \\ \hline Q(\omega 1) \\ \hline Q(\omega 0) \\ \hline \end{array} \qquad y_i = \begin{array}{|c|} \hline 1.021 \\ \hline 1.015 \\ \hline 1 \\ \hline \end{array}$$

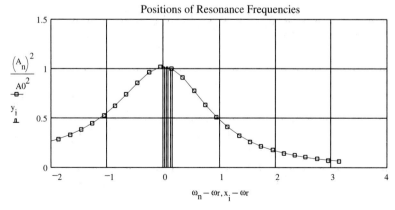

Positions of Resonance Frequencies

How do you explain the gradual change in Q for different values of ω?

The fact that the amplitude of the oscillations of the system is very large at the resonance frequency has both desirable and undesirable effects. In the case of electrical circuits that are used in tuning radios and in certain types of musical instruments such as organ pipes, it is desirable to have a large response for a small driving force. On the other hand, it is very undesirable to have a very large amplitude in mechanical systems, such as in the springs of an automobile or in the spring mounting of an electric motor. There the aim is to minimize the transmission of vibrations.

3.8 ENERGY RESONANCE

In most practical situations involving oscillating systems in nature, the quantity observed experimentally is energy and not amplitude. Also, the total energy of an oscillating system is proportional to the square of the amplitude near resonance; hence we should look for the variation of A^2 versus ω. Once again we assume that ω_0 is fixed. For steady-state motion, the amplitude A is constant, and we may write

$$x = A \cos(\omega t + \phi) \tag{3.111a}$$

and
$$v = \dot{x} = -\omega A \sin(\omega t + \phi) \tag{3.111b}$$

which gives us

$$K(t) = \tfrac{1}{2}mv^2 = \tfrac{1}{2}m\omega^2 A^2 \sin^2(\omega t + \phi) \tag{3.112}$$

$$U(t) = \tfrac{1}{2}kx^2 = \tfrac{1}{2}kA^2 \cos^2(\omega t + \phi) \tag{3.113}$$

$$E(t) = K(t) + U(t)$$

$$= \tfrac{1}{2}A^2[m\omega^2 \sin^2(\omega t + \phi) + k \cos^2(\omega t + \phi)] \tag{3.114}$$

Let us now calculate time averages of $K(t)$, $U(t)$, and $E(t)$ in the case when A changes with ω, and remembering that $\langle \cos^2(\omega t + \phi)\rangle = \langle \sin^2(\omega t + \phi)\rangle = \tfrac{1}{2}$ for an average over one period. Substituting for A from Eq. (3.103) into Eq. (3.112), we get

$$K(t) = \frac{1}{2}m\omega^2 \frac{F_0^2/m^2}{(\omega_0^2 - \omega^2)^2 + 4\gamma^2\omega^2} \sin^2(\omega t + \phi) \tag{3.115}$$

That is,

$$\langle K \rangle = \frac{1}{4}\frac{F_0^2}{m}\frac{\omega^2}{(\omega_0^2 - \omega^2)^2 + 4\gamma^2\omega^2} \tag{3.116}$$

The value of ω for which $\langle K \rangle$ is maximum is obtained by setting

$$\left.\frac{d\langle K\rangle}{d\omega}\right|_{\omega=\omega_k} = 0 \tag{3.117}$$

which gives

$$\omega_k = \omega_0 \tag{3.118}$$

That is, the kinetic energy resonance occurs at the natural free frequency ω_0.

Since $U(t) = \frac{1}{2}kx^2$, the potential energy resonance must occur at the same position as the amplitude resonance; that is, the potential energy resonance frequency ω_u from Eq. (3.106) is

$$\omega_u = \omega_r = (\omega_0^2 - 2\gamma^2)^{1/2} \tag{3.119}$$

It should not be alarming to find that the kinetic and potential energies resonate at different frequencies. This is because we are dealing with a damped oscillator, which is a nonconservative system. The energy is continuously being drained by the damping medium.

The average total energy $\langle E \rangle$ from Eq. (3.114) is

$$\langle E \rangle = \frac{1}{2}A^2 m\omega^2 \langle \sin^2(\omega t + \phi) \rangle + \frac{1}{2}kA^2 \langle \cos^2(\omega t + \phi) \rangle$$

$$= \frac{1}{4}A^2(m\omega^2 + k) = \frac{1}{4}mA^2(\omega^2 + \omega_0^2) \tag{3.120}$$

Substituting for A from Eq. (3.103), we obtain

$$\langle E \rangle = \frac{1}{4}\frac{F_0^2}{m}\frac{\omega^2 + \omega_0^2}{(\omega^2 - \omega_0^2)^2 + 4\gamma^2\omega^2} \tag{3.121}$$

For very weak damping, $\gamma \ll \omega_0$, and we may write

$$\omega^2 + \omega_0^2 \simeq 2\omega_0^2$$

$$\omega^2 - \omega_0^2 = [(\omega + \omega_0)(\omega - \omega_0)] \simeq 2\omega_0(\omega - \omega_0)$$

Using this in Eq. (3.121) and noting that E is a function of ω, we replace $E(t)$ by $E(\omega)$ and simplify

$$\langle E(\omega) \rangle = \frac{1}{8}\frac{F_0^2}{m}\frac{1}{(\omega - \omega_0)^2 + \gamma^2} \tag{3.122}$$

Rewrite this as

$$\langle E(\omega) \rangle \frac{8m}{F_0^2} = \frac{1}{(\omega - \omega_0)^2 + \gamma^2} = L(\omega) \tag{3.123}$$

where the function $L(\omega)$ contains all the necessary frequency dependence of $\langle E(\omega) \rangle$. A plot of function $L(\omega)$ is called a *resonance curve* or *Lorentzian*. Figure 3.13 shows several such plots for different values of γ. Note that for large γ the function is effectively zero except near the resonance frequency ω_0.

The maximum height of the resonance curve occurs at ω_0 and is equal to $1/\gamma^2$. This value will fall to one-half its maximum value when [from Eq. (3.122) or (3.123)]

$$(\omega - \omega_0)^2 = \gamma^2 \tag{3.124}$$

▶ Figure 3.13 _____

Below is the graph of the resonance curve
(or *Lorentzian*), that is, the graph of L(ω)
versus ω for different values of γ.

(a) What causes the change in the values
of L?

(b) What determines the width of the
energy resonance curve?

(c) What does the following analysis
indicate about the location of the
resonance frequency on the ω-axis?

$$n := 0 \dots 200 \qquad \omega_n := n \qquad \omega 0 := 50$$

$$\gamma 1 := 1 \qquad\qquad \gamma 2 := 2 \qquad \gamma 3 := 3$$

$$L1_n := \frac{1}{\left[\left(\omega_n - \omega 0 \right)^2 + \gamma 1^2 \right]}$$

$$L2_n := \frac{1}{\left[\left(\omega_n - \omega 0 \right)^2 + \gamma 2^2 \right]}$$

$$L3_n := \frac{1}{\left[\left(\omega_n - \omega 0 \right)^2 + \gamma 3^2 \right]}$$

$$\max(L1) = 1 \qquad L1_{50} = 1$$

$$\max(L2) = 0.25 \qquad L2_{50} = 0.25$$

$$\max(L3) = 0.111 \qquad L3_{50} = 0.111$$

$$L1_{45} = 0.038 \qquad L1_{55} = 0.038$$

$$L2_{45} = 0.034 \qquad L2_{55} = 0.034$$

$$L3_{45} = 0.029 \qquad L3_{55} = 0.029$$

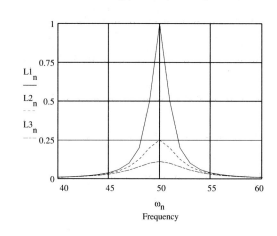

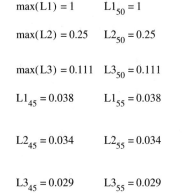

or
$$\omega - \omega_0 = \pm \gamma \qquad\qquad\qquad (3.125)$$

This equation states that the resonance curve drops to half its maximum value at $\omega_+ = \omega_0 + \gamma$
on the higher frequency side of ω_0 and at $\omega_- = \omega_0 - \gamma$ on the lower frequency side of ω_0. Thus
the *full width of the curve at half-maximum, called the resonance width,* is given by $\omega_+ + \omega_- = 2\gamma$. We call $\Delta\omega$ the resonance width; hence

$$\Delta\omega = 2\gamma \qquad\qquad\qquad (3.126)$$

As γ decreases, the width $\Delta\omega$ of the curve also decreases, which means that the resonance curve becomes higher and narrower. This implies that the range of frequencies to which a system will respond becomes narrower; that is, an oscillating system becomes increasingly *selective in frequency* as is obvious in Fig. 3.13.

The frequency-selective property of an oscillating system is characterized by the *quality factor Q*. In Section 3.5, we defined the quality factor as the ratio of the total energy stored in the oscillator to the energy dissipated per radian of oscillation. For a lightly damped oscillator, we showed [Eq. (3.84)] that

$$Q \simeq \frac{\omega_0}{2\gamma} \tag{3.84}$$

When such an oscillator is driven, we get a resonance width $\Delta\omega = 2\gamma$, and we may write

$$Q = \frac{\omega_0}{\Delta\omega} = \frac{\text{resonance frequency}}{\text{frequency width of resonance curve}} \tag{3.127}$$

We can prove that $Q = \omega_0/2\gamma$ by directly applying the energy definition of Q to a driven oscillator.

It becomes clear that systems with a high value of Q have a very narrow resonance width and hence are very highly selective to a frequency response when external driving forces are applied. If the resonance curve is sharp, a system will respond only when the driving frequency is equal to the resonance frequency. As mentioned previously in Section 3.5, Q may vary from 5 to 100 for mechanical systems, to $\simeq 10^8$ for atomic systems, and to $\simeq 10^{14}$ in the case of gas lasers. Such facts have been utilized for defining and making time standards using atomic clocks.

3.9. RATE OF ENERGY DISSIPATION

Finally, let us calculate the rate at which energy is being dissipated, which should be equal to the rate at which work is being done. Starting with the general equation for a forced oscillator, with $\theta_0 = 0$,

$$m\ddot{x} + b\dot{x} + kx = F_0 \cos \omega t \tag{3.87}$$

and multiplying both sides by $\dot{x}$,

$$m\ddot{x}\dot{x} + b\dot{x}^2 + kx\dot{x} = F_0\dot{x} \cos \omega t$$

We may write this as

$$\frac{d}{dt}\left(\frac{m\dot{x}^2}{2} + \frac{m\omega_0^2 x^2}{2}\right) + b\dot{x}^2 = (F_0 \cos \omega t)\dot{x} \tag{3.128a}$$

or

$$\frac{d}{dt}(K + U) + b\dot{x}^2 = (F_0 \cos \omega t)\dot{x} \tag{3.128b}$$

That is,

$$
\begin{pmatrix}
\text{Time rate of} \\
\text{change of } K \\
\text{and } U
\end{pmatrix}
=
\begin{pmatrix}
\text{Rate at which} \\
\text{energy is being} \\
\text{dissipated}
\end{pmatrix}
+
\begin{pmatrix}
\text{Rate at which energy} \\
\text{is being supplied by} \\
\text{the driving force}
\end{pmatrix}
$$

By substituting for x and $\dot{x}$ from any particular solution [Eq. (3.90)] we can prove that the average power $\langle P \rangle$ at which the driving force does work is

$$\langle P \rangle = \left\langle \frac{dW}{dt} \right\rangle = \langle \dot{x} F_0 \cos \omega t \rangle$$

or

$$\langle P \rangle = \tfrac{1}{2} \dot{x}_0 F_0 \sin \phi \tag{3.129}$$

where $\dot{x}_m$ is the maximum velocity, which occurs at $\omega = \omega_0$. Furthermore, we can show that (after transients have died out)

$$\langle P \rangle = \langle b \dot{x}^2 \rangle \tag{3.130}$$

as it should be.

Let us consider the right side of Eq. (3.128b), which, after substituting for $\dot{x}$, may be written as

$$\dot{x}_0 F_0 \cos \omega t = -F_0 \omega A \cos \omega t \sin(\omega t - \phi)$$

$$= F_0 \omega A (\cos^2 \omega t \sin \phi - \cos \omega t \sin \omega t \cos \phi) \tag{3.131}$$

The first term on the right is positive, which means that the driving force is sypplying energy, while the second term is negative, which means that the driving mechanism is receiving energy, that is, the driving force is alternately supplying and absorbing energy. This may appear strange, but that is what actually happens. But it may be pointed out that the average value of $\cos^2 \omega t$ is $\tfrac{1}{2}$, while that of $(\sin \omega t \cos \omega t)$ is zero. Hence, on the whole the driving agent supplies more energy than it absorbs.

PROBLEMS

3.1. Prove the following results for a simple harmonic oscillator; that is, prove Eqs. (3.49) and (3.50):

$$\langle V \rangle_{\text{space}} = \tfrac{1}{6} k A^2, \qquad \langle K \rangle_{\text{space}} = \tfrac{1}{3} k A^2$$

$$\langle E \rangle_{\text{space}} = \langle V \rangle_{\text{space}} + \langle K \rangle_{\text{space}} = \langle E \rangle_{\text{time}}$$

3.2. A liquid in a U tube is in equilibrium. When the liquid is slightly displaced, it executes simple harmonic motion. Calculate the frequency of such oscillations.

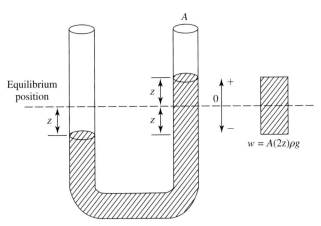

Figure P3.2

3.3. A small mass m is in a box of mass M that is tied to a vertical spring of stiffness constant k. When displaced from its equilibrium position y_0 to y_1 and let go, it executes simple harmonic motion. Calculate the reaction between m and M as a function of time. Does the mass m always stay in contact with the box? If not, what determines that it will not be in contact with M? Calculate the value of h as measured from the equilibrium position for which the contact is lost.

3.4. A wooden block of cross-sectional area A and mass density ρ when floating in water displaces a volume V. If a person of mass m jumps on this block, calculate the period of small oscillations.

3.5. One end of a spring of spring constant k is fixed while the other end is pulled horizontally with a force F for a time t_0, starting from its equilibrium position x_0. Show that

$$x = x_0 + \frac{F}{k}\left(\cos\sqrt{\frac{k}{m}}(t - t_0) - \cos\sqrt{\frac{k}{m}}t\right)$$

3.6. A block of mass M is tied to a horizontal spring of stiffness k. A small mass m is placed on M. The coefficient of friction between M and m is μ. For what value of the amplitude A will the mass m slip away from M? Clearly describe the conditions that lead to this.

3.7. A mass of 0.2 kg is attached to a spring having a spring constant 12 N/cm. The mass is displaced 6 cm and then released. Calculate v_0, T_0, v_{max}, and total energy. Graph these as a function of time.

3.8. In Problem 3.7, if the mass is released with a velocity of 5 cm/s from its displaced position, calculate v_0, T_0, v_{max}, A, the maximum potential energy, the maximum kinetic energy, and the total energy.

3.9. Suppose the motion in Problem 3.7 is taking place in a viscous medium. In 16 seconds the amplitude reduces to half of its initial value. Calculate **(a)** the damping constant, **(b)** the period T_1 and frequency ν_0, and **(c)** the decrement of motion. What are the significant differences in the two situations?

3.10. A pendulum with a time period T spends time Δt between x and $x + \Delta x$. Calculate the value of $\Delta t/T$ and make a plot of $\Delta t/T$ versus x for different values of amplitude A. Calculate the area under each curve. What is the significance of this result? Note:

$$\left|\frac{\Delta t}{T}\right| = \frac{1}{2\pi}\frac{\Delta x}{\sqrt{A^2 - x^2}}$$

3.11. Derive Eq. (3.62) directly from Eq. (3.60).

3.12. Explain why the constants in Eq. (3.60) are complex, while B and C in Eq. (3.61) and A and ϕ in Eq. (3.62) are real?

3.13. An oscillator when undamped has a time period T_0, while its time period when damped is T. Suppose after n oscillations the amplitude of the damped oscillator drops to $1/e$ of its original value. Show that

$$\frac{T}{T_0} = \left(1 + \frac{1}{4\pi^2 n^2} \right)^{1/2} \simeq \left(1 + \frac{1}{8\pi^2 n^2} \right)$$

and

$$\frac{\omega}{\omega_0} \simeq 1 - \frac{1}{8\pi^2 n^2}$$

3.14. Consider a harmonic oscillator of mass m under a restoring force $-kx$ and damping force $-b\dot{x}$. It is displaced a distance $+A$ and then released with zero velocity. Find the equation representing the underdamped, critically damped, and overdamped motions.

3.15. Discuss Problem 3.14 if mass m is released with velocity v_0 from the displaced position.

3.16. In the case of a damped harmonic oscillator, find the position of $x(t)$ when its values are maximum and minimum.

3.17. Discuss the motion of a damped oscillator subject to a constant force F_0.

3.18. In the case of a critically damped oscillator for which the velocity at the equilibrium position is v_0 and that is subject to a force $F_0 \cos \omega t$, calculate the value of $x(t)$ and make a graph for some meaningful values of the constants.

3.19. In the case of a critically damped oscillator for which the velocity at the equilibrium position is v_0 and that is subject to a force $F_0 \sin \omega t$, calculate the value of $x(t)$ and make a graph for some meaningful values of the constants.

3.20. Consider a damped oscillator for which $\gamma = \omega_0/4$, having a velocity v_0 at x_0 at $t = 0$. After a time $t = 2\pi/\omega_0$, a force $F = F_0 \cos \omega t$ is applied. Calculate $x(t)$ and make a graph for some meaningful values of the constants.

3.21. Consider a forced oscillator for which $\dot{x} = 0$ at $t = 0$ and $F = F_0 \cos(\omega t + \theta)$. Find the particular solution and the general solution and make a graph for some meaningful values of the constants.

3.22. Consider a damped oscillator for which $\gamma = \omega_0/4$. If the driving force is given by

$$F = A_1 \cos \omega t + A_3 \cos 3\omega t$$

calculate the value of $x(t)$.

3.23. One end of a massless spring of natural length l_0 is attached to a fixed point, while the other end is attached to a mass m. The whole system is being immersed in a viscous medium that exerts a force on m proportional to the velocity of m ($= kv$). The force constant of the spring is of the form $k^2/4m$. Initially, the spring is stretched to a length x_0 and released. Find the position of m as a function of time; evaluate all the arbitrary constants in terms of the initial conditions. Calculate the rate at which the total energy is changing at time $t = \pi m/k$, and graph this as a function of time.

3.24. A mass of 0.1 kg is attached to a spring of stiffness constant 20 N/cm. The mass is displaced 10 cm and released in a viscous medium. In 12 seconds the amplitude reduces to $1/e$ of its initial value. Calculate **(a)** the damping constant, **(b)** ω, **(c)** the decrement of motion, **(d)** the characteristic time, and **(e)** $x(t)$, and graph it.

3.25. In the case of the forced oscillator considered in the text, using the solution $x = A \cos(\omega t + \theta)$, solve for x_i.

3.26. Prove Eq. (3.124); that is, the maximum height of the resonance curve falls to one-half its value when $\gamma^2 = (\omega - \omega_0)^2$.

3.27. Show that the maximum height of the resonance curve is $1/\gamma^2$.

3.28. Prove Eq. (3.129); that is,

$$\langle P \rangle = \tfrac{1}{2}\dot{x}_m F_0 \sin \phi$$

3.29. In Eq. (3.129), substitute the value of $\sin \phi$ and state the final result in terms of ω, ω_0, and γ.

3.30. Derive Eq. (3.130).

3.31. Consider a damped oscillator with $m = 0.2$ kg, $k = 100$ N/m, and $b = 5$ N-s/m. The oscillator is driven by a force $F = (1.6 \text{ N}) \cos 20t$.
   (a) If the displacement is given by $x = A \cos(\omega t - \delta)$, what are the values of A and δ?
   (b) Calculate the energy dissipated in one cycle.
   (c) What is the average power input?
   (d) Graph the energy and energy dissipated versus time.

SUGGESTIONS FOR FURTHER READING

ARTHUR, W., and FENSTER, S. K., *Mechanics,* Chapter 6. New York: Holt, Rinehart and Winston, Inc., 1969.

BARGER, V., and OLSSON, M., *Classical Mechanics,* Chapter 1. New York: McGraw-Hill Book Co., 1973.

BECKER, R. A., *Introduction to Theoretical Mechanics,* Chapter 7. New York: McGraw-Hill Book Co., 1954.

DAVIS, A. DOUGLAS, *Classical Mechanics,* Chapter 3. New York: Academic Press, Inc., 1986.

FOWLES, G. R., *Analytical Mechanics,* Chapter 3. New York: Holt, Rinehart and Winston, Inc., 1962.

FRENCH, A. P., *Vibrations and Waves,* Chapters 3 and 4. New York: W. W. Norton and Co., Inc., 1971.

HAUSER, W., *Introduction to the Principles of Mechanics,* Chapter 4. Reading, Mass.: Addison-Wesley Publishing Co., 1965.

HOCHSTADT, H., *Differential Equations—A Modern Approach.* New York: Holt, Rinehart and Winston, Inc., 1964.

KITTEL, C., KNIGHT, W. D., and RUDERMAN, M. A., *Mechanics,* Berkeley Physics Course, Volume 1, Chapter 7. New York: McGraw-Hill Book Co., 1965.

KLEPPNER, D., and KOLENKOW, R. J., *An Introduction to Mechanics,* Chapter 10. New York: McGraw-Hill Book Co., 1973.

MARION, J. B., *Classical Dynamics,* 2nd. ed., Chapters 3 and 4. New York: Academic Press, Inc., 1970.

MURPHY, G. M., *Ordinary Differential Equations and Their Solutions.* New York: Van Nostrand Reinhold, 1960.

ROSSBERG, K., *Analytical Mechanics,* Chapter 10. New York: John Wiley & Sons, Inc., 1983.

SLATER, J. C., *Mechanics,* Chapter 2. New York: McGraw-Hill Book Co., 1947.

STEPHENSON, R. J., *Mechanics and Properties of Matter,* Chapter 5. New York: John Wiley & Sons, Inc., 1962.

SYMON, K. R., *Mechanics,* 3rd. ed., Chapter 2. Reading, Mass.: Addison-Wesley Publishing Co., 1971.

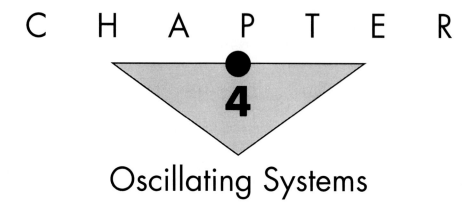

C H A P T E R

4

Oscillating Systems

4.1 INTRODUCTION

We continue our investigation of oscillating systems by dividing our discussion into oscillations in electrical circuits and phase diagrams of nonlinear systems. To start, we extend the results derived for mechanical linear oscillating systems to electrical oscillating systems. There is a complete analogy between electrical and mechanical systems; hence we need not derive all the results. A portion of this chapter will be devoted to the study of nonlinear systems. We shall introduce several techniques for solving oscillating systems, as well.

Chapter 3 was completely devoted to the study of linear oscillating systems. This type of simple harmonic motion, although extensively used in physics and engineering problems, is truly applicable to only a limited number of cases: Nature does not allow such simplicity. Oscillating motion, in general, is nonlinear. A departure from linear motion occurs whenever the restoring force does not have a linear dependence on displacement; furthermore, the damping force may not have a linear dependence on velocity. Thus an equation representing the free motion of an oscillating system may be written as

$$m\ddot{x} + G(\dot{x}) + F(x) = 0 \tag{4.1}$$

where F is nonlinear function of x and G is a nonlinear function of $\dot{x}$.

For arbitrary functions F and G, the general solutions of Eq. (4.1) are not known. Equation (4.1) can be solved only for some particular cases. In general, we must use approximate methods to get some idea of the nature of the oscillations. Each particular situation must be treated as a special case and solved individually. We should keep in mind that the nonlinear nature of oscillating systems is, in general, due to the large amplitudes of oscillating systems.

To get approximate solutions, several techniques have been developed. It is not possible to treat them at any length in this chapter. We shall introduce the principle of superposition and the techniques of Fourier analysis, which are extremely helpful in solving nonlinear systems. We shall employ the method of Green's function whenever the system is acted on by a large force for a short interval of time. We shall extend our discussion to symmetrical and nonsymmetrical nonlinear systems and use the techniques of series expansion and successive approximations in solving such problems. We shall conclude the chapter with a qualitative discussion of these nonlinear systems employing the method of phase diagrams.

4.2 HARMONIC OSCILLATIONS IN ELECTRICAL CIRCUITS

There is a complete analogy between the free, damped, and forced oscillations of a single particle, which we discussed in Chapter 3, and several electrical circuits, which we discuss now. Furthermore, this analogy extends to many physical situations in nature, including atomic, molecular, and nuclear physics. Some time ago, electrical circuits were constructed by analogy with mechanical systems, but recently the situation has been reversed. Electrical circuit designs are now so advanced that mechanical engineers use them extensively in investigating mechanical vibrational problems.

Figure 4.1(i) shows the three mechanical systems that we have been discussing and Fig. 4.1(ii) shows the corresponding electrical oscillating systems. Table 4.1 shows the corresponding mechanical and electrical quantities in the oscillating systems. We shall discuss this analogy presently. To start, let us consider Fig. 4.1(a)(i), which shows a simple oscillator consisting of mass m tied to a spring of spring constant k. The mass moves on a frictionless surface and hence behaves like a free oscillator. Its motion is represented by the equation

$$m\ddot{x} + kx = 0 \quad \text{or} \quad \ddot{x} + \frac{k}{m}x = 0 \tag{4.2}$$

which has a solution

$$x = x_0 \cos \omega_0 t \tag{4.3a}$$

where ω_0 is the free natural frequency given by

$$\omega_0 = \sqrt{\frac{k}{m}} \tag{4.3b}$$

The electrical analogy to this, shown in Fig. 4.1(a)(ii), consists of a capacitor C and an inductor L. Let $Q = Q(t)$ be the charge on the capacitor at any time t and $I = I(t)$ be the current through the inductor. The relations between I and Q are

$$I = \frac{dQ}{dt} = \dot{Q} \tag{4.4a}$$

$$Q = \int I\,dt \tag{4.4b}$$

The voltage drop across the inductor is

$$V_L = L\frac{dI}{dt} \tag{4.5}$$

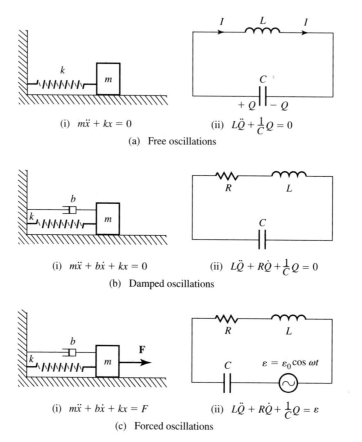

$$\text{(i)} \quad m\ddot{x} + kx = 0 \qquad\qquad \text{(ii)} \quad L\ddot{Q} + \frac{1}{C}Q = 0$$

(a) Free oscillations

$$\text{(i)} \quad m\ddot{x} + b\dot{x} + kx = 0 \qquad \text{(ii)} \quad L\ddot{Q} + R\dot{Q} + \frac{1}{C}Q = 0$$

(b) Damped oscillations

$$\text{(i)} \quad m\ddot{x} + b\dot{x} + kx = F \qquad \text{(ii)} \quad L\ddot{Q} + R\dot{Q} + \frac{1}{C}Q = \varepsilon$$

(c) Forced oscillations

Figure 4.1 Analogy between (i) a mechanical system and (ii) an electrical system: (a) free oscillations, (b) damped oscillations, and (c) forced oscillations.

Table 4.1 Analogy between Mechanical and Electrical Quantities

Mechanical	Electrical	
Displacement	$x \leftrightarrow Q$	Charge
Velocity	$\dot{x} \leftrightarrow \dot{Q} = I$	Current
Mass	$m \leftrightarrow L$	Inductance
Compliance	$1/k \leftrightarrow C$	Capacitance
Damping constant	$b \leftrightarrow R$	Resistance
Applied force	$F \leftrightarrow \varepsilon$	Applied emf

and across the capacitor it is

$$V_C = \frac{Q}{C} = \frac{1}{C} \int I \, dt \qquad (4.6)$$

Thus the sum of the voltages encountered in going around the whole circuit must be zero; that is, by applying Kirchhoff's rule, we get

$$V_L + V_C = 0$$

or

$$L \frac{dI}{dt} + \frac{1}{C} \int I \, dt = 0 \qquad (4.7)$$

Using Eqs. (4.4), we get

$$L\ddot{Q} + \frac{1}{C} Q = 0 \quad \text{or} \quad \ddot{Q} + \frac{1}{LC} Q = 0 \qquad (4.8)$$

This equation is identical to the mechanical system given by Eq. (4.2) provided

$$Q \leftrightarrow x, \quad \ddot{Q} \leftrightarrow \ddot{x}, \quad L \leftrightarrow m, \quad \text{and} \quad C \leftrightarrow 1/k$$

Assuming $Q = Q_0$ at $t = 0$, the solution is

$$Q = Q_0 \cos \omega_0 t \qquad (4.9)$$

with the free natural frequency

$$\omega_0 = \sqrt{\frac{1}{LC}} \qquad (4.10)$$

If we differentiate Eq. (4.9), we may write

$$\frac{dQ}{dt} = I = -\omega_0 Q_0 \sin \omega_0 t = -I_0 \sin \omega_0 t \qquad (4.11)$$

Let us now consider the damped oscillator shown in Fig. 4.1(b)(i). Note that we have added a dashpot filled with viscous fluid to indicate damping with damping constant b. The differential equation representing the motion of a damped oscillator is

$$m\ddot{x} + b\dot{x} + kx = 0 \qquad (4.12)$$

while Fig. 4.1(b)(ii) shows the equivalent electrical circuit, where we have added a resistor R, which is equivalent to the damping or frictional force in mechanical systems. The electrical equation analogous to a mechanical system is

$$L\ddot{Q} + R\dot{Q} + \frac{1}{C} Q = 0 \qquad (4.13)$$

Figure 4.1(c)(i) shows a mechanical driven oscillator represented by the following equation:

$$m\ddot{x} + b\dot{x} + kx = F = F_0 \cos \omega t \qquad (4.14)$$

Figure 4.1(c)(ii) represents an analogous electrical driven oscillator with an emf source given by $\varepsilon = \varepsilon_0 \cos \omega t$. The corresponding equation is

$$L\ddot{Q} + R\dot{Q} + \frac{1}{C}Q = \varepsilon = \varepsilon_0 \cos \omega t \qquad (4.15)$$

To extend our analogy still further we give another example, shown in Fig. 4.2(a) and (b). Suppose we apply a force F to a system consisting of two springs in line [Fig. 4.2(a)(i)]. The net displacement x is given by the sum of the displacement x_1 and x_2 caused by the two springs. Thus

$$x = x_1 + x_2 = \frac{F}{k_1} + \frac{F}{k_2} = F\left(\frac{1}{k_1} + \frac{1}{k_2}\right) = \frac{F}{k}$$

That is, when the *springs* are *in series*, the equivalent k is given by

$$\frac{1}{k} = \frac{1}{k_1} + \frac{1}{k_2} \qquad (4.16)$$

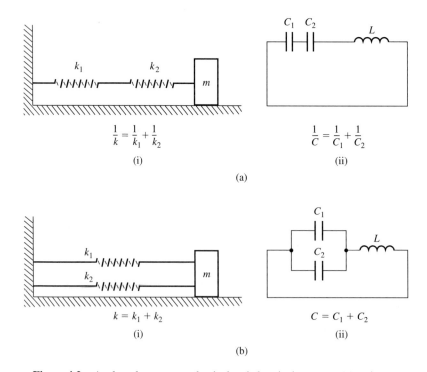

Figure 4.2   Analogy between mechanical and electrical systems: (a) series systems, and (b) parallel systems.

When the *springs* are *in parallel* [Fig. 4.2(b)(i)],

$$k = k_1 + k_2 \tag{4.17}$$

When the *capacitors* are *in series* [Fig. 4.2(a)(ii)], the equivalent capacitance C is

$$\frac{1}{C} = \frac{1}{C_1} + \frac{1}{C_2} \tag{4.18}$$

and when the *capacitors* are *in parallel* [Fig. 4.2(b)(ii)], the equivalent capacitance C is given by

$$C = C_1 + C_2 \tag{4.19}$$

Energy Considerations

For a free oscillator the total energy—the sum of the kinetic and potential energy—is always constant as long as there is no damping and is given by

$$\tfrac{1}{2}m\dot{x}^2 + \tfrac{1}{2}kx^2 = \text{constant} \tag{4.20}$$

Either by analogy with this equation or from Table 4.1, and using Eqs. (4.9) and (4.11), we get

$$\frac{1}{2}LI^2 + \frac{1}{2}\frac{Q^2}{C} = \left(\frac{Q_0^2}{2C}\right) = \text{constant} \tag{4.21}$$

$\tfrac{1}{2}LI^2$ is the energy stored in an inductor and is equivalent to the mechanical kinetic energy $\tfrac{1}{2}mv^2$; $\tfrac{1}{2}Q^2/C$ is the energy stored in the capacitor and is equivalent to the mechanical potential energy $\tfrac{1}{2}kx^2$.

Let us extend this analogy to a damped oscillator. Starting with Eq. (4.13),

$$L\ddot{Q} + R\dot{Q} + \frac{1}{C}Q = 0$$

and multiplying both sides by $\dot{Q}$ yields

$$L\ddot{Q}\dot{Q} + R\dot{Q}^2 + \frac{Q}{C}\dot{Q} = 0$$

or

$$\frac{d}{dt}\left(\frac{1}{2}L\dot{Q}^2\right) + \frac{d}{dt}\left(\frac{1}{2}\frac{Q^2}{C}\right) = -R\dot{Q}^2 \tag{4.22}$$

Since $\dot{Q} = I$, we may write

$$\frac{d}{dt}\left(\frac{1}{2}LI^2 + \frac{1}{2}\frac{Q^2}{C}\right) = -RI^2 \tag{4.23}$$

This equation states that the rate at which the energy is being stored in the inductor and the capacitor is equal to the energy dissipated in the resistor, as it should be. For mechanical systems, Eq. (4.22), after proper substitution, takes the following form:

$$\frac{d}{dt}\left(\frac{1}{2}m\dot{x}^2 + \frac{1}{2}kx^2\right) = -b\dot{x}^2 \tag{4.24}$$

The preceding analogy can be further extended to a driven oscillator, getting its solution and then writing an expression for average power dissipation, resonance, and quality factor.

4.3 PRINCIPLE OF SUPERPOSITION AND FOURIER SERIES

The principle of superposition is used throughout physics. Most students encounter this principle in the following simple form while studying wave motion and optics in general physics.

Principle of Superposition. When two or more waves travel simultaneously through a portion of a medium, each wave acts independently as if the other were not present. The resultant displacement at any point is the vector sum of the displacements of the individual waves.

We now extend this principle to the case of harmonic oscillators and to linear operators in general. If $x_1(t)$, $x_2(2)$, ... are the solutions when the forces acting are $F_1(t)$, $F_2(t)$, ..., respectively, then $x(t) = x_1(t) + x_2(t) + \cdots$ is a solution when the force acting on the system is $F(t) = F_1(t) + F_2(t) + \cdots$. We may further generalize as follows. The second-order linear differential equation describing a forced harmonic oscillator given by

$$\frac{d^2x}{dt^2} + 2\gamma \frac{dx}{dt} + \omega_0^2 x = F(t) \tag{4.25}$$

may be written in general form as

$$\left(\frac{d^2}{dt^2} + a \frac{d}{dt} + b \right) x(t) = F(t) \tag{4.26}$$

We define a *linear operator, L,* as the quantity in the parentheses on the left; that is,

$$L \equiv \left(\frac{d^2}{dt^2} + a \frac{d}{dt} + b \right) \tag{4.27}$$

Thus Eq. (4.25) or (4.26) may be written as

$$Lx(t) = F(t) \tag{4.28}$$

According to the *superposition principle:*

If a set of functions $x_n(t)$, n = 1, 2, 3, . . . , comprises solutions of a linear differential equation

$$Lx_n(t) = F_n(t) \tag{4.29}$$

then the function $x(t)$, which is a linear combination of $x_n(t)$, that is,

$$x(t) = \sum_n C_n x_n(t) \tag{4.30}$$

where C_n are constants, satisfies the differential equation

$$Lx(t) = F(t) \tag{4.31}$$

where

$$F(t) = \sum_n C_n F_n(t) \tag{4.32}$$

We can prove this statement by substituting Eq. (4.30) into Eq. (4.31); that is,

$$Lx(t) = L\left[\sum_n C_n x_n(t)\right] = \sum_n C_n Lx_n(t) = \sum_n C_n F_n(t) = F(t)$$

as it should be.

Let us apply these results to the general case of the driven harmonic oscillator we have discussed in detail. Suppose the individual driving forces $F_n(t)$ have a harmonic dependence of the form $\cos(\omega_n t - \theta_n)$, so that

$$F(t) = \sum_n C_n \cos(\omega_n t - \theta_n) \tag{4.33}$$

When the force was of the form $F_0 \cos(\omega t + \theta_0)$, the steady-state solution was given by Eqs. (3.99) and (3.100). Thus for $F(t)$ given by Eq. (4.33), the steady-state solution is

$$x(t) = \sum_n \frac{C_n}{m} \frac{1}{\sqrt{(\omega_0^2 - \omega_n^2)^2 + 4\gamma^2 \omega_n^2}} \cos(\omega_n t - \theta_n - \phi_n) \tag{4.34}$$

where

$$\phi_n = \tan^{-1}\left(\frac{2\gamma\omega_n}{\omega_0^2 - \omega_n^2}\right) \tag{4.35}$$

The general solution is the sum of the transient and the steady state and is given by

$$x(t) = Ae^{-\gamma t}\cos(\omega_1 t + \theta') + \sum_n \frac{C_n}{m} \frac{\cos(\omega_n t - \theta_n - \phi_n)}{\sqrt{(\omega_0^2 - \omega_n^2)^2 + 4\gamma^2\omega_n^2}} \tag{4.36}$$

where the constants A and θ' are to be determined from initial conditions as usual. Similar results can be obtained if $F(t)$ has a dependence of the form $\sin(\omega_n t - \theta_n)$.

With the help of the Fourier theorem we can extend the preceding type of consideration to the case in which the driving force is periodic (but not harmonic) and is a continuous or piecewise continuous function. $F(t)$ is a periodic function if it satisfies the condition

$$F(t + T) = F(t) \tag{4.37}$$

where $T = 2\pi/\omega$ is the period of the applied force. According to the **_Fourier theorem,_** *any arbitrary periodic function, which is continuous or piecewise continuous, having only a finite number of discontinuities over a time period, can be expressed as a sum of harmonic terms.* Thus any function $F(t)$ that is defined within a time interval $-T/2 < t < T/2$ [or a function $F(x)$ within a time interval $-\pi < x < \pi$] can be expressed as a series of sine and cosine terms as

$$F(t) = \frac{A_0}{2} + A_1 \cos \omega t + A_2 \cos 2\omega t + \cdots + A_n \cos n\omega t + \cdots$$

$$+ B_1 \sin \omega t + B_2 \sin 2\omega t + \cdots + B_n \sin n\omega t + \cdots \tag{4.38a}$$

where A_n and B_n are constants and $n = 1, 2, 3, \ldots, n, \ldots, \infty$. Equation (4.38a) may also be written as

$$F(t) = \frac{A_0}{2} + \sum_{n=1}^{\infty} (A_n \cos n\omega t + B_n \sin n\omega t) \tag{4.38b}$$

The terms in the sum (on the right) form a *Fourier series*. The constants A_n and B_n can be evaluated by integration. For example, to evaluate A_n we multiply both sides of Eq. (4.38b) by $\cos m\omega t$ (m being an integer) and integrate between the limits $-T/2$ and $T/2$:

$$\int_{-T/2}^{+T/2} F(t) \cos m\omega t \, dt = \frac{A_0}{2} \int_{-T/2}^{+T/2} \cos m\omega t \, dt$$

$$+ \sum_{n=1}^{\infty} \left[A_n \int \cos n\omega t \cos m\omega t \, dt + B_n \int \sin n\omega t \cos m\omega t \, dt \right]$$

Since m and n are integers, for all values of m and n, we obtain

$$\int_{-T/2}^{+T/2} \cos n\omega t \cos m\omega t \, dt = 0, \qquad \text{if } m \neq n \text{ and } = T/2 \text{ if } m = n$$

$$\int_{-T/2}^{+T/2} \sin n\omega t \cos m\omega t \, dt = 0, \qquad \text{for all values of } m \text{ and } n$$

$$\int_{-T/2}^{+T/2} \cos m\omega t \, dt = 0, \qquad \text{if } m \neq 0 \text{ and } = T \text{ if } m = 0$$

Thus the values of A_0 and A_n are

$$A_0 = \frac{2}{T} \int_{-T/2}^{+T/2} F(t) \, dt \tag{4.39a}$$

$$A_n = \frac{2}{T} \int_{-T/2}^{+T/2} F(t) \cos n\omega t \, dt \qquad \text{if } n \text{ is an integer} \tag{4.39b}$$

That is, $n = 1, 2, 3, \ldots$. Similarly multiplying Eq. (4.38b) by $\sin m\omega t$ and integrating, we obtain

$$B_n = \frac{2}{T} \int_{-T/2}^{+T/2} F(t) \sin n\omega t \, dt \qquad \text{if } n \text{ is an integer} \tag{4.39c}$$

That is, $n = 1, 2, 3, \ldots$. If necessary we can replace the integration limits $-T/2 \,(= -\pi/\omega)$ to $T/2 \,(= +\pi/\omega)$ by 0 to $T \,(= 2\pi/\omega)$.

    First one has to determine the appropriate number of terms that must be used in the Fourier series to approximate the arbitrary driving force. This is illustrated in two cases: (1) a rectangular function, and (2) a sawtooth function, as we shall discuss shortly. Once we know the series, each term used in the applied force has a corresponding solution. By adding all these solutions we obtain the general solution of a damped harmonic oscillator driven by an arbitrary force. In actual practice, obtaining solutions by this method is quite tedious, but in some situations it is helpful.

Example 4.1

Consider the function $F(\theta)$ given by the equations and graphed below. Find a Fourier series expansion of the function $F(\theta)$, where $\theta\ (=\omega t)$ is a function of time t and angular velocity ω.

Solution

Redefine function $F(\theta)$ in terms of function $f(\theta)$. Graph $f(\theta)$ for the different values of θ given below.

$F(\theta)=-\dfrac{\pi}{2}$ between $-\pi\le\theta\le0$

$F(\theta)=\dfrac{\pi}{2}$ between $0\le\theta\le\pi$

$\theta:=-\pi,-19\cdot\dfrac{\pi}{20}..\pi$ $F(\theta):=\dfrac{\pi}{2}$

$f(\theta):=\text{if}(\theta>0,F(\theta),-F(\theta))$

In order to find the Fourier series expansion that will result in function $f(\theta)$, we evaluate the coefficients A0, A, and B by using Eq. (4.39). Let T be the time period. As t changes from -T/2 to T/2, θ changes from $-\pi$ to π.

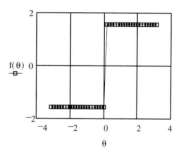

$n:=1,2..10$ $A0:=-\dfrac{1}{2}\cdot\displaystyle\int_{-\pi}^{0}1\,d\theta+\dfrac{1}{2}\cdot\displaystyle\int_{0}^{\pi}1\,d\theta$

$A_n:=-\dfrac{1}{2}\cdot\displaystyle\int_{-\pi}^{0}\cos(n\cdot\theta)\,d\theta+\dfrac{1}{2}\cdot\displaystyle\int_{0}^{\pi}\cos(n\cdot\theta)\,d\theta$ $B_n:=-\dfrac{1}{2}\cdot\displaystyle\int_{-\pi}^{0}\sin(n\cdot\theta)\,d\theta+\dfrac{1}{2}\cdot\displaystyle\int_{0}^{\pi}\sin(n\cdot\theta)\,d\theta$

$A0=0$

The values of the first two are

$A0=0$ $A_n=0$

For B_n even terms are 0, hence the expansion series consists only of odd sine terms.

$F(\theta)=\dfrac{2}{1}\cdot\sin(\theta)+\dfrac{2}{3}\cdot\sin(3\cdot\theta)+\dfrac{2}{5}\cdot\sin(5\cdot\theta)+\ldots\ldots$

As an example, H11, H12, and H13 represent series using n = 1, 2, and 3 terms, respectively.

$H11_n=\dfrac{A0}{2}+\dfrac{2}{1}\cdot\theta_n$ $H12_n=\dfrac{A0}{2}+\dfrac{2}{1}\cdot\sin(\theta_n)+\dfrac{2}{3}\cdot\sin(3\cdot\theta_n)$

$H13_n=\dfrac{A0}{2}+\dfrac{2}{1}\cdot\sin(\theta_n)+\dfrac{2}{3}\cdot\sin(3\cdot\theta_n)+\dfrac{2}{5}\cdot\sin(5\cdot\theta_n)$

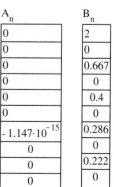

A_n	B_n
0	2
0	0
0	0.667
0	0
0	0.4
0	0
0	0.286
$-1.147\cdot10^{-15}$	0
0	0.222
0	0

Now we change the range number n to the desired value. We plot the functions H (replacing F) using different numbers of terms. H1 means 1 term, H3 means three terms, H5 means 5 terms, and so on. It is clear from the graph below that the H function approaches the F function as the number of terms increases. For example, using n = 15 terms, H almost coincides with the $f(\theta)$ plot. Using more terms will make the graph still closer to the rectangular graph.

$$n := 0..15 \qquad I := 200 \qquad i := 0..I \qquad \theta_i := \frac{1}{20}$$

$$H1_i := \frac{A0}{2} + \left[\sum_{n=1}^{1} \left(A_n \cdot \cos(n \cdot \theta_i) + B_n \cdot \sin(n \cdot \theta_i)\right)\right]$$

$$H3_i := \frac{A0}{2} + \left[\sum_{n=1}^{3} \left(A_n \cdot \cos(n \cdot \theta_i) + B_n \cdot \sin(n \cdot \theta_i)\right)\right]$$

$$H5_i := \frac{A0}{2} + \left[\sum_{n=1}^{5} \left(A_n \cdot \cos(n \cdot \theta_i) + B_n \cdot \sin(n \cdot \theta_i)\right)\right]$$

$$H_i := \frac{A0}{2} + \left[\sum_{n=1}^{10} \left(A_n \cdot \cos(n \cdot \theta_i) + B_n \cdot \sin(n \cdot \theta_i)\right)\right]$$

Fourier Series

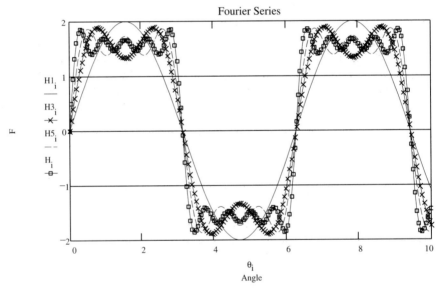

θ_i
Angle

(a) Write H5 in terms of an expansion series.
(b) How will the graph look if A0 is not zero but is constant? Draw the graphs.
(c) Graph for 30 terms and 50 terms. What results do you expect?
(d) How will the plot of H versus θ differ from H versus t?

EXERCISE 4.1 Consider the function shown in Fig. Exer. 4.1 in the interval $-\pi < \theta < \pi$. Find a Fourier series expansion of this function.

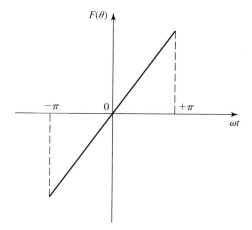

$F(\theta)$

$-\pi$ 0 $+\pi$

ωt

Figure Exer. 4.1

4.4 HARMONIC MOTION AND GREEN'S FUNCTION

Impulsive Force

When a very large force $F(t)$ acts on a system for a very short interval of time Δt, an *impulse* is said to be imparted to the system. It is the same thing when a force $F(t)$ applied to a system has a small value for a short interval of Δt, while almost negligible before and after this interval. By applying the impulse–momentum theorem and the superposition principle to the oscillating system, we can arrive at many interesting and useful results. According to the *impulse–momentum theorem,*

$$p_f - p_i = F \, \Delta t = \int F \, dt \qquad (4.40a)$$

or

$$\Delta p = F \, \Delta t \qquad (4.40b)$$

and

$$\Delta v = \frac{F}{m} \Delta t \qquad (4.40c)$$

We consider the application of this to several situations.

    Case (i)   An Oscillator Initially at Rest: To start, let us assume that we are dealing with an undamped oscillator at rest; that is, $x = 0$ and $\dot{x} = 0$. At time $t = t_0$ an impulse is given to the oscillator so that its velocity right after the impulse is v_0. Thus from Eq. (4.40b),

$$\Delta p = mv_0 = F \, \Delta t \qquad (4.41a)$$

or

$$v_0 = \frac{F}{m} \Delta t \qquad (4.41b)$$

Since at $t = t_0$, $x = 0$ the displacement x of an undamped oscillator is

$$x = A \sin[\omega_0(t - t_0)] \tag{4.42a}$$

where A is to be determined from initial conditions. (Note that we have neglected any short displacement that may result during a short interval Δt when the force is applied.) Differentiating Eq. (4.42a) and substituting $\dot{x} = v_0$ when $t = t_0$, we get

$$\dot{x} = \omega_0 A \cos[\omega_0(t - t_0)] \tag{4.42b}$$

and

$$v_0 = \omega_0 A$$

Therefore,

$$A = \frac{v_0}{\omega_0} = \frac{F \, \Delta t}{m\omega_0} = \frac{\Delta p}{m\omega_0} \tag{4.43}$$

Thus the general solution is

$$x(t) = \begin{vmatrix} 0, & \text{for } t \leqslant t_0 \\ \dfrac{F \, \Delta t}{m\omega_0} \sin[\omega_0(t - t_0)], & \text{for } t \geqslant t_0 \end{vmatrix} \tag{4.44}$$

This procedure can be easily extended to the case of damped harmonic oscillators at rest, for which

$$x = Ae^{-\gamma(t-t_0)} \sin[\omega_1(t - t_0)] \tag{4.45}$$

and the final solution after the impulse has been applied is

$$x(t) = \begin{vmatrix} 0, & \text{for } t < t_0 \\ \dfrac{F \, \Delta t}{m\omega_1} e^{-\gamma(t-t_0)} \sin[\omega_1(t - t_0)], & \text{for } t \geqslant t_0 \end{vmatrix} \tag{4.46}$$

Case (ii)   *An Oscillator Initially Not at Rest*: Once again let us start with an undamped oscillator, but this time initially it has a displacement x_0 and velocity v_0. Thus, starting with

$$x = B \cos \omega_0 t + C \sin \omega_0 t \tag{4.47}$$

and applying the initial conditions at $t = t_0$, $x = x_0$, and $\dot{x} = v_0$, we get

$$x = x_0 \cos \omega_0(t - t_0) + \frac{v_0}{\omega_0} \sin \omega_0(t - t_0) \tag{4.48}$$

Let us now apply a force $F(t)$ at $t = t_0$ for short interval Δt. According to the impulse–momentum theorem,

$$\Delta p = F \, \Delta t \quad \text{or} \quad \Delta v = F \frac{\Delta t}{m} \tag{4.49}$$

where Δv is a small additional velocity given to the system, which already has some velocity v_0 at time t_0. The additional displacement resulting from the impulse may be calculated as if Δv

were the initial velocity; that is, by replacing v_0 in Eq. (4.48) by Δv given by Eq. (4.49), we get the additional displacement x_1 to be

$$x_1 = \frac{F \, \Delta t}{m\omega_0} \sin \omega_0(t - t_0) \tag{4.50}$$

Thus the total displacement is the sum of x and x_1 given by Eqs. (4.48) and (4.50):

$$x(t) = x_0 \cos \omega_0(t - t_0) + \frac{v_0}{\omega_0} \sin \omega_0(t - t_0) + \frac{F \, \Delta t}{m\omega_0} \sin \omega_0(t - t_0) \tag{4.51}$$

We can extend this treatment to the case of a damped harmonic oscillator that at $t = t_0$ has $x = x_0$ and $\dot{x} = v_0$, while an impulse is given at $t = t_0$. (See Problem 4.23.) Before the impulse the motion is described by

$$x = e^{-\gamma t} [B \cos \omega_1 t + C \sin \omega_1 t] \tag{4.52}$$

After the impulse the motion is described by

$$xt = e^{-\gamma(t-t_0)} \left(x_0 \cos \omega_1(t - t_0) + \frac{v_0}{\omega_1} \sin \omega_1(t - t_0) \right) + \frac{F \, \Delta t}{m\omega_1} e^{-\gamma(t-t_0)} \sin[\omega_1(t - t_0)] \tag{4.53}$$

Continuous Arbitrary Force and Green's Function

As long as we are considering a linear oscillator, we can extend the application of the impulse–momentum theorem to the case of an arbitrary force function. The solution is based on the method developed by George Green. According to *Green's method, an arbitrary force function F(t) can be thought of as a series of impulses, each acting for a short interval of time Δt and delivering an impulse $F(t) \, \Delta t$,* as shown in Fig. 4.3. Thus

$$F(t) = \sum_{n=-\infty}^{n=+\infty} F_n(t) \tag{4.54}$$

where

$$F_n(t) \begin{vmatrix} = F(t_n), & \text{if } t_n < t < t_{n+1} \\ = 0, & \text{if } t < 0 \text{ or } t > t_{n+1} \end{vmatrix} \tag{4.55}$$

and $t = n \, \Delta t$. It is clear from Fig. 4.3 that if $\Delta t \to 0$, the sum of the series of impulses $\Sigma \, F_n(t)$ approaches $F(t)$. If the system is linear, we can always apply the principle of superposition. This allows us to write the inhomogeneous part of the differential equation as the sum of the individual impulses. That is, for

$$m\ddot{x} + b\dot{x} + kx = \sum F_n(t) \tag{4.56}$$

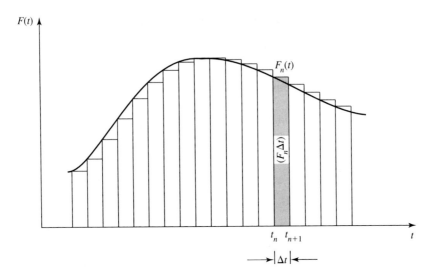

Figure 4.3   Arbitrary force function as a sum of a series of impulses.

the general steady-state solution is the sum of the individual solutions resulting from each $F_n(t)$. The individual solutions are of the type given by Eq. (4.46) for a single impulse. Thus the steady-state solution of Eq. (4.56) is

$$x(t) = \sum_{n=-\infty}^{N} \frac{F(t_n)\,\Delta t}{m\omega_1} e^{-\gamma(t-t_n)} \sin[\omega_1(t - t_n)] \qquad (4.57)$$

which includes all solutions up to and including the Nth impulse. We can replace the summation by integration when $\Delta t \to 0$ and $t_n = t'$. That is,

$$x(t) = \int_{-\infty}^{t} \frac{F(t')}{m\omega_1} e^{-\gamma(t-t')} \sin[\omega_1(t - t')]\, dt' \qquad (4.58)$$

We define *Green's function* $G(t, t')$ as

$$G(t, t') = \frac{e^{-\gamma(t-t')}}{m\omega_1} \sin[\omega_1(t - t')], \qquad \text{for } t \geq t'$$

$$= 0, \qquad\qquad\qquad\qquad\qquad \text{for } t < t' \qquad (4.59)$$

Thus, in terms of Green's function, we may write the steady-state solution [Eq. (4.58)] as

$$x(t) = \int_{-\infty}^{t} F(t')G(t, t')\, dt' \qquad (4.60)$$

The main advantage of this method is that the solution is already adjusted for initial conditions; for example, in this case it is for a damped oscillator that is initially at rest at the equilibrium position. We must add the transient solution to the steady-state solution [Eq. (4.60)] to obtain a complete solution. For different initial conditions, solutions may be obtained by the same procedure. (See Problems 4.24 and 4.25.)

 Example 4.2 _____

A damped oscillator is acted on by the force function below. Using Green's function, graph the response function as well as the applied force.

$F(t) = 0$ if $t < 0$ $F(t) = Fo\ exp(-\gamma t)\ sin(\omega t)$ if $t > 0$

Solution

Let us assume the values of the different variables are

$$i := 0..\,100 \qquad t_i := \frac{i}{5} \qquad F0 := 10 \qquad \gamma := .2$$

$$\omega := 2 \qquad \omega 1 := 1 \qquad M := .5$$

The acting force F(t) may be written as (This means that if $t < 0$, F(t) is zero; otherwise graph the third term in ( ).)

$$F(t) := if\!\left(t<0,0,F0\cdot e^{-\gamma t}\cdot \sin(\omega\cdot t)\right)$$

Using Eq. (4.59), the corresponding Green's function may be written as

$$G(t,t') := if\!\left[t<0,0,F0\cdot\frac{e^{-\gamma(t-t')}}{M\cdot\omega 1}\cdot \sin(\omega\cdot t)\right]$$

The resulting displacement x is given by

$$x_i := \int_0^{t_i} F(t')\cdot G\!\left(t_i,t'\right) dt'$$

The graphs of x and F versus t are as shown.

How do you explain the variation in x with time from the nature of the applied force?

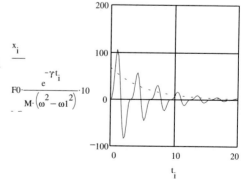

EXERCISE 4.2 Repeat the above example for the case in which the sine term in the force function is replaced by a cosine term. What difference do you observe in the two graphs?

Example 4.3 _____

Consider a damped harmonic oscillator that is suddenly acted on at $t = 0$ by a decaying force

$$F(t) = F_0 e^{-kt}, \quad t > 0$$

Find the general solution by using Green's function. (A typical example of such a force is the decay voltage on a capacitor.)

Solution

Since

$$F(t) = F(t') = F_0 e^{-kt'} \tag{i}$$

using Green's function, Eq. (4.59), the general solution, Eq. (4.60) is

$$x(t) = \int_0^t F(t')G(t, t')\, dt' = \frac{1}{m\omega_1}\int_0^t F_0 e^{-kt'}\, e^{-\gamma(t-t')} \sin[\omega_1(t - t')]\, dt' \tag{ii}$$

If we substitute

$$y = \omega_1(t - t'), \quad dy = -\omega_1\, dt', \quad \text{or} \quad dt' = -\frac{dy}{\omega_1} \tag{iii}$$

then the limits change from $0 \to t$ to $\omega_1 t \to 0$. Therefore,

$$x(t) = \frac{F_0}{m\omega_1^2}\int_0^{\omega_1 t} e^{-kt}\, e^{(k-\gamma)y\gamma/\omega_1} \sin y\, dy \tag{iv}$$

In order to plot x versus t, we first solve this equation for x. Solving for xt and simplifying, the resulting equation is

$$xt = \frac{F0}{m \cdot \omega 1^2} \cdot \int_0^{\omega 1 \cdot t} e^{-k \cdot t} \cdot e^{\frac{k-\gamma}{\omega 1}\cdot \gamma} \cdot \sin(y)\, dy$$

$$x_t = -F0 \cdot \frac{(\omega1 \cdot \cos(\omega1 \cdot t) - \sin(\omega1 \cdot t) \cdot k + \sin(\omega1 \cdot t) \cdot \gamma)}{\left[m \cdot \left[\omega1 \cdot \left(k^2 - 2 \cdot k \cdot \gamma + \gamma^2 + \omega1^2 \right) \right] \right]} \cdot \exp(-t \cdot \gamma) + F0 \cdot \frac{\exp(-k \cdot t)}{\left[m \cdot \left(k^2 - 2 \cdot k \cdot \gamma + \gamma^2 + \omega1^2 \right) \right]}$$

Using $\omega 1$ and A given below, x can be written as

$$\omega 1 = \sqrt{\omega 0^2 - \gamma^2} \qquad A = \frac{\dfrac{F0}{m}}{(k-\gamma)^2 + \omega 1^2} \qquad x_t = A \cdot \left[e^{-k \cdot t} - e^{-\gamma t} \cdot \left(\cos(\omega 1 \cdot t) - \frac{k-\gamma}{\omega 1} \cdot \sin(\omega 1 \cdot t) \right) \right] \tag{v}$$

Below we graph x versus t for 30 values of t and for 3 values of the damping constant γ.

$$N := 30 \qquad t := 0..N \qquad n := 1..3 \qquad k := .4 \qquad F0 := 100 \qquad m := .5$$

$$\omega 0 := \sqrt{\frac{k}{m}}$$

$$b_n := \begin{array}{|c|} \hline .2 \\ \hline .4 \\ \hline 1 \\ \hline \end{array} \qquad \gamma_n := \frac{b_n}{2 \cdot m} \qquad \gamma_n = \begin{array}{|c|} \hline 0.2 \\ \hline 0.4 \\ \hline 1 \\ \hline \end{array} \qquad \omega 1_n := \sqrt{\omega 0^2 - \left(\gamma_n\right)^2}$$

$$\omega 0 = 0.894$$

$$A_n := \frac{\dfrac{F0}{m}}{\left(k - \gamma_n\right)^2 + \left(\omega 1_n\right)^2} \qquad x_{t,n} := A_n \cdot \left[e^{-k \cdot t} - e^{-\gamma_n \cdot t} \cdot \left(\cos\left(\omega 1_n \cdot t\right) - \frac{k - \gamma_n}{\omega 1_n} \cdot \sin\left(\omega 1_n \cdot t\right) \right) \right]$$

x1 for $k > \gamma$

x2 for $k = \gamma$

x3 for $k < \gamma$

(a) If $k \gg \gamma$ and both are small compared to ω_0, $[\omega_1 = (\omega_0^2 - \gamma^2)^{1/2}]$ the term e^{-kt} has an effect only for a short time in the beginning.

(b) When $k = \gamma$ and both are small compared to ω_0, Eq. (v) takes the form

$$x(t) = \frac{F_0}{m\omega_1^2} e^{-\gamma t}(1 - \cos \omega_1 t) \tag{vi}$$

This means that the response function is still oscillatory, but with an exponentially decaying amplitude, as shown.

(c) If $k \ll \gamma$, the forcing function $F(t)$ given by Eq. (i) takes over the oscillatory motion; that is, the amplitude of the oscillations starts decaying exponentially after an initial increase over a short interval of time, as shown.

EXERCISE 4.3 Discuss the example if the decaying force at $t = 0$ is $F(t) = -kt^2$.

4.5 NONLINEAR OSCILLATING SYSTEMS

Before starting with this section, it will be worthwhile to review Section 3.2, where we explained the difference between linear and nonlinear systems. We have already seen that linear systems (systems in which the force is negatively proportional to the displacement) lead to harmonic oscillations (oscillations of one frequency). The condition imposed was that the motion must be limited to a small region near the equilibrium point.

Let us now consider systems in which the motion is not restricted and hence the restoring force is not proportional to the displacement. In general,

$$m\ddot{x} + F(x) = 0 \tag{4.61}$$

where $F(x)$ is the restoring force and is no longer linear. If damping is present, we shall have another free function $G(x)$, which may also be nonlinear. One outstanding characteristic of a nonlinear system is that, unlike linear systems, the time period of nonlinear systems, in general, depends on the amplitude. (There are several good textbooks on nonlinear mechanics; we simply introduce the subject here.)

Consider a system displaced to a position x from its equilibrium position x_0 and under a force $F(x)$. Let us expand $F(x)$ in a Taylor series about x_0; that is,

$$F(x) = F(x_0) + \left(\frac{dF}{dx}\right)_{x_0}(x - x_0) + \frac{1}{2}\left(\frac{d^2F}{dx^2}\right)_{x_0}(x - x_0)^2 + \frac{1}{6}\left(\frac{d^3F}{dx^3}\right)_{x_0}(x - x_0)^3 + \cdots \tag{4.62}$$

where $F(x_0) = 0$ because x_0 is the equilibrium point. Let $x_0 = 0$ be the origin. Define

$$\left(\frac{dF}{dx}\right)_0 = k_1, \quad \frac{1}{2}\left(\frac{d^2F}{dx^2}\right)_0 = k_2, \quad \frac{1}{6}\left(\frac{d^3F}{dx^3}\right)_0 = k_3, \quad \cdots \tag{4.63}$$

and write Eq. (4.62) as

$$F(x) = k_1 x + k_2 x^2 + k_3 x^3 + \cdots \tag{4.64}$$

We need not carry higher terms. If we consider only those forces that lead to stable equilibrium for symmetrical systems, the even terms must vanish; that is, $k_2 = k_4 = \cdots = 0$ and

$$F(x) = k_1 x + k_3 x^3 \tag{4.65}$$

This force is *symmetrical* about the equilibrium $x = 0$; that is, the magnitude of the force exerted on the system is the same for x and $-x$. If we set $k_1 = -k$, where k is positive, and $k_3 = -\epsilon$, we get

$$F(x) = -kx - \epsilon x^3 \tag{4.66}$$

Remember, if $\epsilon > 0$ the system is *hard* and if $\epsilon < 0$ the system is *soft*. For this force, the corresponding symmetrical potential is given by (force $F = -dV/dx$)

$$V(x) = \frac{1}{2}kx^2 + \frac{1}{4}\epsilon x^4 \tag{4.67}$$

On the other hand, if the system is asymmetrical, the force, from Eq. (4.64), after substituting $k_1 = -k$, $k_2 = -\lambda$ and setting $k_3 = 0$, is

$$F(x) = -kx - \lambda x^2 \tag{4.68}$$

Hence the *asymmetrical potential* is

$$V(x) = \tfrac{1}{2} kx^2 + \tfrac{1}{3} \lambda x^3 \tag{4.69}$$

We shall now discuss some situations dealing with symmetrical and asymmetrical potentials.

Symmetrical Nonlinear System

Consider a mass m which is suspended between two identical strings (or springs), as shown in Fig. 4.4(a). The strings are elastic with a force constant k_0 and tied to points A and B. When this system is in position AOB it is in equilibrium and the tension in each string is S_0 as shown. Let us now displace the mass m horizontally through a distance x, as shown in Fig. 4.4(b). The change in length of each string is $(l - l_0)$; hence the restoring force is $k_0(l - l_0)$. The tension S in each string when it is in the displaced position is

$$S = S_0 + k_0(l - l_0) \tag{4.70}$$

We resolve S into components. The vertical components cancel each other, while the sum of the horizontal components is $-2S \sin \theta$. Thus the motion of the mass m is described by the equation

$$m\ddot{x} = -2S \sin \theta \tag{4.71}$$

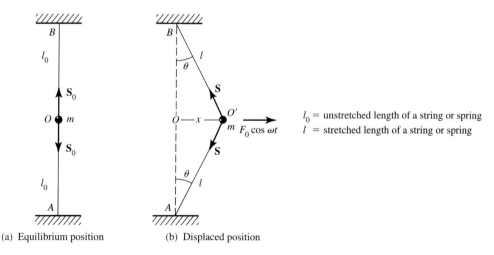

(a) Equilibrium position   (b) Displaced position

l_0 = unstretched length of a string or spring
l = stretched length of a string or spring

Figure 4.4   A mass m tied to two strings (or springs) constitutes a symmetrical nonlinear system: (a) in equilibrium, and (b) in a displaced position.

Remember that we have assumed no damping and also no driving force. (Later we shall apply the force $F_0 \cos \omega t$ and again solve the problem.) From Eqs. (4.70) and (4.71), we obtain

$$m\ddot{x} = -2[S_0 + k_0(l - l_0)] \sin \theta \qquad (4.72)$$

From Fig. 4.4(b), we can calculate the values for l and $\sin \theta$ to be

$$l = (l_0^2 + x^2)^{1/2} = l_0\left(1 + \frac{x^2}{l_0^2}\right)^{1/2} \qquad (4.73a)$$

and

$$\sin \theta = \frac{x}{l} = x(l_0^2 + x^2)^{-1/2} = \frac{x}{l_0}\left(1 + \frac{x^2}{l_0^2}\right)^{-1/2} \qquad (4.73b)$$

Substituting in Eq. (4.72) yields

$$m\ddot{x} = -2\left[S_0 + k_0 l_0\left\{\left(1 + \frac{x^2}{l_0^2}\right)^{1/2} - 1\right\}\right]\frac{x}{l_0}\left(1 + \frac{x^2}{l_0^2}\right)^{-1/2} \qquad (4.74)$$

Since x^2/l_0^2 is a small quantity, we can use the binomial theorem to expand

$$\left(1 + \frac{x^2}{l_0^2}\right)^{\pm 1/2} = 1 \pm \frac{1}{2}\frac{x^2}{l_0^2} + \cdots$$

Substituting these in Eq. (4.74), we get (dropping the x^5 term)

$$m\ddot{x} = -\frac{2S_0}{l_0}x - \left(\frac{k_0}{l_0^2} - \frac{S_0}{l_0^3}\right)x^3 \qquad (4.75)$$

Let

$$\frac{2S_0}{l_0} = k \quad \text{and} \quad \left(\frac{k_0 l_0 - S_0}{l_0^3}\right) = \epsilon \qquad (4.76)$$

Hence we can write Eq. (4.75) as

$$m\ddot{x} = -kx - \epsilon x^3 \qquad (4.77)$$

which is the required equation representing a nonlinear system. Note that ϵ is a small quantity that is positive for a hard spring and negative for a soft spring. Let us assume that the approximate solution of Eq. (4.77) is still sinusoidal as in the linear systems. This should be approximately true because ϵ is a small quantity. Hence,

$$x = A \cos \omega t \qquad (4.78)$$

Substituting for $x = x_1$ and $x^3 = x_1^3$ in the right side of Eq. (4.77), we get a new equation in x_1; that is,

$$m\ddot{x}_1 = -kA \cos \omega t - \epsilon A^3 \cos^3 \omega t \qquad (4.79)$$

Substituting

$$\cos^3 \omega t = \tfrac{1}{4}(3 \cos \omega t + \cos 3\omega t)$$

in Eq. (4.79) and rearranging,

$$m\ddot{x}_1 = -(kA + \tfrac{3}{4}\epsilon A^3) \cos \omega t - \tfrac{1}{4}\epsilon A^3 \cos 3\omega t \qquad \textbf{(4.80)}$$

which on integration (assuming the integration constants to be zero) gives the required solution:

$$x_1 = \frac{1}{m\omega^2}\left(kA + \frac{3}{4}\epsilon A^3\right)\cos \omega t + \frac{\epsilon A^3}{36\omega^2}\cos 3\omega t \qquad \textbf{(4.81)}$$

This is the solution for a first-order approximation.

To find the relation between ω and A we can make use of the assumption that ϵ is small. Thus, substituting a first-order approximation $x = x_1 = A \cos \omega t$ given by Eq. (4.78) in Eq. (4.81) and dropping the last term, or by comparing terms (the second term on the right is zero), we get

$$\omega^2 = \frac{k}{m} + \frac{3}{4}\frac{\epsilon}{m}A^2 \qquad \textbf{(4.82)}$$

which indicates that the natural frequency ω and hence the period $T = 2\pi/\omega$ are functions of the amplitude A. The quantity ω^2 increases or decreases from ω_0^2 by an amount $(3\epsilon/4m)A^2$ depending on the magnitude and the sign of ϵ.

If there were an external driving force $F = F_0 \cos \omega t$ acting on the system, as shown in Fig. 4.4(b), Eq. (4.77) would take the form

$$m\ddot{x} = -kx - \epsilon x^3 + F_0 \cos \omega t \qquad \textbf{(4.83)}$$

Following exactly the outlined procedure, we obtain the following solution (see Problem 4.26):

$$x_1 = \frac{1}{m\omega^2}\left(kA + \frac{3}{4}\epsilon A^3 - \frac{F_0}{m}\right)\cos \omega t + \frac{\epsilon A^3}{36\omega^2}\cos 3\omega t \qquad \textbf{(4.84)}$$

We shall not carry on with the discussion of resonances in this case; although they do occur, they are quite different from those discussed in linear systems.

Asymmetrical Nonlinear System

According to Eqs. (4.68) and (4.69), the asymmetric force and potential representing such a nonlinear system are

$$F(x) = -kx - \lambda x^2 \qquad \textbf{(4.68)}$$

$$V(x) = \tfrac{1}{2}kx^2 + \tfrac{1}{3}\lambda x^3 \qquad \textbf{(4.69)}$$

The differential equation describing such a system without damping is

$$m\ddot{x} + kx + \lambda x^2 = 0 \qquad \textbf{(4.85)}$$

Dividing by m and substituting $k/m = \omega_0^2$ and $\lambda/m = \lambda_1$, we get

$$\ddot{x} + \omega_0^2 x + \lambda_1 x^2 = 0 \tag{4.86}$$

To obtain an approximate solution we use a *perturbation method,* as explained next.

If there were no nonlinear term ($\lambda_1 x^2$), the solution would have been x_0. Since λ_1 is a small quantity, the correct solution of Eq. (4.86) can be obtained by adding a small correction term to x_0; that is,

$$x(t) \simeq x_0 + \lambda_1 x_1 \tag{4.87}$$

[To have higher-order corrections we must write

$$x(t) = x_0 + \lambda_1 x_1 + \lambda_1^2 x_2 + \lambda_1^3 x_3 + \cdots]$$

Substituting Eq. (4.87) in Eq. (4.86), we get

$$(\ddot{x}_0 + \omega_0^2 x_0) + (\ddot{x}_1 + \omega_0^2 x_1 + x_0^2)\lambda_1 + 2x_0 x_1 \lambda_1^2 + x_1^2 \lambda_1^3 = 0 \tag{4.88}$$

Neglecting the higher-order terms in λ_1^2 and λ_1^3 results in

$$(\ddot{x}_0 + \omega_0^2 x_0) + (\ddot{x}_1 + \omega_0^2 x_1 + x_0^2)\lambda_1 = 0 \tag{4.89}$$

For this equation to be valid for any value of λ_1, each term must be zero; that is,

$$(\ddot{x}_0 + \omega_0^2 x_0) = 0 \tag{4.90}$$

$$(\ddot{x}_1 + \omega_0^2 x_1 + x_0^2) = 0 \tag{4.91}$$

Thus the solution can be obtained by first solving Eq. (4.90) for x_0, substituting this in Eq. (4.91), and solving for x_1. Hence the final solution will be

$$x(t) = x_0 + \lambda_1 x_1 \tag{4.92}$$

Suppose the initial conditions are such that we have the following solution for Eq. (4.90):

$$x_0 = A \sin \omega_0 t \tag{4.93}$$

Initial conditions are included in this solution; hence we need not include the transient solution. Substituting Eq. (4.93) in Eq. (4.91) yields

$$\ddot{x}_1 + \omega_0^2 x_1 = -A^2 \sin^2 \omega_0 t = \frac{A^2}{2}(1 - \cos 2\omega_0 t) \tag{4.94}$$

The general solution of this is

$$x_1(t) = B \cos 2\omega_0 t + C \tag{4.95}$$

Substituting in Eq. (4.94) and rearranging, we get

$$\left(\frac{A^2}{2} - 3\omega_0^2 B\right) \cos 2\omega_0 t + \left(-\frac{A^2}{2} + \omega_0^2 C\right) = 0 \tag{4.96}$$

For this to be true for any value of t we must have

$$\frac{A^2}{2} - 3\omega_0^2 B = 0 \quad \text{or} \quad B = \frac{A^2}{6\omega_0^2} \tag{4.97}$$

$$-\frac{A^2}{2} + \omega_0^2 C = 0 \quad \text{or} \quad C = \frac{A^2}{2\omega_0^2} \tag{4.98}$$

Substituting in Eq. (4.95), we get

$$x_1(t) = \frac{A^2}{6\omega_0^2} \cos 2\omega_0 t + \frac{A^2}{2\omega_0^2} \tag{4.99}$$

Thus the general steady-state solution for a first-order approximation in λ ($\lambda_1 = \lambda/m$) is

$$x(t) \simeq x_0 + \lambda_1 x_1 = A \sin \omega_0 t + \frac{\lambda A^2}{6m\omega_0^2}(\cos 2\omega_0 t + 3) \tag{4.100}$$

That is, the solution contains not only the free natural frequency ω_0, but also its higher harmonic $2\omega_0$. This method is not without fault. If we make the next approximation, we obtain term t (a secular term) in the solution, which is physically not acceptable for the present situation.

4.6 QUALITATIVE DISCUSSION OF MOTION AND PHASE DIAGRAMS

Energy Diagram

The following equations [Eqs. (3.2), (3.3), and (3.4)] were obtained in Chapter 3:

$$E = \tfrac{1}{2}m\dot{x}^2 + V(x) \tag{4.101}$$

$$\dot{x} = \pm\sqrt{\frac{2}{m}[E - V(x)]} \tag{4.102}$$

and
$$t_2 - t_1 = \pm\sqrt{\frac{m}{2}}\int_{x_1}^{x_2} \frac{dx}{\sqrt{E - V(x)}} \tag{4.103}$$

Once we know $V(x)$ and E, we can solve Eq. (4.103) to get a relation between x and t. But much can be learned about the qualitative nature of motion without actually solving these equations. This can be achieved in two ways: (1) by plotting $V(x)$ versus x, and (2) by plotting $\dot{x}$ versus x. Before we discuss this, we must note the following:

1. Kinetic energy cannot be negative; this will also assure that v is not imaginary.
2. Potential energy $V(x)$ cannot be greater than the total mechanical energy E of the system. If $V(x) = E$, kinetic energy K must be zero; hence the system must be at rest.

Let us consider an arbitrary potential plot $V(x)$ versus x shown by the boldface curve in Fig. 4.5(a). (A similar situation was discussed in Chapter 2, but now we discuss it by two slightly different methods.) Suppose a particle of mass m can assume different energies, as discussed next.

(0) Particle with energy E_0: This energy corresponds to minimum potential energy V_0 and the particle is at x_0 in stable equilibrium. While at x_0 the kinetic energy is zero; and if the particle is displaced slightly it will return to x_0.

(1) Particle with energy E_1: This energy is greater than the minimum potential energy V_0 and the particle will oscillate between x_1 and x_1'. Since for this low energy E_1 the potential energy between x_1 and x_1' is symmetrical, the oscillations will be simple harmonic. While oscillating, the kinetic energy and the velocity are maximum when in between the points x_1 and x_1'; it has zero velocity at x_1 and x_1'. As the particle approaches x_1 or x_1', its velocity decreases, it comes to a stop, and it then reverses its direction of motion at either of the two points x_1 and x_1'. These points are called the *turning points* and the particle at these points has zero kinetic energy and maximum potential energy. The particle cannot exist in the region for which $x < x_1$ or $x > x_1'$ because this will result in an imaginary velocity.

(2) Particle with energy E_2: The particle can either oscillate between x_2 and x_2' or be at rest and in stable equilibrium, x_2^0. There are two turning points x_2 and x_2'. Since the potential V_{01} ($>V_0$) does not correspond to the lowest energy state (which is V_0), it is called a *metastable state*. Also, the potential $V(x)$ between x_2 and x_2' is asymmetrical; hence the oscillations are nonlinear. Again motion is not permitted in the regions $x < x_2$ and between x_2' and x_2^0.

(3) Particle with energy E_3: There are four turning points, x_3, x_3', and x_3'', x_3'''. Because of the asymmetrical nature of the potential between x_3 and x_3', the oscillations are nonlinear in this valley. On the other hand, the potential between x_3''' is parabolic and hence symmetrical, thereby resulting in linear oscillations in this valley. Once again motion is not permitted in regions for $x < x_3$, $x_3' < x < x_3''$, and $x > x_3'''$.

(4) Particle with energy E_4: When the particle is at x_4^m, the potential energy is maximum (one of the maxima); hence x_4^m is a position of unstable equilibrium. If slightly displaced it could oscillate between x_4 and x_4^m or between x_4^m and x_4'; in both cases the motion is nonlinear because of asymmetrical potentials in both regions. Also, when the particle reaches x_4^m, it could move in either region. Again, for $x < x_4$ and $x > x_4'$, motion is not permitted because it results in negative kinetic energies and hence imaginary velocities.

(5) Particle with energy E_5: There is only one turning point, x_5. The particle traveling from the right with energy E_5 when it reaches x_5 comes to a stop, reverses its direction, and travels back to the right. While coming or going, the particle moves over hills and valleys. As it passes over the hills its velocity decreases; while passing over the valleys the velocity increases. Also, the deeper the valley, the higher the velocity is; and the higher the hill, the lower the velocity. That is, the particle will decelerate as it passes over the hills and accelerate as it passes over the valleys.

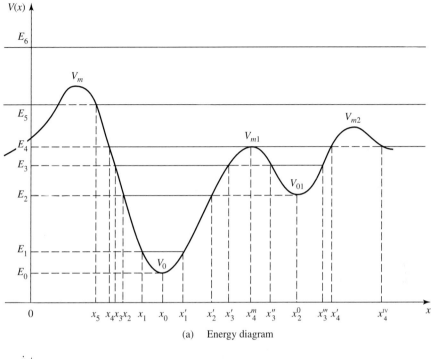

(a)   Energy diagram

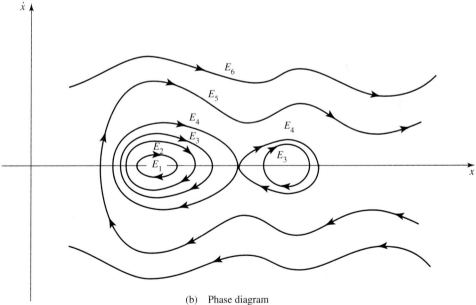

(b)   Phase diagram

Figure 4.5   Motion of a particle with different energies E in an arbitrary potential $V(x)$ represented as (a) an energy diagram in which $V(x)$ is plotted versus x, and (b) a phase diagram in which $\dot{x}$ is plotted versus x.

(6) *Particle with energy E_6:* There are no turning points. The particle simply keeps on moving, slowing down over the hills and speeding up over the valleys.

The preceding discussion is the result of the energy conservation principle given by Eq. (4.101). Also, when the particle has energy E_5 and E_6, the motion of the particle is *unbound*, while in all other cases the motion is *bound*.

Phase Diagrams

To completely specify the state of motion of a one-dimensional oscillator, two quantities must be specified. According to Eq. (4.102), we have

$$\dot{x} = \pm \sqrt{\frac{2}{m} [E - V(x)]} \tag{4.102}$$

If we know $V(x)$ as a function of x, the motion may be represented by plotting $\dot{x}$ versus x. (This is in accord with a second-order differential equation in which two constant quantities are needed to describe motion.) The coordinates $\dot{x}(t)$ and $x(t)$ uniquely describe the state of motion for any time in two dimensions. Any point $P(\dot{x}, x)$ describes the state of the motion in the *phase plane*, and the locus of such points is called the *phase diagram, phase portrait,* or *phase trajectory*. In general, if we are dealing with n-dimensional motion or the system has n degrees of freedom, $2n$ coordinates will be required to describe motion in a $2n$-dimensional phase space. Also, for a constant value E, that is, for a conservative system, the motion in a phase plane is periodic, $x(t + T) = x(t)$ and $\dot{x}(t + T) = \dot{x}(t)$, and the paths are closed curves.

The phase diagram shown in Fig. 4.5(b) for potential energy $V(x)$ and for different values of E in Fig. 4.5(a) can be understood after we discuss the following.

As a first illustration, let us apply the preceding ideas to the case of a one-dimensional simple harmonic oscillator for which $V(x) = \frac{1}{2}kx^2$ Thus, for conservation of energy,

$$\frac{1}{2}m\dot{x}^2 + \frac{1}{2}kx^2 = E \tag{4.104a}$$

or

$$\frac{x^2}{2E/k} + \frac{\dot{x}^2}{2E/m} = 1 \tag{4.104b}$$

which is an equation of an ellipse with $\sqrt{2E/k}$ and $\sqrt{2E/m}$ representing the semimajor and semiminor axes, respectively, and each E representing a unique ellipse. For different values of E we get a family of ellipses, as shown in Fig. 4.6. The same result can be arrived at by starting with the solution of a simple harmonic oscillator; that is,

$$x = A \cos(\omega_0 t + \phi) \tag{4.105}$$

$$\dot{x} = -\omega_0 A \sin(\omega_0 t + \phi) \tag{4.106}$$

$$\frac{x}{A} = \cos(\omega_0 t + \phi)$$

Figure 4.6 _____

Below is the phase diagram for a one-dimensional simple harmonic oscillator for different values of E.

For given values of M, k, b, and the phase angle ϕ, we can calculate the values of $\omega 0$ and γ.

$$M := 1 \qquad\qquad k := 2 \qquad b := .2$$

$$\omega 0 := \sqrt{\frac{k}{M}} \qquad\qquad \gamma := \frac{b}{2 \cdot M} \qquad \phi := 0$$

$$\omega 0 = 1.414 \qquad\qquad \gamma = 0.1$$

As we know the total energy E1, E2, and E3 for three simple harmonic oscillators, we can calculate the corresponding amplitudes.

$$E1 := 15 \qquad\qquad E2 := 7 \qquad E3 := 10$$

$$A1 := \sqrt{\frac{2 \cdot E1}{k}} \qquad A2 := \sqrt{\frac{2 \cdot E2}{k}} \qquad A3 := \sqrt{\frac{2 \cdot E3}{k}}$$

$$A1 = 3.873 \qquad\qquad A2 = 2.646 \qquad A3 = 3.162$$

$$I := 100 \qquad i := 0..I \qquad t_i := \frac{i}{10}$$

With these values, we can now write the expressions for the displacements x and the corresponding velocities v (by differentiating x with respect to t) for the three harmonic oscillators and graph them.

$$x1_i := A1 \cdot \cos(\omega 0 \cdot t_i + \phi) \qquad v1_i := -\omega 0 \cdot A1 \cdot \sin(\omega 0 \cdot t_i + \phi)$$

$$x2_i := A2 \cdot \cos(\omega 0 \cdot t_i + \phi) \qquad v2_i := -\omega 0 \cdot A2 \cdot \sin(\omega 0 \cdot t_i + \phi)$$

$$x3_i := A3 \cdot \cos(\omega 0 \cdot t_i + \phi) \qquad v3_i := -\omega 0 \cdot A3 \cdot \sin(\omega 0 \cdot t_i + \phi)$$

(a) If the phase angle ϕ is $\pi/2$, how do the phase diagrams change?

(b) How does the increase in the values of M, k, b, and γ affect the graphs?

(c) What factors affect the change in the values of the amplitude, frequency, and energy of the oscillating system? Verify your answer by graphing.

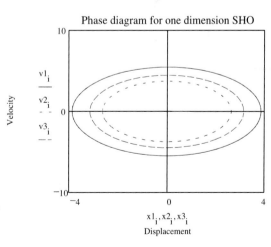

Phase diagram for one dimension SHO

or
$$\frac{\dot{x}}{\omega_0 A} = -\sin(\omega_0 t + \phi)$$

To eliminate t, square and add these two equations; that is,

$$\frac{x^2}{A^2} + \frac{\dot{x}^2}{\omega_0^2 A^2} = 1$$

Using $\omega_0^2 = k/m$ and $E = \frac{1}{2}kA^2$, we get

$$\frac{x^2}{2E/k} + \frac{\dot{x}^2}{2E/m} = 1$$

The first thing we note about the phase paths in Fig. 4.6 is that the motion is clockwise; the reasoning is that, for $x > 0$, $\dot{x}$ is always decreasing, while for $x < 0$, $\dot{x}$ is always increasing. Furthermore, no two phase paths cross each other. Mathematically, the reason is that each solution of the differential equation is unique. Physically, it means that if a particle is capable of changing its path and hence changing its energy, it leads to the nonconservation of energy. But we assume that for a given path a particle has a given energy.

Let us see how the phase diagram will look for an underdamped harmonic oscillator. Equations of x and $\dot{x}$ are

$$x = Ae^{-\gamma t} \cos(\omega_1 t + \phi) \tag{4.107}$$

and
$$\dot{x} = -Ae^{-\gamma t}[\gamma \cos(\omega_1 t + \phi) + \omega_1 \sin(\omega_1 t + \phi)] \tag{4.108}$$

The oscillator is continuously losing energy. Using plane polar coordinates (ρ, θ) we can show that

$$\rho = \omega_1 A e^{(-\gamma/\omega_1)\theta} \tag{4.109}$$

which is an equation of a logarithmic spiral. Without going into any details, we state that the phase path is as shown in Fig. 4.7. An oscillating particle spirals down a potential well and eventually it comes to rest when it reaches $x = 0$.

Finally, we discuss the phase diagram of a nonlinear system, as shown in Fig. 4.8(a). The asymmetric potential shown represents a hard system for $x < 0$ and a soft system for $x > 0$. Using the relation

$$\dot{x} \propto \sqrt{E - V(x)}$$

and the plot of $\dot{x}(x)$ versus x, we get the phase diagram shown in Fig. 4.8(b). The three oval-shaped closed paths correspond to three different energies, assuming that there is no damping. If damping were present, all paths would spiral down to $x = 0$.

The phase diagram in Fig. 4.5(b) should now be easy to understand. This phase diagram corresponds to the energy diagram in Fig. 4.5(a).

Figure 4.7

Below are phase diagrams for a damped harmonic oscillator using polar coordinates and rectangular coordinates.

The values given on the right are good for three damped oscillators with energies E, E1, and E2, but with the same damping constant. The phase diagram below is for the oscillator with energy E. The others can be graphed in the same way.

$$M := .01 \qquad k := 4 \qquad b := .1 \qquad \phi := 0$$

$$\omega 0 := \sqrt{\frac{k}{M}} \qquad \gamma := \sqrt{\frac{b}{2 \cdot M}} \qquad \omega 1 := \sqrt{\omega 0^2 - \gamma^2}$$

$$\omega 0 = 20 \qquad \gamma = 2.236 \qquad \omega 1 = 19.875$$

(a) How will the phase diagrams change if the energy E is increased or decreased? Check this by graphing.

$$E := 1000 \qquad E1 := 20 \qquad E2 := 40$$

$$A := \sqrt{2 \cdot \frac{E}{k}} \qquad A1 := \sqrt{2 \cdot \frac{E1}{k}} \qquad A2 := \sqrt{2 \cdot \frac{E2}{k}}$$

(b) What is the fundamental difference between the two types of phase diagrams?

$$A = 22.361 \qquad A1 = 3.162 \qquad A2 = 4.472$$

(c) Are the plots clockwise or counter-clockwise and why?

$$N := 1000 \qquad n := 0 .. N \qquad t_n := \frac{n}{100}$$

Polar Graph

$$\theta_n := 2 \cdot \pi \cdot \frac{n}{100}$$

$$\rho_n := \omega 1 \cdot A \cdot e^{-\frac{\gamma}{\omega 1} \cdot \theta_n}$$

Rectangular Graph

$$x_n := A \cdot e^{-\gamma t_n} \cdot \cos\left(\omega 1 \cdot t_n + \phi\right)$$

$$v_n := -A \cdot e^{-\gamma t_n} \cdot \left(\gamma \cos\left(\omega 1 \cdot t_n + \phi\right) + \omega 1 \cdot \sin\left(\omega 1 \cdot t_n + \phi\right)\right)$$

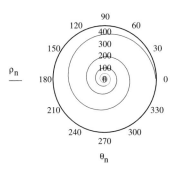

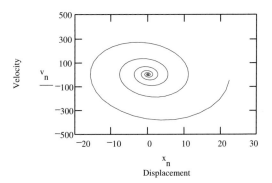

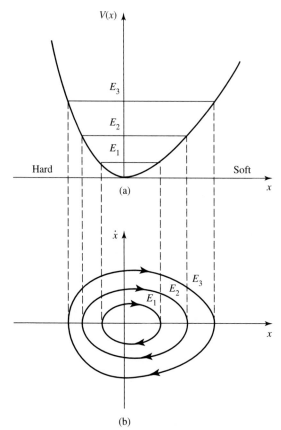

Figure 4.8 Phase and energy diagrams of a nonlinear system with asymmetrical potential.

PROBLEMS

4.1. Draw an equivalent electrical circuit for the mechanical system shown in Fig. P4.1. Set up the equations for describing motion. Calculate different possible frequencies.

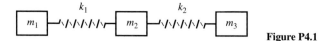

Figure P4.1

4.2. Draw an equivalent electrical circuit for the mechanical system shown in Fig. P4.2. Calculate different possible frequencies.

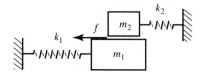

Figure P4.2

4.3. Derive Eqs. (4.17), (4.18), and (4.19).

4.4. For a spring vibrating vertically under the action of gravitational pull, draw the equivalent mechanical and electrical systems and write the necessary equations.

4.5. Discuss the electrical equivalent of the following: **(a)** average power dissipated, and **(b)** the quality factor.

4.6. Consider the electrical system shown in Fig P4.6. Calculate **(a)** the resonance frequency, **(b)** the resonance width, and **(c)** the power absorbed at resonance.

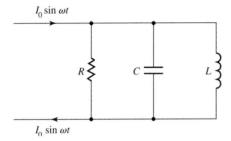

Figure P4.6

4.7. Calculate the oscillation frequency for the circuit in which $R = 150$ ohm (Ω), $C = 20$ microfarad (μF), and $L = 0.1$ henry (H).

4.8. Show that in an RLC circuit for which R is negligibly small the logarithmic decrement of oscillation is $\simeq \pi R(C/L)^{1/2}$.

4.9. A series electrical circuit contains a resistance R, a capacitance C, and an inductor L. An emf $\varepsilon = \varepsilon_0 \cos \omega t$ is applied to the circuit. Solve and discuss the transient and steady-state solutions. Find the steady-state expression for the current. Do this problem in analogy with a forced harmonic oscillator. Calculate the phase angle between the current and the emf. Graph the values to describe motion.

4.10. Consider the RLC circuit discussed in Problem 4.9 and derive an expression for the current as a function of time t. Show that the current decreases to zero as the frequency of the alternating emf goes to zero. Graph this.

4.11. A source of emf is connected to two impedances Z_1 and Z_2 in series. Obtain expressions for the powers dissipated in Z_1 and Z_2.

4.12. Calculate the impedance of the circuit shown in Fig. P4.12. Show that for $\omega = \omega_0 = 1/\sqrt{LC}$ the current flow is minimum. Calculate the Q of the circuit.

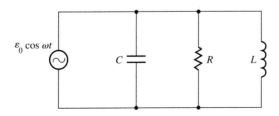

Figure P4.12

4.13. Consider the parallel RLC circuit shown in Fig. P4.13. Calculate Z from the relation

$$1/Z = 1/Z_1 + 1/Z_2$$

where Z_1 is the impedance of R_1 and C in series, while Z_2 is the impedance of R_2 and L in series: hence Z_1 and Z_2 are in parallel. Calculate the value of the total current when both current and emf are in phase. What happens when $R_1 = R_2 = 0$?

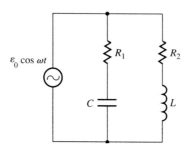

Figure P4.13

4.14. Construct the electrical analog of the system shown in Fig P4.14 and calculate the impedance. b_1 and b_2 are damping parameters resulting from the friction between each mass and surface, ω is the frequency of the driving force, and mass m_2 slides back and forth on mass m_1. Do this problem assuming the absence of spring k_2.

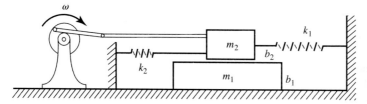

Figure P4.14

4.15. Repeat Problem 4.14 assuming the presence of both springs k_1 and k_2.

4.16. By using a Fourier technique, obtain both the sine and the cosine series in the interval $0 < \theta < \pi$ for the function

$$f(\theta) = \begin{cases} 1, & \text{for } 0 < \theta < \pi/2 \\ 0, & \text{for } \pi/2 < \theta < \pi \end{cases}$$

4.17. Calculate a Fourier series representation (Fourier transform) represented by the following function:

$$F(t) = \begin{cases} 0, & \text{if } nT < t < (n + \frac{1}{2})T \\ F_0 = \text{constant}, & \text{if } (n + \frac{1}{2})T < t < (n + 1)T \end{cases}$$

4.18. Calculate a Fourier series represenation (Fourier transform) represented by the following function:

$$F(t) = \begin{cases} 0, & \text{if } -2\pi/\omega < t < 0 \\ \sin \omega t, & \text{if } 0 < t < 2\pi/\omega \end{cases}$$

4.19. An underdamped oscillator has a natural frequency ω_0. An impulse force function of constant magnitude acts for a time $T = 2\pi/\omega_0$. Calculate the response function and give its physical interpretation by graphing it.

4.20. A linear oscillator is under the influence of the following force function:

$$F(t) = \begin{cases} 0, & \text{if } t < 0 \\ ma \sin \omega t, & \text{if } 0 < t < \pi/\omega \\ 0, & \text{if } t > \pi/\omega \end{cases}$$

Calculate the response function. Make a graph to describe the motion.

4.21. A damped oscillator is acted on by the following force function:

$$F(t) = \begin{cases} 0, & \text{if } t < 0 \\ F_0 e^{-\gamma t} \sin \omega t, & \text{if } t > 0 \end{cases}$$

Using Green's function, calculate the response function. Graph it.

4.22. Derive Eq. (4.46) for the case of a damped harmonic oscillator.

4.23. Derive Eq. (4.53) for the case of a damped harmonic oscillator.

4.24. Extend the results obtained in Eq. (4.51) to the case of multiple impulses and obtain a solution by using Green's function of the form given by Eqs. (4.59) and (4.60).

4.25. Extend the results obtained in Eq. (4.53) to the case of multiple impulses and obtain a solution by using Green's function of the form given by Eqs. (4.59) and (4.60).

4.26. Derive Eq. (4.82) and Eq. (4.84) by the procedure outlined in the text.

4.27. A particle of mass m moving in a resistive medium is acted upon by a series of impulses of force $F(t)$ at times t_1, t_2, and so on, each for a short duration Δt. Find the resulting velocity of the particle.

4.28. Show that for a nonlinear differential equation of the type $\ddot{x} + cx^2 = 0$ the superposition principle does not hold.

4.29. Suppose in Fig. 4.4 that each string must be pulled a distance d before attaching mass m. Show that under such conditions

$$V(x) \simeq \left(k \, \frac{d}{l} \right) x^2 + \left[\frac{k(l-d)}{4l^3} \right] x^4$$

4.30. For Eq. (4.93), assume $x_0 = A \cos \omega_0 t$ and obtain a solution similar to the one leading to Eq. (4.100).

4.31. Consider a particle described by an equation of the form

$$\ddot{x} + \omega_0^2 x - \lambda x^2 = 0$$

Obtain a correct second-order solution by using

$$x(t) = x_0 + \lambda x_1 + \lambda^2 x_2$$

4.32. Suppose in the case of a pendulum that the amplitude of the motion is not small. Show that the horizontal motion of the pendulum may be represented approximately by

$$\ddot{x} + \frac{g}{l} x - \frac{1}{2l^3} x^3 = 0$$

4.33. Consider the case of an overdamped oscillator and draw the phase diagrams for the following cases ($v_0 = \dot{x}_0$): **(a)** $v_0 > 0$, **(b)** $v_0 < 0$ and is small, and **(c)** $v_0 > 0$ and is large.

4.34. For the case of an underdamped oscillator, derive Eq. (4.109) by using $u = \omega_1 x$, $v = \gamma x + \dot{x}$, $\theta = \omega_1 t$, and Eqs. (4.107) and (4.108) in $\rho = (u^2 + v^2)^{1/2}$. Draw the phase diagrams.

4.35. A particle of mass m is under the influence of potential $V = A|x|^n$. Show that the time period for such motion is given by

$$T = \frac{2}{n} \sqrt{\frac{2\pi m}{E}} \left(\frac{E}{A}\right)^{1-n} \frac{\Gamma(1/n)}{\Gamma(1/2 + 1/n)}$$

Calculate the value of T for $n = 2$ and 3.

4.36. Construct the phase diagram for the case $n = 2$ in Problem 4.35.

4.37. Construct the phase diagram for $n = 3$ in Problem 4.35, that is, for $V = -Ax^3$.

4.38. A particle moves under the influence of a constant force F_0 when $x < 0$ and under a constant force $-F_0$ for $x > 0$. Describe the motion by constructing a phase diagram.

SUGGESTIONS FOR FURTHER READING

ARTHUR, W., and FENSTER, S. K., *Mechanics,* Chapter 6. New York: Holt, Rinehart and Winston, Inc., 1969.

BECKER, R. A., *Introduction to Theoretical Mechanics,* Chapter 7. New York: McGraw-Hill Book Co., 1954.

CHURCHILL, R. V., *Fourier Series and Boundary Value Problems.* New York: McGraw-Hill Book Co., 1941.

DAVIS, H. F., *Fourier Series and Orthogonal Functions.* Needham Heights, Mass.: Allyn & Bacon, 1963.

FRENCH, A. P., *Vibrations and Waves,* Chapter 4. New York: W. W. Norton and Co., Inc., 1971.

HAUSER, W., *Introduction to the Principles of Mechanics,* Chapter 4. Reading, Mass.: Addison-Wesley Publishing Co., 1965.

KITTEL, C., KNIGHT, W. D., and RUDERMAN, M. A., *Mechanics,* Berkeley Physics Course, Volume 1, Chapter 7. New York: McGraw-Hill Book Co., 1965.

KU, Y. H., *Analysis and Control of Nonlinear Systems.* New York: Ronald Press, 1958.

MARION, J. B., *Classical Dynamics,* 2nd ed., Chapter 5. New York: Academic Press, Inc., 1970.

MINORSKY, N., *Introduction to Nonlinear Mechanics.* Ann Arbor, Mich.: Edwards, 1947.

SYMON, K. R., *Mechanics,* 3rd, ed., Chapter 2. Reading, Mass.: Addison-Wesley Publishing Co., 1971.

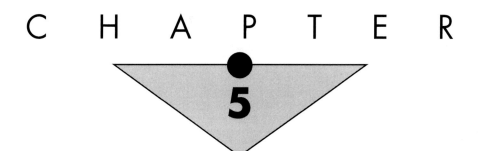

C H A P T E R

5

Vector Analysis, Vector Operators, and Transformations

5.1 INTRODUCTION

Because mathematics is an integral part of physics, it is essential that we have a clear understanding of vectors and are capable of using them frequently and with ease. To make sure that we understand vector terminology and symbolism, we start with a brief review of vector analysis, establishing a thorough understanding of vector operators such as vector differential operators (gradient, divergence, and curl). In the process of doing this, we will make sure to use examples from situations encountered in classical mechanics. Finally we will address the use of matrices in coordinate transformations.

5.2 VECTOR PROPERTIES

Most of the physical quantities we encounter in physics and engineering may be classed as one of two types: scalar or vector. A *scalar quantity* is completely specified by stating its *magnitude only*, together with units, if any. Examples of such quantities are mass, volume, energy, time, and number. These quantities may be treated as ordinary numbers and may be added, subtracted, multiplied, or divided by simple arithmetic rules. A *vector quantity* is completely specified by stating *both magnitude and direction.* Examples include displacement, velocity, acceleration, force, and electric field strength, to name a few. Vector quantities follow the rules of vector algebra.

The vectors may be treated by either a *geometrical* or an *analytical* approach (see Section 5.3); each has its advantages and disadvantages. We will briefly discuss both approaches in this chapter.

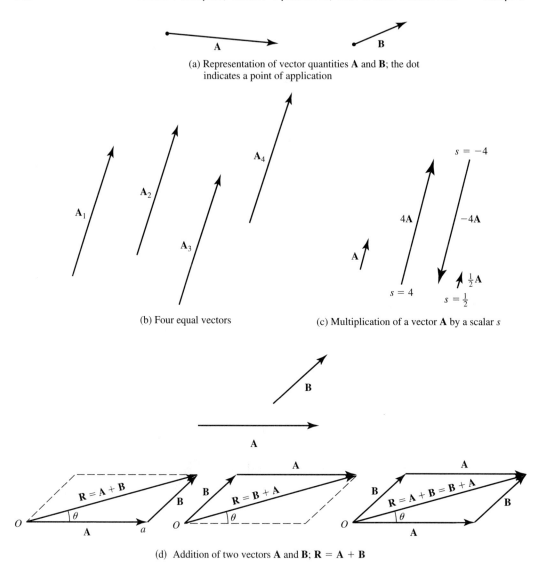

(a) Representation of vector quantities **A** and **B**; the dot
indicates a point of application

(b) Four equal vectors

(c) Multiplication of a vector **A** by a scalar s

(d) Addition of two vectors **A** and **B**; **R** = **A** + **B**

Figure 5.1 Geometrical approach to vectors.

Geometrical Approach

Geometrically, *a vector is represented by a directed line segment with an arrowhead.* In general, we will use boldface letters (**A**, **B**, . . . , **R**) to denote different vectors, as shown in Fig. 5.1(a). If we are concerned with the magnitudes of the vectors, these are represented as A, B, C, . . . , R or $|\mathbf{A}|$, $|\mathbf{B}|$, $|\mathbf{C}|$, . . . , $|\mathbf{R}|$.

Figure 5.1 illustrates the geometrical approach to analyzing the vectors.

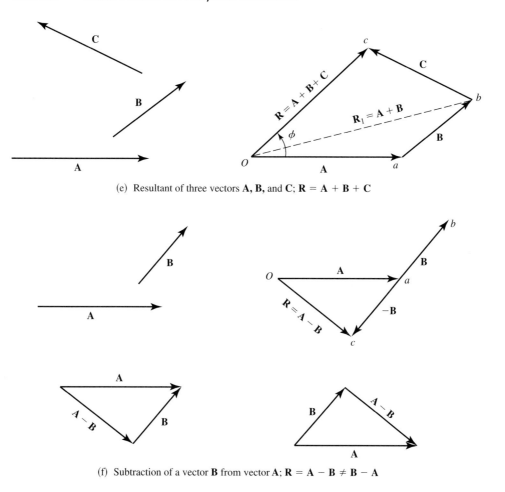

(e) Resultant of three vectors **A, B,** and **C**; **R** = **A** + **B** + **C**

(f) Subtraction of a vector **B** from vector **A**; **R** = **A** − **B** ≠ **B** − **A**

Figure 5.1   (continued)

5.3 VECTOR ADDITION: ANALYTICAL TREATMENT

One advantage of vector formulation as applied to any particular situation is that vectors can be used without reference to any particular coordinate system. But there are two main reasons that eventually compel us to use a proper set of coordinate system: (1) the geometrical method of obtaining the resultant of several vectors is cumbersome and not very accurate, and (2) to make a proper interpretation, it is always helpful to present the results in a suitable set of coordinate system. An alternative is to use an analytical method using the components of vectors.

Again, an analytical approach uses the components of vectors. Figure 5.2 illustrates the components of a vector **R** in two dimensions. The two quantities R_x and R_y are the *rectangular* components of vector **R**, and the magnitudes of these are given by (see Fig. 5.2)

$$R_x = R \cos \theta \qquad \text{and} \qquad R_y = R \sin \theta \qquad\qquad \textbf{(5.1)}$$

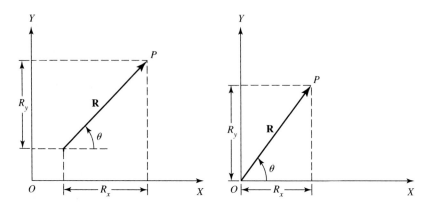

Figure 5.2 Rectangular components R_x and R_y of a vector **R**.

If R and θ are given, we can find R_x and R_y. On the other hand, if R_x and R_y are given, R and θ can be calculated by the relations

$$R = \sqrt{R_x^2 + R_y^2} \qquad \text{and} \qquad \tan \theta = \frac{R_y}{R_x} \tag{5.2}$$

Thus a vector is completely defined in a plane (two dimensions) provided we know R and θ or R_x and R_y.

We shall use the component method for adding several vectors. Suppose we have three vectors $\mathbf{R}_1$, $\mathbf{R}_2$, and $\mathbf{R}_3$ [see Fig. 5.3(a)], making angles θ_1, θ_2, and θ_3, respectively, with the X-

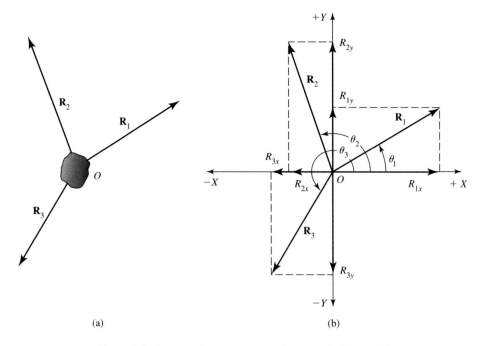

(a)   (b)

Figure 5.3 Rectangular components of vectors $\mathbf{R}_1$, $\mathbf{R}_2$, and $\mathbf{R}_3$.

axis [Fig. 5.3(b)]. We are interested in finding the resultant. First, we draw the XY coordinate [as in Fig. 5.3(b)], and then, at point O, draw $\mathbf{R}_1$, $\mathbf{R}_2$, and $\mathbf{R}_3$ parallel to the vectors in part (a). Next, each vector is resolved into its components as shown. Let R_X and R_Y denote the sum of the X and Y components. Thus

$$R_X = R_{1x} + R_{2x} + R_{3x} = \sum_{i=1}^{3} R_{ix} \tag{5.3a}$$

$$R_Y = R_{1y} + R_{2y} + R_{3y} = \sum_{i=1}^{3} R_{iy} \tag{5.3b}$$

where $R_{1x} = R_1 \cos \theta_1$, $R_{1y} = R_1 \sin \theta_1$, and the summation sign indicates the sum of all the components from $i = 1$ to $i = 3$. Treating R_X and R_Y as single components, we find the resultant to be

$$R = \sqrt{R_x^2 + R_Y^2} \qquad \text{and} \qquad \tan \theta = \frac{R_Y}{R_X} \tag{5.4}$$

It may be pointed out that we need not draw the diagram in Fig. 5.3(b). The calculations are simple, quick, and accurate and may be extended to any number of vectors.

We can extend this procedure to n vectors in three dimensions. As shown in Fig. 5.4 vector $\mathbf{R}$ may be resolved into three components: R_x, R_y, and R_z along the X-, Y-, and Z-axes, respectively. Thus the resultant $\mathbf{R}$ of vectors $\mathbf{R}_1$, $\mathbf{R}_2$, . . . , $\mathbf{R}_n$ may be written as

$$\mathbf{R} = \mathbf{R}_1 + \mathbf{R}_2 + \mathbf{R}_3 + \cdots + \mathbf{R}_n = \sum_{i=1}^{n} \mathbf{R}_i \tag{5.5}$$

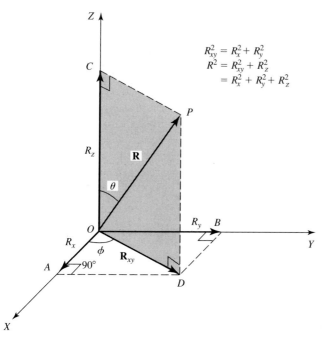

$$R_{xy}^2 = R_x^2 + R_y^2$$
$$R^2 = R_{xy}^2 + R_z^2$$
$$= R_x^2 + R_y^2 + R_z^2$$

Figure 5.4 Rectangular components R_x, R_y, and R_z of a vector $\mathbf{R}$ in three dimensions.

where
$$R_x = R_{1x} + R_{2x} + R_{3x} + \cdots + R_{nx} = \sum_{i=1}^{n} R_{ix} \tag{5.6}$$

with similar expressions for R_Y and R_Z. Referring to Fig. 5.4, we obtain the resultant

$$R = \sqrt{R_x^2 + R_y^2 + R_z^2} \tag{5.7}$$

$$\tan \theta = \frac{R_{xy}}{R_z} = \frac{\sqrt{R_x^2 + R_y^2}}{R_z} \tag{5.8}$$

$$\tan \phi = \frac{R_y}{R_z} \tag{5.9}$$

Since the components of a vector equally define a vector, it should be possible to correlate the two by extending the analytical method to the geometrical method discussed in the previous section. Also,

$$\mathbf{A} = (A_x, A_y) \qquad \text{in two dimensions} \tag{5.10}$$

$$\mathbf{A} = (A_x, A_y, A_z) \qquad \text{in three dimensions} \tag{5.11}$$

Using Eqs. (5.10) and (5.11), we may write the properties of vectors in component form as follows:

Equality of vectors:

$$\mathbf{A} = \mathbf{B} \quad \text{or} \quad [A_x, A_y, A_z] = [B_x, B_y, B_z] \tag{5.12a}$$

which means

$$A_x = B_x, \qquad A_y = B_y, \qquad A_z = B_z \tag{5.12b}$$

Scalar multiplication:

$$s\mathbf{A} = s[A_x, A_y, A_z] = [sA_x, sA_y, sA_z] \tag{5.13}$$

Null vector: The null vector has a zero magnitude and undefined direction

$$\mathbf{0} = [0, 0, 0] \tag{5.14}$$

Vector addition:

$$\mathbf{R} = \mathbf{A} + \mathbf{B} = [R_x, R_y, R_z] \tag{5.15}$$

where

$$R_x = A_x + B_x, \qquad R_y = A_y + B_y, \qquad R_z = A_z + B_z \tag{5.16}$$

Commutative law:

$$\mathbf{A} + \mathbf{B} = \mathbf{B} + \mathbf{A} \tag{5.17}$$

Associative law:

$$A + (B + C) = (A + B) + C \tag{5.18}$$

and similarly,

$$(ns)A = (ns)[A_x, A_y, A_z] = n[sA_x, sA_y, sA_z]$$

$$= n(sA) \tag{5.19}$$

Distributive law:

$$(n + s)A = nA + sA \tag{5.20}$$

and

$$s(A + B) = sA + sB \tag{5.21}$$

5.4 SCALAR AND VECTOR PRODUCTS OF VECTORS

Unlike vector addition and subtraction, there are several ways of defining multiplication of two vectors. Two types of vector products are commonly used and we shall discuss them at some length; they are called (1) the scalar product of two vectors and (2) the vector product of two vectors. A third and fourth product type we shall also introduce are (3) the scalar triple product and (4) the vector triple product.

Scalar Product (or Dot Product)

The scalar or dot product of two vectors **A** and **B** is defined to be a scalar quantity S obtained by taking the magnitude of **A** multiplied by the magnitude of **B** and then multiplied by the cosine of the angle between these two vectors; that is,

$$S = A \cdot B = |A||B| \cos(A, B) = AB \cos \theta, \quad \text{for } 0 < \theta < \pi \tag{5.22}$$

where $A \cdot B$ reads as **A** dot **B** and is defined to be a scalar quantity of magnitude $AB \cos \theta = S$. The angle θ is the smaller of the two angles between the two vectors, as shown in Fig. 5.5. There are two alternatives for defining the scalar products of two vectors, as illustrated in Fig. 5.5(a) and (b): either by projection of **A** on **B** or **B** on **A**. In either case (as shown),

$$A \cdot B = AB \cos \theta \tag{5.23}$$

Let us consider some special cases of the scalar products.

The dot product is zero.   Suppose the scalar or the dot product of two vectors is zero; that is, $A \cdot B = 0$. This is possible if at least one of three conditions is satisfied: $|A| = 0$, $|B| = 0$, or **A** is perpendicular to **B** (that is, $\theta = 90°$). If **A** is perpendicular to **B**, vector **A** is said to be *orthogonal* to vector **B**. Thus it is possible to obtain a scalar product of zero, even though neither of the two vectors is zero.

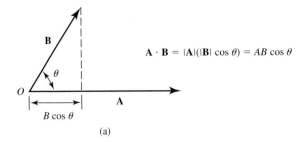

$$\mathbf{A} \cdot \mathbf{B} = |\mathbf{A}|(|\mathbf{B}| \cos \theta) = AB \cos \theta$$

(a)

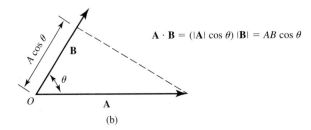

$$\mathbf{A} \cdot \mathbf{B} = (|\mathbf{A}| \cos \theta) |\mathbf{B}| = AB \cos \theta$$

(b)

Figure 5.5   Scalar product of two vectors $\mathbf{A}$ and $\mathbf{B}$; $\mathbf{A} \cdot \mathbf{B}$.

The scalar product is commutative.   It is easy to show that

$$\mathbf{A} \cdot \mathbf{B} = \mathbf{B} \cdot \mathbf{A} \tag{5.24}$$

That is, the order of multiplication is not important:

$$\mathbf{B} \cdot \mathbf{A} = |\mathbf{B}||\mathbf{A}| \cos(-\theta) = AB \cos \theta = \mathbf{A} \cdot \mathbf{B}$$

Two vectors are equal.   Suppose $\mathbf{B} = \mathbf{A}$; then $\cos(\mathbf{A}, \mathbf{B}) = \cos 0° = 1$; hence

$$\mathbf{A} \cdot \mathbf{B} = \mathbf{A} \cdot \mathbf{A} = |\mathbf{A}|^2 = A^2 \tag{5.25}$$

Law of cosines.   The proof of the law of cosines becomes trivial if we make use of the dot product. In Fig. 5.6, let

$$\mathbf{R} = \mathbf{A} + \mathbf{B}$$

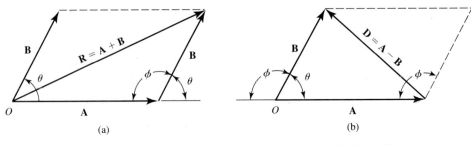

(a)   (b)

Figure 5.6   (a) Law of cosines as applied to $\mathbf{R} = \mathbf{A} + \mathbf{B}$. (b) Law of cosines as applied to $\mathbf{D} = \mathbf{A} - \mathbf{B}$.

$$R^2 = \mathbf{R} \cdot \mathbf{R} = (\mathbf{A} + \mathbf{B}) \cdot (\mathbf{A} + \mathbf{B})$$

$$= \mathbf{A} \cdot \mathbf{A} + \mathbf{B} \cdot \mathbf{B} + 2\mathbf{A} \cdot \mathbf{B}$$

$$= A^2 + B^2 + 2AB \cos(\mathbf{A}, \mathbf{B})$$

That is,

$$R^2 = A^2 + B^2 + 2AB \cos \theta \qquad (5.26)$$

Since $\cos \theta = \cos(\pi - \phi) = -\cos \phi$,

$$R^2 = A^2 + B^2 - 2AB \cos \phi \qquad (5.27)$$

Similarly,

$$\mathbf{D} = \mathbf{A} - \mathbf{B}$$

and

$$D^2 = A^2 + B^2 - 2AB \cos \theta \qquad (5.28)$$

Since $\cos \theta = \cos(\pi - \phi) = -\cos \phi$,

$$D^2 = A^2 + B^2 + 2AB \cos \phi \qquad (5.29)$$

Vector Product (or Cross Product)

The vector or cross product of two vectors $\mathbf{A}$ and $\mathbf{B}$ is defined to be a vector $\mathbf{C}$. The magnitude of this vector is obtained by taking the magnitude of $\mathbf{A}$ multiplied by the magnitude of $\mathbf{B}$ and then multiplied by the sine of the angle between $\mathbf{A}$ and $\mathbf{B}$, while the direction of $\mathbf{C}$ is such that it is perpendicular to both $\mathbf{A}$ and $\mathbf{B}$: that is, $\mathbf{C}$ is perpendicular to the plane containing both $\mathbf{A}$ and $\mathbf{B}$. We denote the cross product as

$$\mathbf{C} = \mathbf{A} \times \mathbf{B} \qquad (5.30)$$

where $\mathbf{A} \times \mathbf{B}$ reads as "A cross B" and is as shown in Fig. 5.7. Thus the magnitude of $\mathbf{C}$ is

$$C = |\mathbf{C}| = |\mathbf{A}||\mathbf{B}|\sin(\mathbf{A}, \mathbf{B}) = AB \sin \theta, \quad \text{for} \quad 0 < \theta < \pi \qquad (5.31)$$

Hence C will be zero if $A = 0$, $B = 0$, or the angle θ is zero. The direction of $\mathbf{C}$ is given by the *right-hand rule* or *right-hand screw*, as illustrated in Fig. 5.7. By drawing simple diagrams, we can draw several conclusions about the cross product.

Commutative and distributive properties.   Because of the right-hand rule convention used in defining the vector product, we may conclude (from Fig. 5.7)

$$\mathbf{A} \times \mathbf{B} = -\mathbf{B} \times \mathbf{A} \qquad (5.32)$$

That is, vector multiplication is not commutative; it is *anticommutative*. From Eq. (5.32), it is also clear that

$$\mathbf{A} \times \mathbf{A} = 0 \qquad (5.33)$$

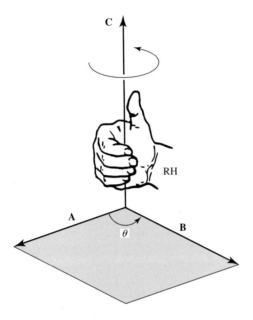

Figure 5.7 Cross product of vectors **A** and **B**, illustrating the direction of **C** by the right-hand rule.

That is, the vector product of any vector with itself is zero. Also, by definition, $\sin \theta = \sin 0° = 0$; that is, $\mathbf{A} \times \mathbf{A} = \mathbf{0}$, and if the angle between **A** and **B** is zero, then $|\mathbf{A} \times \mathbf{B}| = AB \sin 0° = 0$. Note that the right side of Eq. (5.33) is a *null vector*, which obeys the rule

$$\mathbf{A} + \mathbf{0} = \mathbf{A}, \quad \mathbf{A} \times \mathbf{0} = \mathbf{0}, \quad \mathbf{A} \cdot \mathbf{0} = 0 \tag{5.34}$$

The vector product does obey the distributive law:

$$\mathbf{A} \times (\mathbf{B} + \mathbf{C}) = \mathbf{A} \times \mathbf{B} + \mathbf{A} \times \mathbf{C} \tag{5.35}$$

and

$$s(\mathbf{A} \times \mathbf{B}) = (s\mathbf{A}) \times \mathbf{B} = \mathbf{A} \times (s\mathbf{B}) \tag{5.36}$$

Area of a parallelogram.   In many situations it is desirable to know the orientation of the area, and for this purpose the vector product is convenient to use. Consider vectors **B** and **C**, which form the two sides of a triangle *OBC* or two adjacent sides of a parallelogram *OBDC*, as shown in Fig. 5.8. The area A of the parallelogram is

$$A = 2(\text{area of triangle } OBC)$$

$$= 2\left[\tfrac{1}{2} \text{ base} \times \text{height}\right] = 2\left[\tfrac{1}{2} Bh\right]$$

$$= BC \sin \theta = |\mathbf{B}||\mathbf{C}| \sin \theta$$

or, if we represent the area by a vector quantity **A**, we may write

$$A = |\mathbf{A}| = |\mathbf{B} \times \mathbf{C}| = BC \sin(\mathbf{B}, \mathbf{C}) = BC \sin \theta \tag{5.37}$$

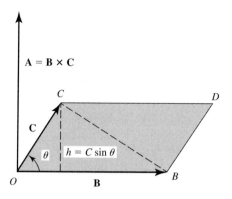

Figure 5.8   Area of a parallelogram.

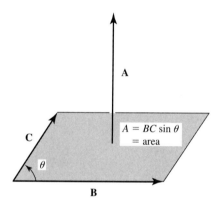

Figure 5.9   Vector representation of the area by **A** that is normal to the plane of vectors **B** and **C**.

Thus the area of a parallelogram of sides **B** and **C** is **B** × **C**. Since the direction of **B** × **C** is normal or perpendicular to the plane of the parallelogram, vector **A** is as shown in Fig. 5.9. That is, the area is assigned a direction of **A**.

Law of sines.   Consider a triangle formed by vectors **A**, **B**, and **C** such that

$$\mathbf{C} = \mathbf{A} + \mathbf{B} \tag{5.38}$$

Take the vector product of both the sides by **A**; that is

$$\mathbf{A} \times \mathbf{C} = \mathbf{A} \times \mathbf{A} + \mathbf{A} \times \mathbf{B} \tag{5.39}$$

Since **A** × **A** = 0, Eq. (5.39) takes the form

$$\mathbf{A} \times \mathbf{C} = \mathbf{A} \times \mathbf{B}$$

or
$$AC \sin(\mathbf{A}, \mathbf{C}) = AB \sin(\mathbf{A}, \mathbf{B})$$

That is,
$$\frac{\sin(\mathbf{A}, \mathbf{C})}{B} = \frac{\sin(\mathbf{A}, \mathbf{B})}{C}$$

or, in general,

$$\frac{\sin(\mathbf{A}, \mathbf{C})}{B} = \frac{\sin(\mathbf{A}, \mathbf{B})}{C} = \frac{\sin(\mathbf{B}, \mathbf{C})}{A} \tag{5.40}$$

which is the law of sines.

Scalar Triple Product: Volume of a Parallelepiped

The quantity $(\mathbf{A} \times \mathbf{B}) \cdot \mathbf{C}$ is a scalar quantity. We can show that if $\mathbf{A}$, $\mathbf{B}$, and $\mathbf{C}$ are the sides of a parallelepiped then such a product represents the volume of the parallelepiped. From Fig. 5.10, the volume V is

$$V = \text{(base area)(height)} = |\mathbf{A} \times \mathbf{B}|h$$

$$= |\mathbf{A} \times \mathbf{B}||\mathbf{C}| \sin[(\mathbf{A} \times \mathbf{B}), \mathbf{C}]$$

or

$$V = (\mathbf{A} \times \mathbf{B}) \cdot \mathbf{C} \tag{5.41}$$

If the vectors in the product follow a cyclic order, we may write V to be

$$V = (\mathbf{A} \times \mathbf{B}) \cdot \mathbf{C} = (\mathbf{B} \times \mathbf{C}) \cdot \mathbf{A} = (\mathbf{C} \times \mathbf{A}) \cdot \mathbf{B} \tag{5.42}$$

or, since a scalar product is commutative,

$$V = \mathbf{A} \cdot (\mathbf{B} \times \mathbf{C}) = \mathbf{B} \cdot (\mathbf{C} \times \mathbf{A}) = \mathbf{C} \cdot (\mathbf{A} \times \mathbf{B}) \tag{5.43}$$

That is, the scalar triple product remains unchanged if the vectors are interchanged in a cyclic order. We may also show that

$$\mathbf{A} \cdot (\mathbf{B} \times \mathbf{C}) = -\mathbf{A} \cdot (\mathbf{C} \times \mathbf{B}) \tag{5.44}$$

Vector Triple Product

There are two different ways in which three vectors may be multiplied so that the resulting product is a vector in each case. One such product is

$$\mathbf{A}(\mathbf{B} \cdot \mathbf{C}) = \mathbf{B}(\mathbf{C} \cdot \mathbf{A}) = \mathbf{C}(\mathbf{A} \cdot \mathbf{B}) \tag{5.45}$$

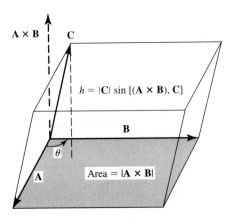

Figure 5.10   Volume of a parallelepiped in terms of sides $\mathbf{A}$, $\mathbf{B}$, and $\mathbf{C}$.

The interpretation of this is: $(\mathbf{B} \cdot \mathbf{C})$ is a scalar; hence $\mathbf{A}(\mathbf{B} \cdot \mathbf{C})$ is vector $\mathbf{A}$ multiplied by a scalar quantity $(\mathbf{B} \cdot \mathbf{C})$. The direction of the new vector is that of $\mathbf{A}$. Similarly, $\mathbf{B}(\mathbf{C} \cdot \mathbf{A})$ is a vector in the direction of $\mathbf{B}$, and $\mathbf{C}(\mathbf{A} \cdot \mathbf{B})$ is a vector in the direction of $\mathbf{C}$.

The second vector triple product is

$$\mathbf{A} \times (\mathbf{B} \times \mathbf{C}) \tag{5.46}$$

$(\mathbf{B} \times \mathbf{C})$ is a vector perpendicular to the plane of $\mathbf{B}$ and $\mathbf{C}$. Therefore, $\mathbf{A} \times (\mathbf{B} \times \mathbf{C})$ is a vector perpendicular to both $\mathbf{A}$ and $(\mathbf{B} \times \mathbf{C})$. Hence $\mathbf{A} \times (\mathbf{B} \times \mathbf{C})$ is a vector in the plane of $\mathbf{B}$ and $\mathbf{C}$. A similar argument applies to other triple products. We may also show that

$$\mathbf{A} \times \mathbf{B} \times \mathbf{C} = \mathbf{B}(\mathbf{A} \cdot \mathbf{C}) - \mathbf{C}(\mathbf{A} \cdot \mathbf{B}) \tag{5.47}$$

5.5 UNIT VECTORS OR BASE VECTORS

As the name indicates, a unit vector is a vector of unit magnitude and has a direction. We shall discuss the concept of a unit vector in this section and use it frequently throughout the text. For a Cartesian or rectangular coordinate system, unit vectors are a set of three mutually perpendicular (or orthogonal) unit vectors, one for each dimension. These are also called *unit coordinate vectors*. For a rectangular coordinate system the unit vectors are denoted by $\hat{\mathbf{i}}, \hat{\mathbf{j}}, \hat{\mathbf{k}}$, where $\hat{\mathbf{i}}$ is along the X-axis, $\hat{\mathbf{j}}$ is along the Y-axis, and $\hat{\mathbf{k}}$ is along the Z-axis (shown in Fig. 5.11), forming a right-handed triad. (Several other notations are used for unit vectors, such as $\hat{\mathbf{x}}, \hat{\mathbf{y}}, \hat{\mathbf{z}}$, or $\hat{\mathbf{x}}_i$, $\mathbf{e}_i$, or $\mathbf{u}_i$, where $i = 1, 2, 3$.) Sometimes unit vectors are represented simply by $\mathbf{i}, \mathbf{j}$, and $\mathbf{k}$. Thus, by definition,

$$\left|\hat{\mathbf{i}}\right| = \left|\hat{\mathbf{j}}\right| = \left|\hat{\mathbf{k}}\right| = 1 \tag{5.48}$$

or in component form

$$\hat{\mathbf{i}} = [1, 0, 0], \qquad \hat{\mathbf{j}} = [0, 1, 0], \qquad \hat{\mathbf{k}} = [0, 0, 1] \tag{5.49}$$

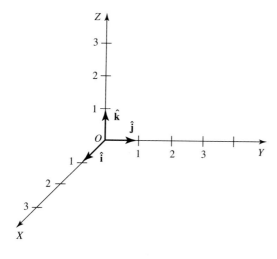

Figure 5.11   Unit vector $\hat{\mathbf{i}}, \hat{\mathbf{j}}, \hat{\mathbf{k}}$ of a rectangular (or Cartesian coordinate) system.

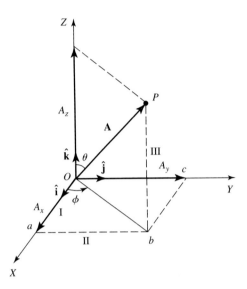

Figure 5.12 Components of a vector $\mathbf{A} = A_x\hat{\mathbf{i}} + A_y\hat{\mathbf{j}} + A_z\hat{\mathbf{k}}$.

Using this definition, we can write vector **A** in component form as

$$\mathbf{A} = [A_x, A_y, A_z] = [A_x, 0, 0] + [0, A_y, 0] + [0, 0, A_z]$$

$$= A_x[1, 0, 0] + A_y(0, 1, 0) + A_z[0, 0, 1]$$

that is, $$\mathbf{A} = A_x\hat{\mathbf{i}} + A_y\hat{\mathbf{j}} + A_z\hat{\mathbf{k}} \tag{5.50}$$

where A_x, A_y, and A_z are the components of **A** along the three axes, as shown in Fig. 5.12. As is clear, the multiplication of $\hat{\mathbf{i}}$ with A_x gives direction to the scalar without changing its magnitude, and similarly for the other two terms, $A_y\hat{\mathbf{j}}$ and $A_z\hat{\mathbf{k}}$. That is, the multiplication of a scalar with a unit vector gives direction to the scalar (the direction of the unit vector) without changing its magnitude. Equation (5.50) has an interesting interpretation, as shown in Fig. 5.12. The tail of the vector is at O, while the head, which is at P, may be located by first going along path I along Oa equal to $A_x\hat{\mathbf{i}}$, followed by path II along ab equal to $A_y\hat{\mathbf{j}}$, and followed by path III along bP equal to $A_z\hat{\mathbf{k}}$. In going from O to P, it is not necessary to follow this particular sequence of paths I, II, III along the axes X, Y, Z; any other sequence—(Y, Z, X) or (Z, X, Y)—will give the same result, that is, will reach point P after starting from point O.

Using the preceding definition of unit vectors and the definition of a scalar product,

$$\mathbf{A} \cdot \mathbf{B} = AB \cos \theta$$

and of a vector product

$$\mathbf{A} \times \mathbf{B} = \mathbf{C} \quad \text{and} \quad |\mathbf{C}| = C = AB \sin \theta$$

we can arrive at the following results, remembering that unit vectors are orthogonal:

$$\hat{\mathbf{i}} \cdot \hat{\mathbf{i}} = \hat{\mathbf{j}} \cdot \hat{\mathbf{j}} = \hat{\mathbf{k}} \cdot \hat{\mathbf{k}} = 1 \tag{5.51}$$

$$\hat{\mathbf{i}} \cdot \hat{\mathbf{j}} = \hat{\mathbf{j}} \cdot \hat{\mathbf{k}} = \hat{\mathbf{k}} \cdot \hat{\mathbf{i}} = \hat{\mathbf{j}} \cdot \hat{\mathbf{i}} = \hat{\mathbf{k}} \cdot \hat{\mathbf{j}} = \hat{\mathbf{i}} \cdot \hat{\mathbf{k}} = 0 \tag{5.52}$$

$$\hat{\mathbf{i}} \times \hat{\mathbf{i}} = \hat{\mathbf{j}} \times \hat{\mathbf{j}} = \hat{\mathbf{k}} \times \hat{\mathbf{k}} = 0 \tag{5.53}$$

$$\hat{\mathbf{i}} \times \hat{\mathbf{j}} = \hat{\mathbf{k}} = -\hat{\mathbf{j}} \times \hat{\mathbf{i}}$$

$$\hat{\mathbf{j}} \times \hat{\mathbf{k}} = \hat{\mathbf{i}} = -\hat{\mathbf{k}} \times \hat{\mathbf{j}} \tag{5.54}$$

$$\hat{\mathbf{k}} \times \hat{\mathbf{i}} = \hat{\mathbf{j}} = -\hat{\mathbf{i}} \times \hat{\mathbf{k}}$$

We shall prove two of these results; the remaining ones can be proved similarly.

$$\hat{\mathbf{i}} \cdot \hat{\mathbf{i}} = |\hat{\mathbf{i}}||\hat{\mathbf{i}}| \cos 0° = (1)(1)(1) = 1$$

$$|\hat{\mathbf{i}} \times \hat{\mathbf{i}}| = |\hat{\mathbf{i}}||\hat{\mathbf{i}}| \sin 0° = (1)(1)(1) = 0$$

Scalar and vector products may be rewritten in a more compact form by making use of unit vector notation and their properties. Using Eq.(5.50), we may write

$$\mathbf{A} = A_x\hat{\mathbf{i}} + A_y\hat{\mathbf{j}} + A_z\hat{\mathbf{k}} \tag{5.55}$$

$$\mathbf{B} = B_x\hat{\mathbf{i}} + B_y\hat{\mathbf{j}} + B_z\hat{\mathbf{k}} \tag{5.56}$$

Addition and substraction of these two vectors may be written as

$$\mathbf{A} + \mathbf{B} = (A_x + B_x)\hat{\mathbf{i}} + (A_y + B_y)\hat{\mathbf{j}} + (A_z + B_z\hat{\mathbf{k}}) \tag{5.57}$$

$$\mathbf{A} - \mathbf{B} = (A_x - B_x)\hat{\mathbf{i}} + (A_y - B_y)\hat{\mathbf{j}} + (A_z - B_z)\hat{\mathbf{k}} \tag{5.58}$$

The scalar product of the two vectors may be evaluated as

$$\mathbf{A} \cdot \mathbf{B} = (A_x\hat{\mathbf{i}} + A_y\hat{\mathbf{j}} + A_z\hat{\mathbf{k}}) \cdot (B_x\hat{\mathbf{i}} + B_y\hat{\mathbf{j}} + B_z\hat{\mathbf{k}})$$

$$= A_xB_x(\hat{\mathbf{i}} \cdot \hat{\mathbf{i}}) + A_yB_y(\hat{\mathbf{j}} \cdot \hat{\mathbf{j}}) + A_zB_z(\hat{\mathbf{k}} \cdot \hat{\mathbf{k}}) + A_xB_y(\hat{\mathbf{i}} \times \hat{\mathbf{j}}) + A_xB_z(\hat{\mathbf{i}} \times \hat{\mathbf{k}})$$

$$+ A_yB_x(\hat{\mathbf{j}} \times \hat{\mathbf{i}}) + A_yB_z(\hat{\mathbf{j}} \cdot \hat{\mathbf{k}}) + A_zB_x(\hat{\mathbf{k}} \cdot \hat{\mathbf{i}}) + A_zB_y(\hat{\mathbf{k}} \cdot \hat{\mathbf{j}})$$

Using the results of Eqs. (5.51) and (5.52), $(\hat{\mathbf{i}} \cdot \hat{\mathbf{i}}) = \cdots = 1$ and $(\hat{\mathbf{i}} \cdot \hat{\mathbf{j}}) = \cdots = 0$, we get

$$\mathbf{A} \cdot \mathbf{B} = A_xB_x + A_yB_y + A_zB_z \tag{5.59}$$

Let us now calculate the cross product:

$$\mathbf{A} \times \mathbf{B} = (A_x\hat{\mathbf{i}} + A_y\hat{\mathbf{j}} + A_z\hat{\mathbf{k}}) \times (B_x\hat{\mathbf{i}} + B_y\hat{\mathbf{j}} + B_z\hat{\mathbf{k}})$$

$$= A_xB_x(\hat{\mathbf{i}} \times \hat{\mathbf{i}}) + A_yB_y(\hat{\mathbf{j}} \times \hat{\mathbf{j}}) + A_zB_z(\hat{\mathbf{k}} \times \hat{\mathbf{k}}) + A_xB_y(\hat{\mathbf{i}} \times \hat{\mathbf{j}}) + A_xB_z(\hat{\mathbf{i}} \times \hat{\mathbf{k}})$$

$$+ A_yB_x(\hat{\mathbf{j}} \times \hat{\mathbf{i}}) + (A_yB_z(\hat{\mathbf{j}} \times \hat{\mathbf{k}}) + A_zB_x(\hat{\mathbf{k}} \times \hat{\mathbf{i}}) + A_zB_y(\hat{\mathbf{k}} \times \hat{\mathbf{j}})$$

Using the results of Eqs. (5.53) and (5.54), $(\hat{\mathbf{i}} \times \hat{\mathbf{i}}) = \cdots = 0$ and $(\hat{\mathbf{i}} \times \hat{\mathbf{j}}) = \mathbf{k} = -(\hat{\mathbf{j}} \times \hat{\mathbf{i}})$, and so on, we get

$$
\begin{aligned}
\mathbf{A} \times \mathbf{B} &= A_x B_x(0) + A_y B_y(0) + A_z B_z(0) + A_x B_y(\hat{\mathbf{k}}) + A_x B_z(-\hat{\mathbf{j}}) \\
&\quad + A_y B_x(-\hat{\mathbf{k}}) + A_y B_z(\hat{\mathbf{i}}) + A_z B_x(\hat{\mathbf{j}}) + A_z B_y(-\hat{\mathbf{i}}) \\
&= \hat{\mathbf{i}}(A_y B_z - A_z B_y) + \hat{\mathbf{j}}(A_z B_x - A_x B_z) + \hat{\mathbf{k}}(A_x B_y - A_y B_x)
\end{aligned}
\tag{5.60}
$$

which may be written in a convenient form as (using the definition of determinants)

$$
\mathbf{A} \times \mathbf{B} = \hat{\mathbf{i}}\begin{vmatrix} A_y & A_z \\ B_y & B_z \end{vmatrix} + \hat{\mathbf{j}}\begin{vmatrix} A_z & A_x \\ B_z & B_x \end{vmatrix} + \hat{\mathbf{k}}\begin{vmatrix} A_x & A_y \\ B_x & B_y \end{vmatrix}
\tag{5.61}
$$

or in a still simpler form as

$$
\mathbf{A} \times \mathbf{B} = \begin{vmatrix} \hat{\mathbf{i}} & \hat{\mathbf{j}} & \hat{\mathbf{k}} \\ A_x & A_y & A_z \\ B_x & B_y & B_z \end{vmatrix}
\tag{5.62}
$$

Note that if $\mathbf{A} = \mathbf{B}$ or $\mathbf{A} = s\mathbf{B}$, where s is a scalar number, s can be taken out of the determinant, thereby making two rows identical; hence the whole determinant will be zero. This is as it should be for $\theta = 0°$, that is, when the two vectors are parallel.

Finally, let us once again consider vector $\mathbf{A}$ as shown in Fig. 5.13. The component of this vector along the X-axis is given by

$$
A_x = A \cos \theta
$$

Another way of writing this is to take the dot product of $\mathbf{A}$ with $\hat{\mathbf{i}}$; that is,

$$
\mathbf{A} \cdot \hat{\mathbf{i}} = |\mathbf{A}||\hat{\mathbf{i}}| \cos \theta = A \cos \theta
$$

which is the same as the preceding equation. Thus

$$
A_x = \mathbf{A} \cdot \hat{\mathbf{i}}
\tag{5.63}
$$

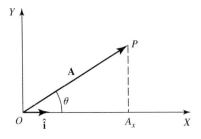

Figure 5.13   Component A_x of vector $\mathbf{A}$ along the X-axis.

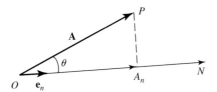

Figure 5.14 Component A_n of vector **A** along the N-axis.

The result of Eq. (5.63) may be written in a general form. Suppose we want to find component A_n of **A** along an arbitrary axis N that has a unit vector $\mathbf{e}_n$ along this axis, as shown in Fig. 5.14. We may write

$$A_n = \mathbf{A} \cdot \mathbf{e}_n \qquad (5.64)$$

5.6 DIRECTIONAL COSINES

Let us start with vector **A** expressed in the form of Eq. (5.50) written in a slightly different form as

$$\mathbf{A} = A_x\hat{\mathbf{i}} + A_y\hat{\mathbf{j}} + A_z\hat{\mathbf{k}} = A\left(\frac{A_x}{A}\hat{\mathbf{i}} + \frac{A_y}{A}\hat{\mathbf{j}} + \frac{A_z}{A}\hat{\mathbf{k}}\right) = A\mathbf{e}_A \qquad (5.65)$$

where $\mathbf{e}_A$ is a unit vector in the direction of **A**. A_x/A is equal to the cosine of the angle between A and the X-axis. Thus

$$\frac{A_x}{A} = \cos(\mathbf{A}, X) = \cos(\mathbf{A}, \hat{\mathbf{i}}) = \alpha \qquad (5.66)$$

$$\frac{A_y}{A} = \cos(\mathbf{A}, Y) = \cos(\mathbf{A}, \hat{\mathbf{j}}) = \beta \qquad (5.67)$$

$$\frac{A_y}{A} = \cos(\mathbf{A}, Z) = \cos(\mathbf{A}, \hat{\mathbf{k}}) = \gamma \qquad (5.68)$$

where α, β, and γ are called the *directional cosines* of the line representing **A**. Thus Eq. (5.65) may be written as

$$\mathbf{A} = A(\alpha\hat{\mathbf{i}} + \beta\hat{\mathbf{j}} + \gamma\hat{\mathbf{k}}) = A\mathbf{e}_A \qquad (5.69)$$

That is,
$$\mathbf{e}_A = \alpha\hat{\mathbf{i}} + \beta\hat{\mathbf{j}} + \gamma\hat{\mathbf{k}} \qquad (5.70)$$

which expresses the unit vector $\mathbf{e}_A$ along **A** in terms of the directional cosines of **A** and the unit vectors. From Eq. (5.65), we may also write

$$\mathbf{A} \cdot \mathbf{A} = A^2\left[\left(\frac{A_x}{A}\right)^2 + \left(\frac{A_y}{A}\right)^2 + \left(\frac{A_z}{A}\right)^2\right] \qquad (5.71)$$

By definition, $\mathbf{A} \cdot \mathbf{A} = A^2$; hence Eq. (5.71) yields

$$\left[\left(\frac{A_x}{A}\right)^2 + \left(\frac{A_y}{A}\right)^2 + \left(\frac{A_z}{A}\right)^2\right] = 1 \qquad (5.72)$$

$$\alpha^2 + \beta^2 + \gamma^2 = 1 \qquad (5.73)$$

That is, the sum of the squares of the directional cosines of any line is equal to 1.

Example 5.1 _____

For the vectors $\mathbf{A} = (6,4,-7)$ and $\mathbf{B} = (2,-2,3)$, calculate the expressions **(a)** to **(j)** given below. The two vectors are expressed in matrix form. Make the calculations and then express the results in terms of unit vectors where possible.

Solution

$$\mathbf{A} := \begin{pmatrix} 6 \\ 4 \\ -7 \end{pmatrix} \quad \mathbf{B} := \begin{pmatrix} 2 \\ -2 \\ 3 \end{pmatrix}$$

(a) $\mathbf{A} + \mathbf{B}$ **(a)** $\mathbf{A} + \mathbf{B} = \begin{pmatrix} 8 \\ 2 \\ -4 \end{pmatrix}$ $\mathbf{A} + \mathbf{B} = 8\cdot\mathbf{i} + 2\cdot\mathbf{j} - 4\cdot\mathbf{k}$

(b) $\mathbf{A} - \mathbf{B}$ **(b)** $\mathbf{A} - \mathbf{B} = \begin{pmatrix} 4 \\ 6 \\ -10 \end{pmatrix}$ $\mathbf{A} - \mathbf{B} = 4\cdot\mathbf{i} + 6\cdot\mathbf{j} - 10\cdot\mathbf{k}$

(c) $4\cdot\mathbf{A} + 2\cdot\mathbf{B}$

  (c) $4\cdot\mathbf{A} + 2\cdot\mathbf{B} = \begin{pmatrix} 28 \\ 12 \\ -22 \end{pmatrix}$ $4\cdot\mathbf{A} + 2\cdot\mathbf{B} = 28\cdot\mathbf{i} + 12\cdot\mathbf{j} - 22\cdot\mathbf{k}$

(d) $4\cdot\mathbf{A} - 2\cdot\mathbf{B}$

  (d) $4\cdot\mathbf{A} - 2\cdot\mathbf{B} = \begin{pmatrix} 20 \\ 20 \\ -34 \end{pmatrix}$ $4\cdot\mathbf{A} - 2 + \mathbf{B} = 20\cdot\mathbf{i} + 20\cdot\mathbf{j} - 34\cdot\mathbf{l}$

(e)   Calculate the magnitudes A, B, and the sum of the squares of A and B.

  (e) $\mathbf{A}\cdot\mathbf{A} = 101$ $\mathbf{B}\cdot\mathbf{B} = 17$

  $\sqrt{\mathbf{A}\cdot\mathbf{A}} = 10.05$ $\sqrt{\mathbf{B}\cdot\mathbf{B}} = 4.123$

  $A := 10.05$ $B := 4.123$

  $A^2 + B^2 = 118.002$

(f)   Calculate the dot product of the two vectors.   **(f)** $\mathbf{A}\cdot\mathbf{B} = -17$

(g)   Calculate the angle θ between the two vectors by making use of the relation

$$\mathbf{A}\cdot\mathbf{B} = AB\cos(\theta)$$

  (g) $\theta := \mathbf{acos}\left(\dfrac{\mathbf{A}\cdot\mathbf{B}}{\sqrt{\mathbf{A}\cdot\mathbf{A}}\cdot\sqrt{\mathbf{B}\cdot\mathbf{B}}} \right)$ $\theta = 1.994\cdot\mathbf{rad}$

  $\theta = 114.221\cdot\mathbf{deg}$

(h)   Calculate the cross product of
the two vectors.

(h)

$$A \times B = \begin{pmatrix} -2 \\ -32 \\ -20 \end{pmatrix}$$

$$A \times B = -2 \cdot i - 32 \cdot j - 20 \cdot k$$

(i)   Calculate the component of **B**
along the direction of vector **A**
(θ is as calculated above).

(i)

$$Ba = B \cdot \cos(\theta)$$

$$Ba := B \cdot \cos(\theta) \qquad\qquad Ba = -1.692$$

Alternate approach:

$$Ba1 := B \cdot \frac{A}{A} \qquad\qquad Ba1 = -1.692$$

(j) Calculate the directional cosines of
A and **B.** (The components are
from the values of the vectors.)

(j)

$$Ax := 6 \qquad Ay := 4 \qquad Az := -7$$

$$A := \begin{pmatrix} Ax \\ Ay \\ Az \end{pmatrix} \qquad A := \sqrt{101}$$

α and β each gives three
directional cosine components
for each vector **A** and **B.**

$$\alpha := \frac{A}{A} \qquad \alpha = \begin{pmatrix} 0.597 \\ 0.398 \\ -0.697 \end{pmatrix}$$

$$Bx := 2 \qquad By := -2 \qquad Bz := 3$$

$$B := \begin{pmatrix} Bx \\ By \\ Bz \end{pmatrix} \qquad B = 4.123$$

As shown, the sum of the squares
of the three directional cosines of
each vector should be unity, and
the sum of the two should be 2.

$$\beta := \frac{B}{B} \qquad \beta = \begin{pmatrix} 0.485 \\ -0.485 \\ 0.728 \end{pmatrix}$$

$$(|\alpha|)^2 = 1 \qquad (|\beta|)^2 = 1$$

$$(|\alpha|)^2 + (|\beta|)^2 = 2$$

EXERCISE 5.1 Repeat the calculations for the vectors $A = (3, 4, -9)$ and $B = (4, -3, 6)$.

5.7 VECTOR CALCULUS

Differentiation of Vectors

Let us consider a vector that is a function of a scalar quantity, say s, where s may be time t, angle θ, or some other quantity. An example is the velocity $\mathbf{v}(t)$ of a particle or the position $\mathbf{r}(\theta)$ of a particle as a function of an angle. The vector $\mathbf{A}$, which is a function of time t, may be written algebraically in component form as

$$\mathbf{A} = \mathbf{A}(t) = A_x(t)\hat{\mathbf{i}} + A_y(t)\hat{\mathbf{j}} + A_z(t)\hat{\mathbf{k}} \tag{5.74}$$

The derivative of $\mathbf{A}$ with respect to t is defined in a manner similar to the derivative of a scalar function. That is,

$$\frac{d\mathbf{A}}{dt} = \lim_{\Delta t \to 0} \frac{\mathbf{A}(t + \Delta t) - \mathbf{A}(t)}{\Delta t} \tag{5.75}$$

This is illustrated in Fig. 5.15, where $\Delta\mathbf{A}$ or $\Delta\mathbf{A}/\Delta t$ becomes tangent to the curve as Δt approaches zero. If $\mathbf{A}$ is replaced by $\mathbf{r}$, then $\Delta\mathbf{r}/\Delta t$ represents the average velocity, while $d\mathbf{r}/dt$ is the instantaneous velocity. We can also write the derivative of a vector in terms of its components as

$$\frac{d\mathbf{A}}{dt} = \left(\frac{dA_x}{dt}, \frac{dA_y}{d}, \frac{dA_z}{dt} \right) = \frac{dA_x}{dt}\hat{\mathbf{i}} + \frac{dA_y}{dt}\hat{\mathbf{j}} + \frac{dA_z}{dt}\hat{\mathbf{k}} \tag{5.76}$$

As an example, if $\mathbf{A} = \mathbf{r}$, the displacement, velocity, and acceleration vectors are

$$\mathbf{r} = x\hat{\mathbf{i}} + y\hat{\mathbf{j}} + z\hat{\mathbf{k}} \tag{5.77}$$

$$\mathbf{v} = \dot{\mathbf{r}} = \dot{x}\hat{\mathbf{i}} + \dot{y}\hat{\mathbf{j}} + \dot{z}\hat{\mathbf{k}} \tag{5.78}$$

$$\mathbf{a} = \ddot{\mathbf{r}} = \dot{\mathbf{v}} = \ddot{x}\hat{\mathbf{i}} + \ddot{y}\hat{\mathbf{j}} + \ddot{z}\hat{\mathbf{k}} \tag{5.79}$$

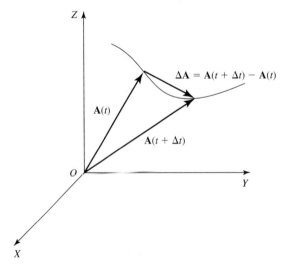

Figure 5.15 Derivative of $\mathbf{A}$ with respect to t.

Note that the magnitudes of the velocity and the acceleration are

$$v = |\mathbf{v}| = \sqrt{\dot{x}^2 + \dot{y}^2 + \dot{z}^2} \tag{5.80}$$

$$a = |\mathbf{a}| = \sqrt{\ddot{x}^2 + \ddot{y}^2 + \ddot{z}^2} \tag{5.81}$$

Thus the differentiation of a vector follows the same procedure used in the differentiation of a scalar function. We can extend these rules to the following particular cases:

$$\frac{d}{ds}(\mathbf{A} \pm \mathbf{B}) = \frac{d\mathbf{A}}{ds} \pm \frac{d\mathbf{B}}{ds} \tag{5.82}$$

$$\frac{d}{ds}[f(s)\mathbf{A}(s)] = \frac{df}{ds}\mathbf{A} + f\frac{d\mathbf{A}}{ds} \tag{5.83}$$

$$\frac{d}{ds}(\mathbf{A} \cdot \mathbf{B}) = \frac{d\mathbf{A}}{ds} \cdot \mathbf{B} + \mathbf{A} \cdot \frac{d\mathbf{B}}{ds} \tag{5.84}$$

$$\frac{d}{ds}(\mathbf{A} \times \mathbf{B}) = \frac{d\mathbf{A}}{ds} \times \mathbf{B} + \mathbf{A} \times \frac{d\mathbf{B}}{ds} \tag{5.85}$$

It is important to note that in the last equation the order of the factors in the cross product must not be changed.

We can use these results to discuss the concepts of relative position, velocity, and acceleration. Let point P_1 be at a distance $\mathbf{r}_1$ and point P_2 at a distance $\mathbf{r}_2$, both with respect to the origin O (Fig. 5.16). The relative position of P_2 with respect to P_1 is $\mathbf{r}_{21}$ and is given by

$$\mathbf{r}_{21} = \mathbf{r}_2 - \mathbf{r}_1 \tag{5.86}$$

Since points P_1 and P_2 are in motion, the relative velocity of P_2 with respect to P_1 is

$$\mathbf{v}_{\text{rel}} = \frac{d\mathbf{r}_{21}}{dt} = \frac{d\mathbf{r}_2}{dt} - \frac{d\mathbf{r}_1}{dt}$$

That is,

$$\mathbf{v}_{21} = \dot{\mathbf{r}}_{21} = \dot{\mathbf{r}}_2 - \dot{\mathbf{r}}_1 \tag{5.87}$$

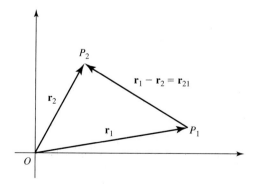

Figure 5.16   Relative position of P_2 with respect to P_1 is $\mathbf{r}_1 - \mathbf{r}_2$.

and, similarly, the relative acceleration of point P_2 with respect to point P_1 is

$$\mathbf{a}_{21} = \dot{\mathbf{v}}_{21} = \ddot{\mathbf{r}}_{21} = \ddot{\mathbf{r}}_2 - \ddot{\mathbf{r}}_1 \tag{5.88}$$

Tangential and Normal Components of Acceleration

By definition, for a particle moving in a curved path, its velocity vector $\mathbf{v}$ is equal to the product of the speed v and a unit vector $\hat{\mathbf{u}}_t$ (in the direction of the tangent); that is,

$$\mathbf{v} = v\hat{\mathbf{u}}_t \tag{5.89}$$

As the particle moves, the speed as well as the direction may change; hence the acceleration of the particle is written as

$$\mathbf{a} = \frac{d\mathbf{v}}{dt} = \frac{d(v\hat{\mathbf{u}}_t)}{dt} = \frac{dv}{dt}\hat{\mathbf{u}}_t + v\frac{d\hat{\mathbf{u}}_t}{dt}$$

or

$$\mathbf{a} = \dot{v}\hat{\mathbf{u}}_t + v\frac{d\hat{\mathbf{u}}_t}{dt} \tag{5.90}$$

Since the unit vector $\hat{\mathbf{u}}_t$ is of constant magnitude, the derivative $d\hat{\mathbf{u}}_t/dt$ means that the direction of $\hat{\mathbf{u}}_t$ is changing with time. As shown in Fig. 5.17(a), initially the particle is at point P and in time Δt it travels a distance Δs reaching a point Q. Let the unit vectors at P and Q be $\hat{\mathbf{u}}_t$ and $\hat{\mathbf{u}}'_t$. As shown in Fig. 5.17(b), the two unit vectors differ by an angle $\Delta\theta$; the magnitude of the diffence in the two unit vectors is

$$\left|\Delta\hat{\mathbf{u}}_t\right| = \left|\hat{\mathbf{u}}'_t - \hat{\mathbf{u}}_t\right| = 2\sin\frac{\Delta\theta}{2}$$

As $\Delta\theta$ approches zero, the right side approaches $\Delta\theta$; hence

$$\lim_{\Delta\theta\to 0}\left|\frac{\Delta\hat{\mathbf{u}}_t}{\Delta\theta}\right| = 1$$

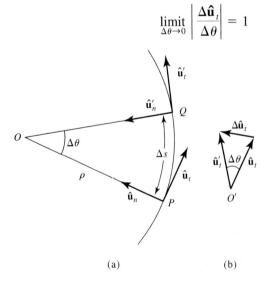

(a)   (b)

Figure 5.17 (a) Unit vector for the motion of a particle along a curved path. (b) The difference $\Delta\mathbf{u}_t$ between unit vectors $\hat{\mathbf{u}}_t$ and $\hat{\mathbf{u}}'_t$ in part (a).

Also in the limit, $\Delta\hat{\mathbf{u}}_t$ becomes perpendicular to $\hat{\mathbf{u}}_t$ and is called the *unit normal vector* $\hat{\mathbf{u}}_n$, as shown in Fig. 5.17(a); that is,

$$\hat{\mathbf{u}}_n = \frac{d\hat{\mathbf{u}}_t}{d\theta} \tag{5.91}$$

Now, using the chain rule, we may write

$$\frac{d\hat{\mathbf{u}}_t}{dt} = \frac{d\hat{\mathbf{u}}_t}{d\theta}\frac{d\theta}{dt} = \hat{\mathbf{u}}_n\frac{d\theta}{ds}\frac{ds}{dt}$$

But $ds/dt = v$ and $ds/d\theta = \rho =$ the radius of curvature of the path; thus

$$\frac{d\hat{\mathbf{u}}_t}{dt} = \hat{\mathbf{u}}_n\frac{v}{\rho} \tag{5.92}$$

Substituting this in Eq. (5.90) yields

$$\mathbf{a} = \dot{v}\hat{\mathbf{u}}_t + \frac{v^2}{\rho}\hat{\mathbf{u}}_n \tag{5.93}$$

Thus the acceleration is described in terms of two components, tangential and normal, where

$$\mathbf{a}_t = \dot{v} = \frac{d^2s}{dt^2} \tag{5.94}$$

$$\mathbf{a}_n = \frac{v^2}{\rho} \tag{5.95}$$

The normal component of the acceleration is always directed toward the concave side of the path and is called the *centripetal acceleration*. The magnitude of the acceleration $\mathbf{a}$ is

$$a = |\mathbf{a}| = \left(\dot{v}^2 + \frac{v^4}{\rho^2}\right)^{1/2} \tag{5.96}$$

Integration of Vectors

In three-dimensional motion, we must distinguish between two types of functions or fields. *Scalar point functions* are of the form

$$f(\mathbf{r}) = f(x, y, z) \tag{5.97}$$

while *vector point functions* are of the form

$$\mathbf{A}(\mathbf{r}) = \mathbf{A}(x, y, z) = [A_x(x, y, z), A_y(x, y, z), A_z(x, y, z)] \tag{5.98}$$

A typical example of a scalar point function is the potential energy function $V(\mathbf{r}) = V(x, y, z)$ of a particle, and that of a vector point function is the electric field intensity $\mathbf{E}(x, y, z)$. These functions may also be functions of time. In mechanics, the corresponding examples are that of density and velocity.

Consider a curve C in space. Suppose a vector point function $\mathbf{A}$ is defined at all the points along this curve. The *line integral* of $\mathbf{A}$ along the curve C is defined as (see Fig. 5.18)

$$\int_c \mathbf{A} \cdot d\mathbf{r} = \text{line integral } \mathbf{A} \text{ along } C \tag{5.99}$$

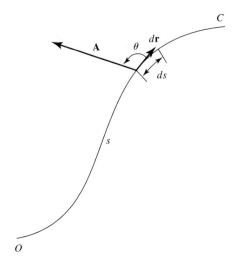

Figure 5.18   Line integral of **A** along the curve C is the sum of the quantities $\mathbf{A} \cdot d\mathbf{r}$.

and if **r** is a position vector,

$$\mathbf{r} = x\hat{\mathbf{i}} + y\hat{\mathbf{j}} + z\hat{\mathbf{k}} \tag{5.100}$$

then

$$d\mathbf{r} = dx\hat{\mathbf{i}} + dy\hat{\mathbf{j}} + dz\hat{\mathbf{k}} \tag{5.101}$$

where dx, dy, and dz are the differences in the two ends of the line segment. Thus, if

$$\mathbf{A} = A_x\hat{\mathbf{i}} + A_y\hat{\mathbf{j}} + A_z\hat{\mathbf{k}} \tag{5.102}$$

we may write the line integral as

$$\int_c \mathbf{A} \cdot d\mathbf{r} = \int_c (A_x dx + A_y dy + A_z dz) \tag{5.103}$$

An alternative way of expressing the line integral is in terms of s, where s is the distance measured along the curve from some fixed point, as shown in Fig. 5.18. Thus, if θ is the angle between **A** and the tangent to the curve at each point,

$$\int \mathbf{A} \cdot d\mathbf{r} = A \cos(\theta)\, ds \tag{5.104}$$

If we know A and $\cos\theta$ as functions of s, the line integral may be evaluated.

It is possible that we know $\mathbf{A}(\mathbf{r})$, but **r** is a known function of parameter s; that is, $\mathbf{r} = \mathbf{r}(s)$, where s is the distance measured along the curve (s could be time t or some other quantity). In such cases, the line integral can be evaluated as

$$\int \mathbf{A} \cdot d\mathbf{r} = \int \left(\mathbf{A} \cdot \frac{d\mathbf{r}}{ds} \right) ds$$

$$= \int \left(A_x \frac{dx}{ds} + A_y \frac{dy}{ds} + A_x \frac{dz}{ds} \right) ds \tag{5.105}$$

This and other methods of evaluating the line integral will be illustrated in subsequent examples.

Example 5.2 _____

Calculate the line integral of $\mathbf{F} = ax\hat{\mathbf{i}} + bxy\hat{\mathbf{j}}$ from $(-R, 0)$ to $(+R, 0)$ along the semicircle shown in Fig. Ex. 5.2 using the parameter θ.

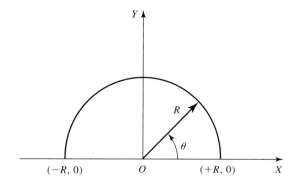

Figure Ex. 5.2

Solution

As the force F is applied, the angle θ changes from π to 0 and the radius vector sweeps out a semicircle as shown. Express x and y in terms of R and θ.
In terms of R and θ, the x and y force components, Fx and Fy ($\mathbf{F} = Fx\mathbf{i} + Fy\mathbf{j}$), take the form

$$\mathbf{F} = a \cdot x \cdot \mathbf{i} + b \cdot x \cdot y \cdot \mathbf{j}$$

$$x(\theta) = R \cdot \cos(\theta) \qquad dx = -R \cdot \sin(\theta) \cdot d\theta$$

$$y(\theta) = R \cdot \sin(\theta) \qquad dy = R \cdot \cos(\theta) \cdot d\theta$$

$$Fx = a \cdot x = a \cdot R \cdot \cos(\theta) \qquad Fy = b \cdot x \cdot y = b \cdot R^2 \cdot \cos(\theta) \cdot \sin(\theta)$$

After substituting for Fx, Fy, dx, and dy, the line integral **A** takes the form

$$\mathbf{A} = \int \mathbf{F} \, d\mathbf{r} = \int Fx \, dx + \int Fy \, dy$$

$$A = \int_{\pi}^{0} \left(-a \cdot R^2 \cdot \cos(\theta) \cdot \sin(\theta) + b \cdot R^3 \cdot \cos(\theta)^2 \cdot \sin(\theta) \right) d\theta \tag{i}$$

Because of the symmetry, we can integrate from $\pi/2$ to 0 and multiply by 2. The resulting integral, which is equal to the work done by the force, is as shown.

$$A = 2 \cdot \left[\int_{\frac{\pi}{2}}^{0} \left(-a \cdot R^2 \cdot \cos(\theta) \cdot \sin(\theta) + b \cdot R^3 \cdot \cos(\theta)^2 \cdot \sin(\theta) \right) d\theta \right] \tag{ii}$$

$$A = a \cdot R^2 - \frac{2}{3} \cdot b \cdot R^3$$

Alternative approach

The alternative approach to the integration above is to substitute $u = \cos(\theta)$ in Eq. (i), as shown below. The integration limits used are from 0 to 1 instead of from -1 to 1.

$\theta1 := 0 \quad \cos(\theta1)^2 = 1 \qquad \cos(\theta1)^3 = 1$

$\theta2 := \dfrac{\pi}{2} \quad \cos(\theta2)^2 = 0 \qquad \cos(\theta2)^3 = 0$

$\theta3 := \pi \quad \cos(\theta3)^2 = 1 \qquad \cos(\theta3)^3 = -1$

$u = \cos(\theta) \qquad du = -\sin(\theta) \cdot d\theta$

$$A = \int_0^1 2 \cdot \left(a \cdot R^2 \cdot u - b \cdot R^3 \cdot u^2 \right) du$$

Integration yields

$$A = a \cdot R^2 - \frac{2}{3} \cdot b \cdot R^3$$

5.8 VECTOR DIFFERENTIAL OPERATORS: GRADIENT, DIVERGENCE, AND CURL

We are now ready to introduce new tools of mathematics that will enable us to study physics in depth. It is not expected that one should grasp these concepts in the first reading; but as they are applied, one becomes familiar with them and can appreciate their usefulness. These new tools are the *vector differential operators,* and we will study them under the headings gradient, divergence, and curl.

The vector differential operator denoted by **grad** or ∇ (del) is not a vector; it is a vector operator. This vector operator, in Cartesian coordinates, is represented by

$$\mathbf{grad} = \nabla \equiv \hat{\mathbf{i}}\,\frac{\partial}{\partial x} + \hat{\mathbf{j}}\,\frac{\partial}{\partial y} + \hat{\mathbf{k}}\,\frac{\partial}{\partial z} \equiv \left(\frac{\partial}{\partial x}, \frac{\partial}{\partial y}, \frac{\partial}{\partial z} \right) \tag{5.106}$$

We will write this operator in other coordinates later. When this operates on a scalar function, it forms a vector. When it operates on a product of functions, it must be treated as a differential operator.

We can perform three different operations with this operator:

1. The gradient operator operates on a scalar function u to form **grad** u or ∇u.
2. When the gradient operator performs a scalar product with another vector function to form $\nabla \cdot \mathbf{A}$ or div $\mathbf{A}$, the result is called the divergence of $\mathbf{A}$. This is a scalar quantity.
3. When the gradient operator performs a vector product with another vector function to form $\nabla \times \mathbf{A}$ or **curl** $\mathbf{A}$, the result is called the curl of $\mathbf{A}$ or rot (meaning "rotation") of $\mathbf{A}$. This is a vector quantity.

The div $\mathbf{A}$ and **curl** $\mathbf{A}$ will be discussed shortly. For the time being, let us concentrate on understanding the meaning and usefulness of the **grad** u.

Gradient Operator (grad $u \equiv \nabla u$)

Consider a scalar function u that is an explicit function of the coordinates x, y, and z, that is, $u = u(x, y, z)$ and this function is continuous and single valued. This scalar function has three components, which may be considered to be the components of a vector called **grad** u or ∇u (del u). That is, even though u is a scalar, **grad** u is a vector with three components given by

$$\text{grad } u = \nabla u = \hat{\mathbf{i}} \frac{\partial u}{\partial x} + \hat{\mathbf{j}} \frac{\partial u}{\partial y} + \hat{\mathbf{k}} \frac{\partial u}{\partial z} = \left(\frac{\partial u}{\partial x}, \frac{\partial u}{\partial y}, \frac{\partial u}{\partial z} \right) \tag{5.107}$$

Physical interpretation of gradient.   Suppose a particle described by a point function $u(x, y, z)$ moves from a point $\mathbf{r} = (x, y, z)$ to a nearby point $\mathbf{r} + d\mathbf{r} = (x + dx, y + dy, z + dz)$. The change in the point function is

$$u(x + dx, y + dy, z + dz) - u(x, y, z)$$

As $dx \to 0$, $dy \to 0$, $dz \to 0$, the differential change du in u takes the form

$$du = \frac{\partial u}{\partial x} dx + \frac{\partial u}{\partial y} dy + \frac{\partial u}{\partial z} dz \tag{5.108}$$

which may be written as

$$du = \left(\frac{\partial u}{\partial x} \hat{\mathbf{i}} + \frac{\partial u}{\partial y} \hat{\mathbf{j}} + \frac{\partial u}{\partial z} \hat{\mathbf{k}} \right) \cdot (dx\,\hat{\mathbf{i}} + dy\,\hat{\mathbf{j}} + dz\,\hat{\mathbf{k}})$$

The first term on the right is **guard** u and the second term is $d\mathbf{r}$. Hence, we may write this equation as

$$du = \text{grad } u \cdot d\mathbf{r} \tag{5.109}$$

Thus this equation allows us to define the change in the function u induced by the changes in its variables. Actually, Eq. (5.109) is the definition of the vector operator gradient. It states that *grad u is a vector such that the change du in u, for an arbitrary small change of position $d\mathbf{r}$, is given by the relation in Eq. (5.109)*. Equation (5.109) may also be written as

$$du = |\text{grad } u||d\mathbf{r}| \cos \theta = |\text{grad } u|\, dr \cos \theta \tag{5.110}$$

du will be maximum when **grad** u and $d\mathbf{r}$ are in the same direction so that $\cos \theta = \cos 0° = 1$, and

$$(du)_{\max} = |\text{grad } u|\, dr \quad \text{when grad } u \text{ is parallel to } d\mathbf{r}$$

Thus
$$|\text{grad } u| = \left(\frac{du}{dr} \right)_{\max} \tag{5.111}$$

Grad u is in the direction in which the change in u is most rapid, and its magnitude is the directional derivative of u, that is, the rate of increase of u per unit distance in that direction.

Let us consider a line path or a surface for which u = constant. Let $d\mathbf{r}$ be directed tangentially along this path of constant u so that $du = 0$. From Eq. (5.109), since neither **grad** u nor $d\mathbf{r}$ is, in general, zero, they must be normal to each other so that cos 90° = 0 gives $du = 0$. *Thus **grad** u is normal to the line (in two dimensions) or to the surface (in three dimensions) for which u = constant.*

Thus the properties of **grad** u may be summarized as follows:

1. **Grad** u is, at any point, normal to the line (in two dimensions) or surface (in three dimensions) for which u is constant.
2. **Grad** u has direction in which u changes most rapidly, and its magnitude is the directional derivative of u.

Once we know **grad** u, the rate of change of u in an arbitrary direction $\hat{\mathbf{n}}$ is given by (the directional derivative)

$$\hat{\mathbf{n}} \cdot \mathbf{grad}\ u = \frac{\partial u}{\partial n} \tag{5.112}$$

Let us illustrate the preceding points with the help of an example. Suppose u stands for a potential function $V(x, y, z)$. Consider two contour lines of constant potential, as shown in Fig. 5.19(a), such that $V_1(x, y, z) = C_1$ and $V_2(x, y, z) = C_2$. Consider a change in $V(x, y, z)$ due to displacement $d\mathbf{r}$ along a contour line. This gives

$$dV = \mathbf{grad}\ u \cdot d\mathbf{r} = \nabla V \cdot d\mathbf{r} \tag{5.113}$$

Since on a contour line V = constant, $dV = 0$, and $\nabla V \cdot d\mathbf{r} = 0$, if $d\mathbf{r}$ is along the line of constant V. Since neither ∇V nor $d\mathbf{r}$ are zero, the vector ∇V must be normal to $d\mathbf{r}$. This is the general result discussed above; that is, at every point in space, ∇V is perpendicular to the constant potential surface (or energy surface) passing through that point.

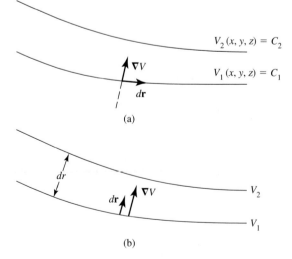

(a)

(b)

Figure 5.19 (a) For displacement $d\mathbf{r}$ along the contour line, $dV = 0$; hence $\nabla V \cdot d\mathbf{r} = 0$, which is possible if ∇V is perpendicular to $d\mathbf{r}$. (b) ∇V is in the direction in which V increases most rapidly, and $d\mathbf{r}$ points in the direction of increasing potential energy.

As shown in Fig. 5.19(b), suppose $d\mathbf{r}$ points in the direction of increasing potential energy. Thus, for change $dV > 0$, $\nabla V \cdot d\mathbf{r} = 0$, which is possible only if ∇V is in the direction in which V increases most rapidly. Furthermore, if $\mathbf{F} = -\nabla V$, then $\mathbf{F}$ is also normal to the constant energy surface everywhere, except that it points from a higher to a lower potential energy. Also, the closer the energy surfaces, the larger will be the gradient; that is, the force is larger where the potential energy is changing rapidly, $\nabla V = (dV/dr)_{\max}$.

Remember that $u(x, y, z)$ is a scalar function and hence defines a scalar field. Examples are temperature and pressure changes in certain volumes of matter. These regions of space may be defined by means of a *temperature gradient* $\nabla T(x, y, z)$ or *pressure gradient* $\nabla P(x, y, z)$. Every point in space may be described by means of one value of temperature if we are dealing with a temperature scalar field or one value of pressure if dealing with a pressure scalar field.

Divergence of a Vector ($\nabla \cdot \mathbf{A} \equiv$ div A)

As stated earlier, when the gradient operator performs a scalar product with a vector point function $\mathbf{A}$, it results in divergence $\mathbf{A}$. That is, in Cartesian coordinates,

$$\text{div } \mathbf{A} \equiv \nabla \cdot \mathbf{A} = \left(\hat{\mathbf{i}} \frac{\partial}{\partial x} + \hat{\mathbf{j}} \frac{\partial}{\partial y} + \hat{\mathbf{k}} \frac{\partial}{\partial z} \right) \cdot (A_x \hat{\mathbf{i}} + A_y \hat{\mathbf{j}} + A_z \hat{\mathbf{k}})$$

$$= \left(\frac{\partial A_x}{\partial x} + \frac{\partial A_y}{\partial y} + \frac{\partial A_z}{\partial z} \right) \tag{5.114}$$

Examples of vector point functions are an electric field vector $\mathbf{E}(x, y, z)$ and a velocity vector $\mathbf{v}(x, y, z)$. The diveregences of such vector point functions describe respective vector fields. Sometimes it is convenient to write (x_1, x_2, x_3) as coordinates instead of (x, y, z), in which case we may write

$$\text{div } \mathbf{A} \equiv \nabla \cdot \mathbf{A} \equiv \left(\frac{\partial A_x}{\partial x} + \frac{\partial A_y}{\partial y} + \frac{\partial A_z}{\partial z} \right) \equiv \left(\frac{\partial A_1}{\partial x_1} + \frac{\partial A_2}{\partial x_2} + \frac{\partial A_3}{\partial x_3} \right)$$

$$= \sum_{i=1}^{3} \frac{\partial A_i}{\partial x_i} \tag{5.115}$$

To fully appreciate the meaning of the divergence of a vector quantity, we consider an example of fluid flow. Consider a region of fluid flow and place a rectangular volume element $dV = dx\, dy\, dz$ located at the origin $O(x, y, z)$ with its faces normal to the three axes, as shown in Fig. 5.20. Let vector $\mathbf{A}$ represent the volume of the flow crossing per unit area per second or the mass of the fluid crossing per unit area (or the momentum carried per unit volume). Without being specific, let $\mathbf{A}$ represent the rate of fluid flow crossing per unit area, the area being normal to $\mathbf{A}$. We consider the volume element to be small enough so that the flow rate $\mathbf{A}$ is constant over each face. Let A_x, A_y, and A_z be the three components of vector $\mathbf{A}$. If the flow rate is $A_x(x)$ at x, then at $(x + dx)$ the flow rate $A_x(x + dx)$ will be, expanding in a Taylor's series,

$$A_x(x + dx) = A_x(x) + \left(\frac{\partial A_x}{\partial x} \right) dx \tag{5.116}$$

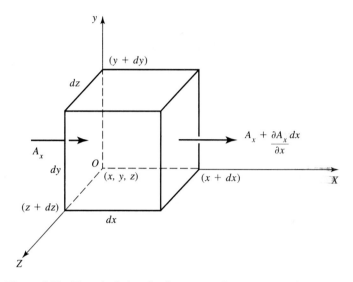

Figure 5.20   For calculating the divergence of a vector quantity.

where we have ignored the higher terms; thus

$$\text{Rate of fluid flowing in at } x = I_x = A_x \, dy \, dz \tag{5.117}$$

$$\text{Rate of fluid flow out at } x + dx = O_x = \left(A_x + \frac{\partial A_x}{\partial x} dx \right) dy \, dz \tag{5.118}$$

The net outward flow parallel to the X-axis through volume element $dV = dx \, dy \, dz$ is

$$O_x - I_x = \left(A_x + \frac{\partial A_x}{\partial x} dx \right) dy \, dz - A_x \, dy \, dz = \frac{\partial A_x}{\partial x} dV \tag{5.119a}$$

with similar expressions for the other two sets of parallel faces:

$$O_y - I_y = \frac{\partial A_y}{\partial y} dV \tag{5.119b}$$

$$O_z - I_z = \frac{\partial A_z}{\partial z} dV \tag{5.119c}$$

Thus the net outward flow rate out of a small rectangular volume element $dV = dx \, dy \, dz$ is obtained by summing these three equations; that is,

$$\text{Net flow rate} = \left(\frac{\partial A_x}{\partial x} + \frac{\partial A_y}{\partial y} + \frac{\partial A_z}{\partial z} \right) dV = (\boldsymbol{\nabla} \cdot \mathbf{A}) \, dV \tag{5.120}$$

Equation (5.120) states that the div $\mathbf{A}$ is the net outward flow rate (volume or mass) per unit volume. But the net flow is also equal to the component of $\mathbf{A}$ normal to the surface area dS; that is,

$$\text{Net flow rate} = \hat{\mathbf{n}} \cdot \mathbf{A} \, dS \tag{5.121}$$

where $\hat{\mathbf{n}}$ is the unit vector normal to the surface pointing outward. From Eqs. (5.120) and (5.121), summing over all the volume and the surface, and replacing by integrations (throughout the whole volume and the entire surface area), we get

$$\iiint_V \boldsymbol{\nabla} \cdot \mathbf{A} \, dV = \iint_S \hat{\mathbf{n}} \cdot \mathbf{A} \, dS \qquad (5.122)$$

which is the mathematical statement of *Gauss's theorem* or the *divergence theorem*.

Gauss's Theorem or the Divergence Theorem. *The divergence of a vector field multiplied by a volume is equal to the net flow of that vector field across the surface bounding that volume.*

That is, since $\hat{\mathbf{n}} \cdot \mathbf{A}$ = the component of $\mathbf{A}$ normal to S, Eq. (5.122) states that

$$\begin{pmatrix} \text{Total amount of } \boldsymbol{\nabla} \cdot \mathbf{A} \\ \text{inside volume } V \end{pmatrix} = \begin{pmatrix} \text{total outward flux} \\ \text{through surface } S \end{pmatrix}$$

Suppose $\mathbf{A} = \mathbf{v}$, the velocity of the moving fluid at any point; then Eq. (5.122) takes the form

$$\iiint_V \boldsymbol{\nabla} \cdot \mathbf{v} \, dV = \iint_S \hat{\mathbf{n}} \cdot \mathbf{v} \, dS \qquad (5.123)$$

which states that

$$\begin{pmatrix} \text{Total volume of the fluid} \\ \text{begin produced within} \\ \text{volume } V \text{ per second} \end{pmatrix} = \begin{pmatrix} \text{volume of the fluid} \\ \text{flowing across } S \\ \text{per second} \end{pmatrix}$$

It is assumed that the fluid is incompressible. Also, $\boldsymbol{\nabla} \cdot \mathbf{v}$ is positive at the source from which the fluid is flowing out, while $\boldsymbol{\nabla} \cdot \mathbf{v}$ is negative at the sink into which the fluid is flowing. On the other hand, if

$$\mathbf{A} = \rho\mathbf{v} = \text{mass flowing through a unit area per second}$$

$$= \text{momentum per unit volume}$$

Gauss's theorem takes the form

$$\iiint_V \boldsymbol{\nabla} \cdot (\rho\mathbf{v}) \, dV = \iint_S \hat{\mathbf{n}} \cdot (\rho\mathbf{v}) \, dS \qquad (5.124)$$

which for an incompressible fluid may be written as

$$\rho\iiint_V \boldsymbol{\nabla} \cdot \mathbf{v} \, dV = \rho\iint_S \hat{\mathbf{n}} \cdot \mathbf{v} \, dS \qquad (5.125)$$

Again depending on the sign of $\nabla \cdot \mathbf{v}$, there will be a source or sink. (A numerical example illustrating the preceding material is given next.) An incompressible fluid must flow out of a given volume element as rapidly as it flows in; that is, there are no sources or sinks and div $\mathbf{A} = 0$, and the flow field is said to be *solenoidal*.

Example 5.3

Consider a box of sides 8 cm by 6 cm by 4 cm, as shown in Fig. Ex. 5.3. Fluid enters the bottom of the box and leaves through the left, right, and top surfaces. No fluid flows out of the front or back surfaces. Calculate the net outward flow through the box for the velocities shown.

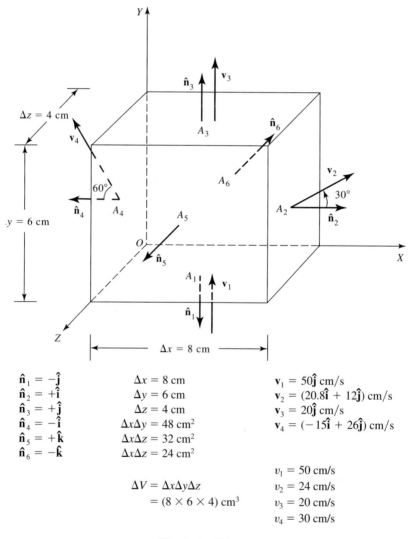

$$\hat{\mathbf{n}}_1 = -\hat{\mathbf{j}}$$
$$\hat{\mathbf{n}}_2 = +\hat{\mathbf{i}}$$
$$\hat{\mathbf{n}}_3 = +\hat{\mathbf{j}}$$
$$\hat{\mathbf{n}}_4 = -\hat{\mathbf{i}}$$
$$\hat{\mathbf{n}}_5 = +\hat{\mathbf{k}}$$
$$\hat{\mathbf{n}}_6 = -\hat{\mathbf{k}}$$

$$\Delta x = 8 \text{ cm}$$
$$\Delta y = 6 \text{ cm}$$
$$\Delta z = 4 \text{ cm}$$
$$\Delta x \Delta y = 48 \text{ cm}^2$$
$$\Delta x \Delta z = 32 \text{ cm}^2$$
$$\Delta x \Delta z = 24 \text{ cm}^2$$

$$\mathbf{v}_1 = 50\hat{\mathbf{j}} \text{ cm/s}$$
$$\mathbf{v}_2 = (20.8\hat{\mathbf{i}} + 12\hat{\mathbf{j}}) \text{ cm/s}$$
$$\mathbf{v}_3 = 20\hat{\mathbf{j}} \text{ cm/s}$$
$$\mathbf{v}_4 = (-15\hat{\mathbf{i}} + 26\hat{\mathbf{j}}) \text{ cm/s}$$

$$\Delta V = \Delta x \Delta y \Delta z$$
$$= (8 \times 6 \times 4) \text{ cm}^3$$

$$v_1 = 50 \text{ cm/s}$$
$$v_2 = 24 \text{ cm/s}$$
$$v_3 = 20 \text{ cm/s}$$
$$v_4 = 30 \text{ cm/s}$$

Figure Ex. 5.3

Solution

Using the divergence theorem as applied to an incompressible fluid flow from Eq. (5.123),

$$\iiint_V \nabla \cdot \mathbf{v} \, dV = \iint_S \hat{\mathbf{n}} \cdot \mathbf{v} \, dS \tag{i}$$

We can solve the problem either by using the left or right side of this equation. We shall do both to show them to be equal. Also note that only normal components of the velocity contribute to the net flow. We start with the right side of Eq. (i) which may be written as

$$\iint_S \hat{\mathbf{n}} \cdot \mathbf{v} \, dS = \sum_i \hat{\mathbf{n}}_i \cdot v_i A_i \tag{ii}$$

where A_i represents the areas of the different faces of the box. Note that the fluid is flowing in only from surface 1.

The unit vectors **i, j,** and **k** are

$$i := \begin{bmatrix} 1 \\ 0 \\ 0 \end{bmatrix} \qquad j := \begin{bmatrix} 0 \\ 1 \\ 0 \end{bmatrix} \qquad k := \begin{bmatrix} 0 \\ 0 \\ 1 \end{bmatrix}$$

Taking into consideration the direction of fluid flow, the unit normal vectors perpendicular to the six surfaces are:

$$n1 := -j \qquad n2 := i \qquad n3 := j \quad n4 := -i \quad n5 := k \qquad n6 := -k$$

$$v1 := 50 \cdot j \cdot \frac{cm}{sec} \qquad\qquad v3 := 20 \cdot j \cdot \frac{cm}{sec}$$

The components of the velocities of the fluid flow, in and out, are

$$|v1| = 50 \cdot cm \cdot sec^{-1} \qquad\qquad |v3| = 20 \cdot cm \cdot sec^{-1}$$

$$v2 := (20.8 \cdot i + 12 \cdot j) \cdot \frac{cm}{sec} \qquad v4 := (-15 \cdot i + 26 \cdot j) \cdot \frac{cm}{sec}$$

$$|v2| = 24.013 \cdot cm \cdot sec^{-1} \qquad\qquad |v4| = 30.017 \cdot cm \cdot sec^{-1}$$

A1, A2, A3, and A4 are the areas of the four surfaces that are needed to calculate the flow rate.

$$x := 8 \cdot cm \qquad\qquad y := 6 \cdot cm \qquad\qquad z := 4 \cdot cm$$

$$A1 := x \cdot z \qquad\qquad A2 := y \cdot z \qquad\qquad A3 := x \cdot z \qquad\qquad A4 := y \cdot z$$

V is the volume of the box. AI is the rate of the fluid flowing into the box. Note that the negative sign indicates the fluid is flowing in, while the positive sign means fluid is flowing out.

$$A1 = 32 \cdot cm^2 \qquad A2 = 24 \cdot cm^2 \qquad A3 = 32 \cdot cm^2 \qquad A4 = 24 \cdot cm^2$$

$$V := x \cdot y \cdot z \qquad\qquad AI := n1 \cdot v1 \cdot A1$$

$$V = 192 \cdot cm^3 \qquad\qquad AI = -1.6 \cdot 10^3 \; \cdot cm^3 \cdot sec^{-1}$$

The amount of fluid flowing out is given by the flowing relation

$$AO := n2 \cdot \left[(20.8 \cdot i + 12 \cdot j) \cdot \frac{cm}{sec} \right] \cdot A2 + n3 \cdot \left(20 \cdot j \cdot \frac{cm}{sec} \right) \cdot A3 + n4 \cdot \left[(-15 \cdot i + 26 \cdot j) \cdot \frac{cm}{sec} \right] \cdot A4$$

or by using the simpler relation:

$$AO := n2 \cdot v2 \cdot A2 + n3 \cdot v3 \cdot A3 + n4 \cdot v4 \cdot A4$$

Thus the outgoing fluid rate AO
and the net flow rate AN from the
box are as shown. The negative
sign means that the box works as
a sink, that is, an amount of fluid is
absorbed.

$$AO = 1.499 \cdot 10^3 \cdot cm^3 \cdot sec^{-1}$$

$$AN := AI + AO \qquad AN = -100.8 \cdot cm^3 \cdot sec^{-1}$$

Alternative approach

The same result can be obtained by using the left side of Eq. (i) rewritten,

$$\int \int \int A \cdot v \, dx \, dy \, dz \equiv \left(\frac{\Delta v1}{\Delta x} + \frac{\Delta v2}{\Delta y} + \frac{\Delta v3}{\Delta z} \right) \cdot \Delta V$$

Using the data given above,
we see that this gives the
same results.

$$\Delta x := 8 \cdot cm \qquad \Delta y := 6 \cdot cm \qquad \Delta z := 4 \cdot cm$$

$$\Delta V := \Delta x \cdot \Delta y \cdot \Delta z \qquad\qquad \Delta V = 192 \cdot cm^3$$

$$\frac{i \cdot (v2 - v4)}{\Delta x} = 4.475 \cdot sec^{-1} \qquad \frac{j \cdot (v3 - v1)}{\Delta y} = -5 \cdot sec^{-1}$$

$$Fnet := \left[\frac{i \cdot (v2 - v4)}{\Delta x} + j \cdot \frac{(v3 - v1)}{\Delta y} \right] \cdot \Delta V$$

$$Fnet = -100.8 \cdot cm^3 \cdot sec^{-1}$$

EXERCISE 5.3: Repeat the calculations if in addition there is a fluid flowing in from the front surface at a speed of 10 cm/s normal to the surface and also leaving from the back surface with a speed of 10 cm/s, making an angle of 45° with the normal.

Curl of a Vector ($\nabla \times A \equiv$ Curl A)

We now take the vector or cross product of the gradient operator with a vector, resulting in a vector called the **curl A** or **rot A** (meaning the rotation of a vector field). Thus (in Cartesian coordinates)

$$\mathbf{curl\ A} \equiv \mathbf{\nabla \times A} = \left(\hat{\mathbf{i}} \frac{\partial}{\partial x} + \hat{\mathbf{j}} \frac{\partial}{\partial y} + \hat{\mathbf{k}} \frac{\partial}{\partial z} \right) \times (A_x \hat{\mathbf{i}} + A_y \hat{\mathbf{j}} + A_z \hat{\mathbf{k}})$$

$$= \hat{\mathbf{i}} \left(\frac{\partial A_z}{\partial y} - \frac{\partial A_y}{\partial z} \right) + \hat{\mathbf{j}} \left(\frac{\partial A_x}{\partial z} - \frac{\partial A_z}{\partial x} \right) + \hat{\mathbf{k}} \left(\frac{\partial A_y}{\partial x} - \frac{\partial A_x}{\partial y} \right) \qquad \textbf{(5.126)}$$

which may also be written in short form as

$$\mathbf{curl\ A} \equiv \boldsymbol{\nabla} \times \mathbf{A} = \begin{vmatrix} \hat{\mathbf{i}} & \hat{\mathbf{j}} & \hat{\mathbf{k}} \\ \dfrac{\partial}{\partial x} & \dfrac{\partial}{\partial y} & \dfrac{\partial}{\partial z} \\ A_x & A_y & A_z \end{vmatrix} \tag{5.127}$$

To understand the physical signficance of the **curl** of a vector quantity, let us consider a fluid flow described by the velocity vector **v**. Place a small paddle wheel in the fluid. In addition to being carried in the fluid, the wheel will tend to rotate in the regions where **curl A** $\neq 0$. A fluid that has a nonvanishing **curl** is said to have a *vortex field*; a fluid that everywhere has a vanishing **curl** has an *irrotational field*.

Consider a vector field that has a nonvanishing **curl.** Let this field be represented by a velocity vector **v** at any point in the field. Suppose the component v_y increases with z, while the component v_z increases with y, as shown in Fig. 5.21. Both $\partial v_y/\partial z$ and $\partial v_z/\partial y$ are positive; but $\partial v_y/\partial z$ results in a negative curling (clockwise rotation) about the X-axis, whereas $\partial v_z/\partial y$ results in a positive curling (counterclockwise rotation) about the X-axis. Thus the X component of **curl v** is

$$(\mathbf{curl\ v})_x = \frac{\partial v_z}{\partial y} - \frac{\partial v_y}{\partial z} \tag{5.128}$$

We can assign similar meaning to the other two components.

The geometrical meaning of **curl A** may also be given by means of *Stokes' theorem*; that is,

$$\iint_S \hat{\mathbf{n}} \cdot (\boldsymbol{\nabla} \times \mathbf{A})\, dS = \int_C \mathbf{A} \cdot d\mathbf{r} \tag{5.129}$$

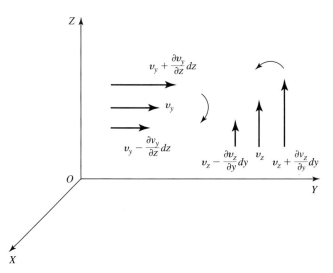

Figure 5.21   For calculating the curl of a velocity vector **v**.

Stokes' Theorem. *The line integral of a vector field along a closed path is equal to the surface integral over an area bounded by the path.*

In Eq. (5.129), C is a curve bounding the surface S in space, $\hat{\mathbf{n}}$ is a unit vector normal to S, and $d\mathbf{r}$ is taken along the path C. Thus the integral on the right is taken around the path, that is, the boundary enclosing any surface that appears on the left. For a small surface ΔS, we may write Stokes' theorem as

$$\hat{\mathbf{n}} \cdot (\nabla \times \mathbf{A})\Delta S = \sum \mathbf{A} \cdot \Delta \mathbf{r} \tag{5.130}$$

as used in the numerical example.

Finally, we may point out that

$$\nabla \cdot \nabla = \text{div grad} = \nabla^2 \tag{5.131}$$

$$= \left(\hat{\mathbf{i}}\,\frac{\partial}{\partial x} + \hat{\mathbf{j}}\,\frac{\partial}{\partial y} + \hat{\mathbf{k}}\,\frac{\partial}{\partial z}\right) \cdot \left(\hat{\mathbf{i}}\,\frac{\partial}{\partial x} + \hat{\mathbf{j}}\,\frac{\partial}{\partial y} + \hat{\mathbf{k}}\,\frac{\partial}{\partial z}\right)$$

$$= \left(\frac{\partial^2}{\partial x^2} + \frac{\partial^2}{\partial y^2} + \frac{\partial^2}{\partial z^2}\right)$$

must be a scalar, and ∇^2, called the *Laplacian operator*, is such a scalar operator. On the other hand, the cross product of ∇ with itself is zero by definition because the two vectors are parallel to each other; that is,

$$\nabla \times \nabla = 0 \tag{5.132}$$

 ## Example 5.4 _____

Consider a 8 cm by 8 cm path $ABCD$ located in a velocity field shown in Fig. Ex. 5.4. Evaluate the left and right side of Stokes' theorem; that is,

$$\iint_S (\nabla \times \mathbf{v}) \cdot \hat{\mathbf{n}}\,dS = \int_C \mathbf{v} \cdot d\mathbf{r} \tag{i}$$

The integration on the right is along the path of the surface used in the left side. We may write Eq. (i) as

$$(\nabla \times \mathbf{v}) \cdot \hat{\mathbf{n}}\Delta S = \sum \mathbf{v} \cdot \Delta \mathbf{r} \tag{ii}$$

Let us consider the left side first.

$$\nabla \times \mathbf{v} = \hat{\mathbf{i}}\left(\frac{\partial v_z}{\partial y} - \frac{\partial v_y}{\partial z}\right) + \hat{\mathbf{j}}\left(\frac{\partial v_x}{\partial z} - \frac{\partial v_z}{\partial x}\right) + \hat{\mathbf{k}}\left(\frac{\partial v_y}{\partial x} - \frac{\partial v_x}{\partial y}\right) = 0 + 0 + \hat{\mathbf{k}}\left(0 - \frac{\partial v_x}{\partial y}\right) \tag{iii}$$

or, in an approximate form, we may write

$$\nabla \times \mathbf{v} = -\hat{\mathbf{k}}\,\frac{\partial v_x}{\partial y} \simeq -\hat{\mathbf{k}}\,\frac{\Delta v_x}{\Delta y} \tag{iv}$$

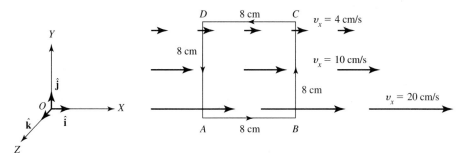

Figure Ex. 5.4

i, j, k are unit vectors while the velocity is in the x-direction only. The surface is in the x-y plane. Thus the resulting change in the velocity is Δv.

$i := 1 \quad j := 1 \quad k := 1$

$\Delta y := 8 \cdot cm \qquad \Delta x := 8 \cdot cm \qquad vy := 0 \qquad vz := 0$

$vx1 := 20 \cdot \dfrac{cm}{sec} \qquad\qquad vx2 := 4 \cdot \dfrac{cm}{sec}$

S is the surface area.
K1 is equal to the product of curl of v and the surface area S, and hence the left side of Eq. (ii) as given in Eq. (iii).
K2 is the right side of Eq. (ii).

$\Delta v := vx2 - vx1 \qquad\qquad \Delta v = -16 \cdot cm \cdot sec^{-1}$

$S := (8 \cdot cm) \cdot (8 \cdot cm) \qquad S = 64 \cdot cm^2$

$K1 := -k \cdot \left(\dfrac{\Delta v}{\Delta y}\right) \cdot S \qquad\qquad K2 := vx1 \cdot \Delta x - vx2 \cdot \Delta y$

Since K1 = K2, this proves the Stokes' theorem.

$K1 = 128 \cdot cm^2 \cdot sec^{-1} \qquad\qquad K2 = 128 \cdot cm^2 \cdot sec^{-1}$

Alternative treatment

We use the same data as above.

$\Delta x := 8 \cdot cm \qquad \Delta y := 8 \cdot cm \qquad k := \begin{pmatrix} 0 \\ 0 \\ 1 \end{pmatrix}$

$\Delta S := \Delta x \cdot \Delta y \qquad \Delta S = 64 \cdot cm^2$

From Eq.(iv)

$curl\_of\_v := \dfrac{-k \cdot \left(4 \cdot \dfrac{cm}{sec} - 20 \cdot \dfrac{cm}{sec}\right)}{\Delta y} \qquad curl\_of\_v = \begin{pmatrix} 0 \\ 0 \\ 2 \end{pmatrix} \cdot sec^{-1}$

From left side of Eq.(ii)

$curl\_of\_v \cdot k \cdot \Delta S = 128 \cdot cm^2 \cdot sec^{-1}$

$i := 0 .. 3 \qquad\qquad \Delta r := 8 \cdot cm \qquad\qquad v_i :=$

From the right side of Eq.(ii), (while going around the surface)

$\displaystyle\sum_{i=0}^{3} v_i \cdot \Delta r = 128 \cdot cm^2 \cdot sec^{-1}$

$20 \cdot \dfrac{cm}{sec}$
$0 \cdot \dfrac{cm}{sec}$
$-4 \cdot \dfrac{cm}{sec}$
$0 \cdot \dfrac{cm}{sec}$

Alternative treatment

Second alternative treatment is

$vfx := 4 \cdot \dfrac{cm}{sec} \qquad vix := 20 \cdot \dfrac{cm}{sec}$

$$\Delta y := 8 \cdot cm \qquad \Delta s := 64 \cdot cm^2$$

$$\frac{vix - vfx}{\Delta y} = 2 \cdot sec^{-1} \qquad \frac{vix - vfx}{\Delta y} \cdot \Delta s = 128 \cdot cm^2 \cdot sec^{-1}$$

5.9 COORDINATE TRANSFORMATIONS

The results of the application of any physical law to a given system must be independent of the coordinate system and the location of the origin of the coordinate system as well. Vectors have this special feature and hence are frequently used in various situations. Thus it becomes relevant to know the procedure by which vectors transfer from one coordinate system to another, that is, to investigate the properties of such transformations. Futhermore, it is convenient and useful to describe these vectors as well as their transformations in matrix notation.

We start with the description of a scalar in different coordinate systems. Suppose mass M is placed at point P and its coordinates are (x, y) in the XY system and (x', y') in the $X'Y'$ system, as shown in Fig. 5.22. The coordinates of the mass are different in the two coordinate systems, but the mass remains constant; that is,

$$M(x, y) \; = \; M(x', y') = \text{constant} \tag{5.133}$$

Quantities that are invariant under a coordinate transformation are called scalars.

Let us now investigate the procedure for coordinate transformation of vectors. Consider a point P that has coordinates (x_1, x_2, x_3) with respect to the $X_1X_2X_3$ coordinate system and (x'_1, x'_2, x'_3) with respect to the $X'_1X'_2X'_3$ coordinate system, as shown in Fig. 5.23. Note that, for convenience, we are using x_1, x_2, x_3 instead of x, y, z.

For simplicity's sake, let us find the relationship between (x'_1, x'_2) and (x_1, x_2). Referring to Fig. 5.24, x'_1 is given by

$$x'_1 = Oa = Ob + bc + ca$$

$$= x_1 \cos \theta + ce \sin \theta + cP \sin \theta$$

$$= x_1 \cos \theta + (ce + cP) \sin \theta$$

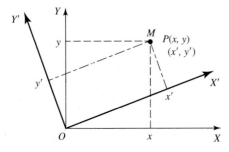

Figure 5.22   Coordinates of a scalar mass M at point P.

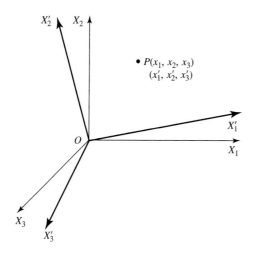

Figure 5.23 Coordinates of point P in two different coordinate systems rotated with respect to each other are (x_1, x_2, x_3) and (x'_1, x'_2, x'_3).

$$= x_1 \cos\theta + eP \sin\theta$$

$$= x_1 \cos\theta + x_2 \sin\theta$$

or

$$x'_1 = x_1 \cos\theta + x_2 \cos\left(\frac{\pi}{2} - \theta\right) \tag{5.134}$$

Similarly,

$$x'_2 = -x_1 \sin\theta + x_2 \cos\theta$$

or

$$x'_2 = x_1 \cos\left(\frac{\pi}{2} + \theta\right) + x_2 \cos\theta \tag{5.135}$$

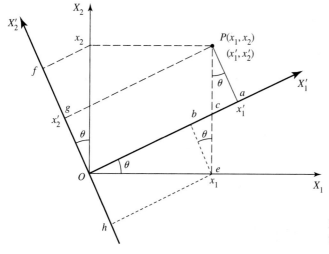

Figure 5.24 To calculate (x'_1, x'_2) in terms of (x_1, x_2).

Thus we have been able to express x_1' and x_2' in terms of x_1, x_2 and the cosines of angle θ. The notation can be simplified by using directional cosines. Let α_1 be the cosine of the angle between the X_1'-axis and X_1-axis or between the unit vectors $\hat{\mathbf{x}}_1'$ and $\hat{\mathbf{x}}_1$; that is,

$$\alpha_1 \equiv \cos(X_1', X_1) = \cos(\hat{\mathbf{x}}_1', \hat{\mathbf{x}}_1) = \hat{\mathbf{x}}_1' \cdot \hat{\mathbf{x}}_1 = \cos\theta$$

Similarly,

$$\alpha_2 \equiv \cos(X_1', X_2) = \cos(\hat{\mathbf{x}}_2', \hat{\mathbf{x}}_2) = \hat{\mathbf{x}}_2' \cdot \hat{\mathbf{x}}_2 = \cos\left(\frac{\pi}{2} - \theta\right) = \sin\theta$$

$$\beta_1 \equiv \cos(X_2', X_1) = \cos(\hat{\mathbf{x}}_2', \hat{\mathbf{x}}_1) = \hat{\mathbf{x}}_2' \cdot \hat{\mathbf{x}}_1 = \cos\left(\frac{\pi}{2} + \theta\right) = -\sin\theta$$

$$\beta_2 \equiv \cos(X_2', X_2) = \cos(\hat{\mathbf{x}}_2', \hat{\mathbf{x}}_2) = \hat{\mathbf{x}}_2' \cdot \hat{\mathbf{x}}_2 = \cos\theta \tag{5.136}$$

Thus Eqs. (5.134) and (5.135) may be written as

$$x_1' = \alpha_1 x_1 + \alpha_2 x_2 \tag{5.137}$$

$$x_2' = \beta_1 x_1 + \beta_2 x_2 \tag{5.138}$$

We may extend these equations to a three-dimensional case.

If we deal with a point P in space with coordinates (x_1, x_2, x_3) in the $X_1 X_2 X_3$ system and (x_1', x_2', x_3') in the $X_1' X_2' X_3'$ system, then

$$x_1' = \alpha_1 x_1 + \alpha_2 x_2 + \alpha_3 x_3 \tag{5.139}$$

$$x_2' = \beta_1 x_1 + \beta_2 x_2 + \beta_3 x_3 \tag{5.140}$$

$$x_3' = \gamma_1 x_1 + \gamma_2 x_2 + \gamma_3 x_3 \tag{5.141}$$

where γ_1 is the cosine of the angle between $\hat{\mathbf{x}}_3'$ and $\hat{\mathbf{x}}_1$. γ_2, γ_3, α_3, and β_3 have similar meanings. The reverse transformation, that is, x_1, x_2, x_3 in terms of x_1', x_2', x_3', may be written as

$$x_1 = \alpha_1 x_1' + \beta_1 x_2' + \gamma_1 x_3' \tag{5.142}$$

$$x_2 = \alpha_2 x_1' + \beta_2 x_2' + \gamma_2 x_3' \tag{5.143}$$

$$x_3 = \alpha_3 x_1' + \beta_3 x_2' + \gamma_3 x_3' \tag{5.144}$$

where α_1 is the cosine of the angle between the X_1-axis and X_1'-axis, and the other directional cosines have similar meaning.

The transformation equations, Eqs. (5.142) to (5.144), may be written in a much neater and more compact form by using the following notation. Let λ_{ij} be the cosine of the angle between the X_i'-axis and X_j-axis; that is, the directional cosine λ_{ij} is

$$\lambda_{ij} \equiv \cos(X_i', X_j) = \hat{\mathbf{x}}_i' \cdot \hat{\mathbf{x}}_j \tag{5.145}$$

and
$$\lambda_{ji} \equiv \cos(X_j, X_1') = \hat{\mathbf{x}}_j \cdot \hat{\mathbf{x}}_i' \qquad (5.146)$$

The coordinates x_1', x_2', x_3' may be expressed in terms of x_1, x_2, x_3 as

$$x_1' = \lambda_{11} x_1 + \lambda_{12} x_2 + \lambda_{13} x_3 \qquad (5.147)$$

$$x_2' = \lambda_{21} x_1 + \lambda_{22} x_2 + \lambda_{23} x_3 \qquad (5.148)$$

$$x_3' = \lambda_{31} x_1 + \lambda_{32} x_2 + \lambda_{33} x_3 \qquad (5.149)$$

while the reverse transformation is

$$x_1 = \lambda_{11} x_1' + \lambda_{21} x_2' + \lambda_{31} x_3' \qquad (5.150)$$

$$x_2 = \lambda_{12} x_1' + \lambda_{22} x_2' + \lambda_{32} x_3' \qquad (5.151)$$

$$x_3 = \lambda_{13} x_1' + \lambda_{23} x_2' + \lambda_{33} x_3' \qquad (5.152)$$

Using summation notation, these transformations may be written as

$$x_i' = \sum_{j=1}^{3} \lambda_{ij} x_j, \qquad i = 1, 2, 3 \qquad (5.153)$$

$$x_i = \sum_{j=1}^{3} \lambda_{ji} x_j', \qquad i = 1, 2, 3 \qquad (5.154)$$

where λ_{ij} may be thought of as elements of a 3 by 3 square matrix $\boldsymbol{\lambda}$ defined as

$$\boldsymbol{\lambda} = \begin{pmatrix} \lambda_{11} & \lambda_{12} & \lambda_{13} \\ \lambda_{21} & \lambda_{22} & \lambda_{23} \\ \lambda_{31} & \lambda_{32} & \lambda_{33} \end{pmatrix} \qquad (5.155)$$

The matrix $\boldsymbol{\lambda}$ is called a *transformation matrix* or a *rotation matrix* and determines the properties of the coordinates of a point under transformation.

According to Eq. (5.155), we need nine quantities λ_{ij} to cause the coordinate transformation of a point. But looking further into the properties of $\boldsymbol{\lambda}$, we find that not all the quantities λ_{ij} are independent. To understand this, we look at two geometrical relations. In Fig. 5.25, the line OP makes angles θ_1, θ_2, and θ_3 with the X_1-, X_2-, and X_3-axes, respectively. Hence directional cosines of the straight line OP are $\cos\theta_1$, $\cos\theta_2$, and $\cos\theta_3$. As discussed in Section 5.6, the sum of the squares of the direction cosines of any line is equal to unity; that is (after replacing α, β, and γ by θ_1, θ_2, and θ_3, respectively)

$$\cos^2\theta_1 + \cos^2\theta_2 + \cos^2\theta_3 = 1 \qquad (5.156)$$

Now with reference to Fig. 5.26, if a line OP makes angles θ_1, θ_2, θ_3 and line OQ makes angles θ_1', θ_2', θ_3' with the axes $X_1 X_2 X_3$, the cosine of the angle between these lines is given by

$$\cos\theta = \cos\theta_1 \cos\theta_1' + \cos\theta_2 \cos\theta_2' + \cos\theta_3 \cos\theta_3' \qquad (5.157)$$

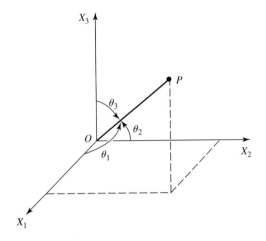

Figure 5.25   The angles θ_1, θ_2, θ_3 that OP makes with the three axes are used for calculating directional cosines.

Let us now consider a set of axes $X_1X_2X_3$. Each of these, when rotated through an angle θ, results in a new set of axes $X_1'X_2'X_3'$. Let us describe the X_1'-axis in the $X_1X_2X_3$ system. Its direction cosines are λ_{11}, λ_{12}, λ_{13}, while for the X_2'-axis in the $X_1X_2X_3$. system, they are λ_{21}, λ_{22}, λ_{23}. Since X_1' is perpendicular to X_2', the angle θ is $\pi2$; when we apply Eq. (5.157), we get

$$\lambda_{11}\lambda_{21} + \lambda_{12}\lambda_{22} + \lambda_{13}\lambda_{23} = \cos \theta = \cos \frac{\pi}{2} = 0 \tag{5.158}$$

which may be written as

$$\sum_{k=1}^{3} \lambda_{1k}\lambda_{2k} = 0 \tag{5.159}$$

In general, applying this to the other two combinations, we get

$$\sum_{k=1}^{3} \lambda_{ik}\lambda_{jk} = 0, \qquad \text{if } i \neq j \tag{5.160}$$

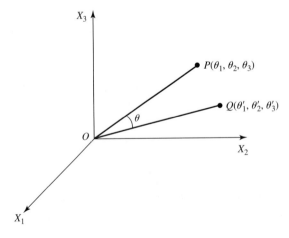

Figure 5.26   For calculating θ between two lines OP and OQ.

Similarly, if we apply Eq. (5.156) to the three axes X_1', X_2', X_3' separately described in the $X_1X_2X_3$ system, we get

$$\lambda_{11}^2 + \lambda_{12}^2 + \lambda_{13}^2 = 1$$

$$\lambda_{21}^2 + \lambda_{22}^2 + \lambda_{23}^2 = 1 \tag{5.161}$$

$$\lambda_{31}^2 + \lambda_{32}^2 + \lambda_{33}^2 = 1$$

which may be written in compact form as

$$\sum_{k=1}^{3} \lambda_{ik}\lambda_{jk} = 1, \quad \text{if } i = j \tag{5.162}$$

Equations (5.160) and (5.162) are called *orthogonality conditions* and apply to any set of coordinate systems in which the coordinate axes are mutually perpendicular; that is, the *systems* are *orthogonal*. Equations (5.160) and (5.162) may be combined into one as

$$\sum_{k=1}^{3} \lambda_{jk}\lambda_{jk} = \delta_{ij} \tag{5.163}$$

where δ_{ij} is called the *Kronecker* delta, which has the properties

$$\delta_{ij} = \begin{cases} 1, & \text{if } i = j \\ 0, & \text{if } i \neq j \end{cases} \tag{5.164}$$

Thus Eq. (5.163) results in six relations between the directional cosines, thereby reducing the number of independent quantities λ_{ij} in the matrix $\boldsymbol{\lambda}$ to only three.

The transformation matrix $\boldsymbol{\lambda}$ described here can be used to describe two different but closely related transformation:

1. *Coordinate transformation:* In this case, point P is fixed, while the base vectors are transformed (say from $X_1 X_2$ to $X_1'X_2'$), causing the coordinates of point P to change, as shown in Fig. 5.27(a). This is the interpretation we have explained here.

2. *Point transformation:* The alternative is to keep the coordinates (or base vectors) fixed and let point P rotate to point P', as shown in Fig. 5.27(b), always keeping the distance from the origin constant.

Let us reconsider Fig. 5.27(a) and (b). In Fig. 5.27(a), axes X_1 and X_2 are fixed and are the reference axes, while axes X_1' and X_2' are obtained by a rotation through an angle θ. Thus the coordinates of point P (x_1', x_2') in the rotated coordinate system are given in terms of the coordinates (x_1, x_2) in the fixed coordinate system as

$$x_1' = x_1 \cos \theta + x_2 \sin \theta \tag{5.165a}$$

$$x_2' = -x_1 \sin \theta + x_2 \cos \theta \tag{5.165b}$$

In this case the transformation acts on the axes and is called a *coordinate transformation*. The same result can be obtained if we keep the axes fixed but rotate point P through an angle θ (in

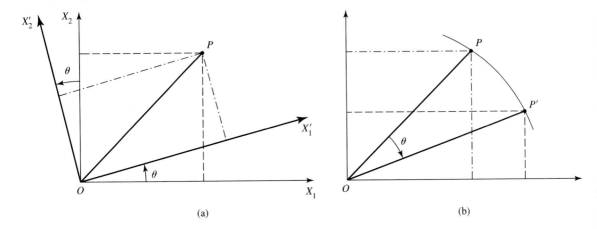

Figure 5.27   (a) Coordinate transformation and (b) point transformation.

the direction opposite to that in which the axes were rotated) to P'. The coordinates of P' are again given by Eqs. (5.165). This type of transformation, which acts on a point, is called a *point transformation*. The two types of transformations are completely equivalent.

Finally, a set of quantities $A_i(A_1, A_2, A_3)$ in an unprimed system may be transformed to a primed system by means of a transformation matrix $\boldsymbol{\lambda}$, resulting in [see Eqs. (5.153)]

$$A_i' = \sum_j \lambda_{ij} A_j \qquad (5.166)$$

The quantities that obey such transformation rules are called vectors; that is, $A_i(A_1, A_2, A_3) \equiv \mathbf{A}$ is a vector quantity.

We have considered a transformation matrix $\boldsymbol{\lambda}$ that is a 3 by 3 square matrix; that is, the number of rows is equal to the number of columns. The matrix may not always be a square. For example, the coordinates (x_1, x_2, x_3) of a point may be represented by a *column matrix x* as

$$x \equiv \begin{pmatrix} x_1 \\ x_2 \\ x_3 \end{pmatrix} \qquad (5.167)$$

or a *row matrix*

$$x \equiv (x_1 \quad x_2 \quad x_3) \qquad (5.168)$$

A common practice is to use the column matrix given by Eq. (5.167) for representing a vector, and we shall use this convention. Thus the coordinates $x_i(x_1, x_2, x_3)$ and $x_i'(x_1', x_2', x_3')$ of point P with respect to the two reference coordinates, the $X_1 X_2 X_3$ and X_1', X_2', X_3' systems, respectively, may be expressed in matrix representation. Thus the transformation equations given by Eq. (5.153)

$$x_i' = \sum_{j=1}^{3} \lambda_{ij} x_j, \quad i = 1, 2, 3 \qquad (5.153)$$

may be written in matrix notation as

$$x' = \lambda x \tag{5.169}$$

This is equivalent to

$$\begin{pmatrix} x_1' \\ x_2' \\ x_3' \end{pmatrix} = \begin{pmatrix} \lambda_1 & \lambda_{12} & \lambda_{13} \\ \lambda_{21} & \lambda_{22} & \lambda_{23} \\ \lambda_{31} & \lambda_{32} & \lambda_{33} \end{pmatrix} \begin{pmatrix} x_1 \\ x_2 \\ x_3 \end{pmatrix} \tag{5.170}$$

which is the same thing as

$$x_1' = \lambda_{11}x_1 + \lambda_{12}x_2 + \lambda_{13}x_3$$

$$x_2' = \lambda_{21}x_1 + \lambda_{22}x_2 + \lambda_{23}x_3 \tag{5.171}$$

$$x_3' = \lambda_{31}x_1 + \lambda_{32}x_2 + \lambda_{33}x_3$$

Note that the multiplication in Eqs. (5.170) or (5.171) is possible only if (1) x and x' are column matrices, and (2) the number of columns in λ must be equal to the number of rows in x. In general, if we want to multiply matrix A with matrix B, the resulting matrix C is given by

$$C = AB \tag{5.172}$$

where the number of columns in matrix A must be equal to the number of rows in B. Any element C_{ij} of matrix C is given by

$$C_{ij} = [AB]_{ij} = \sum_k A_{ik}B_{kj} \tag{5.173}$$

In general, matrix multiplication is not commutative; that is,

$$AB \neq BA \tag{5.174}$$

PROBLEMS

5.1. Prove the following inequalities:
    (a) $|A + B| \leq |A| + |B|$ **(b)** $|A \cdot B| \leq |A||B|$ **(c)** $|A \times B| \leq |A||B|$

5.2 Find the resultant of three forces $\mathbf{F}_1$, $\mathbf{F}_2$, and $\mathbf{F}_3$ in terms of their magnitudes F_1, F_2, and F_3 and angles θ_1, θ_2, and θ_3 between each pair of forces. Also find an expression for the angle α between the resultant force $\mathbf{F}$ and the component force $\mathbf{F}_1$.

5.3. Given the vectors $\mathbf{A} = (4, -2, 6)$ and $\mathbf{B} = (1, 3, -4)$, calculate **(a)** $\mathbf{A} + \mathbf{B}$, **(b)** $\mathbf{A} - \mathbf{B}$, **(c)** $3\mathbf{A} + 2\mathbf{B}$, **(d)** $3\mathbf{A} - 2\mathbf{B}$, **(e)** $A, B, A^2 + B^2$, **(f)** $\mathbf{A} \cdot \mathbf{B}$, **(g)** the angle between $\mathbf{A}$ and $\mathbf{B}$, **(h)** $\mathbf{A} \times \mathbf{B}$, **(i)** the component of $\mathbf{B}$ in the direction of $\mathbf{A}$, and **(j)** the directional cosines of $\mathbf{A}$ and $\mathbf{B}$.

5.4. Given two vectors $\mathbf{A} = 2\hat{\mathbf{i}} + 3\hat{\mathbf{j}} + 4\hat{\mathbf{k}}$ and $\mathbf{B} = -2\hat{\mathbf{i}} - 3\hat{\mathbf{j}} - 4\hat{\mathbf{k}}$, calculate **(a)** $\mathbf{A} + \mathbf{B}$, **(b)** $\mathbf{A} - \mathbf{B}$, **(c)** $3\mathbf{A} + 2\mathbf{B}$, **(d)** $3\mathbf{A} - 2\mathbf{B}$, **(e)** $A, B, A^2 + B^2$, **(f)** $\mathbf{A} \cdot \mathbf{B}$, **(g)** the angle between $\mathbf{A}$ and $\mathbf{B}$, **(h)** $\mathbf{A} \times \mathbf{B}$, **(i)** the component of $\mathbf{B}$ in the direction of $\mathbf{A}$, and **(j)** the directional cosines of $\mathbf{A}$ and $\mathbf{B}$.

5.5. Find the cosine of the angle between vectors $\mathbf{A} = 2\hat{\mathbf{i}} + 3\hat{\mathbf{j}} + 2\hat{\mathbf{k}}$ and $\mathbf{B} = 2\hat{\mathbf{i}} - \hat{\mathbf{j}} + 2\hat{\mathbf{k}}$.

5.6. Prove that the diagonals of an equilateral parallelogram are perpendicular.

5.7. Find a unit vector $\hat{\mathbf{n}}$ that is perpendicular to vectors $\mathbf{A} = \hat{\mathbf{i}} + 2\hat{\mathbf{j}} + 3\hat{\mathbf{k}}$ and $\mathbf{B} = 2\hat{\mathbf{i}} - \hat{\mathbf{j}} + 2\hat{\mathbf{k}}$.

5.8. $\mathbf{A} = 2\hat{\mathbf{i}} + c\hat{\mathbf{j}} + \hat{\mathbf{k}}$ is perpendicular to $\mathbf{B} = \hat{\mathbf{i}} + \hat{\mathbf{j}} + 2\hat{\mathbf{k}}$. What is the value of c?

5.9. Show that vectors $\mathbf{A} = \hat{\mathbf{i}} - 2\hat{\mathbf{j}} - \hat{\mathbf{k}}$ and $\mathbf{B} = 6\hat{\mathbf{i}} + 8\hat{\mathbf{j}} - 10\hat{\mathbf{k}}$ are perpendicular to each other.

5.10. For what values of c will the following two vectors be perpendicular to each other: $\mathbf{A} = c\hat{\mathbf{i}} + 2c\hat{\mathbf{j}} - 4\hat{\mathbf{k}}$ and $\mathbf{B} = 2\hat{\mathbf{i}} - c\hat{\mathbf{j}} + 2c\hat{\mathbf{k}}$?

5.11. Show that the vector $\mathbf{r} = 2\hat{\mathbf{i}} + 5\hat{\mathbf{j}}$ lies in a plane perpendicular to the OZ-axis.

5.12. Vectors $\mathbf{A}$ and $\mathbf{B}$ represent the adjacent sides of a parallelogram. Show that the area of a parallelogram is equal to $|\mathbf{A} \times \mathbf{B}|$.

5.13. $\mathbf{A} = \hat{\mathbf{i}} + \hat{\mathbf{j}}$ and $\mathbf{B} = \hat{\mathbf{i}} + \hat{\mathbf{j}} + \hat{\mathbf{k}}$ are vectors that represent diagonals along the face and through a cube, respectively. Calculate the angle between these two vectors.

5.14. Let $\hat{\mathbf{n}}$ be a unit vector in some fixed direction and $\mathbf{A}$ an arbitrary vector. Show that $\mathbf{A} = (\mathbf{A} \cdot \hat{\mathbf{n}})\hat{\mathbf{n}} + (\hat{\mathbf{n}} \times \mathbf{A}) \times \hat{\mathbf{n}}$.

5.15. For vectors $\mathbf{A}$, $\mathbf{B}$, and $\mathbf{C}$ as given, calculate the following quantities:

$$\mathbf{A} = \hat{\mathbf{i}} + 2\hat{\mathbf{j}} + 3\hat{\mathbf{k}}, \quad \mathbf{B} = 3\hat{\mathbf{i}} - 2\hat{\mathbf{j}} + \hat{\mathbf{k}}, \quad \mathbf{C} = \hat{\mathbf{i}} + \hat{\mathbf{j}} - \hat{\mathbf{k}}$$

(a) $\mathbf{A} + \mathbf{B} + \mathbf{C}$ (b) $\mathbf{A} - \mathbf{B} + \mathbf{C}$ (c) $\mathbf{A} - \mathbf{B} - \mathbf{C}$ (d) $(\mathbf{A} + \mathbf{B}) \cdot \mathbf{C}$

(e) $\mathbf{A} \cdot (\mathbf{B} + \mathbf{C})$ (f) $\mathbf{A} \cdot (\mathbf{B} \times \mathbf{C})$ (g) $(\mathbf{A} \times \mathbf{B}) \cdot \mathbf{C}$ (h) $\mathbf{A} \times \mathbf{B} \times \mathbf{C}$

5.16. Using the fundamental definition of vector diffentiation, prove the results stated in Eqs. (5.82), (5.83), and (5.85).

5.17. For vectors $\mathbf{A}$ and $\mathbf{B}$ as given, calculate the following:

$$\mathbf{A} = ae^{-kt}\hat{\mathbf{i}} + bt\hat{\mathbf{j}} + \hat{\mathbf{k}} \quad \text{and} \quad \mathbf{B} = (c \sin \omega t)\hat{\mathbf{i}} + (d \cos \omega t)\hat{\mathbf{j}}$$

(a) $\dfrac{d\mathbf{A}}{dt}$ $\dfrac{d\mathbf{B}}{dt}$ $\left|\dfrac{d\mathbf{A}}{dt}\right|$ $\left|\dfrac{d\mathbf{B}}{dt}\right|$ (b) $|\mathbf{A}|$ $|\mathbf{B}|$ $\dfrac{d}{dt}|\mathbf{A}|$ $\dfrac{d}{dt}|\mathbf{B}|$

(c) $\dfrac{d}{dt}(\mathbf{A} \cdot \mathbf{B})$ (d) $\dfrac{d}{dt}(\mathbf{A} \times \mathbf{B})$

5.18. Show that a triple scalar product can be written in determinant form as follows:

$$(\mathbf{A} \times \mathbf{B}) \cdot \mathbf{C} = \begin{vmatrix} A_x & A_y & A_z \\ B_x & B_y & B_z \\ C_x & C_y & C_z \end{vmatrix}$$

5.19. Prove the following cyclic relation for a triple scalar product:

$$\mathbf{A} \cdot (\mathbf{B} \times \mathbf{C}) = \mathbf{B} \cdot (\mathbf{C} \times \mathbf{A}) = \mathbf{C} \cdot (\mathbf{A} \times \mathbf{B}) = (\mathbf{A} \times \mathbf{B}) \cdot \mathbf{C}$$

5.20. Prove the following identity:

$$\mathbf{A} \times \mathbf{B} \times \mathbf{C} = (\mathbf{A} \cdot \mathbf{C})\mathbf{B} - (\mathbf{A} \cdot \mathbf{B})\mathbf{C} = \mathbf{B}(\mathbf{A} \cdot \mathbf{C}) - \mathbf{C}(\mathbf{A} \cdot \mathbf{B})$$

5.21. Using the properties of the del operator, ∇, prove the following vector identity: $\mathbf{grad}(uv) = u \, \mathbf{grad} \, v + v \, \mathbf{grad} \, u$.

5.22. Using the properties of the del operator, ∇, prove the following vector identity: $\mathbf{curl} \, (\mathbf{curl} \, \mathbf{A}) = \mathbf{grad} \, (\mathrm{div} \, \mathbf{A}) - \nabla^2\mathbf{A}$.

5.23. Calculate $\mathbf{grad} \, S$, where $S = 1/r^3$ and $r = (x^2 + y^2 + z^2)^{1/2}$.

5.24. Find the gradient of the following functions:

  (a) $f = x + y + z$ (b) $f = xy + xz + yz$ (c) $f = xy^2 + yx^2 + xyz$

5.25. The potential that represents an inverse-square force is $V(r) = k/r$, where $r^2 = x^2 + y^2 + z^2$. Using the definition $\mathbf{F} = -\nabla V$, calculate the components of this force.

5.26. For Problem 5.25, find a unit vector that points in the direction of the maximum increase in V at the position $\mathbf{r} = \hat{\mathbf{i}} + 2\hat{\mathbf{j}} + \hat{\mathbf{k}}$.

5.27. For Problem 5.23, find a unit vector that points in the direction of the maximum increase in S at the position $\mathbf{r} = \hat{\mathbf{i}} + 2\hat{\mathbf{j}} + \hat{\mathbf{k}}$.

5.28. Calculate the divergence of $\mathbf{r}$, div $\mathbf{r} \equiv \nabla \cdot \mathbf{r}$, where:
    (a) $\mathbf{r} = x\hat{\mathbf{i}} + y\hat{\mathbf{j}}$ **(b)** $\mathbf{r} = x\hat{\mathbf{i}} + y\hat{\mathbf{j}} + z\hat{\mathbf{k}}$
    (c) $\mathbf{r} = y\hat{\mathbf{i}} - x\hat{\mathbf{j}} - z\hat{\mathbf{k}}$ **(d)** $\mathbf{r} = 4x\hat{\mathbf{i}} + 2\hat{\mathbf{j}} + 4y\hat{\mathbf{k}}$

5.29. Calculate the divergence of the following vector fields:
    (a) $\mathbf{r} = x^2\hat{\mathbf{i}} + y^2\hat{\mathbf{j}}$ **(b)** $\mathbf{r} = x\hat{\mathbf{i}} - y^2\hat{\mathbf{j}} - z^2\hat{\mathbf{k}}$ **(c)** $\mathbf{r} = xy\hat{\mathbf{i}} - yz\hat{\mathbf{j}} + zx\hat{\mathbf{k}}$

5.30. The gravitational force between two masses may be written as

$$\mathbf{F} = G\frac{Mm}{r^2}\hat{\mathbf{r}} = G\frac{Mn}{r^3}\mathbf{r}$$

Calculate the divergence of $\mathbf{F}$.

5.31. Fluid flowing perpendicular to the six surfaces of a cube of side 10 cm has the magnitude $v_x = 10$ cm/s, $v_{-x} = 20$ cm/s, $v_y = -20$ cm/s, $v_{-y} = -10$ cm/s, $v_z = 50$ cm/s, and $v_{-z} = 50$ cm/s. Calculate the net flud flowing out of the box. Is it a source or sink?

5.32. A cube of side 5 cm is placed in a fluid. For the following velocity values, calculate the average divergence of the flow. Is there a source or sink within this box? The velocities are given in cm/s.

Front	$\mathbf{v}_F = 30\hat{\mathbf{i}} + 60\hat{\mathbf{j}} - 30\hat{\mathbf{k}}$
Right	$\mathbf{v}_R = 20\hat{\mathbf{i}} - 20\hat{\mathbf{j}} + 20\hat{\mathbf{k}}$
Left	$\mathbf{v}_L = -100\hat{\mathbf{i}} + 200\hat{\mathbf{j}} + 300\hat{\mathbf{k}}$
Top	$\mathbf{v}_T = 5\hat{\mathbf{i}} + 50\hat{\mathbf{j}} + 10\hat{\mathbf{k}}$
Bottom	$\mathbf{v}_{Bo} = 0$
Back	$\mathbf{v}_{Ba} = 250\hat{\mathbf{k}}$

Also draw a diagram showing these velocities.

5.33. A box of sides 15 cm by 10 cm by 10 cm is placed in fluid. The flow rates are as shown in Fig. P5.33. Calculate the net flow. Is there a source or sink within the box?

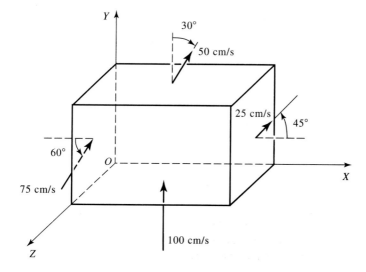

Figure P5.33

5.34. Evaluate the curl of the following vector fields:

    (a) $\mathbf{r} = x\hat{\mathbf{i}} + y\hat{\mathbf{j}}$ **(b)** $\mathbf{r} = y\hat{\mathbf{i}} + x\hat{\mathbf{j}} + z\hat{\mathbf{k}}$

    (c) $\mathbf{r} = x^2\hat{\mathbf{i}} + y^2\hat{\mathbf{j}}$ **(d)** $\mathbf{r} = x^2\hat{\mathbf{i}} - y^2\hat{\mathbf{j}} - z^2\hat{\mathbf{k}}$

    (e) $\mathbf{r} = xy\hat{\mathbf{i}} + yz\hat{\mathbf{j}} + zx\hat{\mathbf{k}}$ **(f)** $4x\hat{\mathbf{i}} - 2\hat{\mathbf{j}} + 4y\hat{\mathbf{k}}$

5.35. Evaluate $\nabla \times \mathbf{F}$ for the following forces:

    (a) $\mathbf{F} = F_x\hat{\mathbf{i}} + F_y\hat{\mathbf{j}} + F_z\hat{\mathbf{k}}$

    (b) $\mathbf{F} = (4abyz^2 - 10bx^2y^2)\hat{\mathbf{i}} + (9abxz^2 - 6bx^3y)\hat{\mathbf{j}} + 8abxyz\hat{\mathbf{k}}$

    (c) $F_x = 6abyz^3 - 20bx^3y^2$

         $F_y = 6abxz^3 - 10bx^4y$

         $F_z = 18abxyz^2$

5.36 Using Stokes' theorem, calculate the average value of the curl of the fluid for a rectangular path 20 cm by 10 cm, as shown in Fig. P5.36.

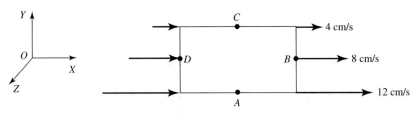

Figure P5.36

5.37 Repeat Problem 5.36 if the velocities (in cm/s) at points A, B, C, and D are

$$\mathbf{v}_A = 10\hat{\mathbf{i}} + 5\hat{\mathbf{j}}, \qquad \mathbf{v}_B = 5\hat{\mathbf{i}} + 10\hat{\mathbf{j}}$$

$$\mathbf{v}_C = 5\hat{\mathbf{i}} + 10\hat{\mathbf{j}}, \qquad \mathbf{v}_D = 10\hat{\mathbf{i}} + 5\hat{\mathbf{j}}$$

5.38. Using Stokes' theorem, calculate the average value of the curl of the fluid for a square of side 10 cm, as shown in Fig. P5.38.

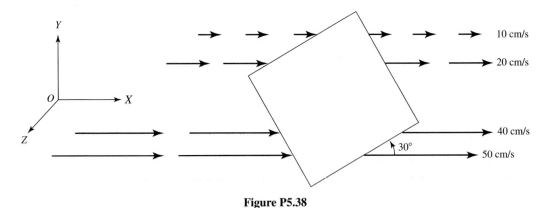

Figure P5.38

5.39. Prove the following relation from trigonometric considerations: $\cos^2 \alpha + \cos^2 \beta + \cos^2 \gamma = 1$.

5.40. A plane formed by the XY-axes is rotated through an angle of 30° about the Z-axis. Find the matrix of rotation.

5.41. A cube is rotated about the Z-axis through $60°$. Find the rotation matrix for the coordinate transformation.

5.42. Consider $\boldsymbol{\lambda}$ to be a two-dimensional transformation matrix. Prove by direct expansion that $|\boldsymbol{\lambda}|^2 = 1$.

5.43. Find the transformation matrix that will cause the rotation of the rectangular coordinates XY system through an angle of $120°$ about the Z-axis.

5.44. Find the transformation matrix that will cause the rotation of the rectangular coordinates XY system through an angle of $120°$ about an axis that makes equal angles with the original three rectangular coordinate axes.

SUGGESTIONS FOR FURTHER READING

ARFKEN, G., *Mathematical Methods for Physicists*. New York: Academic Press, Inc., 1968.

ARTHUR, W., and FENSTER, S. K., *Mechanics,* Chapter 1. New York: Holt, Rinehart and Winston, Inc., 1969.

BECKER, R. A., *Introduction to Theoretical Mechanics,* Chapter 1. New York: McGraw-Hill Book Co, 1954.

DAVIS, A. DOUGLAS, *Classical Mechanics,* Chapter 4. New York: Academic Press, Inc., 1986.

DAVIS, H. F., *Introduction to Vector Analysis*. Needham Heights, Mass.: Allyn & Bacon, 1961.

EISENMAN, R. L., *Matrix Vector Analysis*. New York: McGraw-Hill Book Co., 1963.

FOWLES, G. R., *Analytical Mechanics,* Chapters 1 and 2. New York: Holt, Rinehart and Winston, Inc., 1962.

HAUSER, W., *Introduction to the Principles of Mechanics,* Chapter 1. Reading, Mass.: Addison-Wesley Publishing Co., 1965.

KITTEL, C., KNIGHT, W. D., and RUDERMAN, M. A., *Mechanics,* Berkeley Physics Course, Volume 1, Chapter 2. New York: McGraw-Hill Book Co., 1965.

KLEPPNER, D., and KOLENKOW, R. J., *An Introduction to Mechanics,* Chapters 1 and 5. New York: McGraw-Hill Book Co., 1973.

LINDGREN, B. W., *Vector Calculus*. New York: Macmillan, Inc., 1964.

MARION, J. B., *Classical Dynamics*, 2nd ed., Chapter 1. New York: Academic Press, Inc., 1970.

MARION, J. B., *Principles of Vector Analysis*. New York: Academic Press, Inc., 1965.

ROSSBERG, K., *Analytical Mechanics*, Chapter 1. New York: John Wiley & Sons, Inc., 1983.

SCHEY, H. M., *Div, Grad, Curl and All That*. New York: W. W. Norton and Co., Inc., 1973.

SYMON, K. R., *Mechanics*, 3rd ed., Chapter 3. Reading, Mass.: Addison-Wesley Publishing Co., 1971.

TAYLOR, E. F., *Introductory Mechanics*, Chapter 4. New York: John Wiley & Sons, Inc., 1963.

WYLIE, C. R., Jr., *Advanced Engineering Mathematics*. New York: McGraw-Hill Book Co., 1960.

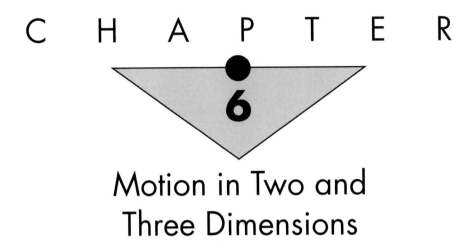

C H A P T E R

6

Motion in Two and Three Dimensions

6.1 INTRODUCTION

So long as we are dealing with one-dimensional motion, rectangular (Cartesian) coordinates have proved quite satisfactory. In describing motion in two and three dimensions, however, rectangular coordinates are frequently not very helpful or convenient, especially in demonstrating physical significance in a given situation. For that purpose we introduce other coordinate systems, such as plane polar, cylindrical, and spherical polar. Furthermore, we will discuss different vector operators in these coordinate systems, and express kinematics and potential energy functions in them. To gain a physical-conceptual understanding of situations such as harmonic oscillators in two and three dimensions and projectile motion, we will need to use these coordinate systems.

6.2 DIFFERENT COORDINATE SYSTEMS

To describe the position and motion of an object or a point in space, it is necessary to have a coordinate system. Some of the commonly used coordinate systems are rectangular coordinates, plane polar coordinates, cylindrical coordinates, and spherical polar coordinates.

Rectangular or Cartesian Coordinates

To start, we choose a two-dimensional rectangular coordinate system. It consists of two mutually perpendicular coordinate axes crossing at the origin O, as shown in Fig. 6.1. As shown, the X- and Y-axes are in the plane of the paper and at 90° to each other. The position of point P is

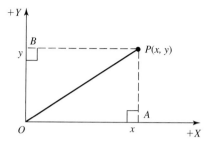

Figure 6.1 Rectangular coordinates (x, y) of a point P in two dimensions.

described by the coordinates (x, y), which are obtained by drawing perpendiculars (or projections) from P to the X- and Y-axes, so that $OA = x$ and $OB = y$. Thus we may write

$$OP^2 = OA^2 + OB^2 = x^2 + y^2 \tag{6.1}$$

Figure 6.2 shows a set of three-dimensional rectangular coordinate axes. Again the X- and Y-axes are in the same plane and at $90°$ to each other, while the Z-axis is perpendicular to this plane. Once again the position of point P is described by the coordinates (x, y, z), and we may write

$$OP^2 = OM^2 + OC^2 = (OA^2 + OB^2) + OC^2$$

or

$$OP^2 = x^2 + y^2 + z^2 \tag{6.2}$$

The three mutually perpendicular axes shown in Fig. 6.2 form a right-handed system.

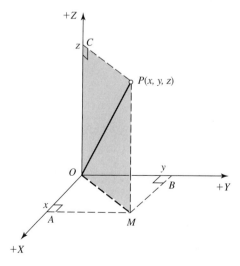

Figure 6.2 Rectangular coordinates (x, y, z) of a point P in three dimensions.

Plane Polar Coordinates

Rectangular coordinate systems are quite useful in describing the motion of an object moving in a straight line. Such coordinates are not always useful when the motion is curved, as in circular motion. For such motion other suitable coordinates are used. A proper choice of a coordinate system can make problem solving much simpler. For example, circular motion in a plane is best described by plane polar coordinates.

Referring to Fig. 6.3, the rectangular coordinates of point P in the XY plane are (x, y). Point P is located at a distance r from the origin, and the line OP makes an angle θ with the X-axis. It is equally acceptable to describe the position of point P by the coordinates (r, θ), called *plane polar coordinates*. The relations between (x, y) and (r, θ) from Fig. 6.3 are

$$x = r \cos \theta, \qquad y = r \sin \theta \tag{6.3}$$

We can express r and θ in terms of x and y by a simple procedure. By squaring and adding Eqs. (6.3), we get

$$x^2 + y^2 = r^2(\cos^2 \theta + \sin^2 \theta) = r^2$$

Also from Eq. (6.3), we get

$$\frac{y}{x} = \frac{r \sin \theta}{r \sin \theta} = \tan \theta$$

That is

$$r = \sqrt{x^2 + y^2} \quad \text{and} \quad \theta = \tan^{-1} \frac{y}{x} \tag{6.4}$$

Thus, in a two-dimensional coordinate system, (x, y) or (r, θ) completely specify the position of a point in a plane. r can have any value between 0 and ∞, while θ can have any value between 0 and 2π radians, with θ increasing counterclockwise.

Further comparison and contrast between rectangular and plane polar coordinates are demonstrated in Figs. 6.4 and 6.5. Figure 6.4 shows the plots of constant x and constant y, while

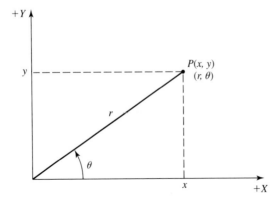

Figure 6.3   Plane polar coordinates (r, θ) of a point P in two dimensions.

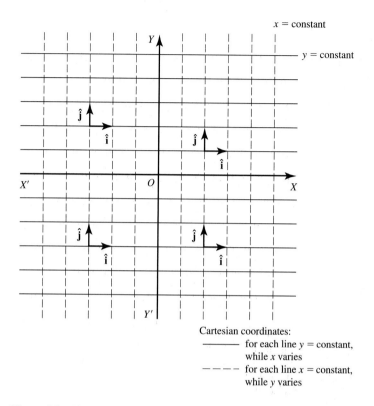

Cartesian coordinates:

———— for each line y = constant, while x varies

– – – – for each line x = constant, while y varies

Figure 6.4   Plots of constant x and constant y. For each continuous line, y = constant while x varies: for each dashed line, x = constant while y varies.

Fig. 6.5 shows the plots of constant r and constant θ. In Fig. 6.4, plots of x = constant are straight lines parallel to the Y-axis, while plots of y = constant are straight lines parallel to the X-axis. These two sets of lines are perpendicular to each other as shown. In Fig. 6.5 plots of θ = constant are straight lines starting from the origin O and directed radially outward, while plots of r = constant are circles with their centers at the origin. Again, the lines of θ = constant and r = constant are perpendicular to each other wherever they cross.

Cylindrical Coordinates

Let us consider a point P located at a distance r from the origin O. Point P can be located by using a set of rectangular coordinates (x, y, z) or cylindrical coordinates (ρ, ϕ, z), as shown in Fig. 6.6 and explained next, or by means of spherical polar coordinates (r, θ, ϕ), as discussed later.

Cylindrical coordinates (ρ, ϕ, z) are shown in Fig. 6.6 and are related to (x, y, z) by the equations

$$x = \rho \cos \phi \qquad\qquad\qquad\qquad\qquad (6.5a)$$

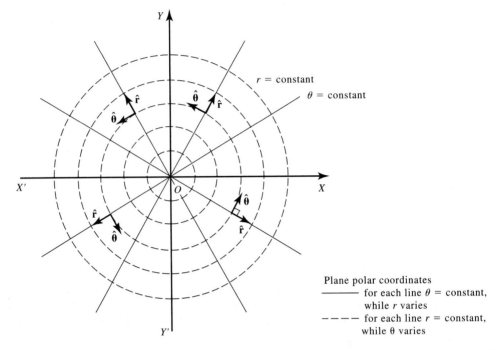

Figure 6.5   Plots of constant r and constant θ. For each continuous line, $\theta =$ constant while r varies; for each dashed line $r =$ constant while θ changes. ($\hat{\mathbf{r}}$ and $\hat{\boldsymbol{\theta}}$ are unit vectors replacing $\hat{\mathbf{i}}$ and $\hat{\mathbf{j}}$.)

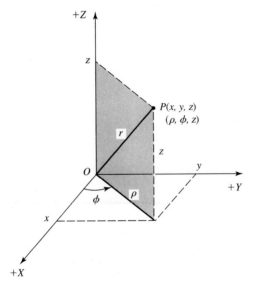

Figure 6.6   Cylindrical coordinates (ρ, ϕ, z) of a point P in space.

$$y = \rho \sin \phi \tag{6.5b}$$

$$z = z \tag{6.5c}$$

while the reverse relations can be obtained from Eqs. (6.5a) and (6.5b) by the procedure used for plane polar coordinates. That is,

$$\rho = \sqrt{x^2 + y^2} \tag{6.6a}$$

$$\phi = \tan^{-1} \frac{y}{x} \tag{6.6b}$$

$$z = z \tag{6.6c}$$

Note that in cylindrical coordinates ρ has replaced r and ϕ has replaced θ of plane polar coordinates.

Spherical Polar Coordinates

Once again consider point P in space located at a distance r from the origin O, as shown in Fig. 6.7. The rectangular coordinates of the point P are (x, y, z), while its spherical polar coordinates are (r, θ, ϕ). To find the relation between these two sets of coordinates, we first resolve $OP = r$ into two components PM and OM, where

$$PM = OC = OP \cos \theta \quad \text{or} \quad z = r \cos \theta$$
$$OM = PC = OP \sin \theta \quad \text{or} \quad OM = r \sin \theta$$

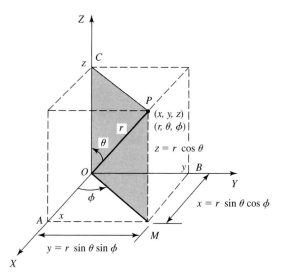

Figure 6.7 Spherical polar coordinates (r, θ, ϕ) of a point P in space.

We further resolve OM into two components, OA and OB, such that

$$OA = OM \cos \phi \quad \text{or} \quad x = r \sin \theta \cos \phi$$

$$OB = OM \cos \phi \quad \text{or} \quad y = r \sin \theta \sin \phi$$

Thus we have the relations

$$x = r \sin \theta \cos \phi \tag{6.7a}$$

$$y = r \sin \theta \sin \phi \tag{6.7b}$$

$$z = r \cos \theta \tag{6.7c}$$

The reverse relations, that is, r, θ, ϕ in terms of x, y, z, may be obtained from Fig. 6.7 as follows:

$$r = OP = \sqrt{OM^2 + OC^2} = \sqrt{(OA^2 + OB^2) + OC^2} = \sqrt{x^2 + y^2 + z^2}$$

$$\tan\theta = \frac{PC}{OC} = \frac{OM}{OC} = \frac{\sqrt{OA^2 + OB^2}}{OC} = \frac{\sqrt{x^2 + y^2}}{z}$$

$$\tan \phi = \frac{OB}{OA} = \frac{y}{x}$$

That is, we have the relations

$$r = \sqrt{x^2 + y^2 + z^2} \tag{6.8a}$$

$$\tan \theta = \frac{\sqrt{x^2 + y^2}}{z} \tag{6.8b}$$

$$\tan \phi = \frac{y}{x} \tag{6.8c}$$

These spherical coordinates will be quite useful in discussing motion in three dimensions.

6.3 KINEMATICS IN DIFFERENT COORDINATE SYSTEMS

We are interested in describing the motion of a particle without regard to the forces that produce such motion. Thus we shall describe the position, velocity, and acceleration of a particle in two and three dimensions. The different coordinate systems that we use in describing motion in detail are plane polar coordinates, cylindrical coordinates, and spherical coordinates.

Cartesian Coordinates

The position of a particle P in the XY plane may be described by the coordinates (x, y), or point P may be described by means of a position vector $\mathbf{r} = (x, y)$, where r is the distance from a specified point called the *origin*. The motion of the point P in the XY plane may be given by describing y as a function of x, or vice versa; that is,

$$y = y(x), \quad x = x(y) \tag{6.9}$$

or it will be still better to give a functional relation between x and y, such as

$$f(x, y) = 0 \tag{6.10}$$

For example, a particle moving in a circular path may be described by

$$x^2 + y^2 = a^2 \tag{6.11}$$

where a is the radius of the circle.

A most convenient way to represent the path of a particle is in terms of some parameter s, such as

$$x = x(s), \qquad y = y(s) \tag{6.12}$$

or

$$r = r(s) \tag{6.13}$$

In such situations, a particle moving in a circular path will be described by

$$x = a \cos \theta, \qquad y = a \sin \theta \tag{6.14}$$

where θ is the parameter in this case.

In general, if a particle moves in a plane XY, its motion is described by

$$x = x(t), \qquad y = y(t) \tag{6.15a}$$

or

$$\mathbf{r} = \mathbf{r}(t) \tag{6.15b}$$

where time t is the parameter in this case. We may write the position vector $\mathbf{r}$, in terms of unit vectors, as

$$\mathbf{r} = \hat{\mathbf{i}}x + \hat{\mathbf{j}}y \tag{6.16}$$

The velocity and the acceleration of the particle and their components are

$$\mathbf{v} = \frac{d\mathbf{r}}{dt} = \hat{\mathbf{i}}\frac{dx}{dt} + \hat{\mathbf{j}}\frac{dy}{dt} = \hat{\mathbf{i}}v_x + \hat{\mathbf{j}}v_y \tag{6.17}$$

$$\mathbf{a} = \frac{d\mathbf{v}}{dt} = \frac{d^2\mathbf{r}}{dt^2} = \hat{\mathbf{i}}\frac{d^2x}{dt^2} + \hat{\mathbf{j}}\frac{d^2y}{dt^2} = \hat{\mathbf{i}}a_x + \hat{\mathbf{j}}a_y \tag{6.18}$$

A three-dimensional motion is represented as

$$\mathbf{r} = \hat{\mathbf{i}}x + \hat{\mathbf{j}}y + \hat{\mathbf{k}}z \tag{6.19}$$

$$\mathbf{v} = \frac{d\mathbf{r}}{dt} = \hat{\mathbf{i}}\frac{dx}{dt} + \hat{\mathbf{j}}\frac{dy}{dt} + \hat{\mathbf{k}}\frac{dz}{dt} = \hat{\mathbf{i}}v_x + \hat{\mathbf{j}}v_y + \hat{\mathbf{k}}v_z \tag{6.20}$$

$$\mathbf{a} = \frac{d\mathbf{v}}{dt} = \frac{d^2\mathbf{r}}{dt^2} = \hat{\mathbf{i}}\frac{d^2x}{dt^2} + \hat{\mathbf{j}}\frac{d^2y}{dt^2} + \hat{\mathbf{k}}\frac{d^2z}{dt^2} = \hat{\mathbf{i}}a_x + \hat{\mathbf{j}}a_y + \hat{\mathbf{k}}a_z \tag{6.21}$$

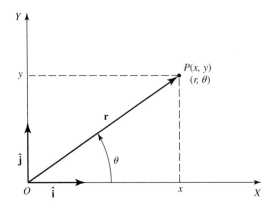

Figure 6.8 Rectangular coordinates (x, y) and plane polar coordinates (r, θ) of a point P. Also shown are the unit vector $\hat{\mathbf{i}}$ and $\hat{\mathbf{j}}$.

Plane Polar Coordinates

In many situations it is convenient to use plane polar coordinates (r, θ) instead of rectangular coordinates (x, y) to describe the motion of a particle. The relations between the two sets of coordinates (Fig. 6.8) are

$$x = r \cos \theta, \qquad y = r \sin \theta \qquad \textbf{(6.22)}$$

while the reverse relations are

$$r = (x^2 + y^2)^{1/2} \qquad \textbf{(6.23a)}$$

$$\theta = \tan^{-1}\left(\frac{y}{x}\right) = \cos^{-1}\left(\frac{x}{\sqrt{x^2 + y^2}}\right) = \sin^{-1}\left(\frac{y}{\sqrt{x^2 + y^2}}\right) \qquad \textbf{(6.23b)}$$

The distance r is measured from the origin, while the polar angle θ is measured counterclockwise from the X-axis, as shown in Fig. 6.8. Unit vector $\hat{\mathbf{i}}$ and $\hat{\mathbf{j}}$ in rectangular coordinates are as shown. We now define two unit vectors in polar coordinates that are perpendicular to each other. These two unit vectors are $\hat{\mathbf{r}}$ and $\hat{\boldsymbol{\theta}}$ and are in the direction of increasing r and θ, as shown in Fig. 6.9. Thus $\hat{\mathbf{r}}$ points in the direction of P along the increasing radial distance r, while $\hat{\boldsymbol{\theta}}$ points

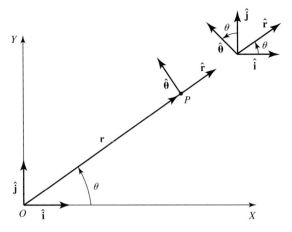

Figure 6.9 Unit vector $\hat{\mathbf{r}}$ and $\hat{\boldsymbol{\theta}}$ in plane polar coordinates.

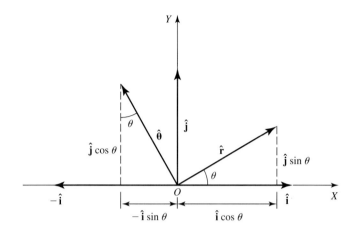

Figure 6.10   Relation between unit vectors $(\hat{\mathbf{r}}, \hat{\boldsymbol{\theta}})$ and $(\hat{\mathbf{i}}, \hat{\mathbf{j}})$.

in the direction that P would move as angle θ increases. Furthermore, both unit vectors are functions of angle θ. Unit vectors $\hat{\mathbf{r}}$ and $\hat{\boldsymbol{\theta}}$ form a new coordinate system called *plane polar coordinates* or simply *polar coordinates*. As shown in Figs. 6.9 and 6.10, unit vectors $\hat{\mathbf{r}}$ and $\hat{\boldsymbol{\theta}}$ are related to $\hat{\mathbf{i}}$ and $\hat{\mathbf{j}}$ by the relation

$$\hat{\mathbf{r}} = \hat{\mathbf{i}} \cos \theta + \hat{\mathbf{j}} \sin \theta \qquad (6.24)$$

$$\hat{\boldsymbol{\theta}} = -\hat{\mathbf{i}} \sin \theta + \hat{\mathbf{j}} \cos \theta \qquad (6.25)$$

Let us differentiate these with respect to θ; that is,

$$\frac{d\hat{\mathbf{r}}}{d\theta} = -\hat{\mathbf{i}} \sin \theta + \hat{\mathbf{j}} \cos \theta = \hat{\boldsymbol{\theta}}$$

$$\frac{d\hat{\boldsymbol{\theta}}}{d\theta} = -\hat{\mathbf{i}} \cos \theta - \hat{\mathbf{j}} \sin \theta = -\hat{\mathbf{r}}$$

Thus we have

$$\frac{d\hat{\mathbf{r}}}{d\theta} = \hat{\boldsymbol{\theta}} \quad \text{and} \quad \frac{d\hat{\boldsymbol{\theta}}}{d\theta} = -\hat{\mathbf{r}} \qquad (6.26)$$

These results can be obtained directly by referring to Fig. 6.11(a) and (b). These figures show the positions of $\hat{\mathbf{r}}$ and $\hat{\boldsymbol{\theta}}$ for a particular angle θ and $\theta + d\theta$. As angle θ increases by $d\theta$, the radial unit vector changes from $\hat{\mathbf{r}}(\theta)$ to $\hat{\mathbf{r}}(\theta + d\theta)$ by an amount $d\hat{\mathbf{r}}$. Similarly, the angular unit vector changes from $\hat{\boldsymbol{\theta}}(\theta)$ to $\hat{\boldsymbol{\theta}}(\theta + d\theta)$ by an amount $d\hat{\boldsymbol{\theta}}$ as shown. Note that $d\hat{\mathbf{r}}$ points in the direction of $\hat{\boldsymbol{\theta}}$, while $d\hat{\boldsymbol{\theta}}$ points in the direction opposite to $\hat{\mathbf{r}}$ that is, in the direction of $-\hat{\mathbf{r}}$ as shown.

The position vector $\mathbf{r}$ in terms of polar coordinates is given by

$$\mathbf{r} = r\hat{\mathbf{r}} = r\hat{\mathbf{r}}(\theta) \qquad (6.27)$$

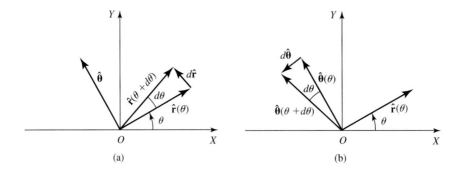

Figure 6.11 For calculating the variation of (a) $\hat{\mathbf{r}}$ with θ and (b) $\hat{\boldsymbol{\theta}}$ with θ.

Note that $\hat{\mathbf{r}} = \hat{\mathbf{r}}(\theta)$; hence expression (6.27) does not contain θ explicitly. The motion of the particle is determined by $r(t)$ and $\theta(t)$ in polar coordinates. Thus the velocity $\mathbf{v}$ is

$$\mathbf{v} = \frac{d\mathbf{r}}{dt} = \frac{d}{dt}(r\hat{\mathbf{r}}) = \frac{dr}{dt}\hat{\mathbf{r}} + r\frac{d\hat{\mathbf{r}}}{dt}$$

Since $\hat{\mathbf{r}} = \hat{\mathbf{r}}(\theta)$ using Eq. (6.26) we must write

$$\frac{d\hat{\mathbf{r}}}{dt} = \frac{d\hat{\mathbf{r}}}{d\theta}\frac{d\theta}{dt} = \hat{\boldsymbol{\theta}}\dot{\theta}$$

That is,

$$\mathbf{v} = \dot{r}\hat{\mathbf{r}} + r\dot{\theta}\hat{\boldsymbol{\theta}} \tag{6.28}$$

We may identify

$$v_r = \dot{r} \quad \text{and} \quad v_\theta = r\dot{\theta} \tag{6.29}$$

where v_r is the component of the velocity along $\hat{\mathbf{r}}$ and is called the *radial velocity*, while v_θ is the component along $\hat{\boldsymbol{\theta}}$ and is called the *angular velocity*.

The acceleration of the system is given by

$$\mathbf{a} = \frac{d\mathbf{v}}{dt} = \frac{d}{dt}(\dot{r}\hat{\mathbf{r}} + r\dot{\theta}\hat{\boldsymbol{\theta}}) = \frac{d\dot{r}}{dt}\hat{\mathbf{r}} + \dot{r}\frac{d\hat{\mathbf{r}}}{d\theta}\frac{d\theta}{dt} + \frac{dr}{dt}\dot{\theta}\hat{\boldsymbol{\theta}} + r\frac{d\dot{\theta}}{dt}\hat{\boldsymbol{\theta}} + r\dot{\theta}\frac{d\hat{\boldsymbol{\theta}}}{d\theta}\frac{d\theta}{dt}$$

$$= \ddot{r}\hat{\mathbf{r}} + \dot{r}(\hat{\boldsymbol{\theta}})\dot{\theta} + \dot{r}\dot{\theta}\hat{\boldsymbol{\theta}} + r\ddot{\theta}\hat{\boldsymbol{\theta}} + r\dot{\theta}(-\hat{\mathbf{r}})\dot{\theta}$$

That is,

$$\mathbf{a} = (\ddot{r} - r\dot{\theta}^2)\hat{\mathbf{r}} + (r\ddot{\theta} + 2\dot{r}\dot{\theta})\hat{\boldsymbol{\theta}} \tag{6.30}$$

Thus the two components of the acceleration $\mathbf{a}$ are the radial acceleration a_r and the angular acceleration a_θ given by

$$a_r = \ddot{r} - r\dot{\theta}^2 \tag{6.31}$$

$$a_\theta = r\ddot{\theta} + 2\dot{r}\dot{\theta} \tag{6.32}$$

A few remarks are in order at this time. The term

$$r\dot{\theta}^2 = r\left(\frac{v_\theta}{r}\right)^2 = \frac{v_\theta^2}{r} \tag{6.33}$$

is the *centripetal acceleration* arising from the motion in the θ direction. Furthermore, if r is held constant in time, $\dot{r} = \ddot{r} = 0$, the path is a circle with centripetal acceleration $a_r = -r\dot{\theta}^2 = -v_\theta^2/r$. The term $2\dot{r}\dot{\theta}$ in a_θ is the *Coriolis acceleration*, and we shall postpone its discussion to Chapter 11.

Cylindrical Polar Coordinates

By adding a Z component to plane polar coordinates we get cylindrical coordinates for describing motion in three dimensions. The three unit vectors $\hat{\boldsymbol{\rho}}$, $\hat{\boldsymbol{\phi}}$, and $\hat{\mathbf{z}}$ in the direction of increasing ρ, ϕ, and z, respectively, are shown in Fig. 6.12. It is important to note that $\hat{\mathbf{z}}$ is constant, while the unit vector $\hat{\boldsymbol{\rho}}$ and $\hat{\boldsymbol{\phi}}$ are functions of ϕ as in the case of plane polar coordinates.

The relations between rectangular coordinates (x, y, z) and cylindrical coordinates (ρ, ϕ, z) are (see Fig. 6.12)

$$x = \rho \cos \phi \tag{6.34a}$$

$$y = \rho \sin \phi \tag{6.34b}$$

$$z = z \tag{6.34c}$$

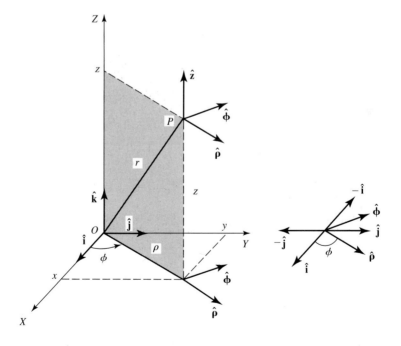

Figure 6.12   Cylindrical coordinates (ρ, ϕ, z) and the corresponding unit vectors $(\hat{\boldsymbol{\rho}}, \hat{\boldsymbol{\phi}}, \hat{\mathbf{z}})$.

while the inverse relations are

$$\rho = (x^2 + y^2)^{1/2} \tag{6.35a}$$

$$\phi = \tan^{-1} \frac{y}{x} = \sin^{-1} \frac{y}{\sqrt{x^2 + y^2}} = \cos^{-1} \frac{x}{\sqrt{x^2 + y^2}} \tag{6.35b}$$

Replacing (r, θ) by (ρ, ϕ) and with an additional Z component, we may write the relations

$$\hat{\boldsymbol{\rho}} = \hat{\mathbf{i}} \cos \phi + \hat{\mathbf{j}} \sin \phi \tag{6.36a}$$

$$\hat{\boldsymbol{\phi}} = -\hat{\mathbf{i}} \sin \phi + \hat{\mathbf{j}} \cos \phi \tag{6.36b}$$

and, as before,

$$\frac{d\hat{\boldsymbol{\rho}}}{d\phi} = \hat{\boldsymbol{\phi}} \tag{6.37}$$

$$\frac{d\hat{\boldsymbol{\phi}}}{d\phi} = -\hat{\boldsymbol{\rho}} \tag{6.38}$$

The position vector $\mathbf{r}$ describing the location of a point P in cylindrical coordinates, shown in Fig. 6.12, is

$$\mathbf{r} = \rho\hat{\boldsymbol{\rho}} + z\hat{\mathbf{z}} \tag{6.39}$$

where ρ gives the distance of P from the Z-axis and ϕ gives its angular rotation from the X-axis, while z gives its elevation above the XY plane. Thus we may write the velocity vector, keeping in mind that $\hat{\boldsymbol{\rho}} = \hat{\boldsymbol{\rho}}(\phi)$, as

$$\mathbf{v} = \frac{d\mathbf{r}}{dt} = \frac{d}{dt}(\rho\hat{\boldsymbol{\rho}} + z\hat{\mathbf{z}}) = \frac{d\rho}{dt}\hat{\boldsymbol{\rho}} + \rho\frac{d\hat{\boldsymbol{\rho}}}{d\phi}\frac{d\phi}{dt} + \frac{dz}{dt}\hat{\mathbf{z}} + z\frac{d\hat{\mathbf{z}}}{dt} = \dot{\rho}\hat{\boldsymbol{\rho}} + \rho(\hat{\boldsymbol{\phi}})\dot{\phi} + \dot{z}\hat{\mathbf{z}} + z(0)$$

where $d\hat{\mathbf{z}}/dt = 0$; hence

$$\mathbf{v} = \dot{\rho}\hat{\boldsymbol{\rho}} + \rho\dot{\phi}\hat{\boldsymbol{\phi}} + \dot{z}\hat{\mathbf{z}} \tag{6.40}$$

Similarly, $$\mathbf{a} = \frac{d\mathbf{v}}{dt} = \frac{d}{dt}(\dot{\rho}\hat{\boldsymbol{\rho}} + \rho\dot{\phi}\hat{\boldsymbol{\phi}} + \dot{z}\hat{\mathbf{z}})$$

can be shown using Eqs. (6.37) and (6.38), to be,

$$\mathbf{a} = (\ddot{\rho} - \rho\dot{\phi}^2)\hat{\boldsymbol{\rho}} + (\rho\ddot{\phi} + 2\dot{\rho}\dot{\phi})\hat{\boldsymbol{\phi}} + \ddot{z}\hat{\mathbf{z}} \tag{6.41}$$

Now we can express any vector $\mathbf{A}$ in terms of three components A_ρ, A_ϕ, and A_z in the directions of the three mutually perpendicular unit vector $\hat{\boldsymbol{\rho}}$, $\hat{\boldsymbol{\phi}}$, and $\hat{\mathbf{z}}$. That is,

$$\mathbf{A} = A_\rho\hat{\boldsymbol{\rho}} + A_\phi\hat{\boldsymbol{\phi}} + A_z\hat{\mathbf{z}} \tag{6.42}$$

where the components depend not only on the vector itself but also on its location in space. This is because both $\hat{\rho}$ and $\hat{\phi}$ depend on ϕ. Thus if $\mathbf{A}$ is a function of a parameter, say t, then in evaluating $d\mathbf{A}/dt$ we must remember the variation of $\hat{\rho}$ and $\hat{\phi}$ as demonstrated next. Using Eq. (6.42),

$$\frac{d\mathbf{A}}{dt} = \frac{dA_\rho}{dt}\,\hat{\rho} + A_\rho \frac{d\hat{\rho}}{d\phi}\frac{d\phi}{dt} + \frac{dA_\phi}{dt}\,\hat{\phi} + A_\phi \frac{d\hat{\phi}}{d\phi}\frac{d\phi}{dt} + \frac{dA_z}{dt}\,\hat{z} + A_z \frac{d\hat{z}}{dt}$$

since $d\hat{z}/dt = 0$, $d\hat{\rho}/d\phi = \hat{\phi}$, and $d\hat{\phi}/d\phi = -\hat{\rho}$, we get, after rearranging,

$$\frac{d\mathbf{A}}{dt} = \left(\frac{dA_\rho}{dt} - A_\phi \dot{\phi} \right)\hat{\rho} + \left(\frac{dA_\phi}{dt} + A_\rho \dot{\phi} \right)\hat{\phi} + \frac{dA_z}{dt}\,\hat{z} \qquad (6.43)$$

Spherical Polar Coordinates

Spherical polar coordinates or spherical coordinates are the most commonly used coordinates in situations of spherical symmetry—for example, in the case of coulomb forces in atoms and gravitational forces. The point P in space is located by the coordinates (r, θ, ϕ), as shown in Fig. 6.13. r is the *radial distance* from the origin O, ϕ is the *azimuthal angle* locating a plane whose angle of rotation is measured from the X-axis, while angle θ is the *polar angle* measured down from the Z-axis. The polar angle θ can have any value between 0 and $\pi/2$, while the azimuthal angle ϕ can have any value between 0 and π.

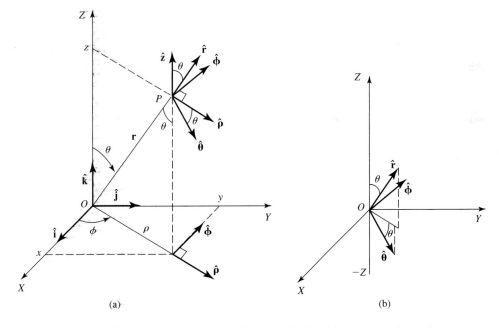

(a)   (b)

Figure 6.13   (a) Spherical polar coordinates (r, θ, ϕ) and the corresponding unit vectors $(\hat{r}, \hat{\theta}, \hat{\phi})$. (b) Orientation of unit vectors $(\hat{r}, \hat{\theta}, \hat{\phi})$ relative to the coordinate system XYZ and polar angle θ.

Rectangular coordinates (x, y, z) are related to spherical polar coordinates (r, θ, ϕ) by the following relations (see Fig. 6.13):

$$x = r \sin \theta \cos \phi \tag{6.44a}$$

$$y = r \sin \theta \sin \phi \tag{6.44b}$$

$$z = r \cos \theta \tag{6.44c}$$

Note that $r \sin \theta = \rho$. The reverse relations are

$$r = (x^2 + y^2 + z^2)^{1/2} \tag{6.45a}$$

$$\theta = \tan^{-1} \frac{\sqrt{x^2 + y^2}}{z} \tag{6.45b}$$

$$\phi = \tan^{-1} \frac{y}{x} \tag{6.45c}$$

The three mutually perpendicular unit vectors used in spherical polar coordinates are $\hat{\mathbf{r}}$, $\hat{\boldsymbol{\theta}}$, and $\hat{\boldsymbol{\phi}}$, as shown in Figs. 6.13(a) and (b). Also shown are the unit vectors $\hat{\mathbf{i}}, \hat{\mathbf{j}}, \hat{\mathbf{k}}, \hat{\mathbf{z}}\ (=\hat{\mathbf{k}})$, and $\hat{\boldsymbol{\rho}}$. The unit vector $\hat{\boldsymbol{\phi}}$ lies in the XY plane, while $\hat{\mathbf{r}}$, $\hat{\boldsymbol{\theta}}$, $\hat{\boldsymbol{\rho}}$, and $\hat{\mathbf{z}}$ all lie in one vertical plane. For fixed r and θ the variation in ϕ, corresponds to rotation about the Z-axis, while for fixed r and ϕ the variation in θ corresponds to the rotation in the plane containing $\hat{\mathbf{r}}$, $\hat{\boldsymbol{\theta}}$, $\hat{\boldsymbol{\rho}}$, and $\hat{\mathbf{z}}$. From Fig. 6.13 we can write the following relations between the unit vectors:

$$\hat{\mathbf{r}} = \hat{\boldsymbol{\rho}} \sin \theta + \hat{\mathbf{z}} \cos \theta = \hat{\mathbf{i}} \sin \theta \cos \phi + \hat{\mathbf{j}} \sin \theta \sin \phi + \hat{\mathbf{k}} \cos \theta \tag{6.46a}$$

$$\hat{\boldsymbol{\theta}} = \hat{\boldsymbol{\rho}} \cos \theta - \hat{\mathbf{z}} \sin \theta = \hat{\mathbf{i}} \cos \theta \cos \phi + \hat{\mathbf{j}} \cos \theta \sin \phi - \hat{\mathbf{k}} \cos \theta \tag{6.46b}$$

$$\hat{\boldsymbol{\phi}} = -\hat{\mathbf{i}} \sin \phi + \hat{\mathbf{j}} \cos \phi \tag{6.46c}$$

Differentiating these equations, we obtain the following relations:

$$\frac{\partial \hat{\mathbf{r}}}{\partial \theta} = \hat{\boldsymbol{\theta}} \qquad \frac{\partial \hat{\mathbf{r}}}{\partial \phi} = \hat{\boldsymbol{\phi}} \sin \theta$$

$$\frac{\partial \hat{\boldsymbol{\theta}}}{\partial \theta} = -\hat{\mathbf{r}}, \qquad \frac{\partial \hat{\boldsymbol{\theta}}}{\partial \phi} = \hat{\boldsymbol{\phi}} \cos \theta \tag{6.47}$$

$$\frac{\partial \hat{\boldsymbol{\phi}}}{\partial \theta} = 0 \qquad \frac{\partial \hat{\boldsymbol{\phi}}}{\partial \phi} = -\hat{\boldsymbol{\rho}} = -\hat{\mathbf{r}} \sin \theta - \hat{\boldsymbol{\theta}} \cos \theta$$

These relations can also be derived from geometrical considerations by drawing figures similar to the ones in the case of plane polar coordinates.

In spherical coordinates, the position of a point P in space is given by the position vector $\mathbf{r}$:

$$\mathbf{r} = r\hat{\mathbf{r}} = r\hat{\mathbf{r}}(\theta, \phi) \tag{6.48}$$

We can now find expressions for velocity and acceleration by making use of the preceding relations. Thus

$$\mathbf{v} = \dot{\mathbf{r}} = \frac{d\mathbf{r}}{dt} = \frac{d}{dt}\left[r\hat{\mathbf{r}}(\theta, \phi)\right] = \frac{dr}{dt}\hat{\mathbf{r}} + r\frac{d\hat{\mathbf{r}}}{dt} = \dot{r}\hat{\mathbf{r}} + r\frac{d\hat{\mathbf{r}}}{dt}$$

Using Eqs. (6.47),

$$\frac{d\hat{\mathbf{r}}}{dt}(\theta, \phi) = \frac{d\hat{\mathbf{r}}}{d\theta}\frac{d\theta}{dt} + \frac{d\hat{\mathbf{r}}}{d\phi}\frac{d\phi}{dt} = \hat{\boldsymbol{\theta}}\dot{\theta} + \hat{\boldsymbol{\phi}}\sin\theta\,\dot{\phi}$$

Hence we obtain

$$\mathbf{v} = \dot{r}\hat{\mathbf{r}} + r\dot{\theta}\hat{\boldsymbol{\theta}} + (r\dot{\phi}\sin\theta)\hat{\boldsymbol{\phi}} \tag{6.49}$$

Similarly,

$$\mathbf{a} = \ddot{\mathbf{r}} = \frac{d\mathbf{v}}{dt} = \frac{d}{dt}\left[\dot{r}\hat{\mathbf{r}} + r\dot{\theta}\hat{\boldsymbol{\theta}} + (r\dot{\phi}\sin\theta)\hat{\boldsymbol{\phi}}\right]$$

which on simplification yields

$$\mathbf{a} = (\ddot{r} - r\dot{\theta}^2 - r\sin^2\theta\,\dot{\phi}^2)\hat{\mathbf{r}} + (r\ddot{\theta} + 2\dot{r}\dot{\theta} - r\sin\theta\cos\theta\,\dot{\phi}^2)\hat{\boldsymbol{\theta}}$$

$$+ (r\sin\theta\,\ddot{\phi} + 2\dot{r}\dot{\phi}\sin\theta + 2r\dot{\theta}\,\dot{\phi}\cos\theta)\hat{\boldsymbol{\phi}} \tag{6.50}$$

Since $\hat{\mathbf{r}}$, $\hat{\boldsymbol{\theta}}$, and $\hat{\boldsymbol{\phi}}$ form a set of mutually perpendicular unit vectors, we may write any vector $\mathbf{A}$ in component form as

$$\mathbf{A} = A_r\hat{\mathbf{r}} + A_\theta\hat{\boldsymbol{\theta}} + A_\phi\hat{\boldsymbol{\phi}} \tag{6.51}$$

where the components depend upon not only on vector $\mathbf{A}$, but also on its location in space. If $\mathbf{A}$ is a function of parameter t, that is, a function of time, we may write

$$\frac{d\mathbf{A}}{dt} = \frac{dA_r}{dt}\hat{\mathbf{r}} + A_r\frac{d\hat{\mathbf{r}}}{dt} + \frac{dA_\theta}{dt}\hat{\boldsymbol{\theta}} + A_\theta\frac{d\hat{\boldsymbol{\theta}}}{dt} + \frac{dA_\phi}{dt}\hat{\boldsymbol{\phi}} + A_\phi\frac{d\hat{\boldsymbol{\phi}}}{dt}$$

As before, using Eqs. (6.47) we may write

$$\frac{d\hat{\mathbf{r}}}{dt} = \frac{d\hat{\mathbf{r}}}{d\theta}\frac{d\theta}{dt} + \frac{d\hat{\mathbf{r}}}{d\phi}\frac{d\phi}{dt} = \hat{\boldsymbol{\theta}}\dot{\theta} + \hat{\boldsymbol{\phi}}\sin\theta\,\dot{\phi}$$

$$\frac{d\hat{\boldsymbol{\theta}}}{dt} = \frac{d\hat{\boldsymbol{\theta}}}{d\theta}\frac{d\theta}{dt} = -\hat{\mathbf{r}}\dot{\theta}$$

$$\frac{d\hat{\boldsymbol{\phi}}}{dt} = \frac{d\hat{\boldsymbol{\phi}}}{d\phi}\frac{d\phi}{dt} = -\hat{\boldsymbol{\rho}}\dot{\phi} = (-\hat{\mathbf{r}}\sin\theta - \hat{\boldsymbol{\theta}}\cos\theta)\dot{\phi}$$

Using these results in the preceding equation, we obtain

$$\frac{d\mathbf{A}}{dt} = \left(\frac{dA_r}{dt} - A_\theta \, \dot{\theta} - A_\phi \sin\theta \, \dot{\phi}\right)\hat{\mathbf{r}} + \left(\frac{dA_\theta}{dt} + A_r \, \dot{\theta} - A_\phi \cos\theta \, \dot{\phi}\right)\hat{\boldsymbol{\theta}}$$

$$+ \left(\frac{dA_\phi}{dt} + A_r \sin\theta \, \dot{\phi} + A_\theta \cos\theta \, \dot{\phi}\right)\hat{\boldsymbol{\phi}} \tag{6.52}$$

6.4 DEL OPERATOR IN CYLINDRICAL AND SPHERICAL COORDINATES

We are now in a position to express the del operator in cylindrical and spherical coordinates by using the definition of gradient. In cylindrical coordinates, a scalar function u is

$$u = u(\rho, \phi, z) \tag{6.53}$$

Therefore, knowing

$$du = \frac{\partial u}{\partial \rho} \, d\rho + \frac{\partial u}{\partial \phi} \, d\phi + \frac{\partial u}{\partial z} \, dz \tag{6.54}$$

and

$$\mathbf{r} = \rho\hat{\boldsymbol{\rho}} + z\hat{\mathbf{z}} \tag{6.55}$$

we may write

$$d\mathbf{r} = d\rho\hat{\boldsymbol{\rho}} + \rho\frac{\partial\hat{\boldsymbol{\rho}}}{\partial\phi} \, d\phi + dz \, \hat{\mathbf{z}} \tag{6.56}$$

Using the relations

$$\frac{\partial\hat{\boldsymbol{\rho}}}{\partial\phi} = \hat{\boldsymbol{\phi}} \qquad \text{and} \qquad \frac{\partial\hat{\boldsymbol{\phi}}}{\partial\phi} = -\hat{\boldsymbol{\rho}}$$

we obtain

$$d\mathbf{r} = \hat{\boldsymbol{\rho}} \, d\rho + \hat{\boldsymbol{\phi}}\rho \, d\phi + \hat{\mathbf{z}} \, dz \tag{6.57}$$

The definition of **grad** u is

$$du = \nabla u \cdot d\mathbf{r} = d\mathbf{r} \cdot \nabla u \tag{6.58}$$

For this relation to yield Eq. (6.54) we must define

$$\nabla \equiv \hat{\boldsymbol{\rho}} \, \frac{\partial}{\partial\rho} + \hat{\boldsymbol{\phi}} \, \frac{1}{\rho}\frac{\partial}{\partial\phi} + \hat{\mathbf{z}} \, \frac{\partial}{\partial z} \tag{6.59}$$

Thus, if we use Eqs. (6.59) and (6.57) in Eq. (6.58), we get Eq. (6.54); that is,

$$dr \cdot \nabla u = (\hat{\boldsymbol{\rho}}\, d\rho + \hat{\boldsymbol{\phi}}\,\rho\, d\phi + \hat{\mathbf{z}}\, dz) \cdot \left(\hat{\boldsymbol{\rho}}\, \frac{\partial}{\partial \rho} + \hat{\boldsymbol{\phi}}\, \frac{1}{\rho} \frac{\partial}{\partial \phi} + z\, \frac{\partial}{\partial z} \right) u$$

$$= \frac{\partial u}{\partial \rho}\, d\rho + \frac{\partial u}{\partial \phi}\, d\phi + \frac{\partial u}{\partial z}\, dz = du$$

Similarly, for spherical coordinates, and using Eq. (6.48),

$$u = u(r, \theta, \phi) \tag{6.60}$$

$$du = \frac{\partial u}{\partial r}\, dr + \frac{\partial u}{\partial \theta}\, d\theta + \frac{\partial u}{\partial \phi}\, d\phi \tag{6.61}$$

$$d\mathbf{r} = \hat{\mathbf{r}}\, dr + \hat{\boldsymbol{\theta}}\, r\, d\theta + \hat{\boldsymbol{\phi}}\, r \sin\theta\, d\phi \tag{6.62}$$

and if we define

$$\nabla \equiv \hat{\mathbf{r}}\, \frac{\partial}{\partial r} + \hat{\boldsymbol{\theta}}\, \frac{1}{r} \frac{\partial}{\partial \theta} + \hat{\boldsymbol{\phi}}\, \frac{1}{r \sin\theta} \frac{\partial}{\partial \phi} \tag{6.63}$$

it will satisfy $du = d\mathbf{r} \cdot \nabla u$, as it should.

6.5 POTENTIAL ENERGY FUNCTION

In considering the motion of a particle in one dimension, we defined a potential energy function as Eq. (2.77),

$$V(x) = \int_x^{x_s} F(x)\, dx = -\int_{x_s}^x F(x)\, dx \tag{6.64}$$

and the corresponding force $F(x)$ is

$$F(x) = -\frac{dV(x)}{dx} \tag{6.65}$$

We can now extend these ideas to the motion of a particle in three dimensions.

Let us consider a particle at $\mathbf{r}(x, y, z)$ that under the action of force $\mathbf{F}$ moves from $\mathbf{r}_1$ to $\mathbf{r}_2$. The work done is given by

$$W = \int_{\mathbf{r}_1}^{\mathbf{r}_2} \mathbf{F}(\mathbf{r}) \cdot d\mathbf{r} \tag{6.66}$$

and to evaluate this integral we must specify a path. As in the case of one dimension, we can introduce a *potential energy function* $V(\mathbf{r}) = V(x, y, z)$ as the work done by the force when the force moves the particle from point $\mathbf{r}$ to some standard reference point $\mathbf{r}_s$. That is,

$$V(\mathbf{r}) = \int_{\mathbf{r}}^{\mathbf{r}_s} \mathbf{F}(\mathbf{r}) \cdot d\mathbf{r} = -\int_{\mathbf{r}_s}^{\mathbf{r}} \mathbf{F}(\mathbf{r}) \cdot d\mathbf{r} \qquad (6.67)$$

Knowing that $V(\mathbf{r})$ must be a function of position alone, this definition is possible only if the preceding integral is independent of the path of integration, that is, only if $\mathbf{F}(r)$ is a *conservative force* so that the work done will be independent of the path; hence the potential energy function may be introduced.

Our task is now to proceed to find the necessary and sufficient conditions for $\mathbf{F}(\mathbf{r})$ to be conservative, and hence justification for the existence of a potential function $V(\mathbf{r})$. Let us say that work done in going from point P to Q is independent of the path. This means that the work done in a closed path (see Fig. 6.14) in going from P to Q and back to P is zero. That is, we find

$$W_{P \to Q \to P} = \oint_{\substack{\text{closed} \\ \text{path}}} \mathbf{F} \cdot d\mathbf{r} = 0 \qquad (6.68)$$

According to Stokes' theorem [Eq. (5.141)], we may write this result as

$$W_{P \to Q \to P} = \oint_{\substack{\text{closed} \\ \text{path}}} \mathbf{F} \cdot d\mathbf{r} = \iint_{\text{surface}} \hat{\mathbf{n}} \cdot (\boldsymbol{\nabla} \times \mathbf{F}) \, dS = 0 \qquad (6.69)$$

This can be true only if the integrand itself $\boldsymbol{\nabla} \times \mathbf{F}$ is zero; that is

$$\boldsymbol{\nabla} \times \mathbf{F} = \mathbf{curl}\ \mathbf{F} = 0 \qquad (6.70)$$

This is a *necessary and sufficient condition for the force to be conservative*; hence a potential energy function given by Eq. (6.67) will exist. Thus Eq. (6.70) states a necessary and sufficient condition for the existence of a potential function. A force $\mathbf{F}$ for which the **curl** $\mathbf{F}$ is zero is called a *conservative force*.

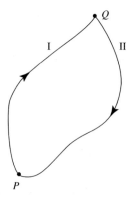

Figure 6.14 The work done in a conservative force field for a closed path in going from P to Q and back to P is zero.

We can now show that the existence of a potential function leads to the conservation of total energy if the force field is conservative. The work done by $\mathbf{F}$ in acting through $\mathbf{r}_1$ to $\mathbf{r}_2$ may be written as

$$W_{1\rightarrow2} = \int_{\mathbf{r}_1}^{\mathbf{r}_2} \mathbf{F}(\mathbf{r}) \cdot d\mathbf{r} = \int_{\mathbf{r}_1}^{\mathbf{r}_s} \mathbf{F}(\mathbf{r}) \cdot d\mathbf{r} + \int_{\mathbf{r}_s}^{\mathbf{r}_2} \mathbf{F}(\mathbf{r}) \cdot d\mathbf{r} = V_1(\mathbf{r}) - V_2(\mathbf{r}) \qquad (6.71)$$

But the work done is also equal to the change in kinetic energy

$$W_{1\rightarrow2} = \int_{\mathbf{r}_1}^{\mathbf{r}_2} \mathbf{F}(\mathbf{r}) \cdot d\mathbf{r} = K_2 - K_1 \qquad (6.72)$$

Combining Eqs. (6.71) and (6.72),

$$K_1 + V_1(\mathbf{r}) = K_2 + V_2(r) \qquad (6.73)$$

That is, if E is the total energy, we obtain

$$K + V = \tfrac{1}{2}m(\dot{x}^2 + \dot{y}^2 + \dot{z}^2) + V(x, y, z) = E \qquad (6.74)$$

which is the energy integral for motion in three dimensions.

Let us consider $\mathbf{F} = \mathbf{F}(\mathbf{r}, t)$, and say at any time t that

$$\nabla \times \mathbf{F}(\mathbf{r}, t) = 0 \qquad (6.75)$$

and we can define a potential function

$$V(\mathbf{r}, t) = \int_{\mathbf{r}}^{\mathbf{r}_s} \mathbf{F}(\mathbf{r}, t) \cdot d\mathbf{r} \qquad (6.76)$$

so that $$\mathbf{F}(\mathbf{r}, t) = -\nabla V(\mathbf{r}, t) \qquad (6.77)$$

But in such cases the sum of the kinetic and potential energy is not constant; hence $\mathbf{F}(\mathbf{r}, t)$ is not a conservative force.

If $\mathbf{F}$ is given and we want to evaluate $V(\mathbf{r})$, we can proceed either directly by evaluating the line integral,

$$V(\mathbf{r}) = -\int_{\mathbf{r}_s}^{\mathbf{r}} \mathbf{F}(\mathbf{r}) \cdot d\mathbf{r} \qquad (6.78)$$

or we can evaluate three ordinary integrals by starting with

$$F_x = -\frac{\partial V}{\partial x}, \qquad F_y = -\frac{\partial V}{\partial y}, \qquad F_z = -\frac{\partial V}{\partial z}$$

to obtain

$$V = -\int F_x \, dx + C_1(y, z)$$

$$V = -\int F_y \, dy + C_2(x, z) \qquad \textbf{(6.79)}$$

$$V = -\int F_z \, dz + C_3(x, y)$$

The resulting potential function must be consistent with all three expressions given in Eqs. (6.79). This is demonstrated in the following example.

Example 6.1

Show that the following forces are conservative and find the corresponding potentials.

(a) $\mathbf{F} = ax\hat{\mathbf{i}} + by\hat{\mathbf{j}} + cz\hat{\mathbf{k}}$

(b) $F_x = 3ayz^3 - 20bx^3y^2$, $F_y = 3axz^3 - 10bx^4y$, $F_z = 9axz^2y$

Solution

In order to prove that the force represents a conservative force field, we must prove that curl of the force vector is zero.

$$i := 1 \qquad\qquad j := 1 \qquad\qquad k := 1$$

$$A \equiv \frac{d}{dx} \centerdot \qquad B \equiv \frac{d}{dy} \centerdot \qquad C \equiv \frac{d}{dz} \centerdot$$

(a)

Let i, j, and k be the unit vectors. A, B, and C represent the differential operators as shown.

$$Sa \equiv \begin{Vmatrix} i & j & k \\ A & B & C \\ a{\cdot}x & b{\cdot}y & c{\cdot}z \end{Vmatrix} \qquad\qquad (i)$$

(a) Let Sa = Curl **F**, which is shown in matrix form. After calculating the absolute value of Sa and then substituting differential operators for A, B, and C, we simplify and find that Sa = 0. Thus **F** in this case is a conservative force field.

$$Sa \equiv i{\cdot}B{\cdot}c{\cdot}z - i{\cdot}C{\cdot}b{\cdot}y - A{\cdot}j{\cdot}c{\cdot}z + A{\cdot}k{\cdot}b{\cdot}y + a{\cdot}x{\cdot}j{\cdot}C - a{\cdot}x{\cdot}k{\cdot}B$$

$$Sa \equiv i{\cdot}\frac{d}{dy}(c{\cdot}z) - i{\cdot}\frac{d}{dz}(b{\cdot}y) - \frac{d}{dx}(j{\cdot}c){\cdot}z + \frac{d}{dx}(k{\cdot}b){\cdot}y + \frac{d}{dz}(a{\cdot}x{\cdot}j) - \frac{d}{dy}(a{\cdot}x{\cdot}k)$$

$$Sa \equiv 0$$

Using Eq. (6.78) or (6.79), we can calculate the potential Va corresponding to this conservative force as shown.

$$Va \equiv \int (-a{\cdot}x) \, dx + \int -b{\cdot}y \, dy + \int -c{\cdot}z \, dz$$

$$Va \equiv \frac{-1}{2}{\cdot}a{\cdot}x^2 - \frac{1}{2}{\cdot}b{\cdot}y^2 - \frac{1}{2}{\cdot}c{\cdot}z^2$$

$$Va \equiv \left(\frac{-1}{2}{\cdot}a{\cdot}x^2 - \frac{1}{2}{\cdot}b{\cdot}y^2 - \frac{1}{2}{\cdot}c{\cdot}z^2\right) + \text{Constant} \qquad (ii)$$

The constant is equal to the sum of three constants Ca = Cx + Cy + Cz.

The three constants Cx, Cy, and Cz, corresponding to the three components of the force, are calculated from the initial conditions.

(b) We follow the same procedure as in (a) and prove that the force is conservative.

$$Sb = \begin{Vmatrix} \begin{bmatrix} i & j & k \\ A & B & C \\ 3 \cdot a \cdot y \cdot z^3 - b \cdot x^3 \cdot y^2 & 3 \cdot a \cdot x \cdot z^3 - 10 \cdot b \cdot x^4 \cdot y & 9 \cdot a \cdot x \cdot z^2 \cdot y \end{bmatrix} \end{Vmatrix} \quad (iii)$$

$Sb = 9 \cdot i \cdot B \cdot a \cdot x \cdot z^2 \cdot y - 3 \cdot i \cdot C \cdot a \cdot x \cdot z^3 + 10 \cdot i \cdot C \cdot b \cdot x^4 \cdot y - 9 \cdot A \cdot j \cdot a \cdot x \cdot z^2 \cdot y + \blacksquare \ldots$
$\quad + 3 \cdot A \cdot k \cdot a \cdot x \cdot z^3 - 10 \cdot A \cdot k \cdot b \cdot x^4 \cdot y + 3 \cdot a \cdot y \cdot z^3 \cdot j \cdot C - 3 \cdot a \cdot y \cdot z^3 \cdot k \cdot B - b \cdot x^3 \cdot y^2 \cdot j \cdot C + b \cdot x^3 \cdot y^2 \cdot k \cdot B$

$Sb = 0$

$$Vx = \int -\left(3 \cdot a \cdot y \cdot z^3 - 20 \cdot b \cdot x^3 \cdot y^2\right) dx \quad Vy = \int -\left(3 \cdot a \cdot x \cdot z^3 - 10 \cdot b \cdot x^4 \cdot y\right) dy \quad Vz = \int -\left(9 \cdot a \cdot x \cdot z^2 \cdot y\right) dz$$

$$Vx = -3 \cdot a \cdot y \cdot z^3 \cdot x + 5 \cdot b \cdot x^4 \cdot y^2 + Cx \qquad Vy = -3 \cdot a \cdot y \cdot z^3 \cdot x + 5 \cdot b \cdot x^4 \cdot y^2 + Cy \qquad Vz = -3 \cdot a \cdot y \cdot z^3 \cdot x + Cz \quad (iv)$$

For a conservative force field, if we assume
Cx = Cy = 0, Vz is short of a term and $\quad Cx = 0 \qquad Cy = 0 \qquad Cz(x,y) = 5 \cdot b \cdot x^4 \cdot y^2$
Cz must be as shown.

Hence for a conservative force, the potential is

$$Vb = -3 \cdot a \cdot y \cdot z^3 \cdot x + 5 \cdot b \cdot x^4 \cdot y^2 + Cx + \left(-3 \cdot a \cdot y \cdot z^3 \cdot x + 5 \cdot b \cdot x^4 \cdot y^2 + Cy\right) + \left(-3 \cdot a \cdot y \cdot z^3 \cdot x + Cz\right)$$

$$Vb = 5 \cdot b \cdot x^4 \cdot y^2 - 3 \cdot a \cdot x \cdot y \cdot z^3$$

or $$Vb = -9 \cdot a \cdot y \cdot z^3 \cdot x + 10 \cdot b \cdot x^4 \cdot y^2 + Cx + Cy + Cz \qquad (v)$$

EXERCISE 6.1 Show that the following forces are conservative and calculate the corresponding potentials.

a) $\mathbf{F} = -yz\hat{\mathbf{i}} - xz\hat{\mathbf{j}} - xy\hat{\mathbf{k}}$
b) $F_x = -6ayzx,\ F_y = az(z^2 - 3x^2),\ F_z = 3ay(z^2 - x^2)$

Example 6.2

A particle of mass m moves from point A to B around a semicircular path of radius R, as shown in Fig. Ex. 6.2. It is attracted toward its starting point A by a force proportional to its distance from A. When it reaches B, the force toward A is F_0. Calculate the work done against this force when the particle moves from A to B in this semicircular path as shown.

Solution

The work done is given by

$$W = \int_A^B \mathbf{F} \cdot d\mathbf{s} \tag{i}$$

and

$$\mathbf{F} = k\mathbf{r} \tag{ii}$$

where k is a constant. But when $r = 2R$ at B (see Fig. Ex. 6.2),

$$F_B = F_0 = k(2R) \quad \text{or} \quad k = \frac{F_0}{2R} \tag{iii}$$

Therefore,

$$\mathbf{F} = \frac{F_0}{2R}\mathbf{r} \tag{iv}$$

To express r in terms of R and θ, we use the law of cosines:

$$r^2 = R^2 + R^2 - 2R^2\cos\theta = 2R^2(1 - \cos\theta)$$

$$F = kr = \frac{F_0}{2R}R\sqrt{2(1 - \cos\theta)}$$

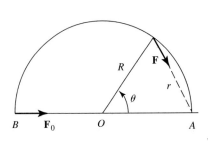 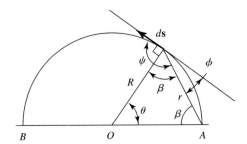

Figure Ex. 6.2

From the Fig. Ex. 6.2, we get the relations between different quantities given here.

$$\mathbf{F} \cdot d\mathbf{s} = F \cdot ds \cdot \cos(\psi) \qquad ds = R \cdot d\theta$$

$$\cos(\psi) = \cos(180 - \phi) = \cos(\phi) = \cos\left(\frac{\theta}{2}\right)$$

The work W done in moving from A to B is

$$W = \int_A^B F\, ds$$

Note that

$$|\mathbf{F}\cdot\mathbf{ds}| = F\cdot r\cdot\cos(\psi)\cdot d\theta = F\cdot r\cdot\cos\left(\frac{\theta}{2}\right) \qquad \text{and} \qquad F = \frac{F0}{2\cdot R}\cdot R\cdot\sqrt{2\cdot(1-\cos(\theta))}$$

Substituting for F, ds, and cos(ψ), the work WF done can be calculated as shown.

$$WF = \int_0^{\pi} \frac{F0}{2\cdot R}\cdot R\cdot\sqrt{2\cdot(1-\cos(\theta))}\cdot R\cdot\cos\left(\frac{\theta}{2}\right) d\theta$$

$$WF = -F0\cdot R$$

EXERCISE 6.2 Solve the problem by the energy conservation method by noting the fact that, for a spring, $\mathbf{F} = k\mathbf{r}$ and potential energy is $\frac{1}{2}kr^2$.

 Example 6.3 _____

A particle moving in an *XY* plane as shown in Fig. Ex. 6.3 is attracted toward the origin by a force $F = k/y$. Calculate the work done when the particle moves **(a)** from *A* to *B* and then to *C*, and **(b)** from *A* to *C* along an elliptical path given by the equations $x = 2a \sin\theta$ and $y = a\cos\theta$.

Solution

(a) In going from *A* to *B* and to *C*, the force is $F = k/y$, and the work done is given by

$$W = \int_A^C \mathbf{F}\cdot d\mathbf{s} = \int_A^B \mathbf{F}\cdot d\mathbf{s} + \int_B^C \mathbf{F}\cdot d\mathbf{s}$$

$$= \int_A^B F\,dr\cos\theta_1 + \int_B^C F\,dr\cos\theta_2 \tag{i}$$

For the path from $A(0, a)$ to $B(2a, a)$,

$$y = a, \qquad dr = dx, \qquad \cos\theta_1 = -\frac{x}{\sqrt{x^2 + a^2}} \tag{ii}$$

For the path from $B(2a, a)$ to $C(2a, 0)$

$$x = 2a, \qquad dr = dy, \qquad \cos\theta_2 = \frac{y}{\sqrt{(2a)^2 + y^2}} \tag{iii}$$

For both the paths

$$F = \frac{k}{y} \tag{iv}$$

The minus sign can also be justified by the fact that the force is opposing the displacement.

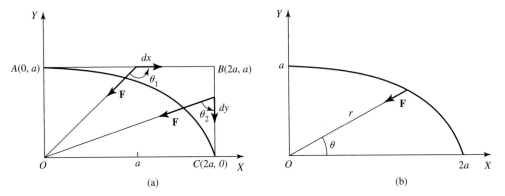

(a)   (b)

Figure Ex. 6.3

Substituting Eqs. (ii), (iii), and (iv) in (i), the work Wa done in going from A to B and then to C is as shown.

$$Wa = \int_0^{2\cdot a} -k \cdot \frac{x}{a \cdot \sqrt{x^2 + a^2}}\, dx + \int_a^0 \frac{-k}{\sqrt{(2 \cdot a)^2 + y^2}}\, dy$$

$$Wa = k \cdot \left[\left(\sqrt{5} - 1 \right) - \ln(2) + \ln\left(\sqrt{5} + 1 \right) \right]$$

(b) We now calculate the work done along the elliptical path. The equation of the ellipse is

$$\frac{x^2}{(2a)^2} + \frac{y^2}{a^2} = 1 \tag{v}$$

and

$$r^2 = x^2 + y^2 \tag{vi}$$

In terms of the inscribed and circumscribed circles and the central angle θ

$$x = 2a \sin \theta \quad \text{and} \quad y = a \cos \theta \tag{vii}$$

The work done is going from A to C along this elliptical path is

$$W = \int \mathbf{F} \cdot d\mathbf{r} = \int F_x\, dx + \int F_y\, dy \tag{viii}$$

where, from Eq. (vii),

$$dx = 2a \cos \theta\, d\theta, \qquad dy = -a \sin \theta\, d\theta \tag{ix}$$

$$F_x = F \cos \phi = F \frac{x}{r} = \frac{k}{y} \frac{x}{r} = -\frac{k}{a \cos \theta} \frac{2a \sin \theta}{\sqrt{(2a \sin \theta)^2 + (a \cos \theta)^2}} \tag{x}$$

$$F_y = F \sin \phi = F \frac{y}{r} = \frac{k}{y} \frac{y}{r} = -\frac{k}{\sqrt{(2a \sin \theta)^2 + (a \cos \theta)^2}} \qquad \textbf{(xi)}$$

Substituting these in Eq. (viii), the work Wb done in going from A to C by the circular path is as shown.

$$Wb = \int_0^{\frac{\pi}{2}} -k \cdot \frac{2 \cdot a \cdot \sin(\theta) \cdot 2 \cdot a \cdot \cos(\theta)}{a \cdot \sqrt{(2 \cdot a \cdot \sin(\theta))^2 + a^2}} \, d\theta + \int_0^{\frac{\pi}{2}} \frac{-k \cdot (-a \cdot \sin(\theta))}{\sqrt{(2 \cdot a)^2 + (a \cdot \cos(\theta))^2}} \, d\theta$$

Which is the same as in Part(a)

$$Wb = -k \cdot \left(\sqrt{5} - 1 + \ln(2) - \ln\left(\sqrt{5} + 1 \right) \right)$$

EXERCISE 6.3 Repeat the problem if the force directed toward the origin is $F = k/x$.

6.6 TORQUE

Let us consider a particle of mass m located at point P at a distance $\mathbf{r}$ from the origin and acted on by a force $\mathbf{F}$, as shown in Fig. 6.15, both $\mathbf{r}$ and $\mathbf{F}$ being in the XY plane. We want to calculate $\boldsymbol{\tau}_0$ about an axis passing through O and perpendicular to the XY plane. The *torque* or *moment of force* about the origin O is defined as the product of the distance $\mathbf{r}$ $(= OP)$ and the component of force $\mathbf{F}$ perpendicular to $\mathbf{r}$; that is, $F \sin \phi = F \sin(\pi - \theta) = F \sin \theta$. Hence

$$\tau_0 = rF \sin \theta \quad \text{or} \quad \boldsymbol{\tau}_0 = \mathbf{r} \times \mathbf{F} \qquad \textbf{(6.80)}$$

The torque $\boldsymbol{\tau}_0$ will be along the $+Z$-axis if $\mathbf{F}$ acts counterclockwise and will be along the $-Z$-axis if $\mathbf{F}$ acts clockwise.

Let us generalize the preceding definition of the torque as applied to a three-dimensional case. As shown in Fig. 6.16, the force $\mathbf{F}$ acts on a particle at P that is at a distance $\mathbf{r}$ from the origin O. We want to calculate the torque or the moment of force of $\mathbf{F}$ acting at P about an axis

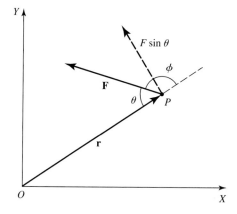

Figure 6.15   Torque τ_0 due to a force $\mathbf{F}$ at a distance $\mathbf{r}$ from the origin O.

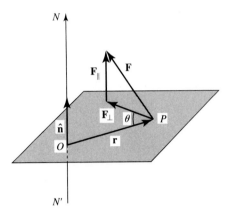

Figure 6.16 Torque $\tau_{NN'}$ (in a three-dimensional case) due to a force $\mathbf{F}$ at a distance $\mathbf{r}$ from the origin O.

NN' passing through O. Let us resolve $\mathbf{F}$ into two vector components: $\mathbf{F}_{\parallel}$, a component parallel to NN' and $\mathbf{F}_{\perp}$, a component perpendicular to NN'. Thus

$$\mathbf{F} = \mathbf{F}_{\parallel} + \mathbf{F}_{\perp} \tag{6.81}$$

Since $\hat{\mathbf{n}}$ is a unit vector along the axis NN', $\hat{\mathbf{n}} \cdot \mathbf{F}$ is the projection of $\mathbf{F}$ along NN'; hence the parallel component is

$$\mathbf{F}_{\parallel} = \hat{\mathbf{n}}(\hat{\mathbf{n}} \cdot \mathbf{F}) \tag{6.82}$$

while the perpendicular component is

$$\mathbf{F}_{\perp} = \mathbf{F} - \mathbf{F}_{\parallel} \tag{6.83}$$

Thus the torque about the axis NN' may be defined as

$$\tau_{NN'} = \pm rF_{\perp} \sin \theta = \pm |\mathbf{r} \times \mathbf{F}_{\perp}| \tag{6.84}$$

The positive sign is for the torque along $\hat{\mathbf{n}}$ and the negative sign for the torque opposite to $\hat{\mathbf{n}}$. On the other hand, the component $\mathbf{F}_{\parallel}$ to NN' does not produce any torque along the NN' axis. Since $\mathbf{r} \times \mathbf{F}_{\parallel}$ is a vector that is perpendicular to $\hat{\mathbf{n}}$, its component parallel to $\hat{\mathbf{n}}$ will be zero; that is,

$$\hat{\mathbf{n}} \cdot (\mathbf{r} \times \mathbf{F}_{\parallel}) = n|\mathbf{r} \times \mathbf{F}_{\parallel}| \cos 90° = 0 \tag{6.85}$$

This result enables us to write $\tau_{NN'}$ given by Eq. (6.84), as

$$\tau_{NN'} = \pm |\mathbf{r} \times \mathbf{F}_{\perp}| = \hat{\mathbf{n}} \cdot (\mathbf{r} \times \mathbf{F}) \tag{6.86}$$

We can prove this result as follows:

$$\hat{\mathbf{n}} \cdot (\mathbf{r} \times \mathbf{F}) = \hat{\mathbf{n}} \cdot (\mathbf{r} \times (\mathbf{F}_{\parallel} + \mathbf{F}_{\perp})]$$

$$= \hat{\mathbf{n}} \cdot (\mathbf{r} \times \mathbf{F}_{\parallel}) + \hat{\mathbf{n}} \cdot (\mathbf{r} \times \mathbf{F}_{\perp})$$

$$= \hat{\mathbf{n}} \cdot (\mathbf{r} \times \mathbf{F}_{\perp})$$

Thus, without considering the orientation of the position vector $\mathbf{r}$ and the applied force $\mathbf{F}$, the torque about the axis NN' with $\mathbf{r}$ being drawn from any point from the axis is given by

$$\tau_{NN'} = \hat{\mathbf{n}} \cdot (\mathbf{r} \times \mathbf{F}) \tag{6.87}$$

This automatically gives the correct sign and takes care of the $\mathbf{F}_{\parallel}$ component, which makes no contribution. We may consider $\tau_{NN'}$ given by Eq. (6.86) to be a component of $\boldsymbol{\tau}_0$, which is defined by

$$\boldsymbol{\tau}_0 = \mathbf{r} \times \mathbf{F} \tag{6.88}$$

where τ_0 is the torque about an axis passing through O and $\mathbf{r}$ is the distance from O to P.

6.7 DYNAMICS IN THREE DIMENSIONS

The general equation describing motion in three dimensions may be written as

$$m \frac{d^2 \mathbf{r}}{dt^2} = \mathbf{F}(\mathbf{r}, \mathbf{v}, t) \tag{6.89}$$

which is a set of three coupled simultaneous second-order differential equations and may be written explicitly as

$$m \frac{d^2 x}{dt^2} = F_x(x, y, z, \dot{x}, \dot{y}, \dot{z}; t) \tag{6.90}$$

with similar expressions for y and z components. Thus, if the initial position $\mathbf{r}_0(x_0, y_0, z_0)$ and the initial velocity $\mathbf{v}_0(v_{0x}, v_{0y}, v_{0z})$ are given, we know the six arbitrary constants x_0, y_0, z_0, v_{0x}, v_{0y}, v_{0z} in the form of initial conditions and we can proceed to solve Eqs. (6.90). Solving these equations is a difficult task and is usually carried out by using numerical analysis.

There are situations where it is possible to solve these equations for two- or three-dimensional motion. For example, if each force component given depends only on the corresponding coordinate and its derivative, Eqs. (6.90) take the form

$$m \frac{d^2 x}{dt^2} = F_x(x, \dot{x}, t) \tag{6.91a}$$

$$m \frac{d^2 y}{dt^2} = F_y(y, \dot{y}, t) \tag{6.91b}$$

$$m \frac{d^2 z}{dt^2} = F_z(z, \dot{z}, t) \tag{6.91c}$$

These are three independent one-dimensional problems and can be solved for $x(t)$, $y(t)$, and $z(t)$ by familiar methods used in previous chapters. A simple example is that of three-dimensional

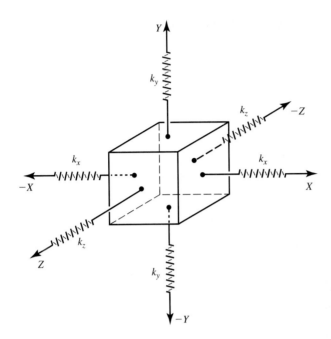

Figure 6.17   Harmonic oscillation in three dimensions with spring constant k_x, k_y, and k_z.

harmonic oscillator, such as the motion of an atom in a crystal lattice with a cubic structure, shown in Fig. 6.17. The forces in this case are

$$F_x = -k_x x, \quad F_y = -k_y y, \quad F_z = -k_z z \tag{6.92}$$

which involve solving separately for three linear harmonic oscillators.

A few remarks about the potential function and the total energy are in order. In the case of one-dimensional motion, if force is a function of position only, that is, $F = F(x)$, we can always define a potential energy function

$$V(x) = -\int_{x_s}^{x} F(x)\, dx \tag{6.93}$$

This is true because, when the particle moves from x_1 to x_2, it has no choice but to return by the same path; hence the work done in a round trip is zero. Thus the total energy is constant, $K + V = E$, and the energy integral can be used to solve the one-dimensional problem.

In the case of three-dimensional motion, even if the force is a function of position only, that is, $\mathbf{F} = \mathbf{F}(\mathbf{r})$, it does not guarantee the existence of a potential energy function $V(\mathbf{r})$. When such a potential function does exist, the conservation of energy theorem still holds, that is, $K + V = E = $ total energy. But unlike the case of one-dimensional motion, the energy integral is no longer sufficient to solve a problem of two- or three-dimensional motion.

6.8 HARMONIC OSCILLATORS IN TWO AND THREE DIMENSIONS

We extend our discussion of the previous section to the motion of two- and three-dimensional harmonic oscillators in this section and projectile motion in the next section.

Harmonic Oscillations in Two Dimensions: Lissajous Figures

A typical example of a two-dimensional anisotropic oscillator is shown in Fig. 6.18. We shall limit our discussion to an isotropic oscillator for which $k_x = k_y = k$. If the restoring force is proportional to the distance,

$$\mathbf{F} = -k\mathbf{r} \tag{6.94}$$

using polar coordinates, we can resolve $\mathbf{F}$ into two components as

$$F_x = -kr \cos \theta = -kx \tag{6.95}$$

$$F_y = -kr \sin \theta = -ky \tag{6.96}$$

Hence the differential equation, Eq. (6.94),

$$m \frac{d^2\mathbf{r}}{dt^2} = -k\mathbf{r} \tag{6.97}$$

may be written as

$$\ddot{x} + \omega^2 x = 0 \tag{6.98}$$

$$\ddot{y} + \omega^2 y = 0 \tag{6.99}$$

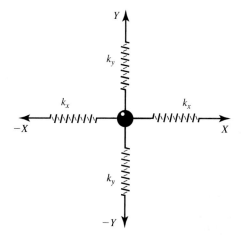

Figure 6.18 Two-dimensional anisotropic harmonic oscillator.

where $\omega^2 = k/m$. The solutions of these equations are

$$x = A \cos(\omega t + \phi_x) \tag{6.100}$$

$$y = B \cos(\omega t + \phi_y) \tag{6.101}$$

where A, B, ϕ_x, and ϕ_y are constants. Both oscillators have the same frequency, but may have different amplitudes and phases. We can obtain the path of the particle by eliminating t from Eqs. (6.100) and (6.101). We may write Eq. (6.101) as

$$\frac{y}{B} = \cos(\omega t + \phi_y) = \cos[\omega t + \phi_x + (\phi_y - \phi_x)]$$

$$= \cos(\omega t + \phi_x) \cos(\phi_y - \phi_x) - \sin(\omega t + \phi_x) \sin(\phi_y - \phi_x)$$

Substituting for $\cos(\omega t + \phi_x)$ and $\sin(\omega t + \phi_x)$ from Eq. (6.100) and rearranging,

$$\frac{y}{B} - \frac{x}{A} \cos(\phi_y - \phi_x) = -\sqrt{1 - \frac{x^2}{A^2}} \sin(\phi_y - \phi_x)$$

Squaring both sides and rearranging, we get

$$\frac{x^2}{A^2} + \frac{y^2}{B^2} - \frac{2xy}{AB} \cos(\phi_y - \phi_x) = \sin^2(\phi_y - \phi_x) \tag{6.102}$$

This is an equation of an ellipse, as shown in Fig. 6.19. The major axis of the ellipse makes an angle ψ with the X-axis and is given by (see Problem 6.30)

$$\tan 2\psi = \frac{2AB \cos(\phi_y - \phi_x)}{A^2 - B^2} \tag{6.103}$$

[Note: Eq. (6.102) is a general equation of the form

$$ax^2 + bxy + cy^2 + dx + ey = f$$

which represents an ellipse if the descriminant $b^2 - 4ac$ is negative, a hyperbola if the discriminant is positive, and a parabola if it is zero.]

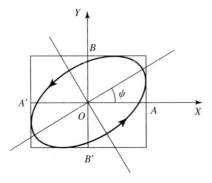

Figure 6.19 The resultant motion of two simple harmonic motions at right angles to each other, in general, is an ellipse.

If, in Eq. (6.102), $\phi_y - \phi_x = \pi/2$, we get

$$\frac{x^2}{A^2} + \frac{y^2}{B^2} = 1 \tag{6.104}$$

which is an equation of an ellipse with its major and minor axes coinciding with the X- and Y-axes, as shown in Fig. 6.20. On the other hand, if $\phi_y - \phi_x = 0$ or π, Eq. (6.102) reduces to

$$y = \pm\frac{B}{A} x \tag{6.105}$$

These equations represent straight lines.

Considering Eq. (6.104) again, if $A = B$, it takes the form

$$x^2 + y^2 = A^2 \tag{6.106}$$

which is an equation of a circle. As a matter of fact, the shape of the orbit changes as the phase difference $\phi = \phi_y - \phi_x$ changes its values, as illustrated in Fig. 6.21 for the case when $A = B$ and $\phi = 0°, 30°, 60°, \ldots$.

Let us now consider the case in which the two frequencies are not equal; hence the solutions of Eqs. (6.98) and (6.99) are of the form

$$x = A \cos(\omega_x t + \phi_x) \tag{6.107}$$

$$y = B \cos(\omega_y t + \phi_y) \tag{6.108}$$

The path of the particle is no longer an ellipse. The paths obtained in such cases are *Lissajous curves* or *figures*. Depending on the ratio of ω_x/ω_y, the curves may be open or closed, as explained next. Figure 6.22 illustrates the Lissajous figures for the case $\omega_y = 2\omega_x$ and $\phi = 0, \pi/4, \pi/2, \ldots$. There are many experimental arrangements, both mechanical and electrical, that demonstrate Lissajous figures.

Harmonic Oscillators in Three Dimensions

As mentioned earlier, the forces acting are such that each component of force is a function of the corresponding single coordinate. A typical example is the motion of an atom in a crystal lattice with a cubic structure. We assume that there is no damping and that the spring constants are k_x, k_y, and k_z. The situation is as shown in Fig. 6.17. Thus the equations describing the motions of three one-dimensional harmonic oscillators are

$$F_x = m\ddot{x} = -k_x x \tag{6.109a}$$

$$F_y = m\ddot{y} = -k_y y \tag{6.109b}$$

$$F_x = m\ddot{z} = -k_z z \tag{6.109c}$$

The corresponding equations resulting in three independent motions are

$$x = A_x \cos(\omega_x t + \phi_x) \tag{6.110a}$$

Figure 6.20 _____

As we know, the resultant motion of two simple harmonic motions at right angles to each other, in general, is an ellipse. If the two amplitudes are equal, the ellipse becomes a circle.

Fx and Fy are components of force F.

$$F = -k \cdot r \qquad Fx = -k \cdot r \cdot \cos(\theta) = -k \cdot x \qquad Fy = -k \cdot r \cdot \sin(\theta) = -k \cdot y$$

The equations describing the two simple harmonic motions are

$$\frac{d^2}{dt^2} x + \omega^2 \cdot x = 0 \qquad \frac{d^2}{dt^2} y + \omega^2 \cdot y = 0 \qquad \omega^2 = \frac{k}{m}$$

$$x = A \cdot \cos(\omega \cdot t + \phi) \qquad y = B \cdot \sin(\omega \cdot t + \phi)$$

The equations describing displacements, velocities, and accelerations for the two oscillators are

$$vx = -A \cdot \sin(\omega \cdot t + \phi) \cdot \omega \qquad vy = B \cdot \cos(\omega \cdot t + \phi) \cdot \omega$$

$$ax = -A \cdot \cos(\omega \cdot t + \phi x) \cdot \omega^2 \qquad ay = -B \cdot \sin(\omega \cdot t + \phi y) \cdot \omega^2$$

$$N := 50 \qquad t := 0 .. N \qquad A := 30 \qquad B := 60 \qquad \omega := .4$$

$$\phi x1 := 0 \qquad \phi y1 := \pi \qquad \phi x2 := 0 \qquad \phi y2 := \pi$$

$$x1_t := A \cdot \cos(\omega \cdot t + \phi x1) \qquad x2_t := A \cdot \cos(\omega \cdot t + \phi x2)$$

$$y1_t := A \cdot \sin(\omega \cdot t + \phi y1) \qquad y2_t := B \cdot \sin(\omega \cdot t + \phi y2)$$

(a) In both graphs the major and minor axes coincide with the x and y– axes. What will cause a shift in the position of the major or minor axis?

(b) In both graphs $\phi y1 - \phi x1 = 0$ or π. If different values are used, how will the graphs differ?

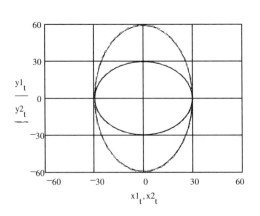

Figure 6.21 _____

More accurately, the resultant motion of two simple harmonic motions at right angles to each other can result in any one of many graphs, depending on the relative amplitudes and the phase differences between the two motions: a circle, an ellipse, an oriented ellipse, or a straight line.

ω = frequency of oscillation, which is the same for both simple harmonic motions.

$N := 50$ $t := 0 .. N$ $i := 0 .. 6$ $\omega := .4$

$A := 30$ $B := 60$ $\phi x_i := 0$ $\phi y_i := i \cdot \dfrac{\pi}{6}$

A and B are the amplitudes of the two simple harmonic motions.

$x_{t,i} := A \cdot \cos\left(\omega \cdot t + \phi x_i\right)$ $y_{t,i} := B \cdot \cos\left(\omega \cdot t + \phi y_i\right)$

ϕx_i and ϕy_i are the phase angles. x and y are the displacements of the two simple harmonic motions.

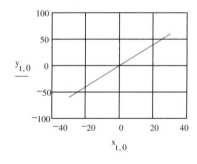

$\phi x_0 = 0$ $\phi y_0 = 0 \cdot deg$

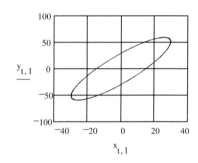

$\phi x_1 = 0$ $\phi y_1 = 30 \cdot deg$

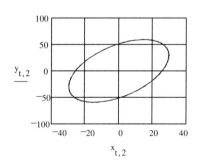

$\phi x_2 = 0$ $\phi y_2 = 60 \cdot deg$

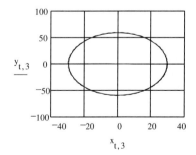

$\phi x_3 = 0$ $\phi y_3 = 90 \cdot deg$

Figure 6.21 (continued)

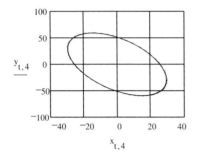

$$\phi x_4 = 0 \qquad \phi y_4 = 120 \cdot deg$$

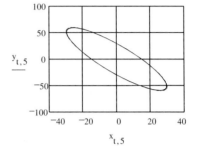

$$\phi x_5 = 0 \qquad \phi y_5 = 150 \cdot deg$$

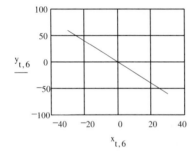

$$\phi x_6 = 0 \qquad \phi y_6 = 180 \cdot deg$$

The following figure shows all the above graphs together.

Which of the above graphs will become circle, if A = B?

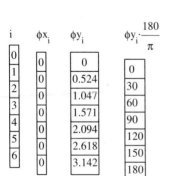

i	ϕx_i	ϕy_i	$\phi y_i \cdot \dfrac{180}{\pi}$
0	0	0	0
1	0	0.524	30
2	0	1.047	60
3	0	1.571	90
4	0	2.094	120
5	0	2.618	150
6	0	3.142	180

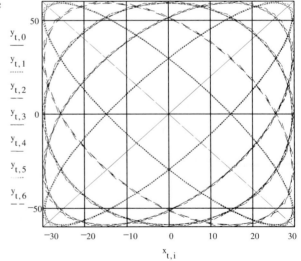

Figure 6.22 _____

A Lissajous figure results from the combination of two simple harmonic motions at right angles, with different frequencies and different amplitudes, as shown here.

$$I := 100 \quad i := 0..I \quad j := 0..4 \quad t_i := \frac{i}{10}$$

$$A := 1 \quad B := 1 \quad \omega x := 1 \quad \omega y := 2$$

$$x_{i,j} := A \cdot \cos\left(\omega x \cdot t_i + \phi x_j\right)$$

$$y_{i,j} := B \cdot \cos\left(\omega y \cdot t_i + \phi y_j\right)$$

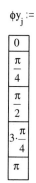

$\phi x_j :=$

0
0
0
0
0

$\phi y_j :=$

0
$\frac{\pi}{4}$
$\frac{\pi}{2}$
$3 \cdot \frac{\pi}{4}$
π

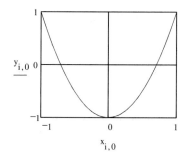

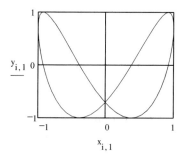

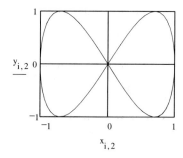

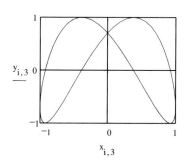

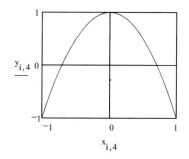

$$y = A_y \cos(\omega_y t + \phi_y) \tag{6.110b}$$

$$z = A_z \cos(\omega_z t + \phi_z) \tag{6.110c}$$

where ω_x, ω_y, and ω_z are the angular frequencies given by

$$\omega_x^2 = \frac{k_x}{m}, \qquad \omega_y^2 = \frac{k_y}{m}, \qquad \omega_z^2 = \frac{k_z}{m} \tag{6.111}$$

The six constants A_x, A_y, A_z, ϕ_x, ϕ_y, and ϕ_z depend on initial conditions: the initial coordinates x_0, y_0, and z_0 and the initial velocities v_{0x}, v_{0y}, and v_{0z}. The resulting motion of the particle is confined to a rectangular box of dimensions $2A_x \times 2A_y \times 2A_z$ placed about the origin.

If the angular frequencies are such that for some set of integers n_x, n_y, and n_z

$$\frac{\omega_x}{n_x} = \frac{\omega_y}{n_y} = \frac{\omega_z}{n_z} \tag{6.112}$$

the frequencies are said to be *commensurable*. The motion of mass m in such cases is *closed*, and the motion repeats itself at regular intervals of time. Furthermore, if the set of integers n_x, n_y, and n_z is such that they have no common integer factor, then the period of motion is given by

$$T = \frac{2\pi n_x}{\omega_x} = \frac{2\pi n_y}{\omega_y} = \frac{2\pi n_z}{\omega_z} \tag{6.113}$$

This means that, in one time period T, coordinate x makes n_x oscillations, coordinate y makes n_y oscillations, and coordinate z makes n_z oscillations. At the end of one period, the particle returns to its initial position and velocity.

If ω_x, ω_y, and ω_z in Eq. (6.112) are *incommensurable*, the curve describing the motion of the particle will never pass through the same point twice with the same velocity. The motion is not periodic. The path fills the entire box of volume $2A_x \times 2A_y \times 2A_z$, and after a sufficiently long time the particle will eventually pass arbitrarily close to every point in the box.

Let us discuss an interesting example of a three-dimensional isotropic harmonic oscillator; that is, all the spring constants are the same and so are the frequencies:

$$k_x = k_y = k_z \quad \text{and} \quad \omega^2 = \frac{k}{m} \tag{6.114}$$

Therefore,
$$x = A_x \cos(\omega t + \phi_x) \tag{6.115a}$$

$$y = A_y \cos(\omega t + \phi_y) \tag{6.115b}$$

$$z = A_z \cos(\omega t + \phi_z) \tag{6.115c}$$

To determine the path or the trajectory of the particle we eliminate t and express z in terms of x and y. The result is

$$z = K_1 x + K_2 y \tag{6.116}$$

(see Problem 6.40) where K_1 and K_2 are constants. This is the equation of a plane. Irrespective of initial conditions, the motion of an isotropic harmonic oscillator is always confined to a plane. The motion is periodic, and each coordinate executes one cycle of oscillation in each period. The path of the particle can be shown to be either an ellipse, a circle, or a straight line.

6.9 PROJECTILE MOTION

To begin, we limit our discussion to two dimensions: without and with air resistance.

No Air Resistance

Consider a projectile of mass m that is launched from the origin of a coordinate system with velocity $\mathbf{v}_0$, making an angle α with the horizontal axis, as shown in Fig. 6.23(a). The only force acting is the downward force of gravity, and the motion of the projectile in the XZ plane is described by the equation

$$m \frac{d^2\mathbf{r}}{dt^2} = -mg\hat{\mathbf{k}} \qquad (6.117)$$

or, in component form,

$$m \frac{d^2x}{dt^2} = 0 \quad \text{and} \quad m \frac{d^2z}{dt^2} = -mg \qquad (6.118)$$

Assume that the starting point is the origin $(0, 0)$, and the initial horizontal and vertical components of velocity $\mathbf{v}_0$ are

$$\dot{x}_0 = v_{x0} = v_0 \cos \alpha \quad \text{and} \quad \dot{z}_0 = v_{z0} = v_0 \sin \alpha \qquad (6.119)$$

Using these initial conditions, the solutions of Eqs. (6.118) are, after integration,

$$\dot{x} = v_x = v_{x0} = \text{constant} \qquad (6.120)$$

$$\dot{z} = v_x = v_{z0} - gt \qquad (6.121)$$

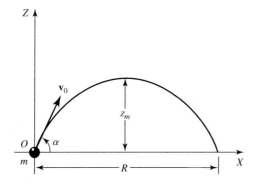

Figure 6.23(a)   A projectile launched from the origin of a coordinate system with velocity $\mathbf{v}_0$ and making an angle α with the horizontal is shown. z_m is the maximum height and R is the range.

and, after a second integration,

$$x = v_{x0}t \tag{6.122}$$

$$z = v_{z0}t - \tfrac{1}{2}gt^2 \tag{6.123}$$

The path of the projectile is obtained by eliminating t between Eqs. (6.122) and (6.123), resulting in the trajectory

$$z = \frac{v_{z0}}{v_{x0}}x - \frac{g}{2v_{x0}^2}x^2 \tag{6.124}$$

which is an equation of a parabola, as shown in Fig. 6.23(b).

Next we proceed to find the range R. Setting $z = 0$ in Eq. (6.124), we obtain two values of x: $x = 0$, corresponding to the starting position, and $x = R$, given by

$$R = \frac{2v_{x0}v_{z0}}{g} = \frac{2v_0^2 \cos\alpha \sin\alpha}{g} = \frac{v_0^2 \sin 2\alpha}{g} \tag{6.125}$$

This equation indicates that, for a given value of v_0, R will be maximum when $\alpha = 45°$.

 Figure 6.23(b) _____

The path of a projectile that is launched with a velocity of 33 m/sec and making an angle of 25 degrees with the horizontal is shown.

$$I := 30 \qquad i := 0 .. I \qquad t_i := \frac{i}{10}\cdot sec$$

$$v0 := 33\cdot\frac{m}{sec} \qquad g := 9.8\cdot\frac{m}{sec^2} \qquad \theta := 25\cdot deg$$

$$vx0 := v0\cdot\cos(\theta) \qquad\qquad vx0 = 29.908\cdot m\cdot sec^{-1}$$

$$vz0 := v0\cdot\sin(\theta) \qquad\qquad vz0 = 13.946\cdot m\cdot sec^{-1}$$

$$x_i := vx0\cdot t_i \qquad\qquad z_i := \frac{vz0}{vx0}\cdot x_i - \frac{g}{2\cdot vx0^2}\cdot\left(x_i\right)^2$$

Note that the projectile travels vertically 9.92 m before landing at a horizontal distance of 89.72 m.

$$x_{30} = 89.724\cdot m$$

$$max(x) = 89.724\cdot m$$

$$max(z) = 9.921\cdot m$$

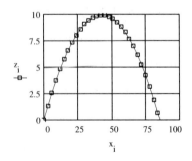

The maximum height z_m reached by the projectile, as shown in Fig. 6.23(a), is obtained by rewriting Eq. (6.124) as (see Problem 6.42)

$$\left(x - \frac{v_{x0}v_{z0}}{g}\right)^2 = -\frac{2v_{x0}^2}{g}\left(z - \frac{v_{z0}^2}{2g}\right) \tag{6.126}$$

Since $z = z_m$, when [from Eq. (6.125)] $x = R/2 = v_{x0}v_{z0}/g$, Eq. (6.126) gives

$$z_m = \frac{v_{z0}^2}{2g} \tag{6.127}$$

Air Resistance as a Function of Velocity

Let us assume that air resistance varies linearly with velocity. Since air resistance always opposes motion, the direction of the resistive force is in the direction opposite to that of **v**. Thus the equation describing the motion is

$$m\frac{d^2\mathbf{r}}{dt^2} = m\mathbf{g} - b\mathbf{v} \tag{6.128}$$

where b is a constant of proportionality for the resistive force and

$$\mathbf{r} = x\hat{\mathbf{i}} + z\hat{\mathbf{k}} \tag{6.129}$$

$$\mathbf{v} = \dot{x}\hat{\mathbf{i}} + \dot{z}\hat{\mathbf{k}} \tag{6.130}$$

$$\mathbf{g} = -g\hat{\mathbf{k}} \tag{6.131}$$

where $\dot{x} = v_x$ and $\dot{z} = v_z$. Thus Eq. (6.128) may be resolved into the following two components:

$$m\ddot{x} = -b\dot{x} \tag{6.132}$$

$$x\ddot{z} = -mg - b\dot{z} \tag{6.133}$$

These equations can now be integrated by methods familiar to us from solving one-dimensional problems. Thus, assuming that at $t = 0$, $(x_0, z_0) = (0, 0)$, and $v_0 = (\dot{x}_0, \dot{z}_0)$ and integrating Eqs. (6.132) and (6.133), we obtain

$$\dot{x} = \dot{x}_0 e^{-bt/m} \tag{6.134}$$

$$\dot{z} = -\frac{mg}{b} + \left(\frac{mg}{b} + \dot{z}_0\right)e^{-bt/m} \tag{6.135}$$

and, integrating again,

$$x = \dot{x}_0\frac{m}{b}(1 - e^{-bt/m}) \tag{6.136}$$

$$z = -\frac{mg}{b}t + \left(\frac{m^2g}{b^2} + \frac{m\dot{z}_0}{b}\right)(1 - e^{-bt/m}) \tag{6.137}$$

Combining Eqs. (6.134) and (6.135), the expression for the velocity of the projectile is

$$\mathbf{v} = \dot{x}\hat{\mathbf{i}} + \dot{z}\hat{\mathbf{k}}$$

$$= \dot{x}_0 e^{-bt/m}\hat{\mathbf{i}} + \left[-\frac{mg}{b} + \left(\frac{mg}{b} + \dot{z}_0\right)e^{-bt/m}\right]\hat{\mathbf{k}} \tag{6.138}$$

For large values of t, this velocity approaches the terminal velocity:

$$v_t = -\frac{mg}{b}, \qquad \text{for large values of } t \tag{6.139}$$

And from Eq. (6.136) for large values of t, we obtain the limiting value of x to be

$$x_t = \frac{\dot{x}_0 m}{b}, \qquad \text{for large values of } t \tag{6.140}$$

Also from Eq. (6.137) for large values of t, z approaches minus infinity. All this leads to the following conclusion: The trajectory has a vertical asymptote, the line $x = \dot{x}_0 m/b$; that is, it has a vertical drop, as shown in Fig. 6.24(a) and (b) and this drop begins above the horizontal plane $z = 0$.

Let us now compare the results obtained in Eqs. (6.136) and (6.137) with those in the case of no air resistance. By using the exponential series

$$e^u = 1 + u + \frac{u^2}{2!} + \frac{u^3}{3!} + \cdots$$

we may write Eqs. (6.136) and (6.137) as

$$x = \dot{x}_0 t - \frac{b\dot{x}_0}{2m}t^2 + \cdots \tag{6.141}$$

$$z = \dot{z}_0 t - \frac{1}{2}gt^2 - \frac{b}{m}\left(\frac{\dot{z}^0 t^2}{2} - \frac{gt^3}{b}\right) + \cdots \tag{6.142}$$

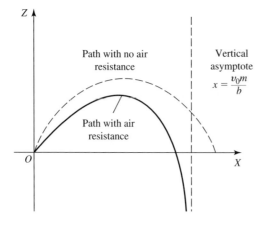

Figure 6.24(a) The trajectory of a projectile with and without air resistance, where the air resistance is proportional to the velocity.

Figure 6.24(b) _____

x0 and z0 when no air resistance
x and z when there is air resistance

$$N := 30 \qquad i := 0..N \qquad t_i := i$$

b = 0.2 is the constant of
proportionality for the resistive force.
The initial velocity in both the x and z
direction is 100 m/sec.

$$b := .2 \qquad m := 5 \qquad vx0 := 100 \qquad vz0 := 100 \qquad g := 9.8$$

$$x_i := vx0 \cdot \left(\frac{m}{b}\right) \cdot \left(1 - e^{\frac{-b \cdot t_i}{m}}\right) \qquad\qquad x0_i := vx0 \cdot t_i$$

The maximum values of x and z are
calculated below.

(a) How does the value of b affect
the maximum values of x and z?

$$z_i := -\left(\frac{m \cdot g}{b}\right) \cdot t_i + \left(\frac{m^2 \cdot g}{b^2} + \frac{m \cdot vz0}{b}\right) \cdot \left(1 - e^{\frac{-b \cdot t_i}{m}}\right)$$

(b) Graph for the values of b to be
0.1 and 0.4 and discuss (a).

$$z0_i := vz0 \cdot t_i - \left(\frac{1}{2}\right) \cdot g \cdot \left(t_i\right)^2$$

(c) If the resistance is proportional to
the square of the velocity, how will
this affect the maximum values?

max(x) = $1.747 \cdot 10^3$

max(x0) = $3 \cdot 10^3$

max(z) = 402.542

max(z0) = 510

max(z0) − max(z) = 107.458

max(x0) − max(x) = $1.253 \cdot 10^3$

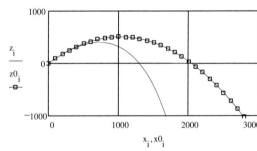

Projectile with, without air resistance

_____

The first term on the right in Eq. (6.141) and the first two terms on the right in Eq. (6.142) are
the same as in the case of no air resistance. The remaining terms in each case are the correction
terms, which are very small for the case in which bt/m is small as compared to unity.

Finally, let us obtain an equation of the trajectory by eliminating t between Eqs. (6.136)
and (6.137). This yields

$$z = \left(\frac{mg}{b\dot{x}_0} + \frac{\dot{z}_0}{\dot{x}_0}\right)x - \frac{m^2g}{b^2}\ln\left(\frac{m\dot{x}_0}{m\dot{x}_0 - bx}\right) \qquad\qquad \textbf{(6.143)}$$

For low air resistance or over short distances, that is, for $(bx/m\dot{x}_0) \ll 1$, we may write the last term as $\ln(1 + u)$, expand it, and obtain

$$z = \frac{\dot{z}_0}{\dot{x}_0} x - \frac{1}{2} \frac{g}{\dot{x}_0^2} x^2 - \frac{1}{3} \frac{bg}{m\dot{x}_0^3} x^3 + \cdots \tag{6.144}$$

This equation without the last term in the equation of a parabola, while the last term is a correction term. That is, Eq. (6.144) represents a parabolic trajectory with a small correction term. As before, we can obtain the maximum range by substituting $x = x_m$ when $z = 0$. We obtain

$$x_m = \frac{2\dot{x}_0\dot{z}_0}{g} - \frac{2}{3} \frac{b}{\dot{x}_0} x_m^2 - \cdots \tag{6.145}$$

The first term on the right is the range R when there is no air resistance, while the second term is the correction term. To solve Eq. (6.145) for x_m we can use the approximate value $x_m = 2\dot{x}_0\dot{z}_0/g$ in the last term and thus obtain

$$x_m \simeq \frac{2\dot{x}_0\dot{z}_0}{g} - \frac{8}{3} \frac{b\dot{x}_0\dot{z}_0^2}{mg^2} \tag{6.146}$$

On the other hand, if air resistance is large and is a major factor in determining the range—that is, if $(b\dot{z}_0/mg) \gg 1$—the maximum range is (see Problem 6.46)

$$x_m \simeq \frac{m\dot{x}_0}{b} \tag{6.147}$$

Projectile Motion in Three Dimensions

For no air resistance and air resistance linearly proportional to the velocity, we do not get any different results from that obtained in the case of two-dimensional projectile motion. An interesting situation is one in which there are crosswinds resulting in a drift force $\mathbf{F}_d$ in the Y direction, as shown in Fig. 6.25. The resulting equations of motion are

$$m \frac{d^2\mathbf{r}}{dt^2} = -mg\hat{\mathbf{k}} - b\mathbf{v} + F_d\hat{\mathbf{j}} \tag{6.148}$$

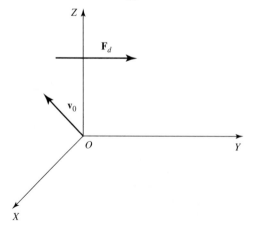

Figure 6.25 Drift force $\mathbf{F}_d$ due to a crosswind in the Y direction acting on a projectile with an initial velocity $\mathbf{v}_0$ in the XZ plane.

which in component form may be written as

$$m\ddot{x} = -b\dot{x} \tag{6.149}$$

$$m\ddot{y} = -b\dot{y} + F_d \tag{6.150}$$

$$m\ddot{z} = -b\dot{z} - mg \tag{6.151}$$

These may be solved as before, yielding almost the same results. We leave this as an exercise (Problem 6.58).

PROBLEMS

6.1. Derive Eq. (6.41), an expression for acceleration in cylindrical coordinates.

6.2. Derive the relations given in Eq. (6.47) from geometrical considerations.

6.3. Starting with Eq. (6.49) for velocity in spherical coordinates, derive an expression for acceleration given by Eq. (6.50).

6.4. The motion of a particle is described by the following equations. Find the velocity and acceleration, giving a geometrical interpretation, if any, in each case.
(a) $\mathbf{r} = a\hat{\mathbf{i}} + bt^2\hat{\mathbf{j}}$
(b) $\mathbf{r} = at\hat{\mathbf{i}} + A\cos\omega t\,\hat{\mathbf{j}}$
(c) $\mathbf{r} = at\hat{\mathbf{i}} + A\cos\omega t\hat{\mathbf{j}} + B\sin\omega t\hat{\mathbf{k}}$
(d) $\mathbf{r} = e^{kt}\hat{\mathbf{u}}_r$, $\theta = at$, in polar coordinates
(e) $\mathbf{r} = a\hat{\mathbf{u}}_r$, $\theta = B\sin\omega t$, $\phi = bt$, in spherical coordinates.

6.5. Consider a river of width w. The speed of water near the banks is zero, but increases linearly and reaches a value v_c at the center of the river. If a boat starts straight across from one bank with a speed v_b, show that when it reaches the other bank, it has drifted downstream by a distance $v_c w/2v_b$.

6.6. Show that the path of a particle is an ellipse having a major axis of $2a$ if its position vector is given by $\mathbf{r} = (a\sin\omega t)\hat{\mathbf{i}} + (b\cos\omega t)\hat{\mathbf{j}}$, where a, b, and ω are constants. Also calculate the speed of the particle.

6.7. A particle moves with a constant speed v in a parabolic path given by $y^2 = 4fx$, where f is a constant. Find its velocity and acceleration components in rectangular and plane polar coordinates. Show that the equation of the given parabola in plane polar coordinates is

$$r = \frac{2f}{1 - \cos\theta} \quad \text{or} \quad r\cos^2\frac{\theta}{2} = f$$

6.8. The acceleration $\mathbf{a}$ of a particle is a function of time t. Find the r and θ components of $d\mathbf{a}/dt$ in plane polar coordinates.

6.9. A vector $\mathbf{A}$ represents the position of a moving particle and is a function of time. Find the components of $d^2\mathbf{A}/dt^2$ in cylindrical polar coordinates.

6.10. A vector $\mathbf{A}$ represents the position of a moving particle and is a function of time. Find the components of $d^3\mathbf{A}/dt^3$ in spherical coordinates. (The time derivative of acceleration is called the *jerk*.)

6.11. The rate of change of acceleration is defined as the *jerk*. Find the magnitude and direction for the jerk of a particle moving in a circle of radius R and angular velocity ω.

6.12. A particle is moving with constant speed, but with continuously changing direction. Show that the acceleration vector is always perpendicular to the velocity vector. In general, if a vector $\mathbf{A}$ has constant length A but its direction in space is changing with time, then the rate of change of $\mathbf{A}$ is another vector that is always perpendicular to vector $\mathbf{A}$.

6.13. Calculate the forces corresponding to the following potential functions: **(a)** $V = ax^2 + by^2 + cz^2$, and **(b)** $V = Kye^{-x}$, where a, b, c, and K are constants.

6.14. Are the following forces conservative? Find the potential energy function if it exists.
   (a) $\mathbf{F} = Kxyz(\hat{\mathbf{i}} + \hat{\mathbf{j}} + \hat{\mathbf{k}})$, where K is a constant.
   (b) $\mathbf{F} = K(x\hat{\mathbf{i}} + y\hat{\mathbf{j}} + z\hat{\mathbf{k}})$, where K is a constant.
   (c) $\mathbf{F} = K(x^2y\hat{\mathbf{i}} + xy^2\hat{\mathbf{j}} + x^2y^2\hat{\mathbf{k}})c^{az}$, where K and a are constants.

6.15. Show that the following force is conservative and find the corresponding potential.

$$F_x = axe^{-u}, \qquad F_y = bye^{-u}, \qquad F_z = cze^{-u}$$

where $u = ax^2 + by^2 + cz^2$ and a, b, and c are constants.

6.16. Which of the following forces are conservative? If any, find the corresponding potential.
   (a) $\mathbf{F} = ax^2\hat{\mathbf{i}} + ay^2\hat{\mathbf{j}} + az^2\hat{\mathbf{k}}$
   (b) $F_x = 3az(x^2 - y^2)$, $F_y = -6azxy$, $F_z = ax(x^2 - 3y^2)$
   (c) $F_x = -ay^2$, $F_y = ayz$, $F_z = -ay^2$

6.17. Determine which of the following forces are conservative. Find the potential energy function for those that are conservative.
   (a) $\mathbf{F} = ax^2yz\hat{\mathbf{i}} + bxy^2z\hat{\mathbf{j}} + cxyz^2\hat{\mathbf{k}}$
   (b) $\mathbf{F} = cx^2yz\hat{\mathbf{i}} + cxy^2z\hat{\mathbf{j}} + cxyz^2\hat{\mathbf{k}}$
   (c) $\mathbf{F} = ax^2y\hat{\mathbf{i}} + ay^2x\hat{\mathbf{j}} + az^3\hat{\mathbf{k}}$

6.18. Show that the following force is conservative and find the corresponding potential.

$$F_r = ar^2\cos\theta, \qquad F_\theta = ar^2\sin\theta, \qquad F_z = 2az^2$$

where a is a constant.

6.19. Show that the following force is conservative and find the corresponding potential.

$$F_r = -2ar\sin\theta\cos\phi, \qquad F_\theta = -ar\cos\theta\cos\phi, \qquad F_\phi = ar\sin\theta\sin\phi$$

where a is a constant.

6.20. Consider a particle of mass m in a uniform gravitational field $\mathbf{F} = -mg\hat{\mathbf{k}}$. Calculate the work done in moving from point A to C along the three paths shown in Fig. P6.20.
   (a) From A to B and then from B to C—path ①.
   (b) From A to O and then From O to C—path ②.
   (c) Along the quadrant of a circle from A to C—path ③.

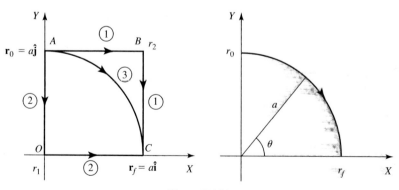

Figure P6.20

6.21. Repeat Problem 6.20 if the particle moves in a force field given by $\mathbf{F} = axy\hat{\mathbf{i}} + bx^2y\hat{\mathbf{j}}$.

6.22. Calculate the work done by a particle of mass m as it moves along a straight line from the origin $(0, 0, 0)$ to the point (x_0, y_0, z_0) under the following force:

$$F_x = ax^2 + bxy + cxz, \qquad F_y = ay^2 + bxy + cyz, \qquad F_z = az^2$$

6.23. A body of mass m can be moved from point A to point C along three different paths shown in Fig. P6.23:

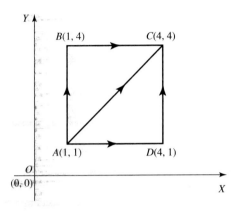

Figure P6.23

(a) From A to B and then from B to C.
(b) From A to D and then from D to C.
(c) From A to C directly.

Calculate the work done in each of the three cases for the following forces:

(i) $\mathbf{F} = ax\hat{\mathbf{i}} + by\hat{\mathbf{j}}$
(ii) $\mathbf{F} = (x + a)\hat{\mathbf{i}} + (y + b)\hat{\mathbf{j}}$
(iii) $\mathbf{F} = x^2y\hat{\mathbf{i}} + y^2x\hat{\mathbf{j}}$
(iv) $\mathbf{F} = x^4\hat{\mathbf{i}} + xy\hat{\mathbf{j}}$

6.24. Calculate the work done in moving a body of mass m from point A to point C along three different paths, as shown in Fig. P6.24:

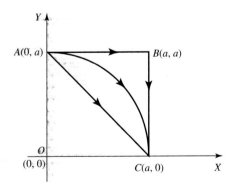

Figure P6.24

(a) From A to B and then from B to C.

(b) Along the straight line path directly from A to C.

(c) Along an arc of a circle from A to C.

Carry out these calculations from each of the following three forces:

(i) $\mathbf{F} = kx\hat{\mathbf{i}} + ky\hat{\mathbf{j}}$

(ii) $\mathbf{F} = \dfrac{k}{x}\hat{\mathbf{i}} + \dfrac{k}{y}\hat{\mathbf{j}}$

(iii) $\mathbf{F} = x^2\hat{\mathbf{i}} + y^2\hat{\mathbf{i}}$

6.25. Calculate the work done in moving a body of mass m from point A to point C along two different paths, as shown in Fig. P6.25:

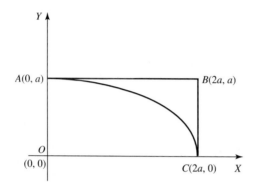

Figure P6.25

(a) From A to B and then from B to C.

(b) Along a segment of an ellipse from A to C. Use the relations $x = 2a \sin \theta$ and $y = a \cos \theta$.

Carry out these calculations for each of the following three forces:

(i) $\mathbf{F} = kx\hat{\mathbf{i}} + ky\hat{\mathbf{j}}$

(ii) $\mathbf{F} = 2y\hat{\mathbf{i}} - 2x\hat{\mathbf{j}}$

(iii) $\mathbf{F} = x^2y\hat{\mathbf{i}} + xy^2\hat{\mathbf{j}}$

6.26. A particle of mass m moves such that

$$x = x_0 + at^2, \qquad y = bt^3, \qquad z = ct$$

Find angular momentum $\mathbf{L}$ as a function of time t. Also find $\mathbf{F}$ and $\boldsymbol{\tau}$ and verify that the angular momentum conservation theorem is satisfied.

6.27. At time $t = 0$, a force $\mathbf{F} = 2a\hat{\mathbf{i}} + 3b\hat{\mathbf{j}}$ is applied to a particle of mass m at rest at the origin. Find its position and velocity as a function of time.

6.28. At time $t = 0$, a force $\mathbf{F} = a\hat{\mathbf{i}} + bt^2\hat{\mathbf{j}} + ct^3\hat{\mathbf{k}}$ is applied to a particle of mass m at rest at $\mathbf{r}_0 = 2\hat{\mathbf{i}} + 3\hat{\mathbf{j}}$. Find its position and velocity as a function of time t.

6.29. At time $t = 0$, a force $\mathbf{F} = a\hat{\mathbf{i}} + be^{kt}\hat{\mathbf{j}}$ is applied to a particle at $\mathbf{r}_0 = 2\hat{\mathbf{i}} + 3\hat{\mathbf{j}}$ and moving with velocity $\mathbf{v}_0 = v_{x0}\hat{\mathbf{i}} + v_{y0}\hat{\mathbf{j}}$. Find the position and velocity of this particle as a function of time t.

6.30. Derive Eq. (6.103).

6.31. Consider two simple harmonic motions:

$$x = A \cos \omega_x t, \qquad y = A \cos(\omega_y t + \phi)$$

Draw Lissajous figures if $\omega_y = 2\omega_x$ and $\phi = 0, \pi/4, \pi/2$, and π.

6.32. Repeat Problem 6.31 if $2\omega_y = 3\omega_x$.

6.33. Two harmonic vibrations at right angles to each other are described by the equations

$$x = 12 \cos(4\pi t), \qquad y = 12 \cos\left(8\pi t + \frac{\pi}{3}\right)$$

Draw Lissajous figures that describe this motion.

6.34. Draw Lissajous figures that describe the following motion:
(a) $x = 4 \sin 2\omega t$ and $y = \cos 2\omega t$
(b) $x = 5 \cos \omega t$ and $y = 5 \cos 2\omega t$

6.35. Draw Lissajous figures for the following motion:
(a) $x = 5 \sin \omega t$ and $y = 5 \cos 2\omega t$
(b) $x = 5 \cos(2\omega t)$ and $y = 5 \cos(2\omega t + \pi/4)$

6.36. Draw Lissajous figures for the case $\phi_x = \phi_y$ and $4\omega_x = 3\omega_y$.

6.37. Draw Lissajous figures for the case $\phi_x = \phi_y$ and $3\omega_x = 4\omega_y$.

6.38. Consider a nonisotropic two-dimensional harmonic oscillator for which the potential energy is

$$V(x, y) = \tfrac{1}{2}x^2 + \tfrac{1}{2}y^2$$

Solve for the equations of motion, considering the particle to have a unit mass. Let $\mathbf{r}_0 = a_2\hat{\mathbf{j}}$ and $\mathbf{v}_0 = a_1\hat{\mathbf{i}}$. Draw an orbit representing this motion.

6.39. Consider a two-dimensional isotropic harmonic oscillator of mass m represented by

$$x(t) = A \cos \omega t \qquad \text{and} \qquad y(t) = B \sin \omega t$$

Show that under the appropriate initial conditions and proper coordinate system, the angular momentum L and total energy E are

$$L = \sqrt{km}\, AB \qquad \text{and} \qquad E = \tfrac{1}{2}k(A^2 + B^2)$$

What is the expression for r in terms of L, E, k, and ω?

6.40. Derive Eq. (6.116) for a three-dimensional isotropic oscillator.

6.41. Discuss the motion of a three-dimensional oscillator for which the motion is confined to a box of dimensions $2A \times 2B \times 2C$.

6.42. Write Eq. (6.124) in the form given by Eq. (6.126).

6.43. Starting with Eqs. (6.132) and (6.133), derive Eqs. (6.134) to (6.137).

6.44. Show that, for large values of t, Eq. (6.138) reduces to Eq. (6.139) and Eq. (6.136) reduces to Eq. (6.140).

6.45. Starting with Eqs. (6.136) and (6.137), derive Eq. (6.143).

6.46. Starting with Eq. (6.143), derive Eqs. (6.146) and (6.147).

6.47. Derive an expression for x_m similar to the one given in Eq. (6.146), but keep an extra term in the expansion arrived at in Eq. (6.145).

6.48. Consider a projectile fired from the origin with velocity $\mathbf{v}_0(v_{0x}, v_{0y}, v_{0z})$ in the presence of a wind velocity $\mathbf{v}_w = v_w\hat{\mathbf{j}}$ and air resistance proportional to the velocity. Solve the equations of motion for the coordinates (x, y, z) as a function of time t. Suppose the projectile returns to the horizontal plane at a point (x_1, y_1). Calculate the position keeping only the first-order term in b. Show that, if we neglect air resistance, the target distance is missed by a fraction $4bv_{z0}/3mg$ of its target distance, and the wind causes an additional miss in the y coordinate by an amount $2bv_w v_{z0}^2/(mg^2)$.

6.49. Using the equations in Problem 6.48, make the appropriate plots using numerical values.

6.50. A projectile is fired from the origin with velocity v_0, making an angle θ with the X-axis in the XZ plane, and hits the target at $(x_0, 0)$. Find the first-order correction to the angle of elevation due to air resistance.

6.51. Using numerical values, make the graphs in Problem 6.50.

6.52. A gun fires a shot with a velocity v_0. Calculate the maximum range in any direction, given that the angle of elevation α of the gun may be varied. From Fig. P6.52 show that

$$r = \frac{2v_0^2}{g \cos^2 \theta} \cos \alpha \sin(\alpha - \theta)$$

and

$$(R_\theta)_{max} = \frac{v_0^2}{g(1 + \sin \theta)}$$

The maximum range R_θ is in a direction making an angle θ with the horizontal.

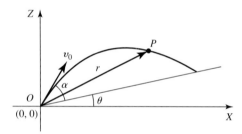

Figure P6.52

6.53. Using numerical values, make the graphs in Problem 6.52.

6.54. A gun mounted at the foot of a hill when fired strikes the hill at a right angle. If the hill makes an angle ϕ with the horizontal, find the angle that the gun barrel makes with the slope of the hill.

6.55. Two projectiles are fired, one with velocity v_{01} making an angle θ_1 and the other with velocity v_{02} making an angle θ_2 $(\theta_1 > \theta_2)$. Show that if they are to collide in midair the time interval between the two firings must be

$$\frac{2v_{01}v_{02} \sin(\theta_1 - \theta_2)}{g(v_{01} \cos \theta_1 + v_{02} \cos \theta_2)}$$

6.56. A projectile is fired with velocity v_0 and passes through two points, both a distance h above the horizontal. Show that, if the angle of the barrel of the gun is adjusted for the maximum range, then the horizontal separation of the two points is

$$d = \frac{v_0}{g} \sqrt{v_0^2 - 4gh}$$

6.57. The motion of a falling spherical drop of liquid of density ρ_0 is opposed by **(a)** a force proportional to the surface area of the drop, **(b)** the density ρ of the medium, and **(c)** the nth power of its velocity. Show that the terminal velocity varies as the nth root of the radius of the drop. Using appropriate values, graph v versus t and v versus x.

6.58. Solve Eqs. (6.149), (6.150), and (6.151).

SUGGESTIONS FOR FURTHER READING

ARTHUR, W., and FENSTER, S. K., *Mechanics*, Chapters 2, 5, and 7. New York: Holt, Rinehart and Winston, Inc., 1969.

DAVIS, A. DOUGLAS, *Classical Mechanics*, Chapters 5 and 6. New York: Academic Press, Inc., 1986.

FOWLES, G. R., *Analytical Mechanics*, Chapters 2 and 4. New York: Holt, Rinehart and Winston, Inc., 1962.

FRENCH, A. P., *Vibrations and Waves*, Chapter 2. New York: W. W. Norton and Co., Inc., 1971.

HAUSER, W., *Introduction to the Principles of Mechanics*, Chapter 2. Reading, Mass.: Addison-Wesley Publishing Co., 1965.

KITTEL, C., KNIGHT, W. D., and RUDERMAN, M. A., *Mechanics*, Berkeley Physics Course, Volume 1, Chapter 5. New York: McGraw-Hill Book Co., 1965.

ROSSBERG, K., *Analytical Mechanics*, Chapters 1 and 2. New York: John Wiley & Sons, Inc., 1983.

SYMON, K. R., *Mechanics*, 3rd ed., Chapter 3. Reading, Mass.: Addison-Wesley Publishing Co., 1971.

C H A P T E R

7

Central Force

7.1 INTRODUCTION

Central force is one of our most important topics and difficult to do it justice in just a few chapters. The importance of central force becomes obvious when one considers the numerous forces in nature that are central forces: gravitational, electric, atomic, and molecular forces to name just a few. To start, we will establish the relation between central force and potential energy. We will describe central force motion as a one-body problem and will show that this leads to the description of the properties of such motion, the most interesting being Kepler's laws of planetary motion. By introducing the concept of effective potential, one can discuss motion qualitatively, thus leading to a clearer understanding. We will extend this approach to describe orbits in an inverse-square central force field, leading to a discussion of Kepler's laws. We will discuss two interesting applications: perturbed circular orbits and orbital transfer, that is, going from one orbit to another.

7.2 CENTRAL FORCE AND POTENTIAL ENERGY

A *central force* acting on a particle is one that is always directed toward a fixed point called the center (or origin) of the force. Furthermore, if interaction between any two objects is represented by a central force, then the force is directed along the line joining the centers of the two objects. Thus a central force acting on a particle at a distance $\mathbf{r}$ from the force center may be represented by

$$\mathbf{F}(\mathbf{r}) = F(\mathbf{r})\hat{\mathbf{r}} \tag{7.1}$$

where $\hat{\mathbf{r}}$ is a radial unit vector. This form of the force law implies that the angular momentum of the particle is conserved or unchanged. On the other hand, if the central force is isotropic, that is,

$$\mathbf{F}(\mathbf{r}) = F(r)\hat{\mathbf{r}} \qquad (7.2)$$

the central force will be conservative; hence the mechanical energy of a particle is constant. In our discussion in this chapter, we shall limit ourselves to this latter form of the central force; hence *both angular momentum and energy will be conserved.* In this case, these conservation laws are the result of radial symmetry. Since the unit vector may be written as $\hat{\mathbf{r}} = \mathbf{r}/r$, we may write Eq. (7.2) as

$$\mathbf{F} = F(r)\frac{\mathbf{r}}{r} \qquad (7.3)$$

The importance of central isotropic forces becomes obvious when one considers the numerous forces of this kind that exist in nature, for example: (1) gravitational forces (attractive) that describe planetary motion; (2) coulomb or electrostatic forces (repulsive and attractive) that, in addition to other applications, lead to the formulation of the Rutherford and Bohr models of the atom; (3) intermolecular long-range forces called van der Walls forces, described by

$$F(r) = \frac{K_1}{r^{13}} - \frac{K_2}{r^7} \qquad (7.4)$$

where K_1 and K_2 are constants [the potential function that results in the force function given by Eq. (7.4) is called Lennard–Jones]; (4) atoms, which in a cubic crystal behave like three-dimensional harmonic oscillators, governed by central forces; and (5) nuclear forces of the Yukawa type represented by

$$F(r) = \left(\frac{K_1}{r} - \frac{K_2}{r^2}\right)e^{-K_3 r} \qquad (7.5)$$

where K_1, K_2, and K_3, as constants.

We stated that central forces are position dependent and conservative. Hence we should be able to introduce a potential energy function $V(r)$ for such central forces. But this is possible only if the **curl** of the force vanishes; that is,

$$\mathbf{curl}\ \mathbf{F} = \nabla \times \mathbf{F} = 0 \qquad (7.6)$$

Writing Eq. (7.3) in component form,

$$\mathbf{F} = \hat{\mathbf{i}}F_x + \hat{\mathbf{j}}F_y + \hat{\mathbf{k}}F_z = \frac{F(r)}{r}(\hat{\mathbf{i}}x + \hat{\mathbf{j}}y + \hat{\mathbf{k}}z) \qquad (7.7)$$

we get

$$F_x = \frac{x}{r}F(r), \qquad F_y = \frac{y}{r}F(r), \qquad F_z = \frac{z}{r}F(r) \qquad (7.8)$$

We may write Eq. (7.6) as

$$\nabla \times \mathbf{F} = \hat{\mathbf{i}}\left(\frac{\partial F_z}{\partial y} - \frac{\partial F_y}{\partial z}\right) + \hat{\mathbf{j}}\left(\frac{\partial F_x}{\partial z} - \frac{\partial F_z}{\partial x}\right) + \hat{\mathbf{k}}\left(\frac{\partial F_y}{\partial x} - \frac{\partial F_x}{\partial y}\right) = 0 \qquad (7.9)$$

For this to be true, each of the three component expressions must be zero. For example,

$$(\nabla \times \mathbf{F})_x = \frac{\partial F_z}{\partial y} - \frac{\partial F_y}{\partial z} \qquad (7.10)$$

must be zero. Using Eq. (7.8), we obtain

$$\frac{\partial F_z}{\partial y} = \frac{\partial}{\partial y}\left(\frac{z}{r}F(r)\right) = z\frac{\partial}{\partial r}\left(\frac{F(r)}{r}\right)\frac{\partial r}{\partial y} = z\frac{\partial r}{\partial y}\frac{\partial}{\partial r}\left(\frac{F(r)}{r}\right) \qquad (7.11a)$$

and, similarly,

$$\frac{\partial F_y}{\partial z} = y\frac{\partial r}{\partial z}\frac{\partial}{\partial r}\left(\frac{F(r)}{r}\right) \qquad (7.11b)$$

Substituting Eqs. (7.11) in Eq. (7.10) yields

$$(\nabla \times \mathbf{F})_x = \left(z\frac{\partial r}{\partial y} - y\frac{\partial r}{\partial z}\right)\frac{\partial}{\partial r}\left(\frac{F(r)}{r}\right) \qquad (7.12)$$

From the relation,

$$r = (x^2 + y^2 + z^2)^{1/2}$$

$$\frac{\partial r}{\partial y} = \frac{y}{r} \quad \text{and} \quad \frac{\partial r}{\partial z} = \frac{z}{r} \qquad (7.13)$$

We substitute these in Eq. (7.12), showing

$$(\nabla \times \mathbf{F})_x = 0 \qquad (7.14a)$$

In the same manner, we can show that

$$(\nabla \times \mathbf{F})_y = 0 \quad \text{and} \quad (\nabla \times \mathbf{F})_z = 0 \qquad (7.14b)$$

Thus, for the central force $\mathbf{F}$,

$$\nabla \times \mathbf{F} = 0 \qquad (7.15)$$

which implies that the central force is conservative, and it is possible to associate with it a potential energy function $V(r)$ such that

$$\mathbf{F}(r) = -\mathbf{grad}\ V(r) = -\nabla V(r) \qquad (7.16)$$

In spherical coordinates, the **gradient** operator $\mathbf{\nabla}$ [Eq. (6.63)] is

$$\mathbf{\nabla} = \hat{\mathbf{r}} \frac{\partial}{\partial r} + \hat{\mathbf{\theta}} \frac{1}{r} \frac{\partial}{\partial \theta} + \hat{\mathbf{\phi}} \frac{1}{r \sin \theta} \frac{\partial}{\partial \phi} \tag{7.17}$$

Since the potential energy function V is a function of the magnitude of r only, that is, $V = V(r)$, θ and ϕ dependence do not enter in Eq. (7.16); therefore,

$$\mathbf{F} = -\mathbf{\nabla}V = -\frac{\partial V}{\partial r} \hat{\mathbf{r}} \tag{7.18}$$

or the magnitude of $\mathbf{F}$ is given by

$$F = -\frac{\partial V}{\partial r} \tag{7.19}$$

We can go a step further and write the reverse relation as

$$V = V(r) = -\int_{r_s}^{r} F(r) \, dr \tag{7.20}$$

7.3 CENTRAL FORCE MOTION AS A ONE-BODY PROBLEM

Consider an isolated system consisting of two bodies that are separated by a distance $r = |\mathbf{r}|$, with the interaction between them described by a central force $F(r)$. If the bodies are spherically symmetric or are point masses, a system consisting of two particles can be described by means of six quantities or coordinates. Thus if $\mathbf{r}_1$ and $\mathbf{r}_2$ are the two radii vectors of particles of masses m_1 and m_2, then the six components of these radii vectors describe the system completely. The equations of motion of the two particles are

$$m_1\ddot{\mathbf{r}}_1 = F(r)\hat{\mathbf{r}} \tag{7.21a}$$

$$m_2\ddot{\mathbf{r}}_2 = -F(r)\hat{\mathbf{r}} \tag{7.21b}$$

where

$$\mathbf{r} = \mathbf{r}_1 - \mathbf{r}_2 \tag{7.22}$$

as shown in Fig. 7.1. The force between the two particles is attractive is $F(r) < 0$ and repulsive if $F(r) > 0$. Equations (7.21) are coupled by the relations of Eq. (7.22); that is, the behavior of $\mathbf{r}_1$ and $\mathbf{r}_2$ depends on the magnitude of $\mathbf{r}$.

Instead of describing the preceding system by the six coordinates of $\mathbf{r}_1$ and $\mathbf{r}_2$, it is convenient to describe the system by an alternate set of six coordinates: three coordinates describing the *center of mass* by $\mathbf{R}$ and three coordinates describing the *relative position* by $\mathbf{r}$. That is,

$$(m_1 + m_2)\mathbf{R} = m_1\mathbf{r}_1 + m_2\mathbf{r}_2 \tag{7.23}$$

and

$$\mathbf{r} = \mathbf{r}_1 - \mathbf{r}_2 \tag{7.24}$$

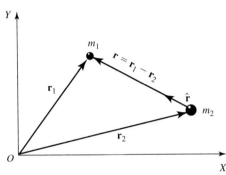

Figure 7.1 System consisting of two particles of mass m_1 and m_2 located at $\mathbf{r}_1$ and $\mathbf{r}_2$ from the origin.

Here, $\mathbf{R}$ describes the motion of the center of mass and $\mathbf{r}$ describes the relative motion of one particle with respect to the other, as shown in Fig. 7.2. Since no external forces are acting on the system, that is, $\ddot{\mathbf{R}} = 0$, the motion of the center of mass is that of uniform translational motion and hence is of no interest.

We now proceed to find the motion described by $\mathbf{r}$. Divide Eq. (7.21a) by m_1, Eq. (7.21b) by m_2, and subtract

$$\ddot{\mathbf{r}}_1 - \ddot{\mathbf{r}}_2 = \left(\frac{1}{m_1} + \frac{1}{m_2}\right)F(r)\hat{\mathbf{r}}$$

which, after rearranging, gives

$$\frac{m_1 m_2}{m_1 + m_2}(\ddot{\mathbf{r}}_1 - \mathbf{r}_2) = F(r)\hat{\mathbf{r}} \qquad (7.25a)$$

or

$$\mu \ddot{\mathbf{r}} = F(r)\hat{\mathbf{r}} \qquad (7.25b)$$

where

$$\mu = \frac{m_1 m_2}{m_1 + m_2} \quad \text{or} \quad \frac{1}{\mu} = \frac{1}{m_1} + \frac{1}{m_2} \qquad (7.26)$$

and μ is called the *reduced mass*. Equation (7.25b) is identical to Eq. (7.21a) or (7.21b), which describe the motion of a single particle m_1 or m_2 under the influence of a central force $F(r)$. In

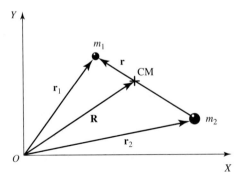

Figure 7.2 Description of a system of two particles by means of six coordinates: three coordinates describing the center of mass by $\mathbf{R}$ and three coordinates describing the relative position by $\mathbf{r}$.

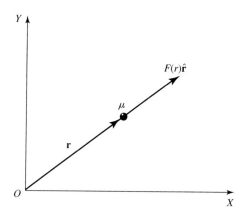

Figure 7.3   System of two bodies described by an equivalent one-body problem of mass μ and relative coordinate **r**.

Eq. (7.25b), m has been replaced by the reduced mass μ. Thus Eq. (7.25b) can be depicted as shown in Fig. 7.3. A two-body problem has been replaced by an equivalent one-body problem in which we have to determine the motion of a particle of mass μ in a central force field $F(r)$. (*Note of caution*: A three-or-more-body problem cannot be reduced to a one-body problem as we have done here. Hence the exact solutions of such problems are not known.)

Now we can use Eq. (7.25b) to find $\mathbf{r} = \mathbf{r}(t)$ and then solve for $\mathbf{r}_1$ and $\mathbf{r}_2$ by using Eqs. (7.23) and (7.24); that is,

$$\mathbf{r}_1 = \mathbf{R} + \frac{m_2}{m_1 + m_2}\mathbf{r} \tag{7.27}$$

and

$$\mathbf{r}_2 = \mathbf{R} - \frac{m_1}{m_1 + m_2}\mathbf{r} \tag{7.28}$$

As we said earlier, the center of mass moves with uniform velocity so that

$$\ddot{\mathbf{R}} = 0 \tag{7.29}$$

which has the solution

$$\mathbf{R} = \mathbf{v}_0 t + \mathbf{R}_0 \tag{7.30}$$

By choosing initial conditions such that at $t = 0$, $\mathbf{v}_0 = 0$, and $\mathbf{R}_0 = 0$, we get $\mathbf{R} \equiv 0$; that is, the origin coincides with the center of mass and Eqs. (7.27) and (7.28) reduce to

$$\mathbf{r}_1 = + \frac{m_2}{m_1 + m_2}\mathbf{r} \tag{7.31}$$

$$\mathbf{r}_2 = - \frac{m_1}{m_1 + m_2}\mathbf{r} \tag{7.32}$$

where $\mathbf{r}_1$ and $\mathbf{r}_2$ are measured from the center of mass, as depicted in Fig. 7.4.

Finally, it may be pointed out that in Eq. (7.26), if the mass of one of the two particles is very large, say $m_2 \gg m_1$, then

$$\frac{1}{\mu} = \frac{1}{m_1} + \frac{1}{m_2} \simeq \frac{1}{m_1}$$

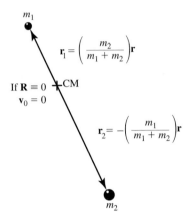

Figure 7.4   Position of two particles in a system from the center of mass at rest at the origin.

and Eq. (7.25b) reduces to

$$m_1 \ddot{\mathbf{r}} = F(r)\hat{\mathbf{r}}$$

which states that the center of mass is almost located at m_2 if $m_2 \gg m_1$ (that is, as if m_2 is infinite as compared to m_1), while $\mathbf{r} = \mathbf{r}_1 - \mathbf{r}_2 \simeq \mathbf{r}_1$. Hence the problem can be treated as a one-body problem. Thus, whenever we use mass m instead of μ, we are indicating that the other mass is very large, whereas the use of μ indicates that either the two masses are comparable or that a better accuracy in calculations is desired.

Once we have calculated $\mathbf{r}(t)$ we can calculate $\mathbf{r}_1(t)$ and $\mathbf{r}_2(t)$ from Eqs. (7.31) and (7.32). Most of the time we are interested in the path or the orbit of one particle with respect to the other particle; hence we need not solve for $\mathbf{r}_1(t)$ and $\mathbf{r}_2(t)$.

7.4 GENERAL PROPERTIES OF MOTION UNDER A CENTRAL FORCE

Equation (7.25b), that is,

$$\mu \ddot{\mathbf{r}} = F(r)\hat{\mathbf{r}} \tag{7.25b}$$

as depicted in Fig. 7.3, describes the motion of a particle of mass μ and can be solved for $\mathbf{r}(t)$ only if we know the form of the central force $F(r)$. Furthermore, it is a vector equation; hence three components must be considered. Much can be learned about the motion of the particle without actually solving these equations if we know that the force is a central force, even though we do not know the actual form of the central force. The properties of the general solution of Eq. (7.25b) are based on the conservation laws.

Central Force Motion Is Confined to a Plane

In the situation under consideration, the central force $F(r)\hat{\mathbf{r}}$ is along $\mathbf{r}$; hence it cannot produce torque $\boldsymbol{\tau}$ on the reduced mass μ. This means that the angular momentum $\mathbf{L}$ of mass μ about an axis passing through the center of force is constant, as we show now. If $\mathbf{p}$ is the linear momentum of a particle of mass μ, the torque $\boldsymbol{\tau}$ about an axis passing through the center of force is

$$\boldsymbol{\tau} = \frac{d\mathbf{L}}{dt} = \frac{d}{dt}(\mathbf{r} \times \mathbf{p}) = \frac{d}{dt}(\mathbf{r} \times m\mathbf{v})$$

$$= \mathbf{r} \times m\frac{d\mathbf{v}}{dt} + \mathbf{v} \times m\mathbf{v} \qquad (7.33)$$

But
$$m\frac{d\mathbf{v}}{dt} = m\mathbf{a} = \mathbf{F} \quad \text{and} \quad \mathbf{v} \times \mathbf{v} = 0$$

Hence
$$\boldsymbol{\tau} = \frac{d\mathbf{L}}{dt} = \mathbf{r} \times \mathbf{F} \qquad (7.34)$$

Since
$$|\mathbf{r} \times \mathbf{F}| = |\mathbf{r}|\,|\mathbf{F}| \sin 0° = 0$$

we have
$$\boldsymbol{\tau} = \frac{d\mathbf{L}}{dt} = 0 \qquad (7.35)$$

That is,
$$\mathbf{L} = \mathbf{r} \times \mathbf{p} = \text{constant} \qquad (7.36)$$

Thus, if the angular momentum $\mathbf{L}$ of mass μ is constant, its magnitude and direction are fixed in space. Hence, by definition of the cross product, if the direction of $\mathbf{L}$ is fixed in space, vectors $\mathbf{r}$ and $\mathbf{p}$ must lie in a plane perpendicular to $\mathbf{L}$. That is, *the motion of particle of mass μ is confined to a plane that is perpendicular to $\mathbf{L}$.*

Thus, if the angular momentum of the particle is constant, the motion of the particle is confined to a single plane, as shown in Fig. 7.5. The problem has been simplified to a motion in

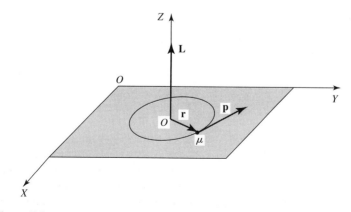

Figure 7.5   Motion of a particle under the influence of a central force is confined to an XY plane perpendicular to $\mathbf{L}$.

two dimensions instead of three dimensions. If the plane is chosen to be an XY plane, the motion can be described by the x, y coordinates. It is convenient to use plane polar coordinates (r, θ), and in such coordinates the equation of motion [Eq. (7.25b)], that is,

$$\mu\ddot{\mathbf{r}} = F(r)\hat{\mathbf{r}} \tag{7.25b}$$

takes the form [after replacing $\ddot{\mathbf{r}}$ by $\mathbf{a}$ and using Eq. (6.30)]

$$\mu(\ddot{r} - r\dot{\theta}^2)\hat{\mathbf{r}} + \mu(r\ddot{\theta} + 2\dot{r}\dot{\theta})\hat{\boldsymbol{\theta}} = F(r)\hat{\mathbf{r}} \tag{7.37}$$

or, equating the coefficients of the $\hat{\mathbf{r}}$ and $\hat{\boldsymbol{\theta}}$ on both sides,

$$\mu(\ddot{r} - r\dot{\theta}^2) = F(r) \tag{7.38}$$

$$\mu(r\ddot{\theta} + 2\dot{r}\dot{\theta}) = 0 \tag{7.39}$$

We shall make use of the preceding equations to arrive at some other important properties of central force motion.

Angular Momentum and Energy Are Constants of Motion

In arriving at the preceding results, we used the fact that the direction of $\mathbf{L}$ is fixed. There are two other constants of central force motion: (1) the magnitude of the angular momentum L, that is, $|\mathbf{L}| = L$; and (2) the total energy E of the system. Each of these constants, L and E, is called the *integral of motion* (or first integral of motion). These constants can be used to solve the problem of central force motion much more easily than solving Eqs. (7.38) and (7.39).

Using Fig. 7.6, the angular momentum of a particle of mass μ at a distance r from the force center is

$$L = rp = r\mu v_\theta = r\mu(r\dot{\theta})$$

That is, since $\mathbf{L}$ is constant,

$$L = \mu r^2\dot{\theta} = \text{constant} \tag{7.40}$$

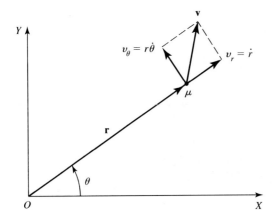

Figure 7.6   Motion of a particle described in plane polar coordinates.

Remember, we said that the center of mass has uniform translational motion; hence it is of no interest. Another way of saying this is that the conservation of linear momentum (even though a constant of motion) does not add much to the description of the motion. The remaining first integral of motion is the conservation of total energy E. Since there are no dissipative systems and central forces are conservative, the total energy is constant. That is,

$$E = K + V(r) = \text{constant} \tag{7.41}$$

where $V(r) = -\int \mathbf{F}(r) \cdot d\mathbf{r}$, and E is evaluated from initial conditions. Using Fig. 7.6,

$$E = \tfrac{1}{2}mv^2 + V(r) = \tfrac{1}{2}\mu(\dot{r}^2 + r^2\dot{\theta}^2) + V(r) \tag{7.42}$$

Substituting for $\dot{\theta} = L/\mu r^2$ from Eq. (7.40) into Eq. (7.42),

$$E = \frac{1}{2}\left(\mu \dot{r}^2 + \frac{L^2}{\mu r^2}\right) + V(r) = \text{constant} \tag{7.43}$$

Note that the expression for E does not contain $\dot{\theta}$.

In the expression for E, the first term is the radial contribution to kinetic energy, and the second term is the contribution of angular momentum to potential energy. We can arrive at the theorem of energy conservation by starting with Eq. (7.38).

Law of Equal Areas

We have shown that for a particle moving under a central force $\mathbf{F}(r)$, which may have any r dependence, the angular momentum $\mathbf{L}$ of the system is constant; that is, it is constant both in magnitude and direction. An important consequence of this fact is that the radius vector $\mathbf{r}$ traces equal areas in equal intervals of time, as shown next.

Consider a mass μ at a distance $\mathbf{r}(\theta)$ at time t from the force center O, as shown in Fig. 7.7. In a time interval dt, the mass moves from P to Q, and when at Q it is at a distance $\mathbf{r}(\theta + d\theta)$ from the force center O. The area dA swept by the radius vector $\mathbf{r}$ in time dt is (assuming ds to be very small, almost a straight line, since $d\theta$ is small) equal to the area of the triangle OPQ; that is,

$$dA = \tfrac{1}{2}r(r\,d\theta) = \tfrac{1}{2}\,r^2\,d\theta \tag{7.44}$$

or

$$\frac{dA}{dt} = \frac{1}{2}\,r^2\,\frac{d\theta}{dt} = \frac{1}{2}\,r^2\dot{\theta}$$

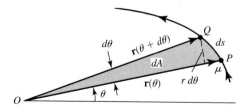

Figure 7.7   Area dA swept by a radius vector $\mathbf{r}$ in time dt.

Substituting for $\dot{\theta} = L/\mu r^2$ from Eq. (7.40), we get

$$\frac{dA}{dt} = \frac{L}{2\mu} = \text{constant} \tag{7.45}$$

which states that the areal velocity is constant, which is true for any particle moving under the influence of a central force. Equation (7.45) is a statement of *Kepler's second law of planetary motion*, also known as the *law of equal areas*. In addition, if the motion is periodic of period T, we may integrate Eq. (7.45) and obtain

$$\int dA = \int_0^T \frac{L}{2\mu} dt$$

or
$$A = \frac{L}{2\mu} T \tag{7.46}$$

Since the law of equal areas is the consequence of the fact that **L** is constant, we may write

$$\mathbf{L} = \mathbf{r}_1 \times \mathbf{p}_1 = \mathbf{r}_2 \times \mathbf{p}_2 \tag{7.47}$$

Applying this to the situation shown in Fig. 7.8, where mass m moves in an orbit around mass M—say, Earth going around the Sun—we may write $p = mv$, where v is the tangential velocity:

$$r_1 v_1 = r_2 v_2 = r_3 v_3 \tag{7.48}$$

For areal velocity to be constant, if r increases, v decreases, as demonstrated in Fig. 7.8, so that

$$A_1 = A_2 = A_3 \tag{7.49}$$

It is important to keep in mind that A remains constant for the orbit of a given object but is different for different objects.

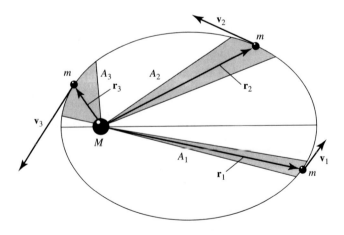

Figure 7.8 Kepler's second law, that is, the law of equal areas: $A_1 = A_2 = A_3$.

7.5 EQUATIONS OF MOTION

Let us go back to Eqs. (7.40) and (7.43), that is,

$$L = \mu r^2 \dot{\theta} = \text{constant} \tag{7.40}$$

and

$$E = \frac{1}{2}\left(\mu \dot{r}^2 + \frac{L^2}{\mu r^2}\right) + V(r) \tag{7.43}$$

If we know $V(r)$, these equations describe the motion of the system and can be solved for $\theta(t)$ and $r(t)$, respectively. The set $[\theta(t), r(t)]$ describes the orbit of the particle in parametric form where time t is the parameter. Very often we are interested in finding the orbit equation in the form of a relation between r and θ, that is, $r = r(\theta)$. We shall set up the equations to get solutions in this form. It may seem that we started to solve Eqs. (7.38) and (7.39) in plane polar coordinates and that somewhere along the way we abandoned them, but this is not true. If you look closely, Eqs. (7.40) and (7.43) are the direct consequences of Eqs. (7.39) and (7.38), respectively.

We may write Eq. (7.38) as

$$m\ddot{r} = F(r) + mr\dot{\theta}^2 \tag{7.50}$$

[Since $\mu = Mm/(M + m)$ and $M \gg m$, $\mu \simeq m$. This approximation is good as long as we do not need high accuracy. Thus we are going to start using m instead of μ unless conditions in the problem dictate otherwise.] Equation (7.50) can be made identical to a one-dimensional case if we can replace $\dot{\theta}$ by L/mr^2 [using Eq. (7.40)] and introduce a term, called the *effective force*, F_{eff},

$$F_{\text{eff}}(r) = F(r) + mr\dot{\theta}^2 = F(r) + \frac{L^2}{mr^3} \tag{7.51}$$

where $mr\dot{\theta}^2 (=L^2/mr^3)$ is treated as another force term—a fictitious force (because it is simply a product of mass and acceleration)—and is usually referred to as *centrifugal force*, F_{cent},

$$F_{\text{cent}} = mr\dot{\theta}^2 = \frac{L^2}{mr^3} \tag{7.52}$$

Thus Eq. (7.50) reduces to

$$m\ddot{r} = F_{\text{eff}}(r) \tag{7.53}$$

and can be treated as an equation in one dimension.

As before, we can use the definition of effective force given by Eq. (7.51) to introduce an *effective potential*, $V_{\text{eff}}(r)$,

$$V_{\text{eff}}(r) = \int_r^{r_s} F_{\text{eff}}(r)\, dr = \int_r^{r_s}\left(F(r) + \frac{L^2}{mr^3}\right) dr = \int_r^{r_s} F(r)\, dr + \frac{L^2}{m}\int_r^{r_s}\frac{dr}{r^3}$$

Assuming r_s to be infinity, we get

$$V_{\text{eff}}(r) = V(r) + \frac{L^2}{2mr^2} \tag{7.54}$$

Thus the effective potential is the sum of the real potential and an additional term called *centrifugal potential* or *centrifugal barrier* defined as

$$V_{\text{cent}} = \frac{L^2}{2mr^2} \tag{7.55}$$

Making Eq. (7.50) look like Eq. (7.53) has a much deeper meaning. Using Eq. (7.53), we are observing radial motion as viewed from a rotating reference frame. In this rotating frame, the force looks to be $F_{\text{eff}}(r)$. It is the central force nature of this force that leads to the conservation of energy given by Eq. (7.43). Solving Eq. (7.43) for $\dot{r}$ (using m instead of μ),

$$\dot{r} = \frac{dr}{dt} = \sqrt{\frac{2}{m}\left(E - V(r) - \frac{L^2}{2mr^2}\right)} \tag{7.56}$$

Integrating this equation, we obtain

$$t = \int_{r_0}^{r} \frac{dr}{\sqrt{\frac{2}{m}\left(E - V(r) - \frac{L^2}{2mr^2}\right)}} \tag{7.57}$$

which gives $t(r)$, but we can rearrange to solve for $r(t)$. Once we have obtained $r(t)$ we can integrate Eq. (7.40) to obtain $\theta(t)$, that is,

$$\dot{\theta} = \frac{d\theta}{dt} = \frac{L}{mr^2} \tag{7.58}$$

On integrating, we get

$$\theta = \theta_0 + \int_0^t \frac{L}{mr^2} \, dt \tag{7.59}$$

We may state that Eqs. (7.57) and (7.59) are the solutions of Eqs. (7.38) and (7.39), respectively. Of course, these solutions are in terms of four constants: L, E, r_0, and θ_0. These constants can be evaluated from initial conditions, that is, the initial position and initial velocity for the motion in the plane.

If we want a relation for $\theta(r)$ or $r(\theta)$, we may proceed as follows:

$$\dot{r} = \frac{dr}{dt} = \frac{dr}{d\theta}\frac{d\theta}{dt} = \frac{dr}{d\theta}\dot{\theta}$$

or

$$d\theta = \frac{\dot{\theta}}{\dot{r}} \, dr$$

Substituting for $\dot{r}$ and $\dot{\theta}$ from Eqs. (7.56) and (7.58), we get

$$d\theta = \frac{(L/mr^2) \, dr}{\sqrt{\frac{2}{m}\left(E - V(r) - \frac{L^2}{2mr^2}\right)}} \tag{7.60}$$

which on integration gives

$$\theta(r) = \int \frac{(L/mr^2)dr}{\sqrt{\frac{2}{m}\left(E - V(r) - \frac{L^2}{2mr^2}\right)}} \tag{7.61}$$

Since L is constant in time, so is $\dot{\theta}$, which will not change signs; hence θ will increase monotonically with time.

Thus to describe motion in a central force problem involves integrating one, two, or all three equations given by Eqs. (7.57), (7.59), and (7.61), yielding $r(t)$, $\theta(t)$, and $r(\theta)$, or $\theta(r)$, respectively. Such integrations are not always possible. Suppose the force law depends on some power of radial distance and is given by

$$F(r) = Kr^n \tag{7.62}$$

where K is a constant. For $n = 1$, the force law corresponds to the case of a harmonic oscillator; for $n = -2$, the force law corresponds to the inverse-square force law representing gravitational force and Coulomb force. These are the two most important cases. One of these, $n = 1$, has already been discussed in detail, while the other, $n = -2$, will be discussed in this and following chapters. For $n = 1$, -2, and -3, the resulting solutions are usually expressed in terms of circular functions. For certain other values of n (integer or fractional), solutions are generally expressed in terms of elliptic integrals.

In most situations we are interested in either (a) finding the orbit of the system (the path of the particle in space) without regard to its time dependence, assuming that the force law $F(r)$ is given, or (b) if the orbit of the system is given, evaluating the form of the force law. An equation that can accomplish either of these can be obtained by starting with Eq. (7.38),

$$F(r) = m\ddot{r} - mr\dot{\theta}^2 \tag{7.38}$$

and making the substitution

$$r = \frac{1}{u} \tag{7.63}$$

From Eq. (7.40)

$$\dot{\theta} = \frac{L}{mr^2} = \frac{L}{m}u^2 \tag{7.64}$$

and from Eq. (7.63), we obtain $\dot{r}$ and $\ddot{r}$ to be

$$\dot{r} = -\frac{L}{m}\frac{du}{d\theta} \tag{7.65}$$

and

$$\ddot{r} = -\frac{L^2}{m^2}u^2\frac{d^2u}{d\theta^2} \tag{7.66}$$

Substituting for r, $\ddot{r}$, and $\dot{\theta}$ in Eq. (7.38), and rearranging, (Problem 7.7) we get

$$F\left(\frac{1}{u}\right) = -\frac{L^2}{m}u^2\frac{d^2u}{d\theta^2} - \frac{L^2}{m}u^3$$

which after rearranging may be written as

or
$$\frac{d^2u}{d\theta^2} = -u - \frac{m}{L^2u^2}F\left(\frac{1}{u}\right) \qquad\qquad\qquad (7.67)$$

This equation is a differential equation that may be solved for $u(\theta)$ and hence for $r(\theta)$, which describes the orbit of a particle moving under a central force $F(r)\hat{\mathbf{r}}$. On the other hand, if the orbit of the particle is given in polar coordinates $r(\theta)$, this differential equation may be solved to find the form of the force law $F(r)$. A special case is in order. For $L = 0$, the above equations do not hold. From Eq. (7.40), $mr^2\dot{\theta} = L = 0$ means that since $m \neq 0$, $r \neq 0$, therefore, $\dot{\theta} = 0$, or $\theta =$ constant, which implies the path of the particle is a straight line passing through the origin.

 Example 7.1 _____

A particle of mass m is observed to move in a spiral orbit given by the equation $r = k\theta$, where k is a constant. Is it possible to have such an orbit in a central force field? If so, determine the form of the force function. Draw the path of the spiral orbit in polar coordinates.

Solution

To determine the force law use Eq. (7.67). Substitute $u = 1/r$ and then calculate the first and the second differential of u with respect to θ as shown.

$$r = k\cdot\theta \qquad u = \frac{1}{r} \qquad u = \frac{1}{k\cdot\theta}$$

$$\frac{d}{d\theta}u = \frac{-1}{\left(k\cdot\theta^2\right)} \qquad \frac{d^2}{d\theta^2}u = \frac{2}{\left(k\cdot\theta^3\right)}$$

Thus Eq. (7.67) takes the form as shown.

$$\frac{d^2}{d\theta^2}u + u = -\frac{M}{L^2\cdot u^2}\cdot Fu \qquad \theta = \frac{r}{k}$$

Substitute for $\theta = r/k$.

$$\frac{2}{\left(k\cdot\theta^3\right)} + \frac{1}{(k\cdot\theta)} = \frac{-M}{L^2}\cdot k^2\cdot\theta^2\cdot Fu$$

Solve this equation for the force Fu.

$$2\cdot\frac{k^2}{r^3} + \frac{1}{r} = \frac{-M}{L^2}\cdot r^2\cdot Fu$$

Thus the force law is central and is a combination of an inverse cube and inverse fifth-power law.

$$Fu = -2\cdot\frac{L^2}{\left(r^5\cdot M\right)}\cdot k^2 - \frac{L^2}{\left(r^3\cdot M\right)}$$

Such a force law gives an almost spiral orbit, as illustrated in the polar graph below. (A pure inverse cube force yields a spiral orbit.) For k = 5, we may write r = 5θ.

$$n := 1..50 \qquad \theta_n := \frac{2 \cdot \pi}{20} \cdot n \qquad r_n := 5 \cdot \theta_n$$

(a) How will the graph look if it is drawn in rectangular coordinates?

(b) How will the graph of Fu versus r look in rectangular or plane polar coordinates?

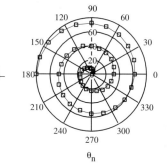

EXERCISE 7.1 Repeat the example for a spiral orbit given by $r = k \cdot \theta^2$

7.6 CENTRAL FORCE FIELD ORBITS AND EFFECTIVE POTENTIAL

In describing the motion of a particle under the action of a central force, we have shown that the motion is limited to two dimensions. Furthermore, by using the conservation of angular momentum and energy, we have reduced a two-dimensional motion to two one-dimensional motions. The conservation of total energy, according to Eq. (7.43), is

$$E = \frac{1}{2} m \dot{r}^2 + \frac{L^2}{2mr^2} + V(r) = K_{rad} + V_{cent}(r) + V(r) \qquad (7.68)$$

where K_{rad} and V_{cent} are the kinetic energies due to the radial and angular motions (V_{cent} may be written as K_{ang}, representing the kinetic energy due to the angular motion), respectively. Since V_{cent} is r dependent and so is $V(r)$, the two terms are combined as the effective potential energy; that is,

$$E = \tfrac{1}{2} m \dot{r}^2 + V_{eff}(r) \qquad (7.69)$$

where
$$V_{eff}(r) = V_{cent}(r) + V(r) = \frac{L^2}{2mr^2} + V(r) \qquad (7.70)$$

The dependence of total energy E on the variables r and $\dot{r}$ is similar to the motion of a particle in one dimension (as discussed in Section 2.5) if we replace x by r, $\dot{x}$ by $\dot{r}$, and $V(x)$ by $V_{eff}(r)$. Hence the energy diagram method discussed in Chapter 2 can be applied here.

Before proceeding further, it is necessary to understand that there are significant differences between the energy method as applied to one-dimensional motion and its application to any two-dimensional motion that has been reduced to two one-dimensional motions. First, in one dimension, energy E alone determines the nature of the motion in a conservative field. In central force motion, the characteristic of the motion depends on both parameters, energy E and angular momentum L. Second, we must not forget that, while the radial distance r is changing with time, so is θ changing with time. That is, as r is changing, the vector $\mathbf{r}$ is also rotating. Hence, even though we solve for r, in order to understand the orbital path of a particle, both r and θ must be taken into consideration. The only exception to this is the case of circular motion in which the magnitude of the vector $\mathbf{r}$ remains constant and is equal to r_0, the radius corresponding to minimum V_{eff}.

We shall now apply the energy diagram method, making plots of V_{eff} versus r, to the two commonly encountered force laws: (1) the isotropic harmonic force law, and (2) the inverse-square force law. The salient features of central force motion can be well understood by first considering the case of a harmonic oscillator, as shown in Fig. 7.9(a). For this case,

$$F(r) = -kr \quad \text{or} \quad V(r) = \tfrac{1}{2}kr^2 \tag{7.71}$$

Hence Eq. (7.70) for effective potential becomes

$$V_{\text{eff}}(r) = V_{\text{cent}}(r) + V(r) = \frac{L^2}{2mr^2} + \frac{1}{2}kr^2 \tag{7.72}$$

The plots of $V(r)$, V_{cent}, and V_{eff} are shown in Fig. 7.9(a). V_{eff} has a minimum at r_0. For a given total energy E (greater than the minimum energy $E_0 = [V_{\text{eff}}(r)]_{\text{min}}$), the particle oscillates between two extreme values of r, $r_1 = r_{\text{min}}$ and $r_2 = r_{\text{max}}$; that is, $r_{\text{min}} < r < r_{\text{max}}$. The two points are the turning points in motion. At these points the radial velocity is zero; that is, $\dot{r} = 0$. These turning points are the roots of the equation [energy conservation, Eq. (7.68)], with $\dot{r} = 0$; that is,

$$E - V(r) - \frac{L^2}{2mr^2} = 0 \tag{7.73}$$

These two radial distances, r_1 and r_2, define two circles of radii r_1 and r_2 about the force center in the plane of the orbit. And it is the angular motion that restricts the motion of the particle within these two circles (discussed later).

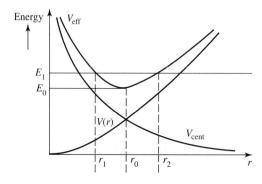

Figure 7.9(a)   Graphs of $V(r)$, V_{cent}, and V_{eff} for an isotropic harmonic force law.

Figure 7.9(b) _____

Below are the graphs of attractive real potential VR, centripetal potential VC, and the total effective potential VE versus distance r. Also shown is a particle with total energy E1 = 212.

The characteristics of the VE plot are clear. No particle exists with energy less than min(VE) = 122.1. Particles with energy slightly greater than this will have harmonic motion. Each particle has two turning points; for example a particle with energy 212 has turning points at r = 5 and r = 16.

$$N := 20 \qquad i := 0 .. N \qquad r_i := i$$

$$k := 1.5 \qquad L := 10 \qquad m := .01$$

$$VC_i := \frac{L^2}{2 \cdot m \cdot (r_i)^2 + .01} \qquad VR_i := \left(\frac{1}{2}\right) \cdot k \cdot (r_i)^2$$

$$VE_i := VR_i + VC_i \qquad E1_i := 212$$

$$VE_5 = 214.828$$

$$VE_{16} = 211.493$$

$$min(VE) = 122.1$$

$$VE_9 = 122.1$$

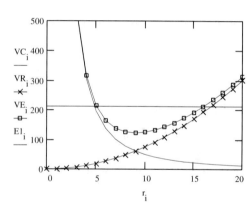

(a) Explain why a particle cannot have energy less than min(VE).

(b) Describe the characteristics of the motion of a particle with energy much greater than min(VE) (for example, 500).

_____

    Figure 7.9(b) is the same as Fig. 7.9(a) for certain values of k, L, m, and E1. It is clear that for the turning points: $(r_{min} = r_5) < (r = r_9) < (r_{max} = r_{16})$. For energy E1 the radii of the two circles that limit the motion of the particle of total energy E1 correspond to points r_5 and r_{16}.

    We now discuss the motion of a particle in the inverse-square force field. Figure 7.10(a) shows an attractive potential V(**r**) versus r, which starts from r = 0, has a very large negative potential, increases with increasing r, and reaches zero as r reaches infinity; that is,

$$V(r) = -\infty \text{ at } r = 0 \quad \text{and} \quad V(r) \to 0 \text{ as } r \to \infty$$

Figure 7.10

$$N := 60 \qquad i := 0..N \qquad j := 1..3 \qquad L_j :=$$

(a) Given are the graphs of VR versus r for an attractive inverse square force law and VC versus r for different values of L.

$$r_i := \frac{i}{15} \qquad m := .1 \qquad K := 37$$

5
8
10

VR = real potential field
VC = centripetal potential for
     three different values of L
VE = effective potential = VR + VC

$$VR_i := -\frac{K}{\left(r_i\right)^2 + .01} \qquad VC_{i,j} := \frac{\left(L_j\right)^2}{2 \cdot m \cdot \left(r_i\right)^2 + .01}$$

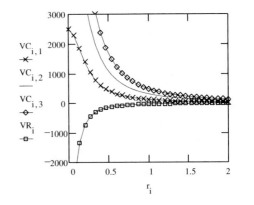

(b) Given are the graphs of VE versus r for different values of L.

$$VE_{i,1} := (-VR)_i + (-VC)_{i,1} \qquad VE_{i,2} := (-VR)_i + (-VC)_{i,2} \qquad VE_{i,3} := (-VR)_i + (-VC)_{i,3}$$

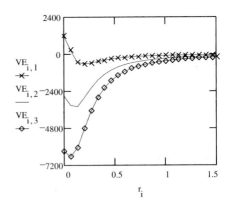

Furthermore, the potential has $1/r$ dependence and the corresponding force has an inverse-square dependence; that is $F(r) = Kr^{-2}$, where K is negative for an attractive force. Shown in Fig. 7.10(a) are the plots of $L^2/2mr^2$ versus r for three different values of angular momentum $L_3 > L_2 > L_1$. The resultant effective potential $V_{eff}(r)$ is obtained by combining the single function $V(r)$ with each of the separate three centrifugal potential curves, as shown in Fig. 7.10(b). All these curves show minima in the potential, and this is due to the fact that $V(r)$ varies as $1/r$ [or $F(r)$ varies as $1/r^2$]. [If $V(r)$ varies as $1/r^2$ or any higher power of r, a minimum in the curve will not occur.]

Now let us discuss the motion of a particle with the help of a typical energy diagram. Figure 7.11 shows the $V_{eff}(r)$ versus r curve for a particular value of L. [This could be any one of the VE versus r plots in Fig. 7.10(b).] If the energy of a particle is less than the minimum energy E_m, no physically meaningful motion is possible because this results in $\dot{r}$ being an imaginary quantity. If the energy of a particle in such that $E = E_m$, there is no radial motion; hence the particle must move in a circle of radius r_0. If the energy of a particle is greater than zero, say $E = E_4$, the motion of the particle is *unbounded*. A particle headed toward the force center can come as close as r_4 and then must turn back, and may go back to infinity. We say there is a single turning point at $r = r_4$. Thus, for a particle with energy $E > 0$ having the assumed form of potential, the motion is unbounded and there is a single turning point.

Now suppose a particle has energy between $E = 0$ and $E = E_m$, say E_1, as shown in Fig. 7.11. The radial motion of the particle will be confined to the values of $r = r_1 = r_{min}$ and $r = r_2 = r_{max}$. The points r_1 and r_2 are the turning points. Actually, the motion is confined between the areas of two circles of radii r_1 and r_2, as shown in Fig. 7.12. The motion is periodic with the radial time period T_r, which is the time the particle takes to go from r_{min} to r_{max} and then back to r_{min}. In Fig. 7.12, T_r is equal to the time in going from A to B. Furthermore, the orbit

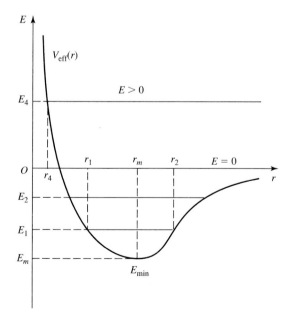

Figure 7.11 Graph of $V_{eff}(r)$ versus r for one particular value of L; the range of motion of particles with different energies.

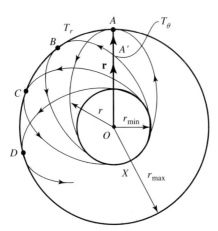

Figure 7.12 Motion of a particle with energy $0 > E > E_m$ is confined between two circles of radii $r_1 = r_{\min}$ and $r_2 = r_{\max}$, resulting in precessing motion. As θ changes for r moving from A to B results in T_r while r moving from A to A' results in T_θ.

must be such that it is tangent to both the circles at the turning points, such as A on the outer circle and X on the inner circle, respectively. The reason is that at the turning points the radial velocity $\dot{r}$ must be zero, while the tangential velocity cannot be zero because of angular momentum. As shown in Fig. 7.12, the orbit at A is tangent to the outer circle, and when it reaches X, it is tangent to the inner circle. The orbit continues and once again becomes tangent to the outer circle at B. The time it takes for the particle motion to go from A to X and then to B is equal to the *radial time period* T_r. Note that vector $\mathbf{r}$ is continuously changing direction. The time it takes to turn through 2π angle is called the *angular time period* T_θ, also called the *characteristic time period* or the *revolution time period*. In Fig. 7.12, to start with, vector $\mathbf{r}$ was equal to OA, and after turning through 2π angle, it is at OA'. The time it takes for vector $\mathbf{r}$ to reach OA' is equal to the period T_θ.

Thus it is clear that, for a particle with energy E such that $0 > E > E_m$, the motion of the particle will be *doubly periodic*, with periods T_r and T_θ. The characteristic of the orbit strongly depends on the ratio of these two periods. If the periods are *commensurable*, that is, the ratio of the two periods T_r/T_θ can be expressed as the ratio of two integers, the particle will ultimately (in time equal to the lowest multiple of T_r and T_θ) come back to exactly the starting position. Such orbits are called *closed orbits*. Another way of stating this fact is to say that the orbit will be closed only if

$$\frac{\Delta\theta}{2\pi} = \frac{a}{b} \tag{7.74}$$

where a and b are integers; that is, $\Delta\theta$ is a rational fraction of 2π. If the orbit is closed, the average frequency of revolution $\bar{f}_{\text{rev}}$ (or angular frequency) may be defined as

$$\bar{f}_{\text{rev}} = \frac{1}{T_{\text{rev}}} = \frac{\text{number of oscillations necessary to close the orbit}}{\text{total time needed to close the orbit}} \tag{7.75}$$

The ratio T_θ/T_r depends on the force law and the values of E and L. If the two periods T_r and T_θ are equal, the closure of the orbit will happen after only one time period; meanwhile θ increases by 2π. This situation happens in the case of an inverse-square law force (gravitational and electrostatic forces). On the other hand, for a situation like the one shown in Fig. 7.12, where T_r is slightly greater than T_θ, the radius vector **r** rotates through greater than 2π, while r completes one revolution from $r_{\min}$ to $r_{\max}$ and back to $r_{\min}$. It will take a large number of revolutions before a particle comes back to its initial position; hence the orbit is nearly closed. Such orbital motion is described as *precessing* motion (discussed in Chapter 13).

If the radial and the angular periods are *incommensurable*, the orbit will never close, that is, it is an *open orbit*. The whole area between the circles will be completely filled by the motion between $r_{\min}$ and $r_{\max}$.

Let us consider the case of a repulsive force $F(r) = K/r^2$ and $V(r) = K/r$, where K is positive for repulsive potential. Hence $V(r)$ is positive and decreases monotonically with increasing r. The resultant effective potential will always be positive; hence there will be no bounded motion. These points are illustrated in Fig. 7.13. For a given value of L, Fig. 7.13(a) shows the plots of $V_{\text{cent}}(r)$ ($= L^2/2mr^2$), $V^+(r)$, which is a repulsive potential, and $V^-(r)$, which is of the same magnitude as $V^+(r)$ except that it is an attractive potential. Figure 7.13(b) shows the plots of

$$V_{\text{eff}}^+ = V^+(r) + \frac{L^2}{2mr^2}, \qquad \text{repulsive force} \qquad (7.76)$$

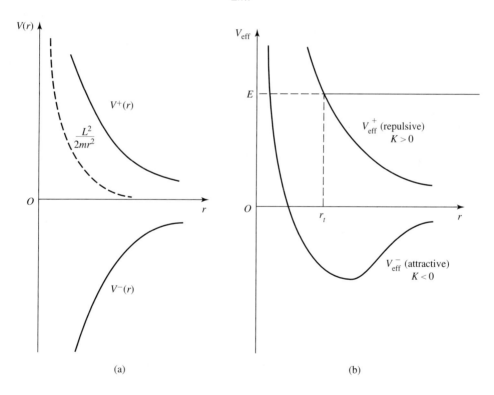

(a)   (b)

Figure 7.13   (a) Plots of attractive potential $V^-(r)$, repulsive potential $V^+(r)$, and V_{cent} ($= L^2/2mr^2$) versus r. (b) Plots of attractive effective potential V_{eff}^- and repulsive effective V_{eff}^+ versus r.

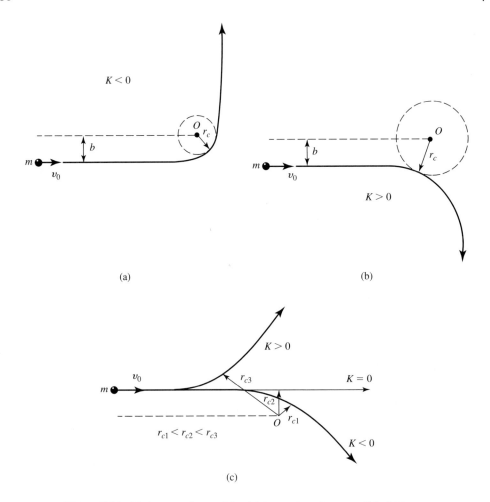

Figure 7.14 Trajectory of a particle: (a) in an attractive potential, (b) in a repulsive potential, and (c) in attractive, repulsive, and no potential. Note that r_{c1} (attractive) $< r_{c2}$ (no potential) $< r_{c3}$ (repulsive).

and

$$V^-_{\text{eff}} = V^-(r) + \frac{L^2}{2mr^2}, \qquad \text{attractive force} \qquad (7.77)$$

It is quite clear that for any energy E (>0) of a particle there is no bound motion, and there is only one turning point at r_t, as shown. If a particle is under the influence of such potentials, V^+_{eff} or V^-_{eff}, the trajectory of the particle will be as shown in Fig. 7.14 and as explained next.

Suppose a particle with energy E is at infinity and is traveling toward the center of force O. At such large distances, both $V(r)$ and $L^2/2mr^2$ are zero; hence the particle travels in a straight line with a speed $v_0 = (2E/m)^{1/2}$. The particle misses the center of force by a distance b, called the *impact parameter*, as shown in Fig. 7.14. Thus the angular momentum L of the particle is

$$L = mv_0 b \qquad (7.78)$$

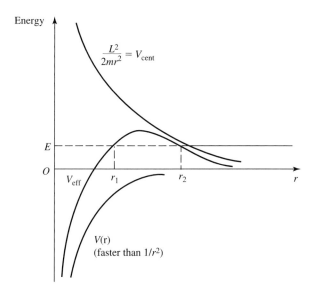

Energy

$$\frac{L^2}{2mr^2} = V_{\text{cent}}$$

V_{eff}

$V(r)$
(faster than $1/r^2$)

Figure 7.15   Plots of $V(r)$, V_{cent} ($=L^2/2mr^2$) and V_{eff} versus r for potentials that fall off faster than $1/r^2$.

The value of v_0 defines E, and the value of b further defines L. Thus, for such given values of E and L (or hence v_0 and b), Fig. 7.14(a) shows the trajectory of a particle in an attractive potential and Fig. 7.14(b) shows a repulsive potential. It is clear that the distance of closest approach r_c (from the force center) or turning point is much smaller for an attractive potential than for a repulsive potential. Figure 7.14(c) shows the distances of the closest approach for the attractive and repulsive potentials, as well as no potential at all ($K = 0$), with the result

$$r_{c1}(\text{attractive}) < r_{c2}(\text{no potential}) < r_{c3}(\text{repulsive})$$

We shall be concerned with these trajectories when we discuss scattering or collision problems.

Finally, let us discuss potentials that fall off more rapidly than $1/r^2$ with increasing r. The resulting effective potential is shown in Fig. 7.15. For a given value of E, a particle has two possible motions: a bound motion if it is between $r = 0$ and $r = r_1$, and an unbound motion if $r > r_2$. The region between r_1 and r_2 is forbidden. This potential is similar to one between an incident proton and an atomic nucleus. According to an energy diagram interpretation, the proton may be trapped between $r = 0$ and $r = r_1$, or it may be free for $r > r_2$, r_2 being the turning point. (It is shown quantum mechanically that the proton has some probability of penetrating the forbidden region $r_1 - r_2$.)

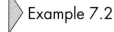 Example 7.2 _____

According to Yukawa's theory of nuclear forces, the attractive force between a neutron and a proton inside the nucleus is represented by a potential function of the form

$$V(r) = \frac{ke^{-ar}}{r} \qquad \text{(i)}$$

where k and a are constants and $k < 0$.

(a) Find the force F(r) and graph it. (b) Discuss the motion of a particle of mass m moving under the influence of such a force by making a graph of Vef. (c) Calculate the time period T, the energy E, and the angular momentum L if the particle moves in a circle of radius r0. (d) Calculate the time period of the small oscillations, that is, for the lightly perturbed circular motion.

Solution

(a) and (b) Given the expression for the potential function, we can find the expression for the force between the two particles as shown.

$$V(r) = \frac{k \cdot e^{-a \cdot r}}{r} \quad (i) \qquad F = \frac{d}{dr} V(r)$$

$$F = -\left(\frac{d}{dr} \frac{k \cdot e^{-a \cdot r}}{r} \right) \quad (ii) \qquad F = k \cdot \exp(-a \cdot r) \cdot \frac{(a \cdot r + 1)}{r^2}$$

From V and F, we can calculate the effective potential Vef = V + Vcent as shown and then make plots of F and Vef. Maximum and minimum values of F and Vef are as shown below.

$$I := 200 \qquad i := 1 .. I \qquad r_i := \frac{i}{200}$$

$$k := -.95 \qquad M := .001 \qquad a := 0.5 \qquad L := .005$$

$$F_i := k \cdot e^{-a \cdot r_i} \cdot \frac{\left(a \cdot r_i + 1\right)}{\left(r_i\right)^2} \qquad\qquad Vef_i := \frac{k \cdot e^{-a \cdot r_i}}{r_i} + \frac{L^2}{2 \cdot M \cdot \left(r_i\right)^2} \quad (iii)$$

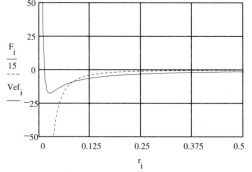

$\min(Vef) = -17.528 \qquad \min(F) = -3.8 \cdot 10^4$

$\max(Vef) = 310.474 \qquad \max(F) = 0$

$Vef_{200} = -0.564 \qquad F_{200} = -0.864$

$r0 := .042$

(c) For the particle to move in a circle, the applied force F [calculated in (a)] must be equal to the centripetal force. By equating the two, we can calculate the velocity in the circular motion as shown.

$$Fc = \frac{-M \cdot v0^2}{r0}$$

$$F(r0) = Fc \qquad k \cdot \exp(-a \cdot r0) \cdot \frac{(a \cdot r0 + 1)}{r0^2} = -M \cdot \frac{v0^2}{r0}$$

$$v0^2 = -k \cdot \exp(-a \cdot r0) \cdot \frac{(a \cdot r0 + 1)}{(M \cdot r0)} \qquad\qquad (iv)$$

Using the velocity, we can calculate the time period.

$$T = \frac{2 \cdot \pi \cdot ro}{v0} \qquad T = 2 \cdot \pi \cdot \frac{ro}{\left[\sqrt{k} \cdot \left(\sqrt{\exp(-a \cdot r0)} \cdot \sqrt{-a \cdot r0 - 1} \right) \right]} \cdot \sqrt{M} \cdot \sqrt{r0}$$

From V0 and V(r), we can
calculate the total energy and
the angular momentum.
E = kinetic energy + potential energy
L = angular momentum

$$E = \frac{1}{2} \cdot M \cdot v^2 + Vr \qquad E = \frac{1}{2} \cdot M \cdot v^2 + k \cdot \frac{\exp(-a \cdot r)}{r}$$

$$E = \frac{1}{2} \cdot M \cdot \left[-k \cdot \exp(-a \cdot r0) \cdot \frac{(a \cdot r0 + 1)}{(M \cdot r0)} \right] + k \cdot \frac{e^{-a \cdot r0}}{r0}$$

or

$$E = \frac{-1}{2} \cdot k \cdot \exp(-a \cdot r0) \cdot \frac{(a \cdot r0 - 1)}{r0} + k \cdot \frac{e^{-a \cdot r0}}{r0} - \quad (v)$$

$$L = M \cdot v0 \cdot r0 \qquad L = \sqrt{M} \cdot \sqrt{r0} \cdot \sqrt{k} \cdot \sqrt{ex'p(-a \cdot r0)} \cdot \sqrt{-a \cdot r0 - 1} \qquad (vi)$$

(d) If the motion is not circular,
we must calculate the time
period by using keff as shown. $Tr = 2 \cdot \pi \cdot \sqrt{\dfrac{M}{keff}}$ $keff = \dfrac{d^2}{d\,r^2} \dfrac{k \cdot e^{-a \cdot r}}{r} + \dfrac{L^2}{2 \cdot M \cdot (r)^2}$ simplifies to

$$keff = \frac{\left(k \cdot a^2 \cdot \exp(-a \cdot r) \cdot M \cdot r^3 + 2 \cdot k \cdot a \cdot \exp(-a \cdot r) \cdot M \cdot r^2 + 2 \cdot k \cdot \exp(-a \cdot r) \cdot M \cdot r + 3 \cdot L^2 \right)}{\left(M \cdot r^4 \right)}$$

$$(vii)$$

$$Tr = 2 \cdot \pi \cdot \sqrt{\frac{M}{keff}} \qquad \text{by substitution, yields}$$

$$Tr = 2 \cdot \pi \cdot \frac{M}{\sqrt{k \cdot a^2 \cdot \exp(-a \cdot r) \cdot M \cdot r^3 + 2 \cdot k \cdot a \cdot \exp(-a \cdot r) \cdot M \cdot r^2 + 2 \cdot k \cdot \exp(-a \cdot r) \cdot M \cdot r + 3 \cdot L^2}} \cdot r^2$$

$$(viii)$$

How do you explain the difference in the expressions for T and Tr?

EXERCISE 7.2 Repeat the example for the case of a potential of the form $V(r) = k \cdot \dfrac{e^{-\alpha \cdot r}}{r^2}$

7.7 ORBITS IN AN INVERSE-SQUARE FORCE FIELD

The force acting on a particle moving in space under an inverse-square force may be written as

$$\mathbf{F}(r) = \frac{K}{r^2}\hat{\mathbf{r}} \quad \text{or} \quad F(r) = \frac{K}{r^2} \tag{7.79}$$

while its potential energy is given by

$$V(r) = -\int_{r_s}^{r} F(r)\,dr = -\int_{r_s}^{r} \frac{K}{r^2}\,dr$$

Assuming $r_s = \infty$ and $V(\infty) = 0$,

$$V(r) = \frac{K}{r} \tag{7.80}$$

where $K < 0$ for an attractive force and $K > 0$ for a repulsive force. Two important cases of an inverse-square force are: (1) gravitational force, which is always attractive, and the quantity K and the constant G, which are

$$K = -Gm_1m_2 \quad \text{and} \quad G = 6.67 \times 10^{-11} \text{ N-m}^2/\text{kg}^2 \tag{7.81}$$

and (2) the coulomb force, for which the quantity K and the constant ϵ_0 are

$$K = \frac{1}{4\pi\epsilon_0}q_1q_2 \tag{7.82}$$

where $$\epsilon_0 = 8.85 \times 10^{-12} \text{ C}^2/\text{N-m}^2$$

ϵ_0 is the permittivity of free space. If q_1 and q_2 have the same signs, the force is repulsive and $K > 0$ (positive), while if q_1 and q_2 have opposite signs, the force is attractive and $K < 0$ (negative).

As before, without actually solving the equations of motion, we can determine the nature of the orbits by discussing the effective potential, which for the inverse-square force is [Eqs. (7.70) and (7.80)]

$$V_{\text{eff}}(r) = \frac{K}{r} + \frac{L^2}{2mr^2} \tag{7.83}$$

The plots of $V_{\text{eff}}(r)$ versus r for four situations ($K < 0, L = 0$; $K < 0, L \neq 0$; $K = 0, L \neq 0$; $K > 0$, $L \neq 0$) are shown in Fig. 7.16 and Fig. 7.17. As discussed earlier for $K > 0$, E is always positive and there is no periodic motion in r. The same is the case for $K = 0$, except that the turning point for a given value of E and L occurs at a smaller value of r than for $K > 0$. The motion in both cases is unbound. For an attractive force ($K < 0$), both bound ($L \neq 0$) and unbound ($L = 0$) motions are possible. The latter corresponds to the one-dimensional motion of a falling body.

For $K < 0$ and $L \neq 0$, then motion of the particle is unbound if $E > 0$, and the turning point occurs at a distance smaller than for the case $K = 0$. The motion is bound and periodic if

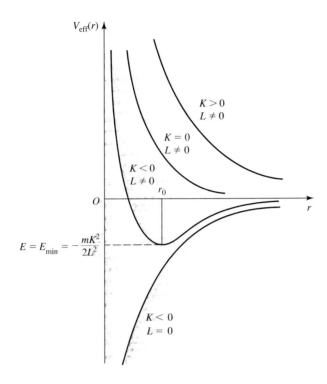

Figure 7.16   Plots of $V_{\text{eff}}(r)$ versus r for different values of K and L. For $E = E_{\min} = -mK^2/2L^2$, the particle moves in a circle of radius r_0.

$E < 0$. The minimum in the effective potential energy curve is given by the condition that at the equilibrium point $dV_{\text{eff}}/dr = 0$:

$$\frac{d}{dr} V_{\text{eff}}(r) = \frac{d}{dt}\left(\frac{K}{r} + \frac{L^2}{2mr^2}\right)\bigg|_{r=r_0} = -\frac{K}{r_0^2} - \frac{L^2}{mr_0^3} = 0$$

That is,

$$r_0 = -\frac{L^2}{mK} \tag{7.84}$$

while the value of V_{eff} at $r = r_0$ is obtained by substituting for r_0 from Eq. (7.84) into Eq. (7.83); that is,

$$V_{\text{eff}}(r_0) = \frac{K}{r_0} + \frac{L^2}{2mr_0^2} = \frac{K}{-L^2/mK} + \frac{L^2}{2m(-L^2/mK)^2}$$

or

$$V_{\text{eff}}(r_0) = -\frac{1}{2}\frac{mK^2}{L^2} \tag{7.85}$$

as shown in Figs. 7.16 and 7.17. Thus for $E = E_0 = E_{\min} = V_{\text{eff}}(r_0)$, as given by Eq. (7.85) a particle moves in a circle of radius $r_0 (= -L^2/mK)$ given by Eq. (7.84). But if the energy is less than 0 but greater than $-mK^2/2L^2$, that is, $-mK^2/2L^2 < E < 0$, the coordinate r oscillates between two turning points, as shown in Fig. 7.18. For all negative values of E and $L \neq 0$, the orbit of the particle is an ellipse. If the value of E is very close to the minimum value of $E_{\min}$, the period

Figure 7.17

Below is an inverse square force field. VR is the graph of an inverse square force law proportional to 1/r with K = KR =–60. VE is the effective potential, which is the sum of VR and L =50.

The bound motion is possible only for potential $VE_{i,3}$ because it has two turning points.

$L := 50$ $m := 10$ $KR := -60$

$j := 1..3$ $N := 30$ $i := 1..N$ $r_i := i$

$K_j :=$

50
0
-60

$$VE_{i,j} := \frac{K_j}{r_i} + \frac{L^2}{2 \cdot m \cdot (r_i)^2} \qquad VR_i := \frac{KR}{r_i}$$

$VE_{5,1} = 15$ $VE_{15,1} = 3.889$

$VE_{5,2} = 5$ $VE_{15,2} = 0.556$

$VE_{5,3} = -7$ $VE_{15,3} = -3.444$

$VR_5 = -12$ $VR_{15} = -4$

$\max(VE) = 175$ $\min(VE) = -7.188$

$\max(VR) = 0$ $\min(VR) = -60$

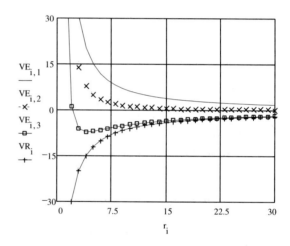

For what values of VE, will the motion of an object almost be a simple harmonic motion?

Which of the potentials lead to bound motions and which to unbound motions?

◁

of small oscillations in *r* is the same as the period of revolution. The orbit is a closed curve (circle) with the origin slightly off center. In the remainder of this section we will continue our discussion of an attractive force.

Now we proceed with an analytical treatment of the attractive inverse square force problem. Starting with Eq. (7.67),

$$\frac{d^2u}{d\theta^2} + u = -\frac{m}{L^2 u^2} F\left(\frac{1}{u}\right) \qquad (7.67)$$

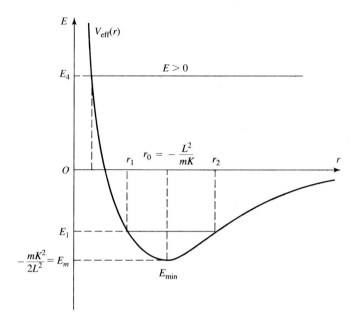

Figure 7.18   For $-mK^2/2L^2 < E < 0$, the coordinate r oscillates between two turning points as shown. For all negative values of E and $L \neq 0$, the orbit of the particle is an ellipse. For E close to E_{min}, the orbit is a closed curve (circle) with the origin slightly off center. r_1 and r_2 are the turning points of a particle with energy E_1. For $E > 0$, the orbit is unbound.

where

$$F\left(\frac{1}{u}\right) = F(r) = \frac{K}{r^2} = Ku^2 \qquad (7.86)$$

when substituted in Eq. (7.67) yields

$$\frac{d^2u}{d\theta^2} + u = -\frac{mK}{L^2} \qquad (7.87)$$

This is a second-order inhomogeneous differential equation similar to the one for a forced harmonic oscillator (see Chapter 3) except that θ plays the role of t. The homogeneous part of the equation,

$$\frac{d^2u}{d\theta^2} + u = 0 \qquad (7.88)$$

which is similar to the equation of a simple harmonic oscillator with $\omega = 1$, has a solution

$$u_h = A \cos(\theta - \phi) \qquad (7.89)$$

where A and ϕ are arbitrary constants. The particular solution of the inhomogeneous Eq. (7.87) is a constant given by

$$u_p = -\frac{mK}{L^2} \tag{7.90}$$

Thus the general solution of Eq. (7.87) is

$$u = u_p + u_h$$

or

$$u = \frac{1}{r} = -\frac{mK}{L^2} + A\cos(\theta - \phi) \tag{7.91}$$

This is an equation of a conic section (ellipse, parabola, or hyperbola) with its focus at $r = 0$. The constant ϕ determines the orientation of the orbit in the plane and can be taken to be zero. The constant A is positive and determines the turning points in the r motion, as discussed next.

Let r_1 and r_2 be minimum and maximum distances of the turning points corresponding to the minimum and maximum values of r in Eq. (7.91). These points can be found by substituting the maximum and minimum values of $A\cos(\theta - \phi)$, which are $+A$ and $-A$, respectively. Thus the turning points (shown in Fig. 7.18) are

$$\frac{1}{r_1} = -\frac{mK}{L^2} + A \quad\text{and}\quad \frac{1}{r_2} = -\frac{mK}{L^2} - A \tag{7.92}$$

A cannot be less than mK/L^2 because that would give a negative value of r, and if A is greater than $-mK/L^2$, then there is only one turning point (the same as for $K > 0$), as shown in Figs. 7.16 and 7.18.

Another way of finding the turning points is to solve the following equation for the case where a particle has energy E:

$$V_{\text{eff}}(r) = \frac{K}{r} + \frac{L^2}{2mr^2} = E \tag{7.93}$$

This is a quadratic in $1/r$, and the two roots are

$$\frac{1}{r_1} = -\frac{mK}{L^2} + \left[\left(\frac{mK}{L^2}\right)^2 + \frac{2mE}{L^2}\right]^{1/2} \tag{7.94}$$

and

$$\frac{1}{r_2} = -\frac{mK}{L^2} - \left[\left(\frac{mK}{L^2}\right)^2 + \frac{2mE}{L^2}\right]^{1/2} \tag{7.95}$$

Comparing Eqs. (7.92) with (7.94) and (7.95), we find the value of A in terms of E and L to be

$$A = \left(\frac{m^2K^2}{L^4} + \frac{2mE}{L^2}\right)^{1/2} \tag{7.96}$$

which is used in Eq. (7.91) to describe the orbit of the particle.

The condition that determines the nature of the orbit can be found by comparing Eq. (7.91) with the standard equation of a conic section. From plane geometry, the general equation of a conic section is

$$r = r_0 \frac{1 + e}{1 + e \cos \theta} \tag{7.97}$$

where e is called the *eccentricity* of the orbit and r_0 is the radius of the circular orbit corresponding to the given values of L, K, and m (or for $e = 0$, $r = r_0$). We now rearrange Eq. (7.91) [with $\phi = 0$ and A given by Eq. (7.96)] to take the form

$$r = -\frac{L^2}{mK} \frac{1}{1 + [-AL^2/mK] \cos \theta} \tag{7.98}$$

Comparing Eqs. (7.97) and (7.98), we obtain

$$e = -\frac{AL^2}{mK} \tag{7.99}$$

and

$$r_0 = -\frac{L^2}{mK} \frac{1}{1 + e} \tag{7.100}$$

The minimum value of r is obtained by letting $\theta = 0$ and $\cos \theta = +1$ in Eqs. (7.97) and (7.98); that is,

$$r_{\min} = r_2 = r_0 \frac{1 + e}{1 + e} = r_0 = -\frac{L^2}{mK} \frac{1}{1 + e} \tag{7.101}$$

while the maximum value of r is obtained from Eqs. (7.97) and (7.98) by substituting $\theta = \pi$ and $\cos \theta = -1$ and using the value of r_0 from Eq. (7.100); that is,

$$r_{\max} = r_1 = r_0 \frac{1 + e}{1 - e} = -\frac{L^2}{mK} \frac{1}{1 - e} \tag{7.102}$$

We can solve Eq. (7.100) to obtain e; that is,

$$e = -\frac{L^2}{mKr_0} - 1 \tag{7.103}$$

Combining Eqs. (7.99) and (7.96), we get

$$e = \sqrt{1 + \frac{2EL^2}{mK^2}} \tag{7.104}$$

Thus we have obtained the values of r_1 (= maximum radius), r_2 (= r_0 = minimum radius), A, and e given by Eqs. (7.102), (7.101), (7.96), and (7.104), respectively. As shown in Fig. 7.19,

Figure 7.19 _____

The value of the eccentricity determines the shape of the orbit.

e = 0 is a circle

e = 0.5 (between 0 and 1)
is an ellipse

e = 1 is a parabola

e = 1.5 is a hyperbola.

(0.002 is added to avoid singularity)

$N := 28$ $n := 0 .. N$ $\theta_n := \dfrac{2 \cdot \pi}{N} \cdot n$ $m := 1 .. 4$ $r0 := 1.5$

$e_m :=$

0
.5
1
1.5

$r_{n,m} := r0 \cdot \left[\dfrac{1 + e_m}{.002 + 1 + e_m \cdot \cos\left[(\pi - \theta)_n \right]} \right]$

(a) What determines that the ellipse will be inside the circle or outside the circle?
(b) Why does the hyperbola have two branches?
(c) Under what conditions will the orbit reduce to a straight line?
(d) Besides e, what other factors determine the shape of the ellipse?
(e) How does the phase angle determine the shape or orientation of an orbit?

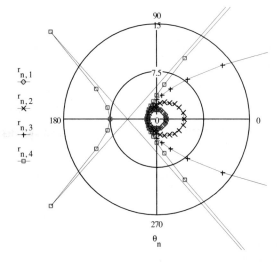

the value of e determines the shape of the orbit. From Fig. 7.18, for $E_m = V_0$, the different orbits are

$V_0 < E < 0,$	$0 < e < 1,$	ellipse
$E = V_0,$	$e = 0,$	circle (special case of ellipse)
$E = 0,$	$e = 1,$	parabola
$E > 0,$	$e > 1,$	hyperbola
$E < V_0,$	$e < 0,$	not allowed

These orbits are shown in Figs. 7.19 and 7.21.

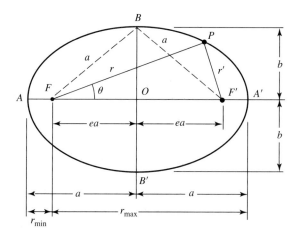

Figure 7.20 Definitions of different quantities in elliptical orbits.

Case (a)   Ellipse: An ellipse is a curve traced by a point so that the sum of its distances from two fixed points F and F' (called *foci*) is constant; that is (Figs. 7.20 and Fig. 7.21),

$$r + r' = 2a$$

Using Fig. 7.20, one obtains the following equation of an ellipse in polar coordinates with the origin at one focus:

$$r = \frac{a(1 - e^2)}{1 - e \cos \theta} \tag{7.105}$$

where a, the semimajor axis, is related to the minimum radius r_0 by the relation

$$r_0 = a(1 - e) \tag{7.106}$$

while the semiminor axis b is given by

$$b = a(1 - e^2)^{1/2} \tag{7.107}$$

where $e < 1$. We can further calculate the length of the major axis in terms of the energy of a particle. According to Eqs. (7.100), (7.101), and (7.102),

$$2a = r_{\min} + r_{\max} = -\frac{2L^2}{mK}\frac{1}{1 - e^2} \tag{7.108a}$$

Substituting for e from Eq. (7.104), we obtain

$$2a = +\frac{K}{E} \tag{7.108b}$$

which states that the length of the major axis is independent of L and that all orbits with the same major axis have the same energy, or vice versa. Furthermore,

$$\frac{r_{\max}}{r_{\min}} = \frac{1 + e}{1 - e} \tag{7.109}$$

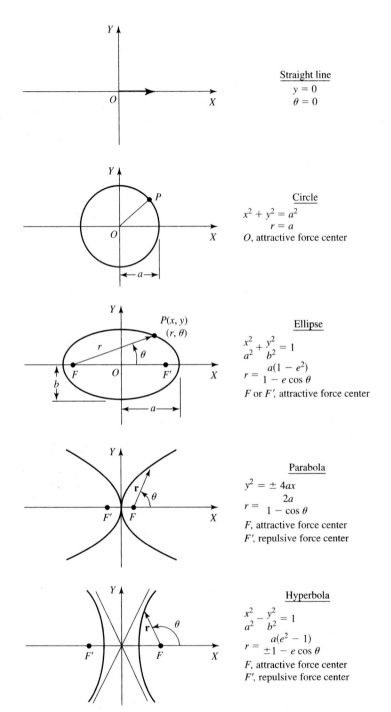

Figure 7.21 Shapes of different orbits and their equations in rectangular and plane polar coordinates.

That is, the shape of the ellipse depends on the value of e and not on r_0; r_0 is simply a scaling factor.

Case (b)   Circle: If $e = 0$, then $a = b$ and the two foci F and F' coincide, while Eq. (7.105) yields

$$r(\theta) = r = a \qquad \qquad \textbf{(7.110)}$$

which is an equation of a circle (see Fig. 7.21).

Case (c)   Hyperbola: A hyperbola is a curve traced by a point such that the difference of its distances from two fixed points F and F', called the *foci*, is constant. The hyperbola has two branches

$$r' - r = 2a, \qquad (+\text{branch to the left})$$
$$r' - r = -2a, \qquad (-\text{branch to the right})$$

For hyperbola $e > 1$, Eq. (7.97) takes the form

$$r = \frac{a(e^2 - 1)}{\pm 1 - e \cos \theta} \qquad \qquad \textbf{(7.111)}$$

These features of a hyperbola are shown in Fig. 7.21. The angle α that the asymptotes make with the axes is obtained by substituting $r = \infty$ in Eq. (7.111) and $\theta = \alpha$; that is,

$$\cos \alpha = \pm \frac{1}{e} \qquad \qquad \textbf{(7.112)}$$

Case (d)   Parabola: A parabola is a curve traced by a point so that its distance from a fixed point, called the *focus F*, is always equal to its distance from a fixed line, called the *directrix*. The resulting equation—which may be obtained from Eq. (7.97) by substituting $e = 1$ and assuming r is minimum when $\theta = \pi$—is

$$r = \frac{2a}{1 - \cos \theta} \qquad \qquad \textbf{(7.113)}$$

where a is the distance from the focus F to the directrix (see Fig. 7.21).

As an alternative, we can write the equation of a conic section in a more general form as

$$\frac{1}{r} = B + A \cos(\theta + \phi) \qquad \qquad \textbf{(7.114)}$$

Comparing the results obtained and keeping in mind that

$b > a$	for an ellipse	**(7.115a)**
$b = a$	for a parabola	**(7.115b)**
$0 < b < a$ or $-a < b < 0$,	for a hyperbola	**(7.115c)**

we can obtain the values of e, A, and B in terms of m, K, L, and E for different orbits (see Problem 7.31).

7.8 KEPLER'S LAWS OF PLANETARY MOTION

After years of analyzing the astronomical data taken by Tycho Brahe, early in the seventeenth century Johannes Kepler announced three laws that described the motion of planets around the Sun. The first two laws were published in 1609 and the third in 1619. These laws, called *Kepler's laws*, may be stated as follows:

    I. *Law of Orbits*: *Planets move in elliptical orbits with the sun at one focus.*

    II. *Law of Areas*: *A line (or position vector) joining any planet to the sun sweeps out equal areas in equal intervals of time.*

    III. *Law of Periods*: *The square of the period of revolution of any planet is proportional to the cube of the semimajor axis of the orbit. (The law of periods is also called harmonic law.)*

    Newton's law of universal gravitation was given shortly after Kepler's laws, and it provided the theoretical description of the motion of planets, consistent with experimental facts. There is a fundamental difference between Newton's laws of motion and Kepler's laws. Newton's laws are about motion and force, in general, and as such implicitly involve an interaction between objects, whereas Kepler's laws describe the motion of a planetary system and do not involve interaction. Newton's laws are *dynamic*, giving relations between force, mass, distance, and time, whereas Kepler's laws are *kinematic*, giving the relation between distance and time. Kepler's laws should apply not only to a solar system but to moons going around planets and to artificial satellites. In addition, Kepler's laws are valid whenever an inverse-square force law is involved.

    We have already seen that the first of these laws, the law of orbits, follows directly from Newton's law of gravitation, that is, from the inverse-square nature of the force of gravitation. The second law, the law of areas, results from the fact that the angular momentum remains constant. We have already shown this to be true (see Section 7.4), because gravitational force is a conservative force; hence angular momentum is conserved. In this section we prove the third law, the law of periods.

    Let T be the time period of an elliptical orbit. Then, according to Eq. (7.46),

$$\text{Area of an ellipse} = \frac{LT}{2\mu} = \pi ab = \pi a^2 \sqrt{1 - e^2} \qquad \textbf{(7.116)}$$

where a is the semimajor axis, b is the semiminor axis, and e is the eccentricity. But using the relation Eq. (7.108a) (note that we are using the reduced mass μ instead of m),

$$a = -\frac{L^2}{\mu k}\frac{1}{1 - e^2}$$

or

$$1 - e^2 = -\frac{L^2}{\mu K a}$$

which, on substituting in Eq. (7.116), and rearranging yields

$$\frac{T^2}{a^3} = \frac{4\pi^2\mu}{-K} = \text{constant} \qquad \textbf{(7.117)}$$

TABLE 7.1. Kepler's Third Law Applied to Planets and Satellites

	e	$a\ (\times 10^7\ \mathrm{km})$	$T\ (\mathrm{yr})$	a^3/T^2
Planets				
Mercury	0.206	5.79	0.24	3.39×10^{18}
Venus	0.007	10.82	0.62	3.31×10^{18}
Earth	0.017	14.96	1.00	3.36×10^{18}
Mars	0.093	22.79	1.88	3.37×10^{18}
Jupiter	0.048	77.83	11.86	3.37×10^{18}
Saturn	0.055	142.7	29.46	3.36×10^{18}
Uranus	0.047	286.9	84.01	3.36×10^{18}
Neptune	0.009	449.8	164.97	3.37×10^{18}
Pluto	0.249	590.0	248.4	3.35×10^{18}
Satellites				
Cosmos 382	0.260	18,117	143	2.91×10^8
ATS 2	0.455	24,123	219.7	2.91×10^8
Explorer 28	0.952	273,740	8.4×10^{13}	2.91×10^8

which is the statement of Kepler's third law. Substituting $K = -GMm$ and $\mu = Mm/(M + m)$, we get

$$T^2 = \frac{4\pi^2}{G(M + m)}\, a^3 \tag{7.118}$$

In all practical situations, m is very small as compared to mass M; hence it may be neglected. For example, even for the largest planet Jupiter, mass $m = (1/1000)M$, where M is the mass of the sun. Thus neglecting m as compared to M, Eq. (7.118) reduces to

$$T^2 \simeq \frac{4\pi^2}{GM}\, a^3 \tag{7.119}$$

The application of Kepler's third law as applied to some planets and some artificial earth satellites is shown in Table 7.1.

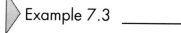

Example 7.3

A satellite of mass m = 2500 kg is going around Earth in an elliptic orbit. At the farthest point from Earth (apogee = da) the altitude is 3600 km, while at the nearest point (perigee = dp) the altiude is 1100 km. Calculate the energy E and angular momentum L of the satellite and its speed va and vp at da and dp, respectively.

The major axis 2a and the constant K = GMm may be calculated as shown.

$$Re := 6400 \cdot 10^3 \qquad da := 3600 \cdot 10^3 \qquad dp := 1100 \cdot 10^3$$

$$m := 2500 \qquad G := 6.673 \cdot 10^{-11} \cdot 10^{-3}$$

$$M := 5.98 \cdot 10^{27} \qquad g := 9.8$$

E = energy of the satellite in an elliptical orbit, from Eq. (7.108b), is E = −K/2a

$$a := \frac{2 \cdot Re + da + dp}{2} \qquad 2 \cdot a = 1.75 \cdot 10^7$$

EI = the energy of the satellite before launching.

EN = Energy needed to put the satellite in the orbit.

K =E 2a

$$K := m \cdot g \cdot Re^2 \qquad K = 1.004 \cdot 10^{18}$$

$$E := \frac{-K}{2 \cdot a} \qquad E = -5.734 \cdot 10^{10}$$

$$EI := -m \cdot g \cdot Re \qquad EI = -1.568 \cdot 10^{11}$$

$$EN := E - EI \qquad EN = 9.946 \cdot 10^{10}$$

$$K := E \cdot 2 \cdot a \qquad K = -1.004 \cdot 10^{18}$$

ec = the eccentricity, may be calculated from Eq. (7.109), da is maximum r and dp is minimum r.

$$ec := \frac{da - dp}{2 \cdot a} \qquad ec = 0.143$$

Using Eq. (7.104), we can calculate the value of L for the given value of ec. There are two roots; we use the positive one.

$$ec^2 = 1 + \frac{2 \cdot E \cdot L^2}{m \cdot K^2} \qquad \begin{bmatrix} \dfrac{-1}{(2 \cdot \sqrt{E})} \cdot \sqrt{m \cdot K} \cdot \sqrt{2} \cdot \sqrt{ec^2 - 1} \\[2ex] \dfrac{1}{(2 \cdot \sqrt{E})} \cdot \sqrt{m \cdot K} \cdot \sqrt{2} \cdot \sqrt{ec^2 - 1} \end{bmatrix}$$

$$L1 := \frac{-1}{(2 \cdot \sqrt{E})} \cdot \sqrt{m \cdot K} \cdot \sqrt{2} \cdot \sqrt{ec^2 - 1} \qquad L1 = 1.466 \cdot 10^{14}$$

$$ra := Re + da \qquad ra = 1 \cdot 10^7$$

ra = the distance at the apogee from the center of Earth

Using the energy relation, we solve for the velocity, va, at the apogee. Once again we get two roots for va. Using the positive root, calculate the velocity va.

$$E = \frac{1}{2} \cdot m \cdot va^2 + \frac{K}{ra} \qquad \begin{bmatrix} \dfrac{-1}{\sqrt{m}} \cdot \sqrt{2} \cdot \dfrac{\sqrt{E \cdot ra - K}}{\sqrt{ra}} \\[2ex] \dfrac{1}{\sqrt{m}} \cdot \sqrt{2} \cdot \dfrac{\sqrt{E \cdot ra - K}}{\sqrt{ra}} \end{bmatrix}$$

$$va := \frac{1}{\sqrt{m}} \cdot \sqrt{2} \cdot \frac{\sqrt{E \cdot ra - K}}{\sqrt{ra}} \qquad\qquad va = 5.866 \cdot 10^3$$

$$Ra := Re + da \qquad\qquad\qquad Rb := Re + dp$$

$$Ra = 1 \cdot 10^7 \qquad\qquad\qquad\qquad Rb = 7.5 \cdot 10^6$$

Using Eq. (7.48), we can
calculate the velocity vp.

$$vp := \frac{Ra}{Rb} \cdot va \qquad\qquad\qquad vp = 7.821 \cdot 10^3$$

EXERCISE 7.3 Repeat the example for a satellite ATS for which $e = 0.455$ and $a = 24,123$ km.

7.9 PERTURBED CIRCULAR ORBITS: RADIAL OSCILLATIONS ABOUT A CIRCULAR ORBIT

We shall divide the discussion in this section into two parts. First, we shall discuss the special case involving only an inverse-square attractive force and, second, a more general case involving a general force law. Consider a satellite of mass m going around Earth in a circular orbit of radius r_0 having energy E_0. Suppose the energy of the satellite increases from E_0 to $E_0 + \Delta E$, where ΔE is a small increment in its energy, while its angular momentum remains the same. This is achieved by firing its engines for a short interval of time radially toward the center of Earth. This situation is shown in Fig. 7.22, where $V_{eff}(r)$ is plotted versus r. For energy E_0, corresponding to the equilibrium point r_0 in the potential curve, the orbit is a circular one of radius r_0 and is described by $r(t) = r_0 =$ constant. We want to find a new orbit corresponding to the energy $E_0 + \Delta E$ as shown. For a small ΔE we use the following procedure, which results in a slight change in the circular orbit. For $E_0 + \Delta E$, the object has two motions: (1) an angular motion corresponding to a circular motion of radius r_0 and energy E_0, and (2) superimposed on it, a radial motion between r_{max} and r_{min}. For small deviations from E_0, the $V_{eff}(r)$ in the vicinity of r_0 can be approximated by a parabolic potential. Thus small displacements about an equilibrium position resulting from radial motion will be simple harmonic about the equilibrium position r_0. We investigate the two motions, radial simple harmonic and angular, as follows.

The effective potential in this case as given by Eq. (7.83) is

$$V_{eff}(r) = \frac{K}{r} + \frac{L^2}{2\mu r^2} \qquad\qquad\qquad\qquad \textbf{(7.83)}$$

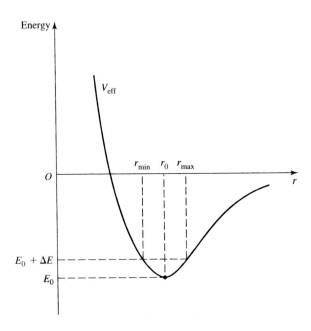

Figure 7.22 For a particle of energy E_0, the orbit is a circular one. A slight increase in energy to $E_0 + \Delta E$ leads to a superposition of radial simple harmonic motion on the circular orbit (angular motion).

where $K = -GMm$ and $\mu = Mm/(M + m)$. Since the slope is zero at the minimum of the potential,

$$\left. \frac{dV_{\text{eff}}}{dr} \right|_{r=r_0} = 0 = -\frac{K}{r_0^2} - \frac{L^2}{\mu r_0^3} \qquad (7.120)$$

gives

$$L = \sqrt{-\mu K r_0} \qquad (7.121)$$

The effective spring constant k_{eff} for simple harmonic motion may be calculated by taking the second derivative of V_{eff}; that is,

$$k_{\text{eff}} = \left. \frac{d^2 V_{\text{eff}}}{dr^2} \right|_{r=r_0} \qquad (7.122a)$$

gives

$$k_{\text{eff}} = \frac{2K}{r_0^3} + \frac{3L^2}{\mu r_0^4} \qquad (7.122b)$$

Using the value of L^2 from Eq. (7.121), we obtain

$$k_{\text{eff}} = -\frac{K}{r_0^3} \qquad (7.123)$$

while the frequency of the radial motion is given by

$$\omega_r = \sqrt{\frac{k_{\text{eff}}}{\mu}} = \sqrt{\frac{-K}{\mu r_0^3}} \tag{7.124a}$$

Substituting for $-K$ from Eq. (7.121) in Eq. (7.124a) gives

$$\omega_r = \sqrt{\frac{-K}{\mu r_0^3}} = \frac{L}{\mu r_0^2} \tag{7.124b}$$

Thus the radial position of an object with the condition $r = r_0$ at $t = 0$ is

$$r = r_0 + A \sin \omega_r t \tag{7.125}$$

which gives $r = r(t)$. But, to describe the orbit, we must know $r = r(\theta)$. This can be done by considering angular motion, which according to Eq. (7.40), for $r = r_0$, is

$$\omega_\theta = \dot{\theta} = \frac{L}{\mu r_0^2} \tag{7.126}$$

Hence

$$\theta = \omega_\theta t = \frac{L}{\mu r_0^2} t \tag{7.127}$$

Comparing Eqs. (7.126) and (7.124b), we find the following surprising result. For small oscillations or for a small perturbation of the circular orbit, the frequency of radial motion is equal to that of rotation; that is,

$$\omega_r = \omega_\theta \tag{7.128}$$

Hence Eq. (7.125), with the help of Eq. (7.127), takes the form

$$r = r_0 + A \sin \theta \tag{7.129}$$

The new orbit is shown by the dashed circle in Fig. 7.23. The orbit is almost a circle except that its center does not coincide with the center of Earth, but is slightly displaced. This same result can be achieved from the general equation of the orbit for $A \ll r_0$ (see Problem 7.42).

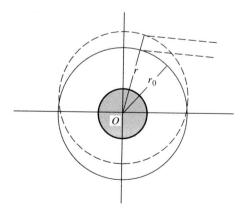

Figure 7.23 Perturbed orbit of a body going around Earth is still a circular one except that its center does not coincide with the center of Earth.

We now generalize the preceding situation to a more general case for which the form of the force law is given by

$$F(r) = Kr^n \tag{7.130}$$

Let us start with the expression for the effective potential:

$$V_{eff} = V(r) + \frac{L^2}{2\mu r^2} \tag{7.131}$$

$$\frac{dV_{eff}}{dr} = \frac{dV}{dr} - \frac{L^2}{\mu r^3} \tag{7.132}$$

or

$$\frac{dV_{eff}}{dr} = -F(r) - \frac{L^2}{\mu r^3} \tag{7.133}$$

Since at the equilibrium $dV_{eff}/dr|_{r=r_0} = 0$, we get, from Eq. (7.133),

$$F(r_0) = -\frac{L^2}{\mu r_0^3} \tag{7.134}$$

Let us now evaluate the spring constant for small radial oscillations. From Eq. (7.133),

$$\frac{d^2V_{eff}}{dr^2} = -\frac{dF}{dr} + \frac{3L^2}{\mu r^4} \tag{7.135}$$

$$k_{eff} = \frac{d^2V_{eff}}{dr^2}\bigg|_{r=r_0} = -F'(r_0) + \frac{3L^2}{\mu r_0^4} = -F'(r_0) - \frac{3F(r_0)}{r_0} \tag{7.136}$$

where the prime indicates derivative with respect to r. Thus

$$\omega_r = \sqrt{\frac{k_{eff}}{\mu}} = \sqrt{\frac{-F'(r_0) - (3/r_0)F(r_0)}{\mu}} \tag{7.137}$$

and the period of radial oscillations is

$$T_r = \frac{2\pi}{\omega_r} = 2\pi\sqrt{\frac{\mu}{-F'(r_0) - (3/r_0)F(r_0)}} \tag{7.138}$$

For angular motion, as before,

$$\dot{\theta} = \frac{L}{\mu r_0^2}$$

Using Eq. (7.134) to eliminate L, we get

$$\omega_\theta = \dot{\theta} = \sqrt{\frac{-F(r_0)}{\mu r_0}} \tag{7.139}$$

As r goes from r_{max} to r_{min}, the angle through which r moves is called the *apsidal angle* Ψ and is given by

$$\Psi = \omega_\theta \left(\tfrac{1}{2} T_r\right) \tag{7.140}$$

Substituting for ω_θ and T_r, we get

$$\Psi = \frac{\pi}{\sqrt{3 + [F'(r_0)/F(r_0)]r_0}} \tag{7.141}$$

This equation, combined with Eq. (7.136), results in some very important results, as discussed next for the general case $F(r) = Kr^n$.

For stable oscillations to occur, $k_{\text{eff}} > 0$. For $k_{\text{eff}} < 0$, there is no restoring force. Any perturbation will increase the radius $r = r_0$ of the circular orbit; hence it will be unstable. Thus, for stable orbits, from Eq. (7.136),

$$k_{\text{eff}} > 0 \quad \text{or} \quad -F'(r_0) - \frac{3}{r_0} F(r_0) > 0 \tag{7.142}$$

Since $F(r_0) = Kr_0^n$, $F'(r_0) = nKr_0^{n-1}$; substituting gives

$$-nKr_0^{n-1} - \frac{3}{r_0} Kr_0^n > 0$$
$$-n - 3 > 0$$

or

$$n > -3 \tag{7.143}$$

Also substituting for $F(r_0)$ and $F'(r_0)$ in Eq. (7.141) yields

$$\Psi = \frac{\pi}{\sqrt{3 + n}} \tag{7.144}$$

For any closed orbit (one that repeats itself) the denominator $\sqrt{3 + n}$ must be an integer. This, together with the condition that $n > -3$, yields $n = -2, 1, 6, 13$, and so on. These being the possible values for n, it is clear that $n = -2$ corresponds to an inverse-square force, while $n = 1$ corresponds to a restoring force in simple harmonic motion. For these values of n, the orbits are stable and limited between r_{min} and r_{max}.

7.10 ORBITAL TRANSFERS: GRAVITATIONAL BOOST AND BREAKING

A few decades ago, sending spacecraft from Earth to other planets was unimaginable, practically speaking. But technology has advanced to the point where some science fiction stories have become reality. In this section, we investigate the method of sending space probes from one planet to another. It is not practical to aim the probe directly inward or outward along the radial line from the Sun. The most efficient way is to let the probe coast in an orbit (elliptical or circular) that joins the orbit of Earth and that of the other planet. To make the problem simple, we may assume that orbits of planets around the Sun are circular—not a bad approximation since most are nearly circular.

Suppose we want to send a space probe from Earth to another planet, say Mars. Let the orbits of these two planets be circle of radii r_E and r_M, with the Sun at their common center. The situation is as in Fig. 7.24, where the transfer orbit is shown dashed. This transfer orbit is tangent to Earth's orbit at E and also tangent to Mars's orbit at M. The length of the major axis of this orbit, as shown, is equal to the sum of the orbital radii of Earth and Mars; that is, $r_t = r_E +$

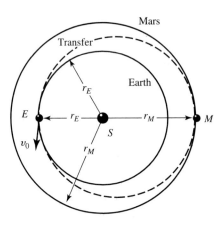

Figure 7.24 The transfer orbit from Earth to Mars (shown dashed) connects the circular orbits of Earth and Mars around the Sun at the center. The transfer orbit is tangent to the circular orbits of Earth and Mars at E and M, respectively.

r_M. Also, $2r_E < 2r_t < 2r_M$; that is, the major axis of the transfer orbit is in between those of the initial and final circular orbits. Since $E = K/2a$, the total energy of the transfer orbit is also in between the energy values of the initial and final circular orbits. Thus, to transfer a space probe from Earth's orbit to Mar's orbit, the probe should be given an acceleration at point E (changing both L and E), and once again an acceleration is given at point M (changing the values to L' and E') so that it can circle Mars. (Note that $E' > E$ and $L' > L$.) The situation would be reversed if we were sending a probe to Venus, as shown in Fig. 7.25. The energy in the final circular orbit around Venus will be less than that in Earth's circular orbit. This means that the probe on Earth will have to be given a retardation so that it can go into transfer orbit and another sudden retardation when arriving at Venus so that it can go into a circular orbit around Venus. We exaplain these steps quantitatively for the case of a space probe transfer from Earth to Mars.

For Earth going in a circular orbit of radius r_E and speed v_0 around the Sun of mass M_S,

$$\frac{mv_0^2}{r_E} = \frac{GM_s m}{r_E^2} \tag{7.145a}$$

Therefore,

$$v_0^2 = \frac{GM_S}{r_E} \tag{7.145b}$$

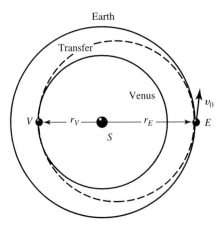

Figure 7.25 The transfer orbit from Earth to Venus connects the circular orbits of Earth and Venus around the Sun at the center. The transfer orbit is tangent to the circular orbits of Earth and Venus at E and V, respectively.

where v_0 is also given by the relation

$$v_0 = \frac{2\pi r_E}{T_E} \qquad (7.146)$$

and T_E is the time period of the orbital motion of Earth. At point E, the space probe is given a thrust so that it will have speed v_1 to be in the transfer orbit. Using Eq. (7.108b), according to which

$$E = \frac{K}{\text{major axis}} = \frac{-GM_S m}{2a} \qquad (7.147)$$

we may write (from Fig. 7.24)

$$E = -\frac{GM_S m}{r_E + r_M} = \frac{1}{2} m v_1^2 - \frac{GM_S m}{r_E} \qquad (7.148)$$

Therefore, using Eq. (7.145b), we get

$$v_1^2 = 2GM_S \frac{r_M}{r_E(r_M + r_E)} = v_0^2 \frac{2r_M}{r_M + r_E} \qquad (7.149)$$

Once the probe leaves Earth orbit and enters the transfer orbit, its speed v_1 is greater than v_0. But when the probe reaches the Mars orbit at M, its speed v_1 is less than the orbital speed of Mars, v_M. That is, the probe is moving slower than Mars when it arrives at the Martian orbit; hence Mars, approaching from behind, will overtake the probe. To an observer on Mars, the probe seems to be approaching from the opposite direction. To avoid this situation, when the probe arrives at point M, its speed is increased from v_1 to $v_2 = v_M$.

Once the space probe is in a transfer orbit, it will coast toward Mars, and it will take one-half of the time period of the transfer orbit. According to Kepler's third law, T^2 is proportional to the cube of the major axis, we may write

$$\frac{T_E^2}{(2r_E)^3} = \frac{T^2}{(r_E + r_M)^3}$$

That is,

$$T = \left(\frac{r_E + r_M}{2r_E}\right)^{3/2} T_E \qquad (7.150)$$

When the probe reaches Mars, it must increase its speed to v_2, given by the following energy conservation equation.

$$E = -\frac{GM_S m}{r_E + r_M} = \frac{1}{2} m v_2^2 - \frac{GM_S m}{r_M} \qquad (7.151)$$

Therefore,

$$v_2^2 = 2GM_S \frac{r_E}{r_M(r_M + r_E)} = v_0^2 \frac{2r_E^2}{r_M(r_M + r_E)} \qquad (7.152)$$

Knowing r_E, r_M, and T_E, we can calculate v_0, v_1, v_2, and T from Eqs. (7.146), (7.149), (7.152), and (7.150) as illustrated in the following example.

The situation is quite different if the probe has to be transferred from Earth to one of the inner planets, say Venus or Mercury. First, to start at point E, the speed is decreased from v_0 to v_1 to put it into a smaller transfer orbit. When it reaches close to point V, it seems to speed up and overtake the planet. To avoid this situation, the probe is slowed down from v_1 to v_2 ($v_1 = v_2$, the orbital speed of Venus). The changes in the speed are achieved by burning rocket fuel, as explained in Chapter 8.

Example 7.4 _____

A spacecraft is launched from Earth into an orbit around Venus. Calculate v0, v1, T, and v2, as defined previously.

Solution

rE = Earth's orbital radius

TE = time period of Earth

$$rE := 1.49 \cdot 10^{11} \cdot m \qquad TE := 3.16 \cdot 10^{7} \cdot sec$$

rV = Venus' orbital radius

$$rV := .72 \cdot rE$$

$$rV = 1.073 \cdot 10^{11} \cdot m$$

The speed v0 of the spacecraft around Earth is calculated by using Eq. (7.146) and the speed v1 of the spacecraft in the transfer orbit, is calculated using Eq. (7.149).

$$v0 := \frac{2 \cdot \pi \cdot rE}{TE} \qquad v0 = 2.963 \cdot 10^{4} \cdot m \cdot sec^{-1}$$

$$v1 := v0 \cdot \sqrt{\frac{2 \cdot rV}{rE + rV}} \qquad v1 = 2.711 \cdot 10^{4} \cdot m \cdot sec^{-1}$$

The decrease in the speed of the aircraft is

$$v0 - v1 = 2.518 \cdot 10^{3} \cdot m \cdot sec^{-1}$$

The time period of the transfer orbit is as shown.

$$T := \left(\frac{rE + rV}{2 \cdot rE} \right)^{\frac{3}{2}} \cdot TE \qquad T = 2.52 \cdot 10^{7} \cdot sec$$

The time it takes to reach the entering point of Venus' orbit is

$$\frac{1}{2} \cdot T = 1.26 \cdot 10^{7} \cdot sec \qquad \frac{\frac{1}{2} \cdot T}{24 \cdot 3600} = 145.845 \cdot sec$$

The speed of the craft when it reaches an orbit around Venus is v2 as calculated by using Eq. (7.152).

$$v2 := v0 \cdot rE \cdot \sqrt{\frac{2}{rV \cdot (rV + rE)}} \qquad v2 = 3.765 \cdot 10^{4} \cdot m \cdot sec^{-1}$$

The orbital speed around Venus is Vv. The decrease in the speed is ΔV.

$$TV := .62 \cdot TE \qquad TV = 1.959 \cdot 10^{7} \cdot sec$$

Thus when arriving at V, the spacecraft must reduce its speed by

$$vV := \frac{2 \cdot \pi \cdot rV}{TV}$$

$$vV = 3.44 \cdot 10^4 \ \cdot m \cdot sec^{-1}$$

$$\Delta V := v2 - vV$$

$$\Delta V = 3.245 \cdot 10^3 \ \cdot m \cdot sec^{-1}$$

EXERCISE 7.4 Repeat the example for a spacecraft launched from Earth into an orbit around Mars.

PROBLEMS

7.1. Calculate the components of force corresponding to the following potential energy functions:

(a) $V = kxyz$

(b) $V = \frac{1}{2}kr^2$

(c) $V = \frac{1}{2}(k_x x^2 + k_y y^2 + k_z z^2)$

7.2. In a hydrogen molecule, the potential energy of an electron is given by

$$V(r) = -\frac{e^2}{r_1} - \frac{e^2}{r_2}$$

where r_1 is the distance of the electron from the point $(-a, 0, 0)$, and r_2 is the distance from the electron to the point $(a, 0, 0)$. Calculate the force on the electron.

7.3. Show that the angular momentum of a particle about an axis through a point O will be zero if and only if one of the following conditions is satisfied: (a) the particle is at O, (b) $\mathbf{p} = 0$, or (c) $\mathbf{p}$ is parallel or antiparallel to $\mathbf{r}$.

7.4. Show that the torque of force $\mathbf{F}$ about an axis through a point O will be zero if and only if one of the following conditions is satisfied: (a) the particle is at O, (b) $\mathbf{F} = 0$, or (c) $\mathbf{F}$ is parallel or antiparallel to $\mathbf{r}$.

7.5. Prove the following statements:

(a) The angular momentum about an axis passing through the origin is not necessarily conserved other than by a central force.

(b) If the mass of the center of force is not larger than the mass of the particle, the coordinate system determined by the central body is not inertial.

7.6. A particle of mass m is moving in an attractive inverse cubed force given by

$$\mathbf{F} = \frac{K}{r^3}\hat{\mathbf{r}}, \quad \text{where} \quad K < 0$$

(a) Discuss the motion qualitatively by graphing the effective potential.

(b) Find E and L when the particle is moving in a circular orbit. Find the time period of the orbit.

(c) If the orbit is perturbed slightly, what will be the period of small oscillations?

7.7. Show that $\mathbf{F} = \hat{\mathbf{r}}F(\mathbf{r})$ is a conservative force by proving that the integral

$$\int_{\mathbf{r}_1}^{\mathbf{r}_2} \mathbf{F} \cdot d\mathbf{r}$$

is independent of the path and depends only on the values r_1 and r_2. Express $\mathbf{F}$ and $d\mathbf{r}$ in spherical coordinates.

7.8. Consider an isotropic harmonic oscillator whose potential is given by $V(r) = \frac{1}{2}kr^2$.
  (a) Calculate the force $F(r)$ and make plots of both $V(r)$ and $F(r)$.
  (b) Make a plot of effective potential for a particle of mass m moving with energy E and angular momentum L. Describe the motion of a particle without solving the equations of motion.
  (c) Find the values of E and L for a circular orbit.
  (d) Calculate the frequency of revolution for the circular orbit and the frequency of small radial oscillations.
  (e) Explain the nature of the orbit that differs slightly from a circular orbit.

7.9. Consider an isotropic harmonic oscillator whose potential is given by $V(r) = \frac{1}{2}kr^2$. Calculate the value of $r(t)$ and $\theta(t)$ for the orbit of a particle. Draw the path of the orbit in such a potential. What conclusions can be drawn from these results?

7.10. In Problem 7.1, find the equations of motion for different situations, solve them for $r(\theta)$, and discuss the nature of the motion.

7.11. A body of mass m is moving in a spiral orbit given by $r = r_n e^{k\theta}$. Show that the force causing such an orbit is a central inverse cube force. Also show that θ varies as the logarithm of time.

7.12. A body of mass m moves in a circular orbit under the influence of a central force such that it passes through the origin of the central field; hence the orbit can be written as $r = r_0 \cos \theta$. Show that the central force is an inverse fifth radial power. Draw the path of such an orbit.

7.13. A particle of mass m describes an orbit $r = a(1 + \cos \theta)$ under the action of a force directed toward the center. Find the law of force.

7.14. The orbit of a particle of mass m under the action of a central force is described by $r = A \sin n\theta$, where A and n are constants. Find the law of force in terms of n, A, m, r, and the angular momentum L. Draw the orbit of the motion of the particle.

7.15. A particle of mass m moves under the action of a central force described by an orbit

$$r = \frac{ab}{(a^2 \cos^2 \theta + b^2 \sin^2 \theta)^{1/2}}$$

where a and b are constants. Find the law of force. Also draw the orbital path of the particle.

7.16. A particle of mass m is moving in a circular orbit of radius r_0 under the action of an attractive force given by $F(r) = k_2/r^2 + k_4/r^4$, where k_2 and k_4 are constants and the center of the force is at the center of the circle. Show that a stable orbit is possible only if $r_0^2 k_2 > k_4$. Calculate the effective potential V_{eff}. Graph $F(r)$ and V_{eff} versus r and discuss the nature of the motion.

7.17. Show that the velocity that a particle at a distance r from the center of Earth must have in order to escape to infinity is equal to $\sqrt{2}$ times the velocity for a stable circular orbit at a distance r.

7.18. A particle of mass m moves under the action of a force whose potential is given by $V(r) = Kr^4$, where $K > 0$.
  (a) Calculate $F(r)$ and make plots of both $F(r)$ and $V(r)$.
  (b) Make a plot of effective potential and discuss the motion of a particle without solving the equations of motion.
  (c) Find the values of E and L and the radius of a circular orbit.
  (d) Calculate the period of this circular motion.
  (e) Calculate the period of small radial oscillations, that is, the period of the motion when a particle is slightly disturbed from a circular orbit.

7.19. Consider a particle that is moving under the influence of a force given by

$$F(r) = -\frac{k}{r^2} + \frac{k'}{r^3}$$

where $k > 0$, while k' can be both positive or negative.
- **(a)** Calculate $V(r)$ and make plots of both $V(r)$ and $F(r)$.
- **(b)** Make plots of effective potential in different cases and discuss the nature of the motion of the particle without actually solving the equations of motion.
- **(c)** Calculate the frequency of a possible circular orbit.
- **(d)** Calculate the frequency of small radial oscillation.

7.20. A particle of mass m under the action of a force law $F = -k/r^2$ is at a distance d from the center of the force.
- **(a)** Show that the time it takes to fall from distance d to the center of force is $\sqrt{m\pi^2 d^3/8k}$.
- **(b)** Show that the time average of the velocity over the first half of the distance traveled to that over the second half is $(\pi - 2)/(\pi + 2)$.

7.21. A system of double stars under the action of their mutual gravitational attraction describes circular orbits around each other with a time period T. If suddenly they are deprived of their velocities, show that they will collide after a time interval of $T/4\sqrt{2}$.

7.22. Prove the following relation for an ellipse:

$$e = \frac{r_{max} - r_{min}}{r_{max} + r_{min}}$$

7.23. Consider the motion of a particle of mass m moving in an attractive central force inversely proportional to the cube of the radius:

$$\mathbf{F(r)} = \frac{K}{r^3}\,\hat{\mathbf{r}}, \qquad \text{where } K < 0$$

- **(a)** Discuss the motion of a particle by the method of plotting effective potential.
- **(b)** Solve the orbit equation $r = r(\theta)$ and show that the solutions can be written in the following forms for different situations:

$$\frac{1}{r} = A\cos[\beta(\theta - \theta_0)]$$

$$\frac{1}{r} = A\cosh[\beta(\theta - \theta_0)]$$

$$\frac{1}{r} = A\sinh[\beta(\theta - \theta_0)]$$

$$\frac{1}{r} = A(\theta - \theta_0)$$

$$\frac{1}{r} = \frac{1}{r_0}e^{\pm\beta\theta}$$

- **(c)** Find the range of values of E and L for which each of the preceding equations holds good. Express the values of constants A and β in terms of E and L.
- **(d)** Sketch a typical orbit in each of the preceding cases.

7.24. A particle of mass m is moving in a repulsive central force of the form $\mathbf{F(r)} = rK\hat{\mathbf{r}}$, where $K > 0$. Show that the path of the particle must be hyperbolic. Graph the path.

7.25. A body of mass m is moving in a repulsive inverse cubed force given by

$$\mathbf{F} = \frac{K}{r^3}\hat{\mathbf{r}}, \qquad \text{where } K > 0$$

Show that the path $r(\theta)$ of the body is given by

$$\frac{1}{r} = A \cos[\beta(\theta - \theta_0)]$$

Find the values of constants A and β in terms of E, L, velocity, and initial position. Draw the path of the orbit.

7.26. If the Sun's mass suddenly reduces to half its value, show that Earth's circular orbit will become parabolic.

7.27 A particle of mass m describes an elliptical path about a center of attractive force at one of its focus given by k/r^2, where k is a constant. Show that the speed v of the particle at any point of the orbit is

$$v^2 = k\left(\frac{2}{r} - \frac{1}{a}\right)$$

where a is the semimajor axis.

7.28. A particle of mass m moves in an elliptical orbit under the influence of an inverse-square central force. Let n be the ratio of the maximum angular velocity to the minimum angular velocity. Show that the eccentricity e of the ellipse is given by

$$e = \sqrt{\frac{n-1}{n+1}}$$

7.29. Show that the product of the maximum and minimum velocities of a particle moving in an elliptical orbit is $(2\pi a/T)^2$.

7.30. For a particle moving in an elliptical orbit of eccentricity e and semimajor axis a, show that $r = r(t)$ and $\theta = \theta(t)$ may be expressed as

$$r(t) = a(1 - e \cos \omega t + \tfrac{1}{2}e^2 - \tfrac{1}{2}e^2 \cos 2\omega t + \cdots)$$

$$\theta(t) = \omega t + 2e \sin \omega t + \tfrac{5}{4}e^2 \sin 2\omega t + \cdots$$

Graph $r(t)$ and $\theta(t)$ versus t.

7.31. Using Eq. (7.114) for the equation of a conic section and Eqs. (7.115), obtain the values of a, b, and e in terms of m, K, L, and E.

7.32. An asteroid is at a distance of 2×10^6 km from Earth and is moving with a speed of 10 km/s. Without gravitational pull, the asteroid would miss Earth by a perpendicular distance of 25,000 km from the center of Earth. What is the closest distance between the asteroid and Earth? Does it hit Earth? If not, will it come back to Earth again?

7.33. A comet with a parabolic path moves in the same plane as that of Earth's orbit. The distance of closest approach of the comet from the Sun is one-third of the radius of Earth's orbit. Using the law of areas, show that the comet cannot stay inside Earth's orbit for more than $\frac{2}{3}\pi$ of a year ($=74.5$ days).

7.34. A particle of mass m when at infinity has velocity v_0 and is headed toward a repulsive inverse cube force. It would miss the force center by a distance b if it were not deflected by the force. Calculate the angular deflection of this incident particle.

7.35. A string passes through a hole in a smooth horizontal table surface and has two equal masses attached to the two ends. One mass hangs vertically, while the other mass describes a circle of radius r_0 on the smooth surface of the table top and has velocity $(gr_0)^{1/2}$. Show that the vertical mass remains at rest; but if slightly disturbed in the vertical plane, the period of small oscillations is $2\pi(2r_0/3g)^{1/2}$.

7.36. Consider a comet that moves in a parabolic orbit in the plane of Earth's orbit. Its distance of closest approach to the Sun is γr_e, where r_e is the radius of Earth's orbit and $\gamma < 1$. Show that the total time that the comet spends within the orbit of Earth is

$$\sqrt{2(1 - \gamma)} \times \frac{1 + 2\gamma}{3\pi} \times 1 \text{ year}$$

7.37. Using the first and second Kepler's laws and Newton's laws of motion, derive the universal law of gravitation.

7.38. Sputnik I had a perigee 227 km above Earth's surface and its speed at this point was 28,710 km/h. Find its apogee distance from the surface of Earth and the time period of revolution.

7.39. A comet is observed at a distance of 10^8 km from the center of the Sun and is traveling with a velocity of 56.6 km/s toward the Sun, making an angle of 45° with the radius of the Sun. Find the orbit of the comet with the Sun at its center and the X-axis passing through the comet. How close to the Sun will it come?

7.40. A comet is observed at a distance D astronomical units from the Sun and is moving with a speed S times Earth's orbital speed. Show that the orbit will be a hyperbola if $DS^2 > 2$, a parabola if $DS^2 = 2$, or an ellipse if $DS^2 < 2$. Graph the path of the comet.

7.41. Halley's comet has an eccentricity of 0.967 and a perihelion distance of 89×10^6 km. Calculate the aphelion distance, the speeds at perihelion and aphelion, and time period of the orbit. Draw the path of the comet.

7.42. Referring to Fig. 7.22, for energy $E + \Delta E$, the equation of an ellipse may be written as

$$r = \frac{r_0}{1 - (A/r_0)\sin\theta}$$

Show that for $A/r_0 \ll 1$, this equation reduces to $r = r_0 + A\sin\theta$ [Eq. (7.129)], which, to first order in A, is an ellipse. Explain.

7.43. A particle of mass m is moving in a circle of radius r_0 under the action of an attractive force $F(r) = (1/r^2)e^{-r/a}$. Show that circular motion is stable only if the radius r_0 of the circle is less than a.

7.44. A particle of mass m moves under the influence of a central force $F(r) = -k/r^n$. Show that, if the orbit of the particle is circular passing through the force center, n must be equal to 5.

7.45. Show that the orbit of a particle of mass m moving under the influence of the following force law is a precessing ellipse.

$$F(r) = -\frac{k_2}{r^2} - \frac{k_3}{r^3}$$

where k_2 and k_3 are constants each greater than 0. Draw graphs showing the precessing ellipse.

7.46. Find the conditions for a stable circular orbit of a particle of mass m moving under the influence of the following force:
  (a) $F(r) = -(k/r^2)e^{-r/a}$
  (b) $F(r) = -k_2/r^2 - k_4/r^4$

7.47. A particle of mass m moving under the influence of a force $F(r) = -(k/r^2)e^{-r/a}$ describes nearly a circular orbit of radius r_0. Calculate the apsidal angle and draw the path of the motion of the particle.

SUGGESTIONS FOR FURTHER READING

ARTHUR, W., and FENSTER, S. K., *Mechanics*, Chapter 8. New York: Holt, Rinehart and Winston, Inc., 1969.

BARGER, V., and OLSSON, M., *Classical Mechanics*, Chapter 4. New York: McGraw-Hill Book Co., 1973.

BECKER, R. A., *Introduction to Theoretical Mechanics*, Chapter 10. New York: McGraw-Hill Book Co., 1954.

DAVIS, A. DOUGLAS, *Classical Mechanics*, Chapter 7. New York: Academic Press, Inc., 1986.

FOWLES, G. R., *Analytical Mechanics*, Chapter 6. New York: Holt, Rinehart and Winston, Inc., 1962.

FRENCH, A. P., *Newtonian Mechanics*, Chapters 8, 11, and 13. New York: W. W. Norton and Co., Inc., 1971.

*GOLDSTEIN, H., *Classical Mechanics*, 2d ed., Chapter 3. Reading, Mass.: Addison-Wesley Publishing Co., 1980.

HAUSER, W., *Introduction to the Principles of Mechanics*, Chapter 7. Reading, Mass.: Addison-Wesley Publishing Co., 1965.

KITTEL, C., KNIGHT, W. D., and RUDERMAN, M. A., *Mechanics*, Berkeley Physics Course, Volume 1, Chapter 9. New York: McGraw-Hill Book Co., 1965.

KLEPPNER, D., and KOLENKOW, R. J., *An Introduction to Mechanics,* Chapter 9. New York: McGraw-Hill Book Co., 1973.

*LANDU, L. D., and LIFSHITZ, E. M., *Mechanics*, Chapter 3. Reading, Mass.: Addison-Wesley Publishing Co., 1960.

MARION, J. B., *Classical Dynamics,* 2d ed., Chapters 2 and 8. New York: Academic Press, Inc., 1970.

McCUSKEY, S. W., *Celestial Mechanics.* Reading, Mass.: Addison-Wesley Publishing Co., 1963.

*MOORE, E. N., *Theoretical Mechanics*, Chapter 4. New York: John Wiley & Sons, Inc., 1983.

ROSSBERG, K., *Analytical Mechanics*, Chapter 6. New York: John Wiley & Sons, Inc., 1983.

STEPHENSON, R. J., *Mechanics and Properties of Matter*, Chapter 4. New York: John Wiley & Sons, Inc., 1962.

SYMON, K. R., *Mechanics,* 3rd ed., Chapter 3. Reading, Mass.: Addison-Wesley Publishing Co., 1971.

TAYLOR, E. F., *Introductory Mechanics,* Chapter 6. New York: John Wiley & Sons, Inc., 1963.

*The asterisk indicates works of an advanced nature.

System of Particles: Conservatiion Laws and Collisions

8.1 INTRODUCTION

In this and subsequent chapters we investigate the motion of a system of particles or body with a large number of particles. Whenever we are dealing with such situations, it is not only convenient but essential to describe the motion in the center of mass coordinates. We must understand the laws of conservation of linear momentum, angular momentum, and energy as applied to such systems. These laws will then be applied to some physical systems of interest, such as the motion of rockets and conveyor belts. We will show that the use of these laws is indispensable in the investigation of scattering or collision problems, both elastic and inelastic. Such investigations lead to an understanding of interactions between microscopic as well as macroscopic systems.

8.2 SYSTEM OF PARTICLES AND CENTER OF MASS

Whenever we are dealing with a system containing a large number of particles, it is, as said, both convenient and essential to describe the motion in the center of mass coordinates. Accordingly, let us consider a system containing N particles labeled $1, 2, \ldots, N$. The masses of these particles are $m_1, m_2, \ldots, m_N$ and they are located at distances $\mathbf{r}_1, \mathbf{r}_2, \ldots, \mathbf{r}_N$ from the origin O, as shown in Fig. 8.1. The velocities of these particles are $\dot{\mathbf{r}}_1, \dot{\mathbf{r}}_2, \ldots, \dot{\mathbf{r}}_N$ (or $\mathbf{v}_1, \mathbf{v}_2, \ldots, \mathbf{v}_N$), while their accelerations are $\ddot{\mathbf{r}}_1, \ddot{\mathbf{r}}_2, \ldots, \ddot{\mathbf{r}}_N$ (or $\mathbf{a}_1, \mathbf{a}_2, \ldots, \mathbf{a}_N$), respectively. For such a system of particles, the center of mass is a point located at a distance $\mathbf{R}(X, Y, Z)$ from the origin and defined by the relation

$$(m_1 + m_2 + \cdots + m_N)\mathbf{R} = m_1\mathbf{r}_1 + m_2\mathbf{r}_2 + \cdots + m_N\mathbf{r}_N$$

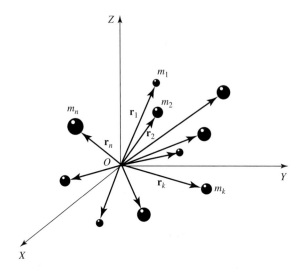

Figure 8.1   System of particles of various masses at different distances from the origin.

or

$$\sum_{k=1}^{N} m_k \mathbf{R} = \sum_{k=1}^{N} m_k \mathbf{r}_k$$

That is,

$$\mathbf{R} = \frac{\Sigma\, m_k \mathbf{r}_k}{\Sigma\, m_k} = \frac{\Sigma\, m_k \mathbf{r}_k}{M} \qquad (8.1)$$

where $M = \Sigma\, m_k$ is the sum of all the masses in the system and the summation Σ is from $k = 1$ to $k = N$. In component form, we may write

$$X = \frac{1}{M} \sum m_k x_k, \qquad Y = \frac{1}{M} \sum m_k y_k, \qquad Z = \frac{1}{M} \sum m_k z_k \qquad (8.2)$$

It should be clear from Eq. (8.1) that the center of mass is a *mass-weighted* average position.

The velocity $\mathbf{V}(=\dot{\mathbf{R}})$ of the center of mass can be obtained by differentiating Eq. (8.1) with respect to t; that is,

$$\mathbf{V} = \dot{\mathbf{R}} = \frac{1}{M} \sum m_k \dot{\mathbf{r}}_k \qquad (8.3)$$

while the components of the velocity of the center of mass may be written as

$$V_X = \dot{X} = \frac{1}{M} \sum m_k \dot{x}_k, \qquad V_Y = \dot{Y} = \frac{1}{M} \sum m_k \dot{y}_k, \qquad V_Z = \dot{Z} = \frac{1}{M} \sum m_k \dot{z}_k \qquad (8.4)$$

The acceleration $\mathbf{A}$ of the center of mass is obtained by differentiating once more; that is,

$$\mathbf{A} = \ddot{\mathbf{R}} = \frac{1}{M} \sum m_k \ddot{\mathbf{r}}_k \qquad (8.5)$$

or, in component form,

$$A_X = \ddot{X} = \frac{1}{M} \sum m_k \ddot{x}_k, \qquad A_Y = \ddot{Y} = \frac{1}{M} \sum m_k \ddot{y}_k, \qquad A_Z = \ddot{Z} = \frac{1}{M} \sum m_k \ddot{z}_k \quad \textbf{(8.6)}$$

In the following sections, we shall find the description of motion of the center of mass coordinates and motion in the center of mass coordinate systems both interesting and useful.

We shall discuss the following three conservation laws in detail as applied to a system of particles:

1. conservation of linear momentum
2. conservation of angular momentum
3. conservation of energy

There are two approaches to this problem: (1) Newton's laws, and (2) symmetry principles.

The conservation laws are the direct consequence of the definitions made in Newtonian mechanics, that is, of Newton's second law of motion. The validity of these conservation laws holds to the extent that Newtonian mechanics provides an adequate description of nature. Furthermore, since there is no such thing as a truly isolated system, these laws can only hold approximately. But ultimately, from a modern point of view, these conservation laws are the consequence of underlying symmetries briefly discussed here and in detail in Chapter 12.

In general, a system is said to have symmetry when some characteristic in the system remains unchanged even though the system is changed in a certain respect. For example, if the system is given a linear displacement, the system remains invariant under linear displacement or translation and the system is said to have *translational symmetry*. Similarly, a system is said to have *rotational symmetry* if it remains invariant under rotation. There is a close relationship between conservation laws and symmetry principles. The conservation of linear momentum is a direct consequence of translational symmetry, that is, the *homogeneity of space*. The law of conservation of angular momentum is the consequence of rotational symmetry, that is, the *isotropy of space*, while the law of conservation of energy leads to the *homogeneity of time*. Actually, we may go a step further and state:

> *Any conservation law is a statement of invariance of some physical property during all physical processes.*

For the time being, we shall investigate the conservation laws from the viewpoint of Newtonian mechanics.

8.3 CONSERVATION OF LINEAR MOMENTUM

For a single particle of mass m moving with velocity **v** and linear momentum **p**, Newton's second law is

$$\mathbf{F} = \frac{d\mathbf{p}}{dt} \tag{8.7}$$

where $\mathbf{F}$ is the net external force acting on mass m and

$$\mathbf{p} = m\mathbf{v} \tag{8.8}$$

If m is constant and does not depend on time,

$$\mathbf{F} = \frac{d\mathbf{p}}{dt} = \frac{d}{dt}(m\mathbf{v}) = m\frac{d\mathbf{v}}{dt} = m\mathbf{a} \tag{8.9}$$

Furthermore, if $\mathbf{F} = 0$, $\mathbf{p} = $ constant, which is the law of conservation of linear momentum for a single particle.

We now extend these ideas to a system of N particles, as shown in Fig. 8.1. Let us consider the motion of the kth particle of mass m_k, which is at a distance $\mathbf{r}_k$ from the origin, has velocity $\dot{\mathbf{r}}_k(= \mathbf{v}_k)$, and acceleration $\ddot{\mathbf{r}}_k$. The total force $\mathbf{F}_k$ acting on the kth particle is the sum of the set of two forces: (1) the sum of the external forces $\mathbf{F}_k^e$ applied to the kth particle, and (2) the sum of the internal force $\mathbf{F}_k^i$ on the kth particle by the remaining $N - 1$ particles in the system. Thus the equation of motion for the kth particle, according to Newton's law, is

$$\mathbf{F}_k = \mathbf{F}_k^e + \mathbf{F}_k^i = m_k\ddot{\mathbf{r}}_k, \qquad k = 1, 2, \dots, N \tag{8.10}$$

where

$$\mathbf{F}_k^i = \sum_{\substack{k=1 \\ k \neq l}}^{N} \mathbf{F}_{kl}^i \tag{8.11}$$

and $\mathbf{F}_{kl}^i$ is the force on the kth particle due to the lth particle. Because of the vector nature of Eq. (8.10), there are $3N$ simultaneous second-order differential equations to be solved. The motion of any particle k at $\mathbf{r}_k$ is obtained by solving such equations in terms of $6N$ arbitrary constants ($3N$ for the initial positions and $3N$ for initial velocities). No general methods are available for solving Eq. (8.10), which is extremely difficult to solve except in some special cases. An alternative approach is to solve these problems by using the center of mass coordinates, as will be explained later.

The momentum of the kth particle is given by

$$\mathbf{p}_k = m_k\mathbf{v}_k = m_k\dot{\mathbf{r}}_k \tag{8.12}$$

Using this, Eq. (8.10) takes the form

$$\frac{d\mathbf{p}_k}{dt} = \mathbf{F}_k = \mathbf{F}_k^e + \mathbf{F}_k^i \tag{8.13}$$

Summing on both sides over all the N particles,

$$\sum_{k=1}^{N} \frac{d\mathbf{p}_k}{dt} = \frac{d}{dt}\sum_{k=1}^{N}\mathbf{p}_k = \sum_{k=1}^{N}\mathbf{F}_k = \sum_{k=1}^{N}\mathbf{F}_k^e + \sum_{k=1}^{N}\mathbf{F}_k^i \tag{8.14}$$

Let $\mathbf{P}$ be the total linear momentum of the system of N particles and $\mathbf{F}$ be the total external force acting on the system; that is,

$$\mathbf{P} = \sum_{k=1}^{N}\mathbf{p}_k = \sum_{k=1}^{N}m_k\dot{\mathbf{r}}_k \tag{8.15}$$

and
$$\mathbf{F} = \sum_{k=1}^{N} \mathbf{F}_k^e \tag{8.16}$$

Furthermore, we shall show that the sum of all the internal forces acting on all the particles of the system is zero; that is

$$\sum_{k=1}^{N} \mathbf{F}_k^i = 0 \tag{8.17}$$

Combining Eqs. (8.15), (8.16), and (8.17) with Eq. (8.14), we obtain

$$\frac{d\mathbf{P}}{dt} = \mathbf{F} \tag{8.18}$$

This is the *momentum theorem* for a system of particles.

> ***Conservation of Linear Momentum.*** *The rate of change of total linear momentum is equal to the total external applied force; thus, if the sum of all the externally applied forces is zero, the total linear momentum* **P** *of the system will be constant.*

That is,
$$\mathbf{P} = \text{constant}, \qquad \text{if } \mathbf{F} = 0 \tag{8.19}$$

In terms of the center of mass coordinates, according to Eqs. (8.3) and (8.15),

$$\mathbf{P} = \sum_{k=1}^{N} m_k \dot{\mathbf{r}} = M\dot{\mathbf{R}} \tag{8.20}$$

which on substituting in Eq. (8.18) yields

$$M\ddot{\mathbf{R}} = \mathbf{F} \tag{8.21}$$

Equations (8.18) and (8.21) are similar in form to Newton's second law as applied to a single particle. Thus, from Eq. (8.21), we may conclude:

> *The center of mass of a system of particles moves like a single particle of mass M (total mass of the system) acted on by a single force* **F** *that is equal to the sum of all the external forces acting on the system.*

All these statements are true only if we can justify Eq. (8.17)—that is, the sum of all internal forces is zero. We now proceed to prove this by two different approaches: (1) Newton's third law, and (2) the principle of virtual work. According to Eq. (8.11),

$$\mathbf{F}_k^i = \sum_{\substack{k=1 \\ k \neq l}}^{N} \mathbf{F}_{kl}^i \tag{8.11}$$

where $\mathbf{F}^i_{kl}$ is the force exerted on the kth particle by the lth particle. According to Newton's third law, the force exerted on the kth particle due to the lth particle is equal and opposite to that exerted on l by k; that is,

$$\mathbf{F}^i_{kl} = -\mathbf{F}^i_{lk} \qquad (8.22)$$

This equation is a statement of *Newton's third law in the weak form* because it simply implies that the two forces are equal and opposite, but not necessarily acting along the line joining the two particles; the *strong form* implies that their line of action should be the same. Using Eq. (8.11), the sum of all the internal forces is

$$\sum_{k=1}^{N} \mathbf{F}^i_k = \sum_{k=1}^{N} \sum_{\substack{l=1 \\ l \neq k}}^{N} \mathbf{F}^i_{kl} \qquad (8.23)$$

The right side contains forces on all pairs of particles. For each pair, the total sum according to Eq. (8.22) is zero; that is $\mathbf{F}^i_{kl} + \mathbf{F}^i_{lk} = 0$. Hence the right side of Eq. (8.23) is zero, thereby proving that in Eq. (8.11) the right side is zero. That is, *the sum of all the internal forces is zero*.

In the preceding proof we had to assume that the internal forces come in pairs. We need not make this assumption if we make use of the *principle of virtual work* or *virtual displacement*. Let us assume that each particle in the system is given a small displacement $\delta\mathbf{r}$. Since each particle in the system is given the same displacement, there is no relative displacement of the system; hence no net work is done by the internal forces. No net total work is done because the internal state of the system has not changed by this virtual or imaginary displacement. The work done by the internal forces $\mathbf{F}^i_k$ in a small virtual displacement $\delta\mathbf{r}$ of the kth particle is

$$\delta W_k = \mathbf{F}^i_k \cdot \delta\mathbf{r} \qquad (8.24)$$

The total work done by all the internal forces is

$$\delta W = \sum_{k=1}^{N} \delta W_k = \sum_{k=1}^{N} (\mathbf{F}^i_k \cdot \delta\mathbf{r}) = \delta\mathbf{r} \cdot \left[\sum_{k=1}^{N} \mathbf{F}^i_k\right] \qquad (8.25)$$

$\delta\mathbf{r}$ has been factored out because it is the same for all the particles. If the total work done by the internal forces is zero for any displacement,

$$\delta\mathbf{r} \cdot \left[\sum_{k=1}^{N} \mathbf{F}^i_k\right] = 0$$

Since $\delta\mathbf{r}$ is not zero, we must have

$$\sum_{k=1}^{N} \mathbf{F}^i_k = \sum_{k=1}^{N} \sum_{\substack{l=1 \\ l \neq k}}^{N} \mathbf{F}^i_{kl} = 0 \qquad (8.26)$$

as required.

8.4 CONSERVATION OF ANGULAR MOMENTUM

The angular momentum of a single particle is defined in terms of a cross product as

$$\mathbf{L} = \mathbf{r} \times \mathbf{p} = \mathbf{r} \times m\dot{\mathbf{r}} = \mathbf{r} \times m\mathbf{v} \tag{8.27}$$

Now we extend this definition to a system of N particles. The total angular momentum $\mathbf{L}$ taken about the origin may be written as a vector sum:

$$\mathbf{L} = \sum_{k=1}^{N} (\mathbf{r}_k \times \mathbf{p}_k) = \sum_{k=1}^{N} (\mathbf{r}_k \times m_k\dot{\mathbf{r}}) \tag{8.28}$$

The total angular momentum could have been taken about any point A instead of the origin O, but in that case we must replace $\mathbf{r}_k$ by $\mathbf{r}_k - \mathbf{r}_A$, where $\mathbf{r}_A$ is the distance of point A from the origin. For simplicity, we shall use expression (8.28). Taking the time derivative of the angular momentum in Eq. (8.28) yields

$$\frac{d\mathbf{L}}{dt} = \sum_{k=1}^{N} (\dot{\mathbf{r}}_k \times m_k\dot{\mathbf{r}}_k) + \sum_{k=1}^{N} (\mathbf{r}_k \times m_k\ddot{\mathbf{r}}) \tag{8.29}$$

The first term on the right vanishes because of the definition of the cross product ($\dot{\mathbf{r}} \times m\dot{\mathbf{r}} = 0$), while $m\ddot{\mathbf{r}}$, from Eq. (8.10), is equal to the total force acting on the particle k; that is, we obtain

$$\frac{d\mathbf{L}}{dt} = \sum_{k=1}^{N} \left[\mathbf{r}_k \times \left(\mathbf{F}_k^e + \sum_{\substack{l=1 \\ l \neq k}}^{N} \mathbf{F}_{kl}^i \right) \right] = \sum_{k=1}^{N} \mathbf{r}_k \times \mathbf{F}_k^e + \sum_{k=1}^{N} \sum_{\substack{l=1 \\ l \neq k}}^{N} \mathbf{r}_k \times \mathbf{F}_{kl}^i \tag{8.30}$$

where, as before, $\mathbf{F}_k^e$ is the total external force acting on the particle k, and $\mathbf{F}_{kl}^i$ is the internal force acting on the kth particle due to the lth particle. We can prove the second term on the right to be zero if we use the *strong form of Newton's third law*; that is, the forces are equal and opposite and their line of action is the same. The second term on the right contains a sum of pairs of torques due to pairs of forces, which according to Newton's third law are equal and opposite. One such pair is

$$(\mathbf{r}_k \times \mathbf{F}_{kl}^i) + (\mathbf{r}_l \times \mathbf{F}_{lk}^i) \tag{8.31}$$

Since $\mathbf{F}_{kl}^i = -\mathbf{F}_{lk}^i$, we may write the expression in Eq. (8.31) as

$$(\mathbf{r}_k - \mathbf{r}_l) \times \mathbf{F}_{kl}^i = \mathbf{r}_{kl} \times \mathbf{F}_{kl}^i \tag{8.32}$$

(see Fig. 8.2). Expression (8.32) is zero if the internal forces are central; that is, the forces acting along the line joining the two particles cause the two particles to either attract or repel each other. Thus the second term on the right in Eq. (8.30) vanishes, and the resulting equation is

$$\frac{d\mathbf{L}}{dt} = \sum_{k=1}^{N} \mathbf{r}_k \times \mathbf{F}_k^e \tag{8.33}$$

Since $\mathbf{r}_k \times \mathbf{F}_k^e$ is the *torque* or the *moment of the external force* $\mathbf{F}_k^e$, the right side of Eq. (8.33) is the total moment (or the total torque) of all the external forces acting on the system. If we denote $\boldsymbol{\tau}_k$ to be the torque on the kth particle and $\boldsymbol{\tau}$ to be the total torque, we may write

$$\frac{d\mathbf{L}}{dt} = \sum_{k=1}^{N} \boldsymbol{\tau}_k = \sum_{k=1}^{N} \mathbf{r}_k \times \mathbf{F}_k^e \tag{8.34}$$

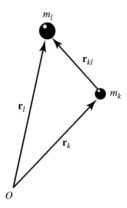

Figure 8.2   Relative distance $\mathbf{r}_{kl}$ between a pair of particles.

and
$$\frac{d\mathbf{L}}{dt} = \boldsymbol{\tau}$$
(8.35)

which states that the time rate of change of the angular momentum of a system is equal to the total torque due to all the net external forces acting on the system. Thus we may state the following principle:

> **Conservation of Angular Momentum.** *For an isolated—one on which no net external forces act—the total torque τ will be zero; hence the angular momentum remains constant both in magnitude and direction.*

That is, if

$$\boldsymbol{\tau} = 0, \qquad \frac{d\mathbf{L}}{dt} = 0$$

and
$$\mathbf{L} = \sum_{k=1}^{N} \mathbf{r}_k \times m_k \mathbf{v}_k = \text{constant}$$
(8.36)

8.5 CONSERVATION OF ENERGY

In many situations, the total force acting on any particle in a system of particles is a function of the positions of the particles in the system. Thus the force $\mathbf{F}_k$ on the kth particle is

$$\mathbf{F}_k = \mathbf{F}_k^e + \mathbf{F}_k^i = \mathbf{F}_k(\mathbf{r}_1, \mathbf{r}_2, \dots, \mathbf{r}_N), \qquad \text{where } k = 1, 2, \dots, N$$
(8.37)

The external forces $\mathbf{F}_k^e$ may depend on the position $\mathbf{r}_k$ of the particle k, while the internal force $\mathbf{F}_k^i$ may depend on the relative positions of the other particles relative to k, that is, $\mathbf{r}_{kl} = (\mathbf{r}_k - \mathbf{r}_l)$, and so on. If the force $\mathbf{F}_k$ satisfies the condition that

$$\boldsymbol{\nabla} \times \mathbf{F}_k = \mathbf{curl}\, \mathbf{F}_k = 0$$
(8.38)

there exists a potential function

$$V = V(\mathbf{r}_1, \mathbf{r}_2, \dots, \mathbf{r}_N)$$
(8.39)

such that

$$F_{kx} = -\frac{\partial V}{\partial x_k}, \qquad F_{ky} = -\frac{\partial V}{\partial y_k}, \qquad F_{kz} = -\frac{\partial V}{\partial z_k}, \qquad \text{where } k = 1, 2, \dots, N \qquad \textbf{(8.40)}$$

Thus, under such conditions, we can derive the law of conservation of energy.

The motion of the kth particle is described by

$$m_k \ddot{\mathbf{r}}_k = m_k \dot{\mathbf{v}}_k = \mathbf{F}_k \qquad \textbf{(8.41)}$$

which, on combining with Eq. (8.40), yields

$$m_k \frac{dv_{kx}}{dt} = -\frac{\partial V}{\partial x_k}, \qquad m_k \frac{dv_{ky}}{dt} = -\frac{\partial V}{\partial y_k}, \qquad m_k \frac{d_{kz}}{dt} = -\frac{\partial V}{\partial z_k} \qquad \textbf{(8.42)}$$

Multiplying the first equation by v_{kx} ($=dx_k/dt$), the second equation by v_{ky} ($=dy_k/dt$), and the third equation by v_{kz} ($=dz_k/dt$), and adding (using $v_k^2 = v_{kx}^2 + v_{ky}^2 + v_{kz}^2$), we get

$$\frac{d}{dt}\left(\frac{1}{2}m_k v_k^2\right) + \frac{\partial V}{\partial x_k}\frac{dx_k}{dt} + \frac{\partial V}{\partial y_k}\frac{dy_k}{dt} + \frac{\partial V}{\partial z_k}\frac{dz_k}{dt} = 0, \qquad \text{where } k = 1, 2, \dots, N \qquad \textbf{(8.43a)}$$

Summing over all values of k gives

$$\frac{d}{dt}\sum_{k=1}^{N}\left(\frac{1}{2}m_k v_k^2\right) + \sum_{k=1}^{N}\left(\frac{\partial V}{\partial x_k}\frac{dx_k}{dt} + \frac{\partial V}{\partial y_k}\frac{dy_k}{dt} + \frac{\partial V}{\partial z_k}\frac{dz_k}{dt}\right) = 0 \qquad \textbf{(8.43b)}$$

where

$$\sum_{k=1}^{N}\left(\frac{1}{2}m_k v_k^2\right) = K \qquad \text{(kinetic energy)} \qquad \textbf{(8.44)}$$

and

$$\sum_{k=1}^{N}\left(\frac{\partial V}{\partial x_k}\frac{dx_k}{dt} + \frac{\partial V}{\partial y_k}\frac{dy_k}{dt} + \frac{\partial V}{\partial z_k}\frac{dz_k}{dt}\right) = \frac{dV}{dt} \qquad \textbf{(8.45)}$$

Hence Eq. (8.43b) takes the form

$$\frac{d}{dt}(K + V) = 0$$

or

$$K + V = E = \text{constant} \qquad \textbf{(8.46)}$$

The total energy E, which is the sum of the kinetic and potential energy, is constant; hence Eq. (8.46) is a statement of the *law of conservation of energy* or the *energy conservation theorem*.

If the external forces are not position dependent, while the internal forces are derivable from a potential function, then the energy conservation theorem takes the form

$$\frac{d}{dt}(K + V^i) = \sum_{k=1}^{N} \mathbf{F}_k^e \cdot \dot{\mathbf{r}}_k \qquad \textbf{(8.47)}$$

Since we have assumed that in this case the internal forces are position dependent and the corresponding potential V^i depends on the relative positions of pairs of particles, that is,

$$V_{kl}^i = V_{kl}^i(\mathbf{r}_{kl}) = V_{kl}^i(\mathbf{r}_k - \mathbf{r}_l) \qquad \textbf{(8.48)}$$

while

$$V^i = \sum_{k=1}^{N} \sum_{l=1}^{k-1} V^i_{kl}(\mathbf{r}_{kl}) \tag{8.49}$$

we may conclude that

$$\mathbf{F}^i_k = -\hat{\mathbf{i}} \frac{\partial V^i}{\partial x_k} - \hat{\mathbf{j}} \frac{\partial V^i}{\partial y_k} - \hat{\mathbf{k}} \frac{\partial V^i}{\partial z_k} \tag{8.50}$$

It is necessary to point out that a potential function exists if the external forces are position dependent and Eq. (8.38) is satisfied. As discussed in previous chapters, that is possible only if the work done by the force between two points is independent of the path. Thus a closed system, that is, one in which no external forces act on the system, leads to the law of the conservation of energy as given by Eq. (8.46).

Suppose a system is such that it has internal frictional forces. Such frictional forces depend on the relative velocities of the particles and are not central forces. Thus the law of conservation of energy, Eq. (8.46), does not hold for such systems.

 Example 8.1 _____

Consider the following three particles of masses m1, m2, and m3 located at distances R1, R2, and R3 from the origin.

m1 = 2 kg   m2 = 3 kg   m3 = 4 kg

$R1 = 2t^2 \cdot i + 3t \cdot j + 4 \cdot k$ $R2 = (1 + t^2) \cdot i + (2 + 5 \cdot t) \cdot j$ $R3 = (1 + 2t^2 + 3t^3) \cdot i + (3t + 4t^2) \cdot k$

Calculate the following quantities at time t = 10 sec. **(a)** The position of the center of mass, **(b)** the velocity of the center of mass, **(c)** the linear momentum of the system, and **(d)** the kinetic energy of the system.

Solution

(a) U1 represents a unit vector matrix. R1, R2, and R3 are expressed in matrix form. R represents the position of the center of mass and may be calculated as shown.

$i := 1$ $j := 1$ $k := 1$

$t := 10$

$m1 := 2$ $m2 := 3$ $m3 := 4$

$$U1 := \begin{pmatrix} i & 0 & 0 \\ 0 & j & 0 \\ 0 & 0 & k \end{pmatrix}$$

$$R1 := \begin{bmatrix} 2 \cdot t^2 & 0 & 0 \\ 0 & 3 \cdot t & 0 \\ 0 & 0 & 4 \end{bmatrix} \qquad R2 := \begin{bmatrix} 1 + t^2 & 0 & 0 \\ 0 & 2 + 5 \cdot t & 0 \\ 0 & 0 & 0 \end{bmatrix} \qquad R3 := \begin{bmatrix} 1 + 2 \cdot t^2 + 3 \cdot t^3 & 0 & 0 \\ 0 & 0 & 0 \\ 0 & 0 & 3 \cdot t + 4 \cdot t^2 \end{bmatrix}$$

$$R := \frac{m1 \cdot U1 \cdot R1 + m2 \cdot U1 \cdot R2 + m3 \cdot U1 \cdot R3}{m1 + m2 + m3} \qquad R = \begin{bmatrix} 1.501 \cdot 10^3 & 0 & 0 \\ 0 & 24 & 0 \\ 0 & 0 & 192 \end{bmatrix}$$

$|R| = 6.916 \cdot 10^6$

(b) Differentiating R1, R2, and R3 with respect to t yield the velocities V1, V2, and V3.

$$V1 = \frac{d}{dt}\left(2 \cdot t^2 \cdot i + 3 \cdot t \cdot j + 4 \cdot k\right) \qquad V2 = \frac{d}{dt}\left[\left(1 + t^2\right) \cdot i + (2 + 5 \cdot t) \cdot j\right] \qquad V3 = \frac{d}{dt}\left[\left(1 + 2 \cdot t^2 + 3 \cdot t^3\right) \cdot i + \left(3 \cdot t + 4 \cdot t^2\right) \cdot k\right]$$

$$V1 = 4 \cdot t \cdot i + 3 \cdot j \qquad V2 = 2 \cdot t \cdot i + 5 \cdot j \qquad V3 = 4 \cdot t \cdot i + 9 \cdot t^2 \cdot i + 3 \cdot k + 8 \cdot k \cdot t$$

$$V1 := \begin{vmatrix} 4 \cdot t & 0 & 0 \\ 0 & 3 & 0 \\ 0 & 0 & 0 \end{vmatrix} \qquad V2 := \begin{vmatrix} 2 \cdot t & 0 & 0 \\ 0 & 5 & 0 \\ 0 & 0 & 0 \end{vmatrix} \qquad V3 := \begin{bmatrix} 4 \cdot t + 9 \cdot t^2 & 0 & 0 \\ 0 & 0 & 0 \\ 0 & 0 & 3 + 8 \cdot t \end{bmatrix}$$

The velocity V of the center of mass is

$$V := \frac{m1 \cdot U1 \cdot V1 + m2 \cdot U1 \cdot V2 + m3 \cdot U1 \cdot V3}{m1 + m2 + m3} \qquad V = \begin{pmatrix} 433.333 & 0 & 0 \\ 0 & 2.333 & 0 \\ 0 & 0 & 36.889 \end{pmatrix} \qquad |V| = 3.73 \cdot 10^4$$

(c) The linear momentum of the system is

$$P := m1 \cdot V1 + m2 \cdot V2 + m3 \cdot V3 \qquad P = \begin{bmatrix} 3.9 \cdot 10^3 & 0 & 0 \\ 0 & 21 & 0 \\ 0 & 0 & 332 \end{bmatrix} \qquad |P| = 2.719 \cdot 10^7$$

(d) The kinetic energy of the system is $\quad K := \frac{1}{2} \cdot \left(m1 \cdot V1^2 + m2 \cdot V2^2 + m3 \cdot V3^2\right)$

$$K = \begin{bmatrix} 1.769 \cdot 10^6 & 0 & 0 \\ 0 & 46.5 & 0 \\ 0 & 0 & 1.378 \cdot 10^4 \end{bmatrix} \qquad |K| = 1.134 \cdot 10^{12}$$

EXERCISE 8.1   Repeat the above example for the following three masses.

m1 = 24 kg   m2 = 3 kg   m3 = 2 kg

$$R1 = \left(1 + 2t^2 + 3t^3\right) \cdot j + \left(3t + 4t^2\right) \cdot k \quad R2 = \left(1 + t^2\right) \cdot i + (2 + 5t) \cdot j \quad R3 = 2t^2 \cdot i + 3t \cdot j + 4 \cdot k$$

8.6 MOTION OF SYSTEMS WITH VARIABLE MASS: ROCKETS AND CONVEYOR BELTS

We will now apply the conservation laws discussed in the previous section to some particular situations. The conservation laws are applicable to any definite system of particles, which may be chosen arbitrarily by including and excluding certain parts so long as it does not exclude the forces acting on the chosen part of the system. Another restriction concerns the law of conservation of kinetic and potential energy. It holds good so long as *no* mechanical energy is converted into other forms of energy, such as heat produced by frictional forces, unless such converted amounts are taken into account.

Rocket Propulsion

Rocket technology is based on the simplest principle of conservation of linear momentum. A rocket is propelled in a forward direction by ejecting mass in a backward direction in the form of gases resulting from the combustion of fuel. Thus the forward force on the rocket is the reaction to the backward force of the ejected gases (burned-out fuel). The problem is to find the velocity of the rocket at any time after launching, or takeoff, from the ground. As shown in Fig. 8.3, at a given time t a rocket of mass m is moving with velocity $\mathbf{v}$ relative to some fixed coordinate system, say Earth. Let the velocity of the exhaust gases from the rocket be $\mathbf{u}$ with respect to the rocket: hence $\mathbf{u} + \mathbf{v}$ with respect to a fixed coordinate system. Let us say that in a time interval between t and $t + dt$ the amount of fuel exhausted is $|dm| = -dm$ (because dm is negative; hence the rate at which the fuel is exhausted is $|dm/dt| = -dm/dt$), while the mass of the rocket is $m + dm$ and its velocity $\mathbf{v} + d\mathbf{v}$.

The momentum of the system at time t is

$$\mathbf{P}(t) = m\mathbf{v} \tag{8.51}$$

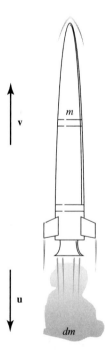

Figure 8.3 Motion of a rocket at some instant t.

and the momentum of the system at time $t + dt$ is

$$\mathbf{P}(t + dt) = \mathbf{P}_{\text{rocket}}(t + dt) + \mathbf{P}_{\text{fuel}}(t + dt) = (m + dm)(\mathbf{v} + d\mathbf{v}) + (-dm)(\mathbf{v} + \mathbf{u}) \quad \textbf{(8.52)}$$

The change in momentum in time interval dt is

$$d\mathbf{P} = \mathbf{P}(t + dt) - \mathbf{P}(t) \simeq m\,d\mathbf{v} - \mathbf{u}\,dm \quad \textbf{(8.53)}$$

where we have dropped the second-order term $dm\,d\mathbf{v}$. Since the rate of change of momentum $d\mathbf{P}/dt$ is equal to the applied external force $\mathbf{F}$, we may write Eq. (8.53) as

$$\frac{d\mathbf{P}}{dt} = \mathbf{F} = m\,\frac{d\mathbf{v}}{dt} - \mathbf{u}\,\frac{dm}{dt} \quad \textbf{(8.54)}$$

Note again that $\mathbf{u}$ is the velocity of the escaping gases. Equation (8.54) may be written as

$$m\,\frac{d\mathbf{v}}{dt} = \mathbf{u}\,\frac{dm}{dt} + \mathbf{F} \quad \textbf{(8.55)}$$

where $\mathbf{F}$ may be a gravitational force, the force of air resistance, or any other external force, $m(d\mathbf{v}/dt)$ is called the *thrust* of the rocket engine. Since dm/dt is negative, the thrust is opposite to the velocity $\mathbf{u}$ of the escaping gases. [The thrust of the rocket engine can be calculated by holding the rocket stationary and burning the fuel at the rate of dm/dt. The force $\mathbf{F}_0$ needed to hold the rocket stationary ($d\mathbf{v}/dt = 0$ and also $\mathbf{F} = 0$),

$$\mathbf{F}_0 = -\mathbf{u}\,\frac{dm}{dt} \quad \textbf{(8.56)}$$

will be the measure of the thrust.]

Let us consider a special case of Eq. (8.55) that prevails when $\mathbf{F} = 0$, that is, when no gravitational force or air resistance is present, which may be the case when the rocket is far in outer space. Equation (8.55) for $\mathbf{F} = 0$ is

$$m\,\frac{d\mathbf{v}}{dt} = \mathbf{u}\,\frac{dm}{dt} \quad \textbf{(8.57)}$$

Multiplying both sides by dt/m and integrating,

$$\int_{v_0}^{v} d\mathbf{v} = \mathbf{u} \int_{m_0}^{m} \frac{dm}{m}$$

$$\mathbf{v} - \mathbf{v}_0 = \mathbf{u}\,\ln m \Big|_{m_0}^{m}$$

Since $m_0 > m$, it is preferable to write

$$\mathbf{v} = \mathbf{v}_0 - \mathbf{u}\,\ln \frac{m_0}{m} \quad \textbf{(8.58)}$$

which states that the change in velocity $\mathbf{v} - \mathbf{v}_0$, or the final velocity $\mathbf{v}$, depends on two factors. A large value of $\mathbf{v}$ results from (1) large values of $\mathbf{u}$, the velocity of the exhaust gases, and

(2) large values of m_0/m, where m_0 is the initial mass of the rocket and its fuel, while m is the final mass when all the fuel has been used up. The final velocity is independent of the rate of burning fuel. Large values of m_0/m mean that we have a large fuel-to-payload ratio. To increase the value of m_0/m by large amounts, *staged rockets* are used for launching satellites and spacecrafts.

Near Earth's surface, we cannot neglect the force of gravitational pull. Thus, substituting $\mathbf{F} = m\mathbf{g}$ in Eq. (8.55), we obtain

$$m \frac{d\mathbf{v}}{dt} = \mathbf{u} \frac{dm}{dt} + m\mathbf{g} \tag{8.59}$$

which on rearranging and integrating,

$$\int_{v_0}^{v} d\mathbf{v} = \mathbf{u} \int_{m_0}^{m} \frac{1}{m} \, dm + \mathbf{g} \int_{0}^{t} dt$$

results in

$$\mathbf{v} = \mathbf{v}_0 - \mathbf{u} \ln \frac{m_0}{m} + \mathbf{g}t \tag{8.60}$$

Assuming that at $t = 0$, $\mathbf{v}_0 = 0$, and since $\mathbf{u}$ is opposite to $\mathbf{v}$, we may write Eq. (8.60) in scalar form as

$$v = u \ln \frac{m_0}{m} - gt \tag{8.61}$$

Initially, the rocket thrust must be large enough to overcome gravitational force $m_0\mathbf{g}$. Subsequently, the preceding equations will describe the motion of the rocket.

A Conveyor Belt

Consider the conveyor belt shown in Fig. 8.4. We are interested in calculating the force $\mathbf{F}$ needed to keep the conveyor belt moving with horizontal uniform speed v, while sand or some other material is continuously dropping on the belt from a stationary hopper at a rate dm/dt. Let M be the mass of the belt and m be the mass of the sand on the belt. The total momentum of the system, the belt, and the sand on the belt is

$$p = (m + M)v \tag{8.62}$$

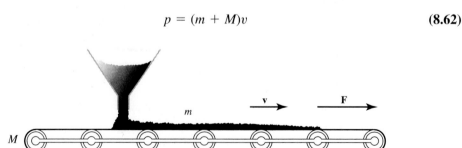

Figure 8.4   Conveyor belt.

Thus, according to the linear momentum theorem, since M and v are constants while m is changing,

$$F = \frac{dp}{dt} = v\frac{dm}{dt} \tag{8.63}$$

where F is the force applied to the belt. The power that must be supplied by the force to keep the belt moving with uniform speed v is

$$\text{Power} = P = Fv = v^2\frac{dm}{dt} = \frac{d}{dt}mv^2 = 2\frac{d}{dt}\left(\frac{1}{2}mv^2\right)$$

$$= 2\frac{d}{dt}\left(\frac{1}{2}(m + M)v^2\right) = 2\frac{dK}{dt} \tag{8.64}$$

That is, the power needed is twice the rate at which the kinetic energy is increasing. This implies that the law of conservation of mechanical energy does not apply here. The missing power is used up in doing work against the friction force, as explained next.[1]

When sand hits the belt, it must accelerate from zero speed to the belt speed over a short distance, during which some sliding must occur between the belt and the sand. To an observer at rest on the belt, the falling sand would appear to have a horizontal motion with speed v in the opposite direction to that of the belt. The belt exerts a horizontal force dF_f on the sand of mass dm to change its speed from $-v$ to 0. It does not matter whether the acceleration time is 1 s or 1/100 s; the power developed by the frictional forces between the belt and the sand is exactly one-half the power supplied.

Example 8.2 _____

A spherical raindrop falling through fog or mist accumulates mass due to condensation at a rate proportional to its cross-sectional area and velocity. (a) Calculate the acceleration of the raindrop in terms of its radius and velocity. The raindrop starts from rest and has almost zero size. (b) Suppose a raindrop falling from a height of 3000 m has a radius of 1 mm and a speed of 10 m/sec when it reaches the surface of Earth. Calculate the time it takes to reach the surface.

Solution

(a) For a spherical raindrop of radius r and density ρ, the cross-sectional area A and mass m are as shown.

$$A = \pi \cdot r^2 \qquad\qquad m = \rho \cdot \frac{4 \cdot \pi \cdot r^3}{3}$$

If k is the constant of proportionality, the rate at which the raindrop gains mass is

$$\frac{d}{dt}m = k \cdot \pi \cdot r^2 \cdot v \tag{i}$$

---

[1] Arom Mu-Shiang Mu, *The Physics Teacher*. April 1986.

The initial momentum pi of a particle of mass m moving with velocity v is

$$pi \equiv m \cdot v$$

The final momentum pf after the drop has gained mass dm and velocity dv, is or

$$pf \equiv (m + dm) \cdot (v + dv)$$

$$pf \equiv m \cdot v + m \cdot dv + dm \cdot v + dm \cdot dv$$

pf after neglecting the product dm dv is

$$pf \equiv m \cdot v + v \cdot dm + m \cdot dv$$

The change in the momentum dp is

$$dp \equiv pf - pi \equiv v \cdot dm + m \cdot dv \tag{ii}$$

The change in the momentum is also

$$dp \equiv F \cdot dt \equiv m \cdot g \cdot dt \tag{iii}$$

Combining the two equations

$$v \cdot dm + m \cdot dv \equiv m \cdot g \cdot dt$$

Substituting for m and dm, rearranging, and noting that a = dv/dt, gives

$$\frac{d}{dt} v \equiv g - \left(\frac{v}{m}\right) \cdot \left(k \cdot \pi \cdot r^2 \cdot v\right)$$

$$a \equiv g - \frac{v^2}{m} \cdot k \cdot \pi \cdot r^2 \qquad \text{or} \qquad a \equiv g - \frac{3}{4} \cdot \frac{v^2}{(\rho \cdot r)} \cdot k$$

Substituting for a = dv/dt and v = dy/dt or writing y in terms of double integral, we can solve for y.

$$\frac{d^2}{dt^2} y \equiv \left[g - \frac{3}{4} \cdot \frac{v^2}{(\rho \cdot r)} \cdot k \right]$$

$$y \equiv \int \int \left[g - \frac{3}{4} \cdot \frac{v^2}{(\rho \cdot r)} \right] dt\, dt \equiv \frac{1}{2} \cdot t^2 \cdot g - \frac{3}{8} \cdot t^2 \cdot \frac{v^2}{(\rho \cdot r)}$$

The resulting equation for y is

$$y \equiv \frac{1}{8} \cdot t^2 \cdot \frac{\left(4 \cdot g \cdot \rho \cdot r - 3 \cdot v^2\right)}{(\rho \cdot r)} \tag{iv}$$

Solving the above equation for t gives two roots. The positive root yields the expression for t given by Eq. (v).

$$\left[\begin{array}{c} -2 \cdot \sqrt{\rho} \cdot \dfrac{\sqrt{r}}{\sqrt{4 \cdot g \cdot \rho \cdot r - 3 \cdot v^2 \cdot k}} \cdot \sqrt{y} \cdot \sqrt{2} \\[4ex] 2 \cdot \sqrt{\rho} \cdot \dfrac{\sqrt{r}}{\sqrt{4 \cdot g \cdot \rho \cdot r - 3 \cdot v^2 \cdot k}} \cdot \sqrt{y} \cdot \sqrt{2} \end{array} \right]$$

(a) What is the significance of the negative root?

(b) Calculate the value of v and a as a function of time t.

$$t \equiv 2 \cdot \sqrt{\rho} \cdot \frac{\sqrt{r}}{\sqrt{4 \cdot g \cdot \rho \cdot r - 3 \cdot v^2 \cdot k}} \cdot \sqrt{y} \cdot \sqrt{2} \tag{v}$$

Given the values for the raindrop, calculate the time t it takes to hit the surface. (As a check, the value of y is also calculated by using Eq. (iv) and equals 3000 m.).

$$r := .001 \cdot m \qquad v := 10 \cdot \frac{m}{sec} \qquad k := .005$$

$$\rho := 1.1 \qquad y := -3000 \cdot m \qquad g := 9.8 \cdot \frac{m}{sec^2}$$

$$t := 2 \cdot \sqrt{\rho} \cdot \frac{\sqrt{r}}{\sqrt{4 \cdot g \cdot \rho \cdot r - 3 \cdot v^2 \cdot k}} \cdot \sqrt{y} \cdot \sqrt{2} \qquad t = 4.257 \cdot sec$$

(c) Calculate the value of v and a just before the raindrop hits the ground.

$$y := \frac{1}{2} \cdot t^2 \cdot g - \frac{3}{8} \cdot t^2 \cdot \frac{v^2}{(\rho \cdot r)} \cdot k$$

$$y = -3 \cdot 10^3 \cdot m$$

EXERCISE 8.2 Repeat the above example for a raindrop that falls from a height of 5000 m and has a radius of 2 mm and a speed of 20 m/sec when it reaches the surface of Earth. Calculate the time it takes to reach Earth's surface. The rest of the constants have the same values as in the above example.

 Example 8.3 _____

Consider a one-stage rocket, assuming constant g, having only a vertical thrust, a constant rate of change of mass k, and a constant exhaust velocity v0. **(a)** Graph the velocity and the altitude as a function of time. **(b)** Calculate the final velocity v and the altitude z.

Solution

After defining constant k and substituting for a and k, we may write the equation for the rocket as

mi = initial mass
mf = final mass

Solving for v, after integration and simplification, gives

$$\frac{d}{dt} m = -k$$

$$m \cdot a = \left(\frac{d}{dt} v\right) \cdot m = k \cdot v0 - m \cdot g$$

$$v = -v0 \cdot \left[\int_{mi}^{mf} \frac{1}{m} \, dm + \frac{g}{k} \cdot \int_{mi}^{mf} 1 \, dm \right]$$

$$v = \frac{(-v0 \cdot k \cdot \ln(mf) + v0 \cdot k \cdot \ln(mi) + g \cdot mf - g \cdot mi)}{k}$$

$$v = -v0 \cdot \ln(mf) + v0 \cdot \ln(mi) + \frac{1}{k} \cdot g \cdot mf - \frac{1}{k} \cdot g \cdot mi$$

Using the definition v = dz/dt, we can write an expression for v, which on integration yields the value of z.

$$z = \int_{mi}^{mf} -v0 \cdot \ln(mf) + v0 \cdot \ln(mi) + \frac{1}{k} \cdot g \cdot mf - \frac{1}{k} \cdot g \cdot mi \, dm$$

$$z = (-v0 \cdot k \cdot \ln(mf) + v0 \cdot k \cdot \ln(mi) + g \cdot mf - g \cdot mi) \cdot \frac{(mf - mi)}{k}$$

Using these expressions and the values of mi, mf, v0, and k, we can calculate the final values of z and v.

$$mi := 60000 \quad mf := 45000 \quad v0 := 6000 \quad k := 150 \quad g := 9.8$$

$$z := (-v0 \cdot k \cdot \ln(mf) + v0 \cdot k \cdot \ln(mi) + g \cdot mf - g \cdot mi) \cdot \frac{(mf - mi)}{k}$$

$$z = -1.119 \cdot 10^7$$

$$v := -v0 \cdot \ln(mf) + v0 \cdot \ln(mi) + \frac{1}{k} \cdot g \cdot mf - \frac{1}{k} \cdot g \cdot mi$$

$$v = 746.092$$

We graph the results.

$v0 := 6000 \quad mi := 1000 \quad k := 150 \quad g := -9.8$

$$n := 1..20 \qquad mf_n := n \cdot \frac{mi}{20} \qquad m_n := \frac{mf_n}{mi}$$

Looking at the graph and the data, answer the following questions.

$$z_n := \left(-v0 \cdot k \cdot \ln\left(mf_n\right) + v0 \cdot k \cdot \ln(mi) + g \cdot mf_n - g \cdot mi \right) \cdot \frac{\left(mf_n - mi\right)}{k}$$

(a) Why do the values of z and v have opposite signs?

$$z_n := \left(v0 \cdot k \cdot \ln\left(\frac{mi}{mf_n}\right) + g \cdot mf_n - g \cdot mi \right) \cdot \frac{\left(mf_n - mi\right)}{k}$$

(b) What effect does the negative value of g have on the value of z and v?

$$v_n := -v0 \cdot \ln\left(mf_n\right) + v0 \cdot \ln(mi) + \frac{1}{k} \cdot g \cdot mf_n - \frac{1}{k} \cdot g \cdot mi$$

$z_1 = -1.713 \cdot 10^7 \qquad v_1 = 1.804 \cdot 10^4$

$z_4 = -7.767 \cdot 10^6 \qquad v_4 = 9.709 \cdot 10^3$

$z_8 = -3.322 \cdot 10^6 \qquad v_8 = 5.537 \cdot 10^3$

$z_{12} = -1.236 \cdot 10^6 \qquad v_{12} = 3.091 \cdot 10^3$

$z_{16} = -2.704 \cdot 10^5 \qquad v_{16} = 1.352 \cdot 10^3$

$z_{20} = 0 \qquad v_{20} = 0$

$m_{20} = 1 \qquad mf_{20} = 1 \cdot 10^3$

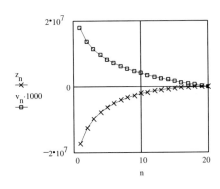

EXERCISE 8.3   Show that the maximum height achieved by the rocket discussed in the example is

$$z_{max} = \frac{v_0^2 (\ln R)^2}{2g} - v_0 t_b \left(\frac{\ln R}{1 - R} - 1 \right)$$

(*Hint*: After burnout, the rocket will continue to climb without power; $mgh = \frac{1}{2} m v_b^2$.)

8.7 ELASTIC COLLISIONS AND CONSERVATION LAWS

When two or more objects come close enough (with or without any physical contact) so that there is some sort of interaction between them, with or without the presence of external forces, we say that a collision has taken place between the objects. After a collision, the velocities of the colliding objects may or may not be the same as before the collision. Very often we are interested in describing the nature of the interaction (or the type of force) between microscopic particles. If the particles are incident on a target, the paths and energies of the interacting particles will change. By measuring the energies and the angular distributions of these scattered particles, we can gain information about their structure and the nature of the forces involved.

By applying conservation laws, many details of a collision can be predicted without knowing much about the nature of the interaction or force. Collisions may be divided into two broad categories: *elastic collisions,* in which both linear momentum and kinetic energy are conserved, and *inelastic collisions,* in which conservation of linear momentum holds good, but kinetic energy is not conserved. Thus, if $\mathbf{P}_i$ and K_i are the initial linear momentum and kinetic energy before collision, while $\mathbf{P}_f$ and K_f are the final linear momentum and kinetic energy after collision, then

$$\text{For elastic collisions:} \qquad \mathbf{P}_i = \mathbf{P}_f \qquad \text{and} \qquad K_i = K_f \qquad \textbf{(8.65)}$$

$$\text{For inelastic collisions:} \qquad \mathbf{P}_i = \mathbf{P}_f \quad \text{and} \quad K_i \neq K_f \qquad \textbf{(8.66)}$$

In this section, we limit our discussion to elastic collisions.

Let us consider an elastic collision between two objects, as shown in Fig. 8.5. An object of mass m_1 moving with a velocity $\mathbf{v}_{1i}$, called the *incident particle,* strikes an object of mass m_2 at rest, called the *target particle,* both being along the X-axis. (Nothing is lost in generality by assuming one of the masses to be at rest. If both masses were moving, we could view the collision from a reference frame that is moving with the same velocity as that of one of the masses, say m_2. In that frame of reference, m_2 will be at rest.) After collision, mass m_1 is moving with velocity $\mathbf{v}_{1f}$, making an angle θ with the X-axis, and mass m_2 is moving with a velocity $\mathbf{v}_{2f}$, making an angle ϕ with the X-axis, as shown in Fig. 8.5(b). Remember that if $\mathbf{v}_{1f}$ is in the XY-plane, then $\mathbf{v}_{2f}$ also must be in the same XY-plane. This is because if $\mathbf{v}_{2f}$ is not in the XY-plane there will be a component of velocity, after collision, in the Z direction; but this cannot happen because there was no Z component of velocity before the collision, hence leading to nonconservation of linear momentum.

The conservation of linear momentum and energy requires

$$\mathbf{P}_i = \mathbf{P}_f \quad \text{and} \quad K_i = K_f \qquad \textbf{(8.67)}$$

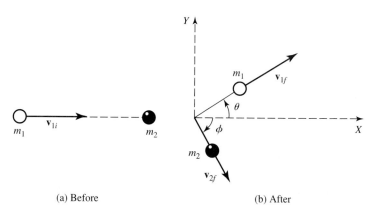

(a) Before   (b) After

Figure 8.5   Elastic collision between two objects: (a) before, and (b) after collision.

where $\mathbf{P}_i$ and $\mathbf{P}_f$ are the initial and final linear momenta, while K_i and K_f are the initial and final kinetic energies. That is,

$$\mathbf{P}_{1i} + \mathbf{P}_{2i} = \mathbf{P}_{1f} + \mathbf{P}_{2f} \tag{8.68}$$

and

$$K_{1i} + K_{2i} = K_{1f} + K_{2f} \tag{8.69}$$

where

$$\mathbf{P}_{1i} = m_1\mathbf{v}_{1i}, \qquad \mathbf{P}_{2i} = 0, \qquad \mathbf{P}_{1f} = m_1\mathbf{v}_{1f}, \qquad \mathbf{P}_{2f} = m_2\mathbf{v}_{2f}$$

$$K_{1i} = \tfrac{1}{2}m_1v_{1i}^2, \qquad K_{2i} = 0, \qquad K_{1f} = \tfrac{1}{2}m_1v_{1f}^2, \qquad K_{2f} = \tfrac{1}{2}m_2v_{2f}^2$$

Using these and writing Eq. (8.68) in component form along the X- and Y-axes with the help of Fig. 8.5, we obtain

$$m_1v_{1i} = m_1v_{1f}\cos\theta + m_2v_{2f}\cos\phi \tag{8.70}$$

$$0 = m_1v_{1f}\sin\theta - m_2v_{2f}\sin\phi \tag{8.71}$$

and, from Eq. (8.69), we get

$$\tfrac{1}{2}m_1v_{1i}^2 = \tfrac{1}{2}m_1v_{1f}^2 + \tfrac{1}{2}m_2v_{2f}^2 \tag{8.72}$$

In most situations, m_1, m_2, and v_{1i} are known, while v_{1f}, v_{2f}, θ, and ϕ are the unknown quantities. Thus we have three equations [(8.70), (8.71), (8.72)] and four unknowns. We can eliminate one of the four unknowns, say ϕ, and find the relations between the other three, v_{1f}, v_{2f}, and θ. We may write Eqs. (8.70) and (8.71) as

$$m_1v_{1i} - m_1v_{1f}\cos\theta = m_2v_{2f}\cos\phi$$

$$m_1v_{1f}\sin\theta = m_2v_{2f}\sin\phi$$

Squaring and adding these equations and dividing by m_1^2 yield

$$v_{1i}^2 + v_{1f}^2 - 2v_{1i}v_{1f}\cos\theta = \left(\frac{m_2}{m_1}\right)^2 v_{2f}^2 \tag{8.73}$$

while from Eq. (8.72) we obtain

$$v_{2f}^2 = \frac{m_1}{m_2}(v_{1i}^2 - v_{1f}^2) \tag{8.74}$$

Substituting for v_{2f}^2 from Eq. (8.74) into Eq. (8.73) resulting in a quadratic equation in v_{1f}/v_{1i}, which when solved gives

$$\frac{v_{1f}}{v_{1i}} = \frac{m_1}{m_1 + m_2}\left[\cos\theta \pm \sqrt{\cos^2\theta - \left(\frac{m_1^2 - m_2^2}{m_1^2}\right)}\right] \tag{8.75}$$

This equation reveals a great deal of information about elastic collisions. In the following discussion we must keep in mind that the quantity under the radical sign cannot be negative because that would yield complex value for v_{1f}, which is physically meaningless.

Case (a) $\theta = 0$:   These are collisions in one dimension; that is, they correspond to a *head-on* collision. Substituting $\theta = 0$ in Eq. (8.75) yields

$$\frac{v_{1f}}{v_{1f}} = 1 \quad \text{or} \quad \frac{v_{1f}}{v_{1f}} = \frac{m_1 - m_2}{m_1 + m_2} \tag{8.76}$$

Substituting these in Eq. (8.74) yields

$$v_{2f} = 0, \qquad \text{if } \frac{v_{1f}}{v_{1f}} = 1 \tag{8.77}$$

which corresponds to no collision; and

$$v_{2f} = \frac{2m_1}{m_1 + m_2} v_{1f} \tag{8.78}$$

if

$$v_{1f} = \frac{m_1 - m_2}{m_1 + m_2} v_{1i} \tag{8.79}$$

Thus Eqs. (8.78) and (8.79) represent head-on collisions, that is, collisions in one direction. Let us consider a few special cases of these two equations for head-on collisions.

(i) Suppose $m_1 = m_2$. Equations (8.78) and (8.79) give

$$v_{1f} = 0 \quad \text{and} \quad v_{2f} = v_{1i} \tag{8.80}$$

That is, the incident particle comes to a stop, while the target particle starts moving with the velocity of the incident particle.

(ii) If $m_1 \ll m_2$, we get

$$v_{1f} \simeq -v_{1i} \quad \text{and} \quad v_{2f} \simeq 0 \tag{8.81}$$

That is, the incident particle is reflected back with the same speed, while the target particle hardly moves.

(iii) If $m_1 \gg m_2$, we get

$$v_{1f} \simeq v_{1i} \quad \text{and} \quad v_{2f} \simeq 2v_{1i} \tag{8.82}$$

That is, the incident particle keeps moving as if nothing happened, while the target particle takes off with twice the velocity of the incident particle.

All the preceding situations are illustrated in Fig. 8.6.

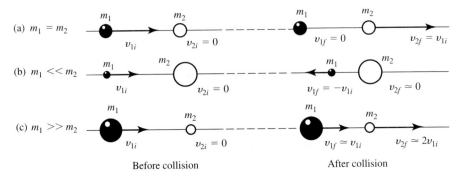

(a) $m_1 = m_2$

(b) $m_1 \ll m_2$

(c) $m_1 \gg m_2$

Before collision   After collision

Figure 8.6   One-dimensional elastic collision between two objects.

Case (b) $m_1 > m_2$:   For v_{1f} to be real, the quantity under the radical sign must be positive; that is,

$$\cos^2\theta \geq \frac{m_1^2 - m_2^2}{m_1^2} \tag{8.83}$$

Furthermore, the quantity under the radical sign will be zero (minimum), say for $\theta = \theta_m$, which according to Eq. (8.83) is

$$\cos^2\theta_m = \frac{m_1^2 - m_2^2}{m_1^2} = 1 - \frac{m_2^2}{m_1^2}, \qquad 0 \leq \theta_m \leq \frac{\pi}{2} \tag{8.84}$$

The scattering angle θ must be less than θ_m, because, if $\theta > \theta_m$ and $\pi/2 \leq \theta \leq \pi$, the quantity under the radical sign will be negative. Thus θ_m represents the maximum angle $= \theta_{\max}$; hence (because $\cos\theta$ decreases with increasing θ)

$$\theta \leq \theta_{\max} \quad \text{and} \quad 0 < \theta_{\max} < \frac{\pi}{2} \tag{8.85}$$

Figure 8.7 shows the plot of maximum scattering angle $\theta_{\max}$ versus m_2/m_1. Note that if $m_1 \gg m_2$, the scattering angle will be very small (a very large mass can hardly be expected to be deflected by a small mass at rest). Furthermore, for $\theta < \theta_{\max}$, there will be two values of v_{1f}/v_{1i}; the larger value corresponds to be *glancing* collision, whereas the smaller value corresponds to a head-on collision.

Case (c) $m_1 < m_2$:   For this case there is no restriction on the value of the scattering angle, which can be anywhere from 0 to π. A situation in which θ is greater than $\pi/2$ is called *backscattering*. If $\theta = 0$, $v_{1f}/v_{1i} = 1$, which corresponds to no collisions. If $\theta = 0$ and $\phi = 0$, we get [as in case (a)]

$$\frac{v_{1f}}{v_{1i}} = \frac{m_1 - m_2}{m_1 + m_2} \quad \text{and} \quad \frac{v_{2f}}{v_{1i}} = \frac{2m_1}{m_1 + m_2} \tag{8.86}$$

Figure 8.7 _____

Consider an elastic collision between a particle of mass m1 moving with velocity v1 and a particle of mass m2 at rest. The graph of the scattering angle θ as a function of the mass ratio m = m2/m1 where m1 > m2 is shown below.

According to Eq. (8.84) $\cos(\theta)^2 = 1 - m^2$ and $0 < \theta < \dfrac{\pi}{2}$ where $m = \dfrac{m2}{m1}$

Solving for θ gives $\cos(\theta) = \sqrt{1 - m^2}$ $\mathrm{acos}\left(\sqrt{1 - m^2}\right) = \theta$

Solve for N = 20 different values of m_i resulting in 20 different values of θ_i .

$N := 20$ $i := 0 .. N$

(a) When m is about 0, that is, m1 >> m2, the value of the scattering angle θ_i is 0 degree.

$m_i := \dfrac{i}{20}$ $\theta_i := -\mathrm{acos}\left[\sqrt{1 - (m_i)^2}\right] \cdot \dfrac{360}{2 \cdot \pi}$

(b) When m1 = m2, and m_i = 1, the scattering angle is maximum, that is, θmax = 90 degree = π/2.

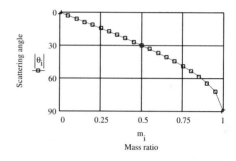

_____

We can also show that for head-on collisions

$$\frac{m_1}{m_2} = \frac{2K_{1i}}{K_{2f}} - 1 \pm \left[\left(\frac{2K_{1i}}{K_{2f}} - 1\right) - 1\right]^{1/2} \qquad (8.87)$$

Case (d) $m_1 = m_2$: By multiplying Eq. (8.70) by cos θ and Eq. (8.71) by sin θ and adding, we get

$$v_{1i} \cos \theta = v_{1f} + v_{2f} \cos(\theta + \phi) \qquad (8.88)$$

Since $m_1 = m_2$, Eq. (8.75) yields

$$v_{1f} = v_{1i} \cos \theta \qquad (8.89)$$

From Eqs. (8.88) and (8.89), we obtain

$$\cos(\theta + \phi) = 0 \quad \text{or} \quad \theta + \phi = \frac{\pi}{2} \qquad (8.90)$$

That is, the two particles leave at right angles to each other. The example of such a collision is observed on a pool table when a cue ball is seen to leave the struck ball at a right angle.

8.8 INELASTIC COLLISIONS

In many situations in both the microscopic and macroscopic worlds, the kinetic energy of the system before collision is not the same as after collision; that is, kinetic energy is not conserved. For example, atoms, molecules, and nuclei possess internal kinetic and potential energies. When such particles collide, kinetic energy may be absorbed or released. Collisions in which the final kinetic energy of the system is less than the initial kinetic energy (that is, energy is absorbed by the system), are called *endoergic* or *first kind reactions* or *collisions*. Collisions in which the final kinetic energy is more than the initial kinetic energy (that is, energy is released), are called *exoergic* or *second kind reactions* or *collisions*. Thus if the initial kinetic energy is K_i and the final kinetic energy is K_f, the *disintegration energy* Q of the reaction is defined as

$$Q = K_f - K_i \tag{8.91}$$

If $Q > 0$ exoergic,	inelastic second kind	**(8.92a)**
If $Q < 0$ endoergic,	inelastic first kind	**(8.92b)**
If $Q = 0$	elastic collision	**(8.92c)**

In all of these cases, the law of the conservation of linear momentum holds good. The law of conservation of energy will hold good only if all internal energies, as well as any other energy, such as heat by friction, are taken into account. Furthermore, in inelastic collisions the nature of the particles after collision may be completely different from those before collision.

Let us consider an inelastic collision between a particle of mass m_1 moving with velocity $\mathbf{v}_{1i}$ with a particle of mass m_2 at rest, as shown in Fig. 8.8. The collision between these two par-

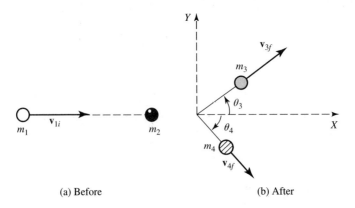

(a) Before   (b) After

Figure 8.8   Inelastic collision between two particles: (a) before, and (b) after collision.

ticles results in two new particles of mass m_3 and m_4 moving with velocities $\mathbf{v}_{3f}$ and $\mathbf{v}_{4f}$, making angles θ_3 and θ_4 with the initial direction of the velocity of the incident particle m_1, which is the X-axis. Let K_1, K_2 (=0 in this case), K_3, and K_4, be the kinetic energies of particles m_1, m_2, m_3, and m_4, respectively, and Q the disintegration energy. From the laws of conservation of momentum and kinetic energy, we may write

$$m_1 v_{1i} = m_3 v_{3f} \cos \theta_3 + m_4 v_{4f} \cos \theta_4 \qquad (8.93)$$

$$0 = m_3 v_{3f} \sin \theta_3 - m_4 v_{4f} \sin \theta_4 \qquad (8.94)$$

and
$$K_1 + Q = K_3 + K_4 \qquad (8.95)$$

θ_4 can be eliminated from Eqs. (8.93) and (8.94) by rearranging, squaring, and adding, resulting in

$$(m_4 v_{4f})^2 = (m_1 v_{1i})^2 + (m_3 v_{3f})^2 - 2m_1 m_3 v_{1i} v_{3f} \cos \theta_3 \qquad (8.96)$$

Combining Eqs. (8.95) and (8.96) and using the relations

$$K_1 = \tfrac{1}{2} m_1 v_{1i}^2, \qquad K_3 = \tfrac{1}{2} m_3 v_{3f}^2, \qquad K_4 = \tfrac{1}{2} m_4 v_{4f}^2$$

we may obtain the following value for Q:

$$Q = K_3 + K_4 - K_1 = K_3\left(1 + \frac{m_3}{m_4}\right) - K_1\left(1 - \frac{m_1}{m_4}\right) - 2\left(\frac{m_1 m_3 K_1 K_3}{m_4^2}\right)^{1/2} \cos \theta_3 \qquad (8.97)$$

Thus, when a particle of mass m_1 and known velocity $\mathbf{v}_{1i}$ collides with mass m_2, Eq. (8.97) allows us to calculate the value of Q by measuring θ_3 and the velocity $\mathbf{v}_{3f}$ of a particle of mass m_3 with additional knowledge of mass m_4. Note that we have eliminated the quantity $\mathbf{v}_{4f}$ because it is usually, especially in nuclear reactions, very hard to measure this quantity.

Consider an inelastic collision in one dimension between two objects. Such collisions are always endoergic, as we will show now. Suppose an object of mass m_1 moving with velocity v_1 collides with an object of mass m_2 at rest, the two objects stick together after the collision (such as a bullet striking a piece of wood and becoming embedded), and now move together with velocity v_2. Thus, according to the law of conservation of momentum,

$$m_1 v_1 = (m_1 + m_2)v_2$$

or
$$v_2 = \frac{m_1 v_1}{m_1 + m_2} \qquad (8.98)$$

Kinetic energy is not conserved in this case; hence

$$Q = K_f - K_i = \tfrac{1}{2}(m_1 + m_2)v_2^2 - \tfrac{1}{2}m_1 v_1^2$$

Substituting for v_2 from Eqs. (8.98), we obtain

$$Q = K_1 \frac{-m_2}{m_1 + m_2} \qquad (8.99)$$

which is a negative quantity; hence the collision is endoergic. The amount of energy changed into heat is equal to $|Q|$. In our discussion we have assumed that no rotational energies are involved. Equation (8.99) also requires that the minimum kinetic energy K_1 needed to start an endoergic reaction be greater than $|Q|$ by a factor $1 + m_1/m_2$. Thus the minimum energy is called the *threshold energy*.

$$(K_1)_{\text{thres}} = \left(1 + \frac{m_1}{m_2}\right)|Q| \tag{8.100}$$

For endoergic reactions in general, K_1 must be $\geq (K_1)_{\text{thres}}$.

Finally, let us define another commonly used term regarding collisions, the *coefficient of restitution*. Consider a head-on elastic collision between two masses, as shown in Fig. 8.9. The laws of conservation of momentum and energy require that

$$m_1 v_{1i} + m_2 v_{2i} = m_1 v_{1f} + m_2 v_{2f} \tag{8.101}$$

$$\tfrac{1}{2} m_1 v_{1i}^2 + \tfrac{1}{2} m_2 v_{2i}^2 = \tfrac{1}{2} m_1 v_{1f}^2 + \tfrac{1}{2} m_2 v_{2f}^2 \tag{8.102}$$

Solving these two equations yields

$$v_{2f} - v_{1f} = v_{1i} - v_{2i} \tag{8.103a}$$

$$(v_{\text{rel}})_f = -(v_{\text{rel}})_i \tag{8.103b}$$

$$\text{Speed of recession} = \text{speed of approach}$$

This results states that *the ratio of the relative velocity after collision to the relative velocity before collision between the two bodies in a head-on collision is constant.* This may be written as

$$v_{2f} - v_{1f} = e(v_{1i} - v_{2i}) \tag{8.104}$$

where e is called the *coefficient of restitution*. As is obvious, if the collision is elastic, $e = 1$, while for a perfect inelastic collision (in which two bodies move as one after collision) $e = 0$. For other inelastic collisions, e varies between 0 and 1.

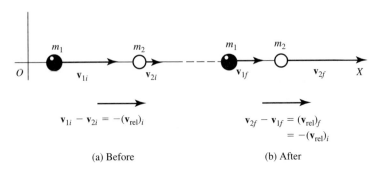

(a) Before   (b) After

Figure 8.9 One-dimensional elastic collision between two masses, m_1 and m_2, showing that $(\mathbf{v}_{\text{rel}})_f = -(\mathbf{v}_{\text{rel}})_i$.

8.9 TWO-BODY PROBLEM IN CENTER-OF-MASS COORDINATE SYSTEM

In many situations we find it quite convenient and useful to describe the motion of a system consisting of two bodies as observed in a center-of-mass coordinate system (CMCS, or CM system) instead of in a laboratory coordinate system (LCS, or LAB system). Furthermore, we shall describe the collisions between two objects as viewed from a CMCS. The advantage of using a CMCS is that under special circumstances the two-body problem can be reduced to two single-body problems described as (1) the motion of the center of mass, and (2) the relative motion (that is, the motion of either particle with respect to the other). The CMCS was discussed in Chapter 7, but we shall consider further details of the system in this chapter.

Let us consider a system consisting of two bodies of mass m_1 and m_2 at distances $\mathbf{r}_1$ and $\mathbf{r}_2$ from the origin O, as shown in Fig. 8.10. Let $\mathbf{F}_1^e$ and $\mathbf{F}_2^e$ be the external forces acting on m_1 and m_2, respectively, while $\mathbf{F}_{12}^i$ is the internal force acting on body m_1 due to m_2, and $\mathbf{F}_{21}^i$ the internal force acting on m_2 due to m_1. According to Newton's third law, force $\mathbf{f}$ may be defined as

$$\mathbf{F}_{12}^i = -\mathbf{F}_{21}^i = \mathbf{f} \tag{8.105}$$

while the total external force acting on the system is

$$\mathbf{F} = \mathbf{F}_1^e + \mathbf{F}_2^e \tag{8.106}$$

According to Newton's second law, the motion of the two bodies in the LAB system may be written as

$$m_1\ddot{\mathbf{r}}_1 = \mathbf{F}_1^e + \mathbf{F}_{12}^i \tag{8.107}$$

$$m_2\ddot{\mathbf{r}}_2 = \mathbf{F}_2^e + \mathbf{F}_{21}^i \tag{8.108}$$

To change now from a LAB system to a CM system, we use the following relations, which were discussed in Chapter 7. The *center-of-mass coordinate* $\mathbf{R}$ is given by (from Section 7.2)

$$\mathbf{R} = \frac{m_1\mathbf{r}_1 + m_2\mathbf{r}_2}{m_1 + m_2} \tag{8.109}$$

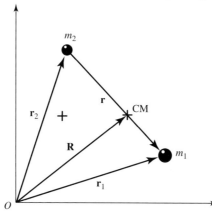

Figure 8.10   Center of mass and relative motion for a system consisting of two particles.

and the *relative coordinate* $\mathbf{r}$ is given by

$$\mathbf{r} = \mathbf{r}_1 - \mathbf{r}_2 \qquad (8.110)$$

while the reverse transformations are given by

$$\mathbf{r}_1 = \mathbf{R} + \frac{m_2}{m_1 + m_2}\mathbf{r} \qquad (8.111)$$

$$\mathbf{r}_2 = \mathbf{R} - \frac{m_1}{m_1 + m_2}\mathbf{r} \qquad (8.112)$$

We want to rewrite the equations of motion of the two bodies m_1 and m_2 in terms of the CM coordinates $\mathbf{R}$ and the relative coordinates $\mathbf{r}$. To do this, we first add Eqs. (8.107) and (8.108); that is,

$$m_1\ddot{\mathbf{r}}_1 + m_2\ddot{\mathbf{r}}_2 = \mathbf{F}_1^e + \mathbf{F}_2^e + \mathbf{F}_{12}^i + \mathbf{F}_{21}^i$$

Using Eqs. (8.105), (8.106), and (8.109), the preceding equation may be written as

$$(m_1 + m_2)\ddot{\mathbf{R}} = \mathbf{F}$$

or
$$M\ddot{\mathbf{R}} = \mathbf{F} \qquad (8.113)$$

where $M = m_1 + m_2$ is the total mass and $\mathbf{F}$ is the total external force acting on the system. This is first of the two equations we are looking for.

Now multiply Eq. (8.107) by m_2 and Eq. (8.108) by m_1 and subtract:

$$m_1 m_2(\ddot{\mathbf{r}}_1 - \ddot{\mathbf{r}}_2) = m_2\mathbf{F}_1^e - m_1\mathbf{F}_2^e + m_2\mathbf{F}_{12}^i - m_1\mathbf{F}_{21}^i$$

Using the result given in Eq. (8.105), we may write this equation as

$$m_1 m_2(\ddot{\mathbf{r}}_1 - \ddot{\mathbf{r}}_2) = m_1 m_2\left(\frac{\mathbf{F}_1^e}{m_1} - \frac{\mathbf{F}_2^e}{m_2}\right) + (m_1 + m_2)\mathbf{f} \qquad (8.114)$$

Let us consider a special case in which either

$$\mathbf{F}_1^e = \mathbf{F}_2^e = 0 \qquad (8.115)$$

or
$$\frac{\mathbf{F}_1^e}{m_1} = \frac{\mathbf{F}_2^e}{m_2} \qquad (8.116)$$

That is, the external forces acting on the objects are proportional to their masses; hence Eq. (8.114) may be written as

$$m_1 m_2(\ddot{\mathbf{r}}_1 - \ddot{\mathbf{r}}_2) = (m_1 + m_2)\mathbf{f} \qquad (8.117)$$

Introducing the quantity *reduced mass* μ, defined as

$$\mu = \frac{m_1 m_2}{m_1 + m_2} \qquad (8.118)$$

and using Eq. (8.110), $\mathbf{r} = \mathbf{r}_1 - \mathbf{r}_2$, we may write Eq. (8.117) as

$$\mu\ddot{\mathbf{r}} = \mathbf{f} \tag{8.119}$$

Thus Eqs. (8.113) and (8.119) are the two required equations. Equation (8.113) is the familiar equation for the motion of the center of mass according to which mass M is acted on by the total external force $\mathbf{F}$, producing an acceleration $\ddot{\mathbf{R}}$, while Eq. (8.119) is the equation of motion of mass μ acted on by an internal force $\mathbf{f} = \mathbf{F}_{12}^i$, producing an acceleration $\ddot{\mathbf{r}}$. Equation (8.119) may also be described as the motion of particle of mass μ at the position of m_1 as viewed from the position of m_2, assuming m_2 to be at rest.

We may also write an expression for the linear momentum $\mathbf{P}$, angular momentum $\mathbf{L}$, and total kinetic energy K in terms of CM coordinates. Using Eqs. (8.109) through (8.112), we may write the center-of-mass velocity $\mathbf{V}$ as

$$\mathbf{V} = \dot{\mathbf{R}} = \frac{m_1\dot{\mathbf{r}}_1 + m_2\dot{\mathbf{r}}_2}{m_1 + m_2} = \frac{m_1\dot{\mathbf{r}}_1 + m_2\dot{\mathbf{r}}_2}{M} \tag{8.120}$$

and the relative velocity $\mathbf{v}$ as

$$\mathbf{v} = \dot{\mathbf{r}} = \dot{\mathbf{r}}_1 - \dot{\mathbf{r}}_2 \tag{8.121}$$

And for the inverse transformation, we may write

$$\mathbf{v}_1 = \dot{\mathbf{r}}_1 = \dot{\mathbf{R}} + \frac{m_2}{m_1 + m_2}\dot{\mathbf{r}} = \dot{\mathbf{R}} + \frac{\mu}{m_1}\dot{\mathbf{r}} \tag{8.122}$$

$$\mathbf{v}_2 = \dot{\mathbf{r}}_2 = \dot{\mathbf{R}} - \frac{m_1}{m_1 + m_2}\dot{\mathbf{r}} = \dot{\mathbf{R}} - \frac{\mu}{m_2}\dot{\mathbf{r}} \tag{8.123}$$

The total linear momentum of the system is

$$\mathbf{P} = m_1\dot{\mathbf{r}}_1 + m_2\dot{\mathbf{r}}_2 = M\dot{\mathbf{R}} \tag{8.124}$$

and the total angular momentum $\mathbf{L}$ of the system is

$$\mathbf{L} = m_1(\mathbf{r}_1 \times \dot{\mathbf{r}}_1) + m_2(\mathbf{r}_2 \times \dot{\mathbf{r}}_2) \tag{8.125}$$

Substituting for $\dot{\mathbf{r}}_1$ and $\dot{\mathbf{r}}_2$ from Eqs. (8.122) and (8.123), we obtain

$$\mathbf{L} = M(\mathbf{R} \times \dot{\mathbf{R}}) + \mu(\mathbf{r} \times \dot{\mathbf{r}})$$

or

$$\mathbf{L} = M(\mathbf{R} \times \mathbf{V}) + \mu(\mathbf{r} \times \mathbf{v}) \tag{8.126}$$

while the total kinetic energy K is given by

$$K = \tfrac{1}{2}m_1\dot{r}_1^2 + \tfrac{1}{2}m_2\dot{r}_2^2 \tag{8.127}$$

Substituting for $\dot{r}_1$ and $\dot{r}_2$, we get

$$K = \tfrac{1}{2}M\dot{R}^2 + \tfrac{1}{2}\mu v^2 \tag{8.128}$$

or
$$K = \tfrac{1}{2} MV^2 + \tfrac{1}{2} \mu v^2 \tag{8.129}$$

This equation states that the kinetic energy of a system is equal to the sum of the kinetic energy of mass M moving with velocity V of the center of mass (kinetic energy of the center of mass) and the kinetic energy of the reduced mass μ moving with a relative velocity v (kinetic energy of relative motion).

8.10 COLLISIONS IN CENTER-OF-MASS COORDINATE SYSTEM

In previous sections, we discussed elastic and inelastic collisions between two objects from the point of view of an observer at rest with respect to the coordinates fixed in a laboratory coordinate system (LCS). In many circumstances, it is convenient to make observations from a coordinate system that is moving with respect to the LCS. One such coordinate system commonly used is the center-of-mass coordinate system (CMCS) as discussed previously. Collisions are observed by an observer at the center of mass, hence moving with the same velocity as the center of mass. We start with the discussion of elastic collisions between two objects as observed from the center of mass.

Suppose at a given instant a particle of mass m_1 at x_1 is moving with velocity v_{1i}, while a particle of mass m_2 at x_2 is at rest, as shown in Fig. 8.11. The center of mass x_c is given by

$$(m_1 + m_2)x_c = m_1 x_1 + m_2 x_2 \tag{8.130}$$

while the velocity of the center of mass obtained by differentiating Eq. (8.130) is

$$(m_1 + m_2)v_c = m_1 \dot{x}_1 + m_2 \dot{x}_2 \tag{8.131}$$

where $v_c = dx_c/dt$, while for the situation shown in Fig. 8.11, $\dot{x}_1 = v_{1i}$ and $\dot{x}_2 = 0$. Thus the velocity of the center-of-mass v_c with respect to the LCS is given by

$$v_c = \frac{m_1 v_{1i}}{m_1 + m_2} = \frac{\mu}{m_2} v_{1i} \tag{8.132}$$

where μ is the reduced mass.

Let the collision between m_1 and m_2 be observed by an observer moving with velocity v_c of the center of mass; that is, the observer is in the CMCS. The velocities of mass m_1 and m_2

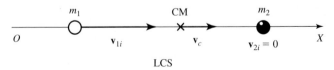

LCS

Figure 8.11   Velocity of m_1 and m_2 and their center of mass in the laboratory coordinate system (LCS).

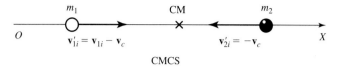

Figure 8.12   Motion of particles m_1 and m_2 in the center-of-mass coordinate system (CMCS).

with respect to the CMCS are v'_{1i} and v'_{2i} (prime indicates that the quantity is described in the CMCS):

$$v'_{1i} = v_{1i} - v_c = v_{1i} - \frac{m_1}{m_1 + m_2} v_{1i} = \frac{m_2}{m_1 + m_2} v_{1i} = \frac{\mu}{m_1} v_{1i} \tag{8.133}$$

$$v'_{2i} = v_{2i} - v_c = 0 - v_c = -\frac{m_1}{m_1 + m_2} v_{1i} = -\frac{\mu}{m_2} v_{1i} \tag{8.134}$$

Figure 8.12 shows the motion of these two particles with respect to the CMCS. The corresponding momentum of each particle before collision in the CMCS is

$$p'_{1i} = m_1 v'_{1i} = \frac{m_1 m_2}{m_1 + m_2} v_{1i} \tag{8.135}$$

$$p'_{2i} = m_2 v'_{2i} = -\frac{m_1 m_2}{m_1 + m_2} v_{1i} \tag{8.136}$$

Thus the total linear momentum of the system in the CMCS before collision is

$$P'_1 = p'_{1i} + p'_{2i} = \frac{m_1 m_2}{m_1 + m_2} v_{1i} - \frac{m_1 m_2}{m_1 + m_2} v_{1i} = 0 \tag{8.137}$$

That the total linear momentum before collision is zero is one of the most important characteristics of the CMCS. This implies that to conserve linear momentum the total linear momentum in the CMCS after collision must also be zero. That is, *as viewed from the CMCS, two particles of mass m_1 and m_2 approach each other in a straight line and, after collision, recede from each other in a straight line with the same initial velocities,* as shown in Fig. 8.13(a). The line joining the receding particles can make any angle θ_c (in CMCS), as shown. For the sake of comparison, Fig. 8.13(b) shows the collision as viewed from the LCS.

We may now look at the following problem. First, how do we get back from the CMCS to the LCS? Second, what is the relation between the angles the particles make after collision with their initial direction in both the LCS and the CMCS?

In the CMCS, the final velocity and direction of the particles after collision are shown in Fig. 8.13(a). To find the final velocities of the particles in the LCS, we may reverse the procedure used for changing from the LCS to the CMCS. This is achieved by adding to the final velocities v'_{1f} $(=v_{1i} - v_c)$ and v'_{2f} $(=v_c)$, the velocity v_c of the center of mass shown in Fig. 8.14. Thus the velocity $\mathbf{v}_{1f}$ and $\mathbf{v}_{2f}$ of m_1 and m_2, respectively, in the LCS are

$$\mathbf{v}_{1f} = \mathbf{v}'_{1f} + \mathbf{v}_c \tag{8.138}$$

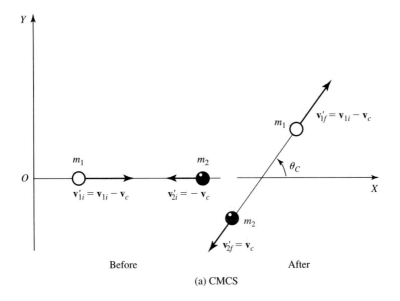

(a) CMCS

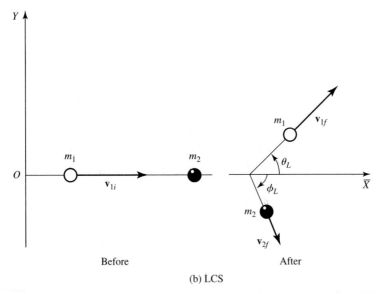

(b) LCS

Figure 8.13   Collision between two particles of mass m_1 and m_2 as viewed from (a) the CMCS and (b) the LCS.

$$\mathbf{v}_{2f} = \mathbf{v}'_{2f} + \mathbf{v}_c \tag{8.139}$$

With the help of Fig. 8.14, we can find the relation between angles θ_L and ϕ_L in the LCS and θ_C in the CMCS. For example, let us consider Eq. (8.138) and the top half of Fig. 8.14. Resolving into components, Eq. (8.138) may be written as

$$v_{1f} \cos \theta_L = v_c + v'_{1f} \cos \theta_C \tag{8.140}$$

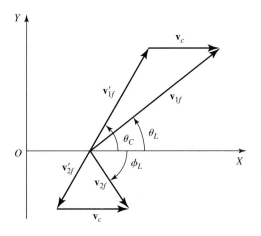

Figure 8.14 Relation between angles θ_L and ϕ_L in the LCS and θ_C in the CMCS after collision.

$$v_{1f} \sin \theta_L = v'_{1f} \sin \theta_C \tag{8.141}$$

Dividing one by the other,

$$\tan \theta_L = \frac{v'_{1f} \sin \theta_C}{v_c + v'_{1f} \cos \theta_C} = \frac{\sin \theta_C}{(v_c/v'_{1f}) + \cos \theta_C} \tag{8.142}$$

or

$$\tan \theta_L = \frac{\sin \theta_C}{\gamma + \cos \theta_C} \tag{8.143}$$

where

$$\gamma = \frac{v_c}{v'_{1f}} = \frac{\text{velocity of the center of mass in LCS}}{\text{velocity of } m_1 \text{ after collision in CMCS}} \tag{8.144}$$

The values of v_c and v'_{1f} are given by Eqs. (8.132) and (8.133). From Eq. (8.132)

$$v_c = \frac{m_1}{m_1 + m_2} v_{1i} = \frac{\mu}{m_2} v_{1i} \tag{8.145}$$

where μ is the reduced mass and v_{1i} is the initial relative velocity ($=v_{1i} - v_{2i} = v_{1i} - 0 = v_{1i}$). v'_{1f} ($=v'_{1i}$), the final relative velocity, from Eq. (8.133) is equal to

$$v'_{1f} = \frac{m_2}{m_1 + m_2} v_{1f} = \frac{\mu}{m_1} v_{1f} \tag{8.146}$$

Thus combining the preceding three equations (and noting that final velocities are equal to initial velocities in the CMCS), we get

$$\gamma = \frac{v_c}{v'_{1f}} = \frac{m_1 v_{1i}}{m_2 v_{1f}} \tag{8.147}$$

For inelastic collisions, $v_{1i} \neq v_{1f}$, Eq. (8.143) becomes

$$\tan \theta_L = \frac{\sin \theta_C}{(m_1 v_{1i}/m_2 v_{1f}) + \cos \theta_C}, \qquad \textit{inelastic collisions} \qquad \textbf{(8.148)}$$

For elastic collisions, $v_{1i} = v_{1f}$, and Eq. (8.148) takes the form

$$\tan \theta_L = \frac{\sin \theta_C}{(m_1/m_2) + \cos \theta_C}, \qquad \textit{elastic collisions} \qquad \textbf{(8.149)}$$

Let us consider some special cases of Eq. (8.149) for elastic collisions.

Case (a):   If $m_1 = m_2$, as is the case in collisions between neutrons and protons, we may write Eq. (8.149) as

$$\tan \theta_L = \frac{\sin \theta_C}{1 + \cos \theta_C} = \frac{2 \sin(\theta_C/2) \cos(\theta_C/2)}{2 \cos^2(\theta_C/2)} = \tan \frac{\theta_C}{2} \qquad \textbf{(8.150)}$$

That is,

$$\theta_L = \frac{\theta_C}{2} \qquad \textbf{(8.151)}$$

Since, in the CMCS, θ_C may have any value of between 0 and π, θ_L can have a maximum value of $\pi/2$, in agreement with the previous discussion.

Case (b):   If $m_2 \gg m_1$, we may write Eq. (8.149) as

$$\tan \theta_L \simeq \frac{\sin \theta_C}{\cos \theta_C} = \tan \theta_C \qquad \textbf{(8.152)}$$

That is,

$$\theta_L \simeq \theta_C \qquad \textbf{(8.153)}$$

which states that, for heavy targets, the scattering angle in the LCS is the same as in the CMCS.

Case (c):   If $m_1 > m_2$, the incident particle is heavier than the target particle. In this case, θ_L must be very small, no matter whatever the value of θ_C. This correctly corresponds to the situation in Eq. (8.85), where it was noted that θ_L cannot be larger than a certain maximum value θ_{max}.

8.11 AN INVERSE-SQUARE REPULSIVE FORCE: RUTHERFORD SCATTERING

Most of our efforts have been devoted to the study of motion of particles in an attractive inverse square force field. There is an important class of physical applications in which the motion of the particle is in an inverse-square repulsive force field. Such situations involve the deflection or scattering of fast-moving atomic particles such as protons and alpha particles by positively

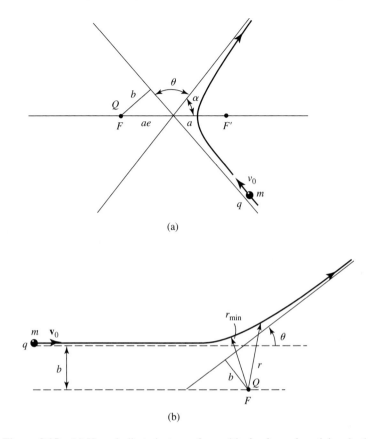

(a)

(b)

Figure 8.15 (a) Hyperbolic trajectory of a positively charged particle q in the field of a positively charged force center Q, that is, in a repulsive force field. Note that θ is the scattering angle and b is the impact parameter. (b) Same as in (a), but also showing the relation between r, r_{min}, b, and θ.

charged nuclei. Paths of such scattered particles are hyperbolic. The first such experiments involving the scattering of alpha particles by nuclei were carried out by Geiger and Marsden (Rutherford's students) and analyzed by Rutherford and will be discussed at some length.

As shown in Fig. 8.15(a) and (b), a positively charged particle of charge q, mass m, and velocity v_0 is incident on a target nucleus of positive charge Q and mass M at rest. The inverse-square repulsive force between the two particles is

$$F = k\frac{Qq}{r^2} = \frac{K}{r^2}$$ (8.154)

where $k = 8.99 \times 10^9$ N-m^2/C^2 and $K = kQq$ is positive; hence F is a repulsive force.

In the particular case of alpha particle scattering by nuclei, $q = 2e$ and $Q = Ze$, where Z is the atomic number of the nuclei and e is the charge of the electron. Since $e = 1.6 \times 10^{-19}$ C

Figure 8.15(c)

The path of the scattered particle is shown using the numerical values for e and a. The calculations for different parameters are given below.

$N := 20 \qquad i := 0..N$

$a := 50 \qquad e := 1.5$

$\theta_i := \dfrac{2 \cdot \pi}{N} \cdot i \qquad r_i := \dfrac{a \cdot (e^2 - 1)}{1 - e \cdot \cos(\theta_i)}$

θ = scattering angle = $\pi - 2\alpha$

$\alpha = \pi - \theta/2$

b = impact parameter

r0 = the distance of closest approach

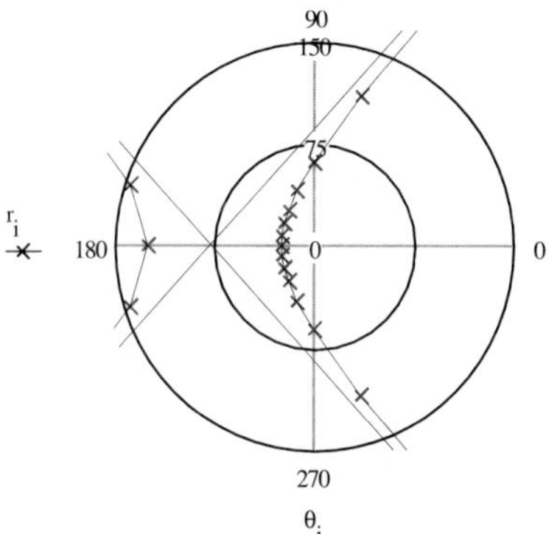

$\alpha := a\cos\left(\dfrac{1}{e}\right) \qquad\qquad \theta := 2 \cdot a\tan\left(\dfrac{1}{\sqrt{e^2 - 1}}\right)$

$\theta 1 := \pi - 2 \cdot \alpha \qquad\qquad r_{20} = -125$

$\alpha = 0.841 \cdot \text{rad} \qquad\qquad \theta = 1.459 \cdot \text{rad} \qquad\qquad \theta 1 = 1.459 \cdot \text{rad} \qquad b := r_{20} \cdot \sin(\theta) \qquad r0 := \min(r)$

$\alpha = 48.19 \cdot \text{deg} \qquad\qquad \theta = 83.621 \cdot \text{deg} \qquad \theta 1 = 83.621 \cdot \text{deg} \quad b = -124.226 \qquad r0 = -292.705$

(C = coulomb),

$$K = kQq = 2kZe^2 = (4.6 \times 10^{-28}\ \text{N-m}^2)Z \tag{8.155}$$

K being positive. Equation (7.106) for the eccentricity e,

$$e = \sqrt{1 + \frac{2EL^2}{mK^2}} \tag{7.106}$$

suggests that $e > 1$; hence the trajectory of an incident alpha particle will be hyperbolic, as shown in Fig. 8.15(a). This is the negative branch trajectory of the hyperbola. The repulsive force center is at F. The scattering angle θ, the angle between the two asymptotes, is

$$\theta = \pi - 2\alpha \tag{8.156}$$

Therefore,

$$\tan \frac{\theta}{2} = \tan\left(\frac{\pi}{2} - \alpha\right) = \cot \alpha \tag{8.157}$$

In the equation for a hyperbola (from Chapter 7) given by

$$r = \frac{a(e^2 - 1)}{1 - e \cos \theta} \tag{7.113}$$

and for the particle at infinity, $r = \infty$, $\theta = \alpha$, the above equation yields

$$\cos \alpha = \frac{1}{e} \qquad (8.158)$$

where e is the eccentricity. Thus, combining Eqs. (8.157) and (8.158),

$$\tan \frac{\theta}{2} = \cot \alpha = \frac{\cos \alpha}{\sin \alpha} = \frac{1/e}{(1 - 1/e^2)^{1/2}}$$

$$\tan \frac{\theta}{2} = \frac{1}{\sqrt{e^2 - 1}} \qquad (8.159)$$

Substituting for e^2 from Eq. (7.106), we obtain

$$\tan \frac{\theta}{2} = \sqrt{\frac{mK^2}{2EL^2}} \qquad (8.160)$$

Referring to Fig. 8.15(b), when the alpha particle is at infinity, its potential energy is $V = K/r = K/\infty = 0$; hence the total energy E is all kinetic and is given by

$$E = \tfrac{1}{2} mv_0^2 \qquad (8.161)$$

Since the alpha particle is headed toward the force center F, in the absence of any force it will not be deflected but will pass the force center at a distance b. This distance b by which the particle misses the force center is called the *impact parameter* for the collision. Also, the angular momentum of the particle is

$$L = mv_0 b \qquad (8.162)$$

which will remain constant during all its motion due to the law of conservation of angular momentum. Substituting for E and L from Eqs. (8.161) and (8.162) into Eq. (8.160),

$$\tan \frac{\theta}{2} = \sqrt{\frac{mK^2}{2(\tfrac{1}{2} mv_0^2)(mv_0 b)^2}}$$

Hence
$$\tan \frac{\theta}{2} = \frac{K}{mv_0^2 b} \qquad (8.163)$$

where $K = kQq$. Relation (8.163) may be written as

$$b = \frac{K}{mv_0^2} \cot \frac{\theta}{2} \qquad (8.164)$$

or
$$\theta = 2 \operatorname{arccot}\left[\left(\frac{mv_0^2}{K}\right)b\right] \qquad (8.165)$$

The scattering angle θ can be measured experimentally; hence the impact parameter b can be calculated from Eq. (8.165). Equation (8.164) also states that as b increases θ decreases—that is, the smaller the impact parameter, the larger the scattering angle. In actual practice, we measure the number of particle $N(\theta)$ scattered at different angles. Hence we must find a way to eliminate b in Eq. (8.165) and find the relation between $N(\theta)$ and θ. This leads us to the concept of a cross section, as discussed next.

A typical experiment setup is shown in Fig. 8.16. A beam of charged particles coming from source S is incident on a thin foil that is the target. The particles are scattered in different directions after colliding with the target nuclei. Suppose the particles with impact parameter b are deflected through an angle θ; then those particles with an impact parameter $b + db$ will be deflected through an angle $\theta + d\theta$, where $d\theta$ is negative, as shown in Fig. 8.17. Suppose there are N particles incident on the target foil and the foil contains n nuclei per unit area; that is, there are n scattering centers per unit area. (The foil is considered thin enough so that nuclei do not hide one behind the other.) Thus the number of alpha particles dN that will be scattered through an angle θ and $\theta + d\theta$ is proportional to the scattering centers n and the number of incident particles N; that is,

$$dN = nN \, d\sigma \qquad\qquad \textbf{(8.166)}$$

where $d\sigma$ is defined as the *cross section* for scattering through an angle θ and $\theta + d\theta$. $d\sigma$ can be thought of as the effective area surrounding each scattering center, which the incident particle must hit in order to be scattered. Thus the total sensitive area for scattering in a unit target area is $n \, d\sigma$; hence the justification for Eq. (8.166). [Note that if the incident particles have im-

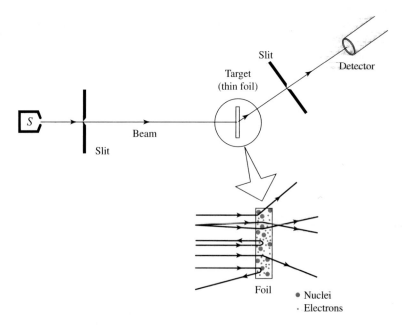

Figure 8.16 Typical experimental setup for investigating the scattering of charged particles from a target of thin foil.

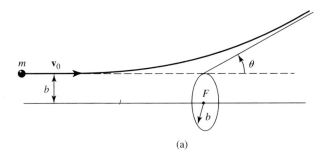

(a)

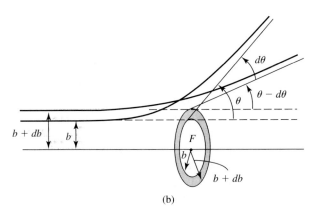

(b)

Figure 8.17   (a) A particle with impact parameter b is scattered through an angle θ. (b) Particles with impact parameters between b and $b + db$ are scattered through angles between θ and $\theta - d\theta$.

pact parameters between 0 and b the particles will be scattered through an angle θ or greater than θ. The cross section in this cases is σ and is equal to the area of a disk of radius b in Fig. 8.17(a) with the center at F:

$$\sigma = \pi b^2 \tag{8.167}$$

hence

$$d\sigma = 2\pi b \, db \tag{8.168}$$

as we shall see next.]

Referring to Fig. 8.17(b), the incident particles approaching the scattering center F have an impact parameter between b and $b + db$. These particles will be scattered through an angle between θ and $\theta - d\theta$ if they hit an area of a ring around F of inner radius b and outer radius $b + db$. Thus the area of the ring is the cross-sectional area $d\sigma$; that is,

$$d\sigma = 2\pi b \, db \tag{8.168}$$

We can express b and db in terms of θ and $d\theta$ by using Eq. (8.164), according to which

$$b = \frac{K}{mv_0^2} \cos \frac{\theta}{2} \tag{8.164}$$

and, differentiating this, we get

$$db = -\frac{K}{2mv_0^2} \frac{1}{\sin^2(\theta/2)} \, d\theta \tag{8.169}$$

We may also use Eq. (8.164) to write b as

$$b = \frac{K}{2mv_0^2} \frac{\sin \theta}{\sin^2(\theta/2)} \qquad (8.170)$$

Substituting for db and b from Eqs. (8.169) and (8.170) into Eq. (8.168), we get, after omitting the negative sign,

$$d\sigma = 2\pi \left(\frac{K}{2mv_0^2}\right)^2 \frac{\sin \theta}{\sin^4(\theta/2)} d\theta \qquad (8.171)$$

Remembering $K = kQq$, we get

$$d\sigma = 2\pi \left(\frac{kQq}{2mv_0^2}\right)^2 \frac{\sin \theta}{\sin^4(\theta/2)} d\theta \qquad (8.172)$$

which is the *Rutherford scattering formula*. $d\sigma$ can be measured experimentally by using Eq. (8.166) and can be compared with the theoretical value calculated by using Eq. (8.172).

Rutherford used the derived formula to make an interpretation of his experiment on the scattering of alpha particles ($q = 2e$) by target nuclei ($Q = Ze$) in the form of thin foils. Expression (8.172) held good as long as the perihelion distance ($a + ae$) was larger than 10^{-14} m. From this he concluded that the positive charge of the nucleus must be concentrated in a sphere with a radius of less than 10^{-14} m. The incident alpha particle can come closest to the nucleus for an impact parameter of $b = 0$. This will result in a minimum distance of the perihelion; and at this distance all the kinetic energy of the incident alpha particle is changed into potential energy, and the particle starts turning back. Thus

$$K = V = \frac{kQq}{r_{\min}} \qquad (8.173)$$

The use of Eq. (8.173) can give some idea about the magnitude of the nuclear radius. Deviations from the Rutherford scattering formula will occur if the kinetic energy K of the incident particle is greater than the minimum potential energy at a distance $r_{\min}$. From such observations, Rutherford concluded that the nuclear radius was 10^{-14} m.

In the preceding discussion, it was assumed that the target was heavy as compared to the incident particle and hence was assumed to be at rest during the collision. If the target nucleus is not heavy, the nucleus itself will move during the collision, as shown in Fig. 8.18. The diffi-

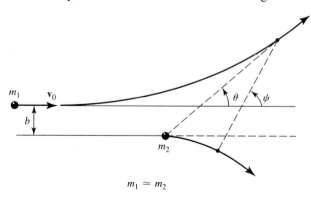

Figure 8.18   Scattering by a target of mass m_2, which is almost equal to the incident particle of mass m_1; that is, $m_1 \simeq m_2$.

culty can be overcome by considering the collision in the CM coordinate system. The final result can be obtained by replacing m by the reduced mass μ $[= mM/(M + m)]$ and θ by θ_C in Eq. (8.171); that is,

$$d\sigma = 2\pi\left(\frac{K}{2\mu v_0^2}\right)^2 \frac{\sin \theta_C}{\sin^4(\theta_C/2)} d\theta_C \qquad (8.174)$$

In the case where $m_1 = m_2$, we have shown, Eq. (8.151), $\theta_C = 2\theta_L = 2\theta$; hence

$$d\sigma = 4\pi\left(\frac{K}{2\mu v_0^2}\right)^2 \frac{\sin 2\theta}{\sin^4 \theta} d\theta \qquad (8.175)$$

PROBLEMS

8.1. Find the center of mass, the velocity of the center of mass, the linear momentum, and the kinetic energy of the following system:

$$m_1 = 1 \text{ kg}, \qquad \mathbf{r}_1 = \hat{\mathbf{i}} + 2\hat{\mathbf{j}} + 3\hat{\mathbf{k}}, \qquad \mathbf{v}_1 = 2\hat{\mathbf{i}} + 3\hat{\mathbf{j}}$$

$$m_2 = 2 \text{ kg}, \qquad \mathbf{r}_2 = \hat{\mathbf{i}} - \hat{\mathbf{j}} + \hat{\mathbf{k}}, \qquad \mathbf{v}_2 = 2\hat{\mathbf{j}} + 3\hat{\mathbf{k}}$$

8.2. Consider the following three particles:

$$m_1 = 1 \text{ kg}, \qquad \mathbf{r}_1 = 2t^2\hat{\mathbf{i}} + 3t\hat{\mathbf{j}} + 4\hat{\mathbf{k}}$$

$$m_2 = 3 \text{ kg}, \qquad \mathbf{r}_2 = (1 + t^2)\hat{\mathbf{i}} + (2 + 5t)\hat{\mathbf{j}}$$

$$m_3 = 5 \text{ kg}, \qquad \mathbf{r}_3 = (1 + 2t^2)\hat{\mathbf{i}} + 4t^2\hat{\mathbf{k}}$$

Calculate the following at $t = 0$ and $t = 10$ s.
(a) The position of the center of mass, (b) the velocity of the center of mass, (c) the linear momentum, and (d) the kinetic energy of the system.

8.3. Find the velocity and acceleration of the center of mass of a system consisting of the following two objects at $t = 0$ and $t = 10$ s.

$$m_1 = 2 \text{ kg}, \qquad \mathbf{r}_1 = 2\hat{\mathbf{i}} + 3t\hat{\mathbf{j}} + 4t^2\hat{\mathbf{k}}$$

$$m_2 = 4 \text{ kg}, \qquad \mathbf{r}_2 = t^2\hat{\mathbf{i}} + 5\hat{\mathbf{j}} + 6t^3\hat{\mathbf{k}}$$

8.4. A projectile of mass m is fired with a velocity of 50 m/s at an angle of 60° with the horizontal. At the top (maximum height), it explodes into two fragments, creating an additional energy E, with the result that one fragment is observed to be moving directly upward. What is the direction of the other fragment? Calculate the velocity of both fragments.

8.5. A projectile of mass M ($=m_1 + m_2$) is fired with velocity v making an angle θ with the horizontal. At the top it explodes into two masses, m_1 and m_2, creating an additional energy E. Show that the two fragments strike the ground at a distance apart equal to

$$\frac{v \sin \theta}{g}\left[2E\left(\frac{1}{m_1} + \frac{1}{m_2}\right)\right]^{1/2}$$

8.6. If a projectile explodes at the top (maximum height) with an additional energy E, under what circumstances will one of the fragments land at the starting position?

8.7. A fire boat draws water from a bay through a vertical inlet and sprays it out at a rate of 10 m/s. The diameter of the nozzle of the fire hose is 20 cm. Calculate the horizontal force from the propellers necessary to keep the boat stationary. The density of water is 1020 kg/m³.

8.8. A bucket of 0.5 kg is placed on a spring scale and water is added to it from a height of 2 m at a rate of 5 ml/s. Find the scale reading as a function of time.

8.9. A chain of length L and mass M is held vertically so that the bottom of the chain just touches the horizontal table top, as shown in Fig. P8.9. If the upper end of the chain is released, determine the force on the table top, as the function of the length of the chain above the table top, while it is falling.

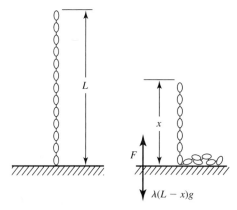

Figure P8.9

8.10. For the falling chain in Fig. P.8.10, show that when all of the chain clears the table the speed of the chain is

$$v = \sqrt{(g/L)(L^2 - a^2)}$$

where $y = a$, when $t = 0$.

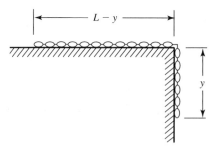

Figure P8.10

8.11. A raindrop as it falls through fog or mist collects mass at a uniform rate. The drop starts from rest with zero radius and remains spherical at all times. Show that the acceleration with which it falls is $g/7$.

8.12. A raindrop of initial mass m_0 is falling under the influence of gravity. Due to condensation from fog or mist, the mass of the drop increases at a rate directly proportional to its instantaneous mass and velocity. Show that eventually the speed of the drop becomes constant. Derive an expression for this terminal velocity. Graph the velocity versus time.

8.13. Consider a spherical raindrop of initial mass m_0 falling through fog or mist. Due to condensation, the raindrop increases in mass at a rate proportional to its mass and velocity. In addition to the force of gravity, the force of friction is present, which is proportional to the velocity and mass of the drop; that is, $F_f = -kmv$. Calculate the velocity of the drop as a function of time. Graph v versus t.

8.14. A raindrop of initial mass m_0 falls vertically through fog or mist. Due to condensation, the mass of the raindrop increases linearly with time; that is, $m = m_0 + \lambda t$. The frictional drag force on the mass m is proportional to its mass and velocity; that is, $F_f = -kmv$. Calculate the velocity of the raindrop as a function of time, assuming the presence of gravitational force and frictional drag. Graph v versus t.

8.15. Calculate the thrust of a test jet engine if it takes in air at a rate of 100 kg/s and exhausts it at a speed of 500 m/s.

8.16. A rocket has an initial mass of 60,000 kg, and the speed of the burned exhaust gases is 6000 m/s. What should be the minimum mass flow rate of the gases to ensure life-off from the surface of Earth?

8.17. A rocket of 60,000-kg mass is burning gases at a rate of 150 kg/s, and the speed of the exhaust gases is 6000 m/s. If the rocket is fired vertically upward from the surface of Earth, what will be its height and speed after 45,000 kg of fuel is expended? Graph the velocity and height as a function of time.

8.18. A rocket propulsion type of car has a mass m_0 without fuel, and its fuel has mass m. The ejecting fuel has a velocity V with respect to the rocket, and the fuel burns at a rate of $k = dm/dt$. Find the acceleration and velocity as a function of time and the velocity when all the fuel has burned out. Graph a and v versus t.

8.19. As a rocket ascends it loses mass at a rate proportional to its instantaneous mass; that is, $dm/dt = bm$, where b is a constant. The motion of the rocket is retarded by air resistance proportional to its velocity; that is, $F_f = -kv$, where k is constant. Find the velocity of the rocket as a function of time. Graph and discuss the outstanding features.

8.20. During the first second of its flight, a rocket exhausts $\frac{1}{50}$ of its mass with a velocity of 2000 m/s. Calculate the acceleration of the rocket. If the rocket exhausts at a constant rate, will it be possible to attain a constant acceleration?

8.21. A rocket of mass $M + m$, where m is the mass of the fuel, rises vertically and ejects gases at a rate of q and with an exhaust velocity of u. Calculate the velocity and the acceleration as a function of time and graph them for the values given next. The initial mass is 4×10^4 kg, $q = 600$ kg/s, and $u = 2000$ m/s. If the fuel burns out in 50 s, calculate the acceleration at $t = 0$ s, 20 s, 40 s, and 50 s.

8.22. A rocket has a mass of m_0 and a mass ratio of R, burns at a rate of dm/dt, and has an exhaust velocity of v_0. Find how long after ignition of the engines it will take the rocket to lift off from the ground. Calculate for the case in which m_0 is 5×10^4 kg, R is 3, the burning rate is 120 kg/s, and the exhaust velocity is 1000 m/s.

8.23. A lunar landing craft is hovering over the Moon's surface. One-third of its mass is fuel, while the exhaust velocity u is 1200 m/s. How long will it take before the craft runs out of fuel? Assume that the acceleration due to gravity on the surface of the Moon is one-sixth of that on Earth.

8.24. Suppose a two-stage rocket starts with a mass m_i. At the end of the first stage, the mass of the rocket is m_1. Before the second stage engines are ignited, some of the mass is discarded, and the starting mass is m_2. The final mass when the engines of the second stage are shut down is m_f. Assuming that the exhaust velocity in both stages is v_0, find the terminal velocity of the second stage.

8.25. An empty truck of mass M starts from rest under an applied force F. At the same time, coal begins to drop into the truck at a rate of $b = dm/dt$. What is the speed of the truck when a mass m of the coal has fallen in?

8.26. An open truck is traveling at a constant speed of 90 km/h and is collecting water from a rainstorm. If it picks up 50 kg of water over a distance of 1000 m, calculate the force and the power required to maintain a constant speed.

8.27. A freight car of mass m contains a mass of coal m. At $t = 0$, a force F is applied. As the car starts rolling, the coal starts dropping at a rate of $b = dm/dt$. What is the speed of the car when all the coal has dropped out?

8.28. A chute discharges sand at the rate of 500 kg/min onto a conveyor belt that is inclined at an angle of 12° to the horizontal and is moving at a rate of 4 m/s. The sand falls at a speed of 5 m/s. Calculate the force necessary to keep the belt moving at a constant speed.

8.29. Consider a conveyor belt inclined at an angle θ from the horizontal so that the belt forms an inclined plane. At the bottom end of the belt, material, deposited at a rate of dm/dt, travels a distance l and then is taken off the upper end of the incline. Calculate the power needed to keep the belt moving at a steady speed v.

8.30. Consider a conveyor belt inclined at an angle θ from the horizontal so that the belt forms an inclined plane. At the top end of the belt, material is deposited at a rate of dm/dt, travels a distance l, and then falls off the lower end of the incline. Assuming a constant force of friction f, calculate the steady speed of the belt.

8.31. Derive Eqs. (8.86) and (8.87).

8.32. In Fig. 8.6(a) if $v_{2i} \neq 0$, show that after the collision $v_{1f} = v_{2i}$ and $v_{2f} = v_{1i}$.

8.33. A neutron of mass m_1 moving with velocity v collides with an atomic nucleus of mass m_2 at rest. Calculate the maximum fractional loss in kinetic energy of the neutron if the atomic nucleus is **(a)** hydrogen, **(b)** carbon, **(c)** iron, and **(d)** lead.

8.34. A particle of mass m_1 and velocity v_{1i} collides with a particle of mass m_2 moving with velocity v_{2i} exactly in the opposite direction. If, after collision, mass m_1 leaves at an angle θ_1 with the initial direction, what is the value of v_{1f}?

8.35. A particle of mass m_1 moving with velocity v_0 collides elasticity with a particle of mass m_2 at rest. At what scattering angle will be momentum of the mass m_1 be half its initial value? What are the restrictions in terms of m_1/m_2?

8.36. A billiard ball of mass m collides with an identical ball at rest. After collision, the two balls leave at angles $\pm\theta$ with the initial direction. Prove that for this to happen the two balls have a rotational kinetic energy of $[1 - (\cos^{-2}\theta)/2]K_i$, where K_i is the initial kinetic energy. Assume that there are no frictional losses in energy.

8.37. Consider a perfect elastic collision between two balls, one of mass m and the other of unknown mass, each moving with a speed v_0 but in opposite directions. After collision, the ball of unknown mass comes to rest. Calculate the unknown mass and the velocity of the ball of mass m.

8.38. A ball of mass m with energy E strikes a ball of mass M at rest. After collision, the ball of mass m is scattered at an angle of 90° from its original direction. Calculate the energy of mass M after collision.

8.39. A particle of mass m_1 moving with velocity v_1 collides with a particle of mass m_2 moving with a velocity v_2, both having the same initial kinetic energy. Find the conditions in terms of v_1/v_2 and m_1/m_2 so that mass m_1 is at rest after collision.

8.40. A particle of mass m moving with velocity v_0 collides with a mass M moving in the opposite direction. After collision, the mass m has velocity $v_0/2$ and moves at right angles to the initial direction, while mass M moves in a direction making an angle of 30° with the initial path of m. Find the ratio m/M.

8.41. A particle of mass m_1 has a head-on collision with a particle of mass m_2 at rest. If the coefficient of restitution is e, calculate the energy loss in this collision.

8.42. A ball of mass m is dropped from a height h onto a horizontal surface. Show that the vertical height through which the ball rises before it stops rebounding is $h(1 + e^2)/(1 - e^2)$, where e is the coefficient of restitution.

8.43. Show that the loss in kinetic energy when two objects collide is $\frac{1}{2}\mu V^2(1 - e^2)$, where μ is the reduced mass, V is the relative speed before collision, and e is the coefficient of restitution.

8.44. A particle of mass m_1 moving at right angles to mass m_2 collides as shown in Fig. P8.44. Calculate the velocity of each particle after collision, assuming that the coefficient of restitution is 0.4, $m_1 =$ 3 kg, $m_2 = 2$ kg, $v_{1i} = 2$ m/s, and $v_{2i} = 3$ m/s.

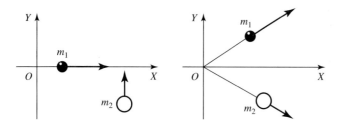

Figure P8.44

8.45. Consider the situation shown in Fig. P8.45. Ball A of mass $2m$ is raised to a height of h so that its string makes an angle of 45° with the vertical, and it is then let go. To what height will ball B of mass m rise if the coefficient of restitution is 0.5?

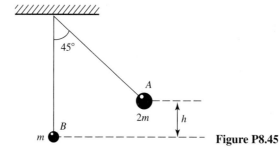

Figure P8.45

8.46. A ball of mass m moving downward with a velocity of $\mathbf{v}$ and making an angle θ with the horizontal strikes a flat surface and rebounds at angle ϕ, as shown in Fig. P8.46. Calculate the velocity of the ball, angle ϕ, and the change in the kinetic energy. Assume that the surface is smooth and the coefficient of restitution is e.

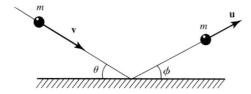

Figure P8.46

8.47. A ball of 1-kg mass moving with a speed of 2 m/s strikes a wooden bar of 2-kg mass moving to the right, with a center-of-mass velocity of 1.5 m/s, as shown in Fig. P8.47. If the coefficient of restitution is 0.4, and the plane is which this collision takes place is smooth, calculate the following quantities just after collision: **(a)** velocity of the ball, and **(b)** linear velocity and angular velocity of the bar.

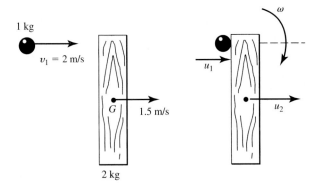

1 kg

$v_1 = 2$ m/s

G 1.5 m/s

2 kg

ω

u_1

u_2

Figure P8.47

8.48. A neutron in a nuclear reactor moving with an initial speed of 120 m/s collides with a deuteron (heavy hydrogen in which the nucleus is made of a proton and a neutron) at rest. The neutron is scattered at an angle of 30°. Calculate the recoil angle for the deuteron and the speed of both the neutron and deuteron after the collision. Draw a diagram showing this collision in the CMCS and the corresponding angles in the CMCS.

8.49. Repeat Problem 8.48 if the deuteron is replaced by a carbon atom with a mass of 12 u.

8.50. An alpha particle of mass 4 u moving with a velocity 2000 m/s collides with a carbon atom of mass 12 u at rest. The alpha particle is scattered through an angle of 30°. Considering the collision to be perfectly elastic, calculate the velocities of both particles after collision and the scattering angle of the recoiling carbon. Describe this collision in the CMCS.

8.51. Derive an expression for the Rutherford scattering cross section in terms of the recoil angle.

8.52. Obtain an expression for the Rutherford scattering cross section for the case in which the mass of the incident particle is very large compared to the mass of the target particle.

8.53. Somewhere in outer space a star of mass m moving with velocity v_0 is headed toward a star of mass $2m$ at rest. The impact parameter in this case is b. Calculate the speeds and the direction of the two stars.

8.54. Show that the differential scattering cross section of mass m from a fixed force center

$$\mathbf{F} = \frac{K}{r^3} \hat{\mathbf{r}}$$

is given by

$$\sigma(\theta) = \frac{k\pi^2(\pi - \theta)}{mv_0^2\theta^2(2\pi - \theta)^2 \sin \theta}$$

8.55. A spaceship of mass m moving with velocity $\mathbf{v}_0$ approaches the Moon ($M \gg m$). The distance of closest approach is b, and the velocity $\mathbf{v}_0$ is perpendicular to the orbital velocity $\mathbf{V}$ of the Moon. Show that if the spaceship passes behind the Moon it gains kinetic energy as it leaves the Moon.

8.56. Obtain an expression for the differential scattering cross section in the CMCS for the case where the target particle is much heavier than the incident particle.

8.57. A particle of mass m moving with velocity v_0 collides with a particle of mass m at rest. M is scattered through an angle θ_C in the CM system. What is the final velocity of m in the LCS? Calculate the fractional loss of kinetic energy of m.

SUGGESTIONS FOR FURTHER READING

ARTHUR, W., and FENSTER, S. K., *Mechanics,* Chapter 11. New York: Holt, Rinehart and Winston, Inc., 1969.

BARGER, V., and OLSSON, M., *Classical Mechanics,* Chapter 3. New York: McGraw-Hill Book Co., 1973.

BECKER, R. A., *Introduction to Theoretical Mechanics,* Chapters 3, 4, and 8. New York: McGraw-Hill Book Co., 1954.

DAVIS, A. DOUGLAS, *Classical Mechanics,* Chapter 8. New York: Academic Press, Inc., 1986.

FOWLES, G. R. *Analytical Mechanics,* Chapter 7. New York: Holt, Rinehart and Winston, Inc., 1962.

FRENCH, A. P., *Newtonian Mechanics,* Chapter 9. New York: W. W. Norton and Co., Inc., 1971.

HAUSER, W., *Introduction to the Principles of Mechanics,* Chapter 8. Reading, Mass.: Addison-Wesley Publishing Co., 1965.

KITTEL, C., KNIGHT, W. D., and RUDERMAN, M. A., *Mechanics,* Berkeley Physics Course, Volume 1, Chapter 6. New York: McGraw-Hill Book Co., 1965.

KLEPPNER, D., and KOLENKOW, R. J., *An Introduction to Mechanics,* Chapters 3 and 4. New York: McGraw-Hill Book Co., 1973.

*LANDU, L. D., and LIFSHITZ, E. M., *Mechanics,* Chapter 4. Reading, Mass.: Addison-Wesley Publishing Co., 1960.

MARION, J. B., *Classical Dynamics,* 2nd. ed., Chapter 9. New York: Academic Press, Inc., 1970.

*MOORE, E. N., *Theoretical Mechanics,* Chapter 1. New York: John Wiley & Sons, Inc., 1983.

ROSSBERG, K., *Analytical Mechanics,* Chapter 7. New York: John Wiley & Sons, Inc., 1983.

STEPHENSON, R. J., *Mechanics and Properties of Matter,* Chapter 3. New York: John Wiley & Sons, Inc., 1962.

SYMON, K. R., *Mechanics,* 3rd. ed., Chapter 4. Reading, Mass.: Addison-Wesley Publishing Co., 1971.

TAYLOR, E. F., *Introductory Mechanics,* Chapters 5, 7, and 10. New York: John Wiley & Sons, Inc., 1963.

*The asterisk indicates works of an advanced nature.

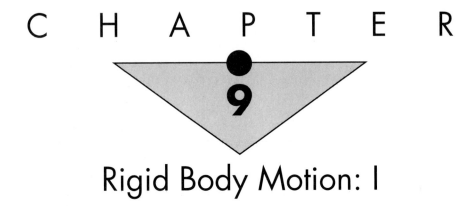

Rigid Body Motion: I

9.1 INTRODUCTION

In previous chapters we have dealt with the motion of a particle or a system of particles under the influence of external forces. In actual, everyday motions, we have to deal with rigid objects of different shapes and sizes which may or may not reduce to equivalent point masses. We will show now that to describe the motion of rigid bodies and apply conservation laws, we must understand the full meanings of center of mass, moment of inertia, and radius of gyration.

Discussion of angular motion is complex in such cases, so simple cases of rotation about a fixed axis will be discussed here, while rotation about an axis passing through a fixed point will be discussed in Chapter 13. Furthermore, in this chapter we will assume that the bodies are rigid and do not deform, which is true in ideal cases only. We will briefly discuss deformable continua in order to understand the elastic properties of objects, which, in turn, is necessary to an understanding of the equilibrium of flexible cables, strings, and solid beams.

9.2 DESCRIPTION OF A RIGID BODY

A *rigid body* is defined as a system consisting of a large number of point masses, called particles, such that the distances between the pairs of point masses remain constant even when the body is in motion or under the action of external forces. This is an idealized definition of a rigid body because (1) there is no such thing as true point masses or particles, and (2) no body of any physical size is strictly rigid; it becomes deformed under the action of applied forces. However, the concept of an idealized rigid body is useful in describing motion, and the resulting deviations are not that significant.

Forces that maintain constant distances between different pairs of point masses are internal forces and are called *forces of constraint*. Such forces come in pairs and obey Newton's third

338

law in the strong form; that is, they are equal and opposite and act along the same line of action. Hence we can apply the laws of conservation of linear momentum and angular momentum to the description of the motion of rigid bodies. Furthermore, in any displacement, the relative distances and the orientations of different particles remain the same with respect to each other; hence no net work is done by the internal forces or the forces of constraint. This implies that for a perfectly rigid body the law of conservation of mechanical energy holds as well.

Our next step is to establish the number of independent coordinates needed to describe the position in space or configuration of a rigid body. Suppose a rigid body consists of N particles. Since the position of each particle is specified by three coordinates, we may be led to conclude that we need $3N$ coordinates to describe the position of the rigid body. This would be true only if the positions and the motions of all particles were independent. But this is not so. The distance r_{kl} between any pair of particles is constant, and there are many such pairs. We are going to show that only *six* independent coordinates are needed to describe the position of a rigid body. Let us consider the rigid body shown in Fig. 9.1. To describe the position of the point mass k we need not specify its distances from all other point masses in the body; we need its distances from three other noncollinear points, such as P_1, P_2, and P_3, as shown in Fig. 9.1. Thus, if the positions of these three points are known, the positions of the remaining points in the body are fixed by the constraints. But P_1, P_2, and P_3 need at the most *nine* coordinates to describe their positions in space. Even these nine coordinates are not all independent. The distances r_{12}, r_{13}, and r_{23} are all constants; that is,

$$r_{12} = d_1, \qquad r_{13} = d_2, \qquad r_{23} = d_3 \qquad\qquad \textbf{(9.1)}$$

where d_1, d_2, and d_3 are constants. These three relations, called the *equations of constraints*, reduce the number of independent coordinates needed to describe the position of the rigid body to six.

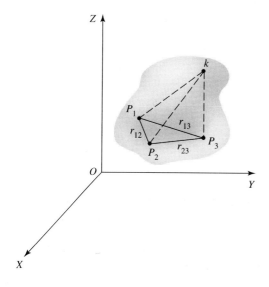

Figure 9.1   The position of any point mass at k in space may be determined by knowing the positions of three non-collinear points P_1, P_2, and P_3.

There is an alternative way of explaining that only six coordinates are needed to establish the positions of the three reference points. The reference point P_1 needs only three coordinates (x_1, y_1, z_1) to specify its position. Once P_1 is fixed, P_2 can be specified by only two coordinates, since it will be constrained to move on the surface of a sphere whose center is at P_1. These two coordinates are (θ_2, ϕ_2). With these two points fixed, point P_3 lies on a circle of radius a whose center lies on an axis joining points P_1 and P_2. Thus only *six* coordinates are needed to locate three noncollinear points P_1, P_2, and P_3 of a rigid body. Once these are fixed, the locations of all other points of the rigid body are fixed; that is, the configuration of a rigid body in space is fixed. If there are other constraints on the body, the number of coordinates needed to specify the position of a rigid body may be less than six.

There are several ways of choosing these six coordinates. One such way is shown in Fig. 9.2. The primed coordinates $X'Y'Z'$, the *body set of axes*, drawn in the rigid body can completely specify the rigid body relative to the external coordinates XYZ, the *space set of axes*. Thus three coordinates are needed to specify the origin of the body set of axes, while the remaining three must specify the orientation of the body set of axes (primed axes) relative to the coordinate axes parallel to the space axes (unprimed axes), as shown in Fig. 9.2. Thus we must know the coordinates of O' with respect to O and the orientation of $X'Y'Z'$ axes relative to the XYZ axes.

Let us consider the motion of a rigid body constrained to rotate about a fixed point. Since there is no translational motion, we are concerned only with the torques that produce rotational motion. But before doing this, we must choose three coordinates that describe the orientation of the body axes relative to the space axes. The choice is not so simple. No simple symmetric set of coordinates can be found that will describe the orientation of the rigid body. We shall postpone the discussion of the rotation of a body about a fixed point until Chapter 13. In this chapter, we limit ourselves to the discussion of simple problems involving rotation about a fixed axis (and not a fixed point).

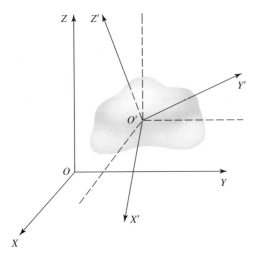

Figure 9.2 The configuration of a rigid body is specified by six independent coordinates: the coordinates of O' with respect to O and the orientation of the $X'Y'Z'$ axes relative to the XYZ axes.

9.3 CENTER OF MASS OF A RIGID BODY

Even a small-sized solid body contains a very large number of atoms and molecules. It is convenient to represent the structure of this body by its *average density*, ρ, defined as mass per unit volume, that is, $\rho = M/V$, where M is the mass and V is the volume, while the *local density* or simply *density* may be defined as

$$\rho = \frac{dM}{dV} \tag{9.2}$$

where dM is the mass of a volume element dV. Since the body is assumed to be continuous over the whole volume, the total mass being given by finite summation over mass particles m_k must now be replaced by an integral over a volume space of infinitesimal masses dM; that is,

$$\sum m_k \rightarrow M = \iiint dM = \iiint \rho\, dV \tag{9.3}$$

For a system containing a discrete number of particles of masses m_k at distances r_k, the center of mass $\mathbf{R}$ was defined in Chapter 8 as

$$\mathbf{R} = \frac{\sum m_k \mathbf{r}_k}{\sum m_k} \tag{9.4}$$

For an extended rigid body, the summation can be replaced by an integration over the whole volume of the body; that is, the center of mass $\mathbf{R}(X, Y, Z)$ is

$$\mathbf{R} = \frac{\iiint \mathbf{r}\, dM}{\iiint dM} = \frac{1}{M} \iiint \mathbf{r}\rho\, dV \tag{9.5}$$

where $dM = \rho\, dV$, and M is the total mass of the body. In component form, the center of mass may be written as

$$X = \frac{1}{M} \iiint x\rho\, dV, \qquad Y = \frac{1}{M} \iiint y\rho\, dV, \qquad Z = \frac{1}{M} \iiint z\rho\, dV \tag{9.6}$$

If a rigid body is in the form of a thin shell, the equation for the center of mass takes the form

$$\mathbf{R} = \frac{1}{M} \iint \mathbf{r}\sigma\, dA \tag{9.7}$$

where σ is the *surface density* defined as the mass per unit area, dA is a small element of area, and the total mass M is given by

$$M = \iint \sigma\, dA \tag{9.8}$$

Similarly, if the body is in the form of a thin wire, the center of mass is

$$\mathbf{R} = \frac{1}{M} \int \mathbf{r}\lambda \, dL \tag{9.9}$$

where λ is the *linear density* defined as mass per unit length, dL is a small element of length, and the total mass M is given by

$$M = \int \lambda \, dL \tag{9.10}$$

If ρ, σ, and λ are constants, they can be taken out of the integration signs, thereby making the problem somewhat simpler.

Suppose a system consists of two or more discrete parts such that the center of mass of M_1 is at $\mathbf{r}_1$, that of M_2 is $\mathbf{r}_2, \ldots$, then the center of mass of the system is

$$\mathbf{R} = \frac{M_1\mathbf{r}_1 + M_2\mathbf{r}_2 + \cdots}{M_1 + M_2 + \cdots} \tag{9.11}$$

In component form,

$$X = \frac{M_1 x_1 + M_2 x_2 + \cdots}{M_1 + M_2 + \cdots} \tag{9.12}$$

with similar expressions for Y and Z. Note that (x_1, y_1, z_1), $(x_2, y_2, z_2), \ldots$, are the coordinates of the center of masses of $M_1, M_2, \ldots$, respectively.

In calculating the center of mass, we should be able to take advantage of symmetry considerations. Suppose a body has a plane of symmetry; that is, every mass m_k has a mirror image of itself m_k' relative to the same plane. Let us assume that the XY plane is the plane of symmetry. In this case,

$$Z = \frac{\Sigma(m_k z_k + m_k' z_k')}{\Sigma(m_k + m_k')} \tag{9.13}$$

But, due to symmetry, $m_k = m_k'$ and $z_k = -z_k'$; that is, $Z = 0$, which implies that the center of mass lies in the XY plane, the plane of symmetry. Similarly, if a rigid body has a line of symmetry, the center of mass lies on this line. Let us discuss some examples to explain the application of the preceding equations.

Center of Mass of a Solid Hemisphere, Hemispherical Shell, and Semicircle

Figure 9.3 shows a solid hemisphere of radius a and uniform density ρ so that its mass $M = (2\pi/3)a^3\rho$. From symmetry considerations, we know that its center of mass lies on the radius that is normal to the plane face. That is, as shown, it lies on the Z-axis. To calculate Z, the center of mass, we consider the volume element shown shaded, so that

$$dV = \pi(a^2 - z^2) \, dz \tag{9.14}$$

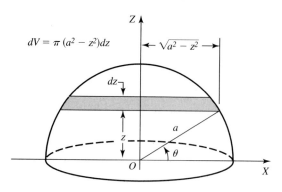

$dV = \pi (a^2 - z^2)dz$

$\leftarrow \sqrt{a^2 - z^2} \rightarrow$

dz

a

θ

O

X

Figure 9.3   Center of mass for a solid hemisphere.

Therefore, the center-of-mass coordinate Z, according to Eq. (9.5), is

$$Z = \frac{\int_0^a z\rho \, dV}{\int_0^a \rho \, dV} = \frac{\int_0^a z\rho\pi(a^2 - z^2) \, dz}{\int_0^a \rho\pi(a^2 - z^2) \, dz} = \frac{3}{8} a \tag{9.15}$$

For a hemispherical shell, the situation is as shown in Fig. 9.4. Again from symmetry considerations, the center of mass is on the Z-axis. A small surface of length $2\pi(a^2 - z^2)^{1/2}$ and width $(a \, d\theta)$, as shown, has an area

$$dA = 2\pi(a^2 - z^2)^{1/2}a \, d\theta \tag{9.16}$$

According to Fig. 9.4,

$$z = a \sin \theta, \qquad dz = a \cos \theta \, d\theta$$

$$d\theta = \frac{dz}{a \cos \theta} = \frac{dz}{(a^2 - z^2)^{1/2}}$$

while the center of mass is $\mathbf{R}(X, Y, Z)$. The translational motion of the body is described by

$$\mathbf{F} = M\ddot{\mathbf{R}} \tag{9.17}$$

where $\mathbf{F}$ is the total external force acting on a body of mass M. The rotational motion of the body is described by the equation

$$\boldsymbol{\tau} = \frac{d\mathbf{L}}{dt} \tag{9.18}$$

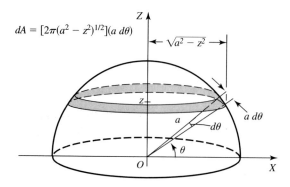

$dA = [2\pi(a^2 - z^2)^{1/2}](a \, d\theta)$

$\leftarrow \sqrt{a^2 - z^2} \rightarrow$

z

$a \, d\theta$

a

$d\theta$

θ

O

X

Figure 9.4   Center of mass for a hemispherical shell.

where $\mathbf{L}$ is the angular momentum and $\boldsymbol{\tau}$ is the total external torque acting about an axis passing through the center of mass. Thus Eqs. (9.17) and (9.18) represent six coupled equations to be solved simultaneously, a hard task to accomplish. But under certain constraints the number of equations can be reduced considerably, as we shall discuss later.

 Example 9.1 _____

Find the center of mass for (a) a solid cone and (b) a frustum of a cone. The two situations are shown in Fig. Ex. 9.1

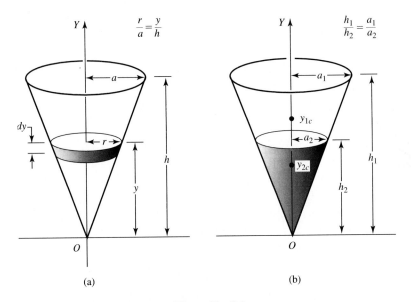

(a)   (b)

Figure Ex. 9.1

Solution

(a) Consider the cone, shown in Fig. Ex. 9.1(a), of radius a, height h, and density ρ. First consider a cone of radius r and thickness dy. Using the relation $r/a = y/h$, we can write the values of r and the mass dm of the cone. Eliminate r by substituting its value as shown.

$$r = \frac{a \cdot y}{h} \qquad dm = \rho \cdot \pi \cdot r^2 \cdot dy \qquad dm = \rho \cdot \pi \cdot \left(\frac{a \cdot y}{h}\right)^2 \cdot dy$$

Using Eq. (9.3), calculate the mass M of the cone by integration.

$$M = \int_0^h \left[\left(\frac{a \cdot y}{h}\right)^2 \cdot \rho \cdot \pi\right] dy \qquad M = \frac{1}{3} \cdot a^2 \cdot h \cdot \rho \cdot \pi$$

Using either of the relations in Eq. (9.5), calculate the center of mass Ycm of the cone as shown.

$$Ycm = \frac{\int_0^h y \cdot \left[\left(\frac{a \cdot y}{h}\right)^2 \cdot \rho \cdot \pi \right] dy}{\int_0^h \left[\left(\frac{a \cdot y}{h}\right)^2 \cdot \rho \cdot \pi \right] dy}$$

$$Ycm = \frac{3}{4} \cdot h$$

(b) The frustum is a cone that has had the lower portion [shaded portion in Figure Ex. 9.1(b)] removed. Treat this as two cones, the whole cone and the lower shaded cone. Find the mass and the center of mass of the frustum by subtracting the values for the lower cone from those for the whole cone as shown.

M1 and M2 are the masses of the two cones, and Y1cm and Y2cm are their center of masses. Using the geometry of the figure, replace the values of h2 by (a2/a1)h1.

$$\frac{h1}{h2} = \frac{a1}{a2} \qquad h2 = \frac{a2}{a1} \cdot h1 \qquad M1 = \frac{1}{3} \cdot a1^2 \cdot h1 \cdot \rho \cdot \pi$$

$$M2 = \frac{1}{3} \cdot a2^2 \cdot h2 \cdot \rho \cdot \pi \qquad M2 = \frac{1}{3} \cdot \frac{a2^3}{a1} \cdot h1 \cdot \rho \cdot \pi$$

$$Y2cm = \frac{3}{4} \cdot h2 \qquad Y2cm = \frac{3}{4} \cdot \frac{a2}{a1} \cdot h1$$

Substituting the values of M1, M2, Y1cm, and Y2cm, and simplifying, we find the center of mass Ycm of the frustum of a cone.

$$Ycm = \frac{M1 \cdot Y1cm - M2 \cdot Y2cm}{M1 - M2}$$

$$Ycm = \frac{3}{4} \cdot (a2 + a1) \cdot (a2^2 + a1^2) \cdot \frac{h1}{\left[a1 \cdot (a2^2 + a1 \cdot a2 + a1^2) \right]}$$

9.4 ROTATION ABOUT AN AXIS

After pure translation motion, the next simplest motion of a rigid body is its rotational motion about a fixed axis. When a body is free to rotate about a fixed axis, it needs only the coordinate to specify its orientation. Let us consider a rigid body that rotates about a fixed Z-axis, as shown in Fig. 9.5(a). The position of the body may be specified by an angle θ, which is between the line OA drawn on the body and the X-axis. Let us consider a particle of mass m_k to be a representative particle located at a distance $\mathbf{R}_k(X_k, Y_k, Z_k)$ from the origin, moving with a velocity $\mathbf{v}_k$ and angular velocity ω. The path of such a particle is a circle of radius $r_k = (x_k^2 + y_k^2)^{1/2}$ with its center on the Z-axis. Let ψ be the angle between the direction of the line OA in the body and the radius r_k from the Z-axis to the mass m_k. Since for a rigid body ψ is constant, as shown in the figure, $\phi = \theta = \psi$ and hence

$$\dot{\phi} = \dot{\theta} = \omega \qquad\qquad (9.19)$$

while

$$v_k = r_k \omega \qquad\qquad (9.20)$$

or in vector notation

$$\mathbf{v}_k = \boldsymbol{\omega} \times \mathbf{r}_k \qquad\qquad (9.21)$$

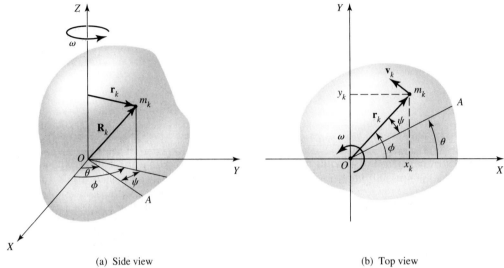

(a) Side view   (b) Top view

Figure 9.5   Rotation of a rigid body about a fixed axis: (a) side view, and (b) top view.

From Eq. (9.21) or from Fig. 9.5(b),

$$\dot{x}_k = -v_k \sin\phi = -\omega y_k \qquad \textbf{(9.22a)}$$

$$\dot{y}_k = v_k \cos\phi = \omega x_k \qquad \textbf{(9.22b)}$$

$$\dot{z}_k = 0 \qquad \textbf{(9.22c)}$$

and
$$v_k = r_k\omega = (\dot{x}_k^2 + \dot{y}_k^2)^{1/2} \qquad \textbf{(9.23)}$$

For further calculations, we can either use the rectangular coordinates (x, y, z) or cylindrical coordinates (r, θ, z). The kinetic energy K of the rotating body about the Z-axis is

$$K = \sum_k \tfrac{1}{2} m_k v_k^2 = \tfrac{1}{2}\left[\sum_k m_k r_k^2\right]\omega^2$$

or
$$K = \tfrac{1}{2}I_z\omega^2 = \tfrac{1}{2}I_z\dot{\theta}^2 \qquad \textbf{(9.24)}$$

where
$$I_z = \sum_k m_k r_k^2 = \sum_k m_k(x_k^2 + y_k^2) \qquad \textbf{(9.25)}$$

The quantity I_z is constant for a given rigid body rotating about a given axis (Z-axis in this case) and is called the *moment of inertia* about that axis. Since the body is continuous, we may replace the summation by the integration, and express I_z as

$$I_z = \iiint r^2\, dm = \iiint r^2 \rho\, dV \qquad \textbf{(9.26)}$$

Let us now calculate the angular momentum of the body about the Z-axis. By definition, the angular momentum of the body about the Z-axis is

$$L = \sum_k r_k(m_k v_k) = \left(\sum_k m_k r_k^2\right)\omega \qquad \textbf{(9.27)}$$

or $$L = I\omega = I\dot{\theta} \tag{9.28}$$

The rate of change of angular momentum for any system is equal to the total external torque (or total moment of force) τ (also written as N). Thus, for a rigid body rotating about the Z-axis, since I is constant,

$$\tau_z = \frac{dL}{dt} = I\frac{d\omega}{dt} = I\ddot{\theta} \tag{9.29}$$

This is an equation of motion for rotation of a rigid body about a fixed axis and is analogous to the equation for the translational motion of a particle along a straight line, that is, Newton's second law.

Similarly, the moment of inertia I is analogous to mass m; that is, I is a measure of the rotational inertia of a body relative to some fixed axis of rotation, just as m is a measure of the translational inertia of a body. Remember, the difference is that the moment of inertia depends on the axis of rotation, while the mass does not depend on its position. Such analogy may be shown between translational and rotational quantities, as well (see Table 9.1).

Furthermore, as an analogy with the translational motion, we may define the rotational potential energy as

$$V(\theta) = -\int_{\theta_s}^{\theta} \tau_z(\theta) \, d\theta \tag{9.30}$$

Table 9.1. Analogy between Rectilinear Motion and
Rotational Motion about an Axis

Rectilinear	Rotational
Position: x	Angular position: θ
Velocity: $v = dx/dt$	Angular velocity: $\omega = d\theta/dt$
Acceleration:	Angular acceleration:
$\quad a = dv/dt = d^2x/dt^2$	$\quad \alpha = d\omega/dt = d^2\theta/dt^2$
$\quad v = v_0 + at$	$\quad \omega = \omega_0 = \alpha t$
$\quad x = v_0 t + \frac{1}{2}at^2$	$\quad \theta = \omega_0 t + \frac{1}{2}\alpha t^2$
Mass: M	Moment of inertia: $I = \Sigma m_k r_k^2$
Linear momentum: $p = mv$	Angular momentum: $L = I\omega$
Force: F	Torque: $\tau = rF\sin\theta$
$\quad\quad F = ma$	$\quad\quad \tau = I\alpha$
$\quad\quad F = dp/dt$	$\quad\quad \tau = dL/dt$
Translational kinetic energy:	Rotational kinetic energy:
$\quad K = \frac{1}{2}mv^2$	$\quad K = \frac{1}{2}I\omega^2$
Potential energy:	Potential energy:
$\quad V(x) = -\int_{x_s}^{x} F(x)\,dx$	$\quad V(\theta) = -\int_{\theta_s}^{\theta} \tau(\theta)\,d\theta$
$\quad F(x) = -\dfrac{dV(x)}{dx}$	$\quad \tau(\theta) = -\dfrac{dV(\theta)}{d\theta}$

and
$$\tau_z = -\frac{dV}{d\theta}$$
(9.31)

Thus the rotational potential energy is the work done against the forces that produce the torque τ_z when the body is rotated from a standard angular position θ_s to a new position θ.

Example 9.2

A stick of mass M and length L is initially at rest in a vertical position on a frictionless table, as shown in Fig. Ex. 9.2. If the stick starts falling, find the speed of the center of mass as a function of the angle that the stick makes with the vertical.

Solution

The situation is as shown in the figure. We can find the speed of the center of mass by using the conservation of energy method. The only force acting on the stick is the gravitational force Mg in the vertical downward direction. Since there is no horizontal force acting on the rod, the center of mass falls vertically downward as shown. Let $\dot{y} = vy$ be the speed of the center of mass.

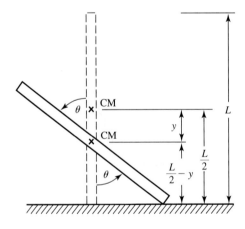

Figure Ex. 9.2

$Ki = 0$

Ei = initial total energy

Ef = final total energy

Vi = initial potential energy .

V = potential energy at different y

E = total energy, which is the sum of the translational kinetic energy (Kt), rotational kinetic energy (Kr), and potential energy V. The energy equation for any position y is

$Ei \equiv K + V \equiv Ef$ $Ei \equiv Vi \equiv M \cdot g \cdot \dfrac{L}{2}$ $V \equiv M \cdot g \cdot \left(\dfrac{L}{2} - y\right)$

$Ei \equiv Kt + Kr + V$

$$Ei \equiv E \equiv \frac{M \cdot vy^2}{2} + \frac{Io \cdot \omega\theta^2}{2} + M \cdot g \cdot \left(\frac{L}{2} - y\right) \qquad \omega\theta \equiv \frac{2 \cdot vy}{L \cdot \sin(\theta)}$$

$$M \cdot g \cdot \frac{L}{2} \equiv \frac{M \cdot vy^2}{2} + \frac{Io \cdot \omega\theta^2}{2} + M \cdot g \cdot \left(\frac{L}{2} - y\right)$$

λ is the linear mass density equal to M/L. Io is the moment of inertia of the rod about the center of mass. Solve for Io and then substitute for λ.

$$Io = \int_{\left(\frac{-L}{2}\right)}^{\frac{L}{2}} x^2 \cdot \lambda \; dx \qquad Io = \frac{1}{12} \cdot L^3 \cdot \lambda \qquad \lambda = \frac{M}{L} \qquad Io = \frac{1}{12} \cdot L^2 \cdot M$$

Substitute the value of Io and $\omega\theta$ in the energy equation and solve the equation for the value of velocity vy.

$$\frac{1}{2} \cdot M \cdot g \cdot L = \frac{1}{2} \cdot M \cdot vy^2 + \frac{1}{2} \cdot Io \cdot \left(\frac{2 \cdot vy}{L \cdot \sin(\theta)}\right)^2 + M \cdot g \cdot \left(\frac{1}{2} \cdot L - y\right)$$

$$\frac{1}{2} \cdot M \cdot g \cdot L = \frac{1}{2} \cdot M \cdot vy^2 + \frac{1}{6} \cdot M \cdot \frac{vy^2}{\sin(\theta)^2} + M \cdot g \cdot \left(\frac{1}{2} \cdot L - y\right)$$

The two resulting solutions for vy are

$$\left[\frac{1}{\left[6 \cdot \left(\frac{-1}{2} \cdot M - \frac{1}{6} \cdot \frac{M}{\sin(\theta)^2} \right) \right]} \cdot \sqrt{2} \cdot M \cdot \sqrt{3 \cdot \sin(\theta)^2 + 1} \cdot \sqrt{g} \cdot \sqrt{y} \cdot \frac{\sqrt{3}}{\sin(\theta)} \right.$$

$$\left. \frac{-1}{\left[6 \cdot \left(\frac{-1}{2} \cdot M - \frac{1}{6} \cdot \frac{M}{\sin(\theta)^2} \right) \right]} \cdot \sqrt{2} \cdot M \cdot \sqrt{3 \cdot \sin(\theta)^2 + 1} \cdot \sqrt{g} \cdot \sqrt{y} \cdot \frac{\sqrt{3}}{\sin(\theta)} \right]$$

Using one of the roots and simplifying, we obtain the final solution for vy. The value of vy may be further simplified as shown.

$$vy = \frac{-1}{\left[6 \cdot \left(\frac{-1}{2} \cdot M - \frac{1}{6} \cdot \frac{M}{\sin(\theta)^2} \right) \right]} \cdot \sqrt{2} \cdot M \cdot \sqrt{3 \cdot \sin(\theta)^2 + 1} \cdot \sqrt{g} \cdot \sqrt{y} \cdot \frac{\sqrt{3}}{\sin(\theta)}$$

$$vy = -i \cdot \sin(\theta) \cdot \frac{\sqrt{2}}{\sqrt{-4 + 3 \cdot \cos(\theta)^2}} \cdot \sqrt{g} \cdot \sqrt{y} \cdot \sqrt{3}$$

(a) What factors determine the velocity of the center of mass?

(b) Explain the changes in the magnitude of V, Kr, and Kt as the stick is falling and just before it hits the floor.

9.5 CALCULATION OF MOMENT OF INERTIA

For a system consisting of masses m_k located at distances r_k from an axis of rotation, the moment of inertia is given by

$$I = \sum_{k=1}^{N} m_k r_k^2 \qquad (9.32)$$

It is important to remember that r_k is the perpendicular distance of m_k from the rotation axis. For an extended, continuous rigid body, the moment of inertia about an axis of rotation is given by

$$I = \int r^2 \, dm \qquad (9.33)$$

where r is the perpendicular distance of the mass element dm from the rotation axis. For a one-dimensional body with a linear mass density λ (mass per unit length), for a two-dimensional body with an area mass density σ (mass per unit area), and for a three-dimensional body with volume mass density ρ (mass per unit volume), the moment of inertia in each case may be written as

$$I = \int r^2 \lambda \, dt \tag{9.34}$$

$$I = \int\int r^2 \sigma \, dA \tag{9.35}$$

$$I = \int\int\int r^2 \rho \, dV \tag{9.36}$$

where dl is the length element, dA is the area element, and dV is the volume element.

The definition of moment of inertia may be extended to the case of a composite body. Thus, if $I_1, I_2, \ldots$, are the moments of inertia of the various parts of the body about a particular axis, then the moment of inertia of the whole body about the same axis is

$$I = I_1 + I_2 + \cdots \tag{9.37}$$

We now calculate the moment of inertia of rigid bodies of different shapes.

Thin Rod

Let us consider a thin rod of length L and mass M, so that the linear mass density will be $\lambda = M/L$. Suppose we want to find the moment of inertia about an axis perpendicular to the rod at one end [Fig. 9.6(a)]. According to Eq. (9.34),

$$I = \int_0^L x^2 \lambda \, dx = \tfrac{1}{3}L^3\lambda = \tfrac{1}{3}ML^2 \tag{9.38}$$

where we have substituted $\lambda = M/L$. If the axis of rotation were at the center of the rod, as shown in Fig. 9.6(b), the moment of inertia would be

$$I = \int_{-L/2}^{+L/2} x^2 \lambda \, dx = \tfrac{1}{12}L^3\lambda = \tfrac{1}{12}ML^2 \tag{9.39}$$

Before proceeding further, it is important at this point to introduce and prove two most important theorems: The parallel axis theorem and the perpendicular axis theorem.

Parallel Axis Theorem

Consider a body rotating about an axis passing through O. There is no loss in generality by assuming this to be a Z-axis. By definition, the moment of inertia about an axis passing through O is

$$I_0 = \sum_k m_k(x_k^2 + y_k^2) = \int\int\int (x^2 + y^2)\rho \, dV \tag{9.40}$$

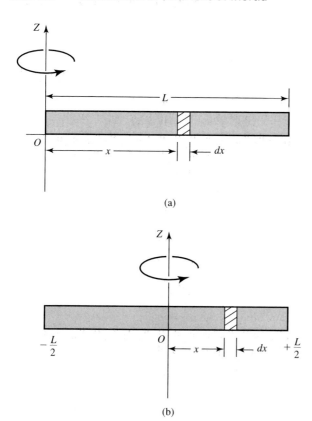

Figure 9.6   The moment of inertia for a thin rod (a) about an axis perpendicular to the rod at one end, and (b) about an axis perpendicular to the rod at the center.

where mass m_k is at a distance $\mathbf{r}_k$ from the origin and $(x_k^2 + y_x^2)^{1/2}$ from the Z-axis. According to Fig. 9.7,

$$\mathbf{r}_k = \mathbf{r}_c + \mathbf{r}_k' \tag{9.41}$$

where $\mathbf{r}_c$ is the distance of the center of mass from the origin O, and $\mathbf{r}_k'$ is the relative coordinate of m_k with respect to the CM. Using Eq. (9.41) (and dropping k),

$$x^2 + y^2 = (x_c + x')^2 + (y_c + y')^2 = x_c^2 + y_c^2 + x'^2 + y'^2 + 2x_c x' + 2y_c y' \tag{9.42}$$

Substituting this in Eq. (9.40), we obtain

$$I_0 = \iiint (x'^2 + y'^2)\rho\, dV + (x_c^2 + y_c^2)\iiint \rho\, dV$$
$$+ 2x_c \iiint x'\rho\, dV + 2y_c \iiint y'\rho\, dV \tag{9.43}$$

where the first term on the right is the moment of inertia about an axis parallel to the Z-axis and passing through the center of mass; that is,

$$I_c = \iiint (x'^2 + y'^2)\rho\, dV \tag{9.44}$$

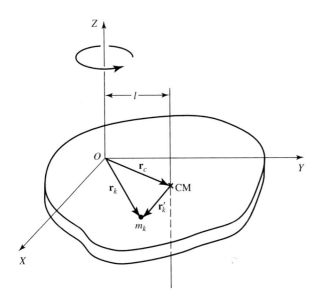

Figure 9.7 Parallel axis theorem.

The second term on the right side of Eq. (9.43) is equal to the mass M of the body multiplied by the square of the distance l between the center of mass and the Z-axis; that is,

$$(x_c^2 + y_c^2)\int\int\int \rho\, dV = (x_c^2 + y_c^2)M = Ml^2 \tag{9.45}$$

The last two terms in Eq. (9.43) are zero by definition of the center of mass; that is, they simply locate the center of mass relative to itself.

$$\int\int\int x'\rho\, dV = \int\int\int y'\rho\, dV = 0 \tag{9.46}$$

Thus, combining Eqs. (9.44), (9.45), and (9.46) with Eq. (9.43), we obtain

$$I_0 = I_c + Ml^2 \tag{9.47}$$

which is the parallel axis theorem and may be stated as follows:

> **Parallel Axis Theorem.** *The moment of inertia of a body about any axis is equal to the sum of the moment of inertia about a parallel axis through the center of mass and the moment of inertia about the given axis for the total mass of the body located at the center of mass.*

Thus, if we know the center of mass of a body and the moment of inertia about the center of mass, then the moment of inertia about any parallel axis can be calculated by using this theorem. This theorem can be applied to composite bodies as well.

Perpendicular Axis Theorem

A body whose mass is concentrated in a single plane is called a *plane lamina*. The perpendicular axis theorem is applicable to a plane lamina of any shape. Let us consider a rigid body in the form of a lamina in the XY-plane, as shown in Fig. 9.8. For rotation about the Z-axis, the moment of inertia about the Z-axis is given by

$$I_Z = \sum m_k(x_k^2 + y_k^2) = \iiint (x^2 + y^2)\rho \, dV \qquad (9.48)$$

If the body were rotating about the X-axis, its moment of inertia about the X-axis would be (for a thin lamina, $z = 0$; hence no z^2 term)

$$I_X = \iiint y^2 \rho \, dV \qquad (9.49)$$

and, similarly, the moment of inertia about the Y-axis would be

$$I_Y = \iiint x^2 \rho \, dV \qquad (9.50)$$

Combining Eqs. (9.49) and (9.50) with Eq. (9.48),

$$I_Z = I_X + I_Y \qquad (9.51)$$

which is the perpendicular axis theorem and may be stated as follows:

> **Perpendicular Axis Theorem.** *The sum of the moments of inertia of a plane lamina about any two perpendicular axes in the plane of the lamina is equal to the moment of inertia about an axis that passes through the point of intersection and perpendicular to the plane of the lamina.*

Let us apply these theorems to different situations.

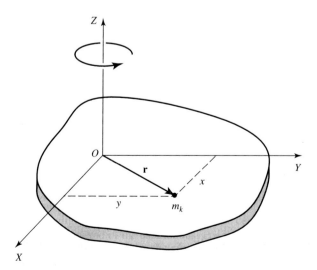

Figure 9.8   Perpendicular axis theorem as applied to a lamina in the XY plane.

Hoop or Cylindrical Shell

Consider a hoop or ring of mass M and radius a, as shown in Fig. 9.9. All the mass M is concentrated at a distance a from the axis. Hence the moment of inertia about the Z-axis is

$$I_Z = Ma^2 \tag{9.52}$$

Now suppose that we want to calculate the moment of inertia about an axis AA' that is parallel to the Z-axis and perpendicular to the plane of the ring, passing through the edge of the ring as shown. The situation is no longer symmetrical, and the direct calculation of the moment of inertia about the axis AA' is no longer trivial. But the application of the parallel axis theorem [Eq. (9.47)] makes such calculations simple; that is,

$$I_0 = I_c + Ml^2$$

When applied to the situation in Fig. 9.9, it gives

$$I_{AA'} = I_Z + Ma^2 = Ma^2 + Ma^2$$
$$= 2Ma^2 \tag{9.53}$$

Next we proceed to calculate the moment of inertia of the ring about an axis in the plane of the ring, such as about an X- or Y-axis. From the symmetry of the situation,

$$I_X = I_Y \tag{9.54}$$

and applying the perpendicular axis theorem, Eq. (9.51), $I_Z = I_X + I_Y$ gives

$$Ma^2 = I_X + I_X = I_Y + I_Y$$

or
$$I_X = I_Y = \tfrac{1}{2}Ma^2 \tag{9.55}$$

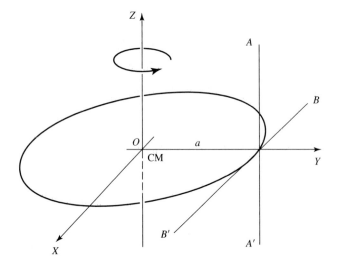

Figure 9.9   Moment of inertia for a hoop or ring about three different axes: Z-axis, AA'-axis, and BB'-axis.

We can now apply the parallel axis theorem to find the moment of inertia about the BB' axis, which is in the plane of the ring and tangent to the edge, as shown in Fig. 9.9. Thus

$$I_{BB'} = I_X + Ma^2 = \tfrac{1}{2}Ma^2 + Ma^2$$
$$= \tfrac{3}{2}\,Ma^2 \tag{9.56}$$

A cylindrical shell is simply a large number of rings piled one upon another. Thus the moment of inertia of a cylindrical shell or hollow cylinder of mass M, radius a, and length l may be calculated in a manner similar to the preceding ring.

Radius of Gyration

It is convenient to express the moment of inertia of a rigid body in terms of a distance k, called the radius of gyration, defined as

$$I = Mk^2, \qquad k = \sqrt{\frac{I}{M}} \tag{9.57}$$

That is, the *radius of gyration* is that distance from the axis of rotation where we may assume all of the mass of the body to be concentrated. Thus, for example, the radius gyration k of a thin rod with the axis of rotation passing through the center is

$$k = \sqrt{\frac{I}{M}} = \sqrt{\frac{(1/12)Ma^2}{M}} = \frac{a}{\sqrt{12}} \tag{9.58}$$

Once we know k for a rigid body rotating about a given axis, the moment of inertia is simply calculated from $I = Mk^2$.

Circular Disk, Solid Cylinder

Let us consider a solid disk of mass M and radius a, rotating about an axis through its center and perpendicular to the plane of the disk, as shown in Fig. 9.10. Let us divide the disk into several concentric rings, such as the one shown shaded in the figure. Thus the moment of inertia of this ring about the given axis is

$$dI = r^2\,dm$$

where r is the radius of the ring. The density per unit area is $\sigma = M/\pi a^2$; hence the mass dm of the ring is

$$dm = \sigma\,dA = \frac{M}{\pi a^2}\,2\pi r\,dr = \frac{2M}{a^2}\,r\,dr$$

Thus the moment of inertia of the disk may be written as

$$I = \int_0^a dI = \int_0^a r^2\,dm = \int_0^a r^2\,\frac{2M}{a^2}\,r\,dr = \frac{2M}{a^2}\int_0^a r^3\,dr = \frac{1}{2}\,Ma^2 \tag{9.59}$$

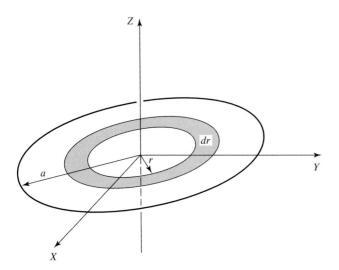

Figure 9.10 Moment of inertia for a disk about an axis perpendicular to the plane of the disk.

The same result can be obtained by using polar coordinates involving double integration. In the expression for the moment of inertia,

$$I = \int \int r^2 \sigma \, dA$$

$dA = r \, d\theta \, dr$ is the area shown in Fig. 9.11 and the mass per unit area is $\sigma = M/\pi a^2$. Substituting this in the preceding expression for I, we get

$$I = \frac{M}{\pi a^2} \int_0^a \int_0^{2\pi} r^3 \, dr \, d\theta = \frac{M}{\pi a^2} 2\pi \int_0^a r^3 \, dr$$

That is,
$$I = \tfrac{1}{2} M a^2$$

the result obtained in Eq. (9.59). To obtain the moment of inertia about different axes we can make use of the parallel and perpendicular axis theorems.

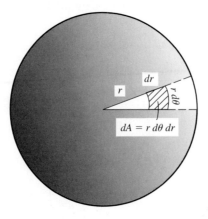

Figure 9.11 Moment of inertia for a disk using plane polar coordinates.

Sphere and Spherical Shell

Let us calculate the moment of inertia of a uniform solid sphere of radius a and mass M about an axis, say the Z-axis, passing through the center, as shown in Fig. 9.12. We can regard the sphere as made of disks, as shown in the figure. Let dI be the moment of inertia of this disk about the Z-axis so that the moment of inertia of the whole sphere will be $I = \int dI$. We calculate dI first. The disk shown has a radius $r = a \sin \theta$ while the density, mass per unit volume, of the material of the disk is

$$\rho = \frac{M}{4\pi a^3/3}$$

and the volume of the disk, with $z = a \cos \theta$, is

$$dV = \pi r^2 \, dz = \pi (a \sin \theta)^2 \, d(a \cos \theta) \tag{9.60}$$

Thus the moment of inertia of the disk, using Eq. (9.59), is

$$dI = \frac{1}{2}r^2 \, dm = \frac{1}{2}r^2 \rho \, dV = \frac{1}{2}(a \sin \theta)^2 \frac{M}{4\pi a^3/3} \pi (a \sin \theta)^2 \, d(a \cos \theta) = \frac{3}{8} M a^2 \sin^5 \theta \, d\theta$$

Hence the moment of inertia of the sphere about the Z-axis is given by

$$I = \int dI = -\tfrac{3}{8}Ma^2 \int_0^{\pi} \sin^5 \theta \, d\theta = \tfrac{2}{5}Ma^2 \tag{9.61}$$

We can obtain the same result by using the rectangular coordinates.

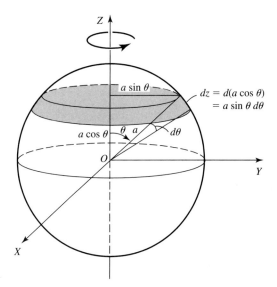

Figure 9.12   Moment of inertia for a solid sphere about the Z-axis passing through its center using spherical coordinates.

Finally, we can calculate the moment of inertia of a thin spherical shell. This can be done either by direct integration, as we have been doing, or alternatively by the application of Eq. (9.56). The final result is

$$I = \tfrac{2}{3}Ma^2 \tag{9.62}$$

9.6 SIMPLE PENDULUM

This is the first of many examples of the treatment of rotational motion. A simple pendulum consists of a mass m suspended from a fixed point O by a massless taut string (or a massless rod) of length l, as shown in Fig. 9.13. The system is treated as a rigid one. When the mass m is displaced from the vertical equilibrium position, it moves back and forth in an arc of a circle as shown. Thus the motion of a pendulum is equivalent to a rotational motion in a vertical plane and about the Z-axis through O, the axis being perpendicular to the plane. Let us apply Eq. (9.29) to this situation:

$$\tau_z = I_z \ddot{\theta} \tag{9.29}$$

where

$$I_z = ml^2 \tag{9.63}$$

and the torque about the Z-axis produced by the force mg is

$$\tau_z = -(mg \sin \theta)l \tag{9.64}$$

The negative sign is taken because the torque acts in such a way as to decrease the angle θ. Substituting Eqs. (9.63) and (9.64) in Eq. (9.31), we obtain

$$\ddot{\theta} + \frac{g}{l} \sin \theta = 0 \tag{9.65}$$

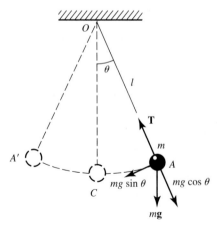

Figure 9.13   Simple pendulum.

This equation is not so easy to solve. But if we assume the angular displacement θ to be very small, that is, $\theta \ll \pi/2$, then $\sin\theta \simeq \theta$ and Eq. (9.65) takes the form

$$\ddot{\theta} + \frac{g}{l}\,\theta \simeq 0 \qquad (9.66)$$

which is the equation for simple harmonic motion and has the solution

$$\theta = \theta_0 \cos(\omega t + \phi) \qquad (9.67)$$

where

$$\omega = 2\pi f = \frac{2\pi}{T} = \sqrt{\frac{g}{l}} \qquad (9.68)$$

θ_0 and ϕ being two arbitrary constants that determine the amplitude and the phase of the oscillations from the initial conditions. Notice that the frequency f and time period T are independent of the amplitude of the oscillations, provided the amplitude is small enough for Eq. (9.66) to hold good. That T is independent of the amplitude for small displacements makes the pendulum well suited for use in clocks to regulate the rate. The exact solution of the pendulum motion, as we shall show, indicates that the time period of the pendulum increases with a slight increase in amplitude.

We now discuss the motion of the pendulum without the restriction that the amplitude be small. Since the motion of the pendulum is under a conservative force, we can solve the pendulum motion problem by the energy integral. The rotational potential energy associated with the torque given by Eq. (9.64) is

$$V(\theta) = -\int_{\theta_s}^{\theta} \tau_z(\theta)\,d\theta = -\int_{\pi/2}^{\theta} (-mgl\sin\theta)\,d\theta = -mgl\cos\theta \qquad (9.69)$$

where we have taken the standard reference angle θ_s to be $\pi/2$. The kinetic energy of mass m is

$$K = \tfrac{1}{2}I\omega^2 = \tfrac{1}{2}\,ml^2\dot{\theta}^2 \qquad (9.70)$$

Thus the energy integral describing the motion is

$$K + V = E = \text{constant}$$

$$\tfrac{1}{2}\,ml^2\dot{\theta}^2 - mgl\cos\theta = E \qquad (9.71)$$

Before solving this equation, we discuss the general features of the motion by drawing an energy diagram. Figure 9.14 shows the graph of $V(\theta)$, $K(\theta)$, and $E(\theta)$ versus θ. The graph of $V(\theta)$ versus θ has the maximum value mgl and the minimum value $-mgl$. For a mass m with energy slightly greater than $-mgl$, the motion will be simple harmonic. For E between $-mgl$ and $+mgl$, the motion is oscillatory and not harmonic. For $E > mgl$, the motion becomes nonoscillatory and the pendulum has enough energy to swing around in a complete circle. But the motion is still periodic, the period being equal to the time it takes to make one revolution, that is, for θ to increase or decrease by 2π.

Figure 9.14 _____

Below is the graph of potential energy V versus θ for a pendulum.

The length L, mass m, amplitude θ0 and the frequency ω of the simple pendulum are as given here. The initial phase φ is assumed to be zero.

$$N := 100 \qquad i := 0..N \qquad t_i := \frac{i}{10} \qquad g := 9.8$$

$$L := 1.5 \qquad m := .5 \qquad \theta 0 := 1 \qquad \omega := \sqrt{\frac{g}{L}} \qquad \theta_i := \omega \cdot t_i$$

Using Eqs.(9.67) through (9.70), we can obtain the expressions for the potential energy V, velocity v, kinetic energy K, and the limits of the potential energy mgL and −mgL as shown.

$$V_i := -m \cdot g \cdot L \cdot \cos\left(\left(\omega \cdot t_i\right)\right) \qquad v_i := -\omega \cdot \theta 0 \cdot L \cdot \sin\left(\omega \cdot t_i\right)$$

$$K_i := \frac{1}{2} \cdot m \cdot \left(v_i\right)^2$$

$$E_i := K_i + V_i \qquad\qquad m \cdot g \cdot L = 7.35$$

max(V) = 7.35   min(V) = −7.35

max(K) = 3.675   min(K) = 0

m·g·L = 7.35   − m·g·L = −7.35

E := max(V) − min(K)   E = 7.35

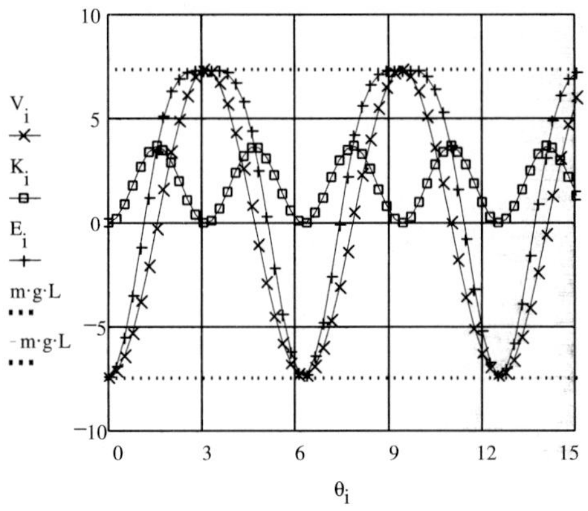

Looking at the graphs explain the variation in the values of K, V, and E at different angles.

From Eq. (9.71), when $\theta = \theta_0$, $\dot{\theta} = 0$, the total energy is

$$E = -mgl \cos \theta_0 \tag{9.72}$$

Substituting this in Eq. (9.71) and rearranging, we obtain

$$\int_{\theta_0}^{\theta} \frac{d\theta}{\sqrt{\cos \theta - \cos \theta_0}} = \sqrt{\frac{2g}{l}} \int_0^t dt \tag{9.73}$$

By assuming θ to be small and using $\cos \theta \simeq 1 - \frac{1}{2} \theta^2$, after integrating, we obtain the same result as obtained for small amplitudes. But we can proceed to solve Eq. (9.73) without a restriction on the amplitude. To obtain an accurate solution, we must transform Eq. (9.73) into a proper form of an elliptic integral. Using the identity $\cos\theta = 1 - 2\sin^2\theta/2$ to write

$$\cos\theta - \cos\theta_0 = 2\left(\sin^2\frac{\theta_0}{2} - \sin^2\frac{\theta}{2}\right)$$

Eq. (9.73) becomes

$$\int_{\theta_0}^{\theta} \frac{d\theta}{\sqrt{\sin^2(\theta_0/2) - \sin^2(\theta/2)}} = 2\sqrt{\frac{g}{l}}\int_0^t dt \tag{9.74}$$

where θ changes between $\pm\theta_0$. Now let us change the variables by substituting

$$\sin\phi = \frac{\sin(\theta/2)}{\sin(\theta_0/2)} = \frac{\sin(\theta/2)}{K} \tag{9.75}$$

where

$$K = \sin\frac{\theta_0}{2} \tag{9.76}$$

As the pendulum swings through a cycle, θ varies between $-\theta_0$ and $+\theta_0$; hence ϕ changes between $-\pi$ and $+\pi$. That is, ϕ runs from 0 to 2π for each cycle. Using Eqs. (9.75) and (9.76), Eq. (9.74) takes the form

$$\int \frac{d\phi}{\sqrt{1 - K^2\sin^2\phi}} = \sqrt{\frac{g}{l}}\int dt \tag{9.77}$$

Let us integrate this equation over one cycle; that is, as ϕ changes from 0 to 2π, t changes from 0 to T. Thus

$$\int_0^{2\pi} \frac{d\phi}{\sqrt{1 - K^2\sin^2\phi}\rho} = \sqrt{\frac{g}{l}}\,T \tag{9.78}$$

This is an *elliptic integral* of the *first kind*, and its value can be obtained from standard tables. However, it is more demonstrative to expand the integrand and then integrate; that is,

$$T = \sqrt{\frac{l}{g}}\int_0^{2\pi}\left(1 + \frac{1}{2}K^2\sin^2\phi + \cdots\right)d\phi = \sqrt{\frac{l}{g}}\left(2\pi + \frac{2\pi}{4}K^2 + \cdots\right) \tag{9.79}$$

$$T = 2\pi\sqrt{\frac{l}{g}}\left(1 + \frac{1}{4}\sin^2\frac{\theta_0}{2} + \cdots\right) \tag{9.80}$$

which clearly states that, as the amplitude θ_0 of the oscillations becomes large, the period becomes slightly larger than for small oscillations. Even for oscillations of small amplitude, the expression $T = 2\pi\sqrt{(l/g)}$, can be improved by assuming

$$\theta_0 \ll 1, \qquad \sin^2(\theta_0/2) \simeq \frac{\theta_0^2}{4}$$

Hence for small oscillations, Eq. (9.80) takes the form

$$T = 2\pi\sqrt{\frac{l}{g}}\,(1 + \tfrac{1}{16}\,\theta_0^2 + \cdots) \tag{9.81}$$

which is a more accurate expression because of the presence of the second term on the right. Expression (9.81) can be experimentally verified by measuring the time periods of two pendulums of the same lengths but with different amplitudes of oscillation.

9.7 PHYSICAL PENDULUM

A rigid body suspended and free to swing under its own weight about a fixed horizontal axis of rotation is known as a *physical pendulum* or *compound pendulum*. The rigid body can be of any shape as long as the horizontal axis does not pass through the center of mass. As shown in Fig. 9.15, the pendulum swings in an arc of a circle about an axis of rotation passing through O, the point O being the point of suspension. The point C is the center of mass of the physical pendulum. The distance between O and C is l. The position of the pendulum is specified by an angle θ between the line OC and the vertical line OA.

The torque τ_0 about the axis of rotation through O produced by the force Mg acting at C is

$$\tau_0 = -Mgl\sin\theta \tag{9.82}$$

If I is the moment of inertia of the body about the axis of rotation through O, the equation of motion

$$\tau_0 = I\ddot{\theta}$$

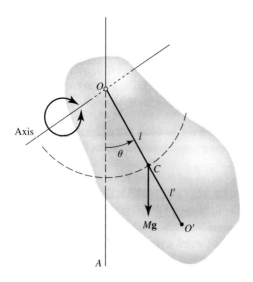

Figure 9.15 Physical pendulum.

takes the form

$$-Mgl \sin \theta = I\ddot{\theta}$$

or
$$\ddot{\theta} + \frac{Mgl}{I} \sin \theta = 0 \tag{9.83}$$

Once again, as in the case of a simple pendulum, for small oscillations we may assume that $\sin \theta \simeq \theta$, and hence

$$\ddot{\theta} + \frac{Mgl}{I} \theta \simeq 0 \tag{9.84}$$

This is the equation of a simple harmonic oscillator and has the solution

$$\theta = \theta_0 \cos(\omega t + \phi) \tag{9.85}$$

where the amplitude θ_0 and the phase angle ϕ are the two arbitrary constants to be determined from the initial conditions. The angular frequency ω is

$$\omega = \sqrt{\frac{Mgl}{I}} \tag{9.86}$$

while the time period T and frequency f are

$$T = \frac{2\pi}{\omega} = \frac{1}{f} = 2\pi\sqrt{\frac{I}{Mgl}} \tag{9.87}$$

If k is the radius of gyration for the moment of inertia about the axis of rotation through O, then

$$I = Mk^2 \tag{9.88}$$

Substituting Eq. (9.88) in Eq. (9.87) gives

$$T = 2\pi\sqrt{\frac{k^2}{gl}} \tag{9.89}$$

which states that *a simple pendulum of length k^2/l will have the same time period as that of a physical pendulum given by Eq. (9.87).*

Let us say that the moment of inertia of the rigid body about an axis passing through the center of mass C and parallel to the axis through O is I_C, and that the corresponding radius of gyration k_c is given by

$$I_C = Mk_c^2 \tag{9.90a}$$

Using the parallel axis theorem, we get the following relation between I and I_c:

$$I = I_C + Ml^2$$

$$Mk^2 = Mk_c^2 + Ml^2$$

or
$$k^2 = k_c^2 = l^2 \tag{9.90b}$$

Thus the time period T given by Eq. (9.89) may be written as

$$T = 2\pi\sqrt{\frac{k_c^2 + l^2}{gl}} \tag{9.91}$$

Let us now change the axis of rotation of this physical pendulum to a different position O' at a distance l' from the center of mass C, as shown in Fig. 9.15. The time period T' of oscillations about this new axis of rotation is

$$T' = 2\pi\sqrt{\frac{k_c^2 + l'^2}{gl'}} \tag{9.92}$$

Furthermore, suppose we assume that O' and l' are adjusted so that the two time periods T and T' are equal; that is,

$$T = T'$$

$$\frac{k_c^2 + l^2}{l} = \frac{k_c^2 + l'^2}{l'} \tag{9.93}$$

which simplifies to

$$k_c^2 = ll' \tag{9.94}$$

The point O' related to O by this relation is called the *center of oscillation* for the point O. Similarly, O is also the center of oscillation for O'.

Substituting Eq. (9.94) into Eq. (9.91) or (9.92) yields

$$T = 2\pi\sqrt{\frac{l + l'}{g}} \tag{9.95}$$

or

$$g = 4\pi^2\frac{l + l'}{T^2} \tag{9.96}$$

Thus, if we know the distance between O and O'—that is, if we know $l + l'$ and measure the time period T—the value of g can be measured very precisely, without knowing the position of the center of mass. Henry Kater used this method for an accurate determination of g. Kater's pendulum, shown in Fig. 9.16, has two knife edges. The pendulum can be suspended from either edge. The position of the edges can be adjusted so that the two time periods are equal. Once this is done, $l + l'$ is measured accurately and, knowing T, the value of g can be calculated from Eq. (9.96).

9.8 CENTER OF PERCUSSION

We shall now discuss some everyday applications of physical pendulum types of problems. Consider the body shown in Fig. 9.17, which is free to rotate about an axis passing through O. Suppose we strike a blow at point O', which is at a distance D from the axis of rotation through O.

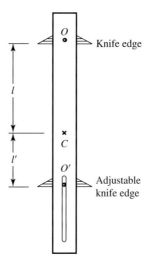

Figure 9.16 Kater's pendulum.

The blow applied is perpendicular to the line OCO', where C is the center of mass of the body. The forces acting on the body during the impact are force F' at the point of impact and another force F that is applied so as to keep the body fixed during the impact. If the body starts rotating with angular frequency ω, a radial force F_r, at O along the line $O'CO$ provides the necessary centripetal force. We want to find the condition under which F will be either zero or minimum. This can be done by the application of the laws of conservation of linear momentum and angular momentum.

By application of the laws of conservation of linear momentum and angular momentum it can be shown that if k_c is the radius of gyration for the momentum of inertia about C and l and l' are, respectively, distances of O and O' from C, then the following relation must be satisfied:

$$ll' = k_c^2 \tag{9.97}$$

Thus, if this relation is satisfied, when a blow is struck at O', no impulse will be felt at O. Such a point O' is called the center of percussion relative to point O. That is, *the point of application of an impulse (or blow) for which there is no reaction at the axis of rotation is known as the center of percussion.* The relation of Eq. (9.97) is exactly the same as for the physical pendulum

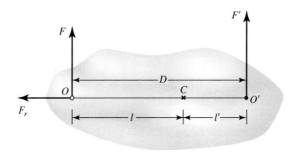

Figure 9.17 Relative positions of the center of oscillation O, the center of percussion O', and the center of mass C for a rigid body.

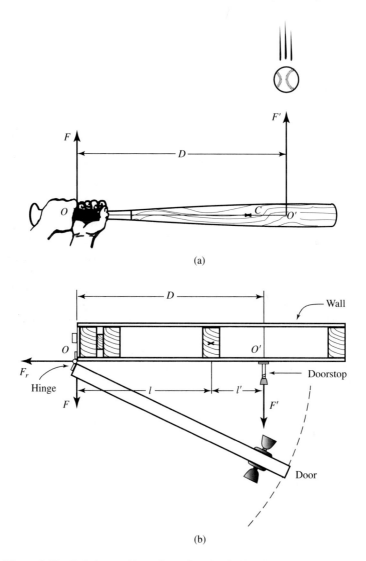

Figure 9.18 Relative position of O, O', and C in the case of (a) a batter hitting a ball, and (b) a door hitting a door stop.

given by Eq. (9.94). Thus the center of oscillation O and the center of percussion O' are identical. That is, the points O and O' are interchangeable.

Let us mention two important everyday applications. For instance, a batter hits a ball with a bat. The batter should try to hit at the center of the percussion at O' relative to his hand on the bat at O. This will minimize the blow to the hand; that is, it will avoid a reaction on the batter's hand, as shown in Fig. 9.18(a). As a second instance, consider a door stop used to prevent tearing door hinges loose. The door stop must be installed in such a way that, as shown in

Fig. 9.18(b), it is at a point of percussion at O' at a distance D from the hinges which are on the axis of rotation of the door.

9.9 DEFORMABLE CONTINUA

In most solids, atoms and molecules are arranged in some order. How rigidly these atoms and molecules are held about their equilibrium positions depends on the relative strength of the short-range electrical forces between them. Even though systems such as a vibrating string consist of a large number of discrete particles, it is advantageous to replace a system of discrete particles with a continuous distribution of matter. So far we have treated such continuous matter as rigid systems. In actual practice, matter is deformable. When under the action of internal and external forces, a change in the size and shape of the body may result. Thus, in this section, we shall be dealing with matter that we assume to be continuous and also deformable, that is, a *deformable continuum*. When external forces are applied to such a system, a distortion results because of the displacement of the atoms from their equilibrium positions and the body is said to be in a state of stress. After the external force is removed the body returns to its equilibrium position, providing the applied force was not too great. The ability of a body to retun to its original shape is called *elasticity*. To reach a quantitative definition of elasticity we must understand the definitions of stress and strain.

Suppose a body with surface area A is acted on by external forces having a resultant $\mathbf{F}$, where $\mathbf{F}$ is neither normal nor tangent to the surface. The *average stress* $\overline{\mathbf{S}}$ acting on an area A is defined as the *force per unit area*

$$\overline{\mathbf{S}} = \frac{\mathbf{F}}{A} \tag{9.98}$$

Let us now consider a small area ΔA, as shown in Fig. 9.19(a), that is acted on by a force $\Delta \mathbf{F}$. Thus the stress at point P is defined as

$$\mathbf{S} = \underset{\Delta A \to 0}{\text{limit}} \frac{\Delta \mathbf{F}}{\Delta A} = \frac{d\mathbf{F}}{dA} \tag{9.99}$$

The magnitude of $\mathbf{S}$ depends on the orientation of plane P in which area ΔA is located. We may resolve stress $\mathbf{S}$ into normal and tangential components by resolving $\Delta \mathbf{F}$ into two components—the normal component force ΔF_n and the tangential component force ΔF_t—which are normal and tangent to plane P. The *normal stress* σ is defined as

$$\sigma = \underset{\Delta A \to 0}{\text{limit}} \frac{\Delta \mathbf{F}_n}{\Delta A} = \frac{d\mathbf{F}_n}{dA} \tag{9.100}$$

while the *shear stress* τ is defined as

$$\tau = \underset{\Delta A \to 0}{\text{limit}} \frac{\Delta F_t}{\Delta A} = \frac{dF_t}{dA} \tag{9.101}$$

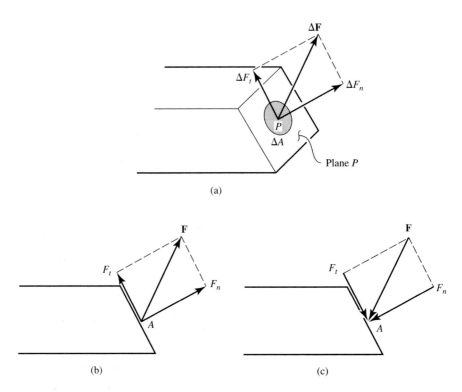

(a)

(b)   (c)

Figure 9.19   (a) Force $\Delta\mathbf{F}$ acting on an area ΔA. (b) Force $\mathbf{F}$ produces a tension and a shear stress. (c) Force $\mathbf{F}$ produces a compression and a shear stress.

If the normal stress is a pull, it is called *tension*, and if the normal stress is a push, it is called *compression*, while the tangential stress is called *shearing stress*. Thus, in Fig. 9.19(a) and (b), the normal force results in tension, while in Fig. 9.19(c) it results in compression, as shown; the tangential components in all three cases result in shear stress. The magnitude of tension and compression in parts (b) and (c) is F_n/A, while the shear stress is F_t/A. (Note: Stresses can result from internal forces as well, but according to Newton's third law, they cancel each other.)

The effect of stress is to cause distortion or a change in size and shape. *A quantity called strain refers to the relative change in size or shape of the body when under the applied stress.* We shall limit our discussion to three types of strain: (1) change in length, (2) change in shape, and (3) change in volume.

Consider a wire or rod of length L_0 and cross-sectional area A subject to a normal a tensile force F. The applied force increases the length. If the final length is L, then the change is $\Delta L = L - L_0$. We define the normal or tensile stress to be

$$\sigma = \frac{F_n}{A} \tag{9.102}$$

while the fractional change in length ϵ, which is called the *longitudinal* or *tensile strain*, is given by

$$\epsilon = \frac{L - L_0}{L_0} = \frac{\Delta L}{L_0} \tag{9.103}$$

It is found experimentally that the ratio of stress to strain is a constant for a given material. This is called the *elastic modulus*. The ratio of the longitudinal stress to the longitudinal strain is called *Young's* (or *stretch*) *modulus, Y;* that is,

$$Y = \frac{\sigma}{\epsilon} = \frac{F_n/A}{\Delta L/L_0} \tag{9.104}$$

Since strain is a dimensionless quantity, the units of Young's modulus are the same as those of stress, that is, N/m^2, lb/in^2, and so on. [The transverse strain (change in length perpendicular to the force) is small and will be considered shortly.]

If the material of the wire obeys Hooke's law, the change in length is proportional to the applied force,

$$F_n = k\Delta L \tag{9.105}$$

where k is the stiffness or spring constant. We may write Eq. (9.104) as

$$F_n = \frac{YA}{L_0} \Delta L \tag{9.106}$$

Comparing these two equations, we get

$$k = \frac{YA}{L_0} \tag{9.107}$$

Since, for a given material Y, A and L_0 are constant, so is k.

Thus elongation increases with increasing force. When the force is removed, the wire returns to its original length. The plot of tensile stress σ versus tensile strain ϵ is shown in Fig. 9.20. The proportionality relation between σ and ϵ holds only if stress is less than a certain maximum value. As shown in Fig. 9.20, this is reached at point A, the *proportional limit* or *yield point*. If the value of the applied stress is between A and B, there is no proportionality; but when the stress is removed, the body does return to its original value. If the applied stress is beyond point C, permanent deformation occurs in the body and eventually, if the applied stress is high, it will break (*fracture point*).

The plot in Fig. 9.20 assumes that, in the expression for stress, the area A remains constant and is equal to the original area before applying any force. But in practical situations there will be lateral contraction. Thus the stress is indicative of the force and the ratio of the force and *true area*. Hence we may define the true strain as follows. Let dL be the infinitesimal amount of elongation when the instantaneous length is L. If L_0 is the initial length and L_f the final length, then the *true strain* is defined as

$$\epsilon_{\text{true}} = \int_{L_0}^{L_f} \frac{dL}{L} = \ln \frac{L_f}{L_0} \tag{9.108}$$

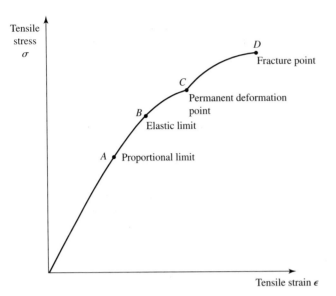

Figure 9.20   Elastic properties of a typical solid under normal stress.

For small elongations, if we expand $\ln (L_f/L_0)$ and neglect higher-order terms, we get ϵ as given by Eq. (9.103); that is, $\epsilon = \Delta L/L_0$.

Let us now consider the rigidity modulus resulting from the application of shear stress. When a pair of equal and opposite forces not acting along the same line of action is applied [Fig. 9.21(a)], the resulting shear stress produces a change in the shape of the body (but no change in length). The resulting strain is called a *shear strain*. It appears that the material consists of layers, and when stress is applied, the layers try to slide over one another. As shown in Fig. 9.21(a), the layer $ABCD$, under the action of a shear stress, has moved to $A'B'C'D'$, while the layer $PQRS$ is not displaced. The shear stress τ as defined in Eq. (9.101) is

$$\tau = \frac{F_t}{A} \tag{9.109}$$

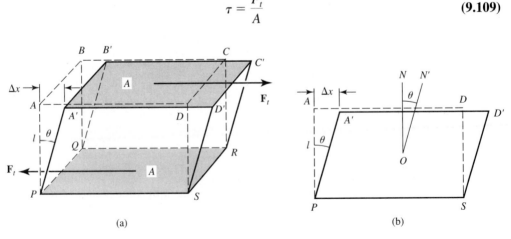

Figure 9.21   (a) A body is under the action of a pair of tangential forces, which results in a shearing stress. (b) Side view of part (a).

The shearing strain γ is defined as the ratio of the displacement Δx and length l, as shown in Fig. 9.21. For small values of Δx, this ratio is equal to the tangent of the angle θ. That is, shearing strain is

$$\gamma = \frac{\Delta x}{l} \simeq \tan \theta \tag{9.110}$$

Thus the *shear modulus*, or *modulus of rigidity*, or *torsion modulus*, η, is defined as

$$\eta = \frac{\tau}{\gamma} = \frac{F_t/A}{\tan \theta} \tag{9.111}$$

In the case of fluids, forces must be applied normal to the surface. Suppose that a fluid of volume V is acted on by a force F_n acting normal to an area A, resulting in a change in volume ΔV. The normal force applied to a fluid is called pressure P. Thus the stress and strain are given by

$$\text{Volume stress} = \sigma = \frac{F_n}{A} = \Delta P \tag{9.112}$$

and
$$\text{Volume strain} = \frac{\Delta V}{V} \tag{9.113}$$

Thus the *volume elasticity* or *bulk modulus*, B, defined as the ratio of the volume stress to volume strain, is given by

$$B = \frac{\Delta P}{-\Delta V/V} = -V\frac{\Delta P}{\Delta V} \tag{9.114}$$

The negative sign indicates that as pressure increases, volume decreases. The reciprocal of the bulk modulus is called *compressibility* β $(\beta = 1/B)$.

Noting that $\Delta P = \sigma = $ normal stress, we may write Eq. (9.114) as

$$\sigma = -B\frac{\Delta V}{V} \tag{9.115}$$

When a stress is applied in one direction, it results in a longitudinal strain as well as a transverse (lateral) strain. In the case of simple tension, the ratio of the lateral strain ϵ_t to the longitudinal strain ϵ_l is called *Poisson's ratio* ν:

$$\nu = \frac{\epsilon_t}{\epsilon_l} \tag{9.116}$$

ν is small for glass $(\simeq 0.25)$, while for rubber it is 0.5. There is a simple relationship among B, Y, and ν which we state here without proof:

$$B = \frac{Y}{3(1 - 2\nu)} \tag{9.117}$$

Note that this relation assumes that the material is uniform, that is, homogeneous and isotropic.

9.10 EQUILIBRIUM OF RIGID BODIES

To start, we shall discuss conditions of equilibrium. We shall apply these conditions to the investigation of equilibrium of flexible strings and cables and then to the equilibrium of solid beams.

Let us consider a body of mass M whose center of mass is at a distance R from a given point O, which is acted on by forces $\mathbf{F}_i$, and has angular momentum L_O about an axis passing through O. The equations of motion describing a rigid body are

$$\sum_i \mathbf{F}_i = M\ddot{\mathbf{R}} \tag{9.118}$$

and
$$\sum_i \boldsymbol{\tau}_{iO} = \frac{d\mathbf{L}_O}{dt} = I\ddot{\boldsymbol{\theta}} \tag{9.119}$$

where $\boldsymbol{\tau}_{iO}$ are the torques about an axis passing through O, I is the moment of intertia, and $\ddot{\boldsymbol{\theta}}$ is the angular acceleration. Once the torques about any one point O are known, the torques about any other point O' may be calculated from the following relation:

$$\sum_i \boldsymbol{\tau}_{iO'} = \sum_i \boldsymbol{\tau}_{iO} + (\mathbf{r}_O - \mathbf{r}_{O'}) \times \sum_i \mathbf{F}_i \tag{9.120}$$

where $\mathbf{r}_O$ and $\mathbf{r}_{O'}$ are the vector distances of points O and O' from the origin. Equation (9.120) states that *the total torque about O' is equal to the sum of two terms: the total torque about O and the total torque taken about O', assuming the total force is acting at O.* The proof is straightforward. Let $\mathbf{F}_i$ be the force acting at a point i, which is at a distance $\mathbf{r}_i$ from the origin. Then, according to the definition, the torque about O' is

$$\sum_i \boldsymbol{\tau}_{iO'} = \sum_i (\mathbf{r}_i - \mathbf{r}_{O'}) \times \mathbf{F}_i$$

$$= \sum_i (\mathbf{r}_i - \mathbf{r}_O + \mathbf{r}_O - \mathbf{r}_{O'}) \times \mathbf{F}_i$$

$$= \sum_i (\mathbf{r}_i - \mathbf{r}_O) \times \mathbf{F}_i + \sum_i (\mathbf{r}_O - \mathbf{r}_{O'}) \times \mathbf{F}_i$$

$$= \sum_i \boldsymbol{\tau}_{iO} + (\mathbf{r}_O - \mathbf{r}_{O'}) \times \sum_i \mathbf{F}_i$$

which is the result stated previously.

For a rigid body to be in translational equilibrium—that is, at rest or moving with uniform velocity—the sum of the forces must be zero. For a body to be in rotational equilibrium—at rest or rotating with uniform velocity, the sum of the external torques must be zero. Thus, from

Eqs. (9.118) and (9.119), with $\ddot{\mathbf{R}} = 0$ and $\ddot{\boldsymbol{\theta}} = \mathbf{0}$, we get

$$\sum_i \mathbf{F}_i = 0 \qquad\qquad (9.121)$$

and

$$\sum_i \boldsymbol{\tau}_{iO} = 0 \qquad\qquad (9.122)$$

Note that if the sum of the torques about any point is zero, then [from Eq. (9.120)] it will be zero about any other point.

Since we notice that the motion of a rigid body is determined by the total forces and total torques, we can make the following statement, which we shall find useful in discussing the equilibrium of rigid bodies. Two systems of forces acting on a rigid body are *equivalent* if they produce the same resultant force and the same total torque about any point.

Here we state the definition of a couple. A *couple* is a system of forces whose sum is zero; that is,

$$\sum_i \mathbf{F}_i = 0 \qquad\qquad (9.123)$$

The total torque resulting from a couple is the same about every point and is given by

$$\sum_i \boldsymbol{\tau}_{iO'} = \sum_i \boldsymbol{\tau}_{iO} = \sum_i \mathbf{r}_{iO} \times \mathbf{F}_i \qquad\qquad (9.124)$$

Thus a couple may be characterized by a vector that is the total torque. This leads to a further statement: *All couples are said to be equivalent if they have the same total torque.*

From this discussion, we can deduce the following very useful result, referred to as the *rigid body theorem*:

> *Every system of forces acting on a rigid body can be reduced to a single force through an arbitrary point and a couple.*

Depending on the type of equilibrium, the resulting force and/or couple, may be zero.

9.11 EQUILIBRIUM OF FLEXIBLE CABLES AND STRINGS

An *ideal flexible cable* or *string* is one that will not support any compression or shearing stress, but there can be tension directed along the tangent to the string at any point. Cables, chains, and light strings used in different structures can be treated as ideal flexible strings. Furthermore, we shall assume that the weight of the cable is negligible compared to the external load acting on it, or that there is no external load and the weight of the cable is the only load. We can divide our discussion into two parts: (1) cables of negligible weight, and (2) cables with loads (or forces) distributed continuously along the length of the cable.

Cable with Concentrated Load

Consider an ideal flexible cable of negligible weight that is suspended between points P_1 and P_2 and is under the action of an external force $\mathbf{F}$ acting at point P_3, as shown in Fig. 9.22. The force $\mathbf{F}$ keeps the cable taut, as shown. Let l_1 be the length of the segment of the string between P_1P_3, l_2 the length between P_2P_3, while l_{12} is the distance between the points P_1 and P_2. Let $\mathbf{T}_1$ and $\mathbf{T}_2$ be the tensions in the two segments of the strings, as shown. By using the law of cosines, the angles α and β are given in terms of l_1, l_2, and l_3 by the following relations:

$$\cos\alpha = \frac{l_1^2 + l_{12}^2 - l_2^2}{2l_1l_{12}} \quad \text{and} \quad \cos\beta = \frac{l_2^2 + l_{12}^2 - l_1^2}{2l_2l_{12}} \tag{9.125}$$

It is assumed that the string does not stretch so that the position of point P_3 is independent of the force $\mathbf{F}$. Since point P_3 is in equilibrium, we may write

$$\mathbf{F} + \mathbf{T}_1 + \mathbf{T}_2 = 0 \tag{9.126}$$

as shown by the triangular relation in Fig. 9.22(b). Using the law of sines, from Fig. 9.22(b) we obtain expressions for $\mathbf{T}_1$ and $\mathbf{T}_2$ in terms of $\mathbf{F}$; that is,

$$\mathbf{T}_1 = \mathbf{F}\frac{\sin(\beta + \gamma)}{\sin(\alpha + \beta)} \quad \text{and} \quad \mathbf{T}_2 = \mathbf{F}\frac{\sin(\gamma - \alpha)}{\sin(\alpha + \beta)} \tag{9.127}$$

     This means that we can find the angles in terms of distances from Eq. (9.125) and then use them in Eq. (9.127) to evaluate $\mathbf{T}_1$ and $\mathbf{T}_2$. But this is not the true answer because we have assumed that the string does not stretch Actually, tension determines the lengths of the segments of the strings, and we must take this into account to evaluate $\mathbf{T}_1$ and $\mathbf{T}_2$. These can be evaluated sufficiently by the method of *iterative approximations* or the *relaxation* method, explained next.

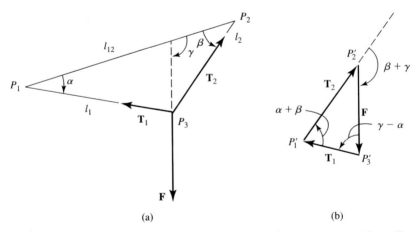

(a)   (b)

Figure 9.22   (a) Ideal flexible cable under the action of an external force $\mathbf{F}$. (b) $\mathbf{F}$, $\mathbf{T}_1$, and $\mathbf{T}_2$ form a closed triangle.

According to Hooke's law, the unstretched lengths l_{10} and l_{20} under tensions T_1 and T_2 become l_1 and l_2, given by

$$l_1 = l_{10}(1 + kT_1) \quad \text{and} \quad l_2 = l_{20}(1 + kT_2) \tag{9.128}$$

where k is a proportionality constant. We now use successive approximation to evaluate different quantities: α, β, T_1, and T_2. For a first approximation, assume that the string does not stretch, so that $l_1 = l_{10}$ and $l_2 = l_{20}$. Using these values in Eqs. (9.125) and (9.127), we evaluate α, β, T_1, and T_2. Now use these values for T_1 and T_2 in Eq. (9.128) to obtain new values for l_1 and l_2. Use these values for l_1 and l_2 in Eq. (9.125) to get new values for α and β, and use these in Eq. (9.127) to get better values for T_1 and T_2. These values for T_1 and T_2 can be used in Eq. (9.128) to get still better values for l_1 and l_2, and then Eqs. (9.125) and (9.127) are used to get better values for α, β, T_1, and T_2. This procedure can be repeated over and over until the values for l_1, l_2, α, β, T_1, and T_2 converge to the correct values. In most situations the amount of stretching is small, and the first few iterations lead to correct values. As stated earlier, this is the method of successive approximation.

Cables under Distributed Loads

Parabolic cables.   Consider a cable AB that is supporting a load that is uniformly distributed horizontally, as shown in Fig. 9.23. The load is denoted by vertical arrows pointing downward. Let this load be **w** per unit length, the length taken to be horizontal. O is the lowest point while $P(x, y)$ is any other point on the cable.

Let us consider the equilibrium of the portion OP of the cable that is horizontally loaded as shown in Fig. 9.24(a). Let $\mathbf{T}_0$ be the tension at the lowest point O that is horizontal, while the tension at P is $\mathbf{T}$, which makes an angle θ with the horizontal. The origin of the coordinate axes is taken to be at O, and XY is the vertical plane. $\mathbf{W}$ is the load acting on a portion of cable OP and is equal to $\mathbf{w}x$. Thus, for static equilibrium [from Fig. 9.24(b)],

$$\mathbf{T} + \mathbf{T}_0 + \mathbf{W} = 0 \tag{9.129}$$

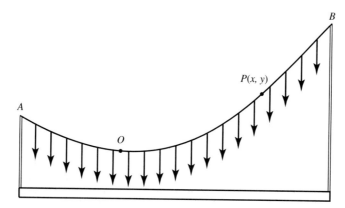

Figure 9.23   Cable AB supports a uniformly distributed load along a horizontal distance.

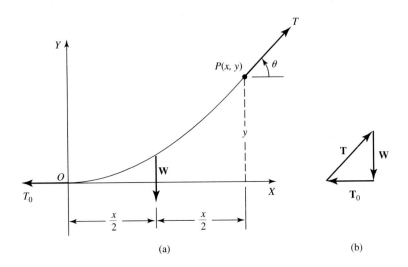

Figure 9.24   (a) Small segment OP of a uniformly loaded string. (b) For static equilibrium, T_0, **T**, and **W** form a closed triangle.

or

$$T \cos \theta = T_0 \tag{9.130}$$

$$T \sin \theta = W = wx \tag{9.131}$$

Solving for T and θ, we get

$$T = \sqrt{T_0^2 + w^2 x^2} \tag{9.132}$$

and

$$\tan \theta = \frac{w}{T_0} x \tag{9.133}$$

Because the load is uniform, W is located at a distance $x/2$, as shown in Fig. 9.24(a). Taking the torque about P, we get

$$W \frac{x}{2} = T_0 y \quad \text{or} \quad wx \frac{x}{2} = T_0 y \tag{9.134}$$

That is,

$$y = \frac{w}{2T_0} x^2 \tag{9.135}$$

which is an equation of a *parabola*; that is, a cable under a uniform horizontal load has a parabolic shape. A cable in a suspension bridge is a typical example.

In Fig. 9.23, if A and B are at the same height, the horizontal distance L between A and B is called a *span*, while the vertical distance h of the lowest point O from A or B is called a *sag*. Substituting $y = h$ and $x = L/2$ in Eq. (9.135), the tension T_0 at the lowest point is

$$T_0 = \frac{wL^2}{8h} \tag{9.136}$$

Suppose point A is (x_A, y_A) and point P is (x_B, y_B). Using these values in Eq. (9.136), we can calculate the span $L = x_B - x_A$ (see Problem 9.46).

Catenary cables.   Let us now consider a cable supporting a load that is distributed uniformly along its length (not along the horizontal distance as in parabolic cables). A typical example is that of a cable supporting its own weight, as shown in Fig. 9.25. Let point P be at a distance s from a point $s = 0$, where the tension T_0 is the supporting force at the end of $s = 0$ and is constant, while the tension at P is $\mathbf{T}(s)$. Let $\mathbf{w}(s)$ be the force per unit length at point s [note that $w(s) \neq w(x)$]. $\mathbf{w}\,ds$ represents the force on a small segment of length ds. Thus, for the portion of the cable AP that is in equilibrium, we must have

$$\mathbf{T}_0 + \mathbf{T}(s) + \int_0^s \mathbf{w}(s)\,ds = 0 \tag{9.137}$$

We can obtain $\mathbf{T}(s)$ by differentiating Eq. (9.137) with respect to s; that is,

$$\frac{d\mathbf{T}}{ds} = -\mathbf{w}(s) \tag{9.138}$$

From Fig. 9.25, θ is the angle that $\mathbf{T}$ makes with the X-axis. The vertical and horizontal components of Eq. (9.138) are

$$\frac{d}{ds}(T \sin \theta) = w \tag{9.139}$$

and

$$\frac{d}{ds}(T \cos \theta) = 0 \tag{9.140}$$

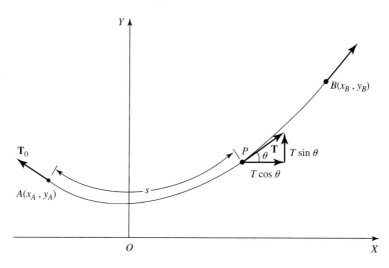

Figure 9.25   Cable supporting a load distributed uniformly along its length (and not its horizontal distance).

Keeping in mind that

$$ds = (dx^2 + dy^2)^{1/2} \tag{9.141}$$

the above equations, on solving, yield

$$y = \frac{C}{w}\cosh\left(\frac{w}{C}x + A\right) + B \tag{9.142}$$

This is the equation of a curve called a *catenary*. *A*, *B*, and *C* are the constants chosen so that *y* has proper values at the end points. Also, if we choose a coordinate system such that $y = 0$ at $x = 0$, then the constant $A = 0$.

9.12 EQUILIBRIUM OF SOLID BEAMS

General Treatment: Bending Moments

Let us consider a horizontal beam that is subject to vertical forces only—that is, the problem of a cantilever. Such a beam is under no compression or tension and there is no torsion about the axis of the beam. In these conditions, the beam bends only in a vertical plane. This is an example of a simple structure under shear forces and bending moments. We can calculate these quantities and the resulting bending as shown text.

Let the vertical forces $\mathbf{F}_1, \mathbf{F}_2, \ldots, \mathbf{F}_n$ be the forces acting on a horizontal beam at distance $x_1, x_2, \ldots, x_n$, as shown in Fig. 9.26. The forces acting vertically upward are taken positive, and those acting downward are taken negative. Draw a plane AA' perpendicular to the beam and at a distance x from the left end of the beam. All the forces acting on the plane AA' due to the portion of the beam to the right of the plane can be reduced to an equivalent single force $\mathbf{F}_t$ through any point in the plane and a couple of torque $\boldsymbol{\tau}$. Since in the present case there are no compression or tension forces, $\mathbf{F}_t$ must be vertical; hence it is a *shearing force* acting from right to left across the plane AA', as shown. We assumed that there was no torsion in the beam. Since all the forces are vertical, the *bending moment* $\boldsymbol{\tau}$ (or torque $\boldsymbol{\tau}$) must be exerted from right to left

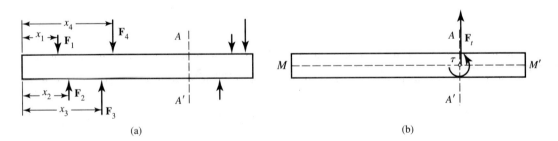

(a)   (b)

Figure 9.26   (a) Horizontal beam under the action of vertical forces. (b) The forces acting on the beam to the right of AA' are equivalent to a shear force $\mathbf{F}_t$ and torque $\boldsymbol{\tau}$.

along a horizontal axis perpendicular to the beam and in the plane AA'. The counterclockwise rotations of the plane about this horizontal axis are taken as positive. (From Newton's third law, force and torque equal and opposite to $\mathbf{F}_t$ and $\boldsymbol{\tau}$ are acting on the plane AA' from the beam on the left.)

We consider the equilibrium of the beam on the left of the plane AA'. Let $\mathbf{w}$ be the weight per unit length of the beam. Thus, from the two conditions of equilibrium, from Fig. 9.26, we get

$$\mathbf{F}_t + \sum_{x_i < x} \mathbf{F}_i - \int_0^x \mathbf{w}\, dx = 0 \tag{9.143}$$

and

$$\boldsymbol{\tau} - \sum_{x_i < x} (x - x_i)\mathbf{F}_i + \int_0^x (x - x')\mathbf{w}\, dx' - \boldsymbol{\tau}_0 = 0 \tag{9.144}$$

where $\mathbf{F}_t$ is the shearing force and $\boldsymbol{\tau}$ is the bending moment acting at a distance x from the left end. The second term on the left of Eq. (9.143) is the sum of the external forces acting on the beam from the left end up to the plane AA', and the third term is the weight of the same portion of the beam and is acting downward. $\boldsymbol{\tau}_0$ is the bending moment exerted by the left end of the beam on its support, providing the beam is fastened or clamped or supported at that end. If all the forces are known, $\mathbf{F}_t$ and $\boldsymbol{\tau}$ acting at x on the beam can be calculated from Eqs. (9.143) and (9.144). If the right end of the beam is free, then $\mathbf{F}_t = 0$ and $\boldsymbol{\tau} = 0$, and the equations can be used to calculate two other forces. Depending on whether the ends are free or supported, we can use these conditions to help solve the preceding equations.

The shearing force $\mathbf{F}_t$ and the bending torque $\boldsymbol{\tau}$ depend on the value of x and may be calculated as a function of x by differentiating Eqs. (9.143) and (9.144):

$$\frac{d\mathbf{F}_t}{dx} = \mathbf{w} \tag{9.145}$$

and

$$\frac{d\boldsymbol{\tau}}{dx} = \sum_{x_i < x} x_i \mathbf{F}_i - \int_0^x \mathbf{w}\, dx' = -\mathbf{F}_t \tag{9.146}$$

PROBLEMS

9.1. Find the center of the mass of the following:
   (a) A thin uniform wire of linear mass density λ bent into an L-shape with both horizontal and vertical lengths equal.
   (b) A thin uniform wire of linear mass density λ bent into a quadrant of a circle of radius R.

9.2. Find the center of mass of the following:
   (a) A thin uniform sheet of metal of surface density σ cut into a semicircle of radius R.
   (b) A thin uniform sheet of metal of surface mass density σ cut into a triangular piece with sides a, a, and b.

(c) A thin uniform sheet of metal of surface density σ cut into an octant of a thin spherical shell of radius R.

9.3. Find the center of mass of an octant of a solid sphere of radius R and uniform density ρ.

9.4. Find the center of mass of a sphere of radius R that is made up of layers of thin spherical shells centered about the center. The variation in density of these shells is **(a)** $\rho = \rho(z) = \rho_0(1 + z/R)$ and **(b)** $\rho = \rho(r) = \rho_0(1 + r/R)$.

9.5. Find the center of mass of a thin sheet in the XY-plane in the form of a parabola $y = ax^2$ and bounded between $y = 0$ and a straight line $y = b$. Calculate for the case when $b = 20$ cm and the surface density is 10 kg/m^2.

9.6. Find the center of mass of a paraboloid $z = a(x^2 + y^2)$ between $z = 0$ and $z = b$, as shown in Fig. P9.6. Calculate for the case when $b = 20$ cm, and the density is $\rho = 8000(1 - 0.5z)$ kg/m^3.

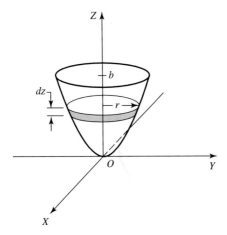

Figure P9.6

9.7. Consider a circular sheet of radius $2R$ having a uniform surface density σ. A circular hole of radius R is made at a distance R from the center of the first circle, as shown in Fig. P9.7. Find the center of mass of the remaining piece.

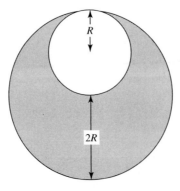

Figure P9.7

9.8. Find the center of the circle plate shown in Fig. P9.8 made up of two semicircular pieces of surface densities densities σ_1 and σ_2.

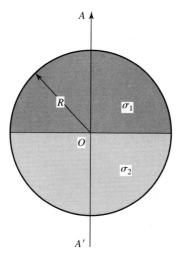

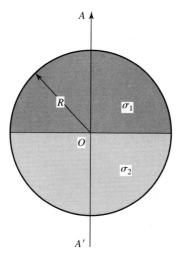

Figure P9.8

9.9. Find the center of mass of a solid hemisphere of radius R whose density varies linearly with distance from the center; that is, $\rho = \rho_0 r/R$.

9.10. Consider a solid sphere of uniform density ρ and radius R and a spherical cavity of radius $R/2$ centered at a distance of $R/2$ from the center. Find the center of mass.

9.11. Find the moment of inertia for a square lamina of mass M and side L, as shown in Fig P9.11, rotating about the following axes:
  (a) Axis AA' passing through the center of mass and perpendicular to the lamina.
  (b) Axis BB' parallel to AA' and at a distance $L/2$.
  (c) Axis CC' parallel to one side of the lamina.

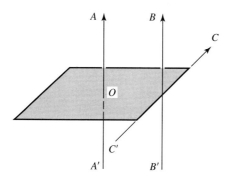

Figure P9.11

9.12. Consider a cube of mass M and side L, as shown in Fig. P9.12. Find the moment of inertia: **(a)** about an axis AA' perpendicular to a face and passing through the center of mass; **(b)** about an axis BB' parallel to the axis in part (a) and parallel to one edge; **(c)** about an axis CC'.

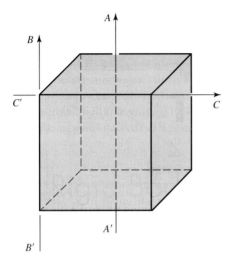

Figure P9.12

9.13. Find the moment of inertia for the following:
 (a) A cylinder of mass M, radius R, and height H rotating about an axis of symmetry.
 (b) Same as part (a), except rotating about an axis parallel to the symmetry axis and tangent to the surface.

9.14. Find the moment of inertia of a solid cone about its symmetry axis.

9.15. Find the moment of inertia for a frustum of a cone of mass M and radii R_1 and R_2 rotating about the symmetry axis.

9.16. Find the moment of inertia about an axis passing through the center O and perpendicular to the plane of a circular disk, as shown in Fig. P9.16. Also calculate I_X and I_Y. (The solid disk was removed from the hollow portion.)

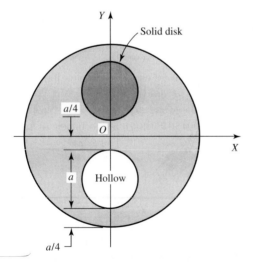

Figure P9.16

9.17. Consider a thin uniform square of side L with its diagonal along the X-axis, as shown in Fig. P9.17. The upper half of the square has a density σ_1 and the lower half σ_2. Find the moment of inertia about an axis passing through the center and perpendicular to the plane of the square. Also calculate I_X and I_Y.

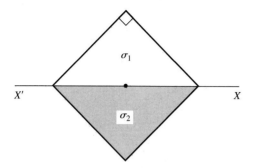

Figure P9.17

9.18. Find the moment of inertia for a sphere of radius R rotating about its axis of symmetry and having a density:

   (a) $\rho_0(r) = \rho_0(kr/R)$:]

   (b) $\rho_0(r) = \rho_0 e^{-kr/R}$

   (c) $\rho_0(r) = \rho_0(1 - kr/R)$

   Discuss the case in which $k \ll 1$.

9.19. Consider a thin disk of radius R and mass M. A small piece of maximum width $R/2$ has been cut off, as shown in Fig. P9.19. Calculate the center of mass and the moment of inertia for rotation about an axis perpendicular to the disk and passing through the center.

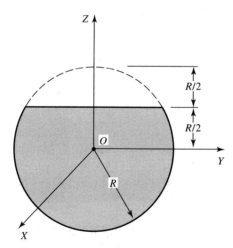

Figure P9.19

9.20. Consider a sphere of radius R that has a portion cut off, simialr to the disk in Fig. P9.19. Calculate the center of mass and the moment of inertia about the symmetry axis of the sphere.

9.21. Find the moment of inertia and the radius of gyration for a uniform rod of mass M and length L that is rotating about an axis through one end, making an angle θ with the rod, as shown in Fig. P9.21.

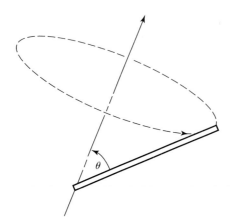

9.22. Show that the moment of inertia for a uniform octant of a sphere of mass m and radius a about an axis along one of the straight edges is $(3/2)ma^2$.

9.23. Calculate the moment of inertia for a parallelepiped about a symmetry axis.

9.24. Show that the moment of inertia for an ellipsoid of principal axes $2a$, $2b$, and $2c$ about the major axis is $(M/5)(b^2 + c^2)$.

9.25. A system consisting of a wheel attached to a fixed shaft is free to rotate without friction. A tape of negligible mass wrapped around the shaft is pulled with a steady constant force F. After a tape of length L has been pulled, the wheel acquires an angular velocity of ω. From these data, calculate the moment of inertia for the wheel.

9.26. If the total force acting on a system of particles is zero, show that the torque on the system is the same about all origins of different coordinate systems.

9.27. If the total linear momentum of a system of particles is zero, show that the angular momentum of the system is the same about all origins of different coordinate systems.

9.28. If τ_z is a funciton of θ alone, then, starting from the equation $dL/dt = I_z\ddot{\theta} = \tau_z$, show that the sum of the kinetic and potential energies is constant.

9.29. As in the case of translational motion, suppose the frictional torque is proportional to the angular velocity, that is, $\tau_f = -k\dot{\theta}$, while the driving torque is $\tau = \tau_0(1 + \alpha \cos \omega_0 t)$. Find the steady-state motion.

9.30. Consider a motor with an armature of 2-kg mass and radius of gyration of 8 cm. Its no-load full speed is when it draws a current of 2 A at 110 V at 1600 rpm. If the frictional torque is proportional to the angular velocity and the electrical efficiency is 75%, calculate the time required to reach a speed of 1200 rpm with no load.

9.31. A homogeneous circular disk of mass M and radius R has a light string wrapped around its circumference. One end of the string is attached to a fixed point. The disk is allowed to fall under gravity with the string unwinding. Find the acceleration of the center of mass.

9.32. A uniform rod of mass M and length L is placed like a ladder against a frictionless wall and frictionless horizontal floor. It is released from rest, making an angle θ with the vertical. Show that the initial reaction of the wall and the floor are (use only one variable to describe motion)

$$R_W = \tfrac{3}{4} mg \cos \alpha \sin \alpha, \qquad R_F = mg(1 - \tfrac{3}{4} \sin^2 \alpha)$$

and that the angle at which the rod will leave the wall is $\cos^{-1}(\tfrac{2}{3} \cos \alpha)$.

9.33. In Problem 9.32, if the coefficient of friction between the rod and the floor is μ, calculate **(a)** the horizontal and vertical components of the reaction as a function of angle θ, **(b)** the angle at which the rod begins to slip, and **(c)** the angular velocity when it hits the ground.

9.34. A uniform rod of length L and mass M is held horizontally with two hands at A and B. If A is suddenly released, at that instant what are **(a)** the torque about B, **(b)** the angular acceleration about B, **(c)** the vertical acceleration of the center of mass, and **(d)** the vertical force at B?

9.35. In the case of a simple pendulum, suppose we carry out correction to the fourth order instead of only the second order (as done in the text). Show that

$$T = 2\pi \sqrt{\frac{1}{g}} \left(1 + \frac{K^2}{4} + \frac{9K^4}{64} + \cdots \right)$$

and that if θ is small, say θ_0, then

$$T \simeq 2\pi \sqrt{\frac{1}{g}} \left(1 + \frac{1}{16} \theta_0^2 + \frac{11}{3072} \theta_0^4 + \cdots \right)$$

9.36. Consider a simple pendulum of 1-m length ($T_0 \simeq 2$s) that has an amplitude of 5 cm. Show that

$$\frac{dT}{T_0} = -\theta_0 \, d\theta_0$$

If θ_0 changes by 10%, that is, $d\theta = \theta_0/10$, calculate dT/T_0 and the resulting error per day.

9.37. In the case of a compound pendulum, we showed that when $T = T'$ the expression for g is given by Eq. (9.96). Suppose $T' = T(1 + \delta)$, where $\delta \ll 1$; find a new expression for g.

9.38. Consider a rod of mass M and length L. A mass m is attached at one end the rod is suspended from the other. If it behaves like a compound pendulum, calculate the time period of oscillations.

9.39. Consider a thin sheet of mass M in the shape of an equilateral triangle with each side of length L. Find the moment of inertia about an axis passing through the vertex and perpendicular to the sheet. If this behaves as a physical pendulum, find the period for small oscillations.

9.40. Consider a homogeneous hemisphere of mass M and radius R. With its flat face up, it rests on a perfectly rough horizontal surface. Find the expression for the length of an equivalent simple pendulum for small oscillations about the equilibrium position. Let the radius of gyration about the horizontal axis passing through the center of mass be k_g.

9.41. Consider a disk of mass m and radius r attached to a rod of length L and mass M. The system is suspended from the other end of the rod and allowed to oscillate. Find the time period of the oscillations. If the disk is mounted in such a way as to be free to spin (mounted on a frictionless bearing), what will be the period of oscillations?

9.42. Suppose a batter lets go of the bat after the ball hits the bat and that the bat starts rotating. For a quarter of a rotation of the bat, describe and sketch the motion of **(a)** the center of mass, and **(b)** the center of percussion; that is, plot $x(t)$ and $y(t)$ for both. Neglect the effect of gravity.

9.43. Consider a square plate of mass M and side L. The plate swings as a compound pendulum with an axis passing through one corner and perpendicular to the plane of the plate. Find the center of percussion and the period of oscillations.

9.44. A physical pendulum is made from a uniform disk of mass M and radius R suspended from a rod of negligible mass. The distance from the center of the disk and the point of oscillations is L. Find the time period of the oscillations. For what value of L will the time period be minimum? Locate the center of percussion.

9.45. Show that for a cube of side a when under a volume stress ΔP the volume strain is given by $\Delta V/V \simeq 3\Delta a/a$.

9.46. If two end points of a cable are at (x_A, y_A) and (x_B, y_B), show that the span L is given by $L = x_B - y_A = 8dT_0/w$, where d is a function of $y_A - y_B$.

9.47. Consider a wire of 1.5-m length and 2-mm diameter. It is clamped at the upper end and a 5-kg mass hangs at the lower end. Young's modulus is 9×10^{10} N/m^2, and its Poisson's ratio is 0.25. Calculate **(a)** the extension of the wire, **(b)** the decrease in the cross-sectional area due to the lateral strain, **(c)** the work done by the stretching force, and **(d)** the potential energy of the stretched wire.

9.48. Calculate the work done per unit volume in the following cases: **(a)** a shearing stress shearing the body through an angle θ, and **(b)** a uniform stress P producing a volume strain V.

9.49. A weightless rod of length L is clamped at two ends in a horizontal position. A weight W is placed at its center. Show that the equation for the shape of the rod is

$$y = \frac{W}{YK^2A}\left(\frac{x^3}{12} - \frac{Lx^2}{16}\right)$$

Calculate the deflection of the center.

SUGGESTIONS FOR FURTHER READING

ARTHUR, W., and FENSTER, S. K., *Mechanics,* Chapter 13. New York: Holt, Rinehart and Winston, Inc., 1969.

BECKER, R. A., *Introduction to Theoretical Mechanics*, Chapter 9. New York: McGraw-Hill Book Co., 1954.

FOWLES, G. R., *Analytical Mechanics*, Chapter 8. New York: Holt, Rinehart and Winston, Inc., 1962.

FRENCH, A. P., *Newtonian Mechanics*, Chapter 14. New York: W. W. Norton and Co., Inc., 1971.

HAUSER, W., *Introduction to the Principles of Mechanics*, Chapter 9. Reading, Mass.: Addison-Wesley Publishing Co., 1965.

KLEPPNER, D., and KOLENKOW, R. J., *An Introduction of Mechanics*, Chapter 6. New York: McGraw-Hill Book Co., 1973.

SLATER, J. C., *Mechanics*, Chapter 12. New York: McGraw-Hill Book Co., 1947.

STEPHENSON, R. J., *Mechanics and Properties of Matter*, Chapter 6. New York: John Wiley & Sons, Inc., 1962.

SYMON, K. R., *Mechanics*, 3rd ed., Chapter 5. Reading, Mass: Addison-Wesley Publishing Co., 1971.

TAYLOR, E. F., *Introductory Mechanics*, Chapter 8. New York: John Wiley & Sons, 1963.

C H A P T E R

10

Gravitational Force and Potential

10.1 INTRODUCTION

In this chapter we shall investigate Newton's universal law of gravitation and its application. We introduce the concepts of gravitational field intensity (or simply gravitational field) **g** and gravitational potential V. We calculate these quantities for different mass distributions by applying Newton's law of gravitation. Gauss's laws will be applied to calculate **g** and V for simple symmetrical mass distributions. Finally, gravitational field equations will be introduced, which are differential equations satisfied by functions such as **g** and V. These equations provide a more general procedure of interest.

10.2 NEWTON'S UNIVERSAL LAW OF GRAVITATION

Newton's universal law of gravitation (which is a law of force), together with Newton's laws of motion, has been applied by physicists to predict and calculate very precisely the motion of the planets, moons, satellites, and other objects in the universe. In 1666, 23-year-old Isaac Newton stated the universal law of gravitation in the following form:

> ***Newton's Universal Law of Gravitation.*** *The gravitational force (or interaction) of attraction between any two objects in the universe is directly proportional to the product of their masses and inversely proportional to the square of the distance between them.*

Thus the magnitude of the force F between any two objects of masses m_i and m_j separated by a distance r_{ij} is given by

$$F = G \frac{m_i m_j}{r_{ij}^2} \tag{10.1}$$

387

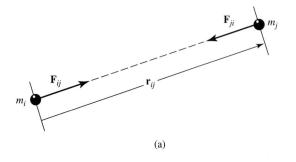

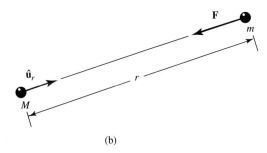

(a)

(b)

Figure 10.1   (a) Gravitational forces between two masses m_i and m_j. (b) Gravitational force on mass m due to mass M.

where G is the *gravitational constant;* its presently accepted value is

$$G = (6.673 \pm 0.003) \times 10^{-11} \text{ N-m}^2/\text{kg}^2 \tag{10.2}$$

Referring to Fig. 10.1(a), we may write the law in vector form as

$$\mathbf{F}_{ij} = G\frac{m_i m_j}{r_{ij}^2}\frac{\mathbf{r}_{ij}}{r_{ij}} = G\frac{m_i m_j}{r_{ij}^3}\mathbf{r}_{ij} \tag{10.3}$$

where $\mathbf{F}_{ij}$ is the gravitational force by which mass m_i is attracted by mass m_j, $\mathbf{r}_{ij} = \mathbf{r}_i - \mathbf{r}_j$ is the distance between the two masses m_i and m_j, and $\mathbf{F}_{ji}$ is the force by which m_j is attracted by mass m_i. According to Newton's third law, we have

$$\mathbf{F}_{ij} = -\mathbf{F}_{ji}$$

$$|\mathbf{F}_{ij}| = |\mathbf{F}_{ji}| = F = G\frac{m_i m_j}{r_{ij}^2} \tag{10.4}$$

From Fig. 10.1(b), mass m is attracted by mass M with a force $\mathbf{F}$; we may write

$$\mathbf{F} = -G\frac{Mm}{r^2}\hat{\mathbf{u}}_r \tag{10.5}$$

where the unit vector $\hat{\mathbf{u}}_r$ is in the direction from M to m. The minus sign indicates that $\mathbf{F}$ is the force of attraction with its line of action passing through a fixed point on the line joining the two masses. Thus the force is directed toward the center of mass M, and the *gravitational force is a central force*. The preceding equations are applicable to the situation in which the masses may be considered point masses. This is possible only if the dimensions of the masses are negligible compared to the distances between them.

Let us consider a point mass m at P attracted by an extended body of mass M, as shown in Fig. 10.2. To calculate the force on m at P, we must assume that the gravitational field is a *linear field*. That is, the force at P may be calculated by the vector addition of the individual forces produced by the interactions between the point particle m and the large number of particles in the extended body. The force $d\mathbf{F}$ between m and a small element of volume dV' of mass dm is

$$d\mathbf{F} = -G\frac{m\, dm}{r^2}\,\hat{\mathbf{u}}_r \tag{10.6}$$

where $dm = \rho(\mathbf{r})\, dV'$, $\rho(\mathbf{r})$ being the density. The force $\mathbf{F}$ acting on m due to the extended body of mass M may be obtained by integrating Eq. (10.6); that is,

$$\mathbf{F} = -\int_{V'} G\frac{m\rho(\mathbf{r})}{r^2}\,\hat{\mathbf{u}}_r\, dV' \tag{10.7}$$

where V' indicates integration over the whole volume. If the extended body is a thin shell that has a surface density or area density σ so that $dm = \sigma\, dA$, we may write

$$\mathbf{F} = -\int_A G\frac{m\sigma(\mathbf{r})}{r^2}\,\hat{\mathbf{u}}_r\, dA \tag{10.8}$$

where A indicates the integration over the whole area. If the extended body is a line source with a linear mass density λ so that $dm = \lambda\, dL$, we may write

$$\mathbf{F} = -\int_L G\frac{m\lambda(\mathbf{r})}{r^2}\,\hat{\mathbf{u}}_r\, dL \tag{10.9}$$

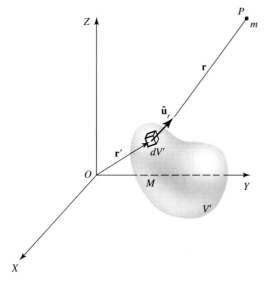

Figure 10.2   Gravitational force on mass m at P due to an extended body of mass M and volume V'.

If the extended body is replaced by a large number of discrete masses $m_1, m_2, m_3, \ldots, m_i$, the force on mass m may be written as

$$\mathbf{F} = -\sum_i G \frac{mm_i}{r_i^2} \hat{\mathbf{u}}_r \tag{10.10}$$

where $\hat{\mathbf{u}}_r$ is the unit vector in the direction along the line joining m_i and m.

   According to Eq. (10.7), the system of forces acting on different portions of the extended body due to mass m at P has a resultant force $\mathbf{F}$ acting along a line through the mass m. According to Newton's third law, the force acting on m is $-\mathbf{F}$, as shown in Fig. 10.3. On this line of action of $\mathbf{F}$, we locate a point CG at a distance r from m at P such that

$$F = G \frac{mM}{r^2} \tag{10.11}$$

Under these conditions, the gravitational force between the body of mass M and the particle of mass m is equivalent to a single resultant force $\mathbf{F}$ acting on M at CG and $-\mathbf{F}$ acting on m at P. The extended body behaves as if all its mass is concentrated at CG. The point CG is called the *center of gravity* of the body of mass M relative to the point mass m at P. If the position of m at P changes, so will the position of CG. In general, CG does not coincide with the center of mass of M; it may not even be on the line joining the center of mass of M with P. The center of gravity will coincide with the center of mass under the following conditions: (1) If the mass m is far away from M, the gravitational field will be uniform, different parts of the body will be acted on by the same force, and the center of gravity will coincide with the center of mass; (2) for a symmetrical body, such as a uniform sphere, its center of gravity coincides with its center of mass.

   We will encounter another complication if the mass m is also an extended body. In such cases Eqs. (10.6) and (10.7) must be rearranged, which will involve integrals of both m and dm.

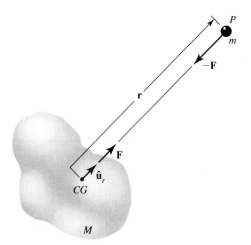

Figure 10.3   Center of gravity CG of an extended body of mass M relative to mass m at point P.

10.3 GRAVITATIONAL FIELD AND GRAVITATIONAL POTENTIAL

As stated before, a gravitational force is a *central force;* that is, it is a purely radial force passing through a given point, the center of force. Furthermore, the gravitational force is *spherically symmetric;* that is, the magnitude of the force depends only on the radial distance from the center of the force and not on its direction. We shall show that *spherically symmetric central forces are conservative;* hence the sum of the kinetic energy and the potential energy is constant. Conversely, *if a central force field is conservative, it must also be spherically symmetric.* (Note of caution: A force that is conservative may or may not be both central and spherically symmetric.)

Suppose a particle of mass *m* is under the action of a spherically symmetric central force **F** with its center of force at *O*, as shown in Fig. 10.4. In this situation, the force **F** has only a radial component F_r, which is a function of **r** only and may be written as

$$F_r = F(\mathbf{r}) \tag{10.12}$$

The work *dW* done by the central force **F** when *m* undergoes a small displacement *ds*, as shown, is

$$dW = \mathbf{F} \cdot d\mathbf{s} = F \, ds \cos \theta \tag{10.13}$$

But $$ds \cos \theta = dr$$

where *dr* is the change of the radial distance from *O* when mass *m* undergoes a displacement *d***s**. Thus

$$dW = F \, dr \tag{10.14}$$

Since the magnitude of the force **F** depends only on *r,* the total work done in going from *A* to *B,* as shown in Fig. 10.4, will be

$$W_{AB} = \int_{r_A}^{r_B} F(r) \, dr \tag{10.15}$$

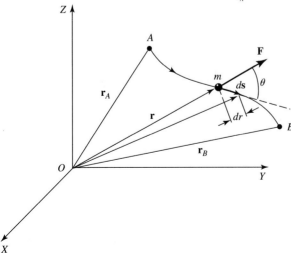

Figure 10.4 Work done by a central force **F** when a mass *m* is displaced from point *A* to point *B*.

Since this integral and hence the work done depend only on the initial and the final values of r (not on the path itself) the spherically symmetric force must be conservative.

Once we know that the force is conservative, we can proceed to define a potential energy function $U(r)$ of an object in such a spherically symmetric central force field. Thus, in going from A to B, the change in potential energy of an object is

$$\Delta U = U_B - U_A = -\int_{r_A}^{r_B} f(r)\, dr \qquad (10.16)$$

From Eqs. (10.15) and (10.16), we obtain

$$W_{AB} = -\Delta U = -(U_B - U_A) \qquad (10.17)$$

But the work done is also equal to change in kinetic energy; that is,

$$W_{AB} = K_B - K_A = -(U_B - U_A) \qquad (10.18)$$

Thus, if E is the total energy, Eq. (10.18) yields

$$K_A + U_A = K_B + U_B = E \qquad (10.19)$$

which is the law of the *conservation of energy*.

Since the gravitational force is an inverse square law force,

$$F(r) = f(r) = \frac{C}{r^2} \qquad (10.20)$$

where C is a constant. Substituting this in Eq. (10.16), we get

$$U_B - U_A = -\int_{r_A}^{r_B} \frac{C}{r^2}\, dr$$

which on integration gives

$$U_B - U_A = C\left(\frac{1}{r_B} - \frac{1}{r_A}\right) \qquad (10.21)$$

As is usually done , we define $U_A = 0$ when $r_A \to \infty$ and $U_B = U(r)$ where $r_B = r$; thus we get

$$U(r) = \frac{C}{r} \qquad (10.22)$$

which states that the potential energy of a particle in a central force field is a function of the distance r from the force center. The constant C is negative for attractive forces and positive for repulsive forces. Since the gravitational force is attractive and has the general form

$$F(r) = -\frac{GMm}{r^2} = \frac{C}{r^2} \quad \text{where } C = GMm \qquad (10.23)$$

the *potential energy* of m in the field of M at a distance r from M is

$$U(r) = -\frac{GMm}{r} \tag{10.24}$$

If M is a continuous mass distribution of arbitrary shape, the potential energy of m at a distance r is

$$U(r) = -\int_{V'} \frac{Gm\rho(\mathbf{r})}{r}\, dV' \tag{10.25}$$

To make the preceding three equations independent of m (the test mass), we introduce the concepts of gravitational field and gravitational potential.

The *gravitational field intensity*, or *gravitational field vector*, or simply *gravitational field*, **g**, is defined as the force per unit mass exerted on a particle in the gravitational field of mass M. That is,

$$\mathbf{g} = \frac{\mathbf{F}}{m} = -\frac{GM}{r^2}\,\hat{\mathbf{u}}_r \tag{10.26}$$

or, for an extended body of mass M, we may write

$$\mathbf{g} = -\int_{V'} \frac{G\rho(\mathbf{r})}{r^2}\,\hat{\mathbf{u}}_r\, dV' \tag{10.27}$$

where **g** has the dimensions of force per unit mass, that is, acceleration. The magnitude of this gravitational acceleration on the surface of Earth is approximately 9.8 m/s^2.

Whenever there is a conservative vector field, as is the gravitational force field, we can always introduce a gravitational potential (which is a scalar quantity) to represent this field, provided certain conditions are satisfied. The condition required is that the **curl** of the vector field **g** must be zero. Since **g** is proportional to $1/r^2$,

$$\mathbf{curl\ g} \equiv \nabla \times \mathbf{g} = 0 \tag{10.28}$$

(as proved in Chapter 6). This condition will also be satisfied if **g** is equal to the gradient of a scalar; that is,

$$\mathbf{g} \equiv -\mathbf{grad}\, V \equiv -\nabla V \tag{10.29}$$

(remembering $\nabla \times \nabla V = 0$), where V is called the *gravitational potential* and has the dimensions of *energy per unit mass*. Since **g** is only r dependent, V will be only r dependent. Substituting for **g** from Eq. (10.26) into Eq. (10.29), we get

$$-\frac{GM}{r^2}\,\hat{\mathbf{u}}_r = -\frac{dV}{dr}\,\hat{\mathbf{u}}_r$$

which on integration gives

$$V(r) = -\frac{GM}{r} \tag{10.30}$$

It is not necessary to have a constant of integration in Eq. (10.30) because we assume that $V(r) \to 0$ as $r \to \infty$.

The gravitational potential due to a continuous distribution of mass M may be written as

$$V(r) = -\int_{V'} \frac{G\rho(\mathbf{r})}{r} \, dV' \tag{10.31}$$

We may summarize this discussion as follows:

Force:

$$\mathbf{F} = -\int_{V'} G \frac{m\rho(\mathbf{r})}{r^2} \, \hat{\mathbf{u}}_r \, dV' \tag{10.7}$$

Potential energy:

$$U(r) = -\int_{V'} \frac{Gm\rho(\mathbf{r})}{r} \, dV' \tag{10.25}$$

Gravitational field:

$$\mathbf{g} = -\int_{V'} \frac{G\rho(\mathbf{r})}{r^2} \, \hat{\mathbf{u}}_r \, dV' \tag{10.27}$$

Gravitational potential:

$$V(r) = -\int_{V'} \frac{G\rho(\mathbf{r})}{r} \, dV' \tag{10.31}$$

Also,

$$\mathbf{F} = m\mathbf{g} \tag{10.32a}$$

$$U = mV \tag{10.32b}$$

$$\mathbf{g} = -\nabla V = -\mathbf{grad}\ V \tag{10.32c}$$

$$\mathbf{F} = -\nabla U = -\mathbf{grad}\ U \tag{10.32d}$$

Whenever a mass m is placed in the field of M, it is conventional to speak of the potential energy of mass m even though such potential energy resides in the field and not in the mass itself.

10.4 LINES OF FORCE AND EQUIPOTENTIAL SURFACES

The lines of force and equipotential lines in two dimensions and equipotential surfaces in three dimensions are very helpful in visualizing a force field. Let us consider a mass M that produces a gravitational field in the surrounding space and that may be described by the gravitational field

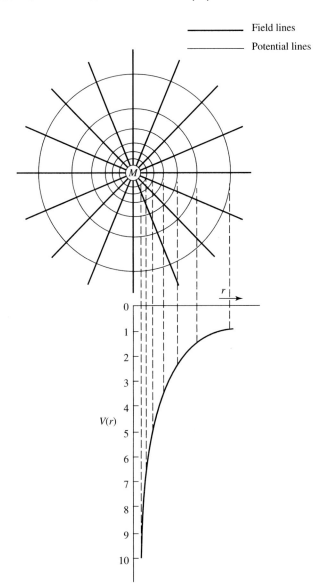

Field lines

Potential lines

Figure 10.5(a)   Gravitational field lines (boldface lines) and equipotential lines due to a sphere of mass M. The graph shows the relative value of $V(r)$ versus r.

vector **g**. We start from an arbitrary point and draw an infinitesimal line element in the direction of the vector **g** at that point. At the end of this line element, we draw another line element in the direction of **g** at this new point. We continue this process, and when we join these small line elements, we obtain a smooth line or curve called the *line of force* or *force field line*. We can draw a large number of such lines in the space surrounding a mass, as shown in Fig. 10.5(a). [See also Fig. 10.5(b).] These lines start from the surface of a mass and extend to infinity. For a single mass point, the force lines are straight lines (or radial) extending to infinity as shown. This is not true in all mass configurations and may be very complicated. For example, Fig. 10.6 will

Figure 10.5(b)

Below is the graph of the gravitational field and gravitational potential versus the distance r from the center of mass M of Earth

$N := 50$ $i := 1 .. N$ $RE := 6.368 \cdot 10^6$

$G := 6.673 \cdot 10^{-11}$ $M := 5.98 \cdot 10^{24}$ $r_i := i \cdot RE$

Which value, g or V, decreases faster with a change in distance r and why?

$$g_i := \frac{-G \cdot M}{\left(r_i\right)^2} \qquad V_i := \frac{-G \cdot M}{r_i}$$

$g_1 = -9.84 \qquad V_1 = -6.266 \cdot 10^7$

$g_{20} = -0.025 \qquad V_{20} = -3.133 \cdot 10^6$

$g_{50} = -0.004 \qquad V_{50} = -1.253 \cdot 10^6$

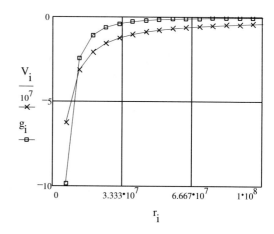

show potential curves resulting from two unequal masses. The force field lines will be perpendicular to potential curves at every point.

    This picture of the lines of force may be used to describe the direction and magnitude of the field vector **g**. A tangent drawn at any point to the field line gives the direction of the force field (of **F** or **g**) at that point. The density of these lines, the number of lines passing through a unit volume (the volume being small, but including the point), gives the magnitude of the vector field **g** at that point. No two field lines cross each other because **g** is a single-valued function; that is, it has only one value at any given point. It may be pointed out that these field lines have no real existence, but give a vivid picture depicting the properties of the force field.

    We now seek to investigate the relation between the force field lines and the gravitational potential lines. Suppose we know the gravitational potential *V* in the space surrounding a mass.

Since the gravitational potential V is defined for each point in space and is a single-valued function, we may write

$$V = V(x, y, z) \tag{10.33}$$

Suppose we join all the points having the same value of gravitational potential V_0. The equation representing these points is

$$V = V(x, y, z) = V_0 = \text{constant} \tag{10.34}$$

This is the equation of a surface, called an *equipotential surface*. We can draw a surface for each different value of V_0, hence resulting in a large number, or a whole family, of equipotential surfaces. In a two-dimensional case instead of equipotential surfaces, we get *equipotential lines*. Once again, since $V(x, y, z)$ is a single-valued function, no two equipotential surfaces or lines will cross each other. Suppose we move a mass m from one point to another point on an equipotential line. By definition, no work will be done. This leads us to the conclusion that *the lines of force are everywhere perpendicular (or orthogonal) to the equipotential lines*. This is true because $\mathbf{g} = -\nabla V$; it means that $\mathbf{g}$ cannot have a component along an equipotential surface because V is constant. Thus every line of force must be normal to the equipotential surface, as shown in Fig. 10.5(a). We shall elaborate on this point shortly. Meanwhile, Fig. 10.6 shows equipotential lines resulting from two masses M_1 and M_2. The equipotential surfaces in this case are defined by the equation

$$V = -G\left(\frac{M_1}{r_1} + \frac{M_2}{r_2} \right) = \text{constant} \tag{10.35}$$

Consider a mass at point P and let it be displaced a distance $d\mathbf{s}$. The change in its potential energy, which is equal to the work done, is given by

$$dU = -\mathbf{F} \cdot d\mathbf{s} = -F_s\, ds \tag{10.36}$$

where F_s is the component of the force in the direction of the displacement $d\mathbf{s}$. Equation (10.36) may be written as

$$F_s = -\frac{dU}{ds} \tag{10.37}$$

This equation states that the *component of* $\mathbf{F}$ *in any direction is equal to the negative rate of change of potential energy with distance in that direction*. The right side of Eq. (10.37) is called the *directional derivative* because its value will depend on the direction of $d\mathbf{s}$ relative to $\mathbf{F}$. For example, consider two equipotential energy lines U_0 and $U_0 + \Delta U$ or two equipotential lines V_0 and $V_0 + \Delta V$, as shown in Fig. 10.7. If we move form P to Q, which is on the same equipotential line, dU/ds will be zero. But if we move from P to R_1, R_2, or R on a different equipotential line, dU/ds will be different for different paths, such that $dU/ds > dU/ds_1$, dU/ds_2, ... In this case, dU/ds is maximum when $d\mathbf{s}$ is the shortest and hence perpendicular to the equipotential line at that point. The particular direction for which dU/ds is maximum in the direction of the

Figure 10.6 _____

Below is a graph of the gravitational potential lines (only three are shown drawn)
due to two nearby unequal masses located at the center of the circles (not shown).
The force field lines, not shown, are perpendicular to potential lines at every point.

$j := 1..10$ $i := 0..200$ $r1_j := 20 \cdot j$ $r2_j := 10 \cdot j$ $x1_i := i - 80$ $x2_i := i - 120$

$$yl_{i,j} := \sqrt{\left(r1_j\right)^2 - \left(x1_i\right)^2}$$
$$y2_{i,j} := \sqrt{\left(r2_j\right)^2 - \left(x2_i\right)^2}$$

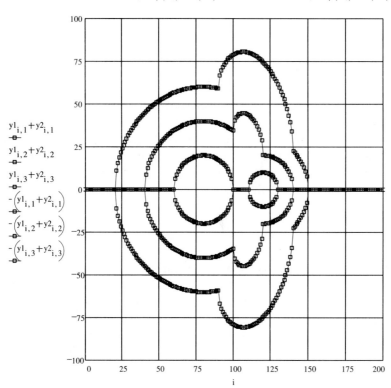

line of force, and the maximum magnitude of dU/ds is the magnitude of the vector force at that
point. The maximum value of dU/ds and its direction is called the *gradient* of the potential energy and is equal to the force **F**; that is,

$$\mathbf{F} = -\mathbf{grad}\ U \qquad\qquad (10.38)$$

Since $\mathbf{F} = m\mathbf{g}$ and $U = mV$, we may write

$$\mathbf{g} = -\mathbf{grad}\ V = -\nabla V \qquad\qquad (10.39)$$

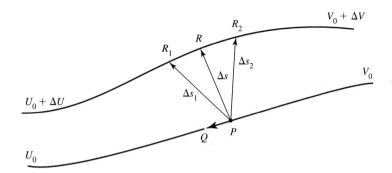

Figure 10.7   Gradient of the potential energy. The magnitude of the gradient is $\Delta U/\Delta s$.

10.5 CALCULATION OF GRAVITATIONAL FORCE AND GRAVITATIONAL POTENTIAL

We shall start by calculating the gravitational force between a uniform spherical shell of mass M and a point mass m. We shall show that any spherical shell may be treated as a point mass located at the center of the shell. Actually, this is true for any uniform spherically symmetric distribution of matter. In any of these situations, instead of calculating the force (which is a vector quantity), it is easier to calculate gravitational potential (which is a scalar quantity). Once the gravitational potential is known, the gravitational force may be calculated from it. We shall elaborate on both these procedures.

Spherical Shell

Consider a thin uniform shell of mass M and radius R, as shown in Fig. 10.8. A particle of mass m is placed outside the shell at point P a distance r ($r > R$) from the center of the shell. We divide the shell into a large number of circular rings like the one shown shaded in the figure. We can calculate the force between one of these rings and mass m and then sum over all the rings. As shown in the figure, the width of the shaded ring is $R \, d\theta$, while the radius of the ring is $R \sin \theta$. The circumference of the ring is $2\pi R \sin \theta$, while the area dA of the circular strip or shaded ring is

$$dA = (2\pi R \sin \theta)R \, d\theta = 2\pi R^2 \sin \theta \, d\theta \tag{10.40}$$

If σ is the density per unit area of the material of the shell, then the mass of the whole spherical shell is

$$M = (4\pi R^2)\sigma, \qquad \sigma = \frac{M}{4\pi R^2} \tag{10.41}$$

while the mass dM of the shaded ring is

$$dM = \sigma \, dA = \sigma 2\pi R^2 \sin \theta \, d\theta = \frac{M}{2} \sin \theta \, d\theta \tag{10.42}$$

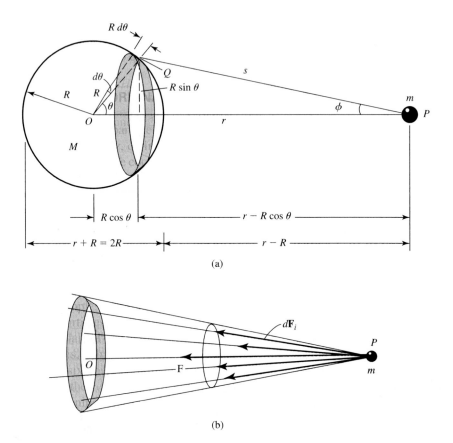

Figure 10.8   Gravitational force between a point mass m and spherical shell of mass M and radius R.

Point Q, or any other point on the shaded ring, is at the same distance s from the point mass m at P. The force dF_i on m due to any small section of this ring, such as at Q, points toward that section [see Fig. 10.8(b)]. This force can be resolved into transverse component $dF_i \sin \phi$, which is perpendicular to PO, and another component $dF_i \cos \phi$, which is parallel to PO. Due to the symmetry of the situation, all the transverse components resulting from considering the whole ring add up to zero, while the force components parallel to PO due to the whole ring add up to give

$$dF = \sum dF_i = \sum \frac{Gm\, dM}{s^2} \cos \phi \qquad (10.43)$$

or, substituting for dM, we have

$$dF = \frac{GMm}{2s^2} \sin \theta\, d\theta \cos \phi \qquad (10.44)$$

The force due to the entire shell is

$$F = \int dF = \int_0^\pi \frac{GMm \sin \theta \, d\theta}{2s^2} \cos \phi$$

or
$$F = GMm \int_0^\pi \frac{\cos \phi \sin \theta \, d\theta}{2s^2} \qquad \text{(10.45)}$$

From triangle OPQ, using the law of cosines, we obtain

$$s^2 = r^2 + R^2 - 2rR \cos \theta \qquad \text{(10.46)}$$

Since r and R are constants, differentiation yields

$$2s \, ds = 2rR \sin \theta \, d\theta \qquad \text{(10.47)}$$

and, similarly, from the same triangle OPQ, we obtain

$$R^2 = s^2 + r^2 - 2sr \cos \phi$$

or
$$\cos \phi = \frac{s^2 + r^2 - R^2}{2sr} \qquad \text{(10.48)}$$

Substituting for $\sin \theta \, d\theta$ and $\cos \phi$ from Eqs. (10.47) and (10.48) into Eq. (10.45) and changing the limits by using Eq. (10.46) from $0 \to \pi$ to $r - R \to r + R$, we obtain

$$F = \frac{GMm}{4r^2R} \int_{r-R}^{r+R} \left(1 + \frac{r^2 - R^2}{s^2}\right) ds \qquad \text{(10.49)}$$

which on integration yields

$$F = \frac{GMm}{r^2} \qquad \text{(10.50)}$$

In vector notation, this may be written as

$$\mathbf{F} = -\frac{GMm}{r^2} \hat{\mathbf{u}}_r, \qquad \text{for } r > R \qquad \text{(10.51a)}$$

and
$$\mathbf{g} = -\frac{GM}{r^2} \hat{\mathbf{u}}_r, \qquad \text{for } r > R \qquad \text{(10.51b)}$$

where $\hat{\mathbf{u}}_r$ is the unit radial vector from the origin O. This result indicates that a uniform spherical shell acts as if the whole mass of the shell were concentrated at the center. A solid uniform spherical body may be assumed to consist of a large number of concentric shells. Each shell may be treated as if its mass is concentrated on the center; hence the mass of the whole sphere may be assumed to be at the center.

To calculate the force on a point mass m placed inside the shell, all we must do is change the lower limit $r - R$ to $R - r$ and the upper limit $r + R$ to $R + r$. Integrating Eq. (10.49) with appropriate limits is

$$F = \frac{GMm}{4r^2R}\left(s + \frac{R^2 - r^2}{s}\right)\Bigg|_{R-r}^{R+r} = 0$$

Thus
$$F = 0 \quad \text{and} \quad g = 0, \quad \text{for } r < R \tag{10.52}$$

It must be kept in mind that this result (for $r < R$) is true only for a spherical shell and not for a solid sphere.

Using the relation given by Eq. (10.16), that is,

$$\Delta U = -\int_{r_A}^{r_B} F(r)\, dr = -\int_{r_A}^{r_B} \mathbf{F} \cdot d\mathbf{r} \tag{10.16}$$

and Eqs. (10.51a) and (10.52), we can calculate the potential energy to be

$$U = -\int\left(-\frac{GMm}{r^2}\right) dr = -\frac{GMm}{r}, \quad \text{for } r > R \tag{10.53}$$

and
$$U = \text{constant} = C_1, \quad \text{for } r < R$$

We can evaluate the constant by substituting $r = R$ in Eq. (10.53), that is,

$$U = -\frac{GMm}{R} = C_1, \quad \text{for } r < R \tag{10.54}$$

while the gravitational potential $V(=U/m)$ is

$$V = -\frac{GM}{r}, \quad \text{for } r > R \tag{10.55}$$

$$V = \text{constant} = C_2 = -\frac{GM}{R}, \quad \text{for } r < R \tag{10.56}$$

The variation is $\mathbf{g}$ and V for this case is shown in Fig. 10.9.

We can obtain the preceding results by first calculating the potential energy $U(r)$ and then calculating $\mathbf{F}(\mathbf{r})$ from the relation $F = -dU/dr$, as shown next.

The potential energy of mass m at P due to the circular ring of mass dM given by Eq. (10.42) at a distance s (each point of the ring is at the same distance s) is (see Fig. 10.8)

$$dU = \frac{Gm\, dM}{s} = -\frac{GMm}{2}\frac{\sin\theta\, d\theta}{s} \tag{10.57}$$

while the total potential energy of m at P is

$$U(r) = -\frac{GMm}{2}\int_0^\pi \frac{\sin\theta\, d\theta}{s} \tag{10.58}$$

Figure 10.9

Below is the graph of the variation in g(r) and V(r) versus r in the case of a spherical shell.

(Because of the large variation in the values of G, M, and R, divide them with appropriate numbers to make the graphs easier to interpret.)

x and y are used for drawing a spherical surface or a circular surface in two dimensions.

The expressions give the values of g and V inside and outside the sphere.

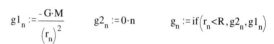

$$G \equiv 6.672 \cdot 10^{-11} \qquad M \equiv 5.98 \cdot 10^{24} \qquad R \equiv 6.37 \cdot 10^6$$

$$G := \frac{6.672 \cdot 10^{-11}}{10^{-11}} \qquad M := 5.98 \cdot \frac{10^{24}}{10^{24}} \qquad R := \frac{6.37 \cdot 10^6}{10^6}$$

$$N := 100 \qquad\qquad n := 1, 5 .. N \qquad\qquad \theta_n := \pi \cdot \frac{n}{50}$$

$$x_n := R \cdot \cos\left(\theta_n\right) \qquad y_n := \left(R \cdot \sin\left(\theta_n\right)\right) \qquad r_n := \frac{n}{5}$$

$$V1_n := -\left(\frac{G \cdot M}{r_n}\right) \qquad V2_n := -\frac{G \cdot M}{R} \cdot \frac{n}{n} \qquad V_n := \text{if}\left(r_n < R, V2_n, V1_n\right)$$

$$g1_n := \frac{-G \cdot M}{\left(r_n\right)^2} \qquad g2_n := 0 \cdot n \qquad g_n := \text{if}\left(r_n < R, g2_n, g1_n\right)$$

min(V) = −6.264

min(g) = −0.916

max(V) = 0

max(g) = 0

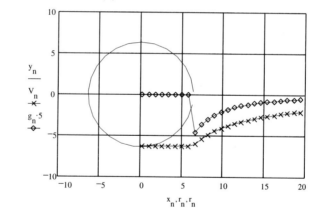

(a) Explain the variations in the values of g and V for the values of r given above.

(b) Since max(V) = 0 and max(g) = 0, what do the variations in V and g mean?

From the triangle *OPQ* in Fig. 10.8, we obtain

$$s^2 = r^2 + R^2 - 2rR \cos \theta \qquad\qquad \textbf{(10.46)}$$

Differentiating, while keeping in mind that *r* and *R* are constants, and rearranging, we get

$$\frac{\sin \theta \, d\theta}{s} = \frac{ds}{Rr}$$

Substituting in Eq. (10.58) yields

$$U(r) = -\frac{GMm}{2Rr} \int ds \qquad (10.59)$$

The limits of integration will depend on the position of the point mass m, as discussed next.

Case (*i*) $r > R$:   That is, the point mass m at P is outside the shell. As before, the limits $0 \to \pi$ change to $s_{min} = r - R \to s_{max} = r + R$. Thus

$$U(r) = -\frac{GMm}{2Rr} \int_{r-R}^{r+R} ds = -\frac{GMm}{2Rr} 2R$$

gives

$$U(r) = -\frac{GMm}{r}, \qquad \text{for } r > R \qquad (10.60)$$

That is, the potential energy varies as $1/r$, while

$$F = -\frac{dU}{dr} = -\frac{d}{dr}\left(-\frac{GMm}{r}\right)$$

gives

$$F = -\frac{GMm}{r^2}, \qquad \text{for } r > R \qquad (10.61)$$

We may also write

$$V(r) = -\frac{GM}{r}, \qquad \text{for } r > R \qquad (10.62)$$

and

$$g(r) = -\frac{GM}{r^2}, \qquad \text{for } r > R \qquad (10.63)$$

Case (*ii*) $r < R$:   That is, the point mass m at P is inside the shell. Hence the limits of integration $0 \to \pi$ change to $s_{min} = R - r \to s_{max} = R + r$. Thus

$$U(r) = -\frac{GMm}{2Rr} \int_{R-r}^{R+r} ds = -\frac{GMm}{2Rr} 2r$$

gives

$$U(r) = -\frac{GMm}{R}, \qquad \text{for } r < R \qquad (10.64)$$

That is, the potential inside the shell is constant, while

$$F = -\frac{dU}{dr} = -\frac{d}{dr}\left(-\frac{GMm}{R}\right) = 0$$

$$F = 0, \qquad \text{for } r < R \qquad (10.65)$$

as expected. We may also write

$$V(r) = -\frac{GM}{R}, \qquad \text{for } r < R \tag{10.66}$$

$$g(r) = 0, \qquad \text{for } r < R \tag{10.67}$$

These results are graphed in Fig. 10.9, as already mentioned.

Solid Sphere

The results derived for a spherical shell may be extended to a solid sphere. The only requirement is that the distribution of matter, that is, the density, be spherically symmetric. Furthermore, the problem becomes simple if the density is uniform.

Case (i) $r > R$:   That is, mass m is at r outside a solid sphere of mass M and radius R. The sphere may be divided into a large number of shells, each behaving as if the mass of the shell were concentrated at the center. Independent of the variation in density with radial distance (that is, symmetric but not necessarily uniform), as in the case of a shell, we obtain

$$F = -\frac{GMm}{r^2}, \qquad \text{for } r > R \tag{10.68}$$

$$g(r) = -\frac{GM}{r^2}, \qquad \text{for } r > R \tag{10.69}$$

$$U(r) = -\frac{GMm}{r}, \qquad \text{for } r > R \tag{10.70}$$

$$V(r) = -\frac{GM}{r}, \qquad \text{for } r > R \tag{10.71}$$

The graphs of $V(r)$ and $\mathbf{g}(r)$ are shown in Fig. 10.10.

Case (ii) $r < R$:   That is, the mass m is inside a solid sphere of mass M. Once again we draw spherical shells. All the shells that are outside a sphere of radius r give zero contribution to the force, while the shells inside r contribute to the force. For convenience, let us assume that the density is uniform; that is, the sphere is homogeneous. The fraction of the mass contained within r is

$$\frac{(4\pi/3)r^3\rho}{(4\pi/3)R^3\rho} = \frac{r^3}{R^3}$$

(see Fig. 10.11) where ρ is the density of the material. Thus the mass concentrated at the center is Mr^3/R^3. Hence the force at r is given by

$$F(r) = -\frac{Gm}{r^2}\left(\frac{Mr^3}{R^3}\right) = -\frac{GMm}{r^3}\,r, \qquad \text{for } r < R \tag{10.72}$$

Figure 10.10 _____

Below is the graph of V(r) and g(r) versus r due to a solid homogeneous sphere of radius R and mass M.

$$G = 6.672 \cdot 10^{-11} \qquad M = 5.98 \cdot 10^{24} \qquad R = 6.37 \cdot 10^{6}$$

$$G := \frac{6.672 \cdot 10^{-11}}{10^{-11}} \qquad M := 5.98 \cdot \frac{10^{24}}{10^{24}} \qquad R := \frac{6.37 \cdot 10^{6}}{10^{6}}$$

Before graphing, we divided the constants by appropriate powers of 10 to make the graphs easier to interpret.

$$G = 6.672 \qquad\qquad M = 5.98 \qquad\qquad R = 6.37$$

$$N := 100 \qquad n := 1 .. N \qquad r_n := \frac{n}{2} \qquad \theta_n := \pi \cdot \frac{n}{50}$$

Graphing y versus x gives a circle.

$$x_n := R \cdot \cos(\theta_n) \qquad\qquad\qquad y_n := \left(R \cdot \sin(\theta_n) \right)$$

$$V1_n := -\left(\frac{G \cdot M}{r_n} \right) \qquad V2_n := -\frac{G \cdot M}{2 \cdot R^3} \cdot \left[3 \cdot R^2 - (r_n)^2 \right] \qquad V_n := if(r_n < R, V2_n, V1_n)$$

$$g1_n := \frac{-G \cdot M}{(r_n)^2} \qquad g2_n := \frac{-G \cdot M}{R^3} \cdot r_n \qquad g_n := if(r_n < R, g2_n, g1_n)$$

min(V) = $^-$9.376

min(g) = $^-$0.944

max(V) = 0

max(g) = 0

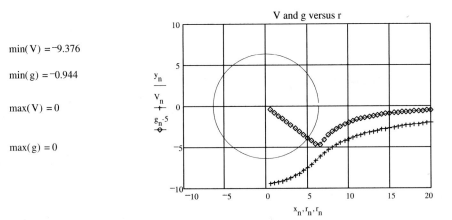

V and g versus r

(a) How do you think the plot for F will differ from these? Explain.

(b) Explain the variations in the values of g and V.

_____

The potential energy $U(r)$ of the mass inside the sphere may be calculated by using Eq. (10.72). For $r < R$, we obtain

$$U(r) - U(R) = -\int_r^R F \, dr = -\int_r^R \frac{GMm}{R^3} r \, dr$$

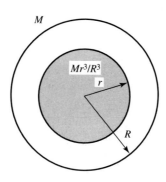

Figure 10.11   Fraction of mass of a sphere of radius r inside a homogeneous sphere of radius R and mass M.

that is,
$$U(r) - U(R) = -\frac{GMm}{2R^3}(R^2 - r^2) \tag{10.73}$$

But at $r = R$, from Eq. (10.70), we obtain

$$U(R) = -\frac{GMm}{R}$$

Substituting this in Eq. (10.73), we get

$$U(r) = -\frac{GMm}{2R^3}(3R^2 - r^2), \qquad \text{for } r < R \tag{10.74}$$

or
$$V(r) = -\frac{GM}{2R^3}(3R^2 - r^2), \qquad \text{for } r < R \tag{10.75}$$

We can calculate $U(r)$ and $V(r)$ at $r = 0$, that is, at the center

$$U(0) = -\frac{3GMm}{2R}, \qquad \text{at } r = 0 \tag{10.76}$$

and
$$V(0) = -\frac{3GM}{2R}, \qquad \text{at } r = 0 \tag{10.77}$$

The graphs of $g(r)$ and $V(r)$ for $r > R$ and $r < R$ are shown in Fig. 10.10.

Shell of Finite Thickness

Consider a shell of finite thickness of inner radius R_1 and outer radius R_2, as shown in Fig. 10.12. We want to calculate the potential at point P at a distance r from the center of the shell. By definition, the potential $V(r)$ at P is

$$V(r) = -G \int_{V'} \frac{\rho(R)}{s}\, dV' \tag{10.78}$$

where dV' is a small volume element at R:

$$dV' = (R \sin\theta\, d\phi)(R\, d\theta)\, dR = R^2 \sin\theta\, dR\, d\theta\, d\phi \tag{10.79}$$

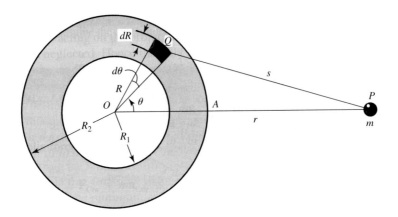

Figure 10.12   Potential and force on a point mass m at P due to a shell of finite thickness.

Because of the symmetry about the line connecting O with P, the azimuthal angle ϕ may be eliminated by integrating over $d\phi$; that is, $\int d\phi = 2\pi$. Also, $\rho(R) = \rho =$ constant for a homogeneous sphere. Thus

$$V(r) = -2\pi G\rho \int_{R_1}^{R_2} R^2 \, dR \int_0^\pi \frac{\sin\theta}{s} \, d\theta \tag{10.80}$$

From triangle OPQ (Fig. 10.12), we obtain

$$s^2 = R^2 + r^2 - 2Rr\cos\theta \tag{10.81}$$

where R and r are constants; hence differentiating gives

$$2s \, ds = 2Rr \sin\theta \, d\theta$$

or
$$\frac{\sin\theta \, d\theta}{s} = \frac{ds}{Rr}$$

Substituting this in Eq. (10.80) yields

$$V(r) = -2\pi G\rho \int_{R_1}^{R_2} R^2 \, dR \int_{s_{\min}}^{s_{\max}} \frac{ds}{Rr} \tag{10.82}$$

From Eq. (10.81), if $\theta = 0$, $s_{\min} = r - R$, and if $\theta = \pi$, $s_{\max} = r + R$. Therefore, for $r > R_2$,

$$V(r) = -\frac{2\pi G\rho}{r} \int_{R_1}^{R_2} R \, dR \int_{r-R}^{r+R} ds \tag{10.83}$$

$$= -\frac{4\pi G\rho}{r} \int_{R_1}^{R_2} R^2 \, dR$$

That is,
$$V(r) = -\frac{4\pi}{3} \frac{G\rho}{r} (R_2^3 - R_1^3) \tag{10.84}$$

Since
$$M = \frac{4\pi}{3}\rho(R_2^3 - R_1^3) \tag{10.85}$$

we get the following expression for the gravitational potential outside the shell:

$$V(r) = -\frac{GM}{r}, \qquad \text{for } r > R_2 \tag{10.86}$$

Thus the gravitational potential at any point outside a shell or sphere with a spherically symmetric mass distribution is independent of the distribution. It behaves as if the whole mass were located at the center.

For a point inside the shell, changing the limits in Eq. (10.83) (as we did in a previous case), we get, for $r < R_1$,

$$V(r) = -\frac{2\pi G\rho}{r}\int_{R_1}^{R_2} R \, dR \int_{R-r}^{R+r} ds \tag{10.87}$$

$$= -4\pi G\rho \int_{R_1}^{R_2} R \, dR$$

which gives

$$V(r) = -2\pi G\rho(R_2^2 - R_1^2) = \text{constant}, \qquad \text{for } r < R_1 \tag{10.88}$$

Thus the potential inside $(r < R_1)$ the shell is constant and is independent of the position.

The potential inside $(R_1 < r < R_2)$ the shell is a little bit tricky to calculate. But an easy approach to this problem is to change the lower limit by replacing R_1 by r in Eq. (10.87) for $V(r)$ for $r < R_1$, and to change the upper limit by replacing R_2 by r in Eq. (10.83) for $V(r)$ for $r > R_2$. Thus, combining the two gives the potential inside the shell:

$$V(R_1 < r < R_2) = -2\pi G\rho(R_2^2 - r^2) - \frac{4\pi G\rho}{3r}(r^3 - R_1^3)$$

That is,

$$V(R_1 < r < R_2) = -4\pi G\rho\left(\frac{R_2^2}{2} - \frac{R_1^3}{3r} - \frac{r^2}{6}\right) \qquad \text{for } R_1 < r < R_2 \tag{10.89}$$

The field intensity vector $\mathbf{g}$ can be calculated from the relation $g = -dV/dr$ for each of the three regions by using Eqs. (10.86), (10.89), and (10.88). That is,

$$g(r) = -\frac{GM}{r^2}, r > R_2 \tag{10.90}$$

$$g(r) = \frac{4\pi G\rho}{3}\left(\frac{R_1^3}{r^2} - r\right) \qquad R_1 < r < R_2 \tag{10.91}$$

$$g(r) = 0, \qquad r < R_1 \tag{10.92}$$

The plots of $V(r)$ using Eqs. (10.86), (10.88), and (10.89) and of $g(r)$ using Eqs. (10.90), (10.91), and (10.92) are shown in Fig. 10.13. Let us make some important observations. The potential function $V(r)$ plotted in Fig. 10.13 is continuous across the points $r = R_1$ and $r = R_2$, and its gradient $dV(r)/dr$, which is force $g(r)$, is also continuous, as shown. If the potential function $V(r)$

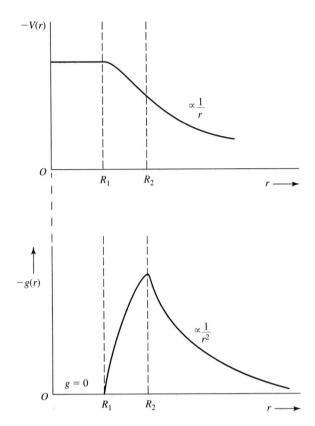

Figure 10.13   Variation of $V(r)$ and $g(r)$ versus r for a shell of finite thickness.

were not continuous, its derivative would be infinite. That is, the force would be infinite, which has no meaning. Thus the potential function must be continuous for the force to have any physical meaning. (Note that the derivative of the force function is not continuous.) That the potential function is continuous may be seen mathematically as follows. In Eq. (10.89), if we substitute $r = R_2$, we get the same result as from Eq. (10.84). Similarly in Eq. (10.89), if we substitute $r = R_1$, we get the same result as from Eq. (10.88).

 Example 10.1 _____

Consider a homogeneous circular disk of radius R, thickness t, and average density ρ (mass per unit volume). Calculate the gravitational potential and gravitational field intensity at a point outside the disk and on the axis of symmetry.

Solution

A circular disk of radius R is shown in Fig. Ex. 10.1(a). Its mass is given by

$$M = \rho V' = \rho(\pi R^2)t \tag{i}$$

where t is the thickness of the disk. Let us consider a ring of radius r and width dr, as shown in Fig. Ex. 10.1(a). As shown in Fig. Ex. 10.1(b), any small element of this ring is at the same distance s from point P. We shall calculate potential at point P due to this ring. The point P is at the same distance from all parts of the ring and lies on the axis of the disk. The mass of the ring is

$$dm = \rho \, dV' = \rho(2\pi r \, dr \, t) \tag{ii}$$

Hence the potential at point P due this ring is

$$dV = -\frac{G \, dm}{s} = -\frac{G2\pi\rho \, tr \, dr}{(z^2 + r^2)^{1/2}} \tag{iii}$$

Refer to two small elements of the ring at A and B in Fig. Ex. 10.1(b). The intensity at P due to these is given by $\mathbf{g}_1$ and $\mathbf{g}_2$ pointing along the lines PA and PB, respectively. When $\mathbf{g}_1$ and $\mathbf{g}_2$ are resolved, the horizontal components cancel. Because of the symmetry of the situation, all the horizontal components cancel and the vertical components add.

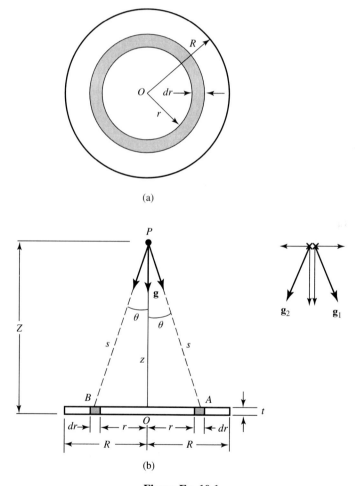

(a)

(b)

Figure Ex. 10.1

Using the value of dV from Eq. (iii), we can calculate V by integrating it as

$$V = \int 1 \, dV$$

$$V = -2 \cdot \pi \cdot \rho \cdot G \cdot t \cdot \int_0^R \frac{r}{\sqrt{z^2 + r^2}} \, dr$$

shown.

$$V = 2 \cdot \pi \cdot \rho \cdot G \cdot t \cdot \left(-\sqrt{z^2 + R^2} + z \right)$$

To calculate g, we have two alternatives: direct integration or the definition $g = dV/dz$.

$$g = \frac{-G \cdot dm}{s^2}$$

$$g = \int_0^R -\frac{2 \cdot \pi \cdot \rho \cdot G \cdot t \cdot z \cdot r}{\left(z^2 + r^2 \right)^{\frac{3}{2}}} \, dr$$

$$g = \frac{d}{dz} V$$

$$g = -\left[\frac{d}{dz} 2 \cdot \pi \cdot \rho \cdot G \cdot t \cdot \left(-\sqrt{z^2 + R^2} + z \right) \right]$$

$$g = 2 \cdot \pi \cdot \rho \cdot G \cdot t \cdot \frac{\left(-\sqrt{z^2 + R^2} + z \right)}{\sqrt{z^2 + R^2}}$$

$$g = 2 \cdot \pi \cdot \rho \cdot G \cdot t \cdot \frac{\left(-\sqrt{z^2 + R^2} + z \right)}{\sqrt{z^2 + R^2}}$$

$$g = 2 \cdot \pi \cdot \rho \cdot G \cdot t \cdot \left(-1 + \frac{1}{\sqrt{z^2 + R^2}} \cdot z \right)$$

How will the graphs of V and g versus z look?

EXERCISE 10.1 Repeat the example for the case of a circular ring of radius R and a linear mass density λ, but with the same mass as the disk. Compare the two results.

 ## Example 10.2 _____

Consider a thin rod of length L and mass M. Calculate the gravitational potential and gravitational field intensity at a point P that is at a distance r from the center of the rod and perpendicular to the rod as shown. Make a plot of F versus r.

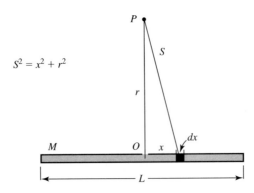

$$S^2 = x^2 + r^2$$

Solution

We can apply Eq. (10.31) using M/L as linear density, and replacing the volume element dV by the length element dx and r by the perpendicular distance from the rod. The gravitational potential energy V is

$$V = \frac{-G \cdot M}{L} \cdot \int_{-\frac{L}{2}}^{\frac{L}{2}} \frac{1}{\sqrt{x^2 + r^2}} \, dx$$

Given the value of V, we can calculate the value of g by differentiating V. Simplifying the expression for g gives the value of F (= mg).

$$V = G \cdot M \cdot \frac{\left(-\ln\left(L + \sqrt{L^2 + 4 \cdot r^2}\right) + \ln\left(-L + \sqrt{L^2 + 4 \cdot r^2}\right)\right)}{L}$$

$$g = -\frac{d}{dr} G \cdot M \cdot \frac{\left(-\ln\left(L + \sqrt{L^2 + 4 \cdot r^2}\right) + \ln\left(-L + \sqrt{L^2 + 4 \cdot r^2}\right)\right)}{L}$$

The graph of F versus r is as shown.

$$g = 8 \cdot G \cdot M \cdot \frac{r}{\left[\sqrt{L^2 + 4 \cdot r^2} \cdot \left[\left(L + \sqrt{L^2 + 4 \cdot r^2}\right) \cdot \left(L - \sqrt{L^2 + 4 \cdot r^2}\right)\right]\right]}$$

$$i := 0..\,10 \qquad r_i := i$$

$$G := 6.672 \cdot 10^{-11} \qquad M := 5 \qquad m := 1 \quad L := 10$$

$$F_i := 8 \cdot G \cdot M \cdot m \cdot \frac{r_i}{\left[-4 \cdot \sqrt{L^2 + 4 \cdot (r_i)^2} \cdot (r_i)^2\right]}$$

How do you explain the variation in the value of F for
(a) r = 0,
(b) r very small as compared to L and
(c) r very large as compared to L?

Graph V and g versus r.

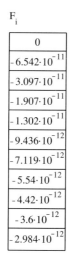

F_i
0
$-6.542 \cdot 10^{-11}$
$-3.097 \cdot 10^{-11}$
$-1.907 \cdot 10^{-11}$
$-1.302 \cdot 10^{-11}$
$-9.436 \cdot 10^{-12}$
$-7.119 \cdot 10^{-12}$
$-5.54 \cdot 10^{-12}$
$-4.42 \cdot 10^{-12}$
$-3.6 \cdot 10^{-12}$
$-2.984 \cdot 10^{-12}$

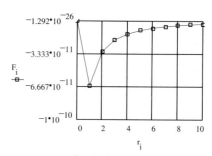

Gravitational force field versus r

$$\min(F) = -6.542 \cdot 10^{-11} \qquad \max(F) = 0$$

EXERCISE 10.2 Repeat the example for the case of a cylindrical rod of radius a, length L, and mass M.

10.6 GAUSS'S LAW

Gauss's law is used extensively in connection with electric fields in electrostatics. Actually, Gauss's law is applicable to any situation that involves the inverse-square force law; it is fair to say that Gauss's law is a compact form of the statement of the inverse-square law. Since gravitational force is an inverse-square force, let us apply Gauss's law and see its usefulness in calculating gravitational field intensity **g** in simple situations.

Let us consider a point mass M. The gravitational field **g** at a distance r from M is given by

$$\mathbf{g(r)} = -\frac{GM}{r^2}\,\hat{\mathbf{u}}_r \tag{10.93}$$

Draw a sphere of radius r with point mass M at the center. We define the radially outward direction as positive. A quantity *flux* ϕ of the gravitational field **g** through the surface of the sphere is defined as, using Eq. (10.93),

$$\phi = 4\pi r^2 g_r = -4\pi GM \tag{10.94}$$

where g_r is the radial component of **g** and $4\pi r^2$ is the surface area of a sphere of radius r. We shall show that the total flux due to any mass is independent of the distance r.

Let us consider a mass M that is completely enclosed by an arbitrarily shaped surface, as shown in Fig. 10.14. Such an arbitrary surface is called a *Gaussian surface* (GS). Let us consider a point P on this surface where the outward normal to the surface makes an angle θ with **r** from M to P. Resolve **g** into two components, a radial (or normal) component and a transverse component (component parallel to the surface). It is only the normal component that contributes to the flux ϕ. Since the normal component of **g** is $-(GM/r^2)\cos\theta$, the flux $d\phi$ through an element of area dA is

$$d\phi = -\frac{GM}{r^2}\cos\theta\,dA = \mathbf{g} \cdot d\mathbf{A} = \hat{\mathbf{n}} \cdot \mathbf{g}\,dA \tag{10.95}$$

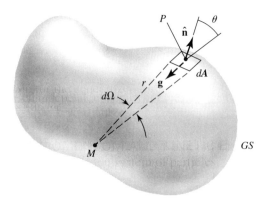

Figure 10.14   Mass M enclosed by a Gaussian surface (GS).

The projection of dA perpendicular to **r** is $dA \cos \theta = \hat{\mathbf{n}} \cdot d\mathbf{A}$, while $dA \cos \theta / r^2 = d\Omega$, where $d\Omega$ is the solid angle subtended by dA at M, as shown. Thus Eq. (10.95) may be written as

$$d\phi = -GM \, d\Omega \qquad (10.96)$$

The total flux ϕ due to **g** is obtained by integrating Eq. (10.96) in which contributions from all the solid angle elements are taken into consideration. Remembering that the complete solid angle is 4π, we get

$$\phi = \int d\phi = -GM \int d\Omega = -4\pi GM \qquad (10.97)$$

This is the same result we obtained using a spherical surface, as in Eq. (10.94).

From Fig. 10.14 and Eqs. (10.95) and (10.96), we may conclude that the flux $d\phi$ is a scalar product of **g** and $d\mathbf{A}$; that is,

$$d\phi = \mathbf{g} \cdot d\mathbf{A} = \hat{\mathbf{n}} \cdot \mathbf{g} \, dA \qquad (10.98)$$

where $\hat{\mathbf{n}} \cdot d\mathbf{A} = dA \cos \theta = $ projection of area dA perpendicular to **g** or **r**.

Let us extend our discussion to a large number of masses $M_1, M_2, M_3, \ldots$, inside an arbitrary closed surface. At any point P on this surface, the total gravitational field is

$$\mathbf{g} = \mathbf{g}_1 + \mathbf{g}_2 + \mathbf{g}_3 + \cdots \qquad (10.99)$$

while the total gravitational flux through the enclosed surface is

$$\phi = \int \mathbf{g} \cdot d\mathbf{A} = \int \hat{\mathbf{n}} \cdot \mathbf{g} \, dA \qquad (10.100)$$

Combining Eqs. (10.99) and (10.100),

$$\phi = \int \mathbf{g}_1 \cdot d\mathbf{A} + \int \mathbf{g}_2 \cdot d\mathbf{A} + \cdots = -4\pi G(M_1 + M_2 + \cdots)$$

If $M_{\text{total}} = M_1 + M_2 + \cdots$, then

$$\phi = -4\pi GM_{\text{total}} \qquad (10.101)$$

Equation (10.101) is a statement of Gauss's law, and its validity is based on the fact that the force is an inverse-square law. Once the flux is calculated by using Eq. (10.101), we can use Eq. (10.100) in simple symmetrical situations to calculate **g**, as illustrated in Example 10.3.

 Example 10.3 _____

By using Gauss's law, calculate the gravitational field intensity at a distance x from an infinite plane sheet having a surface mass density σ.

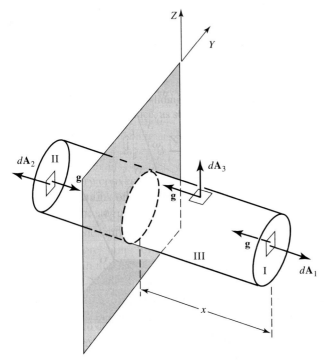

Figure Ex. 10.3

Solution

Draw a Gaussian surface in the form of a right circular cylinder, as shown in Fig. Ex. 10.3. The two end caps of the cylinder each have a surface area A. The total flux through the cylinder is due to the three surface areas (right, left, and curved) and is given by

$$\int \hat{\mathbf{n}} \cdot \mathbf{g} \, dA = \int \mathbf{g} \cdot d\mathbf{A} = \int_{\text{right}} \mathbf{g} \cdot d\mathbf{A} + \int_{\text{left}} \mathbf{g} \cdot d\mathbf{A} + \int_{\text{curved}} \mathbf{g} \cdot d\mathbf{A} \qquad \text{(i)}$$

$\mathbf{g}$ will always be perpendicular to the sheet. One the curved surface of the cylinder, an element like $d\mathbf{A}_3$ (as shown in Fig. Ex. 10.3) will be perpendicular to $\mathbf{g}$; hence there is no flux through the curved surface; that is,

$$\int_{\text{curved}} \mathbf{g} \cdot d\mathbf{A} = \int g \, dA \cos 90° = 0 \qquad \text{(ii)}$$

At the end caps, $\mathbf{g}$ is always antiparallel to the areas as shown; hence

$$\int_{\text{right}} \mathbf{g} \cdot d\mathbf{A} = \int g \, dA \cos 180° = -gA \qquad \text{(iii)}$$

$$\int_{\text{left}} \mathbf{g} \cdot d\mathbf{A} = \int g \, dA \cos 180° = -gA \qquad \text{(iv)}$$

Thus the total flux is

$$\Phi = \int_{\text{total}} \mathbf{g} \cdot d\mathbf{A} = -2gA \tag{v}$$

where A is the area of each cap, which is also equal to the area of the sheet enclosed.

The mass of the enclosed sheet is $M_{\text{total}} = \sigma A$. Hence the total flux according to Gauss's law is

$$\Phi = -4\pi G M_{\text{total}} = -4\pi G(\sigma A) \tag{vi}$$

Combining Eqs. (v) and (vi) gives

$$-2gA = -4\pi G(\sigma A)$$

or

$$g = 2\pi G\sigma \tag{vii}$$

Thus g is independent of the distance from the plane sheet; that is, it is the same everywhere and is directed toward and perpendicular to the sheet.

EXERCISE 10.3 By using Gauss's law, calculate the gravitational field intensity just outside a spherical shell having a mass M, radius R, and surface density σ.

10.7 GRAVITATIONAL FIELD EQUATIONS

We have briefly outlined the procedure for calculating $\mathbf{g}$ and ϕ by using Gauss's law for symmetrical mass distributions and also by the direct application of the inverse-square gravitational force law. A more general procedure of interest will be to find differential equations that are satisfied by the gravitational field intensity $\mathbf{g}(\mathbf{r})$ and the gravitational potential $V(\mathbf{r})$. Thus, from Eq. (10.29), we know the relation between $\mathbf{g}$ and V to be

$$\mathbf{g}(\mathbf{r}) = -\boldsymbol{\nabla} V(\mathbf{r}) \tag{10.102}$$

Taking the curl on both sides, and noting that the curl of a gradient is zero, we get

$$\boldsymbol{\nabla} \times \mathbf{g} = -\boldsymbol{\nabla} \times \boldsymbol{\nabla} V = 0 \tag{10.103}$$

That is,

$$\boldsymbol{\nabla} \times \mathbf{g}(\mathbf{r}) = 0 \tag{10.104}$$

This vector equation is a set of three differential equations giving relations between the components (g_x, g_y, g_z) of $\mathbf{g}$; that is,

$$\frac{\partial g_z}{\partial y} - \frac{\partial g_y}{\partial z} = 0, \qquad \frac{\partial g_x}{\partial z} - \frac{\partial g_z}{\partial x} = 0, \qquad \frac{\partial g_y}{\partial x} - \frac{\partial g_x}{\partial y} = 0 \tag{10.105}$$

These equations are satisfied by any gravitational field. For a unique determination of $\mathbf{g}$, we need a relation between $\mathbf{g}$ and the distribution of mass. This can be achieved as follows.

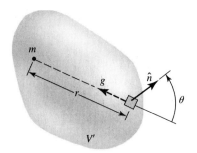

Figure 10.15   Mass m enclosed by a volume V' of surface area A.

As shown in Fig. 10.15, let us consider a mass m enclosed by a volume V' having a surface area A. Thus the flux ϕ through this area is

$$\phi = \iint_A \hat{\mathbf{n}} \cdot \mathbf{g}\, dA \tag{10.106}$$

But ϕ is also given by Eq. (10.97); that is,

$$\phi = -4\pi G M$$

or for a continuous mass distribution

$$\phi = -\iiint_{V'} 4\pi G\, dm = -\iiint_{V'} 4\pi G\rho\, dV' \tag{10.107}$$

where $dm = \rho\, dV'$, ρ being the mass density. Equating the preceding two equations gives

$$\iint_A \hat{\mathbf{n}} \cdot \mathbf{g}\, dA = -\iiint_{V'} 4\pi G\rho\, dV' \tag{10.108}$$

Gauss's divergence theorem applied to any vector $\mathbf{B}$ is (see Chapter 5)

$$\iint_A \hat{\mathbf{n}} \cdot \mathbf{B}\, dA = -\iiint_{V'} \boldsymbol{\nabla} \cdot \mathbf{B}\, dV' \tag{10.109}$$

Applying this to the left side of Eq. (10.108), we obtain

$$\iint_A \hat{\mathbf{n}} \cdot \mathbf{g}\, dA = -\iiint_{V'} \boldsymbol{\nabla} \cdot \mathbf{g}\, dV' \tag{10.110}$$

Substituting this in Eq. (10.108) gives

$$\iiint_{V'} \boldsymbol{\nabla} \cdot \mathbf{g}\, dV' = -\iiint_{V'} 4\pi G\rho\, dV'$$

or
$$\iiint_{V'} (\nabla \cdot \mathbf{g} + 4\pi G\rho)\, dV' = 0 \qquad \textbf{(10.111)}$$

Since this holds for any arbitrary volume V', we may write

$$\nabla \cdot \mathbf{g} + 4\pi G\rho = 0$$

or
$$\nabla \cdot \mathbf{g} = -4\pi G\rho \qquad \textbf{(10.112)}$$

which is the required relation between $\mathbf{g}$ and the mass distribution described by the density $\rho(x, y, z)$. In Cartesian coordinates, Eq. (10.112) may be written as

$$\frac{\partial g_x}{\partial x} + \frac{\partial g_y}{\partial y} + \frac{\partial g_z}{\partial z} = -4\pi G\rho \qquad \textbf{(10.113)}$$

Knowledge of $\rho(x, y, z)$, using Eqs. (10.105) and (10.113) and the boundary condition that $g \to 0$ as $r \to \infty$, will uniquely determine $\mathbf{g}$.

Substituting $\mathbf{g} = -\nabla V$ in Eq. (10.112) yields

$$\nabla \cdot \mathbf{g} = \nabla \cdot (-\nabla V) = -4\pi G\rho$$

or
$$\nabla^2 V = 4\pi G\rho \qquad \textbf{(10.114)}$$

Rewriting in Cartesian coordinates gives

$$\frac{\partial^2 V}{\partial x^2} + \frac{\partial^2 V}{\partial y^2} + \frac{\partial^2 V}{\partial z^2} = 4\pi G\rho \qquad \textbf{(10.115)}$$

This equation is called *Poisson's equation* and uniquely determines the value of $V(r)$ with the boundary conditions that $V(r) \to 0$ as $r \to \infty$. The general solution of Eq. (10.114) is

$$V(r) = \iiint \frac{G\rho(r)}{r}\, dV' \qquad \textbf{(10.116)}$$

where $\rho(r)$ is the density of the volume element dV'. This is in agreement with the value of $V(r)$ given earlier by Eq. (10.31).

In short, Newton's theory of gravitation may be completely summarized by a set of three equations [(10.26), (10.29), (10.114)]; that is,

$$\mathbf{g}(\mathbf{r}) = \frac{\mathbf{F}}{m}, \qquad \mathbf{g} = -\nabla V, \quad \text{and} \quad \nabla^2 V = -4\pi G\rho \qquad \textbf{(10.117)}$$

or, alternatively, Eqs. (10.26), (10.104), and (10.112); that is,

$$\mathbf{g}(r) = \frac{F}{m}, \qquad \nabla \times \mathbf{g} = 0, \quad \text{and} \quad \nabla \cdot \mathbf{g} = -4\pi G\rho \qquad \textbf{(10.118)}$$

PROBLEMS

10.1. Starting with Kepler's laws of planetary motion and Newton's laws of motion, derive Newton's universal law of gravitation.

10.2. Consider a planet of mass M and radius R and with a uniform density ρ. A tunnel is bored through this planet to connect any two points on its surface. An object of mass m is thrown in this tunnel.

     (a) Show that the motion of this object is simple harmonic and calculate the time period of this motion.

     (b) Calculate the time period if the planet were Earth.

     (c) Calculate the time period if the hole were drilled through the Moon.

10.3. In Fig. 10.3, if there are a large number of forces acting on M, the result of combining them should be a single force and a torque. Why is there no torque? (*Hint:* Take an arbitrary point P and show that the forces pass through it.)

10.4. An object of mass m released at a very large distance from Earth falls toward Earth's center. Calculate the time it will take **(a)** to reach halfway to Earth's center, and **(b)** to reach from halfway to Earth's center. Compare the two time intervals. Assume Earth to be a point mass.

10.5. Using the expression $U(r) = -GMm/r$ for $r > R$, show that as a mass m is moved from the surface of Earth to a height h the change in the potential energy is $\approx mgh$.

10.6. A particle of mass m in a certain force field given by $F = -k/x^3$ is moving toward the center of the force. Calculate the time it will take the particle to move from a point at a distance D from the center to the center of the force.

10.7. Suppose an object is dropped from a height h ($h \ll R_E$, where R_E is the radius of Earth). Show that the speed with which it will hit the ground is

$$v = \sqrt{2gh}\left(1 - \frac{1}{2}\frac{h}{R_E}\right)$$

10.8. An object has a free fall in the gravitational field of Earth from infinity to Earth's surface, while another object falls from a height $h = R_E$ with constant acceleration g. Show that they both arrive at Earth's surface with the same speed.

10.9. Draw gravitational field lines and equipotential lines for a thin rod of finite length. What can you say about the equipotential surfaces of the rod?

10.10. The gravitational potential at any point P due to two masses M_1 and M_2 is given by (see Fig. P10.10)

$$V(r) = -\frac{GM_1}{r_1} - \frac{GM_2}{r_2}$$

Suppose $M_1 = nM_2$, where $n = 2$ or 3. Outline a method for drawing equipotential lines, and draw them for these two cases.

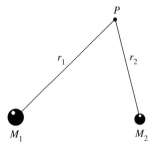

                                                Figure P10.10

10.11. In Problem 10.10, if the masses are not point masses but are spheres of finite sizes, what changes will take place in the equipotential lines?

10.12. Draw lines of force and equipotential lines due to two masses M_1 and M_2 when **(a)** $M_1 = M_2$, and **(b)** $M_1 \gg M_2$.

10.13. Explain the steps necessary to arrive at Eq. (10.89).

10.14. Consider a uniform hemispherical shell of radius R and mass M with its center at $z = 0$. Let the Z-axis be its symmetry axis. Calculate the gravitational potential and field intensity at any point on the Z-axis. Graph $V(r)$ and $g(r)$. How do these compare with those due to a full (solid) shell?

10.15. Consider a uniform solid hemisphere of radius R and mass M with its center at $z = 0$. Let the Z-axis be its symmetry axis. Calculate the gravitational potential and field intensity at any point on the Z-axis and graph the results.

10.16. Consider a planet of radius R_1 and mass M that is surrounded by a cloud of mixed gases with an average density $\bar{\rho}$. Calculate the gravitational potential and the gravitational field intensity in regions I, II, and III (see Fig. P10.16). Graph $V(r)$ and $g(r)$ for different regions.

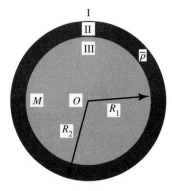

Figure P10.16

10.17. Consider a sphere of radius R having a variable density given by $\rho = \rho_0 e^{-ar}$ for $r < R$, where a is a constant. Calculate the gravitational potential and intensity at some point r inside and outside the sphere.

10.18. In a tunnel at a distance h below Earth's surface, the density of Earth's material is ρ_h. What will be the change in the time period of a clock pendulum, that is, $\Delta T/T$, at this depth? Will the clock run fast or slow? Calculate your answer in terms of h, M, R_E, and ρ_h. If we measure $\Delta T/T$ experimentally, can we determine Earth's mass, knowing the other variables?

10.19. Consider a thin cylindrical rod of length L, radius a, and mass M. Calculate gravitational potential and the gravitational field intensity at a distance r from the center of the rod and in a direction perpendicular to the rod. Graph the results.

10.20. Consider a thin rod of length L and mass M. Calculate the gravitational potential and gravitational field intensity at a point P that is at a distance r $(\gg L)$ from the center of the rod and making an angle θ with the rod, as shown in Fig. P10.20. Calculate only to the second order in L/r. Graph the results and compare them with the results derived in the text.

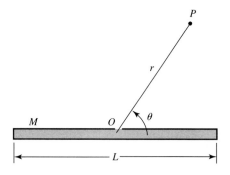

Figure P10.20

10.21. Calculate the gravitational potential and the gravitational field intensity due to a thin circular ring of mass M and radius R at a point in the plane of the ring. For large distances from the ring, expand the expression for the potential and find the first-order correction term. Draw the corresponding graphs.

10.22. Calculate the gravitation potential and the gravitational field intensity due to a thin circular ring of mass M and radius R for a point P on the Z-axis at right angles to the plane of the disk, as shown in Fig. P10.22. Assume $z \gg R$, and expand the expression for the potential, keeping terms only in the second order in R/z. Graph $V(z)$ and $g(z)$.

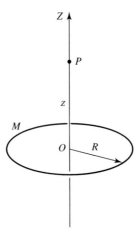

Figure P10.22

10.23. Calculate the gravitational potential and the gravitational field intensity due to a thin circular ring of mass M and radius R at a point P at a distance r from the center of the disk and making an angle θ with the Z-axis, as shown in Fig. P10.23. Assume that $r \gg R$, and expand the expression for the potential, keeping only the second-order terms in R/r. Graph $V(r)$ and $g(r)$.

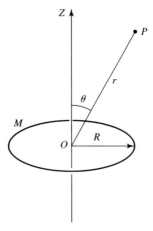

Figure P10.23

10.24. Calculate the gravitational potential and the gravitational field intensity due to a thin circular disk of mass M and radius R for a point in the plane of the ring. For large distances from the ring, expand the expression for the potential and find the first-order correction term. Graph $V(r)$ and $g(r)$.

10.25. Calculate the gravitational potential and the gravitational field intensity due to a thin circular disk of mass M and radius R for point P on the Z-axis at right angles to the plane of the disk. Assume $z \gg R$, and expand the expression for the potential, keeping terms only of the second order in R/z. Graph $V(z)$ and $g(z)$.

10.26. Calculate the gravitational potential and the gravitational field intensity due to a thin circular disk of mass M and radius R at a point P at a distance r from the center of the disk and making an angle θ with the Z-axis, similar to Fig. P10.23. Assume that $r \gg R$, and expand the expression for the potential, keeping only the second-order terms in R/r. Graph $V(r)$ and $g(r)$.

10.27. Consider a body that has a cylindrical symmetry with density $\rho(r, \theta)$ for $r < R$ and $\rho = 0$ for $r > R$. Calculate the gravitational potential at a point (r, θ) far away from the body. (Expand in powers of R/r.)

10.28. Consider a system of binary stars, each of mass M and separated by a distance $2r$. These stars orbit about their common center of mass. A mass m is located at a point P, as shown in Fig. P10.28.
  (a) Calculate the gravitational potential and gravitational field intensity and force at point P.
  (b) Repeat part (a) if $z \gg r$, and $z \ll r$.
  (c) Suppose the mass m is at point O and then is slightly displaced. Show that it executes simple harmonic motion.

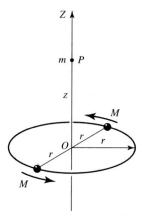

Figure P10.28

10.29. A and B are two thin concentric shells of radii R_1 and R_2 and masses M_1 and M_2, respectively. A point mass m is located at a distance r from O. Calculate the gravitational potential and field intensity in the three regions shown in Fig. P10.29. If a mass m is released from infinity, what will be its speed when it reaches O?

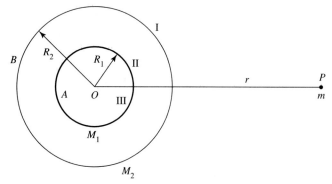

Figure P10.29

10.30. By applying Gauss's law, calculate the gravitational field and gravitational potential due to a homogenous sphere of radius R.

10.31. Calculate the gravitational field intensity and gravitational potential at a distance x from an infinite sheet of surface density σ in the XY plane.

10.32. A mass m is placed at a depth h in a tunnel in Earth. Show by using Gauss's law that the force exerted on this mass is due to the mass of the spherical portion below the tunnel. What will be the force on m at one-half the radius of Earth?

10.33. By using Gauss's law, calculate the gravitational field intensity inside and outside an infinitely long cylindrical shell of radius R and mass M.

10.34. By using Gauss's law, calculate the gravitational field intensity at a distance x from an infinitely long cylindrical rod of mass density σ per unit length.

10.35. Show that if we consider a mass M outside an enclosed surface, then the net flux through this surface is zero.

10.36. Show that the equations $\nabla \times \mathbf{g} = 0$, $\nabla \cdot \mathbf{g} = -4\pi G\rho$, and $\nabla^2 V = 4\pi G\rho$ are all satisfied by the gravitational field intensity and gravitational potential in Problems 10.30 and 10.31.

10.37. Show that **curl g** $= 0$.

SUGGESTIONS FOR FURTHER READING

BARGER, V., and OLSSON, M., *Classical Mechanics,* Chapter 7. New York: McGraw-Hill Book Co., 1973.

FRENCH, A. P., *Newtonian Mechanics,* Chapters 8 and 11. New York: W. W. Norton and Co., Inc., 1971.

KITTEL, C., KNIGHT, W. D., and RUDERMAN, M. A., *Mechanics,* Berkeley Physics Course, Volume 1, Chapter 9. New York: McGraw-Hill Book Co., 1965.

MARION, J. B., *Classical Dynamics,* 2nd ed., Chapter 2. New York: Academic Press, Inc., 1970.

STEPHENSON, R. J., *Mechanics and Properties of Matter,* Chapter 4. New York: John Wiley & Sons, Inc., 1962.

SYMON, K. R., *Mechanics,* 3rd ed., Chapter 6. Reading, Mass.: Addison-Wesley Publishing Co., 1971.

C H A P T E R

11

Noninertial Coordinate Systems

11.1 INTRODUCTION

In previous chapters we found it convenient to describe the motion of a particle or a system of particles in a fixed coordinate system. That is, the equations of motions, say Newton's second law, were described in an inertial system. But in some cases a system has an acceleration; here a reference frame attached to this system will be a *noninertial reference frame*. To describe such a dynamical system we must either modify the usual equations of motion so that these are applicable to a noninertial system or simply ignore the difference. A case in point is Earth, which is rotating (as well as translating) and hence has an acceleration. Thus a reference frame fixed to Earth will be a noninertial system. To describe the motion of a particle or object on Earth's surface, using Newton's laws of motion, an inertial system fixed in space with respect to some stars must be used. An alternative is to modify Newton's equations of motion as they will apply to a frame of reference fixed on Earth, that is, with respect to a noninertial system. We shall take the second approach and seek a modification of Newton's second law as applied to noninertial systems. To start, we give a brief discussion of translating accelerated systems, while a much more detailed discussion will be presented for rotating coordinate systems.

11.2 TRANSLATING COORDINATE SYSTEMS

Let us consider a primed coordinate system $S'(X', Y', Z')$ that is in translational motion with respect to a fixed stationary unprimed coordinate system $S(X, Y, Z)$, as shown in Fig.11.1. At any time t, a point P in space is located by the vectors $\mathbf{r}$ and $\mathbf{r}'$ with respect to the origins O and O',

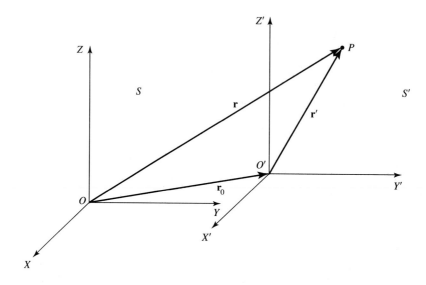

Figure 11.1 Primed coordinate system S' is in translational motion with respect to a fixed stationary unprimed coordinate system S.

respectively, of the two coordinate systems. The origin O' is at a distance $\mathbf{r}_0$ from the origin O. The relation between $\mathbf{r}$ and $\mathbf{r}'$ is given by

$$\mathbf{r} = \mathbf{r}' + \mathbf{r}_0 \tag{11.1}$$

or in component form

$$x = x' + x_0, \qquad y = y' + y_0, \qquad z = z' + z_0 \tag{11.2}$$

We assume that during the motion the axes X', Y', and Z' always remain parallel to the axes, X, Y, and Z. This is called the *translation* of the primed coordinates with respect to the unprimed coordinates. Furthermore, let us assume that the velocity and the acceleration of O' with respect to O are $\mathbf{v}_0$ and $\mathbf{a}_0$. Thus, by differentiating Eq. (11.1), we may write the relations between the velocity and the acceleration of a point P, in Fig. 11.1, with respect to the prime and unprimed coordinate systems S' and S, respectively:

$$\mathbf{v} = \frac{d\mathbf{r}}{dt} = \frac{d\mathbf{r}'}{dt} + \frac{d\mathbf{r}_0}{dt} \tag{11.3}$$

or

$$\mathbf{v} = \mathbf{v}' + \mathbf{v}_0 \tag{11.4}$$

and

$$\mathbf{a} = \frac{d^2\mathbf{r}}{dt^2} = \frac{d^2\mathbf{r}'}{dt^2} + \frac{d^2\mathbf{r}_0}{dt^2} \tag{11.5}$$

or

$$\mathbf{a} = \mathbf{a}' + \mathbf{a}_0 \tag{11.6}$$

Thus, if a particle of mass m at point P is acted on by a force $\mathbf{F}$ in the fixed coordinate system S, we may write Newton's second law as

(for unprimed system)
$$m\frac{d^2\mathbf{r}}{dt^2} = \mathbf{F} \qquad (11.7)$$

while in the moving coordinate system S', from Eqs. (11.5) and (11.7),

(for primed system)
$$m\frac{d^2\mathbf{r'}}{dt^2} = \mathbf{F} - m\mathbf{a}_0 \qquad (11.8)$$

The form of Newton's second law, as is clear from Eqs. (11.7) and (11.8), is different in the two coordinate systems. If the primed system S' is moving with uniform velocity with respect to S, Eq. (11.8) takes the form

$$m\frac{d^2\mathbf{r'}}{dt^2} = \mathbf{F}, \qquad \text{if } \mathbf{a}_0 = 0 \qquad (11.9)$$

which is similar in form to Eq. (11.7). Thus we may conclude that the form of Newton's second law is the same in all reference systems that are moving with uniform relative velocities. This is the *Newtonian principle of relativity* and asserts that there is no unique reference frame; that is, all reference frames moving with uniform relative velocities are equivalent. All such reference frames are called *inertial systems*. Another way of stating this is to say that *Newton's equations of motion remain invariant in form.*

Thus, according to Eq. (11.8), if the coordinate system S' has uniform acceleration $\mathbf{a}_0$ with respect to S, Newton's second law takes the form

$$m\frac{d^2\mathbf{r'}}{dt^2} = \mathbf{F'} \qquad (11.10)$$

where
$$\mathbf{F'} = \mathbf{F} - m\mathbf{a}_0 = m\mathbf{a'} \qquad (11.11)$$

Equation (11.10), $\mathbf{F'} = m\mathbf{a'}$, in the moving coordinate system S' is similar to Eq. (11.7), $\mathbf{F} = m\mathbf{a}$, in the fixed coordinate system S, but $\mathbf{F} \neq \mathbf{F'}$; $\mathbf{F'}$ must be replaced by $\mathbf{F} - m\mathbf{a}_0$ as given by Eq. (11.11). The term $-m\mathbf{a}_0$ is called a *noninertial force* or *fictitious force*. Force $\mathbf{F}$ is called a real force because $\mathbf{F}$ acting on m depends on the positions and the velocities of the objects with which it interacts. On the other hand, from the viewpoint of classical mechanics, the term $-m\mathbf{a}_0$ is not a force at all, but simply a product of mass and acceleration. This noninertial force depends on the acceleration of the moving coordinate system with respect to the fixed coordinate system. (On the contrary, in the general theory of relativity, such terms as $-m\mathbf{a}_0$ are regarded as legitimate forces in moving coordinate systems.) The fixed coordinate is called an *inertial system* or *inertial reference frame,* while the accelerated moving system is a *noninertial system.* In a noninertial system, Newton's second law takes the form given by Eqs. (11.8), (11.10), or (11.11).

Example 11.1 _____

A wheel of radius R is rolling with a uniform linear velocity $\mathbf{v}_0$ along a straight line path, as shown in Fig. Ex. 11.1. What are the position, velocity, and acceleration of a particle fixed to the rim of the wheel according to **(a)** a reference frame fixed with the wheel, and **(b)** a stationary observer on the ground?

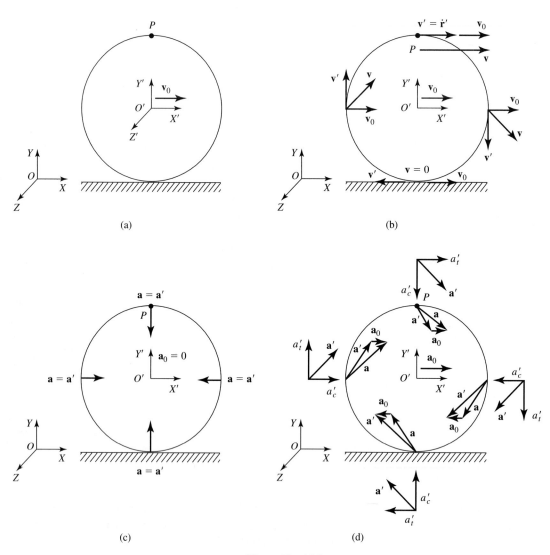

Figure Ex. 11.1

Solution

Let $O'X'Y'Z'$ be the coordinate system that is translating with the wheel, with origin O' at the center of the wheel. The stationary reference frame on the ground is represented by $OXYZ$, as shown in Fig. Ex. 11.1. Let us consider the motion of point P fixed to the rim of the wheel, as shown in Fig. Ex. 11.1.(a).

(a) According to the observer in the moving coordinate system $X'Y'Z'$, the angular speed of the wheel is

$$\omega = \frac{v_0}{R} \quad \text{or} \quad v_0 = R\omega \tag{i}$$

To the moving observer, the position, velocity, and acceleration of point P are

$$\mathbf{r}' = R\hat{\mathbf{j}}' \tag{ii}$$

$$\dot{\mathbf{r}}' = \mathbf{v}' = R\omega\hat{\mathbf{i}}' = v_0\hat{\mathbf{i}}' \tag{iii}$$

$$\ddot{\mathbf{r}}' = \mathbf{a}' = -\frac{v_0^2}{R}\hat{\mathbf{j}}' = -R\omega^2\hat{\mathbf{j}}' \tag{iv}$$

(b) Now we look at point P from the stationary reference frame XYZ. We orient the two coordinate systems so that $\hat{\mathbf{i}} = \hat{\mathbf{i}}', \hat{\mathbf{j}} = \hat{\mathbf{j}}'$, and $\hat{\mathbf{k}} = \hat{\mathbf{k}}'$. The position of O' with respect to O is

$$\mathbf{r}_0 = v_0 t\hat{\mathbf{i}} + R\hat{\mathbf{j}} \tag{v}$$

Thus, using Eqs. (11.1), (11.4), and (11.6), the motion of point P according to the stationary reference frame XYZ is

$$\mathbf{r} = \mathbf{r}' + \mathbf{r}_0, \qquad \mathbf{r} = v_0 t\hat{\mathbf{i}} + 2R\hat{\mathbf{j}} \tag{vi}$$

$$\mathbf{v} = \mathbf{v}' + \mathbf{v}_0, \qquad \dot{\mathbf{r}} = \mathbf{v} = 2v_0\hat{\mathbf{i}} \tag{vii}$$

$$\mathbf{a} = \mathbf{a}' + \mathbf{a}_0, \qquad \ddot{\mathbf{r}} = \mathbf{a} = -\frac{v_0^2}{R}\hat{\mathbf{j}} = -R\omega^2\hat{\mathbf{j}} \tag{viii}$$

Note that $\mathbf{a}_0 = 0$. These results are demonstrated in Fig. Ex. 11.1.(b), (c), and (d), where part (b) shows velocity vectors for various positions of the particle, part (c) shows acceleration vectors for various positions for a constant velocity of O', that is, $\mathbf{a}_0 = 0$, and part (d) shows acceleration vectors for various positions for the case of a forward acceleration of the wheel.

EXERCISE 11.1 Repeat the example for the case when the wheel is rolling down along an inclined plane making an angle θ with the ground.

11.3 ROTATING COORDINATE SYSTEMS

Let us consider a fixed reference frame S with coordinate axes XYZ and a rotating frame S' with coordinate axes $X'Y'Z'$. Thus S is an inertial system, while S' is a noninertial system. The origins of the two coordinate systems always coincide. The unit vectors in the two coordinate systems are $\hat{\mathbf{i}}, \hat{\mathbf{j}}$, and $\hat{\mathbf{k}}$ and $\hat{\mathbf{i}}', \hat{\mathbf{j}}'$, and $\hat{\mathbf{k}}$, respectively, as shown in Fig. 11.2. Let the position of

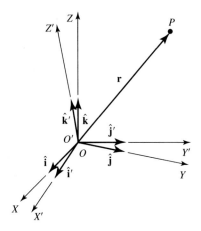

Figure 11.2   Coordinates X', Y', and Z' of system S' are rotating with respect to the stationary coordinates X, Y, and Z of system S.

the point P in space shown in Fig. 11.2 be represented by a vector $\mathbf{r}$. Since the origins of the two coordinate systems coincide, $\mathbf{r}$ is the same in both systems; only the components are different along different axes. (Also, $\mathbf{r}$ does not have to be measured from the origin.) Thus $\mathbf{r}$ in terms of components in either coordinate system may be written as

$$\mathbf{r} = x\hat{\mathbf{i}} + y\hat{\mathbf{j}} + z\hat{\mathbf{k}} \tag{11.12}$$

$$\mathbf{r} = x'\hat{\mathbf{i}}' + y'\hat{\mathbf{j}}' + z'\hat{\mathbf{k}}' \tag{11.13}$$

The relation between the components in the two coordinate systems may be obtained by using the dot product. Thus

$$
\begin{aligned}
x &= \mathbf{r} \cdot \hat{\mathbf{i}} = x'(\hat{\mathbf{i}}' \cdot \hat{\mathbf{i}}) + y'(\hat{\mathbf{j}}' \cdot \hat{\mathbf{i}}) + z'(\hat{\mathbf{k}}' \cdot \hat{\mathbf{i}}) \\
y &= \mathbf{r} \cdot \hat{\mathbf{j}} = x'(\hat{\mathbf{i}}' \cdot \hat{\mathbf{j}}) + y'(\hat{\mathbf{j}}' \cdot \hat{\mathbf{j}}) + z'(\hat{\mathbf{k}}' \cdot \hat{\mathbf{j}}) \\
z &= \mathbf{r} \cdot \hat{\mathbf{k}} = x'(\hat{\mathbf{i}}' \cdot \hat{\mathbf{k}}) + y'(\hat{\mathbf{j}}' \cdot \hat{\mathbf{k}}) + z'(\hat{\mathbf{k}}' \cdot \hat{\mathbf{k}})
\end{aligned} \tag{11.14}
$$

These equations may also be written as

$$
\begin{aligned}
x &= \lambda_{11}x' + \lambda_{21}y' + \lambda_{31}z' \\
y &= \lambda_{12}x' + \lambda_{22}y' + \lambda_{32}z' \\
z &= \lambda_{13}x' + \lambda_{23}y' + \lambda_{33}z'
\end{aligned} \tag{11.15}
$$

where λ_{11}, λ_{21}, $\ldots$, λ_{33} are the cosines of the angles between the prime and the unprimed unit vectors. For example, $(\hat{\mathbf{i}}' \cdot \hat{\mathbf{i}}) = |1||1|\cos\theta_{11} = \cos\theta_{11} = \lambda_{11}$, where θ_{11} is the angle between $\hat{\mathbf{i}}'$ and $\hat{\mathbf{i}}$. Similarly, $(\hat{\mathbf{j}}' \cdot \hat{\mathbf{i}}) = |1||1|\cos\theta_{21} = \lambda_{21}$, where θ_{21} is the angle between $\hat{\mathbf{j}}'$ and $\hat{\mathbf{i}}$, with similar meanings for the other terms. Since the prime axes are rotating with respect to the unprimed axes, the cosines of the angles are a function of time.

Let us now consider a vector $\mathbf{A}$ in space. Since the coordinates are rotating, if $\mathbf{A}$ is constant in time in one coordinate system, it will not be constant in another coordinate system. That is, the time derivatives of a vector will be different in the two coordinate systems. This is because, by the time $\mathbf{A}(t)$ changes to $\mathbf{A}(t + \Delta t)$ in one system in time Δt, $\mathbf{A}(t)$ has changed in the other coordinate system due to the rotation. Let $\mathbf{A}$ have the components (A_x, A_y, A_z) and (A'_x, A'_y, A'_z) in the unprimed and the primed coordinates respectively. Thus

$$\mathbf{A} = A_x\hat{\mathbf{i}} + A_y\hat{\mathbf{j}} + A_z\hat{\mathbf{k}} \tag{11.16}$$

$$\mathbf{A} = A'_x\hat{\mathbf{i}}' + A'_y\hat{\mathbf{j}}' + A'_z\hat{\mathbf{k}}' \tag{11.17}$$

Let d/dt denote the derivative with respect to the unprime (fixed) coordinate system and d'/dt with respect to the prime (rotating) coordinate system. From Eqs. (11.16) and (11.17), we obtain (by denoting the derivative by an overdot)

$$\frac{d\mathbf{A}}{dt} = \dot{A}_x\hat{\mathbf{i}} + \dot{A}_y\hat{\mathbf{j}} + \dot{A}_z\hat{\mathbf{k}} \tag{11.18}$$

$$\frac{d'\mathbf{A}}{dt} = \dot{A}'_x\hat{\mathbf{i}}' + \dot{A}'_y\hat{\mathbf{j}}' + \dot{A}'_z\hat{\mathbf{k}}' \tag{11.19}$$

To obtain a relation between $d\mathbf{A}/dt$ and $d'\mathbf{A}/dt$, we proceed as follows:

$$\frac{d\mathbf{A}}{dt} = \frac{d}{dt}(A'_x\hat{\mathbf{i}}' + A'_y\hat{\mathbf{j}}' + A'_z\hat{\mathbf{k}}')$$

$$= \dot{A}'_x\hat{\mathbf{i}}' + \dot{A}'_y\hat{\mathbf{j}}' + \dot{A}'_z\hat{\mathbf{k}}' + A'_x\frac{d\hat{\mathbf{i}}'}{dt} + A'_y\frac{d\hat{\mathbf{j}}'}{dt} + A'_z\frac{d\hat{\mathbf{k}}'}{dt} \tag{11.20}$$

Using the result of Eq. (11.19) in Eq. (11.20),

$$\frac{d\mathbf{A}}{dt} = \frac{d'\mathbf{A}}{dt} + A'_x\frac{d\hat{\mathbf{i}}'}{dt} + A'_y\frac{d\hat{\mathbf{j}}'}{dt} + A'_z\frac{d\hat{\mathbf{k}}'}{dt} \tag{11.21}$$

Thus we are left with evaluating the last three terms. To do this, let us suppose that the primed coordinate system is rotating about some axis ON through an origin with an angular velocity $\boldsymbol{\omega}$, as shown in Fig. 11.3. Let a vector $\mathbf{B}$ making an angle θ with the axis ON be at rest in the primed coordinate system so that its prime derivative $d'\mathbf{B}/dt = 0$, while we calculate the unprimed derivative $d\mathbf{B}/dt$. (Note that the time derivative of $\mathbf{B}$ depends on the component $\mathbf{B}$ along the axis, and not on the position of $\mathbf{B}$ in space.) At time t, vector $\mathbf{B}(t)$ is along OP. At a later time $t + \Delta t$, $\mathbf{B}(t + \Delta t)$ is along OQ. Thus $\mathbf{B}(t + \Delta t) - \mathbf{B}(t) = \Delta\mathbf{B}$. The magnitude of $\Delta\mathbf{B}$ from Fig. 11.3. is

$$\Delta B = RQ\,\Delta\phi = (B\sin\theta)(\omega\,\Delta t)$$

$$\frac{\Delta B}{\Delta t} = \omega B\sin\theta \tag{11.22}$$

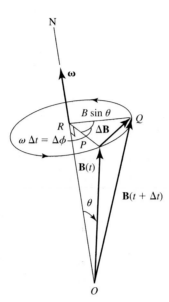

Figure 11.3 Primed coordinate system with vector **B** in it is rotating about an axis *ON* with an angular velocity **ω**.

or, in the limit $\Delta t \to 0$, we get

$$\frac{dB}{dt} = \omega B \sin \theta \tag{11.23}$$

or

$$\frac{d\mathbf{B}}{dt} = \boldsymbol{\omega} \times \mathbf{B} \tag{11.24}$$

This equation correctly gives the direction of $d\mathbf{B}/dt$ by using the definition of the cross product. The direction of $d\mathbf{B}/dt$ is perpendicular to the plane containing $\boldsymbol{\omega}$ and $\mathbf{B}$. We can make use of Eq. (11.24) to evaluate the last three terms in Eq. (11.21). Each of the three terms takes the form

$$\frac{d\hat{\mathbf{i}}'}{dt} = \boldsymbol{\omega} \times \hat{\mathbf{i}}', \qquad \frac{d\hat{\mathbf{j}}'}{dt} = \boldsymbol{\omega} \times \hat{\mathbf{j}}', \qquad \frac{d\hat{\mathbf{k}}'}{dt} = \boldsymbol{\omega} \times \hat{\mathbf{k}}' \tag{11.25}$$

Substituting these in Eq. (11.21), we obtain

$$\frac{d\mathbf{A}}{dt} = \frac{d'\mathbf{A}}{dt} + A'_x(\boldsymbol{\omega} \times \hat{\mathbf{i}}') + A'_y(\boldsymbol{\omega} \times \hat{\mathbf{j}}') + A'_z(\boldsymbol{\omega} \times \hat{\mathbf{k}}')$$

or

$$\frac{d\mathbf{A}}{dt} = \frac{d'\mathbf{A}}{dt} + \boldsymbol{\omega} \times \mathbf{A} \tag{11.26}$$

Thus, knowing the time derivative of **A** in the unprimed (fixed) coordinate system, we can calculate the time derivative of **A** in the prime (rotating) coordinate system by using Eq. (11.26). In general, we may write the relation between d/dt and d'/dt as

$$\frac{d}{dt} = \frac{d'}{dt} + \boldsymbol{\omega} \times \tag{11.27}$$

When applied to a position vector **r**, this gives

$$\frac{d\mathbf{r}}{dt} = \frac{d'\mathbf{r}}{dt} + \boldsymbol{\omega} \times \mathbf{r} \tag{11.28}$$

or
$$\mathbf{v} = \mathbf{v}' + \boldsymbol{\omega} \times \mathbf{r} \tag{11.29}$$

Equation (11.28) or (11.29) gives the relation between **v** in a fixed coordinate system and velocity **v**′ in a rotating coordinate system. This relation applies even when **ω** is changing in both magnitude and direction. We are interested in getting the second derivative of **r**. This is achieved by differentiating Eq. (11.28) once again. That is,

$$\frac{d^2\mathbf{r}}{dt^2} = \frac{d}{dt}\left(\frac{d\mathbf{r}}{dt}\right) = \frac{d}{dt}\left(\frac{d'\mathbf{r}}{dt} + \boldsymbol{\omega} \times \mathbf{r}\right) \tag{11.30}$$

Substituting for d/dt from Eq. (11.27), we have

$$\frac{d^2\mathbf{r}}{dt^2} = \left(\frac{d'}{dt} + \boldsymbol{\omega} \times \right)\left(\frac{d'\mathbf{r}}{dt} + \boldsymbol{\omega} \times \mathbf{r}\right)$$

$$= \frac{d'^2\mathbf{r}}{dt^2} + \boldsymbol{\omega} \times \frac{d'\mathbf{r}}{dt} + \frac{d'\boldsymbol{\omega}}{dt} \times \mathbf{r} + \boldsymbol{\omega} \times \frac{d'\mathbf{r}}{dt} + \boldsymbol{\omega} \times \boldsymbol{\omega} \times \mathbf{r}$$

or
$$\frac{d^2\mathbf{r}}{dt^2} = \frac{d'^2\mathbf{r}}{dt^2} + \boldsymbol{\omega} \times \boldsymbol{\omega} \times \mathbf{r} + 2\boldsymbol{\omega} \times \frac{d'\mathbf{r}}{dt} + \frac{d'\boldsymbol{\omega}}{dt} \times \mathbf{r} \qquad \text{Coriolis theorem} \qquad \textbf{(11.31)}$$

Also, we may replace $d'\omega/dt$ by $d\omega/dt$ because the prime and unprime derivatives of any vector parallel to the axis of rotation are the same. We can verify this by using Eq. (11.26), in which we substitute **A** = **ω**; that is,

$$\frac{d\boldsymbol{\omega}}{dt} = \frac{d'\boldsymbol{\omega}}{dt} + \boldsymbol{\omega} \times \boldsymbol{\omega} = \frac{d'\boldsymbol{\omega}}{dt}, \qquad \text{that is, } \dot{\boldsymbol{\omega}} = \dot{\boldsymbol{\omega}}' \tag{11.32}$$

since $|\boldsymbol{\omega} \times \boldsymbol{\omega}| = \omega^2 \sin 0° = 0$; **ω** × **ω** is a null vector.

In summary, we may say that if O and O' remain fixed, then any point in space in located by a position vector **r**, which is the same in both coordinate systems, while the velocity and acceleration are given by Eqs. (11.28) and (11.31), respectively.

Equation (11.31) is a statement of the *Coriolis theorem*. We briefly discuss each term of this equation.

$$\frac{d^2\mathbf{r}}{dt^2} \equiv \text{acceleration relative to the unprimed or fixed coordinate system}$$

$$\frac{d'^2\mathbf{r}}{dt^2} \equiv \text{acceleration relative to the prime coordinate system}$$

$$\boldsymbol{\omega} \times \boldsymbol{\omega} \times \mathbf{r} \equiv \text{centripetal acceleration of a point in rotation about an axis}$$

$2\boldsymbol{\omega} \times \dfrac{d'\mathbf{r}}{dt} \equiv$ Coriolis acceleration, which is present when a particle is moving in the prime (rotating) coordinate system

$\dfrac{d\boldsymbol{\omega}}{dt} \times \mathbf{r} \equiv$ nonuniform rotation, which vanishes if $\boldsymbol{\omega}$ is constant about a fixed axis

If we assume that Newton's second law is valid in the unprimed (fixed) coordinate system, using Eq. (11.31) we obtain

$$m \frac{d^2\mathbf{r}}{dt^2} = \mathbf{F} = m \frac{d'^2\mathbf{r}}{dt^2} + m\boldsymbol{\omega} \times \boldsymbol{\omega} \times \mathbf{r} + 2m\boldsymbol{\omega} \times \frac{d'\mathbf{r}}{dt} + m \frac{d\boldsymbol{\omega}}{dt} \times \mathbf{r} \qquad (11.33)$$

while the effective force $\mathbf{F}'$, as observed in the rotating coordinate system acting on m, is given by

$$m \frac{d'^2\mathbf{r}}{dt^2} \equiv \mathbf{F}' = \mathbf{F} - m\boldsymbol{\omega} \times \boldsymbol{\omega} \times \mathbf{r} - 2m\boldsymbol{\omega} \times \frac{d'\mathbf{r}}{dt} - m \frac{d\boldsymbol{\omega}}{dt} \times \mathbf{r} \qquad (11.34)$$

where

$-m\boldsymbol{\omega} \times \boldsymbol{\omega} \times \mathbf{r} \equiv$ centrifugal force acting away from the center $(11.35a)$

as we will explain shortly.

$-2m\boldsymbol{\omega} \times \dfrac{d'\mathbf{r}}{dt} \equiv$ Coriolis force $(11.35b)$

$-m \dfrac{d\boldsymbol{\omega}}{dt} \times \mathbf{r} \equiv$ transverse force for the case of a nonuniform rotation, which is zero because we shall deal with uniform rotation only $(11.35c)$

In Eq. (11.33), $|\boldsymbol{\omega} \times \boldsymbol{\omega} \times \mathbf{r}| = a_c$ is the centripetal acceleration because, as shown in Fig. 11.4, it is directed toward the center and perpendicular to the axis of rotation. As shown, $\mathbf{v} = \boldsymbol{\omega} \times \mathbf{r}$ or $v = \omega r \sin \theta$, where v is the speed of the circular motion and $r \sin \theta$ is the distance from the axis. From Fig. 11.5, using $\omega = v/(r \sin \theta)$, we get

$$a_c = |\boldsymbol{\omega} \times \boldsymbol{\omega} \times \mathbf{r}| = \omega^2 r \sin \theta = \frac{v^2}{r \sin \theta} \qquad (11.36)$$

The quantity $-m\boldsymbol{\omega} \times \boldsymbol{\omega} \times \mathbf{r}$ is called the *centrifugal force* and is equal to $-m\omega^2 r$ in the case where $\boldsymbol{\omega}$ is normal to the radius vector. The negative sign means that the centrifugal force is directed outward or away from the center of rotation, as shown in Fig. 11.5. According to classical mechanics, the centrifugal force is not a real force; it is a fictitious or noninertial force.

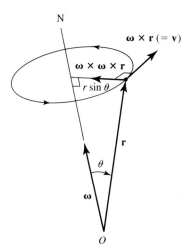

Figure 11.4   Centripetal acceleration $a_c = |\boldsymbol{\omega} \times \boldsymbol{\omega} \times \mathbf{r}|$ resulting from the rotation of the primed coordinate system.

This force is present only if we refer to moving coordinates in space. Thus, for example, a particle moving in a circle has no centrifugal force acting on it. A force that is acting toward the center, producing centripetal acceleration, is present. On the other hand, if we observe this moving particle from a reference frame that is moving with the particle, the particle will be at rest in this system. There is a force acting toward the center, but the particle does not fall toward the center. This is possible only if the force toward the center is balanced by an outward force, the centrifugal force.

The term $-2m\boldsymbol{\omega} \times (d'\mathbf{r}/dt)$ is called the *Coriolis force* and results from the motion of a particle in a rotating coordinate system. This force is directly proportional to a velocity $\mathbf{v}'$ and

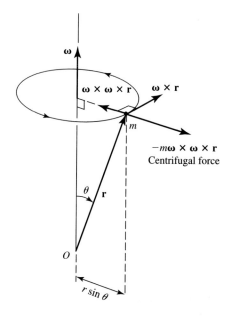

Figure 11.5   Centrifugal force resulting from rotational motion is shown directed away from the center.

will disappear if there is no motion. Once again, according to classical mechanics, it is not a real force; it is a noninertial force (or fictitious force).

It is essential to note that both the centripetal force and Coriolis force have been introduced for only one purpose: to write an equation that is similar to Newton's second law and that is still applicable to noninertial systems including rotational coordinate systems. Thus, if we write Newton's second law in the unprime (or fixed) coordinates as

$$\mathbf{F} = m \frac{d^2 \mathbf{r}}{dt^2} = m\mathbf{a}$$

then the force $\mathbf{F}'$ acting on m in a rotating coordinate system will be

$$\mathbf{F}' = m \frac{d'^2 \mathbf{r}}{dt^2} = m\mathbf{a}'$$

$$= \mathbf{F} + \text{noninertial forces}$$

$$= \mathbf{F} + \mathbf{F}_{\text{cent}} + \mathbf{F}_{\text{Cor}} + \mathbf{F}_{\text{trans}} \tag{11.37}$$

where the noninertial (or fictitious) forces are the centrifugal and Coriolis forces. This modification serves a very useful purpose, as we shall see in the following sections.

If the prime coordinate system has both translational and rotational motion, the following equations give the relations between the displacement, velocity, and acceleration vectors in the two systems. If $\mathbf{r}_0$ is the distance of O' from O,

$$\mathbf{r} = \mathbf{r}' + \mathbf{r}_0 \tag{11.38}$$

$$\frac{d\mathbf{r}}{dt} = \frac{d'\mathbf{r}'}{dt} + \boldsymbol{\omega} \times \mathbf{r}' + \frac{d\mathbf{r}_0}{dt} \tag{11.39}$$

$$\frac{d^2 \mathbf{r}}{dt^2} = \frac{d'^2 \mathbf{r}'}{dt^2} + \boldsymbol{\omega} \times \boldsymbol{\omega} \times \mathbf{r}' + 2\boldsymbol{\omega} \times \frac{d'\mathbf{r}'}{dt} + \frac{d\boldsymbol{\omega}}{dt} \times \mathbf{r}' + \frac{d^2 \mathbf{r}_0}{dt^2} \tag{11.40}$$

The centrifugal and Coriolis forces are not due to any physical interaction but are a result of kinematics; hence such forces are called *noninertial* or *fictitious forces*. For example, real forces always decrease with distance, whereas the centrifugal force increases with distance. It is true that it is convenient to use a rotational coordinate system to describe rotational motion of an object, but one must remember that the noninertial fictitious forces must be used only in a noninertial system or rotational coordinate system and not in an inertial system. For example, when a stone tied to a string is whirled in a circle, we feel as if a force is pulling the stone outward; we call this centrifugal force. For an observer in a rotating coordinate system with the stone, the stone is stationary and the outward centrifugal force balances the inward tension in the string. But in an inertial system there is no centrifugal force, only the tension in the string that causes the radial acceleration. Description in either coordinate system is correct provided the proper forces are taken into consideration. Similarly, when a car is going around a curve too fast, it will skid outward. According to an observer in an inertial system, the sideways force exerted by the road on the tires of a car is not sufficient to keep the car turning with the road. To an observer in the car (in the noninertial system), it may feel as if the car is being pushed outward by a centrifugal force.

The most important application of the above discussion is presented in the next two sections.

11.4 DESCRIPTION OF MOTION ON ROTATING EARTH

The results derived in the previous section can now be applied to describe motion in a coordinate system that is fixed on Earth and hence moving with the rotating Earth; that is, we describe motion in a noninertial coordinate system. The angular velocity ω of Earth with a radius vector relative to the sun is 2π radians per day. This value of angular velocity, when corrected to give the angular velocity with respect to fixed stars, is

$$\omega = \frac{2\pi}{24 \times 3600 \text{ s}} \frac{366.5}{365.5} \simeq 7.292 \times 10^{-5} \text{ rad/s} \qquad (11.41)$$

where 366.5 is the number of sidereal days in one year and 365.5 is the solar days in a year. This angular velocity, although very small, has profound effects. Some of the quantities that may be cited are as follows:

1. It is the spinning motion of Earth that causes the equatorial bulge; that is, Earth is flattened at the North and South Poles, resulting in an equatorial radius of $\simeq 21$ km ($\simeq 13$ miles) greater than the polar radius.
2. It is the Coriolis force on moving masses that produces a counterclockwise circulation of winds in the Northern Hemisphere. It affects the course of the trade winds and the Gulf Stream.
3. It is necessary to take into account the Coriolis force to accurately compute the trajectories of long-range projectiles and missiles.
4. The motion of the Foucault pendulum is the result of the Coriolis force.

Suppose a particle of mass m at a distance $\mathbf{r}$ from the center of Earth is subjected to a gravitational force $m\mathbf{g}$ and some other nongravitational force $\mathbf{F}$, such as friction. The equation of motion of this particle relative to coordinates fixed in space is

$$m \frac{d^2\mathbf{r}}{dt^2} = \mathbf{F} + m\mathbf{g} \qquad (11.42)$$

To obtain an equation of motion for this particle with respect to the coordinate system fixed with Earth, which has an angular velocity ω, we use the results obtained in Eq. (11.33) or (11.34). Keeping in mind that $d\omega/dt = 0$ for constant angular velocity, we obtain the equation of motion of the particle to be

$$m \frac{d^2\mathbf{r}}{dt^2} = \mathbf{F} + m\mathbf{g} = m \frac{d'^2\mathbf{r}}{dt^2} + m\boldsymbol{\omega} \times \boldsymbol{\omega} \times \mathbf{r} + 2m\boldsymbol{\omega} \times \frac{d'\mathbf{r}}{dt} \qquad (11.43)$$

or
$$\mathbf{F'} = m\frac{d'^2\mathbf{r}}{dt^2} = \mathbf{F} + m\mathbf{g} - m\boldsymbol{\omega} \times \boldsymbol{\omega} \times \mathbf{r} - 2m\boldsymbol{\omega} \times \frac{d'\mathbf{r}}{dt} \tag{11.44}$$

We divide our discussion into two parts: (1) static effects (plumb line), and (2) dynamic effects (motion of a projectile and free fall).

Static Effects (Plumb Line)

We are interested in measuring the gravitational force, both magnitude and direction, on a body of mass m at rest on the surface of Earth. Let us apply Eq. (11.44) to this situation. In this case, $d'\mathbf{r}/dt = 0$, and in the absence of frictional or any other force, $\mathbf{F} = 0$; therefore, we may write

$$\mathbf{F'} = m\frac{d'^2\mathbf{r}}{dt^2} = m\mathbf{g} - m\boldsymbol{\omega} \times \boldsymbol{\omega} \times \mathbf{r} = m\mathbf{g}_e \tag{11.45}$$

where
$$\mathbf{g}_e = \mathbf{g} - \boldsymbol{\omega} \times \boldsymbol{\omega} \times \mathbf{r} \tag{11.46}$$

The quantity $-\boldsymbol{\omega} \times \boldsymbol{\omega} \times \mathbf{r}$, as shown in Fig. 11.6, points radically outward from Earth's axis. It is the combination of this term and $\mathbf{g}$ that gives the resultant $\mathbf{g}_e$, the effective acceleration of gravity. As is clear from the figure, $\mathbf{g}_e$ at any point to the north of the equator points slightly to the south of Earth's center. Any object released near the surface of Earth will follow the direction of $\mathbf{g}_e$. This is also the direction of a plumb line. If we had a bucket of rotating liquid, the surface of the rotating liquid will settle in such a way that $\mathbf{g}_e$ is perpendicular to the surface at every point. This is also the reason that Earth is in the form of an oblate ellipsoid, resulting in a flattening at the poles. It is the combined effect of the centripetal acceleration and the flattening of Earth that has resulted in the polar radius's being ≈ 21 km (≈ 13 miles) smaller than the equatorial radius; $r_e \simeq r_p + 21$ km. This difference in the radius gives the value of $\mathbf{g}_e$ to be about 0.53% less at the equator than at the poles.

In general, the value of $\mathbf{g}_e$ changes with latitude λ and may be calculated as shown next. Referring to Fig. 11.6(b), the magnitude of the centrifugal acceleration of any point mass m is

$$\rho\omega^2 = (r\cos\lambda)\omega^2 \tag{11.47}$$

where ρ is the distance of the mass m from the axis and λ is the latitude of the place at P on Earth. The magnitude of the centrifugal force is

$$m\rho\omega^2 = (mr\cos\lambda)\omega^2 \tag{11.48}$$

The angle ϵ by which $\mathbf{g}_e$ deviates from $\mathbf{g}$ (the direction of the gravitational force) or the angle ϵ by which a plumb line deviates is [from Fig. 11.6(c)]

$$\frac{\sin\epsilon}{mr\omega^2\cos\lambda} = \frac{\sin\lambda}{mg_e} \tag{11.49}$$

Since ϵ is very small, we may write

$$\sin\epsilon \simeq \epsilon = \frac{r\omega^2}{g_e}\sin\lambda\cos\lambda = \frac{r\omega^2}{2g_e}\sin 2\lambda \tag{11.50}$$

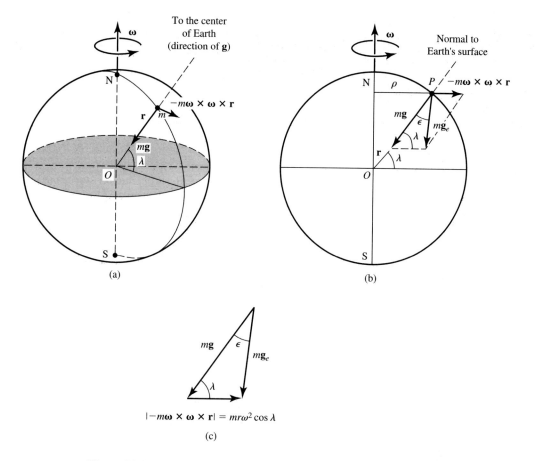

Figure 11.6   Rotating Earth: (a) centrifugal force $-m\boldsymbol{\omega} \times \boldsymbol{\omega} \times \mathbf{r}$ on an object of mass m at rest on Earth's surface, (b) direction of a plumb line is along $\mathbf{g}_e$, and (c) deviation ϵ from the vertical.

At the equator ($\lambda = 0°$) and at the poles ($\lambda = 90°$), ϵ will be zero, while the maximum deviation of a plumb line from the true vertical is at $\lambda = 45°$, where

$$\epsilon_{max} = \frac{r\omega^2}{2g_e} \sin(2 \times 45°) \simeq 1.7 \times 10^{-3} \text{ rad} = \frac{1}{10} \text{ degree} \qquad \textbf{(11.51)}$$

The direction of a plumb line is normal to the surface of Earth at every point.

Dynamic Effects (Motion of a Projectile and Free Fall)

We shall divide our discussion into two parts:

1. Deflection of a projectile fired horizontally
2. Deflection from the vertical of a freely falling body

Again, we apply Eq. (11.44) to these situations. We assume that there are no frictional or other external forces acting on the projectile; that is, $\mathbf{F} = 0$. Also, the term $m\boldsymbol{\omega} \times \boldsymbol{\omega} \times \mathbf{r}$ is very small and may be neglected. Hence the only contribution due to the rotational motion of Earth is the Coriolis force, and the working equation is

$$m\frac{d'^2\mathbf{r}}{dt^2} = m\mathbf{g} - 2m\boldsymbol{\omega} \times \frac{d'\mathbf{r}}{dt} \tag{11.52}$$

where the Coriolis force is

$$\mathbf{F}_{\text{Cor}} = m\mathbf{a}_{\text{Cor}} = -2m\boldsymbol{\omega} \times \frac{d'\mathbf{r}}{dt} = -2m\boldsymbol{\omega} \times \mathbf{v}' \tag{11.53}$$

Let us refer to Fig. 11.7 and discuss two special cases. $\boldsymbol{\omega}$ is along the axis passing through the North Pole. From Fig. 11.7(b), we obtain

$$\omega_x = 0, \qquad \omega_y = \omega \cos \lambda, \qquad \omega_z = \omega \sin \lambda \tag{11.54}$$

where λ is the latitude of a given place on the surface of Earth.

Deflection of a projectile fired horizontally.   According to Eq. (11.53), the Coriolis force acting on any particle is perpendicular to both $\boldsymbol{\omega}$ and $\mathbf{v}'$. In the Northern Hemisphere, $\boldsymbol{\omega}$ points out of the ground. If the projectile is shot horizontally along Earth's surface, that will be the direction of $\mathbf{v}'$. Thus the Coriolis force $-2m\boldsymbol{\omega} \times \mathbf{v}'$ will tend to deflect the projectile to the right of its direction of travel, as shown in Fig. 11.8. The direction of deflection is opposite in the Southern Hemisphere. At the equator $\boldsymbol{\omega}$ (being along the Y-axis) is horizontal, as shown in Fig. 11.7(b), $\boldsymbol{\omega}$ has no component along the Z-axis at the equator; hence the deflection is zero. We can calculate this deflection at other places as shown next.

Let us assume that the projectile is fired horizontally in the Northern Hemisphere. The Coriolis acceleration from Eq. (11.53) is

$$a_{\text{Cor}} = |2\boldsymbol{\omega} \times \mathbf{v}'| = 2\omega_z v' = 2\omega v' \sin \lambda \tag{11.55}$$

The linear deflection s $(=y)$ after time t will be

$$s = \tfrac{1}{2}a_{\text{Cor}}\, t^2 = \tfrac{1}{2}(2\omega v' \sin \lambda)t^2 = \omega v' t^2 \sin \lambda \tag{11.56}$$

If R is the horizontal range for a time of flight t, $R = v't$ or $t = R/v'$. Thus

$$y = s = \omega v'\left(\frac{R}{v'}\right)^2 \sin \lambda = \frac{\omega}{v'} R^2 \sin \lambda \tag{11.57}$$

Thus, as shown in Fig. 11.8(b), the angular deflection θ is

$$\theta = \frac{\text{linear deflection}}{\text{distance of travel}} = \frac{\omega v' t^2 \sin \lambda}{v' t} \tag{11.58a}$$

$$\theta = \omega t \sin \lambda \tag{11.58b}$$

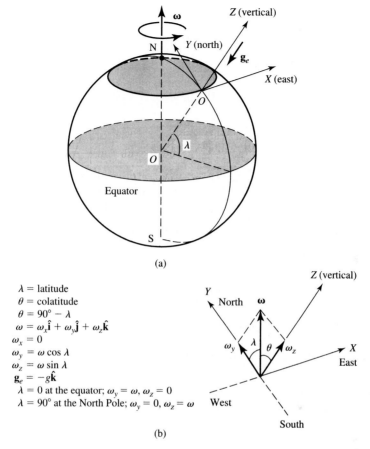

λ = latitude
θ = colatitude
$\theta = 90° - \lambda$
$\boldsymbol{\omega} = \omega_x \hat{\mathbf{i}} + \omega_y \hat{\mathbf{j}} + \omega_z \hat{\mathbf{k}}$
$\omega_x = 0$
$\omega_y = \omega \cos \lambda$
$\omega_z = \omega \sin \lambda$
$\mathbf{g}_e = -g\hat{\mathbf{k}}$
$\lambda = 0$ at the equator; $\omega_y = \omega$, $\omega_z = 0$
$\lambda = 90°$ at the North Pole; $\omega_y = 0$, $\omega_z = \omega$

(b)

Figure 11.7   (a) Coordinates XYZ, $\boldsymbol{\omega}$, and $\mathbf{g}_e$ and latitude λ needed to describe the motion of a projectile, and (b) components of $\boldsymbol{\omega}$.

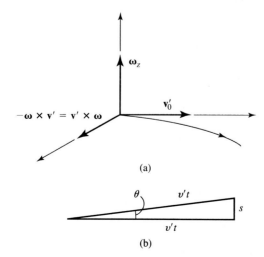

Figure 11.8   In the Northern Hemisphere, a projectile fired horizontally is deflected to the right due to the Coriolis force.

At the North Pole, $\lambda = 90°$ and $\sin 90° = 1$; hence the maximum value of a_{Cor} from Eq. (11.55) is

$$a_{Cor} = 2\omega v' = (1.5 \times 10^{-4}/s)v'$$

which, for $v' = 10^3$ m/s ($\simeq 2000$ mph), gives

$$a_{Cor} \simeq 0.15 \text{ m/s}^2 \simeq (0.015)g$$

The deflection θ at the pole, from Eq. (11.58b), is

$$\theta = \omega t \qquad\qquad\qquad \textbf{(11.59)}$$

which is the angle Earth will rotate in time t. This result implies that when a projectile is fired from the North Pole, its trajectory in the inertial frame is a straight line; the apparent deflection is due to Earth rotating under it. From Eq. (11.59), for $t = 100$ s,

$$\theta \simeq 7 \times 10^{-5} \text{ rad/s} \times 100 \text{ s} \simeq 7 \times 10^{-3} \text{ rad} \simeq 0.4°$$

Although θ is small, it becomes important in launching long-range rockets and missiles.

Deflection from the vertical of a freely falling body.   Referring to Fig. 11.7, since the velocity of the body is almost vertical along the Z-axis, and $\boldsymbol{\omega}$ lies in the north-south vertical plane, that is, in the YZ plane, the Coriolis force, which is now the deflecting force $-2m\boldsymbol{\omega} \times \mathbf{v}' = 2m\mathbf{v}' \times \boldsymbol{\omega}$, will be in the east-west direction. Thus in the Northern Hemisphere a freely falling body will be deflected to the east (the X-axis). The equation of motion in the X-direction is

$$m\frac{d^2x}{dt^2} = -2m(\boldsymbol{\omega} \times \mathbf{v}')_x$$

$$= -2m\omega v'_z \sin\theta \qquad\qquad \textbf{(11.60)}$$

where $\theta = 90° - \lambda$ is the *colatitude* angle.
[We can evaluate the $\boldsymbol{\omega} \times \mathbf{v}'$ term as follows:

$$\boldsymbol{\omega} \times \mathbf{v}' = \begin{vmatrix} \hat{\mathbf{i}} & \hat{\mathbf{j}} & \hat{\mathbf{k}} \\ \omega_x & \omega_y & \omega_z \\ v'_x & v'_y & v'_z \end{vmatrix} = \begin{vmatrix} \hat{\mathbf{i}} & \hat{\mathbf{j}} & \hat{\mathbf{k}} \\ 0 & \omega\cos\lambda & \omega\sin\lambda \\ 0 & 0 & v'_z \end{vmatrix}$$

$$= \hat{\mathbf{i}}\omega v'_z \cos\lambda + 0 + 0$$

$$= (\boldsymbol{\omega} \times \mathbf{v}')_x = \hat{\mathbf{i}}\omega v'_z \cos(90° - \theta) = \hat{\mathbf{i}}\omega v'_z \sin\theta]$$

The contribution of the Coriolis force to the velocity v'_z is very small; hence the contribution to the horizontal deflection x will be negligible. We may calculate v'_z as if there were no contribution by the Coriolis force; that is,

$$v'_z = -gt \qquad\qquad\qquad \textbf{(11.61)}$$

and $$z = \frac{1}{2}gt^2 \quad \text{or} \quad t = \sqrt{\frac{2z}{g}} \qquad\qquad \textbf{(11.62)}$$

Substituting for v_z' from Eq. (11.61) in Eq. (11.60) gives

$$\frac{d^2x}{dt^2} = -2\omega(-gt)\sin\theta = 2\omega gt\sin\theta \tag{11.63}$$

Integrating this twice, with the initial conditions that at $t = 0$, $\dot{x} = 0$, and $x = 0$, we get

$$x = \tfrac{1}{3}\omega gt^3\sin\theta \tag{11.64}$$

Substituting for t from Eq. (11.62), and $\theta = 90° - \lambda$,

$$x = \frac{\omega}{3}\sqrt{\frac{8z^3}{g}}\sin\theta = \frac{\omega}{3}\sqrt{\frac{8z^3}{g}}\cos\lambda \tag{11.65}$$

Thus at the equator $\lambda = 0°$ ($\theta = \pi/2$), and if $z = 100$ m, Eq. (11.65) gives $x \simeq 2.30$ cm. At the pole, $\lambda = 90°$, gives $x = 0$; while at $\lambda = 45°$, $x \simeq 1.55$ cm. Thus the deflection is maximum at the equator. The drift is always toward the right (or southward) in the Northern Hemisphere. This deflection from the vertical is comparable to such deflections as those due to friction and air currents. Such small effects can be more easily seen in the Foucault pendulum (see Section 11.5) and in rotational and vibrational levels of molecules.

Dynamic Effects (Motion of a Projectile): Alternative Treatment

In this alternative treatment, we shall start with the general equation of a projectile and then specialize to the two cases discussed previously. The general equation of motion of the projectile is [Eq. (11.52)]

$$m\frac{d'^2\mathbf{r}}{dt^2} = m\mathbf{g} - 2m\boldsymbol{\omega} \times \frac{d'\mathbf{r}}{dt} \tag{11.66}$$

where

$$\mathbf{g} = -g\hat{\mathbf{k}} \tag{11.67}$$

and

$$\boldsymbol{\omega} = \omega_x\hat{\mathbf{i}} + \omega_y\hat{\mathbf{j}} + \omega_z\hat{\mathbf{k}} \tag{11.68}$$

while (note that we are not using the prime notation any longer)

$$\boldsymbol{\omega} \times \dot{\mathbf{r}} = \begin{vmatrix} \hat{\mathbf{i}} & \hat{\mathbf{j}} & \hat{\mathbf{k}} \\ \omega_x & \omega_y & \omega_z \\ v_x & v_y & v_z \end{vmatrix} = \begin{vmatrix} \hat{\mathbf{i}} & \hat{\mathbf{j}} & \hat{\mathbf{k}} \\ 0 & \omega\cos\lambda & \omega\sin\lambda \\ v_x & v_y & v_z \end{vmatrix}$$

$$= \hat{\mathbf{i}}(\omega v_z\cos\lambda - \omega v_y\sin\lambda) + \hat{\mathbf{j}}(\omega v_x\sin\lambda) + \hat{\mathbf{k}}(-\omega v_x\cos\lambda) \tag{11.69}$$

Hence Eq. (11.66) may be written in component form as

$$\ddot{x} = -2\omega(\dot{z}\cos\lambda - \dot{y}\sin\lambda) \tag{11.70}$$

$$\ddot{y} = -2\omega(\dot{x}\sin\lambda) \tag{11.71}$$

$$\ddot{z} = -g + 2\omega(\dot{x}\cos\lambda) \tag{11.72}$$

Solving these equations (by integrating twice and neglecting terms in ω^2), we get Eqs. (11.56) or (11.57) and (11.65) for x and y, respectively.

 Example 11.2 _____

A ballistic projectile is launched, with an initial speed v of 500 m/s southward, at an angle θ of 30° from the horizontal from a point at a latitude λ of 60° north as shown in Fig. Ex. 11.2. Calculate the point of impact.

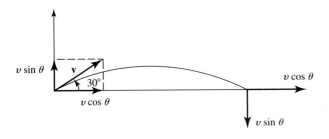

Figure Ex. 11.2

Solution

Neglecting the effect of the centrifugal force, let us assume that the value of g remains the same throughout the path of the projectile. First we solve the problem of finding the range by ignoring the rotation of Earth, that is, ignoring the Coriolis force. The projectile makes an angle θ with the horizontal and is shot southward. Thus the vertical component of the velocity is $v_v = v \sin\theta$, while the horizontal component is $v_h = v \cos\theta$. The time of flight of the projectile is

$$t_t = \frac{2v \sin\theta}{g} \tag{i}$$

while the range R is

$$R = \frac{v^2 \sin 2\theta}{g} \tag{ii}$$

For $g = 9.8$ m/s², $v = 500$ m/s, and $\theta = 30°$ (θ is not a colatitude angle),

$$R = 22092.5 \text{ m} = 22.09 \text{ km} \tag{iii}$$

Let us now calculate the effect of the Coriolis force in changing the position of the point of impact. The Coriolis force is given by [from Eq. (11.60)]

$$\mathbf{F}_{\text{Cor}} = -2\,m\boldsymbol{\omega} \times \mathbf{v} \tag{iv}$$

or

$$F_{\text{Cor}} = 2m\omega v \cos\lambda \tag{v}$$

If the projectile is moving upward, that is, v is upward, this expression gives the force; hence the deflection is toward the west. On the other hand, if the projectile is moving downward, that is, v is downward, the force F_{Cor} is toward the east; hence the deflection is toward the east. In addition to these two deflections, if the projectile has a velocity $v_{\text{N--S}}$ in the north-south direction, there are two more deflections; that is [from Eq. (11.55)],

$$F_{\text{Cor}} = 2m\omega v_{\text{N--S}} \sin\lambda$$

or

$$F_{\text{Cor}} = 2m\omega v_{\text{N}} \sin\lambda, \qquad \text{toward the east} \tag{vi}$$

$$F_{\text{Cor}} = 2m\omega v_{\text{S}} \sin\lambda, \qquad \text{toward the west} \tag{vii}$$

Thus there are four deflections, two from Eq. (v) and two from Eqs. (vi) and (vii). Even if there were an east-west motion, it would not contribute to any deflection because an east-west direction is perpendicular to $\boldsymbol{\omega}$.

We could use the above equations to do the calculations or take a slightly different approach as shown.

Alternative approach

We divide our discussion in two parts: (a) flight without taking into account the Coriolis force and (b) the effect of the Coriolis force that results in a change in the position of impact.

(a) Flight without Coriolis force.
θ, λ and g are given. Since the projectile is fired southward, the only components of velocity are in the y-direction, v0y, and z-direction (vertical), v0z.

$$\theta := \frac{\pi}{6} \qquad\qquad \theta = 0.524 \cdot rad \qquad \theta = 30 \cdot deg$$

$$\lambda := \frac{\pi}{2} - \theta \qquad\qquad \lambda = 1.047 \cdot rad \qquad \lambda = 60 \cdot deg$$

$$v0 := 500 \cdot \frac{m}{sec} \qquad ge := 9.8 \cdot \frac{m}{sec^2}$$

$$v0z := v0 \cdot \sin\left(\frac{\pi}{6}\right) \qquad v0z = 250 \cdot m \cdot sec^{-1}$$

$$v0y := v0 \cdot \cos\left(\frac{\pi}{6}\right) \qquad v0y = 433.013 \cdot m \cdot sec^{-1}$$

Time of flight tf is

$$tf := \frac{2 \cdot v0 \cdot \sin(\theta)}{ge} \qquad tf = 51.02 \cdot sec$$

The maximum range R of the projectile is

$$R := \frac{v0^2 \cdot \sin(2 \cdot \theta)}{ge} \qquad R = 2.209 \cdot 10^4 \cdot m$$

(b) Effect of Coriolis force on the position of the impact. Since the components of velocity are in the x- and z-direction the Coriolis force, hence the accelerations, are only in these directions: vertical up and down motion, causing the projectile to deflect westward (explain why), and horizontal southward velocity, also causing southward deflection.

The general expression for the Coriolis force is $\mathbf{F} = -2m\ \boldsymbol{\omega}\mathbf{x}\mathbf{v}$
Force acting on the vertical motion due to z-component of the Coriolis force is Fzcor = 2mω v cos(λ), where the angular velocity is

$$\omega := 7.292 \cdot 10^{-5} \cdot \frac{rad}{sec}$$

$$ac\equiv(2 \cdot \omega) \times v$$

$$\omega \equiv \begin{bmatrix} 0 \cdot \dfrac{rad}{sec} \\ \omega \cdot \cos(\lambda) \\ \omega \cdot \sin(\lambda) \end{bmatrix} \quad v \equiv \begin{bmatrix} 0 \cdot \dfrac{m}{sec} \\ \omega \cdot \cos(\theta) \\ v0 \cdot \sin(\theta) - g \cdot t \end{bmatrix} \quad ac \equiv \left| 2 \cdot \left[\begin{bmatrix} 0 \cdot \dfrac{rad}{sec} \\ \omega \cdot \cos(\lambda) \\ \omega \cdot \sin(\lambda) \end{bmatrix} \times \begin{bmatrix} 0 \cdot \dfrac{m}{sec} \\ \omega \cdot \cos(\theta) \\ v0 \cdot \sin(\theta) - g \cdot t \end{bmatrix} \right] \right|$$

simplifies to

$$ac \equiv \left| 2 \cdot \omega \cdot \cos(\lambda) \cdot vo \cdot \sin(\theta) - 2 \cdot \omega \cdot \cos(\lambda) \cdot g \cdot t - 2 \cdot \omega^2 \cdot \sin(\lambda) \cdot \cos(\theta) \right|$$

$voz \equiv vo \cdot \sin(\theta)$ $voy \equiv vo \cdot \cos(\theta)$

$Fzc \equiv 2 \cdot m \cdot \omega \cdot v \cdot \cos(\lambda)$ $Fyc \equiv 2 \cdot m \cdot \omega \cdot v \cdot \sin(\lambda)$

$vz \equiv vo \cdot \sin(\theta) - g \cdot t$

$vy \equiv vo \cdot \cos(\theta)$

$aw1 \equiv 2 \cdot \omega \cdot (vo \cdot \sin(\theta) - g \cdot t) \cdot \cos(\lambda)$

$aw2 \equiv 2 \cdot \omega \cdot vo \cdot \cos(\theta) \cdot \sin(\lambda)$

$vw1 \equiv \displaystyle\int 2 \cdot \omega \cdot (vo \cdot \sin(\theta) - g \cdot t) \cdot \cos(\lambda)\ dt$

$vw2 \equiv \displaystyle\int 2 \cdot \omega \cdot vo \cdot \cos(\theta) \cdot \sin(\lambda)\ dt$

$vw1 \equiv 2 \cdot \omega \cdot \cos(\lambda) \cdot t \cdot vo \cdot \sin(\theta) - \omega \cdot \cos(\lambda) \cdot t^2 \cdot g$

$vw2 \equiv 2 \cdot \omega \cdot vo \cdot \cos(\theta) \cdot \sin(\lambda) \cdot t$

$xw1 \equiv \displaystyle\int \left(2 \cdot \omega \cdot \cos(\lambda) \cdot t \cdot vo \cdot \sin(\theta) - \omega \cdot \cos(\lambda) \cdot t^2 \cdot g\right)\ dt$

$xw2 \equiv \displaystyle\int 2 \cdot \omega \cdot vo \cdot \cos(\theta) \cdot \sin(\lambda) \cdot t\ dt$

$xw1 \equiv \omega \cdot \cos(\lambda) \cdot t^2 \cdot vo \cdot \sin(\theta) - \dfrac{1}{3} \cdot \omega \cdot \cos(\lambda) \cdot t^3 \cdot ge$

$xw2 \equiv \omega \cdot vo \cdot \cos(\theta) \cdot \sin(\lambda) \cdot t^2$

Substituting the value for tf

$xw1 \equiv \dfrac{4}{3} \cdot vo^3 \cdot \cos(\lambda) \cdot \omega \cdot \sin(\theta) \cdot \dfrac{\left(1 + \cos(\theta)^2\right)}{ge^2}$

$xw2 \equiv 4 \cdot \omega \cdot vo \cdot \cos(\theta) \cdot \sin(\lambda) \cdot vo^2 \cdot \dfrac{\sin(\theta)^2}{ge^2}$

$xw1 := \dfrac{4}{3} \cdot vo^3 \cdot \cos(\lambda) \cdot \omega \cdot \sin(\theta) \cdot \dfrac{\left(1 + \cos(\theta)^2\right)}{ge^2}$

$xw2 := 4 \cdot \omega \cdot vo \cdot \cos(\theta) \cdot \sin(\lambda) \cdot vo^2 \cdot \dfrac{\sin(\theta)^2}{ge^2}$

$xw1 = 55.363 \cdot m$

$xw2 = 71.181 \cdot m$

$xw := xw1 + xw2$ $xw = 126.544 \cdot m$ $R = 2.209 \cdot 10^4\ \cdot m$ $R + xw = 2.222 \cdot 10^4\ \cdot m$

Using the appropriate equations, we graph the projectile motion with and without Coriolis force.

$N := 550$ $i := 0 .. N$

$t_i := i \cdot sec$ $g0 := 9.8 \cdot \dfrac{m}{sec^2}$

$X0_i := v0 \cdot \cos(\theta) \cdot t_i$

$Z0_i := v0 \cdot \sin(\theta) \cdot t_i - \dfrac{1}{2} \cdot g0 \cdot \left(t_i\right)^2$

$xw1_i := \dfrac{1}{2} \cdot 2 \cdot \omega \cdot v0 \cdot \sin(\theta) \cdot \cos(\lambda) \cdot \left(t_i\right)^2$

$xw2_i := \dfrac{1}{2} \cdot 2 \cdot \omega \cdot v0 \cdot \cos(\theta) \cdot \sin(\lambda) \cdot \left(t_i\right)^2$

$X_i := X0_i + xw1_i + xw2_i$

$X \equiv X0 + xw1 + xw2$

$\max(X) = 2.492 \cdot 10^5\ \cdot m$

$\max(X0) = 2.382 \cdot 10^5\ \cdot m$

$\max(xw1) = 2.757 \cdot 10^3\ \cdot m$

$\max(xw2) = 8.272 \cdot 10^3\ \cdot m$

$\max(X) - \max(X0) = 1.103 \cdot 10^4\ \cdot m$

$\max(xw1) + \max(xw2) = 1.103 \cdot 10^4\ \cdot m$

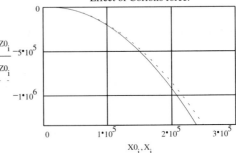

Effect of Coriolis force.

What is the significance of the above calculations?

EXERCISE 11.2 Repeat the example for a projectile launched at a latitude λ of 60° south.

11.5 FOUCAULT PENDULUM

In 1851, Jean Leon Foucault (1819–1868) devised a system, called the Foucault pendulum, that demonstrates that *Earth is a noninertial system.* In other words, as we shall show, he demonstrated the presence of the Coriolis force due to the rotation of Earth. The Foucault pendulum consisted of a heavy bob of 28-kg mass hanging from a long wire of 67-m length mounted from the dome of the Pantheon in Paris. The pendulum swung freely in any vertical plane. Once the pendulum was set swinging in a definite vertical plane, it was observed that the plane of the swing precessed about the vertical. The plane precessed almost a centimeter on each swing, thereby directly demonstrating that Earth is rotating. (Such a pendulum is exhibited at the Smithsonian Institution in Washington, D.C.)

We analyze the problem of the motion of the Foucault pendulum with the help of a diagram, as in Fig. 11.9. The necessary conditions are that the string is very long, the mass of the bob is very large, and the support is frictionless so that the pendulum can swing freely. As shown, the origin O of the coordinate system $OXYZ$ is directly below the point of support and is the equilibrium point of the pendulum. The Z-axis is the local vertical, and the plane defined by the X and Y axes is the horizontal plane. We are interested in the motion of the bob in this horizontal plane. To achieve this, the motion of the pendulum is limited to oscillations of very small amplitude. Under these conditions, the displacement $\mathbf{r}$ of the bob from the equilibrium point is almost horizontal; hence $\dot{z}\,(=v_z)$ is very small as compared to $\dot{x}\,(=v_x)$ or $\dot{y}\,(=v_y)$. If $\mathbf{T}$ is the tension in the string and $\mathbf{v} = \dot{\mathbf{r}} = d\mathbf{r}/dt$ is the velocity of the bob, the equation of motion of the mass m may be written as

$$m\,\frac{d^2\mathbf{r}}{dt^2} = \mathbf{T} + m\mathbf{g}_e - 2m\boldsymbol{\omega} \times \frac{d\mathbf{r}}{dt} \tag{11.73}$$

The last term in this equation is the Coriolis force and would have been absent if Earth were not rotating. Equation (11.73) without the last term is simply an equation of a simple pendulum in a nonrotating Earth.

For small velocities, say $v \simeq 10$ km/h, the Coriolis force is about 0.1% of the gravitational force $m\mathbf{g}_e$; hence the vertical component of the Coriolis force is negligible as compared to $m\mathbf{g}_e$. On the other hand, the horizontal component of the Coriolis force that is perpendicular to $\dot{\mathbf{r}}$ and is in the XY plane has an appreciable effect on the motion of the pendulum because no other forces are acting in this plane. This component of the Coriolis force makes it impossible for the pendulum to continue to swing in a fixed vertical plane and results in a precession or rotation of this plane about the vertical axis. We calculate the frequency of the precession as follows.

The differential equation of motion of mass m given by Eq. (11.73) may be written as

$$m\ddot{\mathbf{r}} = \mathbf{T} + m\mathbf{g}_e - 2m\boldsymbol{\omega} \times \dot{\mathbf{r}} \tag{11.74}$$

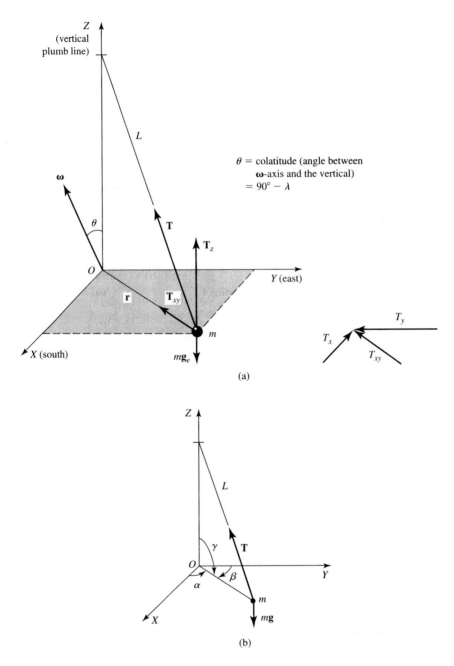

Figure 11.9   Forces acting on a bob of mass m of a Foucault pendulum.

The components of $\boldsymbol{\omega} \times \dot{\mathbf{r}}$ as calculated before are

$$\boldsymbol{\omega} \times \dot{\mathbf{r}} = \begin{vmatrix} \hat{\mathbf{i}} & \hat{\mathbf{j}} & \hat{\mathbf{k}} \\ \omega_x & \omega_y & \omega_z \\ \dot{x} & \dot{y} & \dot{z} \end{vmatrix} = \begin{vmatrix} \hat{\mathbf{i}} & \hat{\mathbf{j}} & \hat{\mathbf{k}} \\ 0 & \omega \cos \lambda & \omega \sin \lambda \\ \dot{x} & \dot{y} & \dot{z} \end{vmatrix}$$

$$= \hat{\mathbf{i}}(\omega \dot{z} \cos \lambda - \omega \dot{y} \sin \lambda) + \hat{\mathbf{j}}(\omega \dot{x} \sin \lambda) + \hat{\mathbf{k}}(-\omega \dot{x} \cos \lambda) \qquad \textbf{(11.75)}$$

The components of T, from Fig. 11.9(b) (where, α, β, and γ are the directional cosines) are

$$T_x = -T \cos \alpha = -\frac{x}{L} T \qquad \textbf{(11.76a)}$$

$$T_y = -T \cos \beta = -\frac{y}{L} T \qquad \textbf{(11.76b)}$$

$$T_z = T \cos \gamma = \frac{z}{L} T \qquad \textbf{(11.76c)}$$

Using these equations, we may write the differential equation Eq. (11.74) in component form as

$$m\ddot{x} = -\frac{x}{L} T - 2m\omega(\dot{z} \cos \lambda - \dot{y} \sin \lambda) \qquad \textbf{(11.77)}$$

$$m\ddot{y} = -\frac{y}{L} T - 2m\omega \dot{x} \sin \lambda \qquad \textbf{(11.78)}$$

$$m\ddot{z} = T_z - mg + 2m\omega \dot{x} \cos \lambda \qquad \textbf{(11.79)}$$

In our case the displacement of the pendulum from the vertical is very small, so that $z \simeq L$ and from Eq. (11.76c) $T_z \simeq T \simeq mg$. Also, in this case $\dot{z}$ is almost zero as compared to $\dot{x}$ and $\dot{y}$. Thus the motion in the XY plane of mass m with $\dot{z} = 0$ and $T = mg$ is described by Eqs. (11.77) and (11.78); that is,

$$m\ddot{x} = -\frac{x}{L} mg + 2m\omega \dot{y} \sin \lambda \qquad \textbf{(11.80)}$$

$$m\ddot{y} = -\frac{y}{L} mg - 2m\omega \dot{x} \sin \lambda \qquad \textbf{(11.81)}$$

Let us define

$$\omega' = \omega \sin \lambda \qquad \textbf{(11.82)}$$

and Eqs. (11.80) and (11.81) take the form

$$\ddot{x} = -\frac{g}{L} x + 2\omega' \dot{y} \qquad \textbf{(11.83)}$$

$$\ddot{y} = -\frac{g}{L} y - 2\omega' \dot{x} \qquad \textbf{(11.84)}$$

These are called *coupled equations* because the equation for $\ddot{x}$ contains $\dot{y}$ and the equation for $\ddot{y}$ contain $\dot{x}$. Instead of trying to solve these equations to have some physical interpretation, it is easier to take an alternative approach, as discussed next.

The motion described by the differential equations (11.83) and (11.84) can be visualized by transforming to a new set of coordinate axes $O'X'Y'$ rotating in the XY plane with constant angular speed $-\omega'$ ($= -\omega \sin \lambda$) relative to the OXY axes, as shown in Fig. 11.10. Note that the rotation is about the Z'-axis, which coincides with the Z-axis. The equations of transformation are

$$x = x' \cos \omega' t + y' \sin \omega' t \qquad\qquad \textbf{(11.85)}$$

$$y = -x' \sin \omega' t + y' \cos \omega' t \qquad\qquad \textbf{(11.86)}$$

Calculate the values of $\ddot{x}$ and $\dot{y}$ from Eqs. (11.85) and (11.86); drop the terms containing ω'^2. Substitute the values of x, $\dot{y}$, and $\ddot{x}$ in Eq. (11.83) and, after rearranging, we get

$$\left(\ddot{x}' + \frac{g}{L} x'\right) \cos \omega' t + \left(\ddot{y}' + \frac{g}{L} y'\right) \sin \omega' t = 0 \qquad\qquad \textbf{(11.87)}$$

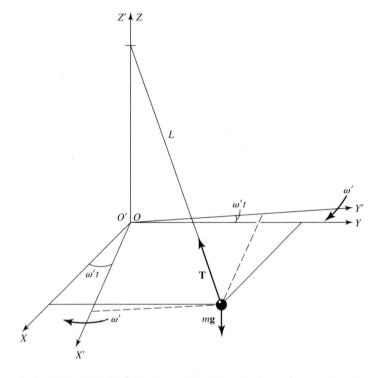

Figure 11.10   Motion of the Foucault pendulum as viewed by transformation to a new set of coordinate axes $O'X'Y'$ rotating in the XY plane with constant angular velocity $-\omega'$ ($= -\omega \sin \lambda$) relative to the OXY axes.

For this equation to hold good for any value of t, the coefficients of $\cos \omega' t$ and $\sin \omega' t$ must be zero. That is,

$$\ddot{x}' + \frac{g}{L}x' = 0 \tag{11.88}$$

$$\ddot{y}' + \frac{g}{L}y' = 0 \tag{11.89}$$

Let us define

$$\omega_x^2 = \frac{g}{L} \quad \text{and} \quad \omega_y^2 = \frac{g}{L} \tag{11.90}$$

and Eqs. (11.88) and (11.89) take the form

$$\ddot{x}' + \omega_x^2 x' = 0 \tag{11.91}$$

$$\ddot{y}' + \omega_y^2 y' = 0 \tag{11.92}$$

Differential equations (11.91) and (11.92) reveal that the path of the particle of mass m in the coordinate system $O'X'Y'$ is elliptical (due to the two simple harmonic motions at right angles to each other), as shown in Fig. 11.11(a). In the $O'X'Y'$ coordinates, the major axis of the ellipse has a fixed orientation, but not when this major axis is viewed from the coordinate system $O'XY$, because $O'X'Y'$ is rotating with respect to OXY with a constant angular velocity $-\omega'$. Thus, when viewed from the OXY coordinate system, the major axis undergoes a steady precession in a clockwise direction (in the Northern Hemisphere) with an angular velocity $\omega' = \omega \sin \lambda$. That is, the vertical plane precesses about the local vertical with an angular velocity ω' when referred to the OXY system, as shown in Fig. 11.11(b). The period of precession T_p is given by

$$T_p = \frac{2\pi}{\omega'} = \frac{2\pi}{\omega \sin \lambda} = \frac{T_0}{\sin \lambda} = \frac{24 \text{ h}}{\sin \lambda} \tag{11.93}$$

Thus, if

$$\lambda = 45° \qquad\qquad T_p \simeq 34 \text{ h} \tag{11.94a}$$

$$\omega = 90° \text{ (north pole)} \qquad T_p = 24 \text{ h} \tag{11.94b}$$

$$\lambda = 0° \text{ (equator)} \qquad T_p = \infty \tag{11.94c}$$

At the North or South Pole, $\omega' = \pm\omega$; that is, the pendulum swings in a fixed vertical plane (fixed with respect to an inertial system such as defined by space or the stars), while Earth underneath turns with a period of 24 h. At the equator $\omega' = 0$; hence the plane of the pendulum does not precess.

Figure 11.11 _____

Below the precession of the plane of the pendulum is illustrated.

$g := 9.8$	$L := 20$	$\omega e := .01$
$\lambda := \dfrac{\pi}{4}$	$\omega 0 := \sqrt{\dfrac{g}{L}}$	$\omega 1 := \omega e \cdot \sin(\lambda)$
$\lambda = 0.785$	$\omega 0 = 0.7$	$\omega 1 = 0.007$
$A := 10$	$B := 5$	$7.292 \cdot 10^{-5}$
$N := 500$	$i := 0 .. N$	$t_i := \dfrac{i}{10}$

The values used in the calculations are as shown.

(a) An elliptical path of the bob of a Foucault pendulum as viewed from the rotating coordinates $O'X'Y'$ (primed system). Note that $\omega 1 = 0$ in the prime system.

(b) Steady precession of the major axis of the ellipse when viewed from the nonrotating coordinate OXY (unprimed) system is a precessing ellipse. Note that if $\omega 1 = 0$, it gives the graph in (a).

$$x1_i := A \cdot \cos\left(\omega 0 \cdot t_i\right)$$

$$y1_i := B \cdot \sin\left(\omega 0 \cdot t_i\right)$$

$$x_i := A \cdot \cos\left(\omega 0 \cdot t_i\right) \cdot \cos\left(\omega 1 \cdot t_i\right) + B \cdot \left(\sin\left(\omega 0 \cdot t_i\right) \cdot \sin\left(\omega 1 \cdot t_i\right)\right)$$

$$y_i := A \cdot \cos\left(\omega 0 \cdot t_i\right) \cdot \sin\left(\omega 1 \cdot t_i\right) + B \cdot \sin\left(\omega 0 \cdot t_i\right) \cdot \cos\left(\omega 1 \cdot t_i\right)$$

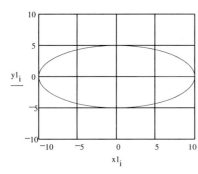

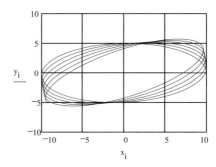

(a) Is it possible to have equal values of A and B? Explain how this will influence the shape of the precessing ellipse.

(b) How will the precession look for the same value of altitude in the Southern Hemisphere? Explain.

(c) How will these plots be affected by the frequency of precession?

_____

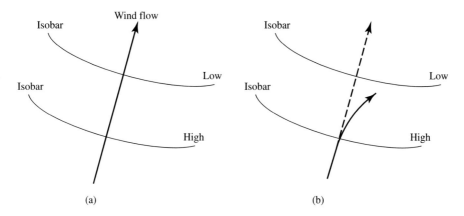

Figure 11.12   Wind direction from a high to low pressure (a) in the absence of the Coriolis force, and (b) in the presence of the Coriolis forces.

11.6 HORIZONTAL WIND CIRCULATIONS: WEATHER SYSTEMS

Coriolis force is one of the many factors that affect the motion of the winds and hence the weather system. We give here a very simplified picture of this situation. Suppose, because of differential heating of air, a region of low pressure results in the atmosphere. The horizontal curves in Fig. 11.12 and the closed curves in Figs. 11.13 and 11.14 represent lines of constant pressure and are called *isobars*. Due to the pressure gradient, a force results on each element of air, and in the absence of any other forces, the mass of air tends to move from region of high pressure to a region of low pressure, and the pressure difference will be quickly equalized. Such flow of air is called a *pressure gradient flow*. In the vertical direction, the pressure gradient is balanced by the gravitational force. Hence air masses flow only in a horizontal plane and are called *winds*.

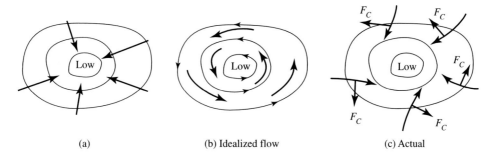

Figure 11.13   Wind flow about a low pressure in the Northern Hemisphere (a) without the Coriolis force, (b) idealized in the presence of the Coriolis force F_C, and (c) actual.

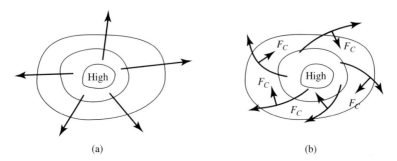

Figure 11.14 Wind flow about a high pressure in the Northern Hemisphere (a) without the Coriolis force, and (b) actual with the Coriolis and other forces.

The presence of the Coriolis force, which is comparable to the pressure gradient forces, changes the situation. In the absence of the Coriolis forces, the wind direction will be perpendicular to the isobars, as shown in Fig. 11.12(a). The Coriolis force deflects them to the right, as shown in Fig. 11.12(b). Such deflection of air masses to the right continues until the wind vector is parallel to the isobar, and the Coriolis force just balances the pressure gradient force. Thus the wind continues parallel to the isobars circulating in a counterclockwise direction about a low pressure in the Northern Hemisphere as shown in Fig. 11.13, where part (a) shows no Coriolis force present, part (b) shows an idealized flow in the presence of the Coriolis force, and part (c) shows the actual flow (modification resulting from friction and other forces). On the other hand, winds circulate in a clockwise direction about regions of high pressure in the Northern Hemisphere as shown in Fig. 11.14. [Formation of high winds (hurricanes) about a high-pressure region is not very common, as we shall explain later.] The directions of wind flow in the Southern Hemisphere about low- and high-pressure regions are opposite to those in the Northern Hemisphere. Such wind directions are called *geostrophic winds* or *cyclonic winds*. At the equator, there is no Coriolis force, and hence the absence of circular winds leads to more uniform weather near the equator.

We now analyze this motion from an analytical point of view. Let us consider a mass of air of volume $V = A\,\Delta r$, where A is the cross-sectional area and Δr is the thickness (see Fig. 11.15). This mass of air is at a distance r from the center of low pressure. The pressure force on the inner surface is PA and that on the outer surface is $(P + \Delta P)A$. Thus the net inward pressure force F_P is ΔPA. The Coriolis force F_C is $2m\omega v \sin \lambda$, where $m = \rho V = \rho A\,\Delta r$ is the mass of the air, ρ is the density, and v is its velocity. Since the air mass is rotating counterclockwise, the Coriolis force is acting outward. The centripetal acceleration being v^2/r, we may write the equation of motion for this mass of air as

$$\frac{mv^2}{r} = \Delta PA - 2m\omega v \sin \lambda \qquad (11.95)$$

Substituting $m = \rho A\,\Delta r$ and taking the limit $\Delta r \to 0$, we get

$$\frac{v^2}{r} = \frac{1}{\rho}\frac{dP}{dr} - 2\omega v \sin \lambda \qquad (11.96)$$

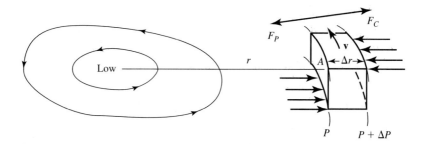

Figure 11.15   Forces acting on a mass of air at a distance r from the center of a low pressure.

Near the center of a low pressure, dP/dr is very large and hence the wind velocity is very large. One the other hand, far away from the center of the low, we can neglect the term v^2/r and write

$$\frac{1}{\rho}\frac{dP}{dr} = 2\omega v \sin \lambda \tag{11.97}$$

$$v = \frac{1}{2\omega \sin \lambda}\frac{1}{\rho}\frac{dP}{dr} \tag{11.98}$$

Wind flow around low- and high-pressure regions together with the directions of the Coriolis force F_C and the force F_P due to the pressure gradient is shown in Fig. 11.16.

In the case of a low-pressure region for low-pressure gradients (1 to 4 millibars/m), we get a wind velocity on the order of $\leqslant 80$ km/h. If the pressure gradient is ≈ 30 mbars/m, the wind velocities reach ≈ 160 km/h and are called hurricanes. These calculations are shown in Example 11.3.

In the case of a high-pressure region, the directions of F_P and F_C are opposite to those of a low-pressure region, as shown in Fig. 11.16(b), and after rewriting Eq. (11.96), we get

$$v = r\omega \sin \lambda - \left[(r\omega \sin \lambda)^2 - \frac{r}{\rho}\left|\frac{dP}{dr}\right|\right]^{1/2} \tag{11.99}$$

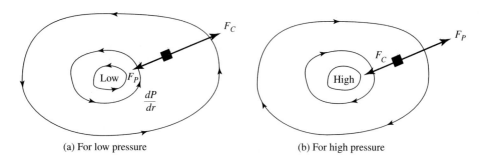

(a) For low pressure   (b) For high pressure

Figure 11.16   Flow and direction of the forces around (a) low-pressure and (b) high-pressure regions.

Thus, if

$$\frac{r}{\rho}\left|\frac{dP}{dr}\right| > (r\omega \sin \lambda)^2 \tag{11.100a}$$

or

$$\frac{1}{\rho}\left|\frac{dP}{dr}\right| > r(\omega \sin \lambda)^2 \tag{11.100b}$$

that is, $F_P > F_C$, the high pressure cannot provide a centripetal acceleration to form a hurricane. This is because the Coriolis force is too weak to supply the needed centripetal force. Thus we conclude that storms and hurricanes are always low-pressure systems, and not high-pressure systems.

　　　Air masses moving with high velocities are called *hurricanes*. As shown in Example 11.3, when v is large, we cannot ignore v^2/r term. If F_P exceeds F_C, it can provide centripetal acceleration.

 Example 11.3 _____

Suppose at a place 20°N latitude the atmospheric pressure is 105 N/m² and the air density is 1.3 kg/m³. **(a)** Calculate the wind velocity if the pressure gradient is 1.5 millibars/m from the center of a low pressure area over a distance of about 100 km. **(b)** Calculate the wind velocity if the pressure gradient is 30 millibars/m over the same distance.

Solution

(a) In this case, as the pressure gradient PG = dP/dr is small, we may neglect the centripetal acceleration term and use Eq. (11.98) to calculate the wind velocity v.

For PG = 1.5 mb/m, the wind velocity is 23.123 m/sec

$$\rho := 1.3 \cdot \frac{\text{kg}}{\text{m}^3} \qquad PG := 1.5 \cdot 10^{-3} \cdot \frac{\text{newton}}{\text{m}^3} \qquad \omega := 7.292 \cdot 10^{-5} \cdot \frac{\text{rad}}{\text{sec}}$$

$$\lambda := 20 \cdot \text{deg} \qquad v := \frac{1}{2 \cdot \omega \cdot \sin(\lambda)} \cdot \frac{1}{\rho} \cdot PG \qquad v = 23.132 \cdot \text{m} \cdot \text{sec}^{-1}$$

$$n := 10 \qquad i := 1 .. n \qquad PG_i := PG \cdot i$$

$$v_i := \frac{1}{2 \cdot \omega \cdot \sin(\lambda)} \cdot \frac{1}{\rho} \cdot PG_i \qquad v_1 = 23.132 \cdot \text{m} \cdot \text{sec}^{-1}$$

(b) In this case the pressure gradient PG = dP/dt is *not* small and we cannot neglect the centripetal acceleration term. We use Eq. (11.99) to calculate the wind velocity vc (including the effect of the centripetal force) at different pressure gradients Pg.

Using these equations we can calculate the wind velocity vc for different pressure gradients.

The graphs for (a) and (b) are as shown. It is clear that at a low PG one may neglect the effect of the centripetal force.

$$r := 100 \cdot 10^3 \cdot \text{m} \qquad PG1 := 30 \cdot 10^{-3} \cdot \frac{\text{newton}}{\text{m}^3}$$

$$PG1_i := PG1 \cdot i$$

$$vc_i := \left[(r \cdot \omega \cdot \sin(\lambda))^2 + \frac{r}{\rho} \cdot PG1_i \right]^{\frac{1}{2}}$$

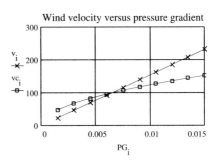

$v_1 = 23.132 \cdot m \cdot sec^{-1}$ $vc_1 = 48.103 \cdot m \cdot sec^{-1}$

$v_5 = 115.662 \cdot m \cdot sec^{-1}$ $vc_5 = 107.446 \cdot m \cdot sec^{-1}$

$v_{10} = 231.323 \cdot m \cdot sec^{-1}$ $vc_{10} = 151.931 \cdot m \cdot sec^{-1}$

(c) What is the significance of the point of the PG gradient where velocities v and vc have the same value.

(d) At what Pg will the error in predicting the hurricanes be more than 10 percent, if the wrong equation is used ?

(e) How will the results change if we have a high-pressure gradient, that is, if the center is a high pressure region?

EXERCISE 11.3 Recalculate parts (a) and (b) for a pressure gradient from a high-pressure region. What do you conclude in part (b)?

PROBLEMS

11.1. A wheel of radius a is rolling in mud with a velocity v_0 along a straight-line path and the mud is thrown as the wheel rolls. Calculate **(a)** the maximum height above the ground that the mud particles can attain, and **(b)** the point on the rim where the mud leaves the rim.

11.2. Consider a cart moving with an acceleration **a** to the right along the X-axis. A shaft OA is mounted on the cart. A rod CD of length $2L$ is pivoted on this shaft and is free to rotate in the horizontal XY plane, as shown in Fig. P11.2. Masses m_1 and m_2 are mounted at the two ends of the rod. Discuss the motion of the rod with respect to a moving coordinate system attached to the car. If the rod vibrates or oscillates, find the time period.

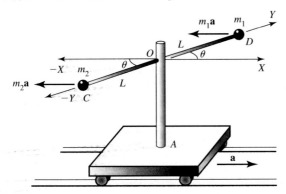

Figure P11.2

11.3. A mass m is tied to a spring of spring constant k, and the spring is fastened to a support S, as shown in Fig. P11.3. S moves with simple harmonic motion of frequency ω and amplitude A as observed from a stationary reference system $OXYZ$. If mass m also moves along the X-axis, what is the displacement and the time period as observed from the reference system attached to the support S?

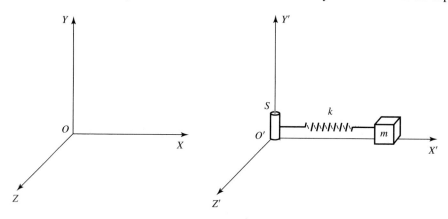

Figure P11.3

11.4. A bicycle tire of radius r is moving with a speed v_0 **(a)** in a straight line, and **(b)** in a circular path of radius R. What is the acceleration of the air valve with respect to the ground when the air valve is at the bottom?

11.5. A mass m is tied to a string of length L and hung from the ceiling of a train. Find the tension in the string and the deflection from the vertical if **(a)** the train is moving with a constant acceleration **a** on a straight track, and **(b)** if the train is going around a curve of radius R at a constant speed v.

11.6. In an unprimed coordinate system an expression called a *jerk* is defined as $\dddot{\mathbf{r}} = d^3\mathbf{r}/dt^3 = d\mathbf{a}/dt$. Find an expression for the jerk in a primed coordinate system.

11.7. We may write $\mathbf{a}_{PS} = \mathbf{a}_{PE} + \mathbf{a}_{ES}$, where $\mathbf{a}_{PE}$ is the acceleration of a point P on the surface of Earth with respect to the center of Earth and $\mathbf{a}_{ES}$ is the acceleration of Earth with respect to the Sun, both accelerations being due to their rotations (see Fig. P11.7). Show that the acceleration $\mathbf{a}_{PS}$ of the point P with respect to the Sun in small as compared to g.

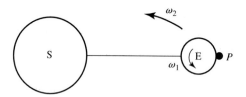

Figure P11.7

11.8. A merry-go-round is revolving with frequency ω. A boy standing at a distance d from the center O is holding a small mass m tied to a string of length L. The mass oscillates with simple harmonic motion with a small amplitude. Calculate the period of oscillations.

11.9. A rod is fixed on a turntable in a radial direction starting at the center O. A bug starts crawling from O toward A radially outward. The bug can hold onto the rod with a maximum force equal to its weight on Earth. If the turntable is rotating with an angular frequency ω, how far can the bug crawl before it loses its grip?

11.10. On a turntable with a 12-in. diameter and rotating at 33 rpm, a bug starts crawling from the center toward the rim. Assuming the coefficient of friction to be 0.2, will the bug ever reach the edge?

11.11. The effect on the weight of a body due to Earth's spin is greatest at the poles. Calculate the contribution due to the spinning of Earth to the weight of a body. Show that when measuring weight by a spring even this effect is negligible.

11.12. Show that the angular deviation ϵ of a plumb line from the true vertical at a point on Earth's surface at a latitude λ is

$$\epsilon = \frac{R_E \omega^2 \sin \lambda \cos \lambda}{g_0 - r_0 \omega^2 \cos^2 \lambda}$$

where R_E is the radius of Earth.

11.13. We showed in the text that $\mathbf{g}_e = \mathbf{g}_0 + \mathbf{a}$, where $\mathbf{g}_0$ is directed vertically down to the center of Earth, while $\mathbf{a}$ is the centrifugal acceleration. Neglecting higher-order terms in ω, show that

$$g_e = g_0 \left(1 - \frac{\omega^2 R}{g_0} \cos^2 \theta \right)$$

where R is the radius of Earth, θ is the colatitude, and ω is the angular velocity of Earth.

11.14. A small mass m attached to the midpoint of a spring is held between the end points of a track along the X'-axis (Fig. P11.14). The track is fixed on a table that rotates about a vertical axis with angular velocity ω. If T_0 is the time period when the table does not rotate, calculate the time period when the table does rotate.

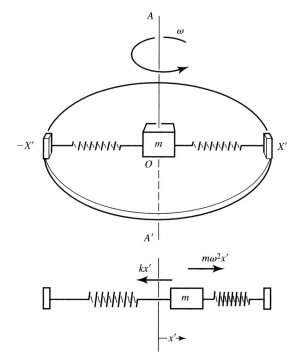

Figure P11.14

11.15. A particle's position is given by a vector $\mathbf{r} = x\hat{\mathbf{i}} + y\hat{\mathbf{j}} + z\hat{\mathbf{k}}$ that is rotating around the Z-axis with an angular velocity $\boldsymbol{\omega} \simeq \omega\hat{\mathbf{k}}$. What is the direction of the centripetal acceleration? What is the acceleration if the point lies on the Z-axis?

11.16. The equation for rotational motion about the center of mass is $d\mathbf{L}/dt = \boldsymbol{\tau}$, where $\mathbf{L}$ is the angular momentum and $\boldsymbol{\tau}$ is the torque about the point at the center of mass, which is at a distance R from the stationary origin O. What form will the preceding equation take if observed from an accelerated system?

11.17. A jet is circling in a holding pattern over an airport with a speed of 250 km/h and a radius of 6 km. Inside the jet, a mass of 0.5 kg drops from a height of 1.5 m from rest. What will be the lateral displacement of the point of impact? Graph the path of the falling object.

11.18. Find the magnitude and direction of the centripetal force on a 2-metric-ton jet sitting on the ground at 45° latitude. How does this compare with the Coriolis force acting on it when cruising at a speed of 750 km/h?

11.19. Find the magnitude of the Coriolis force on a 1200-kg automobile that is traveling north at 90 km/h at a latitude of 60° north.

11.20. A particle of mass m is dropped from rest at a height of 500 m. Where does it hit the ground?

11.21. A bug is crawling in a circular path on a rotating turntable. Show that, if to an observer watching from outside the bug seems to be stationary, the centripetal force must be equal to the Coriolis force.

11.22. A particle of mass m located in the Northern Hemisphere at latitude λ moves with a velocity $\mathbf{v}$ with components $\mathbf{v} = \mathbf{v}_e + \mathbf{v}_s + \mathbf{v}_v$, where e, s, and v stand for east, south, and vertically upward from the center of Earth. Calculate F_e, F_s, and F_v, which are the components of a Coriolis force $\mathbf{F}_{\text{Cor}}$.

11.23. A jet is flying across the North Pole at a speed of 900 km/h, following a meridian of longitude that rotates with Earth. Calculate the angle between the direction of a plumb line hanging freely in the jet as it passes over the pole and one hanging freely on the surface of Earth over the North Pole.

11.24. Repeat Problem 11.23 for the same jet flying over a point with latitude of 45°N.

11.25. Consider two children standing at the two ends of the 6-m diameter of a merry-go-round that is turning at 12 revolutions per minute. One child throws a basketball of 0.5-kg mass to the other child with a speed of 6 m/s. Calculate **(a)** the Coriolis force on the basketball while it is in the air, and **(b)** the distance by which it misses the mark due to this force.

11.26. A body is thrown upward with a velocity of 15 m/s from a point on the surface of Earth at 45°N latitude. Calculate the displacement of the point of impact due to the Coriolis force.

11.27. A bird of mass m is flying at a latitude λ with a speed of v_0. Calculate the Coriolis force on this bird as a function of the latitude. What does the bird have to do to make sure that it will fly in a straight line? Perform a numerical calculation for a typical case in which $m = 0.5$ kg, $v = 30$ kg/h, and $\lambda = 45°$N.

11.28. A satellite at a distance R from the center of Earth is in a circular orbit around Earth with an angular velocity of $\boldsymbol{\omega}$. Let the system of coordinates fixed with the satellite have its X-axis along the line joining the satellite with the center of Earth and its Y-axis along the direction of its motion. Show that a particle of mass m near the satellite will experience a force of $\mathbf{F}_{\text{inertial}} = 3m\omega^2 x \hat{\mathbf{i}} - 2m\boldsymbol{\omega} \times \mathbf{r}$.

11.29. A particle of mass m is projected from the North Pole in a path that is close to Earth. Obtain an expression for the angular deviation of this particle. What is the angular deviation for a missile that travels a distance of 1200 km in 3 min? Calculate the "miss distance" if the missile were aimed directly at the target.

11.30. Calculate the difference in the apparent values of the acceleration due to gravity at the equator and the poles. Assume Earth to be spherical in shape.

11.31. Derive the familiar expression for velocity in plane polar coordinates by examining the motion of a particle in a rotating coordinate system in which the velocity is instantaneously radial. Remember in an inertial coordinate system that $\mathbf{v} = \dot{r}\hat{\mathbf{r}} + r\dot{\theta}\hat{\boldsymbol{\theta}}$.

11.32. Assume plane polar coordinates are attached to a merry-go-round. For such coordinates, describe the motion and the acceleration of an onlooker standing nearby. Explain his acceleration as measured from the rotating frame in terms of the centrifugal and Coriolis accelerations.

11.33. A pendulum of length l is rigidity fixed on an axle AA'. The supports S and S' are attached to a platform P that is rotating with an angular velocity ω, as shown in Fig. P11.33. The pendulum can swing with a very small amplitude in a plane perpendicular to the axis AA'. Calculate the frequency and the time period of such a pendulum.

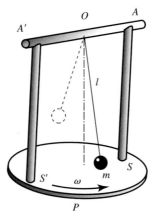

Figure P11.33

11.34. A 500-ton train at a latitude of 30°N is running at 120 km/h due south. What is the magnitude and direction of the force on the tracks?

11.35. A body is dropped from a height of 100 m above the ground at a latitude of 45°N. Find the deviation of the body from the point vertically beneath the initial position, knowing that $\omega = 7.29 \times 10^{-5}$ rad/s.

11.36. A body is thrown vertically upward to a height h and parallel to a plumb line at a north latitude of λ. Where will it strike the ground?

11.37. A tidal current is running northward along a channel of width d at a north latitude λ. Show that the height of the water on the East Coast exceeds that on the West Coast by $(2dv\omega \sin \lambda)/g$, where v is the velocity of the water and ω is the angular velocity of Earth.

11.38. A circular platform of 5-m radius rotates with an angular velocity of $\omega = 0.25$ rad/s. A child of 50-kg mass walks with constant velocity $v' = 1$ m/s along the diameter of the platform. Let $t = 0$ when he is at the center and $t = 5$ s when he jumps off the edge of the platform. Write expressions for the centripetal force and the Coriolis force felt by the child. Sketch these quantities during the time interval $t = 0$ and $t = 5$ s.

11.39. A projectile is fired due east from a point on the surface of Earth at a north latitude of λ, with a velocity v_0 and at angle of inclination α with the horizontal. Show that the lateral deflection when the projectile hits the ground is (neglect terms of ω^2)

$$\Delta x = \frac{2v_0^3}{g^2} \omega \sin 2\alpha \sin \alpha \sin \lambda$$

11.40. In Problem 11.39, if the range of the projectile is R for the case $\omega = 0$, show that the change in the range ΔR due to the rotation of Earth is (neglecting terms in ω^2)

$$\Delta R = \sqrt{\frac{2R^3}{g}} \, \omega \cos \lambda \left[\cot^{1/2} \alpha - \frac{1}{3} \tan^{3/2} \alpha \right]$$

11.41. A mass m is dropped from a small height above the surface of Earth at a north latitude of λ. The air resistance is proportional to the velocity, with the constant of proportionally being k. The time of fall of the mass is t_0. Show that when the mass strikes the ground the deviation that results as compared to the nonrotating Earth is

$$\Delta D = \frac{g\omega \cos \lambda}{3} t_0^3 \left(1 - \frac{k}{4m} t_0\right)$$

We may assume that the translational acceleration of Earth is negligible and terms of the order of k^2 are neglected.

11.42. A spherical pendulum undergoes small oscillations about a conical angle ϕ. If the pendulum is at a latitude of 45°N, for what values of ϕ will the precession owning to Earth's rotation just cancel the pendulum's natural precession?

11.43. A ball of mass m slides on a frictionless horizontal plane on Earth's surface. Show that the motion of the ball is similar to that of a Foucault pendulum provided the mass remains near the point of tangency. Find the time period and frequency of the oscillations.

11.44. A bob of mass m of a pendulum of length L is set in motion so as to swing in a circle. Calculate the contribution due to the Coriolis force; that is, show that its precessional frequency is approximately $-\omega \sin \lambda$. Note that the vertical component of the Coriolis force may be neglected because on the average it is zero.

11.45. Suppose at a place 20°N latitude the atmospheric pressure is 10^5 N/m^2 and the air density is 1.3 kg/m^3. Calculate v for a pressure gradient of 3 millibars/m over 100 km from the center of a low-pressure region. Graph the velocity, with and without centripetal acceleration, versus the pressure gradient.

11.46. Repeat Problem 11.45 if the pressure gradient is from a high-pressure region.

11.47. Repeat Problem 11.45 if the latitude is 40°N and the pressure gradient is 1.5 millibars/m.

11.48. Repeat Problem 11.45 if the latitude is 45°N and the pressure gradient is 35 millibars/m. What will happen if the center is a high-pressure region?

SUGGESTIONS FOR FURTHER READING

ARTHUR, W., and FENSTER, S. K., *Mechanics,* Chapter 9. New York: Holt, Rinehart and Winston, Inc., 1969.

BARGER, V., and OLSSON, M., *Classical Mechanics,* Chapter 6. New York: McGraw-Hill Book Co., 1973.

BECKER, R. A., *Introduction to Theoretical Mechanics,* Chapter 11. New York: McGraw-Hill Book Co., 1954.

DAVIS, A. DOUGLAS, *Classical Mechanics,* Chapter 6. New York: Academic Press, Inc., 1986.

FOWLERS, G. R., *Analytical Mechanics,* Chapter 5. New York: Holt, Rinehart and Winston, Inc., 1962.

FRENCH, A. P., *Newtonian Mechanics,* Chapter 12. New York: W. W. Norton and Co., Inc., 1971.

HAUSER, W., *Introduction to the Principles of Mechanics,* Chapter 3. Reading, Mass.: Addison-Wesley Publishing Co., 1965.

KLEPPNER, D., and KOLENKOW, R. J., *An Introduction to Mechanics,* Chapter 8. New York: McGraw-Hill Book Co., 1973.

MARION, J. B., *Classical Dynamics,* 2nd ed., Chapter 11. New York: Academic Press, Inc., 1970.

SYMON, K. R., *Mechanics,* 3rd ed., Chapter 7. Reading, Mass: Addison-Wesley Publishing Co., 1971.

TAYLOR, E. F., *Introductory Mechanics,* Chapter 9. New York: John Wiley & Sons, Inc., 1963.

Lagrangian and Hamiltonian Dynamics

12.1 INTRODUCTION

In previous chapters, we clearly demonstrated and established the importance of Newton's laws. By using *Newton's second law* and given *initial conditions*, we were able to obtain the equations of motion of a given system and describe the motion of the system. Newton's laws can be used only if all the forces acting on the system are known; that is, the *dynamical conditions* are known. Furthermore, we used rectangular coordinates, with an occasional use of polar, cylindrical, or spherical coordinates.

In most situations, the problems are not that simple to solve by means of dynamical and initial conditions, for example, a mass that is constrained to move on a spherical surface or a bead that slides on a wire. In these situations, not only the unknown form of the forces of constraints makes the problem difficult to solve, but using the rectangular or other commonly used coordinates may make it impossible to tackle the problem (even if the forces of constraints were known). Two different methods, *Lagrange's equations and Hamilton's equations*, have been developed to handle such problems. These two techniques are not the result of new theories. They are derived from Newton's second law and they offer much ease in handling very difficult problems of a physical nature. First, these techniques use *generalized coordinates.* That is, instead of being limited to the use of rectangular or polar coordinates and the like, any suitable quantity, such as velocity, linear momentum, angular momentum, or (length)^2, is used in solving problems. Such generalized coordinates are usually denoted by q_k, where q_1 may be v, q_2 may be x, q_3 may be the angle θ, and so on. Furthermore, these techniques use an energy approach, having the primary advantage of dealing with *scalars*, rather than vectors. We shall discuss these in detail in the following sections. We may briefly mention the difference between Lagrange's and Hamilton's method. In Lagrange's formalism the generalized coordinates used are position and velocity, resulting in second-order linear differential equations. In Hamilton's formalism the generalized coordinates used are position and momentum, resulting in first-order linear differential equations. These methods are not only helpful in solving the equations of motion describing the system, but also can be used to calculate the constraint and reaction forces.

12.2 GENERALIZED COORDINATES AND CONSTRAINTS

To locate the position of a particle, we need three coordinates. These coordinates could be Cartesian coordinates x, y, and z, cylindrical coordinates r, θ, and z, spherical coordinates r, θ, and ϕ, or any other three suitable coordinates. If there are some restrictions or constraints on the motion of the particle, we need less than three coordinates. For example, if a particle is constrained to move on a plane surface, only two coordinates are sufficient, while if the particle is constrained to move in a straight line, only one coordinate is sufficient to describe the motion of the particle.

Let us consider a mechanical system consisting of N particles. To specify the position of such a system at any given time, we need N vectors, while each vector can be described by three coordinates. Thus, in general, we need $3N$ coordinates to describe a given mechanical system. If there are constraints, the total number of coordinates needed to specify the system will be reduced. As an example, suppose the system is a rigid body, and as we know, the distances between different particles are fixed. These fixed distances can be expressed in the form of equations. As we explained in Chapter 9, a rigid body can be completely described by only six coordinates; that is, only six coordinates are needed to specify the configuration of a rigid body system. Of these six, three coordinates give the position of some convenient reference point in the body, usually the center of mass with respect to the origin of some chosen coordinate system, and the remaining three coordinates describe the orientation of the body in space.

We are interested in finding the minimum number of coordinates needed to describe a system of N particles. Usually, the constraints on any given system are described by means of equations. Suppose there are m number of such equations that describe the constraints. The minimum number of coordinates, n, needed to completely describe the motion or the configuration of such a system at any given time is given by

$$n = 3N - m \tag{12.1}$$

where n is the number of *degrees of freedom* of the system. It is not necessary that these n coordinates should be rectangular, cylindrical, or any other curvilinear coordinates. As a matter of fact, n could be any parameter, such as length, (length)^2, angle, energy, a dimensionless quantity, or any other quantity, as long as it completely describes the configuration of the system. The name *generalized coordinates* is given to *any set of quantities that completely describes the state or configuration of a system*. These n generalized coordinates are customarily written as

$$q_1, q_2, q_3, \ldots, q_n \tag{12.2a}$$

or　　　　　　　　　　q_k,　where　$k = 1, 2, 3, \ldots, n$ $\tag{12.2b}$

These n generalized coordinates are not restricted by any constraints. If each coordinate can vary independently of the other, the system is said to be *holonomic*. In a *nonholonomic* system, the coordinates *cannot* vary independently. Hence in such systems the number of degrees of freedom is less than the minimum number of coordinates needed to specify the configuration of the system. As an example, a sphere constrained to roll on a perfectly rough plane surface needs only five coordinates to specify its configuration, two for the position of its center of mass and three for its orientation. But these five coordinates cannot all vary independently. When the sphere rolls, at least two coordinates must change. Hence this is a nonholonomic system. The investigation and description of nonholonomic systems are involved and will not be considered here. We shall limit ourselves to the discussion of holonomic systems for the time being.

A suitable set of generalized coordinates of a system is that which results in equations of motion leading to any easy interpretation of the motion. These q_n generalized coordinates form a configuration space, with each dimension represented by a coordinate q_k. The path of the system is represented by a curve in this configuration space. The path in the configuration space does not lend itself to the same interpretation as a path in ordinary three-dimensional space. In analogy with Cartesian coordinates, we may define the derivatives of q_k, that is $\dot{q}_1, \dot{q}_2, \ldots,$ or $\dot{q}_k$ as *generalized velocities*.

Let us consider a single particle whose rectangular coordinates x, y, and z are a function of the generalized coordinates q_1, q_2, and q_3; that is

$$x = x(q_1, q_2, q_3) = x(\mathbf{q}_k)$$

$$y = y(q_1, q_2, q_3) = y(\mathbf{q}_k)$$

$$z = z(q_1, q_2, q_3) = z(\mathbf{q}_k) \tag{12.3}$$

Suppose the system changes from an initial configuration given by (q_1, q_2, q_3) to a neighborhood configuration given by $(q_1 + \delta q_1, q_2 + \delta q_2, q_3 + \delta q_3)$. We can express the corresponding changes in the Cartesian coordinates by the following relations:

$$\delta x = \frac{\partial x}{\partial q_1} \delta q_1 + \frac{\partial x}{\partial q_2} \delta q_2 + \frac{\partial x}{\partial q_3} \delta q_3 = \sum_{k=1}^{n} \frac{\partial x}{\partial q_k} \delta q_k \tag{12.4}$$

with similar expression for δy and δz, where n is equal to three and the partial derivatives $\partial x/\partial q_k, \ldots,$ are functions of q's. The value of n depends on the degrees of freedom. For example, if there were no constraints, $m = 0$, and from Eq. (12.1) for $N = 1$, $n = 3$, as we have used above, n would be less than 3 if there were constraints on the system.

Let us consider a more general case in which a mechanical system consists of a large number of particles having n degrees of freedom. The configuration of the system is specified by the generalized coordinates $q_1, q_2, \ldots, q_n$. Suppose the configuration of the system changes from $(q_1, q_2, \ldots, q_n)$ to a new configuration $(q_1 + \delta q_1, q_2 + \delta q_2, \ldots, q_n + \delta q_n)$. The Cartesian coordinates of a particle i change from (x_i, y_i, z_i) to $(x_i + \delta x_i, y_i + \delta y_i, z_i + \delta z_i)$. This displacements δx_i, δy_i, and δz_i can be expressed in terms of the generalized coordinates q_k as

$$\delta x_i = \frac{\partial x_i}{\partial q_1} \delta q_1 + \frac{\partial x_i}{\partial q_2} \delta q_2 + \cdots + \frac{\partial x_i}{\partial q_k} \delta q_k = \sum_{k=1}^{n} \frac{\partial x_i}{\partial q_k} \delta q_k \tag{12.5}$$

with similar expression for δy_i and δz_i. Once again the partial derivatives are functions of the generalized coordinates q_k.

It is essential at this point to distinguish between two types of displacements: an actual displacement $d\mathbf{r}_i$ and a virtual (not in actual fact or name) displacement $\delta\mathbf{r}_i$. Suppose a mass m_i is acted on by an external force $\mathbf{F}_i$ and causes the mass m_i to move from $\mathbf{r}_i$ to $\mathbf{r}_i + d\mathbf{r}_i$ in a time interval dt. This displacement must be consistent with both the equations of motion and the equations of constraints that describe this mass system; hence such displacements are *actual displacements*. On the other hand, *virtual displacements* are consistent with the equations of the constraints but do not satisfy the equations of motion or time. For example, the bob of a pendulum of length l may be moved from (l, θ) to $(l, \theta + \delta\theta)$ in any arbitrary time interval as long as the bob remains on the arc of a circle of radius l. Thus $\delta\mathbf{r}_i$ and δq_i are the virtual displacements. We shall make use of the principle of virtual work in the following. We shall cause a virtual displacement $\delta\mathbf{r}$, resulting in virtual work δW. Basically, in such displacements, the relative orientations and distances between the particles remain unchanged.

12.3 GENERALIZED FORCES

Single Particle

Consider a force $\mathbf{F}$ that is acting on a single particle of mass m and produces a virtual displacement $\delta\mathbf{r}$ of the particle. The work done δW by this force is given by

$$\delta W = \mathbf{F} \cdot \delta\mathbf{r} = F_x\,\delta x + F_y\,\delta y + F_z\,\delta z \tag{12.6}$$

where F_x, F_y, and F_z are the rectangular components of $\mathbf{F}$. We can express the displacements δx, δy, and δz in terms of the generalized coordinates q_k. Making use of Eqs. (12.4) and (12.6), we may write

$$\delta W = \sum_{k=1}^{n}\left(F_x\frac{\partial x}{\partial q_k} + F_y\frac{\partial y}{\partial q_k} + F_z\frac{\partial z}{\partial q_k}\right)\delta q_k = \sum_{k=1}^{n} Q_k\,\delta q_k \tag{12.7}$$

where

$$Q_k = F_x\frac{\partial x}{\partial q_k} + F_y\frac{\partial y}{\partial q_k} + F_z\frac{\partial z}{\partial q_k} \tag{12.8}$$

Q_k is called the *generalized force* associated wth the generalized coordinate q_k. The dimensions of Q_k depend on the dimensions of q_k. The dimensions of $Q_k\,\delta q_k$ are that of work. If the increment δq_k has the dimensions of distance, Q_k will have the dimensions of force; if δq_k has the dimensions of angle θ, Q_k will have dmensions of torque τ_θ. It may be pointed out that the quantity δq_k and the quantities δx, δy, and δz are called *virtual displacements* of the system because it is not necessary that such displacements represent any actual displacements.

A System of Particles

Let us apply the preceding ideas to a general case of a system consisting of N particles acted on by forces $\mathbf{F}_i$ ($i = 1, 2, \ldots, N$). The total work done δW for a virtual displacement $\delta\mathbf{r}_i$ of the system is

$$\delta W = \sum_{i=1}^{N}\mathbf{F}_i \cdot \delta\mathbf{r}_i = \sum_{i=1}^{N} F_{x_i}\,\delta x_i + F_{y_i}\,\delta y_i + F_{z_i}\,\delta z_i \tag{12.9}$$

Once again, expressing the virtual displacements in terms of the generalized coordinates, using Eq. (12.5), we get

$$\delta W = \sum_{i=1}^{N}\left[\sum_{k=1}^{n}\left(F_{x_i}\frac{\partial x_i}{\partial q_k} + F_{y_i}\frac{\partial y_i}{\partial q_k} + F_{z_i}\frac{\partial z_i}{\partial q_k}\right)\delta q_k\right] \tag{12.10a}$$

Interchanging the order of summation, we get

$$\delta W = \sum_{k=1}^{n}\left[\sum_{i=1}^{N}\left(F_{x_i}\frac{\partial x_i}{\partial q_k} + F_{y_i}\frac{\partial y_i}{\partial q_k} + F_{z_i}\frac{\partial z_i}{\partial q_k}\right)\delta q_k\right] \tag{12.10b}$$

or

$$\delta W = \sum_{k=1}^{n} Q_k\,\delta q_k \tag{12.11}$$

where

$$Q_k = \sum_{i=1}^{N}\left(F_{x_i}\frac{\partial x_i}{\partial q_k} + F_{y_i}\frac{\partial y_i}{\partial q_k} + F_{z_i}\frac{\partial z_i}{\partial q_k}\right) \tag{12.12}$$

Q_k is called the *generalized force* associated with the generalized coordinate q_k. Once again, the dimensions of the generalized force Q_k depend on the dimensions of q_k but the product $Q_k q_k$ is always work.

Conservative Systems

Let us write an expression for the generalized forces that are conservative. Suppose a conservative force field is represented by a potential function $V = V(x, y, z)$. The rectangular components of a force acting on a particle are given by

$$F_x = -\frac{\partial V}{\partial x}, \qquad F_y = -\frac{\partial V}{\partial y}, \qquad F_z = -\frac{\partial V}{\partial z} \tag{12.13}$$

Expression Q_k for a generalized force given by Eq. (12.8) becomes

$$\begin{aligned}
Q_k &= F_x\frac{\partial x}{\partial q_k} + F_y\frac{\partial y}{\partial q_k} + F_z\frac{\partial z}{\partial q_k} \\
&= -\left(\frac{\partial V}{\partial x}\frac{\partial x}{\partial q_k} + \frac{\partial V}{\partial y}\frac{\partial y}{\partial q_k} + \frac{\partial V}{\partial z}\frac{\partial z}{\partial q_k}\right)
\end{aligned} \tag{12.8}$$

The expression in the parentheses is the partial derivative of the function V with respect to q_k. That is,

$$Q_k = -\frac{\partial V}{\partial q_k} \tag{12.14}$$

This expresses the relation between a generalized force and the potential representing a conservative system.

Example 12.1 _____

Consider the motion of a particle of mass m moving in a plane. Using the plane polar coordinates (r, θ) as generalized coordinates, calculate (**a**) the displacement δx and δy, and (**b**) the generalized forces for a particle acted on by a force $\mathbf{F} = \hat{\mathbf{i}}F_x + \hat{\mathbf{j}}F_y + \hat{\mathbf{k}}F_z$.

Solution

Since the plane polar coordinates (r, θ) are the generalized coordinates

$$q_1 = r \quad \text{and} \quad q_2 = \theta$$

$$x = x(r, \theta) = r \cos \theta \quad \frac{\partial x}{\partial r} = \cos \theta, \quad \frac{\partial x}{\partial \theta} = -r \sin \theta \tag{i}$$

$$y = y(r, \theta) = r \sin \theta \quad \frac{\partial y}{\partial r} = \sin \theta, \quad \frac{\partial y}{\partial \theta} = r \cos \theta \tag{ii}$$

(**a**) The changes in the Cartesian coordinates are

$$\delta x = \frac{\partial x}{\partial r}\, \delta r + \frac{\partial x}{\partial \theta}\, \delta \theta = \cos \theta\, \delta r - r \sin \theta\, \delta \theta \tag{iii}$$

$$\delta y = \frac{\partial y}{\partial r}\, \delta r + \frac{\partial y}{\partial \theta}\, \delta \theta = \sin \theta\, \delta r + r \cos \theta\, \delta \theta \tag{iv}$$

(**b**) From the definition of generalized forces,

$$Q_k = F_x \frac{\partial x}{\partial q_k} + F_y \frac{\partial y}{\partial q_k} + F_z \frac{\partial z}{\partial q_k}$$

we get

$$Q_r = F_x \frac{\partial x}{\partial r} + F_y \frac{\partial y}{\partial r} = F_x \cos \theta + F_y \sin \theta = F_r \tag{v}$$

$$Q_\theta = F_x \frac{\partial x}{\partial \theta} + F_y \frac{\partial y}{\partial \theta} = -rF_x \sin \theta + rF_y \cos \theta$$

$$= r(-F_x \sin \theta + F_y \cos \theta) = rF_\theta \tag{vi}$$

EXERCISE 12.1 Consider the motion of a particle of mass m moving in space. Using the generalized coordinates (r, θ, z), calculate (**a**) the displacement δx, δy, and δz, and (**b**) the generalized forces for the particle acted on by a force $\mathbf{F} = \hat{\mathbf{i}}F_x + \hat{\mathbf{j}}F_y + \hat{\mathbf{k}}F_z$.

12.4 LAGRANGE'S EQUATIONS OF MOTION FOR A SINGLE PARTICLE

We are interested in describing the motion of a single particle by means of equations written in terms of generalized coordinates. This leads us to Lagrange's equations. We could start with Newton's second law, $\mathbf{F} = m\mathbf{a}$. But it is convenient to start with an expression for kinetic energy

T in terms of Cartesian coordinates and then write T in terms of generalized coordinates. (Note that we are using T instead of K for kinetic energy.) Let x, y, and z be Cartesian coordinates, while $q_1, q_2, \ldots, q_n$ are generalized coordinates. The kinetic energy of the particle in Cartesian coordinates is

$$T = \tfrac{1}{2}m(\dot{x}^2 + \dot{y}^2 + \dot{z}^2) \tag{12.15}$$

since

$$x = x(q_1, q_2, \ldots, q_n) = x(q) \tag{12.16}$$

similarly

$$y = y(q), \quad z = z(q) \tag{12.17}$$

we can evaluate $\dot{x}$ in terms of q_k by the following procedure:

$$\dot{x} = \frac{\partial x}{\partial q_1}\frac{\partial q_1}{\partial t} + \frac{\partial x}{\partial q_2}\frac{\partial q_2}{\partial t} + \cdots + \frac{\partial x}{\partial q_n}\frac{\partial q_n}{\partial t}$$

$$= \sum_{k=1}^{n} \frac{\partial x}{\partial q_k}\frac{\partial q_k}{\partial t} = \sum_{k=1}^{n} \frac{\partial x}{\partial q_k}\dot{q}_k = \dot{x}(q, \dot{q}) \tag{12.18}$$

Thus we can describe the different components of velocity in terms of the generalized coordinates q_k and generalized velocities $\dot{q}_k$; that is,

$$\dot{x} = \dot{x}(q, \dot{q}), \quad \dot{y} = \dot{y}(q, \dot{q}), \quad \dot{z} = \dot{z}(q, \dot{q}) \tag{12.19}$$

We may now write Eq. (12.15) for kinetic energy as

$$T = \tfrac{1}{2}m[\dot{x}^2(q, \dot{q}) + \dot{y}^2(q, \dot{q}) + \dot{z}^2(q, \dot{q})] \tag{12.20}$$

Taking the derivative with respect to the generalized velocity $\dot{q}_k$,

$$\frac{\partial T}{\partial \dot{q}_k} = m\left(\dot{x}\frac{\partial \dot{x}}{\partial \dot{q}_k} + \dot{y}\frac{\partial \dot{y}}{\partial \dot{q}_k} + \dot{z}\frac{\partial \dot{z}}{\partial \dot{q}_k}\right) \tag{12.21}$$

Using Eq. (12.18), we may write

$$\frac{\partial \dot{x}}{\partial \dot{q}_k} = \frac{\partial x}{\partial q_k} \tag{12.22}$$

Note that $\partial x/\partial q_k$ is the coefficient of $\dot{q}_k$ in the expression of $\dot{x}$ in Eq. (12.18). Substituting this and similar expressions for other terms in Eq. (12.21),

$$\frac{\partial T}{\partial \dot{q}_k} = m\left(\dot{x}\frac{\partial x}{\partial q_k} + \dot{y}\frac{\partial x}{\partial q_k} + \dot{z}\frac{\partial z}{\partial q_k}\right) \tag{12.23}$$

Now differentiate both sides of this equation with respect to t:

$$\frac{d}{dt}\left(\frac{\partial T}{\partial \dot{q}_k}\right) = m\ddot{x}\frac{\partial x}{\partial q_k} + m\ddot{y}\frac{\partial y}{\partial q_k} + m\ddot{z}\frac{\partial z}{\partial q_k} + m\dot{x}\frac{d}{dt}\left(\frac{\partial x}{\partial q_k}\right) + m\dot{y}\frac{d}{dt}\left(\frac{\partial y}{\partial q_k}\right)$$

$$+ m\dot{z}\frac{d}{dt}\left(\frac{\partial z}{\partial q_k}\right) \tag{12.24}$$

To simplify the last three terms on the right side, we use the fact that d/dt and $\partial/\partial q_k$ are interchangeable.

$$\frac{d}{dt}\left(\frac{\partial x}{\partial q_k}\right) = \frac{\partial}{\partial q_k}\left(\frac{dx}{dt}\right) = \frac{\partial \dot{x}}{\partial q_k} \tag{12.25}$$

Thus the fourth term on the right of Eq. (12.24) may be written as

$$m\dot{x}\frac{d}{dt}\left(\frac{\partial x}{\partial q_k}\right) = m\dot{x}\frac{\partial \dot{x}}{\partial q_k} = \frac{\partial}{\partial q_k}\left(\frac{1}{2}m\dot{x}^2\right) \tag{12.26}$$

with similar expressions for other terms. Also note that

$$F_x = m\ddot{x}, \quad F_y = m\ddot{y}, \quad F_z = m\ddot{z} \tag{12.27}$$

Combining Eqs. (12.25) and (12.26) with Eq. (12.24), we obtain

$$\frac{d}{dt}\left(\frac{\partial T}{\partial \dot{q}_k}\right) = F_x\frac{\partial x}{\partial q_k} + F_y\frac{\partial y}{\partial q_k} + F_z\frac{\partial z}{\partial q_k} + \frac{\partial}{\partial q_k}\left[\frac{1}{2}m(\dot{x}^2 + \dot{y}^2 + \dot{z}^2)\right] \tag{12.28}$$

Using the definition of generalized force and kinetic energy given by Eqs. (12.8) and (12.20),

$$Q_k = F_x\frac{\partial x}{\partial q_k} + F_y\frac{\partial y}{\partial q_k} + F_z\frac{\partial z}{\partial q_k} \tag{12.8}$$

$$T = \tfrac{1}{2}m[\dot{x}^2(q, \dot{q}) + \dot{y}^2(q, \dot{q}) + \dot{z}^2(q, \dot{q})] \tag{12.20}$$

in Eq. (12.28) gives

$$\frac{d}{dt}\left(\frac{\partial T}{\partial \dot{q}_k}\right) = Q_k + \frac{\partial T}{\partial q_k} \tag{12.29}$$

These differential equations in generalized coordinates describe the motion of a particle and are known as *Lagrange's equations* of motion.

Lagrange's equations take a much simpler form if the motion is in a conservative force field so that

$$Q_k = -\frac{\partial V}{\partial q_k} \tag{12.30}$$

which on substituting in Eq. (12.29) yields

$$\frac{d}{dt}\left(\frac{\partial T}{\partial \dot{q}_k}\right) = \frac{\partial T}{\partial q_k} - \frac{\partial V}{\partial q_k} \tag{12.31}$$

Let us define a *Lagrangian function L* as the difference between the kinetic energy and potential energy; that is,

$$L = T - V \quad \text{or} \quad L(q, \dot{q}) = T(q, \dot{q}) - V(q) \tag{12.32}$$

It is important to know that, if V is a function of the generalized coordinates and not of the generalized velocities, then

$$V = V(q) \quad \text{and} \quad \frac{\partial V}{\partial \dot{q}_k} = 0 \tag{12.33}$$

[If V is not independent of velocity $\dot{q}$, then $V = V(q, \dot{q})$ will lead to a tensor force, which we will not discuss here.] Thus we may write

$$\frac{\partial L}{\partial \dot{q}_k} = \frac{\partial}{\partial \dot{q}_k}(T - V) = \frac{\partial T}{\partial \dot{q}_k}$$

$$\frac{\partial L}{\partial q_k} = \frac{\partial}{\partial q_k}(T - V) = \frac{\partial T}{\partial q_k} - \frac{\partial V}{\partial q_k}$$

Substituting these results in Eq. (12.31) yields

$$\frac{d}{dt}\left(\frac{\partial L}{\partial \dot{q}_k}\right) - \frac{\partial L}{\partial q_k} = 0 \tag{12.34}$$

which are *Lagrange's equations describing the motion of a particle in a conservative force field.* To solve these equations, we must know the Lagrangian function L in the appropriate generalized coordinates. Since energy is a scalar quantity, the Lagrangian L is a scalar function. Thus the Lagrangian L will be invariant with respect to coordinate transformations. This means that *the Lagrangian gives the same description of the system under given conditions no matter which generalized coordinates are used.* Thus Eq. (12.34) describes the motion of a particle moving in a conservative force field in terms of any generalized coordinates.

 Example 12.2 _____

Consider a particle of mass m moving in a plane and subject to an inverse-square attractive force. Find the equations of motion and expressions for the generalized forces.

Solution

Let the plane polar coordinates (r, θ) be the generalized coordinates to be used in this problem. The polar coordinates (r, θ) and the Cartesian coordinates (x, y) are related by

$$x = r \cos \theta \quad \text{and} \quad y = r \sin \theta \tag{i}$$

Using these relations, we obtain the following expressions for the kinetic and potential energy:

$$T = \tfrac{1}{2}mv^2 = \tfrac{1}{2}m(\dot{x}^2 + \dot{y}^2) = \tfrac{1}{2}m(\dot{r}^2 + r^2\dot{\theta}^2) \tag{ii}$$

$$V = -\frac{k}{(x^2 + y^2)^{1/2}} = -\frac{k}{r} \tag{iii}$$

Thus the Lagrangian in the coordinates (r, θ) is

$$L = T - V = \frac{1}{2}m(\dot{r}^2 + r^2\dot{\theta}^2) + \frac{k}{r} \tag{iv}$$

In Lagrange's equations,

$$\frac{d}{dt}\left(\frac{\partial L}{\partial \dot{q}_k}\right) - \frac{\partial L}{\partial q_k} = 0$$

let us substitute $q_1 = r$ and $q_2 = \theta$ so that

$$\frac{d}{dt}\left(\frac{\partial L}{\partial \dot{r}}\right) - \frac{\partial L}{\partial r} = 0 \qquad\qquad \textbf{(v)}$$

and

$$\frac{d}{dt}\left(\frac{\partial L}{\partial \dot{\theta}}\right) - \frac{\partial L}{\partial \theta} = 0 \qquad\qquad \textbf{(vi)}$$

From Eq. (iv),

$$\frac{\partial L}{\partial \dot{r}} = m\dot{r}, \qquad \frac{d}{dt}\left(\frac{\partial L}{\partial \dot{r}}\right) = m\ddot{r}, \qquad \text{and} \qquad \frac{\partial L}{\partial r} = mr\dot{\theta}^2 - \frac{k}{r^2}$$

Substituting these in Eq. (v), we obtain

$$m\ddot{r} - mr\dot{\theta}^2 = -\frac{k}{r^2} \qquad\qquad \textbf{(vii)}$$

Since the particle is moving in a conservative field, we may write

$$F(r) = -\frac{\partial V(r)}{\partial r} = -\frac{\partial}{\partial r}\left(-\frac{k}{r}\right) = -\frac{k}{r^2} \qquad\qquad \textbf{(viii)}$$

and Eq. (vii) takes the form, $F(r) = F_r$.

$$m\ddot{r} = mr\dot{\theta}^2 + F_r \qquad\qquad \textbf{(ix)}$$

Once again from Eq. (iv)

$$\frac{\partial L}{\partial \dot{\theta}} = mr^2\dot{\theta}, \qquad \frac{\partial L}{\partial \theta} = 0, \qquad \text{and} \qquad \frac{d}{dt}\left(\frac{\partial L}{\partial \dot{\theta}}\right) = 2mr\dot{r}\dot{\theta} + mr^2\ddot{\theta}$$

Hence Lagrange's Equation [Eq. (vi)] takes the form

$$2mr\dot{r}\dot{\theta} + mr^2\ddot{\theta} = 0 \qquad\qquad \textbf{(x)}$$

or

$$\frac{d}{dt}(mr^2\dot{\theta}) = \frac{dJ}{dt} = 0 \qquad\qquad \textbf{(xi)}$$

where J, which may be identified as the angular momentum, is constant. That is, the integration of Eq. (xi) yields

$$J = mr^2\dot{\theta} = \text{constant} \qquad\qquad \textbf{(xii)}$$

Thus we may conclude that in a conservative force field the angular momentum **J** is a constant of motion. Also, as from the previous example,

$$Q_r = F_r \quad \text{and} \quad Q_\theta = rF_\theta$$

we may arrive at the following using Eq. (12.33),

$$\frac{d}{dt}\left(\frac{\partial T}{\partial \dot{r}}\right) - \frac{\partial T}{\partial r} = Q_r = F_r = m\ddot{r} - mr\dot{\theta}^2 \tag{xiii}$$

$$\frac{d}{dt}\left(\frac{\partial T}{\partial \dot{\theta}}\right) - \frac{\partial T}{\partial \theta} = rF_\theta = \frac{d}{dt}(mr^2\dot{\theta}) \tag{xiv}$$

That is,

$$F_r = m\ddot{r} - mr\dot{\theta}^2$$

and

$$Q_\theta = \tau = rF_\theta = \frac{d}{dt}(mr^2\dot{\theta}) = \frac{dJ}{dt} = 0$$

where $Q_\theta = \tau$ is the torque and is equal to zero.

EXERCISE 12.2 Repeat the example for the case of a repulsive inverse-square force. How does the situation in this exercise differ from the one in the example?

 Example 12.3 _____

Consider an Atwood machine consisting of a single pulley of moment of inertia I about an axis through its center and perpendicular to its plane. The length of the inextensible string connecting the two masses and going over the pulley is l. Calculate the acceleration of the system.

Solution

Let x be the variable vertical distance from the pulley to the mass m_1, while the mass m_2 is at a distance $l - x$ from the pulley, as shown in Fig. Ex. 12.3. Thus there is only one degree of freedom x that represents the configuration of the system. The velocity of the two masses and the angular velocity of the disk may be written as

$$v_1 = \frac{dx}{dt} = \dot{x}, \qquad v_2 = \frac{d(l - x)}{dt} = -\dot{x} \tag{i}$$

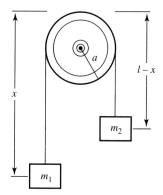

Figure Ex. 12.3

and
$$\omega = \frac{v}{a}, \quad \text{where } v = |v_1| = |v_2| = \dot{x} \tag{ii}$$

Therefore, the total kinetic energy of the system is

$$T = \frac{1}{2}m_1v_1^2 + \frac{1}{2}m_2v_2^2 + \frac{1}{2}I\omega^2 = \frac{1}{2}m_1\dot{x}^2 + \frac{1}{2}m_2\dot{x}^2 + \frac{1}{2}I\frac{\dot{x}^2}{a^2} \tag{iii}$$

while the potential energy of the system is

$$V = -m_1gx - m_2g(l - x) \tag{iv}$$

The Lagrangian of the system is

$$L = T - V = \frac{1}{2}\left(m_1 + m_2 + \frac{1}{a^2}\right)\dot{x}^2 + g(m_1 - m_2)x + m_2gl \tag{v}$$

There is only one degree of freedom, which is the generalized coordinate $q = x$. Lagrange's equation is

$$\frac{d}{dt}\left(\frac{\partial L}{\partial \dot{x}}\right) - \frac{\partial L}{\partial x} = 0 \tag{vi}$$

From Eq. (v)

$$\frac{\partial L}{\partial \dot{x}} = \left(m_1 + m_2 + \frac{I}{a^2}\right)\dot{x}$$

$$\frac{d}{dt}\left(\frac{\partial L}{\partial \dot{x}}\right) = \left(m_1 + m_2 + \frac{I}{a^2}\right)\ddot{x} \quad \text{and} \quad \frac{\partial L}{\partial x} = g(m_1 - m_2)$$

Substituting these in Eq. (vi), Lagrange's equation takes the form

$$\left(m_1 + m_2 + \frac{I}{a^2}\right)\ddot{x} = g(m_1 - m_2)$$

$$\ddot{x} = \frac{g(m_1 - m_2)}{(m_1 + m_2 + I/a^2)} \tag{vii}$$

If $I = 0$, $\ddot{x} = [(m_1 - m_2)/(m_1 + m_2)]g$.
If $m_1 > m_2$, mass m_1 descends with a constant acceleration.
If $m_1 < m_2$, mass m_1 ascends with a constant acceleration.

EXERCISE 12.3 Consider a double Atwood machine, as shown in Fig. Exer. 12.3. Assuming frictionless pulleys, that is, $I_1 = I_2 = 0$, calculate the accelerations of the masses. Assume two degrees of freedom x_1 and x_2, as shown.

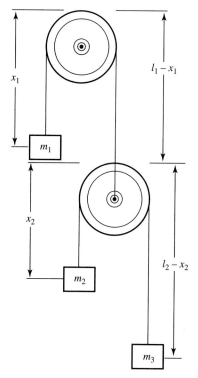

Figure Exer. 12.3

Example 12.4

Consider Atwood's machine discussed in Example 12.3. Assume that the pulley is frictionless, and calculate the tension S in the string, as shown in Fig. Ex. 12.4.

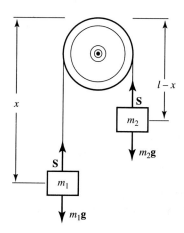

Figure Ex. 12.4

Solution

In Example 12.3, while discussing Atwood's machine, we were interested only in the motion of the system; hence the coordinate l was constrained to have a constant value. To find the tension in the string, the length l must be included as a coordinate. Since the pulley is frictionless, there is no rotational kinetic energy. Hence the expression for the kinetic energy is given by

$$T = \frac{1}{2} m_1 \left(\frac{dx}{dt} \right)^2 + \frac{1}{2} m_2 \left[\frac{d}{dt} (l - x) \right]^2 = \frac{1}{2} m_1 \dot{x}^2 + \frac{1}{2} m_2 (\dot{l} - \dot{x})^2 \qquad \text{(i)}$$

The two forces acting on the system are the tension S in the string and the force of gravity, as shown. The work done when x increase to $x + \delta x$, while l remains constant, is

$$\delta W = (m_1 g - S) \, \delta x + (m_2 g - S) \, \delta(l - x)$$

$$= (m_1 g - S) \, \delta x + (m_2 g - S)(- \delta x)$$

$$= (m_1 - m_2)g \, \delta x$$

Comparing with

$$\delta W = Q_x \, \delta x$$

we get
$$Q_x = (m_1 - m_2)g \qquad \text{(ii)}$$

The work done when l increases to $l + \delta l$, while x remains constant, is

$$\delta W = (m_1 g - S) \, \delta x + (m_2 g - S) \, \delta(l - x)$$

$$= 0 + (m_2 g - S) \, \delta l$$

Comparing with

$$\delta W = Q_l \, \delta l$$

we get
$$Q_l = m_2 g - S \qquad \text{(iii)}$$

Note that the generalized force Q_x does not contain S, while Q_l depends on S. To solve for S, we must solve the following two Lagrange equations.

Lagrangian equations for the coordinates x and l, given by Eq. (12.29), and the generalized

forces given by Eqs. (ii) and (iii), are

$$\frac{d}{dt}\frac{d}{dvx}T - \frac{d}{dx}T = (m1 - m2) \cdot g \qquad \text{(iv)} \qquad \frac{d}{dt}\left[\left(\frac{d}{dvl} T \right) - \frac{d}{dl} T = (m2 \cdot g - S) \right] \qquad \text{(v)}$$

We solve these equations by substituting for T and simplifying

$$\left[\frac{d}{dt}(m1 \cdot vx - m2 \cdot (vl - vx)) \right] = (m1 - m2) \cdot g \qquad \frac{d}{dt} m2 \cdot (vl - vx) = (m2 \cdot g - S)$$

Substituting for v1 = 0 and a1 = 0, we get the two resulting equations

$$m1 \cdot ax + m2 \cdot ax = (m1 - m2) \cdot g \qquad \text{(vi)} \qquad - m2 \cdot ax = m2 \cdot g - S \qquad \text{(vii)}$$

We solve Eqs. (vi) and (vii) for S and ax, as shown below.

Given

$$m1 \cdot ax + m2 \cdot ax = (m1 - m2) \cdot g$$

$$- m2 \cdot ax = m2 \cdot g - S$$

$$\text{Find}(S, ax) \rightarrow \begin{bmatrix} 2 \cdot m2 \cdot g \cdot \dfrac{m1}{(m1 + m2)} \\[2ex] -g \cdot \dfrac{(-m1 + m2)}{(m1 + m2)} \end{bmatrix}$$

$$S = 2 \cdot m2 \cdot g \cdot \frac{m1}{(m1 + m2)} \quad \text{(viii)} \qquad ax = (m1 - m2) \cdot \frac{g}{(m1 + m2)} \quad \text{(ix)}$$

The value of S is given by Eq. (viii), while the value of x is calculated from Eq. (ix), as shown below.

Let ax = (vf–v0)/t1, and using this in Eq. (ix) we get

$$vf - v0 = g \cdot \left[(m1 - m2) \cdot \frac{t1}{(m1 + m2)} \right]$$

Integrating from x0 to x and t = 0 to t0 = t

$$\int_{x0}^{x} 1 \, dx1 = \int_{0}^{t} v0 + g \cdot (m1 - m2) \cdot \frac{t1}{(m1 + m2)} \, dt1$$

We get the displacement x as

$$x - x0 = t \cdot v0 + \frac{1}{2} \cdot \frac{(m1 - m2)}{(m1 + m2)} \cdot t^2 \cdot g$$

EXERCISE 12.4 Repeat the example assuming that the pulley is not frictionless and that it has a moment of inertia *I*.

12.5 LAGRANGE'S EQUATIONS OF MOTION FOR A SYSTEM OF PARTICLES

Let us extend the procedure outlined in the previous section to a more general case consisting of *N* particles. Thus the kinetic energy of such a system is

$$T = \sum_{i=1}^{N} \left[\frac{1}{2} m_i (\dot{x}_i^2 + \dot{y}_i^2 + \dot{z}_i^2) \right] \tag{12.35}$$

Instead of using the coordinates *x*, *y*, and *z*, we represent the Cartesian coordinates by x_i. Since each particle has three degrees of freedom, the total number of x_i coordinates needed to represent *N* particles will be 3*N*. Hence we may write the kinetic energy of the system as

$$T = \sum_{i=1}^{3N} \frac{1}{2} m_i \dot{x}_i^2 \tag{12.36}$$

where the Cartesian coordinates x_i are functions of the generalized coordinates q_k. It is possible that the relationship between x_i and q_k may involve time explicitly. Hence, we may write

$$x_i = x_i(q_1, q_2, \ldots, q_n, t) = x_i(q, t) \tag{12.37}$$

$$\frac{dx_i}{dt} = \dot{x}_i = \frac{\partial x_i}{\partial q_1} \dot{q}_1 + \frac{\partial x_i}{\partial q_2} \dot{q}_2 + \cdots + \frac{\partial x_i}{\partial q_n} \dot{q}_n + \frac{\partial x_i}{\partial t}$$

That is,

$$\dot{x}_i = \sum_k \frac{\partial x_i}{\partial q_k} \dot{q}_k + \frac{\partial x_i}{\partial t} \tag{12.38}$$

where $i = 1, 2, \ldots, 3N,$ N being the number of particles in the system
$k = 1, 2, \ldots, n,$ n being the generalized coordinates or degrees of freedom of the system

From Eq. (12.38), we may conclude that T is a function of the generalized coordinates q_k, of the generalized velocities $\dot{q}_k$, and of time t; thus

$$T = T(q, \dot{q}, t) \tag{12.39}$$

Differentiating T with respect to $\dot{q}$, we get

$$\frac{\partial T}{\partial \dot{q}_k} = \frac{\partial}{\partial \dot{q}_k} \left(\sum_{i=1}^{3N} \frac{1}{2} m_i \dot{x}_i^2 \right) = \sum_i \frac{\partial}{\partial \dot{q}_k} \left(\frac{1}{2} m_i \dot{x}_i^2 \right) = \sum_i \left(m_i \dot{x}_i \frac{\partial \dot{x}_i}{\partial \dot{q}_k} \right)$$

From Eq. (12.22),

$$\frac{\partial \dot{x}}{\partial \dot{q}_k} = \frac{\partial x}{\partial q_k} \tag{12.22}$$

which on substituting in the preceding equation gives

$$\frac{\partial T}{\partial \dot{q}_k} = \sum_i \left(m_i \dot{x}_i \frac{\partial x_i}{\partial q_k} \right) \tag{12.40}$$

Differentiating with respect to t, we get

$$\frac{d}{dt} \left(\frac{\partial T}{\partial \dot{q}_k} \right) = \sum_i m_i \ddot{x}_i \frac{\partial x_i}{\partial q_k} + \sum_i m_i \dot{x}_i \frac{d}{dt} \left(\frac{\partial x_i}{\partial q_k} \right) \tag{12.41}$$

The expression for the generalized force Q_k given by Eq. (12.8),

$$Q_k = F_x \frac{\partial x}{\partial q_k} + F_y \frac{\partial y}{\partial q_k} + F_z \frac{\partial z}{\partial q_k} \tag{12.8}$$

takes the following form for a system of particles:

$$Q_k = \sum_i F_{x_i} \frac{\partial x_i}{\partial q_k} = \sum_i m_i \ddot{x}_i \frac{\partial x_i}{\partial q_k} \tag{12.42}$$

We also extend the results of Eq. (12.26) to the present case for a system of particles; that is,

$$\sum_i m_i \dot{x}_i \frac{d}{dt}\left(\frac{\partial x_i}{\partial q_k}\right) = \sum_i \frac{\partial}{\partial q_k}\left(\frac{1}{2}m_i\dot{x}_i^2\right) = \frac{\partial T}{\partial q_k} \tag{12.43}$$

Combining Eqs. (12.41), (12.42), and (12.43), we obtain the following Lagrange's equations for a system of particles:

$$\frac{d}{dt}\left(\frac{\partial T}{\partial \dot{q}_k}\right) = Q_k + \frac{\partial T}{\partial q_k}, \qquad k = 1, 2, \ldots, n \tag{12.44}$$

The number of equations is equal to the number of degrees of freedom n of the system.

If the system is conservative so that there exists a potential function $V(q)$, we may write

$$Q_k = -\frac{\partial V}{\partial q_k} \tag{12.45}$$

and, as before, we may define the Lagrangian function $L = T - V$ and write Lagrange's equations of motion for a system of particles as

$$\frac{d}{dt}\left(\frac{\partial L}{\partial \dot{q}_k}\right) - \frac{\partial L}{\partial q_k} = 0, \qquad k = 1, 2, \ldots, n \tag{12.46}$$

Of the generalized forces Q_k, suppose some of these, say Q'_k, are not conservative and cannot be derived from a potential function, while the remaining forces are conservative. (Frictional forces are a typical example of nonconservative forces.) In such cases, we can still define the Lagrangian function as $L = T - V$, while Eqs. (12.45) and (12.46) take the form

$$Q_k = Q'_k + \left(-\frac{\partial V}{\partial q_k}\right) \tag{12.47}$$

and

$$\frac{d}{dt}\left(\frac{\partial L}{\partial \dot{q}_k}\right) - \frac{\partial L}{\partial q_k} = Q'_k, \qquad k = 1, 2, \ldots, n \tag{12.48}$$

These equations can be applied to the motion of a single particle as well.

We are now in a position to illustrate the use of Lagrange's method for obtaining and solving equations for simple systems. It is convenient to do this if we use the following procedure as a guide.

1. Select a proper set of generalized coordinates to represent the configuration of the system.
2. Express the kinetic energy T of the system in terms of generalized coordinates and their time derivatives (velocities).
3. If the system is conservative, express the potential energy V as a function of the generalized coordinates; otherwise, find an expression for the generalized forces Q_k.
4. Finally, using the preceding information, write Lagrange's equations of motion.

For systems with constraints and for finding the forces of constraints or reactions, a few more steps are needed, as will be discussed in the next section.

 Example 12.5 _____

An inclined plane of mass M is sliding on a smooth horizontal surface, while a particle of mass m is sliding on a smooth inclined surface, as shown in Fig. Ex. 12.5. Find equations of motion of the particle and the inclined plane.

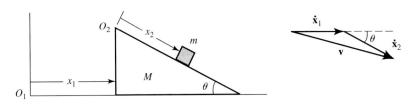

Figure Ex. 12.5

Solution

The system has two degrees of freedom; hence we need two generalized coordinates to describe the configuration of the system. Let the two coordinates x_1 and x_2, as shown in the figure, represent the displacements of M and m from the origins O_1 and O_2, respectively. The velocity of M with respect to O_1 is $\dot{x}_1$, while that of m with respect to O_2 is $\dot{x}_2$. The velocity $\mathbf{v}$ of m with respect to O_1, as shown in the insert, is $v = \dot{x}_1 + \dot{x}_2 = v_1 + v_2$.

Different quantities used are

$$v = \left(\frac{d}{dt}x1\right) + \frac{d}{dt}x2$$

v1 = velocity of M with respect to O₁
v2 = velocity of m with respect to O₂
v = velocity of m with respect to O₁
T = kinetic energy
V = potential energy
L = Lagrangian

$$v = \sqrt{v1^2 + v2^2 + 2 \cdot v1 \cdot v2 \cdot \cos(\theta)}$$

$$T = \frac{1}{2} \cdot M \cdot v1^2 + \frac{1}{2} \cdot m \cdot v^2$$

$$T = \frac{1}{2} \cdot M \cdot v1^2 + \frac{1}{2} \cdot m \cdot \left(v1^2 + v2^2 + 2 \cdot v1 \cdot v2 \cdot \cos(\theta)\right)$$

$$V = -m \cdot g \cdot x2 \cdot \sin(\theta)$$

$$L = T - V$$

The two Lagrangian equations for the coordinates x1 and x2 are

$$\frac{d}{dt}\frac{d}{dv1}L + \frac{d}{dx1}L = 0 \qquad \frac{d}{dt}\frac{d}{dv2}L + \frac{d}{dx2}L = 0$$

The resulting two equations are

$$m \cdot a2 + m \cdot a1 \cdot \cos(\theta) = 0 \qquad (M \cdot a1 + m \cdot a1 + m \cdot a2 \cdot \cos(\theta)) - m \cdot g \cdot \sin(\theta) = 0$$

Let us solve for a1 and a2, the two accelerations

Given

$$m \cdot a2 + m \cdot a1 \cdot \cos(\theta) = 0$$

$$(M \cdot a1 + m \cdot a1 + m \cdot a2 \cdot \cos(\theta)) - m \cdot g \cdot \sin(\theta) = 0$$

$$\text{Find}(a1, a2) \rightarrow \begin{bmatrix} \dfrac{-1}{\left(m \cdot \cos(\theta)^2 - M - m\right)} \cdot m \cdot g \cdot \sin(\theta) \\[2em] \dfrac{1}{\left(m \cdot \cos(\theta)^2 - M - m\right)} \cdot m \cdot g \cdot \sin(\theta) \cdot \cos(\theta) \end{bmatrix}$$

$$a1 = \dfrac{-1}{\left(m \cdot \cos(\theta)^2 - M - m\right)} \cdot m \cdot g \cdot \sin(\theta) \qquad a2 = \dfrac{1}{\left(m \cdot \cos(\theta)^2 - M - m\right)} \cdot m \cdot g \cdot \sin(\theta) \cdot \cos(\theta)$$

Knowing the initial conditions, we can solve these equations for the velocities and displacements by integrating a1 and a2.

EXERCISE 12.5 Solve the example for the situation shown in Fig. Exer. 12.5.

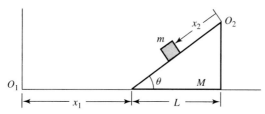

Figure Exer. 12.5

12.6 LAGRANGE'S EQUATIONS OF MOTION WITH UNDETERMINED MULTIPLIERS AND CONSTRAINTS

In the beginning of this chapter, we briefly mentioned the difference between holonomic systems (in which the coordinates can very independently) and nonholonomic systems. In this section, we describe holonomic and nonholonomic constraints in more detail and their usefulness in solving for the forces or reactions of constraints. *Holonomic constraints* can be expressed as algebraic relations among the coordinates, such as

$$f_l(x_i, t) = 0, \qquad \text{where } l = 1, 2, \ldots, m \tag{12.49}$$

where m is the number of constraints. In such cases it is always possible to find a set of proper generalized coordinates in terms of which the equations of motion can be written and are free from explicit reference to the constraints. On the contrary, in *nonholonomic constraints*, the constraints are expressed as relations among the velocities of the particles of the system; that is,

$$f_l(x_i, \dot{x}_i, t) = 0, \qquad \text{where } l = 1, 2, \ldots, m \tag{12.50}$$

If these equations of nonholonomic constraints can be integrated to yield relations among the coordinates, then the constraints become holonomic and the usual procedure for solving the problem can be carried out. Equations representing inequalities, such as a molecule moving anywhere in a cube, are examples of nonholonomic constraints, because in this case the constraint on the motion of the molecule is that it can be anywhere as long as $x \leqslant L$, $y \leqslant L$, and $z \leqslant L$. Let us illustrate these points with the help of a few examples.

As an example of a holonomic constraint, consider the motion of a particle constrained to move over a spherical surface of radius a with its center at the origin. In rectangular coordinates, the equation of constraint is

$$x^2 + y^2 + z^2 = a^2 \tag{12.51}$$

The displacements are related by the equations

$$x\,dx + y\,dy + z\,dz = 0 \tag{12.52}$$

The differential Eq. (12.52) can be integrated to obtain Eq. (12.51). Thus Eqs. (12.51) and (12.52) form only one equation of constraint. Not all three coordinates x, y, and z are independent. Since there is one equation of constraint, only two coordinates will be enough to describe the position of the particle. Similarly, if using spherical coordinates r, θ, and ϕ,

$$r = a = \text{constant}$$

and θ and ϕ are sufficient to describe the position of the particle. Suppose we use directional cosines as generalized coordinates to describe the position of the particle; that is,

$$q_1 = \frac{x}{a}, \quad q_2 = \frac{y}{a}, \quad \text{and} \quad q_3 = \frac{z}{a} \tag{12.53}$$

q_1, q_2, and q_3 are not all independent, and from Eq. (12.51)

$$\left(\frac{x}{a}\right)^2 + \left(\frac{y}{a}\right)^2 + \left(\frac{z}{a}\right)^2 = 1$$

or
$$q_1^2 + q_2^2 + q_3^2 = 1 \tag{12.54}$$

That is,
$$q_3 = \sqrt{1 - q_1^2 - q_2^2} \tag{12.55}$$

Since q_3 can be expressed in terms of the coordinates q_1 and q_2, there are only two independent coordinates.

As a second example, let us consider a circular disk of radius a rolling over a perfectly rough (absolutely no slipping allowed) horizontal plane XY, as shown in Fig. 12.1. The plane of the disk is vertical at all times (that is, $z = a$ and $\alpha = \pi/2$, where α is the angle between the plane of the disk and the horizontal plane XY). Thus, to describe the configuration of the disk at any instant, we need four coordinates: x, y, θ and ϕ. The coordinates x, y of the center of the disk (with $z = a$) describe the translational motion of the disk and locate the point of contact of the disk with the plane. Angle θ describes the rotational motion of the disk about the center of mass; that is, θ is the angle between a fixed radius in the disk and the vertical. Angle ϕ deter-

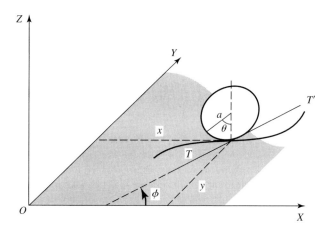

Figure 12.1   Circular disk of radius a rolling over a perfectly rough (absolutely no slipping) horizontal plane XY. Tangent TT' makes an angle ϕ with the X-axis.

mines the orientation of the plane of the disk with respect to the X-axis; that is, it gives the instantaneous direction of the motion. These coordinates are not all independent. Because of the constraints, the velocity $\mathbf{v}$ of the center of mass and $\dot{\theta}$ are related by

$$v = a\dot{\theta} \qquad (12.56)$$

which results in the velocity components

$$\dot{x} = v\cos\phi = a\dot{\theta}\cos\phi$$
$$\dot{y} = -v\sin\phi = -a\dot{\theta}\sin\phi \qquad (12.57)$$

These yield the following two equations of constraints:

$$dx = a\,d\theta\cos\phi$$
$$dy = -a\,d\theta\sin\phi \qquad (12.58)$$

Neither of these differential equations can be integrated to obtain two relations between x, y, and ϕ. Such constraints in which differential equations are not integratable are called *nonholonomic constraints*. A system containing such constraints is called a *nonholonomic system*. Thus the disk in this example has four degrees of freedom and we need four coordinates to solve the problem.

What happens if the plane is not rough, that is, slipping takes place? In such cases, the Eqs. (12.58) of constraints do not hold and the system becomes holonomic: again four coordinates are needed to describe the motion. We need to known $\dot{x}$, $\dot{y}$, $\dot{\theta}$, and $\dot{\phi}$. On the other hand, when the disk is allowed to *roll only*, if $\dot{\theta}$ is given, and any one of the remaining three $\dot{x}$, $\dot{y}$, and $\dot{\phi}$ is also given, then the remaining two can be calculated from Eqs. (12.57). The rolling disk has two degrees of freedom. The disk is free to roll and rotate.

As an alternative, suppose the disk was constrained to roll along a prescribed curve. Let s measure the length of the path along this curve. In this case, Eq. (12.56) takes the form

$$ds = a\,d\theta \qquad (12.59)$$

which may be integrated

$$s - a\theta = \text{constant} \tag{12.60}$$

We have a condition representing a holonomic constraint, hence a holonomic system.

Finally, let us consider an example of a disk rolling down an inclined plane, as shown in Fig. 12.2. The disk is rolling without slipping. The position can be located by two coordinates s and θ. The velocities $\dot{s}$ and $\dot{\theta}$ are related by

$$\dot{s} = a\dot{\theta} \tag{12.61}$$

$$ds = a\,d\theta \tag{12.62}$$

which can be integrated to give

$$s - a\theta = \text{constant } C \tag{12.63}$$

Even though initially the constraint is expressed in terms of velocities [Eq. (12.61)] it can be integrated to give the relation between the coordinates [Eq. (12.63)]. Thus the system is holonomic with one equation of constraint, and only one coordinate is needed to describe the system.

Thus we conclude: *If in a certain system, the constraint on the velocities can be integrated to give a relationship between the coordinates, then the constraint is holonomic.*

Let us pursue our discussion still further. Consider a constraint relation of the form

$$\sum_i A_i \dot{x}_i + B = 0, \qquad i = 1, 2, 3, \ldots \tag{12.64}$$

In general, this is nonintegrable; hence it represents a nonholonomic constraint. But if A_i and B have the following form,

$$A_i = \frac{\partial f}{\partial x_i}, \qquad B = \frac{\partial f}{\partial t}, \qquad f = f(x_i, t) \tag{12.65}$$

then Eq. (12.64) may be written in the form

$$\sum_i \frac{\partial f}{\partial x_i}\frac{dx_i}{dt} + \frac{\partial f}{\partial t} = 0 \tag{12.66}$$

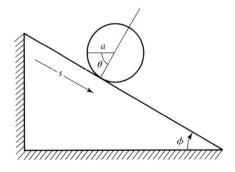

Figure 12.2 Disk rolling down, without slipping, an inclined plane.

which gives

$$\frac{df}{dt} = 0 \tag{12.67}$$

and may be integrated to give

$$f(x_i, t) = \text{constant} \tag{12.68}$$

Hence a constraint given by Eq. (12.64) is actually holonomic. Thus, in general, a constraint expressed as

$$\sum_{i=1}^{m} \frac{\partial f_l}{\partial q_k} \frac{dq_k}{dt} + \frac{\partial f_l}{\partial t} = 0 \tag{12.69}$$

or

$$\sum_{i=1}^{m} \frac{\partial f_l}{\partial q_k} \dot{q}_k + \frac{\partial f_l}{\partial t} = 0 \tag{12.70}$$

is entirely equivalent to

$$f_l = f_l(q_k, t) = 0 \tag{12.71}$$

There are certain advantages in expressing constraints in a differential form, rather than as algebraic expressions. In these situations, the constraint relations can be directly incorporated (without first integrating them) into Lagrange's equations by means of the Lagrange undetermined multipliers. Suppose the constraints are expressed as

$$\sum_{l} \frac{\partial f_l}{\partial q_k} dq_k = 0 \tag{12.72}$$

where $l = 1, 2, \ldots, m$ and $k = 1, 2, \ldots, n$; then Lagrange's equations take the form

$$\frac{d}{dt}\left(\frac{\partial L}{\partial \dot{q}_k}\right) - \frac{\partial L}{\partial q_k} = \sum_{l} \lambda_l(t) \frac{\partial f_l}{\partial q_k} \tag{12.73}$$

$\lambda_l(t)$ are the *undetermined multipliers* and they simply represent the forces of constraints. There are the same number of λ_l as the number of equations of constraints. The following examples illustrate the use of the equations of constraints and the method of undetermined multipliers for calculating the forces of constraints.

 Example 12.6 _____

Discuss the motion of a disk that is rolling down an inclined plane without slipping. Also, find the forces of constraints by using the method of undetermined multipliers.

Solution

The situation is shown in Fig. Ex. 12.6. Use y and θ as the two generalized coordinates. Thus the total kinetic energy, which is the sum of the translational energy and rotational energy, may be written as, noting that the moment of inertia of the disk is $I = \frac{1}{2}MR^2$,

$$T = \frac{1}{2}M\dot{y}^2 + \frac{1}{2}I\dot{\theta}^2 = \frac{1}{2}M\dot{y}^2 + \frac{1}{4}MR^2\dot{\theta}^2 \tag{i}$$

while the potential energy is, assuming potential energy at the bottom to be zero,

$$V = Mg(l - y)\sin\phi \tag{ii}$$

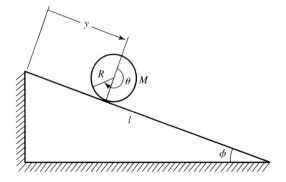

Figure Ex. 12.6

Thus the Lagrangian of the system is

$$L = T - V = \tfrac{1}{2}M\dot{y}^2 + \tfrac{1}{4}MR^2\dot{\theta}^2 - Mg(l - y)\sin\phi \qquad \textbf{(iii)}$$

The equation of the holonomic constraint giving the relation between the coordinates y and θ is

$$f(y, \theta) = y - R\theta = 0 \qquad \textbf{(iv)}$$

Thus, if the disk rolls down without slipping, the preceding constraint relation must hold good. Hence, instead of two degrees of freedom y and θ, we have only one degree of freedom. One of the two coordinates y and θ may be eliminated from Eq. (iii) by using the relation given by Eq. (iv); then the equations of motion may be obtained by using Lagrange's equations (see Exercise 12.6).

Alternatively, we can use both y and θ as generalized coordinates and the method of undetermined multipliers. This method yields much more information, as we see now.

The different quantities are

$$a\theta \equiv \ddot{\theta} \qquad ay \equiv \ddot{y} \qquad v\theta \equiv \dot{\theta} \qquad vy \equiv \dot{y}$$

The equations of holonomic constraints, which after differentiation give the relation between the acceleration ay and $a\theta$, are given in Eq. (i)

$$y - R \cdot \theta \equiv 0 \qquad\qquad ay - R \cdot a\theta \equiv 0 \qquad \text{(i)}$$

$$T \equiv \frac{1}{2} \cdot M \cdot vy^2 + \frac{1}{2} \cdot I \cdot v\theta^2 \equiv \frac{1}{2} \cdot M \cdot vy^2 + \frac{1}{4} \cdot M \cdot R^2 \cdot v\theta^2$$

$$V \equiv M \cdot g \cdot (1 - y) \cdot \sin(\phi)$$

The expressions for T, V, and L are

$$L \equiv T - V \equiv \frac{1}{2} \cdot M \cdot vy^2 + \frac{1}{4} \cdot M \cdot R^2 \cdot v\theta^2 - M \cdot g \cdot (1 - y) \cdot \sin(\phi)$$

f = constraint function
λ = undetermined multiplier

$$f \equiv y - R \cdot \theta \qquad \frac{df}{dy} \equiv 1 \qquad \frac{df}{d\theta} \equiv -R$$

The resulting Lagrangian equations for y and θ are as shown.

$$\frac{d}{dt}\left(\frac{d}{dvy}L\right) - \frac{d}{dy}L - \lambda \cdot \frac{d}{dy}f \equiv 0 \qquad \text{(ii)} \qquad\qquad \frac{d}{dt}\left(\frac{d}{dv\theta}L\right) - \frac{d}{d\theta}L - \lambda \cdot \frac{d}{d\theta}f \equiv 0 \qquad \text{(iii)}$$

Substituting the values of L and f, we get the following two equations.

$$M \cdot ay - M \cdot g \cdot \sin(\phi) - \lambda \equiv 0 \qquad \text{(iv)} \qquad\qquad \frac{-1}{2} \cdot M \cdot R^2 \cdot a\theta - \lambda \cdot R \equiv 0 \qquad \text{(v)}$$

Using Eqs. (i), (iv), and (v), we can solve for the three unknowns ay, $a\theta$, and λ.

Given

$$ay - R \cdot a\theta = 0$$

$$\frac{-1}{2} \cdot M \cdot R^2 \cdot a\theta - \lambda \cdot R = 0$$

$$M \cdot ay - M \cdot g \cdot \sin(\phi) - \lambda = 0$$

$$S = Find(ay, a\theta, \lambda) \rightarrow S = \begin{bmatrix} \dfrac{2}{3} \cdot g \cdot \sin(\phi) \\[2mm] \dfrac{2}{(3 \cdot R)} \cdot g \cdot \sin(\phi) \\[2mm] \dfrac{-1}{3} \cdot M \cdot g \cdot \sin(\phi) \end{bmatrix} \qquad \begin{bmatrix} ay \\[2mm] a\theta \\[2mm] \lambda \end{bmatrix} = \begin{bmatrix} \dfrac{2}{3} \cdot g \cdot \sin(\phi) \\[2mm] \dfrac{2}{3 \cdot R} \cdot g \cdot \sin(\phi) \\[2mm] -\dfrac{M}{3} \cdot g \cdot \sin(\phi) \end{bmatrix}$$

The expressions for ay, $a\theta$, and λ reveal that these are constants for this situation. λ gives the magnitude of the force of constraint resulting from a frictional force.

EXERCISE 12.6 (a) Find Lagrange's equation for the example using θ as the independent coordinate. Solve the equation and interpret the results. (b) Find Lagrange's equation for the example using y as the independent coordinate. Solve the equation and interpret the results.

12.7 GENERALIZED MOMENTA AND CYCLIC (OR IGNORABLE) COORDINATES

In previous sections, we have shown how to describe a system by means of Lagrange's equations. For a system with n degrees of freedom, we need n generalized coordinates. The Lagrangian L was described in terms of generalized coordinates q_k and generalized velocities $\dot{q}_k$. Furthermore, if a Lagrangian were an explicit function of time, we wrote

$$L(q, \dot{q}, t) = L(q_1, q_2, \ldots, q_n; \dot{q}_1, \dot{q}_2, \ldots, \dot{q}_n; t) \tag{12.74}$$

As we know, Lagrange's formalism leads to n second-order differential equations.

An alternative to Lagrange's formalism is Hamilton's formalism. Hamilton's formalism is carried out in terms of generalized coordinates and generalized momenta. That is, if the Hamiltonian H is an explicit function of time, then

$$H = H(q, p, t) = H(q_1, q_2, \ldots, q_n; p_1, p_2, \ldots, p_n; t) \tag{12.75}$$

Such formalism for a system of n degrees of freedom leads to $2n$ first-order differential equations. These $2n$ first-order equations are much easier to solve than n second-order differential equations, as in the case of Lagrange's formalism.

To begin, let us define the generalized momentum. As a simple example, let us consider the motion of a single particle moving with velocity $\dot{x}$ along the X-axis. The kinetic energy of such a particle is

$$T = \tfrac{1}{2} m \dot{x}^2 \tag{12.76}$$

Usually one defines momentum p as $p = m\dot{x}$. But in the present formalism, we define the momentum p to be

$$p = \frac{\partial T}{\partial \dot{x}} \tag{12.77}$$

If we substitute the value of T from Eq. (12.76) into (12.77), we get $p = m\dot{x}$. Furthermore, if V is not a function of the velocity $\dot{x}$, that is, $V = V(x)$, then the momentum p may also be written as

$$p = \frac{\partial L}{\partial \dot{x}} \tag{12.78}$$

We now utilize these concepts to define the generalized momentum. For a system described by a set of generalized coordinates $q_1, q_2, \ldots, q_k, \ldots, q_n$, and the corresponding generalized momenta $p_1, p_2, \ldots, p_k, \ldots, p_n$, we define the *generalized momentum* p_k corresponding to generalized coordinate q_k as

$$p_k = \frac{\partial L}{\partial \dot{q}_k} \tag{12.79}$$

The generalized momentum p_k is also called the *conjugate momentum* p_k (conjugate to coordinate q_k). It may be pointed out that the generalized momenta do not always convey the same physical concept as we are used to in introductory physics using rectangular coordinates.

For a conservative system, Lagrange's equations are

$$\frac{d}{dt}\left(\frac{\partial L}{\partial \dot{q}_k}\right) - \frac{\partial L}{\partial q_k} = 0 \tag{12.80}$$

and, from Eq. (12.79),

$$\dot{p}_k = \frac{d}{dt}\frac{\partial L}{\partial \dot{q}_k} \tag{12.81}$$

Hence Lagrange's equations take the form

$$\dot{p}_k = \frac{\partial L}{\partial q_k} \tag{12.82}$$

Let us now see the connection between the generalized momenta and the constants of motion. *Quantities that are functions of the coordinates and velocities that remain constant in time are called* **constants of motion.** Suppose in the expression for the Lagrangian L of a system a certain coordinate q_λ does not occur explicitly. In such cases,

$$\frac{\partial L}{\partial q_\lambda} = 0 \tag{12.83}$$

Therefore, Eq. (12.81) together with Eq. (12.82) takes the form

$$\dot{p}_\lambda = \frac{d}{dt}\left(\frac{\partial L}{\partial \dot{q}_\lambda}\right) = 0 \tag{12.84}$$

which on integration yields

$$p_\lambda = \frac{\partial L}{\partial \dot{q}_\lambda} = \text{a constant} = C_\lambda \tag{12.85}$$

That is, whenever the Lagrangian function does not contain a coordinate q_λ explicitly, the corresponding generalized momentum p_λ is a constant of the motion. The coordinate q_λ is said to be *cyclic* or *ignorable*. Thus we may conclude that *the generalized momentum associated with an ignorable coordinate is a constant of motion of the system.*

As an example, let us consider the motion of a particle in a central force field. In polar coordinates, the Lagrangian L is

$$L = T - V = \tfrac{1}{2}m(\dot{r}^2 + r^2\dot{\theta}^2) - V(r) \tag{12.86}$$

Since L does not contain the coordinate θ, θ is cyclic (*or ignorable*), and the generalized momentum corresponding to θ is

$$p_\theta = \frac{\partial L}{\partial \dot{\theta}} = mr^2\dot{\theta} = \text{constant} \tag{12.87}$$

where p_θ is the angular momentum and is a constant of motion. A note of caution: p_θ is a constant of motion; it does not mean that $\dot{\theta}$ is constant. For example, in an elliptic orbit resulting from an inverse-square attractive force, $\dot{\theta}$ varies, but r also varies in such a way as to make p_θ remain constant.

12.8 HAMILTONIAN FUNCTION: CONSERVATION LAWS AND SYMMETRY PRINCIPLES

A system that does not interact with anything outside the system is called a *closed system.* There may or may not be any interactions between the particles of the closed system. In either case, for such a closed system there are always *seven constants or integrals of motion:* (1) the linear momentum, which has three components, (2) the angular momentum, which also has three components, and (3) the total energy. In this section we shall investigate the process by which these constants are arrived at from the considerations of the Lagrangian of a closed system.

Conservation of Linear Momentum

Let us consider a Lagrangian of a closed system in an inertial reference frame. An outstanding property of an inertial system is that *space is homogeneous in an inertial system or frame;* that is, a closed system is unaffected by a translation of the entire system in space. This implies that *the Lagrangian of a closed system in an inertial frame remains unaffected or is invariant.* For simplicity, let us consider a single particle with a Lagrangian $L(q, \dot{q})$. The variation in L due to the variation in the generalized coordinates must be zero. That is, L is invariant:

$$\delta L = \sum_k \frac{\partial L}{\partial q_k}\delta q_k + \sum_k \frac{\partial L}{\partial \dot{q}_k}\delta \dot{q}_k = 0 \tag{12.88}$$

Since we are causing only the displacement of the system, δq_i are not functions of time; hence

$$\delta \dot{q}_k = \delta\left(\frac{dq_k}{dt}\right) = \frac{d}{dt}(\delta q_k) = 0 \tag{12.89}$$

and Eq. (12.88) reduces to

$$\delta L = \sum_k \frac{\partial L}{\partial q_k}\delta q_k = 0 \tag{12.90}$$

Each displacement δq_k is independent or arbitrary. Hence δL in Eq. (12.90) will be zero only if each of the partial derivatives of L is zero; that is,

$$\frac{\partial L}{\partial q_k} = 0 \tag{12.91}$$

Thus Lagrange's equation, Eq. (12.80), reduces to

$$\frac{d}{dt}\left(\frac{\partial L}{\partial \dot{q}_k}\right) = 0 \tag{12.92}$$

Hence
$$\frac{\partial L}{\partial \dot{q}_k} = \text{constant} \tag{12.93}$$

Since $L = T(\dot{q}_k) - V(q_k)$, we may write Eq. (12.93) as

$$\frac{\partial L}{\partial \dot{q}_k} = \frac{\partial}{\partial \dot{q}_k}(T - V) = \frac{\partial}{\partial \dot{q}_k}\left(\frac{1}{2}m\sum_k \dot{q}_k^2\right) = m\dot{q}_k = p_k = \text{constant} \tag{12.94}$$

Equation (12.94) states that if space is homogeneous then the linear momentum p_k of a closed system is constant. Since the motion of a single particle can be described by three Cartesian coordinates (or any other three generalized coordinates), there will be three constants of motion, $p_x, p_y,$ and p_z, which are the three components of a linear momentum vector $\mathbf{p}_k$. In more general terms, we may make the following statement:

> *If the Lagrangian of a system, closed or otherwise, is invariant with respect to a translation in a certain direction, then the linear momentum of the system in that direction is constant in time.*

Conservation of Angular Momentum

Another outstanding property of an inertial system is that *space is isotropic in an inertial frame*; that is, a closed system is unaffected by the orientation or rotation of the entire system. This implies that *the Lagrangian of a closed system remains invariant if the system is rotated through an infinitesimal angle*. Once again consider a system consisting of a single particle. The change in the Lagrangian as given by Eq. (12.88) is

$$\delta L = \sum_k \frac{\partial L}{\partial q_k} \delta q_k + \sum_k \frac{\partial L}{\partial \dot{q}_k} \delta \dot{q}_k = 0 \qquad \textbf{(12.88)}$$

By definition,

$$p_k = \frac{\partial L}{\partial \dot{q}_k} \qquad \textbf{(12.79)}$$

Hence Lagrange's equations, Eq. (12.80), may be written as

$$\dot{p}_k = \frac{\partial L}{\partial q_k} \qquad \textbf{(12.95)}$$

Combining the preceding equations with Eq. (12.88), we may write

$$\delta L = \sum_k \dot{p}_k \, \delta q_k + \sum_k p_k \, \delta \dot{q}_k = 0 \qquad \textbf{(12.96)}$$

Let us apply these results to the case shown in Fig. 12.3. A point particle is at a distance **r** from the origin O. The system is rotated through an angle $\delta\theta$ about an axis. The value of **r** changes so that

$$\delta \mathbf{r} = \delta\boldsymbol{\theta} \times \mathbf{r} \qquad \textbf{(12.97)}$$

This leads to a change in velocity given by

$$\delta \dot{\mathbf{r}} = \delta\boldsymbol{\theta} \times \dot{\mathbf{r}} \qquad \textbf{(12.98)}$$

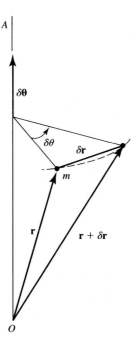

Figure 12.3 Rotation about an axis OA of a point particle of mass m at a distance **r** from the origin O.

Applying Eq. (12.96) to this case, where $p_k \equiv \mathbf{p}$ and $\dot{q}_k \equiv \dot{\mathbf{r}}$, we get ($k = 1, 2, 3$, the three components of a vector)

$$\delta L = \dot{\mathbf{p}} \cdot \delta \mathbf{r} + \mathbf{p} \cdot \delta \dot{\mathbf{r}} = 0 \qquad (12.99)$$

Using Eqs. (12.97) and (12.98) in Eq. (12.99),

$$\dot{\mathbf{p}} \cdot (\delta \boldsymbol{\theta} \times \mathbf{r}) + \mathbf{p} \cdot (\delta \boldsymbol{\theta} \times \dot{\mathbf{r}}) = 0 \qquad (12.100)$$

From the properties of a scalar triple product, Eq. (12.100) takes the form

$$\delta \boldsymbol{\theta} \cdot (\mathbf{r} \times \dot{\mathbf{p}}) + \delta \boldsymbol{\theta} \cdot (\dot{\mathbf{r}} \times \mathbf{p}) = 0 \qquad (12.101)$$

or

$$\delta \boldsymbol{\theta} \cdot [(\mathbf{r} \times \dot{\mathbf{p}}) + (\dot{\mathbf{r}} \times \mathbf{p})] = 0 \qquad (12.102)$$

or

$$\delta \boldsymbol{\theta} \cdot \left(\frac{d}{dt} (\mathbf{r} \times \mathbf{p}) \right) = 0 \qquad (12.103)$$

But

$$\mathbf{r} \times \mathbf{p} = \mathbf{J} \qquad (12.104)$$

where $\mathbf{J}$ is the angular momentum about the given axis. Therefore,

$$\delta \boldsymbol{\theta} \cdot \frac{d\mathbf{J}}{dt} = 0 \qquad (12.105)$$

Since $\delta \boldsymbol{\theta}$ is arbitrary, we must have

$$\frac{d\mathbf{J}}{dt} = 0 \qquad (12.106)$$

That is,

$$\mathbf{J} = \mathbf{r} \times \mathbf{p} = \text{constant} \qquad (12.107)$$

where $\mathbf{J}$ has three components. In general:

> *If the Lagrangian remains invariant under rotation about an axis, then the angular momentum of the system about this axis will remain constant in time.*

Suppose a system is acted on by a force and has a symmetry about this force field. This means that the Lagrangian of this system will be invariant about this axis of symmetry. Hence the angular momentum $\mathbf{J}$ of the system about this axis will remain constant in time.

Conservation of Energy and the Hamiltonian Function

Still another outstanding property of an inertial frame is that *the time is homogeneous within an inertial reference frame.* This implies that the Lagrangian of a closed system cannot be an explicit function of time. That is, in the total differential of L,

$$dL = \sum_k \frac{\partial L}{\partial q_k} dq_k + \sum_k \frac{\partial L}{\partial \dot{q}_k} d\dot{q}_k + \frac{\partial L}{\partial t} dt \qquad (12.108)$$

we must have $$\frac{\partial L}{\partial t} = 0 \tag{12.109}$$

Hence the total time derivative of L reduces to

$$\frac{dL}{dt} = \sum_k \frac{\partial L}{\partial q_k} \frac{dq_k}{dt} + \sum_k \frac{\partial L}{\partial \dot{q}_k} \frac{d\dot{q}_k}{dt} = 0$$

or $$\frac{dL}{dt} = \sum_k \frac{\partial L}{\partial q_k} \dot{q}_k + \sum_k \frac{\partial L}{\partial \dot{q}_k} \ddot{q}_k = 0 \tag{12.110}$$

From Lagrange's equations,

$$\frac{d}{dt}\left(\frac{\partial L}{\partial \dot{q}_k}\right) = \frac{\partial L}{\partial q_k} \tag{12.80}$$

Substituting for $\partial L/\partial q_k$ in Eq. (12.110), we get

$$\frac{dL}{dt} = \sum_k \dot{q}_k \frac{d}{dt}\left(\frac{\partial L}{\partial \dot{q}_k}\right) + \sum_k \frac{\partial L}{\partial \dot{q}_k} \ddot{q}_k = \sum_k \frac{d}{dt}\left(\dot{q}_k \frac{\partial L}{\partial \dot{q}_k}\right) = 0$$

That is,

$$\frac{d}{dt}\left(\sum_k \dot{q}_k \frac{\partial L}{\partial \dot{q}_k} - L\right) = 0 \tag{12.111}$$

Thus the quantity in parentheses must be constant in time. This constant is denoted by H, called the *Hamiltonian H*, and is given by [using the definition of generalized momentum in Eq. (12.79)]

$$H = \sum_k \dot{q}_k \frac{\partial L}{\partial \dot{q}_k} - L = \sum_k p_k \dot{q}_k - L = \text{constant} \tag{12.112}$$

Hence H is *a constant of motion if L is not an explicit function of time t*; that is, $\partial L/\partial t = 0$. H takes a special form if we make the following two assumptions:

(i) The potential energy V is independent of the velocity coordinate so that

$$\frac{\partial L}{\partial \dot{q}_k} = \frac{\partial(T - V)}{\partial \dot{q}_k} = \frac{\partial T}{\partial \dot{q}_k} \tag{12.113}$$

(ii) If the equations representing the transformation of coordinates do not contain time explicitly, then the kinetic energy T will not only be the quadratic function of the generalized velocity, but will also be homogeneous in all its terms. [Caution: The transformation equations between the coordinates, say from a rectangular to generalized, will contain time explicitly if q's are rotating with respect to the inertial coordinates (say rectangular) or if the constraints are functions of time.]

Now, according to Euler's theorem for a homogeneous function $f(q_1, q_2, \ldots, q_k, \ldots, q_n)$ of order N in variables $(q_1, q_2, \ldots, q_k, \ldots, q_n)$,

$$\sum_{k=1}^{N} q_k \frac{\partial f}{\partial q_k} = Nf \qquad (12.114)$$

Thus, if the kinetic energy T is a homogeneous quadratic function, that is, of the order $N = 2$, from Eq. (12.114) we get

$$\sum_{k=1}^{N} \dot{q}_k \frac{\partial f}{\partial \dot{q}_k} = 2f \qquad (12.115)$$

Combining Eqs. (12.113) and (12.115) with Eq. (12.112), we obtain

$$H = 2T - (T - V) = T + V = E = \text{constant} \qquad (12.116)$$

where E is the total energy and is constant. That is, under the assumptions given above, (1) $V = V(q_k)$ and (2) T is a homogeneous quadratic function; the constant of motion, the Hamiltonian H, is equal to the total energy E of the system. It is very important to keep in mind that H is not always equal to E. The different possibilities are as follows:

H is *constant* and is *equal* to the total energy E.

H is *constant* but is *not equal* to the total energy E.

H is *not constant* but is *equal* to the total energy E.

H is *not constant* and is *not equal* to the total energy E.

The conservation laws derived here may be summarized as shown in Table 12.1. It is important to note that the *invariance* of physical quantities results from the *symmetry* properties of the system and is not limited only to the three cases discussed. The preceding type of reasoning is frequently used in arriving at different conservation laws in modern theories of elementary particles and fields.

TABLE 12.1   SYMMETRY PROPERTIES AND CONSERVATION LAWS

Property in inertial frame	Conserved quantity	Restrictions on Lagrangian L
Homogeneous space	Linear momentum	L is invariant to translational; $\delta L = 0$
Isotropic space	Angular momentum	L is invariant to rotation; $\delta L = 0$
Homogeneous time	Total energy	L is not an explicit function of time t; $\partial L/\partial t = 0$

12.9 HAMILTONIAN DYNAMICS: HAMILTON'S EQUATIONS OF MOTION

Hamilton's equations of motion, also called the *canonical equations* of motion, will be derived here. The Lagrangian L is a function of generalized coordinates and generalized velocities and may be an explicit function of time; that is,

$$L = L(q_1, q_2, \ldots, q_n; \dot{q}_1, \dot{q}_2, \ldots, \dot{q}_n; t) \tag{12.117}$$

The differential of L is

$$dL = \sum_{k=1}^{n} \left(\frac{\partial L}{\partial q_k} dq_k + \frac{\partial L}{\partial \dot{q}_k} d\dot{q}_k \right) + \frac{\partial L}{\partial t} dt \tag{12.118}$$

Using the following relations, proved already by the definition of generalized momentum and the Lagrange equation,

$$\dot{p}_k = \frac{\partial L}{\partial q_k} \quad \text{and} \quad \frac{\partial L}{\partial \dot{q}_k} = p_k \tag{12.119}$$

we obtain

$$dL = \sum_{k=1}^{n} (\dot{p}_k \, dq_k + p_k \, d\dot{q}_k) + \frac{\partial L}{\partial t} dt \tag{12.120}$$

Add $\dot{q}_k \, dp_k$ to both sides of this equation and, after rearranging, we get

$$d\left(\sum_{k=1}^{n} p_k \dot{q}_k - L \right) = \sum_{k=1}^{n} (\dot{q}_k \, dp_k - \dot{p}_k \, dq_k) - \frac{\partial L}{\partial t} dt \tag{12.121}$$

As before, we define the Hamiltonian function H to be

$$H = \sum_{k=1}^{n} p_k \dot{q}_k - L(q_1, \ldots, \dot{q}_k; \dot{q}_1, \ldots, \dot{q}_k; t) \tag{12.122}$$

and Eq. (12.121) takes the form

$$dH = \sum_{k=1}^{n} (\dot{q}_k \, dp_k - \dot{p}_k \, dq_k) - \frac{\partial L}{\partial t} dt \tag{12.123}$$

Let us concentrate on Eq. (12.122) for a moment. L is an explicit function of $(q_1, \ldots, q_n; \dot{q}_1, \ldots, \dot{q}_n; t)$. In many cases it is possible to express H as an explicit function of $(q_1, \ldots, q_n; p_1, \ldots, p_n; t)$. This can be done by using the relation defining generalized momenta, that is, $\partial L / \partial \dot{q}_k = p_k$; hence we can express $\dot{q}_k$ in terms of p_k. When this is possible, we can write

$$H = H(q_1, \ldots, q_n; p_1, \ldots, p_n; t) \tag{12.124}$$

That is, H is expressed as a function of the generalized coordinates, generalized momenta, and t. Thus, using Eq. (12.124), we may write the differential of H as

$$dH = \sum_{k=1}^{n} \left(\frac{\partial H}{\partial q_k} dq_k + \frac{\partial H}{\partial p_k} dp_k \right) + \frac{\partial H}{\partial t} dt \qquad (12.125)$$

Comparing Eqs. (12.125) and (12.123), we obtain

$$\dot{q}_k = \frac{\partial H}{\partial p_k} \qquad (12.126)$$

$$-\dot{p}_k = \frac{\partial H}{\partial q_k} \qquad (12.127)$$

and

$$\frac{\partial H}{\partial t} = -\frac{\partial L}{\partial t} \qquad (12.128)$$

Substituting Eqs. (12.126) and (12.127) in Eq. (12.125) also yields $dH/dt = \partial H/\partial t$. Equations (12.126) and (12.127) are *Hamilton's equations of motion;* and because of their symmetrical nature, they are also called *canonical equations* of motion. This procedure of describing motion by these equations is called *Hamilton's dynamics.* These $2n$ first-order differential equations are much easier to solve than the n second-order differential equations in Lagrangian formalism. It must be clear from Eq. (12.127), that if any coordinate q_λ is ignorable (that is, it is not contained in the Hamiltonian H) then the corresponding conjugate momentum p_λ is a constant of the motion.

Let us consider the case in which L, and hence H, do not contain time explicitly. In such conditions, $\partial H/\partial t = 0$, and Eq. (12.125) reduces to

$$\frac{dH}{dt} = \sum_{k=1}^{n} \left(\frac{\partial H}{\partial q_k} \dot{q}_k + \frac{\partial H}{\partial p_k} \dot{p}_k \right) \qquad (12.129)$$

Using Hamilton's equations (12.126) and (12.127), we obtain

$$\frac{dH}{dt} = \sum_{k=1}^{n} \left(\frac{\partial H}{\partial q_k} \frac{\partial H}{\partial p_k} - \frac{\partial H}{\partial p_k} \frac{\partial H}{\partial q_k} \right) = 0 \qquad (12.130)$$

Hence H *is a constant of motion if it does not contain t explicitly.* Furthermore, as we showed earlier, H is identical to E, if (1) the equations describing the transformation of generalized coordinates do not contain time explicitly, and (2) the potential energy is not a function of the generalized velocity.

Example 12.7 _____

A particle of mass m is attracted toward a given point by a force of magnitude k/r^2, where k is a constant. Derive an expression for the Hamiltonian and Hamilton's equations of motion.

Solution

Using plane polar coordinates (r, θ),

$$T = \tfrac{1}{2} m(\dot{r}^2 + r^2 \dot{\theta}^2) \tag{i}$$

$$V = -\int \mathbf{F} \cdot d\mathbf{r} = \int \frac{k}{r^2}\, dr = -\frac{k}{r} \tag{ii}$$

Therefore,

$$L = L(r, \dot{r}, \dot{\theta}) = T - V = \frac{1}{2} m(\dot{r}^2 + r^2\dot{\theta}^2) + \frac{k}{r} \tag{iii}$$

$\dot{r}$ and $\dot{\theta}$ in Eq. (i) must be replaced by p_r and p_θ by using Eq. (iii). That is,

$$p_r = \frac{\partial L}{\partial \dot{r}} = m\dot{r} \quad \text{or} \quad \dot{r} = \frac{p_r}{m} \tag{iv}$$

$$p_\theta = \frac{\partial L}{\partial \dot{\theta}} = mr^2\dot{\theta} \quad \text{or} \quad \dot{\theta} = \frac{p_\theta}{mr^2} \tag{v}$$

Thus the kinetic energy in Eq. (i) may be written as

$$T = \frac{1}{2} m\left[\left(\frac{p_r}{m}\right)^2 + r^2\left(\frac{p_\theta}{mr^2}\right)^2 \right] = \frac{p_r^2}{2m} + \frac{p_\theta^2}{2mr^2} \tag{vi}$$

Hence the Hamiltonian H becomes

$$H = H(r, p_r, p_\theta) = \frac{1}{2m}\left(p_r^2 + \frac{p_\theta^2}{r^2}\right) - \frac{k}{r} \tag{vii}$$

The equations of motion can now be found from the canonical Eqs. (12.131) and (12.132):

$$\frac{\partial H}{\partial q_k} = -\dot{p}_k \quad \text{and} \quad \frac{\partial H}{\partial p_k} = \dot{q}_k$$

The generalized coordinates are r, θ, p_r, and p_θ. Thus

$$-\dot{p}_r = \frac{\partial H}{\partial r} = -\frac{p_\theta^2}{mr^3} + \frac{k}{r^2} \quad \text{or} \quad \dot{p}_r = \frac{p_\theta^2}{mr^3} - \frac{k}{r^2} \tag{viii}$$

$$-\dot{p}_\theta = \frac{\partial H}{\partial \theta} = 0 \quad \text{or} \quad \dot{p}_\theta = 0 \quad \text{or} \quad p_\theta = \text{constant} \tag{ix}$$

$$\dot{r} = \frac{\partial H}{\partial p_r} = \frac{p_r}{m} \quad \text{or} \quad p_r = m\dot{r} \tag{x}$$

$$\dot{\theta} = \frac{\partial H}{\partial p_\theta} = \frac{p_\theta}{mr^2} \quad \text{or} \quad p_\theta = mr^2\dot{\theta} \tag{xi}$$

Note that Eqs. (x) and (xi) duplicate Eqs. (iv) and (v), respectively, while Eq. (ix) (since H does not contain θ) gives the familiar constant of motion, $p_\theta = mr^2\dot\theta = $ constant.

EXERCISE 12.7 Repeat the example in rectangular coordinates.

 Example 12.8 _____

Describe the motion of a particle of mass m constrained to move on the surface of a cylinder of radius a and attracted toward the origin by a force that is proportional to the distance of the particle from the origin.

Solution

The motion of a particle of mass m in Fig. Ex. 12.8 may be described by the Cartesian coordinates x, y, and z or the cylindrical coordinates r, θ, and z. The equation of constraint is

$$x^2 + y^2 = a^2 \tag{i}$$

while the attractive force is

$$\mathbf{F} = -k\mathbf{r} \tag{ii}$$

where k is the force constant and $r^2 = x^2 + y^2 + z^2$. Using cylindrical coordinates, the kinetic energy T is

$$T = \tfrac{1}{2}mv^2 = \tfrac{1}{2}m(\dot r^2 + r^2\dot\theta^2 + \dot z^2) \tag{iii}$$

Since in the present situation $r = a$, $\dot r = 0$; hence

$$T = \tfrac{1}{2}m(a^2\dot\theta^2 + \dot z^2) \tag{iv}$$

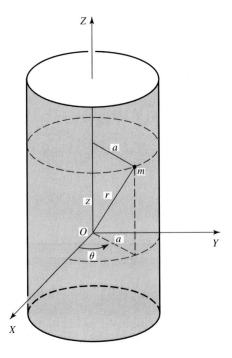

Figure Ex. 12.8

while the potential energy is given by, using Eq. (i),

$$V = \tfrac{1}{2}kr^2 = \tfrac{1}{2}k(x^2 + y^2 + z^2) = \tfrac{1}{2}k(a^2 + z^2) \tag{v}$$

Therefore,

$$L = L(z, \dot{\theta}, \dot{z}) = T - V = \tfrac{1}{2}m(a^2\dot{\theta}^2 + \dot{z}^2) - \tfrac{1}{2}k(a^2 + z^2) \tag{vi}$$

We have to replace $\dot{\theta}$ and $\dot{z}$ by p_θ and p_r. Thus

$$p_\theta = \frac{\partial L}{\partial \dot{\theta}} = ma^2\dot{\theta} \quad \text{or} \quad \dot{\theta} = \frac{p_\theta}{ma^2} \tag{vii}$$

$$p_z = \frac{\partial L}{\partial \dot{z}} = m\dot{z} \quad \text{or} \quad \dot{z} = \frac{p_z}{m} \tag{viii}$$

Substituting these in Eq. (iv),

$$T = \frac{1}{2}m\left[a^2\left(\frac{p_\theta}{ma^2}\right)^2 + \left(\frac{p_z}{m}\right)^2\right] = \frac{p_\theta^2}{2ma^2} + \frac{p_z^2}{2m}$$

Therefore,

$$H = H(z, \theta, p_\theta, p_z) = T + V = \frac{p_\theta^2}{2ma^2} + \frac{p_z^2}{2m} + \frac{1}{2}kz^2 \tag{ix}$$

where we have dropped the constant term $\tfrac{1}{2}ka^2$. Thus, using the canonical equations,

$$\frac{\partial H}{\partial q_k} = -\dot{p}_k \quad \text{and} \quad \frac{\partial H}{\partial p_k} = \dot{q}_k$$

and the generalized coordinates z, θ, p_z, and p_θ, we obtain

$$\dot{z} = \frac{\partial H}{\partial p_z} = \frac{p_z}{m} \quad \text{or} \quad p_z = m\dot{z} \tag{x}$$

$$\dot{\theta} = \frac{\partial H}{\partial p_\theta} = \frac{p_\theta}{ma^2} \quad \text{or} \quad p_\theta = ma^2\dot{\theta} \tag{xi}$$

$$-\dot{p}_z = \frac{\partial H}{\partial z} = kz \quad \text{or} \quad \dot{p}_z = -kz \tag{xii}$$

$$-\dot{p}_\theta = \frac{\partial H}{\partial \theta} = 0 \quad \text{or} \quad \dot{p}_\theta = 0 \quad \text{or} \quad p_\theta = \text{constant} \tag{xiii}$$

Combining Eqs. (x) and (xii), we get

$$m\ddot{z} + kz = 0 \tag{xiv}$$

which states that the motion of the particle in the Z direction is simple harmonic with the frequency ω given by

$$\omega = \sqrt{\frac{k}{m}} \tag{xv}$$

while from Eqs. (xi) and (xiii) we obtain

$$p_\theta = \text{constant} = ma^2\dot\theta \qquad\qquad \textbf{(xvi)}$$

That is, the angular momentum about the Z-axis is a constant of motion, as it should be, since the Z-axis is the axis of symmetry.

EXERCISE 12.8 Repeat the example of a sliding mass (Example 12.5) using Hamilton's dynamics by finding the cyclic coordinate and the constant of motion.

PROBLEMS

12.1. Write the Lagrangian and derive the equations of motion for the four systems shown in Fig. P12.1. **(a)** A particle of mass m is shot vertically upward in a uniform gravitational field. **(b)** A mass m tied to a spring of spring constant k is allowed to vibrate vertically in a uniform gravitational field. **(c)** A simple pendulum swinging in a vertical plane. **(d)** A uniform rod of mass m and length L pivoted at O, at a distance h from the center of mass, swings in a vertical plane in a uniform gravitational field.

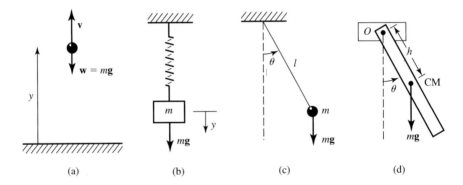

(a)　　　　(b)　　　　(c)　　　　(d)

Figure P12.1

12.2. A uniform rod of mass M, length L, and radius R rolls without slipping (translates as well as rotates) on a horizontal plane under the action of a horizontal force $\mathbf{F}$ applied at its center, as shown in Fig. P12.2. Derive Lagrange's equations of motion.

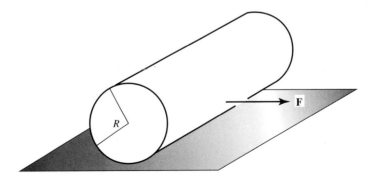

Figure P12.2

12.3. Two identical masses are tied to two identical springs as shown in Fig. P12.3 and move on a smooth horizontal plane. Derive the equations of motion for the system by using Lagrange's method.

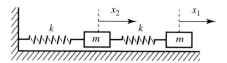

Figure P12.3

12.4. A particle of mass m moves on the outer surface of a hoop of radius R, as shown in Fig. P12.4. Determine the position of the mass as a function of time t. At what point does the particle leave the hoop?

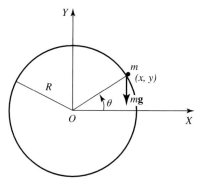

Figure P12.4

12.5. Two masses m_1 and m_2 connected by a massless string of length l are allowed to slide on two frictionless inclined planes under the influence of gravity, as shown in Fig. P12.5. Write an expression for the generalized force using **(a)** l as the generalized coordinate, and **(b)** l_1 as the generalized coordinate. Under what conditions will the system be in equilibrium? Find this position of equilibrium.

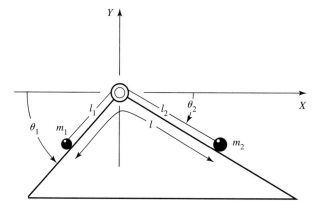

Figure P12.5

12.6. A point mass m is constrained to move on a circle of radius R and acted on by a force $\mathbf{F}$, as shown in Fig. P12.6. Calculate the generalized force and the virtual work done considering **(a)** s as a generalized coordinate, and **(b)** θ as a generalized coordinate.

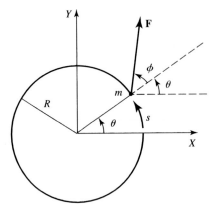

Figure P12.6

12.7. An oscillating pendulum, as shown in Fig. P12.7, consists of a mass m attached to a massless spring of spring constant k and of relaxed length l_0. Find Lagrange's equations of motion. Under what conditions will this system reduce to **(a)** a simple pendulum, and **(b)** a linear harmonic oscillator?

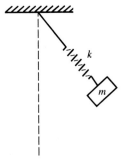

Figure P12.7

12.8. A point mass m slides down a wedge of mass M, as shown in Fig. P12.8. The wedge slides on a frictionless surface with velocity v. Write Lagrange's equations for the system if **(a)** the wedge is an inclined plane, and **(b)** the surface of the wedge is a quadrant of a circle of radius R.

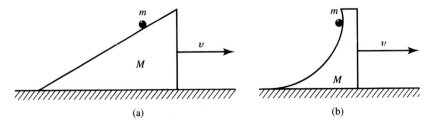

(a)   (b)

Figure P12.8

12.9. **(a)** Consider a particle of mass m subject to a force with spherical components F_r, F_θ, and F_ϕ. Set up Lagrange's equations of motion for the particle in spherical coordinates r, θ and ϕ.

(b) Find the equations of motion for the particle in part (a) if the system in spherical coordinates is rotating with an angular velocity ω about the Z-axis. Can you identify the generalized centrifugal and Coriolis forces by comparing the results in part (a)?

12.10. The point of support of a simple pendulum of length l is being elevated at constant acceleration a. Find the time period.

12.11. Two masses m and M are connected by a light inextensible string, as shown in Fig. P12.11. If the surface is frictionless, set up the equations of motion and find the acceleration of the system.

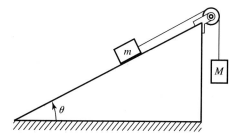

Figure P12.11

12.12. A particle of mass m slides on a smooth inclined plane whose inclination angle θ is increasing at a constant rate ω, starting from $\theta = 0$, at $t = 0$. Write equations describing the subsequent motion of the particle.

12.13. A uniform rod of length L and mass M is pivoted at O and can swing in a vertical plane (Fig. P12.13). The other end of the rod is pivoted to the center of a thin disk of mass M and radius R. (The rod swings, while the disk swings as well as rotates.) Using Lagrange's equations, derive the equations of motion for the system and describe the motion.

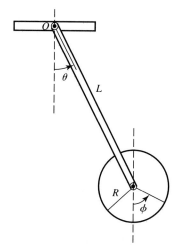

Figure P12.13

12.14. Using Lagrange's equations, derive the equations of motion of the planar double pendulum shown in Fig. P12.14. Assume the amplitude of the oscillations to be very small.

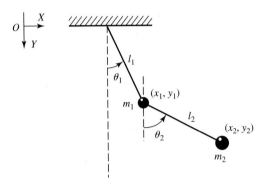

Figure P12.14

12.15. Discuss the double-pendulum problem for the special cases **(a)** $m_1 \gg m_2$, and **(b)** $m_1 \ll m_2$.

12.16. Analyze the motion of a spherical pendulum by using Lagrange's equations. Use the coordinates R, θ, and ϕ as shown in Fig. P12.16.

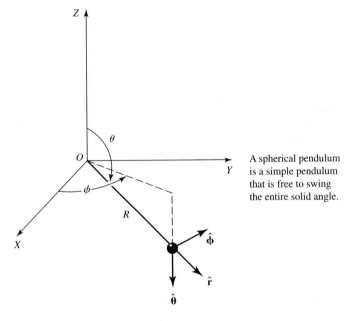

A spherical pendulum is a simple pendulum that is free to swing the entire solid angle.

Figure P12.16

12.17. Two point masses m_1 and m_2 are connected by a massless cord of length l that passes through a hole in a horizontal table (Fig. P12.17). Mass m_1 moves without friction on the table top in a circular path, while mass m_2 oscillates as a bob of a simple pendulum. Set up and solve Lagrange's equations of motion. Solve these equations for different accelerations and also calculate the frequency of the oscillations.

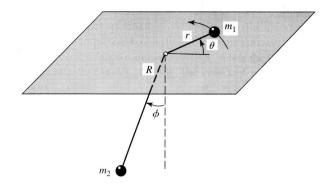

Figure P12.17

12.18. In Problem 12.17, if masses m_1 and m_2 simply move in horizontal and vertical straight lines in a uniform gravitational field, set up and solve Lagrange's equations.

12.19. The elliptical coordinates (u, v) of a point mass m are defined by

$$u = A \cosh u \cos v \quad \text{and} \quad y = A \sinh u \sin v$$

where A is a constant. Calculate the kinetic energy of the particle and its Lagrange's equations.

12.20. Suppose the coordinate transformations between the Cartesian coordinates and plane polar coordinates contain time explicitly given by

$$x = r \cos \theta + vt \quad \text{and} \quad y = r \sin \theta + at^2$$

where v and a are constants. Show that the kinetic energy in plane polar coordinates is not a homogeneous quadratic function of the velocities $\dot{r}$ and $\dot{\theta}$. Find Lagrange's equations of motion.

12.21. The coordinates (u, v) are related to the plane polar coordinates (r, θ) by

$$u = \ln \frac{r}{a} - \theta \cot \phi \quad \text{and} \quad v = \ln \frac{r}{a} - \theta \tan \phi$$

where a and ϕ are constants. Calculate the kinetic energy of a particle of mass m in terms of the coordinates u and v. Calculate the forces Q_u and Q_v. Solve the equations of motion.

12.22. *Double Atwood machine:* Consider a system of masses and pulleys as shown in Fig. P12.22. Masses m_1 and m_2 are suspended from a string of length l_1; and m_3 and m_4 are suspended by a string of length l_2. Pulleys A and B are hung from the ends of a string of length l_3 over a third fixed pulley C. Set up Lagrange's equations, and find the accelerations and the tensions in the strings. Find the acceleration if $m_1 = m$, $m_2 = 2m$, $m_3 = 3m$, and $m_4 = 4m$.

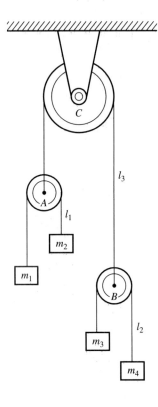

Figure P12.22

12.23. A ladder of mass M and length L rests against a smooth wall and makes an angle θ with the floor. The ladder starts sliding both on the floor and the wall. Set up Lagrange's equations of motion, assuming the ladder remains in contact with the wall. At what angle will the ladder leave the wall?

12.24. A mass m is suspended by a string of length l from a support S and oscillates as a pendulum in a vertical plane containing the X-axis and making an angle θ with the vertical, as shown in Fig. P12.24. The support S slides back and forth along a horizontal X-axis according to the equation $x = a \cos \omega t$. Set up Lagrange's equations. Show that for small values of θ the system behaves as a forced harmonic oscillator.

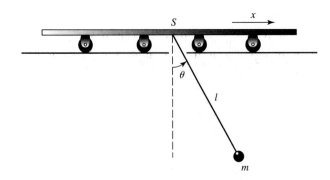

Figure P12.24

12.25. By using Lagrange's method, calculate the acceleration of a solid sphere rolling down a perfectly rough inclined plane at an angle θ with the horizontal.

12.26. Show that Lagrange's method yields the correct equations of motion for a particle moving in a plane in a rotating coordinate system $OX'Y'$.

12.27. A particle of mass m is confined to move on the inside surface of a smooth cone of half-angle α. The vertex is at the origin and the axis of the cone is vertical. What is the angular velocity ω of the particle so that it can describe a horizontal circle at a height h above the vertex?

12.28. A hoop of mass m and radius r rolls without slipping down an inclined plane of mass M with an angle of inclination ϕ. The plane slides without friction on the horizontal surface. Find and solve Lagrange's equations for the system.

12.29. Two masses m_1 and m_2 are connected by a string of length l. One mass is placed on a smooth horizontal surface, while the other mass hangs over the side after the string passes over a solid pulley of mass M and radius R. Find the Lagrangian and the equations of motion.

12.30. A bead of mass m subject to no external force is constrained to move on a straight wire rotating at constant angular velocity about an axis through O and perpendicular to the wire (Fig. P12.30). Using Lagrange's equations, set up the equations of motion in both Cartesian and polar coordinates. Calculate the constraining force.

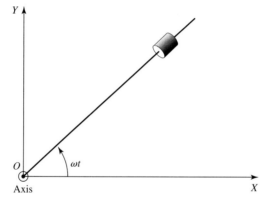

Figure P12.30

12.31. Consider a smooth wire bent so as to form a helix, the equations of which, in cylindrical coordinates, are $z = k\theta$ and $r = a$, where k and a are constants. The origin is the center of an attractive force that varies directly with distance. By using Lagrange's equations, discuss the motion of a bead that is free to slide.

12.32. In Problem 12.31, find the components of the reaction of the wire in the r, θ, and z directions.

12.33. A small sphere of mass m and radius r slides down a smooth stationary large sphere of radius R under the action of gravity from a position of rest at the top. Use Lagrange's equations to find the reaction of the sphere on the particle at any value θ, θ being the angle between the vertical diameter of the sphere and the normal to the sphere that passes through the particle. Find the value of θ at which the particle falls off.

12.34. A particle of mass m is constrained to move along a smooth wire bent into the form of a horizontal circle of radius a. Initially, the particle has a velocity v_0. The motion is subject to air resistance proportional to the square of the velocity. Using Lagrange's method, find the angular position of the particle as a function of time. Calculate the reaction of the wire on the particle. Ignore the effect of gravity.

12.35. In Problem 12.34, if the wire is rough and the coefficient of friction is μ, what will be the resultant reaction of the wire on the particle?

12.36. Write Hamilton's equations for a point mass moving in a straight line.

12.37. Two masses m_1 and m_2 are attached to the ends of a massless spring of spring constant k and relaxed length l_0. The system oscillates and rotates in a plane but moves freely through space.

   (a) Find the number of degrees of freedom.

   (b) Derive Hamilton's equations.

   (c) Identify the cyclic coordinates and state the corresponding conservation laws.

12.38. Derive Hamilton's equations for the oscillating pendulum in Problem 12.1(b) (mass–spring pendulum).

12.39. A point mass m is subjected to a central isotropic force. Using plane polar coordinates, compare the Hamiltonian H of this point mass relative to a reference frame S fixed in space with the Hamiltonian H' of the same point mass relative to frame S' rotating about the force center with constant angular speed $\omega(\dot{\phi}' = \dot{\phi} - \omega)$.

12.40. For Problem 12.39, set up Hamilton's equations for the particle in terms of r' and θ'. Also write these equations in terms of r and θ.

12.41. Set up and solve Hamilton's canonical equations for **(a)** a projectile in two dimensions and **(b)** a simple pendulum.

12.42. Using Hamilton's method, derive an expression that describes the motion of a particle executing simple harmonic motion.

12.43. Write the Hamiltonian of a simple pendulum and derive its equation of motion.

12.44. Find Hamilton's equations for a harmonic oscillator for which $V = \frac{1}{2}kx^2 + \frac{1}{4}\epsilon x^4$, where k and ϵ are constants.

12.45. A particle of mass m moves under the influence of gravity along the spiral $z = k\theta$, $r = a$, where k and a are constants and Z is the vertical axis. Obtain Hamilton's equations of motion.

12.46. A pendulum of mass m and length l is suspended from a car of mass M that moves with velocity v on a frictionless horizontal overhead rail. The pendulum swings only in the vertical plane containing the rail.

   (a) Set up the Lagrangian L and the Hamiltonian H.

   (b) Show that there is one cyclic (ignorable) coordinate. Eliminate this coordinate and discuss the motion.

12.47. A mass m tied to a vertical spring of spring constant k and vertical length l_0 has a constant upward acceleration a_0. The acceleration due to gravity is g. Find the Lagrangian and the equations of motion. Calculate the Hamiltonian function, and write Hamilton's equations. What is the time period of the motion?

12.48. A particle of mass m moves in a central field attractive force of magnitude $[k/r^2]e^{-\alpha t}$, where k and α are constants. Find the Lagrangian and Hamiltonian functions. Obtain Hamilton equations of motion. Is H constant? Is H the total energy?

12.49. A mass m is suspended by means of a string of length L that passes through a hole in a table. The string is pulled up at a constant rate k (cm/s); that is, $dL/dt = -k$. The suspension point remains constant, and the mass oscillates as a pendulum. Calculate the Lagrangian and Hamiltonian functions. Compare the Hamiltonian and the total energy, and discuss the conservation of energy for the system. Is it a constant of motion? Is it the total energy? Obtain Hamilton's equations of motion.

12.50. A particle of mass m moves in one dimension under the influence of a force

$$F(x, t) = \frac{k}{x^2} \, e^{-(t/\tau)}$$

where k and τ are positive constants. Calculate the Lagrangian and Hamiltonian functions. Is the Hamiltonian a constant of motion? Is it the total energy? Derive and solve the equations of motion.

SUGGESTIONS FOR FURTHER READING

ARTHUR, W., and FENSTER, S. K., *Mechanics,* Chapter 14. New York: Holt, Rinehart and Winston, Inc., 1969.

BECKER, R. A., *Introduction to Theoretical Mechanics,* Chapter 13. New York: McGraw-Hill Book Co., 1954.

CORBEN, H. C., and STEHLE, P., *Classical Mechanics,* Chapters 6, 7, and 10. New York: John Wiley & Sons, Inc., 1960.

DAVIS, A. DOUGLAS, *Classical Mechanics,* Chapter 10. New York: Academic Press, Inc., 1986.

FOWLES, G. R., *Analytical Mechanics,* Chapter 10. New York: Holt, Rinehart and Winston, Inc., 1962.

*GOLDSTEIN, H., *Classical Mechanics,* 2nd ed., Chapters 1 and 2. Reading, Mass.: Addison-Wesley Publishing Co., 1980.

HAUSER, W., *Introduction to the Principles of Mechanics,* Chapters 5 and 6. Reading, Mass.: Addison-Wesley Publishing Co., 1965.

MARION, J. B., *Classical Dynamics,* 2nd ed., Chapters 6 and 7. New York: Academic Press, Inc., 1970.

*MOORE, E. N., *Theoretical Mechanics,* Chapters 2 and 3. New York: John Wiley & Sons, Inc., 1983.

ROSSBERG, K., *Analytical Mechanics,* Chapter 8. New York: John Wiley & Sons, Inc., 1983.

SLATER, J. C., *Mechanics,* Chapter 4. New York: McGraw-Hill Book Co., 1947.

SOMMERFELD, A., *Mechanics,* Chapters 2 and 6. New York: Academic Press. Inc., 1952.

STEPHENSON, R. J., *Mechanics and Properties of Matter,* Chapter 7. New York: John Wiley & Sons, Inc., 1962.

SYMON, K. R., *Mechanics,* 3rd ed., Chapter 9. Reading, Mass.: Addison-Wesley Publishing Co., 1971.

SYNGE, J. L., and GRIFFITH, B. A., *Principles of Mechanics,* Chapter 15. New York: McGraw-Hill Book Co., 1959.

*The asterisk indicates works of an advanced nature.

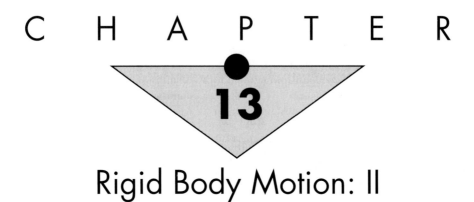

C H A P T E R

13

Rigid Body Motion: II

13.1 INTRODUCTION

We continue our discussion of rigid body motion started in Chapter 9. We briefly summarize the properties of rigid bodies as discussed there. A *rigid body* may be defined as a collection of discrete point particles for which the distance between any pair of particles is constrained to remain constant with time. Actually, these point particles are atoms and molecules that are always in constant vibrational motion. But these vibrations are on a microscopic scale and may be neglected. A perfectly rigid body will have no elastic deformation, and a mechanical pulse signal (a blow) will travel with infinite velocity. In actual practice, all rigid bodies have elastic properties and transmission velocities are $\simeq 10^3$ m/s. In most situations we shall ignore elastic deformation.

The motion of a rigid body can be described by using two coordinate systems, an inertial coordinate system and a body coordinate system, that is, a coordinate system fixed with respect to the body. Furthermore, to specify the position of the body, six coordinate must be specified. Three of these are usually taken to be the coordinates of the center of mass of the rigid body (usually the origin of the body coordinate system is taken to coincide with the center of mass), and the other three coordinates are taken to be the angles that describe the orientation of the body coordinate axes with respect to the inertial (or fixed) coordinate axes. One set of three commonly used independent angles are the Eulerian angles, as will be described in this chapter.

13.2 ANGULAR MOMENTUM AND KINETIC ENERGY

Let us consider a rigid body B as shown in Fig. 13.1. The body is rotating about an axis passing through a single fixed point O, while the coordinate system $OXYZ$ is fixed in the body with its

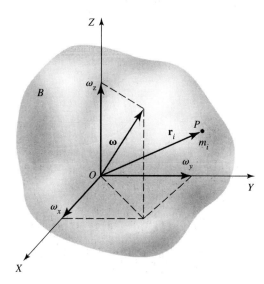

Figure 13.1   Rigid body B rotating with angular velocity $\boldsymbol{\omega}(\omega_x, \omega_y, \omega_z)$ about an axis passing through a single fixed point O.

origin at O. The instantaneous translational velocity $\mathbf{v}_i$ of particle P of mass m_i, which is at a distance $\mathbf{r}_i$ from the origin O, is

$$\mathbf{v}_i = \boldsymbol{\omega} \times \mathbf{r}_i \tag{13.1}$$

where $\boldsymbol{\omega}$ is the angular velocity of the body with its components $(\omega_x, \omega_y, \omega_z)$ as shown. The angular momentum $\mathbf{L}$ relative to origin O, for a system of particles m_i may be defined as

$$\mathbf{L} = \sum_{i=1}^{n} m_i \mathbf{r}_i \times \mathbf{v}_i \tag{13.2}$$

Substituting for $\mathbf{v}_i$ from Eq. (13.1),

$$\mathbf{L} = \sum_{i=1}^{n} m_i \mathbf{r}_i \times (\boldsymbol{\omega} \times \mathbf{r}_i) \tag{13.3}$$

Using the identity for a triple cross product,

$$\mathbf{A} \times (\mathbf{B} \times \mathbf{A}) = A^2 \mathbf{B} - \mathbf{A}(\mathbf{A} \cdot \mathbf{B}) \tag{13.4}$$

we may write

$$\mathbf{r}_i \times (\boldsymbol{\omega} \times \mathbf{r}_i) = r_i^2 \boldsymbol{\omega} - \mathbf{r}_i(\mathbf{r}_i \cdot \boldsymbol{\omega})$$

$$= (x_i^2 + y_i^2 + z_i^2)(\hat{\mathbf{i}}\omega_x + \hat{\mathbf{j}}\omega_y + \hat{\mathbf{k}}\omega_z)$$

$$- (\hat{\mathbf{i}}x_i + \hat{\mathbf{j}}y_i + \hat{\mathbf{k}}z_i)(x_i\omega_x + y_i\omega_y + z_i\omega_z) \tag{13.5}$$

Combining this result with Eq. (13.3) and rearranging,

$$\mathbf{L} = \hat{\mathbf{i}}L_x + \hat{\mathbf{j}}L_y + \hat{\mathbf{k}}L_z$$

$$= \hat{\mathbf{i}}\left[\omega_x \sum_{i=1}^{n} m_i(y_i^2 + z_i^2) - \omega_y \sum_{i=1}^{n} m_i x_i y_i - \omega_z \sum_{i=1}^{n} m_i x_i z_i \right]$$

$$+ \hat{\mathbf{j}}\left[-\omega_x \sum_{i=1}^{n} m_i x_i y_i + \omega_y \sum_{i=1}^{n} m_i(x_i^2 + z_i^2) - \omega_z \sum_{i=1}^{n} m_i y_i z_i \right]$$

$$+ \hat{\mathbf{k}}\left[-\omega_x \sum_{i=1}^{n} m_i x_i z_i - \omega_y \sum_{i=1}^{n} m_i y_i z_i + \omega_z \sum_{i=1}^{n} m_i(x_i^2 + y_i^2) \right] \qquad (13.6)$$

We may obtain the same result by using a matrix expansion

$$\mathbf{r}_i \times (\boldsymbol{\omega} \times \mathbf{r}_i) = \begin{vmatrix} \hat{\mathbf{i}} & \hat{\mathbf{j}} & \hat{\mathbf{k}} \\ x_i & y_i & z_i \\ (\omega_y z_i - \omega_z y_i) & (\omega_z x_i - \omega_x z_i) & (\omega_x y_i - \omega_y x_i) \end{vmatrix} \qquad (13.7)$$

which on simplification and combining with Eq. (13.3) gives the same result as Eq. (13.6).
    In short, we may write Eq. (13.6) as

$$\mathbf{L} = \hat{\mathbf{i}}L_x + \hat{\mathbf{j}}L_y + \hat{\mathbf{k}}L_z$$

$$= \hat{\mathbf{i}}[\omega_x I_{xx} - \omega_y I_{xy} - \omega_z I_{xz}] + \hat{\mathbf{j}}[-\omega_x I_{yx} + \omega_y I_{yy} - \omega_z I_{yz}]$$

$$+ \hat{\mathbf{k}}[-\omega_x I_{zx} - \omega_y I_{zy} + \omega_z I_{zz}] \qquad (13.8)$$

where the quantities I_{xx}, I_{yy}, and I_{zz} involve the sums of the squares of the coordinates and are called the *moments of inertia* of the body about the coordinate axes; that is (the summation is taken from $i = 1$ to n),

$$I_{xx} = \sum m_i(y_i^2 + z_i^2) = \sum m_i(r_i^2 - x_i^2) = \text{moment of inertia about the } X\text{-axis}$$

$$I_{yy} = \sum m_i(x_i^2 + z_i^2) = \sum m_i(r_i^2 - y_i^2) = \text{moment of inertia about the } Y\text{-axis}$$

$$I_{zz} = \sum m_i(x_i^2 + y_i^2) = \sum m_i(r_i^2 - z_i^2) = \text{moment of inertia about the } X\text{-axis} \qquad (13.9)$$

The quantities I_{xy}, I_{xz}, . . . , involve the sums of the products of the coordinates and are called the *products of inertia*; that is,

$$I_{xy} = I_{yx} = \sum m_i x_i y_i, \qquad xy \text{ product of inertia} \qquad (13.10a)$$

$$I_{yz} = I_{zy} = \sum m_i y_i z_i, \qquad yz \text{ product of inertia} \qquad \textbf{(13.10b)}$$

$$I_{zx} = I_{xz} = \sum m_i z_i x_i, \qquad zx \text{ product of inertia} \qquad \textbf{(13.10c)}$$

It is clear from Eq. (13.8) that $\mathbf{L}$ is not necessarily always in the same direction as the instantaneous axis of rotation; that is $\mathbf{L}$ is not always in the same direction as $\boldsymbol{\omega}$. For example, if the Z-axis is the direction of rotation, $\boldsymbol{\omega} = (0, 0, \omega)$; that is, $\omega_x = \omega_y = 0$ and $\omega_z = \omega$, then from Eq. (13.8)

$$L_x = -I_{xz}\omega, \quad L_y = -I_{yz}\omega, \quad L_z = +I_{zz}\omega$$

That is, $\mathbf{L}$ has a component $L_z = I_{zz}\omega$ in the direction of rotation, but also has two other components in the directions at right angles to the direction of rotation. Thus $\mathbf{L}$ and $\boldsymbol{\omega}$ are not in the same direction. This point is further illustrated in Example 13.1.

The components of $\mathbf{L}$ given by Eq. (13.8) may be written in a compact form as

$$\mathbf{L}_k = \sum_{i=1}^{3} \omega_l I_{kl} \qquad \textbf{(13.11)}$$

where $k = 1, 2, 3$ and $l = 1, 2, 3$; that is x, y, and z have been replaced by 1, 2, and 3.

Now we are in a position to derive a general expression for the rotational kinetic energy of a body. In simple cases, the axis of rotation always remains normal to a fixed plane. This need not be the case, as we demonstrate now. Let us calculate the kinetic energy of a rigid body that is rotating about an axis passing through a fixed point with an angular velocity $\boldsymbol{\omega}$. A particle of mass m_i at a distance $\mathbf{r}_i$ has a velocity $\mathbf{v}_i$.

$$\mathbf{v}_i = \boldsymbol{\omega} \times \mathbf{r}_i \qquad \textbf{(13.1)}$$

Thus the kinetic energy of the whole body is given by

$$T = \sum_{i=1}^{n} \frac{1}{2} m_i v_i^2 = \sum_{i=1}^{n} \frac{1}{2} m_i \mathbf{v}_i \cdot \mathbf{v}_i = \frac{1}{2} \sum_{i=1}^{n} [(\boldsymbol{\omega} \times \mathbf{r}_i) \cdot (m_i \mathbf{v}_i)] \qquad \textbf{(13.12)}$$

But in a triple scalar product, the dot and cross may be interchanged; that is,

$$(\mathbf{A} \times \mathbf{B}) \cdot \mathbf{C} = \mathbf{A} \cdot (\mathbf{B} \times \mathbf{C}) \qquad \textbf{(13.13)}$$

or
$$(\boldsymbol{\omega} \times \mathbf{r}_i) \cdot m_i \mathbf{v}_i = \boldsymbol{\omega} \cdot (\mathbf{r}_i \times m_i \mathbf{v}_i) \qquad \textbf{(13.14)}$$

For kinetic energy T, we may write Eq. (13.12) as

$$T = \frac{1}{2} \sum_{i=1}^{n} \boldsymbol{\omega} \cdot (\mathbf{r}_i \times m_i \mathbf{v}_i) \qquad \textbf{(13.15a)}$$

Since $\boldsymbol{\omega}$ is the same for all particles, and from the definition of angular momentum given by Eq. (13.2), we may write

$$T = \frac{1}{2} \boldsymbol{\omega} \cdot \left[\sum_{i=1}^{n} (\mathbf{r}_i \times m_i \mathbf{v}_i) \right] \qquad \textbf{(13.15b)}$$

or
$$T = \frac{1}{2}\boldsymbol{\omega} \cdot \mathbf{L} \tag{13.16}$$

It may be pointed out that unlike $\mathbf{L}$, which is a vector and has three components, the rotational kinetic energy T is a scalar (a dot product of $\frac{1}{2}\boldsymbol{\omega}$ and $\mathbf{L}$). Also, this expression for T is analogous to the expression for the translational kinetic energy T_{tran} given by

$$T_{tran} = \frac{1}{2}\mathbf{v}_c \cdot \mathbf{p}_c \tag{13.17}$$

where $\mathbf{v}_c$ is the velocity of the center of mass and $\mathbf{p}_c$ is the linear momentum of the system. Using the expression

$$\boldsymbol{\omega} = \hat{\mathbf{i}}\omega_x + \hat{\mathbf{j}}\omega_y + \hat{\mathbf{k}}\omega_z \tag{13.18}$$

and Eq. (13.8) for $\mathbf{L}$ in Eq. (13.16), we may write

$$T = \frac{1}{2}\boldsymbol{\omega} \cdot \mathbf{L} = \frac{1}{2}\omega_x L_x + \frac{1}{2}\omega_y L_y + \frac{1}{2}\omega_z L_z$$

$$= \frac{1}{2}\omega_x^2 I_{xx} + \frac{1}{2}\omega_y^2 I_{yy} + \frac{1}{2}\omega_z^2 I_{zz} - \omega_x\omega_y I_{xy} - \omega_y\omega_z I_{yz} - \omega_z\omega_x I_{zx} \tag{13.19}$$

Instead of using (x, y, z), we may use $k = 1, 2, 3$ and $l = 1, 2, 3$ and write T in a compact form as

$$T = \frac{1}{2}\sum_{\substack{k=1 \\ l=1}}^{3} \omega_k \omega_l I_{kl} = \frac{1}{2}\boldsymbol{\omega} \cdot \mathbf{L} \tag{13.20}$$

In many practical situations, a rigid body consists of continuous mass with density ρ, which may not be constant. In such cases, summation must be replaced by volume integration. Thus the moment of inertia and the product of inertia may be written as

$$I_{xx} = \int_V \rho(y^2 + z^2)\, dx\, dy\, dz$$

$$I_{yy} = \int_V \rho(x^2 + z^2)\, dx\, dy\, dz$$

$$I_{zz} = \int_V \rho(x^2 + y^2)\, dx\, dy\, dz \tag{13.21a}$$

$$I_{xy} = \int_V \rho xy\, dx\, dy\, dz$$

$$I_{yz} = \int_V \rho yz\, dx\, dy\, dz$$

$$I_{zx} = \int_V \rho zx\, dx\, dy\, dz \tag{13.21b}$$

Example 13.1 _____

Two point masses of equal mass m are connected by a massless rigid rod of length $2a$ forming a dumb-bell. The dumbbell is constrained to rotate with a constant angular velocity $\boldsymbol{\omega}$ about an axis that makes an angle ϕ with the rod. Calculate the magnitudes and the directions of the angular momentum and the torque that is applied to the system.

Solution

As shown in Fig. Ex. 13.1(a), let the dumbbell rotate with an angular velocity $\boldsymbol{\omega}$ about an axis AOA' passing through O and lying in the inertial coordinate system. (AOA' is also the direction of the axle and the bearings are at O.) The point O is the origin of the coordinate system. The angular momentum of the system due to the two masses is

$$\mathbf{L} = \mathbf{L}_1 + \mathbf{L}_2 = m\mathbf{r}_1 \times (\boldsymbol{\omega} \times \mathbf{r}_1) + m\mathbf{r}_2 \times (\boldsymbol{\omega} \times \mathbf{r}_2) \qquad \textbf{(i)}$$

Note that both $\mathbf{L}_1$ and $\mathbf{L}_2$ point in the same direction as does $\mathbf{L}$, as shown in Fig. Ex. 13.1(a). It is quite clear that $\mathbf{L}$ is not in the same direction as $\boldsymbol{\omega}$. As shown in part (b), if $\mathbf{L}$ is resolved into two components, only $\mathbf{L}_{\parallel}$ is in the direction of $\boldsymbol{\omega}$, while $\mathbf{L}_{\perp}$, although in a plane at right angles to $\boldsymbol{\omega}$, is not zero. The magnitude of the angular momentum is

$$L = ma^2\omega \sin \phi + ma^2\omega \sin \phi = 2ma^2\omega \sin \phi = I\omega \sin \phi \qquad \textbf{(ii)}$$

where I is the moment of inertia of the dumbbell about an axis perpendicular to the length of the connecting rod.

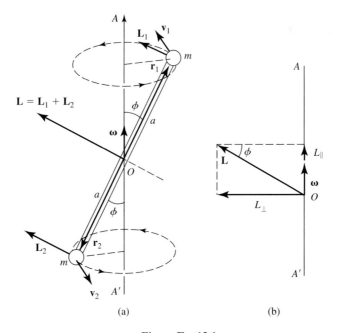

(a)   (b)

Figure Ex. 13.1

Furthermore, the angular momentum vector **L** is continuously changing direction as it rotates about **ω**. Thus **L** is not constant, and it is necessary to apply a torque **τ** to maintain this motion. By definition

$$\boldsymbol{\tau} = \frac{d\mathbf{L}}{dt} = \dot{\mathbf{L}} \tag{iii}$$

where $\dot{\mathbf{L}}$ is a vector in the direction in which the tip (or head) of vector **L** is moving. In analogy to the relation $\dot{\mathbf{r}} = \boldsymbol{\omega} \times \mathbf{r}$, we may write

$$\dot{\mathbf{L}} = \boldsymbol{\omega} \times \mathbf{L} \tag{iv}$$

Thus the magnitude of the applied torque is [substituting for **L** from Eq. (ii)]

$$|\boldsymbol{\tau}| = |\dot{\mathbf{L}}| = \omega L \sin(90° - \phi) = 2ma^2\omega^2 \sin \phi \cos \phi \tag{v}$$

and the direction of the torque, from Eq. (iv), is perpendicular to the plane containing **ω** and **L** at any instant. If—rather than having one dumbbell as in Fig. Ex. 13.1(a)—there are two dumbbells moving symmetrically, by drawing a simple diagram we can show that **L** and **ω** will be in the same direction.

EXERCISE 13.1 Discuss the motion of the double dumbbells if the masses of the one dumbbell are different from those of the other (say twice).

13.3 INERTIA TENSOR

We proceed to write expressions for kinetic energy and angular momentum in tensor notation. Once again, we are considering a rigid body rotating about an axis passing through a fixed point located inside or outside the body. We shall use i, j for running indexes referring to the particles, while k, l, and s will be used to refer to the coordinate axes. The expression for rotational kinetic energy is

$$T = T_{\text{rot}} = \frac{1}{2} \sum_{i=1}^{n} m_i(\boldsymbol{\omega} \times \mathbf{r}_i)^2 \tag{13.22}$$

Making use of the vector identity

$$(\mathbf{A} \times \mathbf{B})^2 = (\mathbf{A} \times \mathbf{B}) \cdot (\mathbf{A} \times \mathbf{B}) = A^2B^2 - (\mathbf{A} \cdot \mathbf{B})^2 \tag{13.23}$$

In Eq. (13.22), we may write

$$T = T_{\text{rot}} = \frac{1}{2} \sum_{i=1}^{n} m_i[\omega^2 r_i^2 - (\boldsymbol{\omega} \cdot \mathbf{r}_i)^2] \tag{13.24}$$

The vector $\mathbf{r}_i$ has components x_{is}, that is (x_{i1}, x_{i2}, x_{i3}), and **ω** has components $\omega_k(\omega_1, \omega_2, \omega_3)$. Thus

$$T = \frac{1}{2} \sum_{i=1}^{n} m_i \left[\left(\sum_{k=1}^{3} \omega_k^2 \right) \left(\sum_{s=1}^{3} x_{is}^2 \right) - \left(\sum_{k=1}^{3} \omega_k x_{ik} \right) \left(\sum_{i=1}^{3} \omega_l x_{il} \right) \right] \tag{13.25}$$

Making use of the relation

$$\omega_k = \sum_{l} \omega_l \delta_{kl}$$

where [from Eq. (5.176)], we write $\delta_{kl} = 1$ if $k = l$ and $\delta_{kl} = 0$ if $k \neq l$, we may write Eq. (13.25) as

$$T = \frac{1}{2} \sum_i \sum_{k,l} m_i \left[\omega_k \omega_l \delta_{kl} \left(\sum_s x_{is}^2 \right) - \omega_k \omega_l x_{ik} x_{il} \right]$$

Since all points in a rigid body have the same angular velocity, we may factor these out and write

$$T = \frac{1}{2} \sum_{k,l} \omega_k \omega_l \sum_i m_i \left[\delta_{kl} \sum_s x_{is}^2 - x_{ik} x_{il} \right] \tag{13.26}$$

If we define I_{kl} to be the klth element of the sum over i, that is,

$$I_{kl} = \sum_{i=1}^{n} m_i \left[\delta_{kl} \sum_s x_{is}^2 - x_{ik} x_{il} \right] \tag{13.27a}$$

or, noting that $x_{i1}^2 + x_{i2}^2 + x_{i3}^2 = r_i^2$, we may write

$$I_{kl} = \sum_{i=1}^{n} m_i \left[\delta_{kl} r_i^2 - x_{ik} x_{il} \right] \tag{13.27b}$$

Then Eq. (13.26) for rotational kinetic energy becomes

$$T = \frac{1}{2} \sum_{k,l} I_{kl} \omega_k \omega_l \tag{13.28}$$

I_{kl}, given by Eq. (13.27), has *nine* components and constitutes the elements of a quantity I, called the *moment of inertia tensor* or simply an *inertia tensor* of a rigid body relative to a body coordinate system. I is very similar in form to a 3×3 matrix and, as we shall see shortly, it is a factor of proportionality between L and ω and also between T and $\omega\omega$ (a quantity called *dyadic*, discussed in Section 13.7). The dimensions of I are (mass) $\times$ (length)^2. The elements of I can be obtained from Eq. (13.27) and may be written in a 3×3 array.

$$\mathbf{I} = \begin{pmatrix} \sum_i m_i(x_{i2}^2 + x_{i3}^2) & -\sum_i m_i x_{i1} x_{i2} & -\sum_i m_i x_{i1} x_{i3} \\ -\sum_i m_i x_{i2} x_{i1} & \sum_i m_i(x_{i1}^2 + x_{i3}^2) & -\sum_i m_i x_{i2} x_{i3} \\ -\sum_i m_i x_{i3} x_{i1} & -\sum_i m_i x_{i3} x_{i2} & \sum_i m_i(x_{i1}^2 + x_{i2}^2) \end{pmatrix} \tag{13.29}$$

which for a single point mass m reduces to

$$\mathbf{I} = m \begin{pmatrix} x_2^2 + x_3^2 & -x_1 x_2 & -x_1 x_3 \\ -x_2 x_1 & x_1^2 + x_3^2 & -x_2 x_3 \\ -x_3 x_1 & -x_3 x_2 & x_1^2 + x_2^2 \end{pmatrix} \tag{13.30}$$

or, in general,

$$\mathbf{I} = I_{kl} = \begin{pmatrix} I_{11} & I_{12} & I_{13} \\ I_{21} & I_{22} & I_{23} \\ I_{31} & I_{32} & I_{33} \end{pmatrix} \tag{13.31}$$

The diagonal elements I_{11}, I_{22}, and I_{33}, that is,

$$I_{kk} = \sum_{i=1}^{n} m_i(r_i^2 - x_{ik}^2) \tag{13.32}$$

are called the *moment of inertia* about the k-axis. The off-diagonal elements given by

$$I_{kl} = I_{lk} = -\sum_{i=1}^{n} m_i x_{ik} x_{il} \tag{13.33}$$

are called the *product of inertia.* Since the off-diagonal element are symmetric, $I_{kl} = I_{lk}$, the inertia tensor is a *symmetric tensor.* Hence only six elements of $\mathbf{I}$ are independent. Furthermore, the tensor $\mathbf{I}$ has a positive definite form.

Let us consider a particular element I_{11}; that is,

$$I_{11} = \sum_{i=1}^{n} m_i(r_i^2 - x_{i1}^2) = \sum_{i=1}^{n} m_i(x_{i1}^2 + x_{i2}^2 + x_{i3}^2 - x_{i1}^2) = \sum_{i=1}^{n} m_i(x_{i2}^2 + x_{i3}^2) \tag{13.34}$$

$(x_{i2}^2 + x_{i3}^2)$ is the square of the distance from the ith mass point to the X_1-axis; hence I_{11} is always positive or zero. In general, we may conclude that the diagonal elements I_{kk} are always positive or zero. I_{kk} is zero only if all the masses lie on the kth axis. On the other hand, the off-diagonal elements I_{kl} may be positive, negative, or zero.

Another property of the inertia tensor is the additive property of the elements. That is, the inertia tensor for a body can be considered to be the sum of the tensors for the various portions for the body. Thus, for a continuous distribution, we may write, using Eq. (13.27),

$$I_{kl} = \int \int_{V} \int \rho(\mathbf{r}) \left[\delta_{kl} \sum_{s} x_s^2 - x_k x_l \right] dV = \int \int_{V} \int \rho(\mathbf{r}) [\delta_{kl} r^2 - x_k x_l] \, dV \tag{13.35}$$

where the volume elements $dV = dx_1 \, dx_2 \, dx_3$, $\rho(\mathbf{r})$ is the density, and the integration is taken over the whole volume. Note that the indexes for the mass of the particles are not needed.

We may arrive at the same expression for the inertia tensor by starting with the expression for angular momentum. That is, by definition

$$\mathbf{L} = \sum_i m_i \mathbf{r}_i \times \mathbf{v}_i = \sum_i m_i \mathbf{r}_i \times (\boldsymbol{\omega} \times \mathbf{r}_i) \tag{13.36}$$

Using the vector identity

$$\mathbf{A} \times (\mathbf{B} \times \mathbf{A}) = A^2 \mathbf{B} - \mathbf{A}(\mathbf{A} \cdot \mathbf{B}) \tag{13.37}$$

we get

$$\mathbf{L} = \sum_{i=1}^{n} m_i[r_i^2 \boldsymbol{\omega} - \mathbf{r}_i(\mathbf{r}_i \cdot \boldsymbol{\omega})] \tag{13.38}$$

Unlike T, the angular momentum is a vector quantity, and hence for the kth component we may write

$$L_k = \sum_i m_i \left[\omega_k \sum_s x_{is}^2 - x_{ik} \sum_l x_{il} \omega_l \right] = \sum_i m_i \sum_l \left[\omega_l \delta_{kl} \sum_s x_{is}^2 - \omega_l x_{ik} x_{il} \right]$$

$$= \sum_l \omega_l \sum_i m_i \left[\delta_{kl} \sum_s x_{is}^2 - x_{ik} x_{il} \right] \tag{13.39}$$

As before, I_{kl} is defined as

$$I_{kl} = \sum_{i=1}^{n} m_i \left[\delta_{kl} \sum_{s} x_{is}^2 - x_{ik} x_{il} \right]$$ (13.27a)

and we may write

$$L_k = \sum_{l} I_{kl} \omega_l$$ (13.40)

or in tensor notation

$$\mathbf{L} = \mathbf{I} \cdot \boldsymbol{\omega}$$ (13.41)

As mentioned earlier and shown in Example 13.1, $\mathbf{L}$ and $\boldsymbol{\omega}$ are not in the same direction.

The relation between $\mathbf{L}$ and T may be arrived at in the following manner. Multiplying both sides of Eq. (13.40) by $\frac{1}{2} \omega_k$ and summing over k,

$$\frac{1}{2} \sum_{k} \omega_k L_k = \frac{1}{2} \sum_{k,l} I_{kl} \omega_k \omega_l = T$$

or

$$T = \frac{1}{2} \sum_{k} \omega_k L_k = \frac{1}{2} \boldsymbol{\omega} \cdot \mathbf{L}$$ (13.42)

Substituting for $\mathbf{L}$ from Eq. (13.41),

$$T = \tfrac{1}{2} \boldsymbol{\omega} \cdot \mathbf{L} = \tfrac{1}{2} \boldsymbol{\omega} \cdot \mathbf{I} \cdot \boldsymbol{\omega}$$ (13.43)

From Eq. (13.41), we may conclude that a product of a tensor and a vector is a vector; while from Eq. (13.43) we conclude that the product of two vectors and a tensor is a scalar.

 Example 13.2 _____

Calculate the components of a moment of inertia tensor for the following configuration. Point masses of 1,2,3, and 4 units are located at (1,0,0), (1,1,0), (1,1,1), and (1,1,–1).

Solution

n = 4, the number of point masses
m_i = mass of the ith particle
r_i = distance of the ith particle from the origin
x_{i1}, x_{i2}, and x_{i3} are the coordinates for the particles i = 1,2,3,4. All quantities are in arbitrary units. The masses and the coordinates of the particles are shown in the column matrices

$n := 4$ $i := 1 .. n$ $k := 1 .. 3$ $j := 1 .. 3$

$m_i :=$ $x_{i,1} :=$ $x_{i,2} :=$ $x_{i,3} :=$

m_i	$x_{i,1}$	$x_{i,2}$	$x_{i,3}$
1	1	0	0
2	1	1	0
3	1	1	1
4	1	1	-1

Calculating distance r_i from the origin and the definition of $\delta_{k,j}$ function,

$$r_i := \sqrt{\left(x_{i,1}\right)^2 + \left(x_{i,2}\right)^2 + \left(x_{i,3}\right)^2} \qquad \delta_{k,j} := \text{if}(k \equiv j, 1, 0)$$

r_i	$\left(r_i\right)^2$
1	1
1.414	2
1.732	3
1.732	3

$$\delta = \begin{pmatrix} 1 & 0 & 0 \\ 0 & 1 & 0 \\ 0 & 0 & 1 \end{pmatrix}$$

Using Eq. (13.27b), we can calculate the moment of inertia tensor as shown.

$$I_{k,j} := \sum_{i=1}^{n} m_i \left[\delta_{k,j} \cdot \left(r_i\right)^2 - x_{i,k} \cdot x_{i,j}\right]$$

$$I = \begin{pmatrix} 16 & -9 & 1 \\ -9 & 17 & 1 \\ 1 & 1 & 19 \end{pmatrix} \qquad I = \begin{pmatrix} 16 & -9 & 1 \\ -9 & 17 & 1 \\ 1 & 1 & 19 \end{pmatrix}$$

EXERCISE 13.2 Calculate **I** for point masses 4, 3, 2, and 1 units and located at $(1,1,-1)$, $(1,1,1)$, $(1,1,0)$ and $(1,0,0)$.

 Example 13.3 _____

Consider a homogeneous cube of density ρ, mass M, and side L. For origin O at one corner and an axis directed along the edges as shown in Fig. Ex. 13.3, evaluate the elements of the inertia tensor.

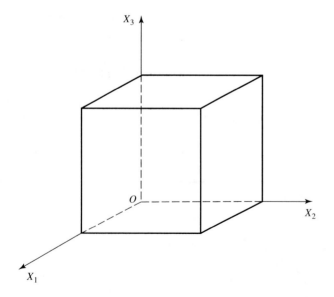

Figure Ex. 13.3

Solution

Calculate the elements of inertia tensor by using Eq. (13.35).

$$I_{k,1} = \int \int \int \rho(r) \cdot \left[\delta_{k,1} \cdot \sum_i \left(x_i \right)^2 - x_k \cdot x_1 \right] dx\, dy\, dz$$

Since the cube is homogenous, ρ is constant.

$$\delta_{k,1} = \text{if}(k=1,1,0) \qquad M = \rho \cdot L^3$$

The diagonal elements of the inertia tensor are all equal and calculated as shown. (Integrate and then substitute for ρ.)

$$I11 = \rho \cdot \int_0^L \int_0^L \int_0^L (y^2 + z^2)\, dz\, dy\, dx$$

$$I11 = \frac{2}{3} \cdot \rho \cdot L^5 = \frac{2}{3} \cdot M \cdot L^2$$

Because of symmetry, all the off-diagonal elements are equal and are calculated as shown.

$$I12 = \rho \cdot \int_0^L \int_0^L \int_0^L x \cdot y\, dz\, dy\, dx$$

$$I12 = \frac{-1}{4} \cdot \rho \cdot L^5 = \frac{-1}{4} \cdot M \cdot L^2$$

All diagonal elements = I11
All off diagonal elements = I12

$$k := 1 .. 3 \qquad 1 := 1 .. 3 \qquad M := 1 \qquad L := 1 \qquad \rho := 1$$

$$m := 1 .. 3 \qquad n := 1 .. 3$$

Using different given values, we can calculate the moment of inertia tensor. Note that each element must be multiplied γ.

$$I11 := \frac{2}{3} \cdot M \cdot L^2 \qquad I12 := \frac{-1}{4} \cdot M \cdot L^2$$

$$I_{m,n} := \text{if}(m=n, I11, I12)$$

$$I = \begin{pmatrix} 0.667 & -0.25 & -0.25 \\ -0.25 & 0.667 & -0.25 \\ -0.25 & -0.25 & 0.667 \end{pmatrix} \qquad I = \begin{pmatrix} 0.667 & -0.25 & -0.25 \\ -0.25 & 0.667 & -0.25 \\ -0.25 & -0.25 & 0.667 \end{pmatrix} \cdot \gamma \qquad \gamma = M \cdot L^2$$

EXERCISE 13.3 Repeat the example for a rectangular body of homogeneous density ρ and sides $2a$, a, and a.

13.4 MOMENT OF INERTIA FOR DIFFERENT BODY SYSTEMS (STEINER THEOREM)

We have seen that if we choose a body coordinate system whose origin coincides with the center of mass, it is possible to express the kinetic energy as a sum of the translational and rotational kinetic energy. Hence, it is convenient to know the relationship between inertia tensors

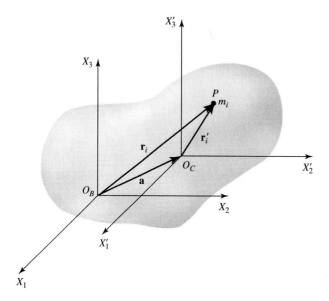

Figure 13.2 Body coordinate system with origin at O_B having its axes oriented parallel to the center-of-mass coordinate axes with origin at O_C.

expressed in different body coordinate systems. Let **I** be the inertia tensor defined in a body coordinate system with the origin fixed at O_B and **I′** be the inertia tensor defined in a center-of-mass coordinate system with its origin at the center-of-mass O_C, as shown in Fig. 13.2. Furthermore, it is assumed that the Cartesian coordinate axes in the two systems are parallel to each other as shown; that is, they have the same orientation. We wish to find the relation between **I** and **I′**. The components I_{kl} of the inertia tensor **I**, from Eq. (13.27a) are

$$I_{kl} = \sum_{i=1}^{n} m_i \left[\delta_{kl} \sum_{s} x_{is}^2 - x_{ik} x_{il} \right] \tag{13.27}$$

while the components I'_{kl} of the inertia tensor **I′** are

$$I'_{kl} = \sum_{i=1}^{n} m_i \left[\delta_{kl} \sum_{s} x_{is}'^2 - x'_{ik} x'_{il} \right] \tag{13.44}$$

Referring to Fig. 13.2, if the center of mass O_C is at a distance **a** from the origin O_B, the relation between **r** and **r′** is

$$\mathbf{r}_i = \mathbf{r}'_i + \mathbf{a} \tag{13.45a}$$

or, in component form,

$$x_{is} = x'_{is} + a_s, \quad s = 1, 2, 3 \tag{13.45b}$$

Substituting this in Eq. (13.27) and rearranging,

$$I_{kl} = \sum_i m_i \left[\delta_{kl} \sum_s (x'_{is} + a_s)^2 - (x'_{ik} + a_k)(x'_{il} + a_l) \right]$$

$$= \sum_i m_i \left[\delta_{kl} \sum_s x'^2_{is} - x'_{ik}x'_{il} \right] + \sum_i m_i \left[\delta_{ki} \sum_s a_s^2 - a_k a_l \right]$$

$$+ 2 \sum_i m_i [\delta_{kl} x'^2_{is} a_s] - \sum_i m_i x'_{ik} a_l - \sum_i m_i x'_{il} a_k \qquad (13.46)$$

Each of the last three terms on the right side is zero because of the definition of the center of mass with the origin at O_C. That is,

$$\sum_i m_i \mathbf{r}'_i = 0 \quad \text{or} \quad \sum_i m_i x'_{is} = 0$$

Thus Eq. (13.46) with the help of Eq. (13.44) takes the form

$$I_{kl} = I'_{kl} + \sum_i m_i \left[\delta_{kl} \sum_s a_s^2 - a_k a_l \right] \qquad (13.47)$$

If, instead of discrete particles, we had an extended rigid body, we would obtain the following relation

$$I_{kl} = I'_{kl} + (\delta_{kl} a_s^2 - a_k a_l) \int \int \int \rho \, dV \qquad (13.48)$$

In either situation, the mass M is given by

$$M = \sum_i m_i \quad \text{or} \quad M = \int \int \int_v \rho \, dV$$

and Eq. (13.47) or (13.48) takes the form

$$I_{kl} = I'_{kl} + M(a^2 \delta_{kl} - a_k a_l) \qquad (13.49)$$

which is the required relation. It states that the difference in the elements $I_{kl} - I'_{kl}$ is equal to the mass M of the body multiplied by the square of the distance $(a^2 \delta_{kl} - a_k a_l)$.

As a special case, let us find the relation between the diagonal elements; that is,

$$I_{kk} = I'_{kk} + M(a^2 - a_k^2) = I'_{kk} + M d_k^2 \qquad (13.50a)$$

where d_k is the shortest distance from the axis of rotation in the body system to the center of mass. The relation of Eq. (13.50a) is the statement of *Steiner's theorem*.

The moment of inertia of a rigid body in a body coordinate system about a given axis is equal to the moment of inertia of the body in the center-of-mass coordinate system about an axis parallel to the given axis plus the moment of inertia of M located at the center of mass about the given body axis.

Note that if O_B and O_C coincide, $d_k = 0$, which implies that the body will have a minimum moment of inertia in the center-of-mass coordinate system.

Let us consider a relation between the diagonal elements.

$$I_{11} = I'_{11} + M(a_1^2 + a_2^2 + a_3^2 - a_1^2) = I'_{11} + M(a_2^2 + a_3^2) = I'_{11} + Md_1^2 \qquad \textbf{(13.50b)}$$

where $d_1^2 = a_2^2 + a_3^2$. Equation (13.50b) states that the difference between elements $I_{11} - I'_{11}$ is equal to the mass M of the body multiplied by the square of the distance between the parallel axes. This is the special case of Steiner's theorem and is called the *parallel axes theorem.*

Perpendicular Axis Theorem

As explained earlier, a *plane lamina* is a rigid body whose mass is distributed in a single plane; that is, it has almost zero thickness. Suppose this plane lamina lies in the $X_1 - X_2$ plane; hence $x_3 = 0$. Let σ be the mass per unit area of this body. Let dA be a small area element. In such situations, the diagonal elements of the inertia tensor $\mathbf{I}$ of a plane lamina are

$$I_{11} = \int\int_A \sigma x_2^2 \, dA$$

$$I_{22} = \int\int_A \sigma x_1^2 \, dA$$

$$I_{33} = \int\int_A \sigma(x_1^2 + x_2^2) \, dA \qquad \textbf{(13.51)}$$

From these relations, we may conclude

$$I_{33} = I_{11} + I_{22} \qquad \textbf{(13.52)}$$

which is the statement of the *perpendicular axis theorem*:

If for a certain rigid body in the form of a plane lamina the moments of inertia about the X_1 and X_2 axes are I_{11} and I_{22}, the moment of inertia about the X_3 axis is equal to $I_{11} + I_{22}$.

▷ Example 13.4 _____

Consider the homogenous cube of density ρ, mass M, and side L discussed in Example 13.3. For a coordinate system with its origin at the center of mass of the cube as shown in Fig. Ex. 13.4, evaluate the elements of the inertia tensor.

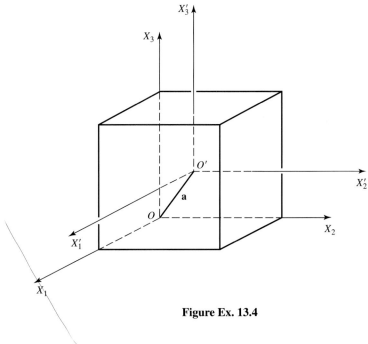

Figure Ex. 13.4

Solution

To transform the elements of an inertia tensor from one coordinate system to another, we make use of the relation given by Eq. (13.49).

$I_{k,l}$ are the elements of the inertia tensor I evaluated in Example 13.3. Thus

$$I_{k,l} = (I')_{k,l} + M\left(a^2 \cdot \delta_{k,l} - a_k \cdot a_l\right)$$

$$I_{1,1} = I_{2,2} = I_{3,3} = \frac{2}{3} \cdot M \cdot L^2$$

$$I_{1,2} = I_{1,3} = I_{2,3} = I_{3,1} = I_{2,1} = I_{3,2} = \frac{-1}{4} \cdot M \cdot L^2$$

$$I := \begin{bmatrix} \dfrac{2}{3} & -\dfrac{1}{4} & -\dfrac{1}{4} \\[2mm] -\dfrac{1}{4} & \dfrac{2}{3} & -\dfrac{1}{4} \\[2mm] -\dfrac{1}{4} & -\dfrac{1}{4} & \dfrac{2}{3} \end{bmatrix}$$

$$I_{k,l} = I \cdot \gamma$$

$$\gamma = M \cdot L^2$$

The center of mass of the cube is at (L/2, L/2, L/2) in the X-system and the components of the vector **a** are

$$a_1 = a_2 = a_3 = \frac{L}{2}$$

Assuming L = 1 and M = 1, we can assign values to I and a as shown.

$$k := 0..2 \qquad l := 0..2 \qquad n := 0..2 \qquad M := 1$$

$$I := \begin{bmatrix} \dfrac{2}{3} & -\dfrac{1}{4} & -\dfrac{1}{4} \\[2mm] -\dfrac{1}{4} & \dfrac{2}{3} & -\dfrac{1}{4} \\[2mm] -\dfrac{1}{4} & -\dfrac{1}{4} & \dfrac{2}{3} \end{bmatrix}$$

$$a_k :=$$

$$\begin{bmatrix} 1 \\ 2 \\ 1 \\ 2 \\ 1 \\ 2 \end{bmatrix} \qquad \begin{bmatrix} a_l \\ \hline 0.5 \\ 0.5 \\ 0.5 \end{bmatrix}$$

$$a_l := a_k$$

$$|a| = 0.866025$$

Using the values of the inertia
tensor I, we can calculate the
components of It.

$$\delta_{k,1} := \text{if}(k=1,1,0)$$

$$It_{k,1} := M \cdot \left[\left| (a_k)^2 + (a_1)^2 \right| \cdot \delta_{k,1} - a_k \cdot a_1 \right]$$

$$It = \begin{pmatrix} 0.25 & -0.25 & -0.25 \\ -0.25 & 0.25 & -0.25 \\ -0.25 & -0.25 & 0.25 \end{pmatrix}$$

The transfer matrix It' components
may be calculated as shown. Note
that the diagonal elements are the
only elements that are not zero.

$$It' := I - It$$

$$It' = \begin{pmatrix} 0.416667 & 0 & 0 \\ 0 & 0.416667 & 0 \\ 0 & 0 & 0.416667 \end{pmatrix}$$

Thus the transfer matrix It' may be
written by using the matrix Iu as
shown.

$$It' = \frac{1}{6} \cdot M \cdot L^2 \cdot Iu \qquad Iu := \frac{It'}{0.417} \qquad Iu = \begin{pmatrix} 1 & 0 & 0 \\ 0 & 1 & 0 \\ 0 & 0 & 1 \end{pmatrix}$$

EXERCISE 13.4 Repeat the calculations for the case discussed in Exercise 13.3.

13.5 PRINCIPAL MOMENT OF INERTIA AND PRINCIPAL AXES

We have described the inertia tensor of a rigid body with respect to a set of coordinate systems
with the origin fixed in the body. A particular set of coordinate axes can be chosen such that the
product of inertia elements will be zero in such a set. A set of axes possessing this property is
called the *principal axes*. We shall find such a set of axes very useful in many situations in order
to understand the description of motion of a rigid body.

Three mutually orthogonal coordinate axes meeting at a point O are said to form a set of
principal axes provided all the product of inertia elements I_{xy}, I_{yz}, and I_{zx} of the rigid body are
zero as expressed in terms of these axes. The point O, the origin of these principal axes, is called
the *principal point*. Three coordinate planes, each of which passes through two principal axes,
are called *principal planes* at point O.

If the product of inertia elements is zero, then the inertia tensor consists only of diagonal
elements; that is,

$$I = \begin{pmatrix} I_1 & 0 & 0 \\ 0 & I_2 & 0 \\ 0 & 0 & I_3 \end{pmatrix} \tag{13.53}$$

or in compact form we may write

$$I_{kl} = I_k \delta_{kl} \tag{13.54}$$

Furthermore, the use of the principal axes leads to a considerable simplification in the expression for L and T. Thus

$$L_k = \sum_l I_{kl} \omega_l = \sum_l I_k \delta_{kl} \omega_l = I_k \omega_k \tag{13.55}$$

$$T = \frac{1}{2} \sum_{k,l} I_{kl} \omega_k \omega_l = \frac{1}{2} \sum_{k,l} I_k \delta_{kl} \omega_k \omega_l = \frac{1}{2} \sum_k I_k \omega_k^2 \tag{13.56}$$

Before proceeding to understand the mathematical procedure for finding the principal axes so that the resulting moment of inertia will be diagonal, we shall present a physical description of the process and some particular situations of common interest.

For some situations in rigid body dynamics, the principal axes may be determined by examining the symmetry of the body. For example, consider a plane laminar body in the XY plane so that $z = 0$ for every particle. Thus

$$I_{yz} = I_{zx} = 0$$

Furthermore, suppose the lamina has an axis of symmetry, say the X-axis as shown in Fig. 13.3, such that the xy product in $\iiint \rho xy\, dV$ consists of two parts of equal magnitudes but of opposite sign. This results in $I_{xy} = 0$. Thus the inertia tensor is diagonal and the three coordinate axes in this case are the principal axes for a laminar rigid body. This also leads to the fact (using the definition of the product of inertia) that the coordinate axes will be the principal axes if the coordinate planes are planes of symmetry. (A note of caution: A body does not have to be symmetrical for its product of inertia elements to be zero.) But symmetry of a rigid body is helpful in determining the principal axes by inspection. For example, a cylindrical rod (which is a solid of revolution) has one principal axis along the symmetry axis, say the Z-axis through the center line of the cylindrical rod, and the two other principal axes are in a plane perpendicular to the symmetry axis, as shown in Fig. 13.4. The placement of these two principal axes in the XY plane is arbitrary.

Let us consider the relation between **L** and $\boldsymbol{\omega}$ for a rigid body when the coordinate axes are the principal axes. In such situations, **L** takes the form

$$\mathbf{L} = \hat{\mathbf{i}} I_x \omega_x + \hat{\mathbf{j}} I_y \omega_y + \hat{\mathbf{k}} I_z \omega_z \tag{13.57}$$

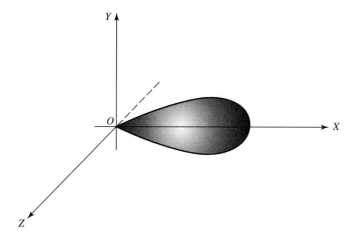

Figure 13.3   Plane laminar body in the XY-plane with the axis of symmetry along the X-axis, which has elements of the product of inertia to be zero.

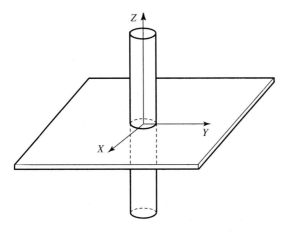

Figure 13.4   Cylindrical rod with one principal axis along the Z-axis and the other two principal axes in a plane perpendicular to the Z-axis.

where I_x, I_y, and I_z are the principal moments of inertia. Let the body rotate about the Z-axis such that $\omega_z = \omega$, while $\omega_x = \omega_y = 0$. Thus Eq. (13.57) takes the form

$$\mathbf{L} = \hat{\mathbf{k}} I_z \omega_z \qquad\qquad (13.58)$$

which states that the angular momentum is parallel to the axis of rotation; that is, **L** is in the same direction as $\boldsymbol{\omega}$. Thus we may conclude: If **L** and $\boldsymbol{\omega}$ are in the same direction (that is, the direction of rotation), then the axis of rotation is the principal axis. **L** and $\boldsymbol{\omega}$ will be in different directions if the rotation axis is not the principal axis. An important application of this principal is described next.

Dynamic Balancing

Consider a rotating device such as a fan blade or flywheel. This device will be *statically balanced* if the center of mass lies on the axis of rotation. If the device is *dynamically balanced*, the axis of rotation must be the principal axis; hence **L** and $\boldsymbol{\omega}$ will lie along this axis of rotation. If the rotational axis is not a principal axis, the angular momentum varies in direction, as shown in Fig. 13.5. Such variations require that there must be a torque acting on the body, that is, $\boldsymbol{\tau} = \dot{\mathbf{L}} = d\mathbf{L}/dt$, and the direction of this torque is at right angles to the direction of rotation. This leads a rigid body, a rotor in this case, to be dynamically unbalanced, resulting in the vibrations and wobbling of the whole system.

Figure 13.5   **L** not being in the same direction as $\boldsymbol{\omega}$ results in a rotating device such as a fan blade being dynamically unbalanced.

Determination of Principal Axes

We are given the moment of inertia and the product of inertia elements of a rigid body in terms of arbitrarily chosen coordinate axes through point O. We wish to find the principal axes about the origin at O. The process is called *diagonalizing* the matrix tensor. We make use of the fact that if the rotation axis is the principal axis then both the angular momentum $\mathbf{L}$ and the angular velocity $\boldsymbol{\omega}$ are directed along this axis and hence are proportional to each other. If I is the moment of inertia about the axis, we may write

$$\mathbf{L} = I\boldsymbol{\omega} = I\omega_x\hat{\mathbf{i}} + I\omega_y\hat{\mathbf{j}} + I\omega_z\hat{\mathbf{k}} \tag{13.59}$$

Thus, using Eq. (13.40), we may write

$$L_x = I\omega_x = I_{xx}\omega_x + I_{xy}\omega_y + I_{xz}\omega_z$$

$$L_y = I\omega_y = I_{yx}\omega_x + I_{yy}\omega_y + I_{yz}\omega_z$$

$$L_z = I\omega_z = I_{zx}\omega_x + I_{zy}\omega_y + I_{zz}\omega_z \tag{13.60}$$

or, after rearranging,

$$(I_{xx} - I)\omega_x + I_{xy}\omega_y + I_{xz}\omega_z = 0$$

$$I_{yx}\omega_x + (I_{yy} - I)\omega_y + I_{yz}\omega_z = 0$$

$$I_{zx}\omega_x + I_{zy}\omega_y + (I_{zz} - I)\omega_z = 0 \tag{13.61}$$

For these equations to have nontrivial solutions, the determinants of the coefficients must vanish; that is,

$$\begin{vmatrix} I_{xx} - I & I_{xy} & I_{xz} \\ I_{yx} & I_{yy} - I & I_{yz} \\ I_{zx} & I_{zy} & I_{zz} - I \end{vmatrix} = 0 \tag{13.62}$$

Equation (13.62), called a *secular* or *characteristic equation*, is cubic in I of the form

$$-I^3 + AI^2 + BI + C = 0 \tag{13.63}$$

where A, B, and C are constants and depend on the values of the moment of inertia and product of inertia elements. Each of the three roots I_x, I_y, and I_z (or I_1, I_2, and I_3) corresponds to the moment of inertia about one of the principal axes. These values of I_x, I_y, and I_z are called the *principal moments of inertia*. The direction of any one principal axis is determined by substituting for I equal to one of the three roots I_x, I_y, or I_z, say I_x, in Eq. (13.61) and determining the ratio of the components of the angular velocity $\boldsymbol{\omega}$, that is, to find $\omega_x : \omega_y : \omega_z$. Hence we can determine the directional cosines of the axis about which the moment of inertia is I_x. A similar procedure can be followed for finding the directions of the principal axes corresponding to principal moments of inertia I_y and I_z. This procedure results in the directions of the axes. The magnitude of the angular velocity is arbitrary and we are free to assume any value. The elements of the principal moment of inertia are generally called the *eigenvalues* or *characteristic values* of the inertia tensor. The directions of the principal axes are the *eigenvectors* or the *characteristic vectors*.

In most situations in rigid body dynamics, the body has some regular shape and the principal axes may be determined by determining the symmetry of the body. The axis of symmetry is the principal axis. Furthermore, if the body is a *solid of revolution* and has a moment of inertia I_x along the symmetry axis, then $I_y = I_z$; hence the secular equation has two distinct roots. Similarly, if the secular equation has a triple root, that is, $I_x = I_y = I_z$, it is called a *spherical top*; it is called an *asymmetrical top* if all the roots are distinct, that is, $I_x \neq I_y \neq I_z$. A body is a *rotor* if $I_x = 0$ and $I_y = I_z$, such as a dumbbell and diatomic molecules.

From symmetry properties or otherwise, if one of the principal axes is known, then the other two can be determined by the following procedure. Suppose one of the principal axes is the Z-axis; then the other two principal axes must lie in the *XY* plane. Since the Z-axis is the principal one, we must have

$$I_{zx} = I_{zy} = 0 \tag{13.64}$$

and the first two equations in (13.61) take the form

$$(I_{xx} - I)\omega_x + I_{xy}\omega_y = 0$$
$$I_{xy}\omega_x + (I_{yy} - I)\omega_y = 0 \tag{13.65}$$

Let us define

$$\tan \phi = \frac{\omega_y}{\omega_x} \tag{13.66}$$

where ϕ is the angle between the principal axis and the *X*-axis. Substituting from Eq. (13.66) into Eq. (13.65) and eliminating I from the resulting two equations,

$$\tan 2\phi = \frac{2I_{xy}}{I_{yy} - I_{xx}} \tag{13.67}$$

This equation gives two values of ϕ between $0°$ and $180°$, and these are directions of the two principal axes in the *XY* plane.

 Example 13.5 _____

Consider a homogeneous cube of density ρ, mass M, and side L, as discussed in Example 13.3. Evaluate the principal axes and their associated moment of inertia.

Solution

The moment of inertia tensor
of the cube with the axes directed
along the edges, as evaluated in
Example 13.3, is

$$I = \begin{bmatrix} \dfrac{2}{3} & -\dfrac{1}{4} & -\dfrac{1}{4} \\[2mm] -\dfrac{1}{4} & \dfrac{2}{3} & -\dfrac{1}{4} \\[2mm] -\dfrac{1}{4} & -\dfrac{1}{4} & \dfrac{2}{3} \end{bmatrix}$$

To evaluate the principal moment of inertia, we must solve the secular equation Eq. (13.62),

$$\left\| \begin{bmatrix} \dfrac{2}{3} - I & -\dfrac{1}{4} & -\dfrac{1}{4} \\[2mm] -\dfrac{1}{4} & \dfrac{2}{3} - I & -\dfrac{1}{4} \\[2mm] -\dfrac{1}{4} & -\dfrac{1}{4} & \dfrac{2}{3} - I \end{bmatrix} \right\| = 0$$

Simplifying this equation gives

Solving for I gives three roots

$$\frac{121}{864} - \frac{55}{48} \cdot I + 2 \cdot I^2 - I^3 = 0$$

Now for each of these roots and the value of I given above, we use the secular equation Eq. (13.61), which gives the principal axes corresponding to each root as
    I1 = 1/6   I2 = I3 = 11/12

$$\begin{bmatrix} \dfrac{1}{6} \\[2mm] \dfrac{11}{12} \\[2mm] \dfrac{11}{12} \end{bmatrix}$$

Substitute for the first root I1 = 0.167 = 1/6 in the secular equation, Eq. (13.62). S = 0 gives three equations, which when solved give the three values ω11, ω21, and ω31.

$$S = \left\| \begin{bmatrix} \left(\dfrac{2}{3} - \dfrac{1}{6}\right) \cdot \omega 11 & -\dfrac{1}{4} \cdot \omega 21 & -\dfrac{1}{4} \cdot \omega 31 \\[2mm] -\dfrac{1}{4} \cdot \omega 11 & \left(\dfrac{2}{3} - \dfrac{1}{6}\right) \cdot \omega 21 & -\dfrac{1}{4} \cdot \omega 31 \\[2mm] -\dfrac{1}{4} \cdot \omega 11 & -\dfrac{1}{4} \cdot \omega 21 & \left(\dfrac{2}{3} - \dfrac{1}{6}\right) \cdot \omega 31 \end{bmatrix} \right\|$$

Given

$$\left(\frac{2}{3} - \frac{1}{6}\right) \cdot \omega 11 + \left(-\frac{1}{4} \cdot \omega 21 + -\frac{1}{4} \cdot \omega 31\right) = 0$$

$$-\frac{1}{4} \cdot \omega 11 + \left[\left(\frac{2}{3} - \frac{1}{6}\right) \cdot \omega 21 + -\frac{1}{4} \cdot \omega 31\right] = 0$$

$$-\frac{1}{4} \cdot \omega 11 + -\frac{1}{4} \cdot \omega 21 + \left(\frac{2}{3} - \frac{1}{6}\right) \cdot \omega 31 = 0$$

Thus the first eigenvector ω1 will have all equal components. The resulting eigenvector ω1 is as shown.

$$\text{Find}(\omega 11, \omega 21, \omega 31) \rightarrow \begin{pmatrix} \omega 11 \\ \omega 11 \\ \omega 11 \end{pmatrix} \qquad \boldsymbol{\omega 1 = i + j + k}$$

Repeat the above procedure for the second root I2 = 11/12 = 0.917 to obtain the eigenvector. Note that the third root I3 is also 11/12 = 0.917.

Given

$$\left(\frac{2}{3} - \frac{11}{12}\right) \cdot \omega 12 + \left(-\frac{1}{4} \cdot \omega 22 + -\frac{1}{4} \cdot \omega 32\right) = 0$$

Thus the eigenvector ω2 has two equal
components and a third that is equal
to the negative of the sum of the other two.

$$-\frac{1}{4}\cdot\omega12+\left[\left(\frac{2}{3}-\frac{11}{12}\right)\cdot\omega22+-\frac{1}{4}\cdot\omega32\right]=0$$

$$-\frac{1}{4}\cdot\omega12+-\frac{1}{4}\cdot\omega22+\left(\frac{2}{3}-\frac{11}{12}\right)\cdot\omega32=0$$

$$\text{Find}(\omega12,\omega22,\omega32)\rightarrow\begin{pmatrix}\omega12\\\omega22\\-\omega12-\omega22\end{pmatrix}\qquad\qquad\omega2=\mathbf{i}-2\cdot\mathbf{j}+\mathbf{k}$$

Since the two roots ω2 and ω3 are equal, the corresponding eigenvectors must lie in the same plane. (Note that all the three roots are interchangeable, that is, naming them 1, 2, 3, is arbitrary.)

Alternate Direct Treatment

The moment of inertia tensor of the cube
with the origin at the corner and axes
directed along the edges as evaluated
in Exercise 13.3 is (without a constant γ)

$$I:=\begin{bmatrix}\frac{2}{3}&-\frac{1}{4}&-\frac{1}{4}\\-\frac{1}{4}&\frac{2}{3}&-\frac{1}{4}\\-\frac{1}{4}&-\frac{1}{4}&\frac{2}{3}\end{bmatrix}\qquad c:=\text{eigenvals}(I)$$

c = eigenvalues   c has three values, two
v = eigenvector   of which are equal

$$c=\begin{pmatrix}0.917\\0.917\\0.167\end{pmatrix}$$

$$v=I\cdot c$$

$$v1:=\text{eigenvec}\left(I,c_1\right)\qquad v2:=\text{eigenvec}\left(I,c_2\right)\qquad v3:=\text{eigenvec}\left(I,c_3\right)$$

The three eigenvectors
are as shown

$$v1=\begin{pmatrix}-0.615\\-0.158\\0.773\end{pmatrix}\qquad v2=\begin{pmatrix}-0.615\\-0.158\\0.773\end{pmatrix}\qquad v3=\begin{pmatrix}0.577\\0.577\\0.577\end{pmatrix}$$

$$\mathbf{v1}=\mathbf{v2}=-0.62\cdot\mathbf{i}+-0.16\cdot\mathbf{j}+0.77\cdot\mathbf{k}\qquad\qquad \mathbf{v3}=0.58\cdot\mathbf{i}+0.58\cdot\mathbf{j}+0.58\cdot\mathbf{k}$$

Thus the eigenvectors v1 and v2 are in the same plane, while v3 is perpendicular to them.

This implies that the principal axis corresponding to I_3 must lie along the diagonal of the cube, that is, along *OA* as shown in Fig. Ex. 13.5. Since $I_1=I_2$, this means that the remaining two principal axes must lie in a plane normal to the axis *OA*. This plane is shown shaded in the figure. Thus the second principal axis can be picked in any direction in this plane, while the third one will be perpendicular to the second but in the same plane.

EXERCISE 13.5 Evaluate the principal axes and their associated moment of inertia for the inertia tensor obtained in Exercise 13.3.

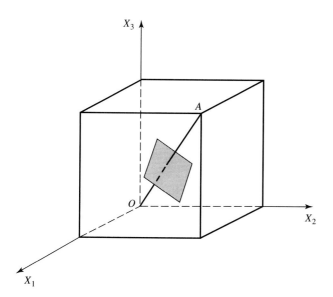

X_3

A

O

X_2

X_1

13.6 INERTIA ELLIPSOID

The inertia ellipsoid is helpful in visualizing the inertia tensor geometrically, thereby enabling us to predict some inertial properties of rigid bodies without going deeply into mathematical details. The motion of a rigid body depends on three numbers: I_1, I_2, and I_3, the principal moments of inertia. Bodies that have the same principal moments of inertia will move in exactly the same manner independent of their shape and size (provided we ignore the effects of frictional force and other forces that may be functions of the shape of the body). We show in the following that *the simplest geometrical shape of a body having three given principal moments of inertia is a homogeneous ellipsoid.* Hence, we may conclude that *the motion of any rigid body can be represented by the motion of an equivalent ellipsoid.*

Consider an arbitrary axis of rotation OA passing through a body, as shown in Fig. 13.6. Let P be a point on the axis such that the distance OP is numerically equal to the reciprocal of the square root of the moment of inertia I about OA. That is,

$$OP = \frac{1}{\sqrt{I}} \tag{13.68}$$

If the coordinates of P are x, y, and z and the directional cosines of line OP are l, m, and n, then

$$l = \frac{x}{OP} = x\sqrt{I}, \qquad m = \frac{y}{OP} = y\sqrt{I}, \qquad n = \frac{z}{OP} = z\sqrt{I} \tag{13.69}$$

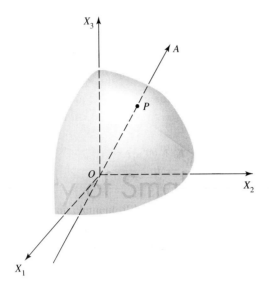

Figure 13.6   Axis of rotation OA through an arbitrarily shaped rigid body and passing through a point P.

Thus the moment of inertia of a rigid body about any line in terms of the directional cosines of that line with its inertia elements for some coordinate system with its origin on the line is

$$I = l^2 I_{xx} + m^2 I_{yy} + n^2 I_{zz} + 2nm I_{yz} + 2ln I_{zx} + 2ml I_{xy} \tag{13.70}$$

Substituting for l, m, and n, Eq. (13.70), after rearranging, takes the form

$$x^2 I_{xx} + y^2 I_{yy} + z^2 I_{zz} + 2yz I_{yz} + 2zx I_{zx} + 2xy I_{xy} = 1 \tag{13.71}$$

This is an equation of a surface (the locus of points P) as the direction of axis OA is varied. It is the equation of a general quardratic surface in three dimensions, and the surface is bounded; hence it must be an *ellipsoid*. If the coordinate axes are the principal axes. Eq. (13.70) takes the form

$$I = l^2 I_{xx} + m^2 I_{yy} + n^2 I_{zz} \tag{13.72}$$

and the inertia ellipsoid, Eq. (13.71), takes the form

$$x^2 I_1 + y^2 I_2 + z^2 I_3 = 1 \tag{13.73}$$

where I_1, I_2, and I_3 (which have replaced I_{xx}, I_{yy}, and I_{zz}) are the principal moments of inertia. The two inertia ellipsoids given by Eqs. (13.71) and (13.73) are shown in Fig. 13.7(a) and (b). Note that the semiaxes of the ellipsoid in Fig. 13.7(a) are

$$\frac{1}{\sqrt{I_{11}}}, \quad \frac{1}{\sqrt{I_{22}}}, \quad \text{and} \quad \frac{1}{\sqrt{I_{33}}}$$

and the semiaxes of the ellipsoid in Fig. 13.7(b) are

$$\frac{1}{\sqrt{I_1}}, \quad \frac{1}{\sqrt{I_2}}, \quad \text{and} \quad \frac{1}{\sqrt{I_3}}$$

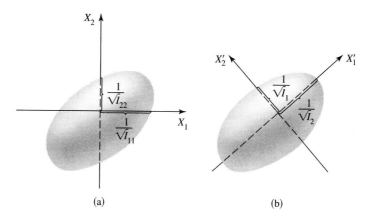

Figure 13.7   (a) Inertia ellipsoid in a nonprincipal coordinate axes system given by Eq. (13.71), and (b) inertia ellipsoid in a principal coordinate axes system given by Eq. (13.73).

That is, the semiaxes are

$$x_k = \frac{1}{\sqrt{I_k}} \tag{13.74}$$

Also, the ellipsoid in Fig. 13.7(b) can be obtained from Fig. 13.7(a) by causing proper rotations.

If two of the I_k are equal, the inertia ellipsoid has rotational symmetry about the third axis. Suppose $I_1 = I_2$, then the intersection of the inertia ellipsoid with the $X_1 - X_2$ planes may be drawn, all having the same moments of inertia. If $I_1 = I_2 = I_3$, the inertia ellipsoid reduces to a sphere, and the moments of inertia about any axis passing through the origin are equal.

13.7 MORE ABOUT THE PROPERTIES OF THE INERTIA TENSOR

We start with the definition of an inertia tensor and tensors in general. Then we introduce a slightly different way of defining an inertia tensor by means of a dyad product. And, finally, we see the similarity in the treatment of tensors as matrices.

The relation between the quantities $\mathbf{L}$ and $\boldsymbol{\omega}$ may be written as

$$\mathbf{I} = \frac{\mathbf{L}}{\boldsymbol{\omega}} \tag{13.75}$$

where $\mathbf{I}$ is the quotient of two vector quantities. In general, the quotient of two vector quantities is not necessarily a member of the same class as that of the two dividing factors. Hence, we do not expect the ratio of the two dividing vectors to be a vector. As a matter of fact, it is altogether a different quantity, called a *tensor of the second rank*.

In a Cartesian three-dimensional space, a Cartesian tensor $\mathbf{T}$ of the Nth rank may be defined as (1) a quantity that has 3^N components $T_{ijk\ldots N}$, and (2) under orthogonal transformation of coordinates it obeys the following rule:

$$T'_{ijk\ldots}(x') = a_{il}a_{jm}a_{kn}\cdots T_{imn\ldots}(x) \tag{13.76}$$

where $a_{il}, a_{jm}, \ldots$ are the elements of transformation. Since we shall not be using any other coordinates except Cartesian, we shall simply use the term tensor **T** instead of Cartesian tensor **T**. Thus from this definition, for $N = 0$, $3° = 1$. That is, a tensor of *zero rank* has only *one component*; hence this quantity will be invariant under an orthogonal transformation. We may say that a *scalar is a tensor of zero rank* and has only *one component*. On the other hand, if $N = 1$, $3^1 = 3$; the tensor of *first rank* will have *three components*. These components transfer, according to Eq. (13.76), as

$$T'_i = a_{ij}T_j \tag{13.77}$$

which is similar to the transformation equation for a vector. Thus a *vector is a tensor of the first rank*, and has three components. For $N = 2$, a tensor of *second rank* will have *nine components*, which will transfer as

$$T'_{ij} = a_{il}a_{jm}T_{lm} \tag{13.78}$$

This transformation is similar to a 3×3 square matrix, except for one fundamental difference between the two. Unlike a tensor of second rank, a matrix transformation is not limited only to orthogonal transformation. In spite of these differences, we shall make use of the properties of matrices in tensors.

Another way of representing a tensor **I** is as a *dyadic*. We start with the definition of angular momentum, Eq. (13.38):

$$\mathbf{L} = \sum_{i=1}^{n} m_i[r_i^2\boldsymbol{\omega} - \mathbf{r}_i(\mathbf{r}_i \cdot \boldsymbol{\omega})] \tag{13.38}$$

Or we may write this as

$$\mathbf{L} = \left(\sum_{i=1}^{n} m_i r_i^2\right)\boldsymbol{\omega} - \left(\sum_{i=1}^{n} m_i\mathbf{r}_i\mathbf{r}_i\right) \cdot \boldsymbol{\omega} \tag{13.79}$$

The second term on the right has no meaning because we have not yet defined quantities of the form $\mathbf{r}_i\mathbf{r}_i$. We define a *dyad* as a simple pair of two vectors written as **AB**. The quantity **AB** has meaning only when it operates on other quantities. Thus we define the scalar dot product of a dyad with a vector **C** as a vector quantity given by

$$(\mathbf{AB}) \cdot \mathbf{C} = \mathbf{A}(\mathbf{B} \cdot \mathbf{C}) \tag{13.80}$$

or

$$\mathbf{C} \cdot (\mathbf{AB}) = (\mathbf{C} \cdot \mathbf{A})\mathbf{B} \tag{13.81}$$

where $\mathbf{B} \cdot \mathbf{C}$ is a scalar $(= b_1c_1 + b_2c_2 + b_3c_3)$; hence $(\mathbf{AB}) \cdot \mathbf{C}$ is a vector. Similarly, $\mathbf{C} \cdot (\mathbf{AB})$ is a vector. But the two vectors given by Eqs. (13.80) and (13.81), in general, will not be equal. That is, *dyad scalar multiplication is not commutative*. If we let

$$\mathbf{T} = \mathbf{AB} \tag{13.82}$$

then we may write

$$\mathbf{T} \cdot \mathbf{C} = \mathbf{A}(\mathbf{B} \cdot \mathbf{C}) \tag{13.83}$$

$$\mathbf{C} \cdot \mathbf{T} = (\mathbf{C} \cdot \mathbf{A})\mathbf{B} \tag{13.84}$$

Also,
$$\mathbf{T} \cdot (\mathbf{C} + \mathbf{D}) = \mathbf{T} \cdot \mathbf{C} + \mathbf{T} \cdot \mathbf{D} \tag{13.85}$$

$$\mathbf{T} \cdot (c\mathbf{C}) = c(\mathbf{T} \cdot \mathbf{C}) \tag{13.86}$$

where c is a constant.

A linear polynomial of dyads is called a *dyadic*, such as $\mathbf{AB} + \mathbf{CD} + \cdots$. Actually, any dyad may be expressed as a dyadic if we express the vectors $\mathbf{A}$ and $\mathbf{B}$ in terms of unit vectors. Thus, if

$$\mathbf{C} = c_1\hat{\mathbf{i}} + c_2\hat{\mathbf{j}} + c_3\hat{\mathbf{k}}$$

$$\mathbf{A} = a_1\hat{\mathbf{i}} + a_2\hat{\mathbf{j}} + a_3\hat{\mathbf{k}}$$

$$\mathbf{B} = b_1\hat{\mathbf{i}} + b_2\hat{\mathbf{j}} + b_3\hat{\mathbf{k}}$$

then the dyad $\mathbf{AB}$ may be written as a dyadic:

$$\mathbf{T} = \mathbf{AB} = a_1b_1\hat{\mathbf{i}}\hat{\mathbf{i}} + a_1b_2\hat{\mathbf{i}}\hat{\mathbf{j}} + a_1b_3\hat{\mathbf{i}}\hat{\mathbf{k}} + a_2b_1\hat{\mathbf{j}}\hat{\mathbf{i}} + a_2b_3\hat{\mathbf{j}}\hat{\mathbf{k}}$$
$$+ a_3b_1\hat{\mathbf{k}}\hat{\mathbf{i}} + a_3b_2\hat{\mathbf{k}}\hat{\mathbf{j}} + a_3b_3\hat{\mathbf{k}}\hat{\mathbf{k}} \tag{13.87}$$

Thus, in matrix notation, we may write

$$\mathbf{T} = \begin{pmatrix} T_{11} & T_{12} & T_{13} \\ T_{21} & T_{22} & T_{23} \\ T_{31} & T_{32} & T_{33} \end{pmatrix} = \begin{pmatrix} a_1b_1 & a_1b_2 & a_1b_3 \\ a_2b_1 & a_2b_2 & a_2b_3 \\ a_3b_1 & a_3b_2 & a_3b_3 \end{pmatrix} \tag{13.88}$$

Any given component of $\mathbf{T}$ is written as T_{ij}.

In component form, we may write

$$\mathbf{C} = \sum_{j=1}^{3} c_j\hat{\mathbf{u}}_j \tag{13.89}$$

where $\hat{\mathbf{u}}_j = (\hat{\mathbf{u}}_1, \hat{\mathbf{u}}_2, \hat{\mathbf{u}}_3)$ are the unit vectors; hence

$$(\mathbf{T} \cdot \mathbf{C})_i = \sum_{j=1}^{3} T_{ij}c_j \tag{13.90}$$

$$(\mathbf{C} \cdot \mathbf{T})_i = \sum_{j=1}^{3} c_j T_{ji} \tag{13.91}$$

$$T_{ij} = \hat{\mathbf{u}}_i \cdot (\mathbf{T} \cdot \hat{\mathbf{u}}_j) = (\hat{\mathbf{u}}_i \cdot \mathbf{T}) \cdot \hat{\mathbf{u}}_j \tag{13.92}$$

while
$$\mathbf{T} = \sum_{i,j=1}^{3} T_{ij}\hat{\mathbf{u}}_i\hat{\mathbf{u}}_j \tag{13.93}$$

Now we may define a *unit dyadic* **1** as

$$\mathbf{1} = \hat{\mathbf{i}}\hat{\mathbf{i}} + \hat{\mathbf{j}}\hat{\mathbf{j}} + \hat{\mathbf{k}}\hat{\mathbf{k}} \tag{13.94}$$

and **1** behaves exactly like a unit matrix, giving the results

$$\mathbf{1} \cdot \mathbf{A} = \mathbf{A} \cdot \mathbf{1} = \mathbf{A} \tag{13.95}$$

We may also write **1** as a unit tensor such that

$$\mathbf{1} = \begin{pmatrix} 1 & 0 & 0 \\ 0 & 1 & 0 \\ 0 & 0 & 1 \end{pmatrix} \tag{13.96}$$

Finally, we take full advantage of the fact that a tensor of second rank is very similar to a 3×3 square matrix in its representation. Hence transformation properties in orthogonal Cartesian coordinates may be directly utilized here. Let us start with a vector **L** in space or fixed in an inertial coordinate system so that

$$L_k = \sum_l I_{kl}\omega_l \tag{13.97}$$

In a body coordinate system that is simply rotated with respect to space coordinates, the angular momentum **L**' must have an analogous form:

$$L_i' = \sum I_{ij}'\omega_j' \tag{13.98}$$

Using the transformation properties of vectors, we may write the transformation of **L** and **ω** as [note that from Eq. (5.166)

$$x_i = \sum_j \lambda_{ji}x_j' \tag{5.166}$$

where λ_{ji} is an element of the transformation matrix **λ**]

$$L_k = \sum_m \lambda_{mk}L_m' \tag{13.99}$$

and

$$\omega_l = \sum_j \lambda_{jl}\omega_j' \tag{13.100}$$

Substituting these in Eq. (13.97), we obtain

$$\sum_m \lambda_{mk}L_m' = \sum_l I_{kl}\sum_j \lambda_{jl}\omega_j' \tag{13.101}$$

Multiplying both sides by λ_{ik} and summing over k,

$$\sum_m \left(\sum_k \lambda_{ik}\lambda_{mk} \right) L'_m = \sum_j \left(\sum_{k,l} \lambda_{ik}\lambda_{kl}I_{kl} \right) \omega'_j \tag{13.102}$$

This left side may be written as

$$\sum_m \left(\sum_k \lambda_{ik}\lambda_{mk} \right) L'_m = \sum_m \delta_{im} L'_m = L'_i$$

That is,

$$L'_i = \sum_j \left(\sum_{k,l} \lambda_{ik}\lambda_{jl}I_{kl} \right) \omega'_j \tag{13.103}$$

But this must be identical to Eq. (13.98). Comparing the two yields

$$I'_{ij} = \sum_{k,l} \lambda_{ik}\lambda_{jl}I_{kl} \tag{13.104}$$

Thus each element I_{kl} of inertia tensor **I** in a fixed coordinate system can be transformed into rotated (body) coordinates resulting in elements I'_{ij} of inertia tensor **I**'. The preceding result may be written as

$$I'_{ij} = \sum_{k,l} \lambda_{ik}I_{kl}\lambda^t_{lj} \tag{13.105}$$

where λ^t_{lj} are the elements of a transposed matrix $\boldsymbol{\lambda}^t$. Just as in matrix notation, we may write

$$\mathbf{I}' = \boldsymbol{\lambda}\mathbf{I}\boldsymbol{\lambda}^t \tag{13.106}$$

Since for orthogonal transformations, $\boldsymbol{\lambda}^t = \boldsymbol{\lambda}^{-1}$, where $\boldsymbol{\lambda}^{-1}$ is the inverse matrix, we may write

$$\mathbf{I}' = \boldsymbol{\lambda}\mathbf{I}\boldsymbol{\lambda}^{-1} \tag{13.107}$$

which is the *similarity transformation* (**I**' is similar to **I**).

These results indicate the method for transferring an inertia tensor from one system to another rotated system by using the rotation matrix. Furthermore, we may utilize this method to find the principal axes by determining the eigenvalues from the secular equation

$$\left| I_{ml} - I\delta_{ml} \right| = 0 \tag{3.108}$$

That is,

$$\begin{vmatrix} I_{11} - I & I_{12} & I_{13} \\ I_{21} & I_{22} - I & I_{23} \\ I_{31} & I_{32} & I_{33} - I \end{vmatrix} = 0 \tag{13.109}$$

which is the same as Eq. (13.62). These points are illustrated in the following example.

▷ Example 13.6 _____

Diagonalize the inertia tensor of a cube by rotating the coordinate axes.

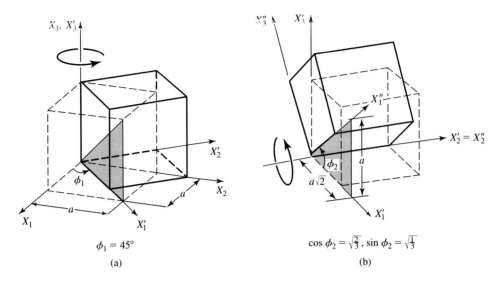

$$\cos\phi_2 = \sqrt{\tfrac{2}{3}}, \ \sin\phi_2 = \sqrt{\tfrac{1}{3}}$$

$\phi_1 = 45°$

(a)

(b)

Figure Ex. 13.6

Solution

As before, let the origin be at one corner of the cube. We have to perform a rotation in such a way that the X_1-axis will coincide with the diagonal of the cube. This can be achieved by means of two rotations: (a) perform a rotation through an angle $\phi_1 = 45°$ about the X_3-axis, and (b) perform a rotation through an angle $\phi_2 = \cos^{-1}(\sqrt{2/3})$ about the X_2'-axis, as shown in Fig. Ex. 13.6.

(a)

$\phi1$ = first angle of rotation about X3-axis

$$\phi1 := 45 \cdot \deg \qquad\qquad \phi1 = 0.785 \cdot \text{rad}$$

$\lambda1$ = the matrix of first rotation

$$\lambda1 := \begin{bmatrix} \cos(\phi1) & \sin(\phi1) & 0 \\ -\sin(\phi1) & \cos(\phi1) & 0 \\ 0 & 0 & 1 \end{bmatrix} \quad \lambda1 = \begin{pmatrix} 0.707 & 0.707 & 0 \\ -0.707 & 0.707 & 0 \\ 0 & 0 & 1 \end{pmatrix}$$

(b)

$\phi2$ = second angle of rotation about X2'-axis

$$\phi2 := a\cos\left(\sqrt{\frac{2}{3}}\right) \qquad \phi2 = 0.615$$

$\lambda2$ = matrix of second rotation

$$\lambda2 := \begin{bmatrix} \cos(\phi2) & 0 & \sin(\phi2) \\ 0 & 1 & 0 \\ -\sin(\phi2) & 0 & \cos(\phi2) \end{bmatrix} \quad \lambda2 = \begin{pmatrix} 0.816 & 0 & 0.577 \\ 0 & 1 & 0 \\ -0.577 & 0 & 0.816 \end{pmatrix}$$

λ = total matrix rotation

$$\lambda := \lambda2 \cdot \lambda1$$

λ^T = the inverse or transform matrix

$$\lambda = \begin{pmatrix} 0.577 & 0.577 & 0.577 \\ -0.707 & 0.707 & 0 \\ -0.408 & -0.408 & 0.816 \end{pmatrix} \quad \lambda^T = \begin{pmatrix} 0.577 & -0.707 & -0.408 \\ 0.577 & 0.707 & -0.408 \\ 0.577 & 0 & 0.816 \end{pmatrix}$$

I is from Example 13.3.

$$I = \begin{Vmatrix} \dfrac{2}{3}\cdot\gamma & \dfrac{-1}{4}\cdot\gamma & \dfrac{-1}{4}\cdot\gamma \\[2mm] \dfrac{-1}{4}\cdot\gamma & \dfrac{2}{3}\cdot\gamma & \dfrac{-1}{4}\cdot\gamma \\[2mm] \dfrac{-1}{4}\cdot\gamma & \dfrac{-1}{4}\cdot\gamma & \dfrac{2}{3}\cdot\gamma \end{Vmatrix}$$

$$I' = \lambda \cdot I \cdot \lambda^T$$

We can now calculate the inverse trasform matrix by substituting the values of λ and λ^T and I.

$$I' = \begin{bmatrix} .5761773 & .5756886 & .5802 \\ -.7068 & .7074 & 0 \\ -.41043348 & -.41008536 & .8145 \end{bmatrix} \cdot \begin{bmatrix} \dfrac{2}{3}\cdot\gamma & \dfrac{-1}{4}\cdot\gamma & \dfrac{-1}{4}\cdot\gamma \\[2mm] \dfrac{-1}{4}\cdot\gamma & \dfrac{2}{3}\cdot\gamma & \dfrac{-1}{4}\cdot\gamma \\[2mm] \dfrac{-1}{4}\cdot\gamma & \dfrac{-1}{4}\cdot\gamma & \dfrac{2}{3}\cdot\gamma \end{bmatrix} \cdot \begin{bmatrix} 0.576 & -0.707 & -0.41 \\ 0.576 & 0.707 & -0.41 \\ 0.58 & 0 & 0.815 \end{bmatrix}$$

It is clear that we get the same matrix as in Example 13.5.

$$I' = \begin{bmatrix} .1666\cdot\gamma & 0 & 0 \\ 0 & .9165\cdot\gamma & 0 \\ 0 & 0 & .9168\cdot\gamma \end{bmatrix}$$

$$\frac{1}{6} = 0.167 \qquad \frac{11}{12} = 0.917$$

The alternate treatment

The procedure is self-explanatory.

$$A := \begin{bmatrix} \dfrac{2}{3} & -\dfrac{1}{4} & -\dfrac{1}{4} \\[2mm] -\dfrac{1}{4} & \dfrac{2}{3} & -\dfrac{1}{4} \\[2mm] -\dfrac{1}{4} & -\dfrac{1}{4} & \dfrac{2}{3} \end{bmatrix}$$

$$E := \text{eigenvals}(A) \qquad E = \begin{pmatrix} 0.917 \\ 0.917 \\ 0.167 \end{pmatrix}$$

$$\text{diag}(E) = \begin{pmatrix} 0.917 & 0 & 0 \\ 0 & 0.917 & 0 \\ 0 & 0 & 0.167 \end{pmatrix}$$

The three columns are the three eigenvectors.

$$V := \text{eigenvecs}(A) \qquad V = \begin{pmatrix} 0.711 & -0.401 & 0.577 \\ -0.009 & 0.816 & 0.577 \\ -0.703 & -0.416 & 0.577 \end{pmatrix}$$

$$V = \begin{bmatrix} 0.711 & -0.401 & 0.577 \\ -0.009 & 0.816 & 0.577 \\ -0.703 & -0.416 & 0.577 \end{bmatrix}$$

which gives

$$\omega 1 = 0.577 \cdot \mathbf{i} + 0.577 \cdot \mathbf{j} + 0.577 \cdot \mathbf{k}$$
$$\omega 2 = -0.401 \cdot \mathbf{i} + 0.816 \cdot \mathbf{j} + -0.46 \cdot \mathbf{k}$$
$$\omega 3 = -0.71 \cdot \mathbf{i} - 0.01 \cdot \mathbf{j} + 0.71 \cdot \mathbf{k}$$

EXERCISE 13.6 Diagonalize the inertia tensor of the rectangular body discussed in Exercise 13.3 by rotating the coordinate axes.

13.8 EULERIAN ANGLES

We are interested in evaluating a matrix that will enable us to cause transformation from one co-ordinate system to another. Let us say we want to transfer from coordinates $\mathbf{X}'$ of a fixed or inertial coordinate system to coordinates $\mathbf{X}$ of a body coordinate system. The transformation may be represented by a matrix equation

$$\mathbf{X} = \boldsymbol{\lambda}\mathbf{X}' \tag{13.110}$$

where the rotation matrix $\boldsymbol{\lambda}$ completely specifies the relative orientation of the two systems. Such a rotation matrix should contain three independent angles. Of the several possible choices for these angles, the most common and convenient ones to use are the *Eulerian angles* represented as ϕ, θ, and ψ.

To go from an X' system to an X system, the following sequence of rotations of three angles is followed, as demonstrated in Fig. 13.8:

1. The first rotation is counterclockwise through an angle ϕ about the X_3'-axis and in the X_1'-X_2' plane, transforming the axes $X_i' \rightarrow X_i''$, as shown in Fig. 13.8(a). The transformation matrix for this rotation in the X_1'-X_2' plane is

$$\mathbf{R}_\phi = \begin{pmatrix} \cos\phi & \sin\phi & 0 \\ -\sin\phi & \cos\phi & 0 \\ 0 & 0 & 1 \end{pmatrix} \tag{13.111}$$

   The angle ϕ is called the *precession angle*.

2. The second rotation is counterclockwise through an angle θ about the X_1''-axis and in the X_2''-X_3'' plane, transforming the axes $X_i'' \rightarrow X_i'''$, as shown in Fig. 13.8(b). The transformation matrix for this rotation in the X_2''-X_3'' plane is

$$\mathbf{R}_\theta = \begin{pmatrix} 1 & 0 & 0 \\ 0 & \cos\theta & \sin\theta \\ 0 & -\sin\theta & \cos\theta \end{pmatrix} \tag{13.112}$$

   The angle θ is called the *nutation angle*.

3. The third rotation is counterclockwise through an angle ψ about the X_3'''-axis and in the X_1'''-X_2''' plane, transforming the axes $X_i''' \rightarrow X_i$, as shown in Fig. 13.8(c). The transformation matrix for this rotation is

$$\mathbf{R}_\psi = \begin{pmatrix} \cos\psi & \sin\psi & 0 \\ -\sin\psi & \cos\psi & 0 \\ 0 & 0 & 1 \end{pmatrix} \tag{13.113}$$

   The angle ψ is called the *body angle*.

The line NN' formed by the intersection of the planes containing the X_1-X_2 axes of the body system and the X_1'-X_2' axes of the fixed system is called the *line of nodes*. The transforma-

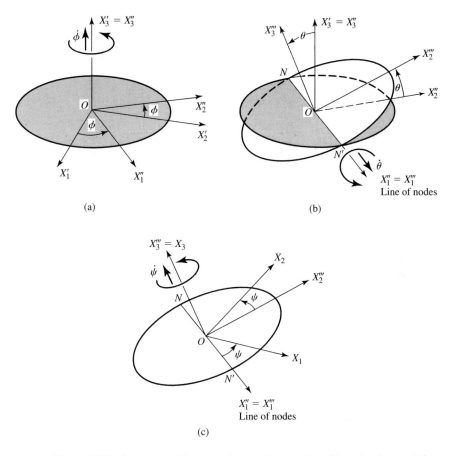

(a)

(b)

(c)

Figure 13.8   Sequence of three angular rotations employed in going from an X' system to an X system.

tion from the fixed coordinate system X_i' to the body coordinate system X_i is given by the rotation matrix $\boldsymbol{\lambda}$ obtained by the product of the three individual matrices $\mathbf{R}_\phi$, $\mathbf{R}_\theta$, and $\mathbf{R}_\psi$ given previously. That is,

$$\boldsymbol{\lambda} = \mathbf{R}_\psi \mathbf{R}_\theta \mathbf{R}_\phi \tag{13.114}$$

$$\boldsymbol{\lambda} = \begin{pmatrix} \lambda_{11} & \lambda_{12} & \lambda_{13} \\ \lambda_{21} & \lambda_{22} & \lambda_{23} \\ \lambda_{31} & \lambda_{32} & \lambda_{33} \end{pmatrix}$$

$$\boldsymbol{\lambda} = \begin{pmatrix} \cos\psi\cos\phi & \cos\psi\cos\phi & \sin\psi\sin\theta \\ -\cos\theta\sin\phi\sin\psi & +\cos\theta\cos\phi\sin\psi & \\ -\sin\psi\cos\phi & -\sin\psi\sin\phi & \cos\psi\sin\theta \\ -\cos\theta\sin\phi\cos\psi & +\cos\theta\cos\phi\cos\psi & \\ \sin\theta\sin\phi & -\sin\theta\cos\phi & \cos\theta \end{pmatrix} \tag{13.115}$$

All infinitesimal rotations can be represented by vector notation. This enables us to represent the three time derivatives of rotation, that is, $\dot{\phi}$, $\dot{\theta}$, and $\dot{\psi}$, as the components of an angular velocity vector $\boldsymbol{\omega}(\omega_\phi, \omega_\theta, \omega_\psi)$. The three components of $\boldsymbol{\omega}$ are not all either along the fixed axes or the body axes. Actually,

$$\omega_\phi = \dot{\phi} \quad \text{is directed along the } X_3' \text{ (fixed) axis}$$

$$\omega_\theta = \dot{\theta} \quad \text{is along the line of nodes}$$

$$\omega_\psi = \dot{\psi} \quad \text{is directed along the } X_3 \text{ (body) axis} \tag{13.116}$$

It is not very convenient to use these components to describe the motion of a rigid body. Rigid body equations of motion are described in terms of a body coordinate system. Thus we must calculate the angular velocity vector $\boldsymbol{\omega}(\omega_1, \omega_2, \omega_3)$ in the body coordinate system. To do this, we must first resolve $\dot{\phi}$, $\dot{\theta}$, and $\dot{\psi}$ along the body axes; that is,

$$\dot{\phi}_1 = \dot{\phi} \sin\theta \sin\psi \qquad\qquad \text{along } X_1\text{-axis}$$

$$\dot{\phi}_2 = \dot{\phi} \sin\theta \cos\psi \qquad\qquad \text{along } X_2\text{-axis}$$

$$\dot{\phi}_3 = \dot{\phi} \cos\theta \qquad\qquad\qquad \text{along } X_3\text{-axis} \tag{13.117}$$

$$\dot{\theta}_1 = \dot{\theta} \cos\psi \qquad\qquad\qquad \text{along } X_1\text{-axis}$$

$$\dot{\theta}_2 = -\dot{\theta} \sin\psi \qquad\qquad\quad \text{along } X_2\text{-axis}$$

$$\dot{\theta}_3 = 0 \qquad\qquad\qquad\qquad \text{along } X_3\text{-axis} \tag{13.118}$$

$$\dot{\psi}_1 = 0 \qquad\qquad\qquad\qquad \text{along } X_1\text{-axis}$$

$$\dot{\psi}_2 = 0 \qquad\qquad\qquad\qquad \text{along } X_2\text{-axis}$$

$$\dot{\psi}_3 = \dot{\psi} \qquad\qquad\qquad\qquad \text{along } X_3\text{-axis} \tag{13.119}$$

Using these results, we get the components of $\boldsymbol{\omega}$ to be

$$\omega_1 = \dot{\phi}_1 + \dot{\theta}_1 + \dot{\psi}_2 = \dot{\phi} \sin\theta \sin\psi + \dot{\theta} \cos\psi$$

$$\omega_2 = \dot{\phi}_2 + \dot{\theta}_2 + \dot{\psi}_2 = \dot{\phi} \sin\theta \cos\psi - \dot{\theta} \sin\psi$$

$$\omega_3 = \dot{\phi}_3 + \dot{\theta}_3 + \dot{\psi}_3 = \dot{\phi} \cos\theta + \dot{\psi} \tag{13.120}$$

These equations are called *Euler's geometrical equations.* We can make use of these to describe rigid body motion using body axes.

It is important to emphasize that the angular displacements and other rotational quantities may be represented as vectors only if these quantities are infinitesimally small; then they obey the law of vector addition. The exception is the case in which the rotations are in the same plane.

13.9 EULER'S EQUATIONS OF MOTION FOR A RIGID BODY

The translational motion of a rigid body is described by the equations

$$\mathbf{F} = \frac{d\mathbf{P}}{dt}, \qquad \text{where } \mathbf{P} = M\mathbf{V} \tag{13.121}$$

$\mathbf{F}$ is the resultant force acting on the body, $\mathbf{P}$ is its linear momentum, M is its mass, and $\mathbf{V}$ is the velocity of its center of mass. The rotational motion of a body is described by

$$\boldsymbol{\tau} = \frac{d\mathbf{L}}{dt}, \qquad \text{where } \mathbf{L} = \mathbf{I}\boldsymbol{\omega} \tag{13.122}$$

$\boldsymbol{\tau}$ is the net torque acting on the rigid body, $\mathbf{L}$ is its angular momentum, $\boldsymbol{\omega}$ is its angular velocity, and $\mathbf{I}$ is the inertia tensor. The methods used for solving equations of translational motion can be directly extended to those for rotational motion only for the special case in which the rotational motion is restricted about a *fixed* axis. For a general case, this is not true.

Let us now proceed to obtain Euler's equations of motion for a rigid body in a force field. Equation (13.22), which describes the motion of a rigid body as viewed from a fixed, inertial, or laboratory coordinate system (LCS), may be written as

$$\frac{d\mathbf{L}}{dt} = \dot{\mathbf{L}} = \frac{d}{dt}(\mathbf{I} \cdot \boldsymbol{\omega}) = \boldsymbol{\tau} \tag{13.123}$$

Note that $\mathbf{I}$ changes as the body rotates. To overcome this difficulty, we refer Eq. (13.123) to a set of axes that are fixed with the rotating body. Let d'/dt be the time derivative with respect to the coordinate axes fixed in the body. Using the results given in Chapter 11 [Eq. (11.26)], Eq. (13.123) takes the form

$$\frac{d\mathbf{L}}{dt} = \frac{d'\mathbf{L}}{dt} + \boldsymbol{\omega} \times \mathbf{L} = \boldsymbol{\tau} \tag{13.124}$$

Since $\mathbf{L} = \mathbf{I} \cdot \boldsymbol{\omega}$, where $\mathbf{I}$ is constant relative to the body axes, we may substitute in Eq. (13.124) to obtain

$$\frac{d'(\mathbf{I} \cdot \boldsymbol{\omega})}{dt} + \boldsymbol{\omega} \times (\mathbf{I} \cdot \boldsymbol{\omega}) = \boldsymbol{\tau}$$

$$\mathbf{I} \cdot \frac{d'\boldsymbol{\omega}}{dt} + \frac{d'\mathbf{I}}{dt} \cdot \boldsymbol{\omega} + \boldsymbol{\omega} \times (\mathbf{I} \cdot \boldsymbol{\omega}) = \boldsymbol{\tau} \tag{13.125}$$

But

$$\frac{d'\mathbf{I}}{dt} = 0 \quad \text{and} \quad \frac{d'\boldsymbol{\omega}}{dt} = \frac{d\boldsymbol{\omega}}{dt} \tag{13.126}$$

which when substituted in Eq. (13.125) yields

$$\mathbf{I} \cdot \frac{d\boldsymbol{\omega}}{dt} + \boldsymbol{\omega} \times (\mathbf{I} \cdot \boldsymbol{\omega}) = \boldsymbol{\tau} \tag{13.127}$$

For convenience, choose the body axes to be the principal axes so that

$$\mathbf{L} = \mathbf{I} \cdot \boldsymbol{\omega} = \hat{\mathbf{i}} L_1 + \hat{\mathbf{j}} L_2 + \hat{\mathbf{k}} L_3$$

Also

$$L_i = I_i \omega_i$$

That is,

$$L_1 = I_1 \omega_1, \qquad L_2 = I_2 \omega_2, \quad \text{and} \quad L_3 = I_3 \omega_3$$

Using these relations, Eq. (13.127) may be written in component form as

$$I_1 \dot{\omega}_1 + (I_3 - I_2)\omega_3 \omega_2 = \tau_1 \tag{13.128a}$$

$$I_2 \dot{\omega}_2 + (I_1 - I_3)\omega_1 \omega_3 = \tau_2 \tag{13.128b}$$

$$I_3 \dot{\omega}_3 + (I_2 - I_1)\omega_2 \omega_1 = \tau_3 \tag{13.128c}$$

These are known as *Euler's dynamical equations* or simply *Euler's equations* for the motion of a rigid body in a force field. In the absence of a torque, Eqs. (13.128), take the form

$$I_1 \dot{\omega}_1 + (I_3 - I_2)\omega_3 \omega_2 = 0 \qquad \tau_1 = 0 \tag{13.129a}$$

$$I_2 \dot{\omega}_2 + (I_1 - I_3)\omega_1 \omega_3 = 0 \qquad \tau_2 = 0 \tag{13.129b}$$

$$I_3 \dot{\omega}_3 + (I_2 - I_1)\omega_2 \omega_1 = 0 \qquad \tau_3 = 0 \tag{13.129c}$$

Furthermore, for the net zero external torque, the angular momentum must remain constant in both magnitude and direction. For this, Euler's equations require that if $\omega_1 \neq 0$, then $\omega_2 = \omega_3 = 0$, and if $\omega_2 \neq 0$, then $\omega_1 = \omega_3 = 0$. These results imply that for no net external torque, only rotations about the body's principal axes are possible.

As is clear from our discussion, the three principal moment of inertia elements I_1, I_2, and I_3 determine the motion of a rigid body. Any two rigid bodies that have the same principal moment of inertia will have the same behavior regardless of their structure and shape. Motions of such bodies are described by means of an equivalent ellipsoid constructed with principal moment of inertia elements, as discussed in Section 13.6.

13.10 FORCE FREE MOTION OF A SYMMETRICAL TOP

We can solve Euler's equations, Eqs. (13.128), for the special case in which $\tau = 0$. Furthermore, we limit our discussion to the case in which the body is symmetrical, that is, a symmetrical top in which two of the principal moments of inertia are the same. We choose the X_3-axis to be the symmetry axis so that $I_1 = I_2 = I_{12} \neq I_3$. Thus Eqs. (13.128) reduce to

$$I_{12} \dot{\omega}_1 + (I_3 - I_{12})\omega_2 \omega_3 = 0 \tag{13.130}$$

$$I_{12} \dot{\omega}_2 + (I_{12} - I_3)\omega_1 \omega_3 = 0 \tag{13.131}$$

$$I_3 \dot{\omega}_3 = 0 \tag{13.132}$$

Before solving these equations, two points must be made clear. First, since the motion is force free, the center of mass of the body is at rest or moving with uniform velocity. There will be no loss of generality if we assume that the center of mass is at rest and is located at the origin of a fixed or laboratory coordinate system. Second, assume that the angular velocity $\boldsymbol{\omega}$ does *not* lie along one of the principal axes of the body coordinate system, because if it does the problem will be a trivial one.

From Eq. (13.132), since $I_3 \neq 0$,

$$\dot{\omega}_3 = 0$$

which on integration gives

$$\omega_3(t) = \text{constant} \tag{13.133}$$

This equation states that for any rigid body rotating with angular velocity $\boldsymbol{\omega}$ the component of angular velocity along the symmetry axis, ω_3, remains constant. (If $\boldsymbol{\omega}$ were along the X_3-axis, the principal axis, the entire angular velocity $\boldsymbol{\omega}$ would remain constant.)

Equations (13.130) and (13.131) may be written as

$$\dot{\omega}_1 + \frac{I_3 - I_{12}}{I_{12}} \omega_3 \omega_2 = 0 \tag{13.134}$$

and

$$\dot{\omega}_2 - \frac{I_3 - I_{12}}{I_{12}} \omega_3 \omega_1 = 0 \tag{13.135}$$

Let us define Ω (or Ω_B to be more specific),

$$\Omega_B \equiv \frac{I_3 - I_{12}}{I_{12}} \omega_3 = \gamma \omega_3 \tag{13.136}$$

and rewrite Eqs. (13.134) and (13.135) as

$$\dot{\omega}_1 + \Omega_B \omega_2 = 0 \tag{13.137}$$

$$\dot{\omega}_2 - \Omega_B \omega_1 = 0 \tag{13.138}$$

These are two first-order coupled equations and can be solved by the usual procedure for such equations. Multiply the second equation by i and add to the first, that is,

$$(\dot{\omega}_1 + i\dot{\omega}_2) - i\Omega_B(\omega_1 + i\omega_2) = 0 \tag{13.139}$$

Substitute

$$\eta \equiv \omega_1 + i\omega_2 \tag{13.140}$$

and

$$\dot{\eta} = \dot{\omega}_1 + i\dot{\omega}_2 \tag{13.141}$$

into Eq. (13.139), resulting in

$$\dot{\eta} - i\Omega_B \eta = 0 \tag{13.142}$$

Assuming that the phase angle $\delta = 0$ when $t = 0$, the solution of Eq. (13.142) is

$$\eta(t) = Ae^{i\Omega_B t} \tag{13.143}$$

or

$$\omega_1 + i\omega_2 = A \cos \Omega_B t + iA \sin \Omega_B t \tag{13.144}$$

where A is an arbitrary constant. Comparing the two sides,

$$\omega_1(t) = A \cos \Omega_B t \tag{13.145}$$

$$\omega_2(t) = A \sin \Omega_B t \tag{13.146}$$

Squaring the two equations and adding,

$$\omega_1^2 + \omega_2^2 = A^2 \tag{13.147}$$

That is, the sum of the squares of the angular velocity components ω_1 and ω_2 is constant and is equal to A^2. Furthermore, according to Eq. (13.133), ω_3 is constant; therefore, the magnitude of ω is also constant; that is,

$$\omega = |\boldsymbol{\omega}| = \sqrt{\omega_1^2 + \omega_2^2 + \omega_3^2} = \sqrt{A^2 + \omega_3^2} = \text{constant} \tag{13.148}$$

Equations (13.145) and (13.146) are parametric equations of a circle, and ω_1 and ω_2 are the components of $\boldsymbol{\omega}$ in the X_1X_2 body plane. Thus the components ω_1 and ω_2 of $\boldsymbol{\omega}$ trace out a circle with time in the X_1X_2 plane, which implies that the angular velocity vector $\boldsymbol{\omega}$ precesses in a cone about the X_3-axis (the body symmetry axis) with a constant angular frequencey Ω_B, as shown in Fig. 13.9, while ω_3 remains constant around the symmetry axis. The net result is *to an*

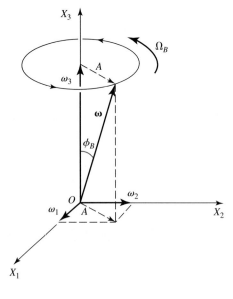

Figure 13.9(a)   The angular velocity vector $\boldsymbol{\omega}$ precesses in a cone about the X_3-axis, the body symmetry axis, with a constant angular frequency Ω_B. ϕ_B is the half-angle of the body cone.

Figure 13.9(b) _____

Assuming arbitrary values, we can show the angular velocity precessing in a cone.

$n := 1 .. 26$ $m := 1$ $I1 := 2$ $I2 := 2$ $I12 := 2$ $I3 := 3$ $\omega := 3$

$\omega z_n := 1$ $\Omega B_m := \dfrac{I3 - I12}{I12} \cdot \omega z_n \cdot m$ $\Omega B_m = 0.5$ $A_n := \sqrt{\omega^2 - \left(\omega z_n\right)^2}$

$\omega x_{n,m} := A_n \cdot \cos\left(\Omega B_m \cdot \dfrac{n}{2}\right)$ $\omega y_{n,m} := A_n \cdot \sin\left(\Omega B_m \cdot \dfrac{n}{2}\right)$ $Z_{n,m} := \omega z_n \cdot \pi \cdot 2$

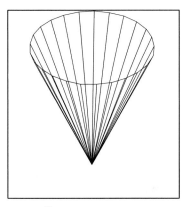

$\omega x, \omega y, Z$
Precessing cone

observer in the body coordinate system, $\boldsymbol{\omega}$ *traces out a cone around the body symmetric axis.* This is called the *body cone*, and in the body reference frame its half angle, ϕ_B, is

$$\tan \phi_B = \frac{(\omega_1^2 + \omega_2^2)^{1/2}}{\omega_3} = \frac{A}{\omega_3} \qquad (13.149)$$

as shown in Fig. 13.9(a) and (b).

Remember, we have been considering the force free motion of a rigid body. As viewed from the inertial system, there should be two constants of motion, the angular momentum and kinetic energy. Thus, as viewed from the fixed, LCS, or inertial coordinate system,

$$\mathbf{L}(t) = \text{constant} \qquad (13.150)$$

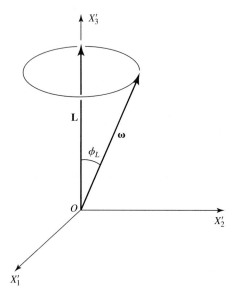

Figure 13.10 As viewed from a fixed, LCS, or inertial coordinate system, the angular vector **ω** moves in such a way that its projection on the angular momentum vector **L** or the X_3'-axis is constant. ϕ_L is the half-angle of the space cone.

and is fixed about the X_3'-axis, as shown in Fig. 13.10. Since the center of mass is fixed, the kinetic energy is all rotational and constant; that is,

$$T_{\text{rot}} = \tfrac{1}{2}\boldsymbol{\omega} \cdot \mathbf{L} = \text{constant} \tag{13.151}$$

We know that **L** is constant; T_{rot} will be constant only if **ω** moves in such a way that its projection on the angular momentum vector **L** or the X_3'-axis is constant. As shown in Fig. 13.10, the angle ϕ_L between **ω** and **L** is given by [using the definition of the dot product and Eq. (13.151)]

$$\cos \phi_L = \frac{\boldsymbol{\omega} \cdot \mathbf{L}}{\omega L} = \frac{2T_{\text{rot}}}{\omega L} = \text{constant} \tag{13.152}$$

Angle ϕ_L remains constant and is the half-angle of the *laboratory* or *space cone*. This cone is the result of precession of **ω** about the constant angular momentum **L** as viewed from the inertial or LCS reference frame, **L**, **ω**, and the X_3(body)-axis all lie in one plane, and since **L** has been designated to be along the X_3'(LCS)-axis, it has resulted in **ω** precessing around the X_3'-axis when viewed in the LCS or inertial coordinate system. On the other hand, when viewed from the body coordinate system, **ω** precesses around the X_3(body or symmetry)-axis. The situation is shown in Fig. 13.11(a) and (b) and may be described as one cone rolling on another; that is, the body cone is rolling without slipping around the LCS cone and the line of contact is the direction of the angular velocity **ω**, which precesses around the X_3-axis when viewed from the body reference frame and around the X_3'-axis when viewed from the LCS frame. The angular frequency of precession of **ω** about the X_3-axis (the symmetry axis), as stated earlier, is Eq. (13.136)

$$\Omega_B \equiv \frac{I_3 - I_{12}}{I_{12}} \omega_3 = \gamma \omega_3 \tag{13.136}$$

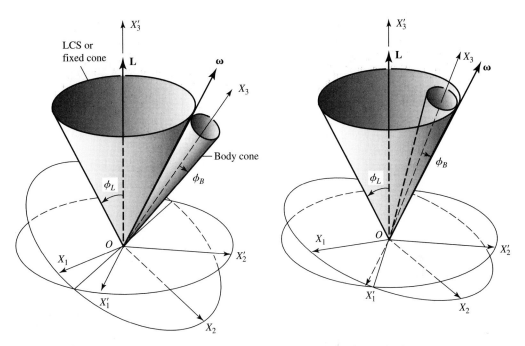

Figure 13.11   Body cone rolling around a LCS cone without slipping. Depending on the values of I_{12} and I_3, the body cone may roll (a) outside or (b) inside the LCS cone.

and the angular frequency of precession of $\boldsymbol{\omega}$ about the X_3'-axis (or **L**) is

$$\Omega_L = \gamma \omega_3 \frac{\sin \phi_B}{\sin \phi_L} \qquad \textbf{(13.153)}$$

Depending on the values of I_{12} and I_3, the body cone may roll outside or inside the LCS cone, as shown in Fig. 13.11.

One striking example is the application of the above theory to the rotating Earth. Earth is known to be slightly flattened near the poles, resulting in an oblate spheroid shape. This gives $I_3 \simeq I_{12}$ and $I_3 > I_{12}$, resulting in

$$\Omega = \frac{I_3 - I_{12}}{I_{12}} \omega_3$$

being very small as compared to ω_3, such that $\Omega \simeq \omega_3/300$. Since the period of Earth's rotation is $(1/\omega) = 1$ day and $\omega_3 \simeq \omega$, we get $(1/\Omega \simeq 300$ days$)$

$$T_P = \frac{2\pi}{\Omega} = \frac{2\pi I_{12}}{\omega_3(I_3 - I_{12})} = \frac{1 \text{ day}}{0.00327} = 305 \text{ days} \qquad \textbf{(13.154)}$$

The measured value is $\simeq 440$ days. The disagreement is not due to lack of knowledge of I_3 or I_{12}, but to the fact that Earth is not a perfect rigid body nor an oblate spheroid in shape. Actually, the shape of Earth resembles a flattened pear. Thus Earth's rotation axis precesses about the North Pole in a circle with a radius of $\simeq 10$ m and with a period of about 430 days. Since latitude is dependent on the rotation axis, a measurable change in latitude results. Such changes in latitude are called the *Chandler wobble* and were discovered by S. C. Chandler in 1891.

Another and more familiar precession, Earth's axis about a cone with a half-angle $23.5°$, is the result of the external gravitational torques due to the Sun and Moon. (That is, the rotational axis is inclined at $23.5°$ to the plane of Earth's orbit around the Sun.) This results in a slow precession of Earth'a axis. The period of such precessional motion is 26,000 years. This means that, as time passes, different stars become the polar star. Today the North Star (Polaris) is the polar star; in 3000 B.C., Thuban was the polar star; in 14,000 A.D., Vega will be the polar star. This is due to the precession of the rotational axis of Earth resulting from the gravitational forces of the Sun and Moon.

13.11 MOTION OF A SYMMETRICAL TOP WITH ONE POINT FIXED (THE HEAVY TOP)

A rigid body rotating about some fixed point O under the influence of a torque produced by its weight (in the gravitational force field) is called a *heavy top*. We shall limit our discussion to a special case of a symmetrical top in which $I_3 > I_1 = I_2 \, (=I_{12})$. Furthermore, the fixed point O does not coincide with the center of mass, but still lies on the symmetry axis. Such a situation is shown in Fig. 13.12. The fixed point is O, which coincides with the origins of the fixed and body coordinate systems. Because of the coincidence of the origins of the body and fixed sys-

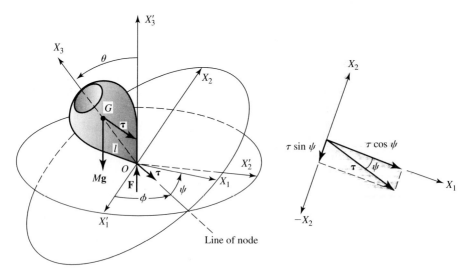

Figure 13.12 Heavy symmetrical top with one point fixed.

tems, the translational kinetic energy will be zero because $v = \dot{r} = 0$. The X_3'(fixed)-axis corresponds to the vertical, and the X_3(body)-axis is the symmetry axis of the top. The tip of the top is stationary at O. The only two forces acting are the reaction force **F** passing through point O, which does not produce any torque, while the gravitational force $M\mathbf{g}$ produces a torque $\boldsymbol{\tau}$ that is parallel to the line of nodes, as shown. We can use the Euler angles to describe the motion of the symmetrical top. The torque $\boldsymbol{\tau}$ on the symmetrical top is

$$\boldsymbol{\tau} = \mathbf{r} \times M\mathbf{g} \tag{13.155}$$

thus

$$\tau \simeq Mgl \sin \theta \tag{13.156}$$

The torque $\boldsymbol{\tau}$ that is along the lines of node may be resolved along the body axes, as shown in the insert in Fig. 13.12, resulting in

$$\tau_1 = Mgl \sin \theta \cos \psi \tag{13.157}$$

$$\tau_2 = Mgl \sin \theta \sin \psi \tag{13.158}$$

$$\tau_3 = 0 \tag{13.159}$$

Using the values of ω_1, ω_2, and ω_3 from Eqs. (13.120) and the Euler equations given by Eqs. (13.128), we obtain the following Euler's equations for a symmetrical top:

$$I_{12} \frac{d}{dt} (\dot{\phi} \sin \theta \sin \psi + \dot{\theta} \cos \psi) + (I_3 - I_{12})(\dot{\phi} \cos \theta + \dot{\psi})$$
$$\times (\dot{\phi} \sin \theta \cos \psi - \dot{\theta} \sin \psi) = Mgl \sin \theta \cos \psi \tag{13.160}$$

$$I_{12} \frac{d}{dt} (\dot{\phi} \sin \theta \cos \psi - \dot{\theta} \sin \psi) - (I_3 - I_{12})(\dot{\phi} \cos \theta + \dot{\psi})$$
$$\times (\dot{\phi} \sin \theta \sin \psi + \dot{\theta} \cos \psi) = -Mgl \sin \theta \sin \psi \tag{13.161}$$

$$I_3 \frac{d}{dt} (\dot{\psi} + \dot{\phi} \cos \theta) = 0 \tag{13.162}$$

In principle, Euler's equations can be solved to obtain three first integrals (two angular momenta and one energy); hence the three Euler angles. Since this is quite cumbersome, we will simply summarize the results.

The kinetic energy of the symmetrical top is

$$T = \frac{1}{2} \sum_i I_i \omega_i^2 = \frac{1}{2} I_{12}(\omega_1^2 + \omega_2^2) + \frac{1}{2} I_3 \omega_3^2 \tag{13.163}$$

or

$$T = \frac{1}{2} I_{12}(\dot{\phi}^2 \sin^2 \theta + \dot{\theta}^2) + \frac{1}{2} I_3(\dot{\phi} \cos \theta + \dot{\psi})^2 \tag{13.164}$$

while the potential energy is

$$V = Mgl \cos \theta \tag{13.165}$$

Thus the Lagrangian L is

$$L = L(\theta, \dot{\phi}, \dot{\theta}, \dot{\psi}) = T - V \tag{13.166}$$

We notice that ϕ and ψ are ignorable or cyclic coordinates. Therefore, the momenta conjugate to these coordinates are constants of motion. The cyclic coordinates are angles; the conjugate momenta are angular momenta. Thus

$$p_\phi = \frac{\partial L}{\partial \dot{\phi}} = I_{12} \dot{\phi} \sin^2 \theta + I_3 \cos \theta (\dot{\phi} \cos \theta + \dot{\psi}) = \text{constant} \tag{13.167}$$

$$p_\psi = \frac{\partial L}{\partial \dot{\psi}} = I_3 (\dot{\phi} \cos \theta + \dot{\psi}) = \text{constant} \tag{13.168}$$

These are the two first integrals of motion. Another first integral is the total energy E. Since the symmetrical top is in the gravitational force field, which is conservative, the total energy is a constant of motion and may be written as

$$E = T + V = \tfrac{1}{2} I_{12}(\omega_1^2 + \omega_2^2) + \tfrac{1}{2} I_3 \omega_3^2 + Mgl \cos \theta$$

$$= \tfrac{1}{2} I_{12}(\dot{\phi}^2 \sin^2 \theta + \dot{\theta}^2) + \tfrac{1}{2} I_3(\dot{\phi} \cos \theta + \dot{\psi})^2 + Mgl \cos \theta = \text{constant} \tag{13.169}$$

From Eq. (13.168),

$$p_\psi = I_3(\dot{\phi} \cos \theta + \dot{\psi}) = I_3 \omega_3 = \text{constant} \tag{13.170}$$

and

$$\frac{1}{2} I_3 \omega_3^2 = \frac{p_\psi^2}{2 I_3} = \text{constant} \tag{13.171}$$

Thus not only E, but $E' = E - \tfrac{1}{2} I_3 \omega_3^2$ is also a constant of motion. Substituting the values of $\dot{\phi}$ from Eq. (13.75) in Eq. (13.168) and after rearranging, we get

$$E' = \tfrac{1}{2} I_{12} \dot{\theta}^2 + V(\theta) \tag{13.172}$$

where

$$V(\theta) = \frac{p_\phi - p_\psi \cos \theta}{2 I_{12} \sin^2 \theta} + Mgl \cos \theta \tag{13.173}$$

and $V(\theta)$ is called the *effective potential*. From Eq. (13.172),

$$\dot{\theta} = \left(\frac{2}{I_{12}} [E' - V(\theta)] \right)^{1/2} \tag{13.174}$$

which on integration gives

$$t(\theta) = \int \frac{d\theta}{\sqrt{(2/I_{12})[E' - V(\theta)]}} \tag{13.175}$$

This equation, in principle, can be solved to obtain $\theta(t)$. These values of $\theta(t)$ can be used to yield the values of $\phi(t)$ and $\psi(t)$. Thus we have all three Eulerian angles that specify the orientation of a rigid body. Hence, the problem at hand is completely solved. Unfortunately, the integration of these equations involves an elliptic integral and the procedure becomes complicated. Hence, it becomes essential to limit ourselves to a qualitative discussion, similar to the one used in describing the motion of a particle in a central force field.

Steady Precession

Figure 13.13 shows the plot of effective potential $V(\theta)$ [given by Eq. (13.173)] versus θ between the physically acceptable range of $0 \le \theta \le \pi$. This energy diagram with a minimum effective potential is similar to the diagram for the central force field. For any energy value $E' = E_1'$, the motion is limited between two extreme values, which are similar to the turning points, that is, between $\theta = \theta_1$ and $\theta = \theta_2$, as shown. This implies that the symmetrical axis OX_3 of the rotating top can vary its inclination θ to the vertical between $\theta_1 \le \theta \le \theta_2$. If the energy of the top is such that $E' = E_0' = V_{\min}$, the value of θ is limited to a single value of $\theta = \theta_0$, as shown. The resulting motion is a steady precession at a fixed angle of inclination θ_0. This is an interesting special case of steady precession in which the axis of the gyroscope or heavy top describes a right

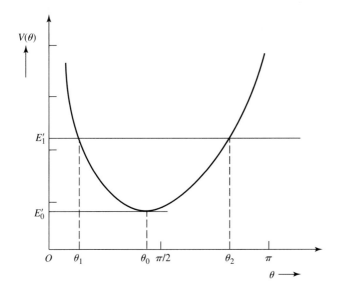

Figure 13.13 Energy diagram: plot of effective potential $V(\theta)$ versus θ ($0 \le \theta \le \pi$) for a heavy symmetrical top.

circular cone about the vertical (X_3'-axis). Before discussing the general situation, we shall discuss this case in some detail. We can evaluate the value θ_0 by setting the derivative of the effective potential $V(\theta)$ equal to zero at θ_0. (Note that, in general, $\tau(\theta) = dV(\theta)/d\theta$.) Hence, from Eq. (13.173),

$$\tau(\theta_0) = \left| -\frac{\partial V}{\partial \theta} \right|_{\theta = \theta_0}$$

$$= \frac{(p_\phi - p_\psi \cos\theta_0)^2 \cos\theta_0 - (p_\phi - p_\psi\cos\theta_0) p_\psi \sin^2\theta_0}{I_{12}\sin^3\theta_0} + Mgl\sin\theta_0 = 0 \qquad \textbf{(13.176)}$$

Let us define

$$\gamma \equiv p_\phi - p_\psi\cos\theta_0 \qquad \textbf{(13.177)}$$

and rewrite Eq. (13.176) as

$$(\cos\theta_0)\gamma^2 - (p_\psi\sin^2\theta_0)\gamma + (MglI_{12}\sin^4\theta_0) = 0 \qquad \textbf{(13.178)}$$

This is a quadratic in γ, which has two values. For the given value of $\theta = \theta_0$, let us discuss the value of $\dot\phi$. The precessional angular velocity $\dot\phi_0$ has two possible values, one for each value of γ given by solving Eq. (13.178). A large value of γ results in fast precession and a small value in slow precession; that is,

$$\dot\phi_0(+) = \dot\phi_{0f} \rightarrow \text{fast precession}$$

$$\dot\phi_0(-) = \dot\phi_{0s} \rightarrow \text{slow precession} \qquad \textbf{(13.179)}$$

It is this slow precessional angular velocity, that is usually observed in gyroscopes.

Thus, for the symmetry axis at $\theta = \theta_0$ and less than $\pi/2$, the top is rotating about the symmetry axis at frequency ω_3 and the symmetry axis can precess about the fixed axis with two possible frequencies $\dot\phi$. A special case is in order. If the top is spinning sufficiently fast and is in the vertical position, the axis of the top will remain fixed in the vertical direction. This condition is called *sleeping,* and the top is a *sleeping top.* If the top slows down due to friction or other causes, the top starts undergoing a nutation (as discussed later) and eventually will topple over.

Let us now discuss the case in which $\theta_0 > \pi/2$. In such a case, the fixed tip of the top is at a position above the center of mass. The symmetrical top is hanging with its axis below the horizontal. Furthermore, the values of $\dot\phi_{0f}$ and $\dot\phi_{0s}$ have opposite signs. That is, for $\theta_0 > \pi/2$, the fast precession $\dot\phi_{0f}$ is in the same direction as that for $\theta_0 < \pi/2$, while the slow precession $\dot\phi_{0s}$ takes place in the opposite sense.

θ Motion: Nutation

As discussed earlier in connection with the effective potential $V(\theta)$ versus θ plot in Fig. 13.13, the motion of the symmetrical axis is limited between $\theta_1 < \theta < \theta_2$ for any given energy E' of the top. As θ varies between these limits, the value of $\dot\phi$ may or may not change sign. If there is no change in the sign of $\dot{\phi}$, the top precesses monotonically around the fixed, inertial, or LCS X_3'-axis, while the X_3(symmetry)-axis of the body oscillates between $\theta = \theta_1$ and $\theta = \theta_2$. This motion of the top is called *nutation*. The path desribed by the body symmetry axis when projected on a unit sphere in the fixed system is shown in Fig. 13.14(a). On the other hand, if $\dot{\phi}$ does change sign between the limiting values of θ, the precessional angular velocity must have opposite signs at $\theta = \theta_1$ and $\theta = \theta_2$. In this situation the nutational-precessional motion results in a looping motion of the symmetry axis, as shown in Fig. 13.14(b), which is a projection of the symmetry axis on a unit sphere. Note that the changes in $\dot{\phi}$ are not only due to the values of p_ϕ and p_ψ. If these values are such that at $\theta = \theta_1$,

$$(p_\phi - p_\psi \cos \theta)\big|_{\theta = \theta_1} = 0 \tag{13.180}$$

then
$$\dot{\phi}\big|_{\theta = \theta_1} = 0, \qquad \dot{\theta}\big|_{\theta = \theta_1} = 0, \quad \text{and} \quad \dot{\psi} = \omega_3 \tag{13.181}$$

The resulting motion of the projection of the symmetry axis on a sphere is cusplike, as shown in Fig. 13.14(c). (a) and (c) are redrawn for arbitrary values in Fig. 13.14(d).

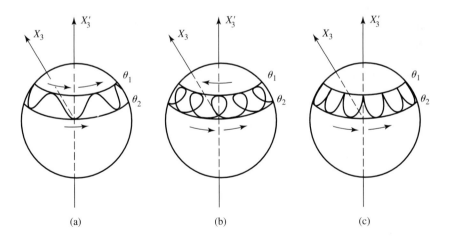

(a)   (b)   (c)

Figure 13.14 θ motion (nutation), the motion of the symmetrical axis limited between $\theta_1 < \theta < \theta_2$. The diagrams show the path of a body symmetry axis as projected on a unit sphere in a fixed system. (a) monotonic precession, (b) looping motion, and (c) cusplike motion of the symmetry axis about a fixed axis.

Figure 13.14(d) _____

Assuming arbitrary values, we can show the angular velocity precessing in a cone. The two cases (a) and (c) above are redrawn here.

$$n := 1 .. 60 \qquad m := 1 \qquad\qquad I1 := 2 \quad I2 := 2 \quad I12 := 3.5 \quad I3 := 4 \quad \omega := 40$$

$$\omega z := 5 \qquad \Omega B_m := \frac{I3 - I12}{I12} \cdot \omega z \qquad \Omega B_m = 0.714 \qquad A := \sqrt{\omega^2 - (\omega z)^2}$$

$$\omega x_{n,m} := A \cdot \sin\left(\Omega B_m \cdot m\right) \cdot \sin\left(n \cdot \frac{2 \cdot \pi}{60}\right) \qquad\qquad \omega y_{n,m} := A \cdot \sin\left(\Omega B_m \cdot m\right) \cdot \cos\left(n \cdot \frac{2 \cdot \pi}{60}\right)$$

$$Za_{n,m} := 4 \cdot \cos\left(\Omega B_m \cdot n\right) + 60 \qquad\qquad Zc_{n,m} := -4 \cdot \left|\cos\left(\Omega B_m \cdot n\right)\right| + 60$$

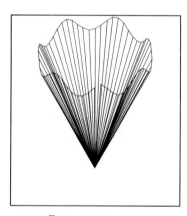

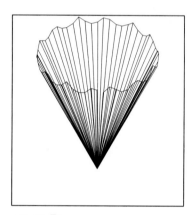

$\omega x, \omega y, Za$ $\omega x, \omega y, Zc$

       Precessing cone   Precessing cone

Compare these with (a) and (c) and explain the difference.

PROBLEMS

13.1. Combine Eq. (13.7) with Eq. (13.3) to obtain the results in Eq. (13.6).

13.2. Prove the following identities.
   (a) $\mathbf{A} \times (\mathbf{B} \times \mathbf{A}) = A^2\mathbf{B} - \mathbf{A}(\mathbf{A} \cdot \mathbf{B})$
   (b) $(\mathbf{A} \times \mathbf{B}) \cdot (\mathbf{C} \times \mathbf{D}) = (\mathbf{A} \cdot \mathbf{C})(\mathbf{B} \cdot \mathbf{C}) - (\mathbf{B} \cdot \mathbf{C})(\mathbf{A} \cdot \mathbf{D})$
   (c) $(\mathbf{A} \times \mathbf{B})^2 = (\mathbf{A} \times \mathbf{B}) \cdot (\mathbf{A} \times \mathbf{B}) = (\mathbf{A} \cdot \mathbf{A})(\mathbf{B} \cdot \mathbf{B}) - (\mathbf{B} \cdot \mathbf{A})(\mathbf{A} \cdot \mathbf{B}) = A^2B^2 - (\mathbf{A} \cdot \mathbf{B})^2$

13.3. Find the angular momentum and kinetic energy for the rotation of a uniform square lamina of side L and mass M about a diagonal with an angular velocity $\boldsymbol{\omega}$.

13.4. Consider a uniform rectangular lamina of sides a and b and surface density (mass per unit area) σ. It is rotating about a diagonal with constant angular velocity ω. Find the magnitude and the direction of the angular momentum about an axis passing through **(a)** its center, and **(b)** one of its corners. Also calculate the rotational kinetic energy.

13.5. A uniform disk of radius R and mass M is rotating with uniform angular velocity ω about an axis that makes an angle θ with the axis of the disk. Calculate **(a)** the angular momentum (magnitude as well as direction), and **(b)** the total rotational kinetic energy.

13.6. A particle of mass m is rotating in a vertical plane with angular velocity ω, which lies in the XY plane and makes an angle of 45° with the X-axis, as shown in Fig. P13.6. Calculate velocity $\mathbf{v}$, angular momentum $\mathbf{L}$, and the rotational kinetic energy. Are ω and $\mathbf{L}$ in the same direction? What does this imply?

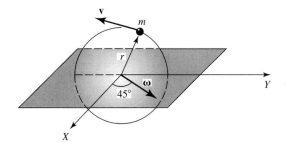

Figure P13.6

13.7. A thin uniform disk of mass m, radius r, and thickness h rolls without slipping about the Z-axis. It is supported by an axle of length R through its center (as shown in Fig. P13.7) and circles around the Z-axis with angular velocity Ω. Calculate the instantaneous angular velocity ω of the disk and its angular momentum $\mathbf{L}$. Are ω and $\mathbf{L}$ parallel to each other?

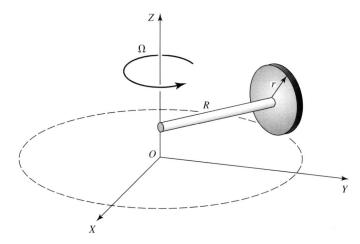

Figure P13.7

13.8. The axis of symmetry of a misaligned armature makes an angle ϕ with the rotation axis, as shown in Fig. P13.8. If the rotational angular velocity of the armature is $\dot{\theta}$, what are the reactions at the bearings at A and B? The disk has a radius of R, thickness h, and mass M.

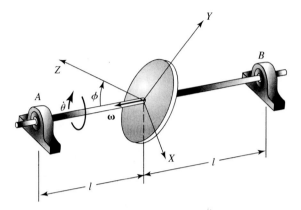

Figure P13.8

13.9. Calculate the components of a moment of inertia tensor for the following configuration. Point masses 2, 3, 1, and 5 units located at $(0, 0, 1)$, $(0, 1, 1)$, $(1, -1, 1)$, and $(1, -1, -1)$.

13.10. Calculate the components of a moment of inertia tensor for the following configuration. Point masses 2, 3, 6, and 8 units located at $(0, 1, 2)$, $(0, 2, 2)$, $(2, 2, 2)$, and $(2, -2, -2)$.

13.11. Find the elements of the inertia tensor of a rod of mass M and length l. The origin of the coordinate system is at its center, the X-axis is along the length of the rod, and the Z-axis is perpendicular to the rod.

13.12. Find the inertia tensor for a square lamina of length L and mass M for a coordinate system whose origin is located **(a)** at one corner, and **(b)** at the center of the lamina.

13.13. Find the inertia tensor of a rectangular lamina of sides L and W and mass M for a coordinate system whose origin is **(a)** at one corner, and **(b)** at the center of the lamina.

13.14. Consider a homogeneous sphere of mass M and radius R. Find a coordinate system whose origin is at the center of the sphere, calculate the moments of inertia I_1, I_2, and I_3.

13.15. Show that for any homogeneous regular polyhedron the principal moments of inertia will all be equal for a coordinate system whose origin is at the center of the polyhedron. Find the radius of a homogeneous solid sphere of the same mass that has the same moment of inertia elements.

13.16. Find the inertia tensor of a rectangular block of mass M and dimension $a \times b \times c$. The origin of the axes coincides with the center of mass, the Z-axis is parallel to the thickness c, and the Y-axis is parallel to a diagonal of rectangular $a \times b$, as shown in Fig. P13.16. Find the relation between the coordinates axes and the principal coordinate axes.

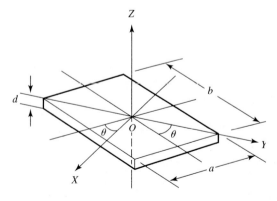

Figure P13.16

13.17. Consider a right triangular solid of mass M and length L along the X_1-, X_2-, and X_3-axes, as shown in Fig. P13.17. Calculate the elements of the inertia tensor for these axes.

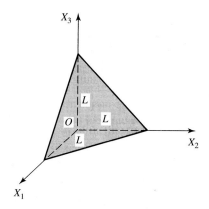

Figure P13.17

13.18. Six particles, each of mass m, are fixed at the end of massless rods of length $2l$. The rods are perpendicular to each other, as shown in Fig. P13.18. For the axis along the three rods, calculate the elements of the inertia tensor for this configuration. Show that these axes are the principal axes.

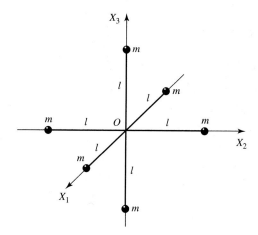

Figure P13.18

13.19. Show that any one of the three principal moments of inertia cannot be greater than the sum of the other two.

13.20. Consider a uniform density object of mass M in the shape of an ellipsoid whose equation is

$$\frac{x^2}{a^2} + \frac{y^2}{b^2} + \frac{z^2}{c^2} = 1$$

where the axes $2a > 2b > 2c$ are the dimensions of the solid. Find the principal moments of inertia I_1, I_2, and I_3 and the principal axes.

13.21. Consider a homogeneous cone of mass M, height h, and radius R. Let the origin be at the apex of the cone and the X_3-axis be along the axis of symmetry of the cone. Calculate the elements of the inertia tensor. Now make the transformation so that the center of mass of the cone is also the origin of the coordinate axes. Find the principal moment of inertia about this new coordinate system.

13.22. Show that in a plane rectangular lamina the direction of the principal axis at a corner is given by

$$\tan 2\phi = \frac{2(Mab/4)}{(Ma^2/3) - (Mb^2/3)} = \frac{3ab}{2(a^2 - b^2)}$$

13.23. For the following two cases, draw a sketch of the ellipsoid of inertia: **(a)** a uniform circular disk of radius R and mass M; **(b)** a solid rectangular parallelepiped of mass M and sides a, $2a$, and $4a$.

13.24. Draw a sketch of the ellipsoid of inertia of a solid right circular cylinder of radius R and length L ($ = 8R$). What should be the ratio R/L so that the ellipsoid of inertia at the center is a sphere?

13.25. Find the principal moment of inertia and principal axes for the right triangular solid discussed in Problem 13.17 and shown in Fig. P13.17. What types of rotation and rotation matrices are needed to go from the given axes to the principal axes?

13.26. Consider a right elliptical cylinder of mass M. The cylinder is bounded by plane ends with $Z = -c$ and $Z = +c$, while its wall is an elliptical surface defined by $(x/a)^2 + (y/b)^2 = 1$. Calculate the moment of inertia for rotation of the cylinder about the Z-axis.

13.27. In Problem 13.12, transfer from **(a)** to **(b)** by a proper rotation matrix.

13.28. In Problem 13.13, transfer from **(a)** to **(b)** by a proper rotation matrix.

13.29. The trace of a tensor $\mathbf{I}$ is defined as the sum of the diagonal elements; that is, trace $\mathbf{I} \equiv \Sigma_k I_{kk}$. By performing a similarity transformation, show that the trace is invariant under a coordinate transformation; that is, trace $\mathbf{I} \equiv$ trace $\mathbf{I}'$, where $\mathbf{I}'$ is the tensor in a coordinate system rotated with respect to the coordinate system of $\mathbf{I}$.

13.30. Show that the determinant of the elements of a tensor is invariant under different coordinate systems rotated with respect to each other.

13.31. Consider a tensor $\mathbf{T} = \mathbf{AB} + \mathbf{BA}$, where $\mathbf{A} = 10\hat{u}_1 - 6\hat{u}_2 + 4\hat{u}_3$ and $\mathbf{B} = \hat{u}_2 + 2\hat{u}_3$. Transform this tensor into a coordinate system rotated $45°$ about the X_3-axis. Diagonalize the resulting tensor.

13.32. Diagonalize the following tensor and find the principal axes:

$$\mathbf{T} = \begin{pmatrix} 7 & \sqrt{6} & -\sqrt{3} \\ \sqrt{6} & 2 & -5\sqrt{2} \\ -\sqrt{3} & -5\sqrt{2} & -3 \end{pmatrix}$$

13.33. Consider a thin homogeneous plate of mass M and of dimensions l and w that lies in the X_1-X_2 plane. Show that its inertia tensor has the following form:

$$\mathbf{T} = \begin{pmatrix} A & -C & 0 \\ -C & B & 0 \\ 0 & 0 & A+B \end{pmatrix}$$

Calculate the value of A, B, and C in terms of M, l, and w.

Now rotate the coordinate axis through angle θ about the X_3-axis. Show that the new inertia tensor takes the form

$$\mathbf{T}' = \begin{pmatrix} A' & -C' & 0 \\ -C' & B' & 0 \\ 0 & 0 & A' + B' \end{pmatrix}$$

Calculate the values of A', B', and C' in terms of A, B, and C and θ.

Show that if the angle of rotation θ is given by the expression

$$\theta = \frac{1}{2} \tan^{-1} \frac{2C}{B - A}$$

then the X_2- and X_3-axes will be the principal axes.

13.35. Obtain the components of angular velocity $\boldsymbol{\omega}$ directly from the transformation matrix $\boldsymbol{\lambda}$ given by Eq. (13.115).

13.36. Obtain the inverse transformation matrix of $\boldsymbol{\lambda}$ given in Eq. (13.115), and then obtain the components of $\boldsymbol{\omega}'$.

13.37. By using Fig. 13.8, obtain the components of $\boldsymbol{\omega}$ along the fixed X_i'-axes; that is, calculate ω_1', ω_2', and ω_3'.

13.38. Derive Eq. (13.153).

13.39. Find the componets of a tensor that corresponds to a rotation by an angle θ about the Z-axis and followed by a rotation by an angle ϕ about the Y-axis.

13.40. Consider a homogeneous sphere with moments of inertia $I_1 = I_2 = I_3$. Find the equations of motion of the sphere by using Euler's equations.

13.41. A uniform rod of length l and mass m is mounted on a horizontal frictionless axle through its center. The axle is mounted on a platform that is rotating with angular velocity Ω, as shown in Fig. P13.41. The axis of the platform passes through the center of the rod. Using Euler's equations, calculate the angle θ that the rod makes with the horizontal as a function of time. Show that for a small θ the motion of the rod is simple harmonic with angular frequency $[(I_3 - I_2)/I_1]^{1/2}\Omega$.

Figure P13.41

13.42. Derive Eqs. (13.198) and (13.199).

13.43. Consider the force free motion of a symmetrical top and show that the angular velocity $\boldsymbol{\omega}$, the angular momentum $\mathbf{L}$ about the fixed (space) X_3'-axis, and the body X_3-axis are coplanar.

13.44. A circular disk of mass M and radius R is rotating freely under no external torque. The angle between the axis of symmetry of the disk and the angular velocity $\boldsymbol{\omega}$ is ϕ. Calculate the time in which the axis of symmetry describes a cone about the direction of $\mathbf{L}$, that is, about the invariable line.

13.45. Consider the force free rotation of a plane lamina. By using Euler's equations, show that the component of the angular velocity in the plane of the lamina is constant in magnitude. Under what conditions will the component of the angular velocity normal to the plane of the lamina be constant?

13.46. Consider a symmetrical rigid body moving freely in space and powered by two jet engines that are symmetrically placed with respect to the symmetry body axis (that is, X_3-axis) and supply a constant torque $\boldsymbol{\tau}$ about the symmetry axis. Find the general expression for the angular velocity $\boldsymbol{\omega}$ as a function of time. Show that $\boldsymbol{\omega}$ increases in magnitude with time, and its components perpendicular to the X_3-axis describe a constant ellipse.

13.47. Consider a rigid body with three different principal moments of inertia, $I_1 > I_2 > I_3$, rotating freely about its center of mass. Show by using Euler's equations that the rotational motion of the body is stable about either the axis of greatest moments of inertia or the axis of least moment of inertia.

13.48. A symmetrical rigid body rotates freely about a fixed point free of any external torque. Let θ be the angle between the axis of rotation and the axis of the system. The moment of inertia I_s about the symmetry axis is greater than the moment of inertia I_n about an axis normal to the symmetry axis. Show that the angle between the axis of rotation and the invariable line (the $\mathbf{L}$ vector) is

$$\tan^{-1}\left[\frac{(I_s - I_n)\tan\theta}{I_s + I_n\tan^2\theta}\right]$$

What is the maximum possible value for this angle?

13.49. A flywheel (a disk of mass M and radius R) is mounted with its axis vertical in a truck and works as a stabilizing gyroscope. Suppose the disk is rotating at full speed ω. Show that the torque needed to make it precess in a vertical plane is $\tau \simeq \frac{1}{2}MR^2\omega\Omega$, where Ω is the frequency of precession.

13.50. Suppose a heavy top is spinning in a stable configuration. What is the effect of friction on the motion as friction gradually reduces the value of ω_3?

13.51. A simple gyroscope consists of a disk of 0.2-kg mass and has a radius of 0.06 m; it is mounted at the center of a light rod of length 0.12 m. It is set spinning such that precessional frequency is 0.2 revolution per second. Calculate approximately the spinning frequency.

13.52. A symmetrical rigid body rotates with an angular velocity $\boldsymbol{\omega}$ in three-dimensional motion about its center of mass. If there is a frictional torque $-b\boldsymbol{\omega}$ due to air drag, show that the component of $\boldsymbol{\omega}$ in the direction of the symmetry axis decreases exponentially with time.

13.53. To investigate the turning points of the nutational motion of a symmetrical top, we substitute $\dot{\theta} = 0$. Show that the resulting equation is a cubic in $\cos\theta$ with two real roots and one imaginary root for θ.

SUGGESTIONS FOR FURTHER READING

ARTHUR, W., and FENSTER, S. K., *Mechanics*, Chapter 12. New York: Holt, Rinehart and Winston, Inc., 1969.

BARGER, V., and OLSSON, M., *Classical Mechanics*, Chapter 6. New York: McGraw-Hill Book Co., 1973.

BECKER, R. A., *Introduction to Theoretical Mechanics*, Chapter 12. New York: McGraw-Hill Book Co., 1954.

CORBEN, H. C., and STEHLE, P., *Classical Mechanics*, Chapter 9. New York: John Wiley & Sons, Inc., 1960.

DAVIS, A. DOUGLAS, *Classical Mechanics*, Chapter 9. New York: Academic Press, Inc., 1986.

FOWLES, G. R., *Analytical Mechanics*, Chapter 9. New York: Holt, Rinehart and Winston, Inc., 1962.

*GOLDSTEIN, H., *Classical Mechanics*, 2nd ed., Chapters 4 and 5. Reading, Mass.: Addison-Wesley Publishing Co., 1980.

HAUSER, W., *Introduction to the Principles of Mechanics*, Chapter 9. Reading, Mass.: Addison-Wesley Publishing Co., 1965.

KITTEL, C., KNIGHT, W. D., and RUDERMAN, M. A., *Mechanics*, Berkeley Physics Course, Volume 1, Chapter 8. New York: McGraw-Hill Book Co., 1965.

KLEPPNER, D., and KOLENKOW, R. J., *An Introduction to Mechanics*, Chapter 7. New York: McGraw-Hill Book Co., 1973.

*LANDU, L. D., and LIFSHITZ, E. M., *Mechanics*, Chapter 6. Reading. Mass.: Addison-Wesley Publishing Co., 1960.

MARION, J. B., *Classical Dynamics*, 2nd ed., Chapter 12. New York: Academic Press, Inc., 1970.

*MOORE, E. N., *Theoretical Mechanics*, Chapters 5 and 6. New York: John Wiley & Sons, Inc., 1983.

ROSSBERG, K., *Analytical Mechanics*, Chapter 9. New York: John Wiley & Sons, Inc., 1983.

SCARBOROUGH, J. B., *They Gyroscope: Theory and Applications*. New York: Wiley-Interscience, 1958.

SLATER, J. C., *Mechanics*, Chapters 5 and 6. New York: McGraw-Hill Book Co., 1947.

SYMON, K. R., *Mechanics*, 3rd ed., Chapters 10 and 11. Reading, Mass.: Addison-Wesley Publishing Co., 1971.

SYNGE, J. L., and GRIFFITH, B. A., *Principles of Mechanics*, Chapter 14. New York: McGraw-Hill Book Co., 1959.

*The asterisk indicates works of an advanced nature.

C H A P T E R

14

Theory of Small Oscillations and Coupled Oscillators

14.1 INTRODUCTION

In Chapters 3 and 4, we discussed the oscillatory motion of undamped, damped, and forced oscillators. As in the previous chapters where we extended the motion of single particles to the motion of rigid bodies, we now investigate the oscillatory motion of system of particles. One of the most deeply investigated concepts in modern physics is that of oscillatory motion of atoms in the field of molecular physics and solids in the field of solid state physics.

We will describe the oscillatory motion of many coupled oscillators in terms of normal coordinates and normal frequencies. Theory of small oscillations in analyzing coupled oscillatory motion uses methods of Lagrange's equations together with matrix tensor formulation. We will close the chapter with the discussion of vibrations and beats in the vibrating systems. Also, we will briefly touch the topic of dissipative systems under forced oscillators.

14.2 EQUILIBRIUM AND POTENTIAL ENERGY

To understand the general theory of vibrations, it is essential to know the relation between potential energy and equilibrium that leads to the conditions of stable or unstable equilibrium of a given system. To start, let us consider a system with n degrees of freedom, and let its configuration be specified by the generalized coordinative: $q_1, q_2, \ldots, q_n$. Furthermore, let us assume that the system is conservative; hence the potential energy V is a function of the generalized coordinates; that is,

$$V = V(q_1, q_2, \ldots, q_n) \tag{14.1}$$

The generalized forces Q_k are given by

$$Q_k = -\frac{\partial V}{\partial q_k}, \qquad k = 1, 2, \ldots, n \tag{14.2}$$

If the system is in such a configuration that it is in equilibrium, it implies that all the generalized forces Q_k must be zero. Thus the condition for an equilibrium configuration is

$$Q_k = -\frac{\partial V}{\partial q_k} = 0 \tag{14.3}$$

The system will remain at rest in this configuration if no external force is applied. Now let us displace this system slightly from its equilibrium configuration. After displacement, the system may or may not return to its equilibrium configuration. If after a small displacement the system does return to its original equilibrium configuration, the system is said to be in a *stable equilibrium*. If the system does not return to its equilibrium configuration, it is in an *unstable equilibrium*. On the other hand, if the system is displaced and it has no tendency to move toward or away from the equilibrium configuration, the system is said to be in *neutral equilibrium*.

We are interested in finding a relation between the potential energy function V and the stability of the system. Suppose, when the system is in an equilibrium configuration, it has kinetic energy T_0 and potential energy V_0. Now the system is given a small displacement (by a small impulsive force), and at any subsequent time the system has kinetic energy T and potential energy V. Since total energy is conserved, we may write

$$T_0 + V_0 = T + V$$
$$T - T_0 = -(V - V_0) \tag{14.4}$$

Let us assume an arbitrary form of a potential function V versus q, as shown in Fig. 14.1. The points A and B, where $\partial V/\partial q$ is zero, are equilibrium points. Let us consider the nature of stability at these points.

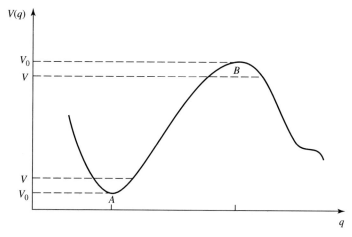

Figure 14.1   Arbitrary form of a potential function V versus q.

Suppose initially the system is in equilibrium corresponding to the configuration at B where the potential energy V_0 is maximum. Any displacement from this equilibrium will lead to a potential energy V that is less than V_0. Thus $V - V_0$ is negative, and from Eq. (14.4) $T - T_0$ will be positive; that is, T increases. Since T increases with displacement, the system never returns to the equilibrium point B; hence B is a position of *unstable equilibrium*. Now let us consider point A, where the equilibrium potential V_0 is minimum. If the system is displaced slightly, the potential energy V_0 increases to V; hence $V - V_0$ is positive. From Eq. (14.4), $T - T_0$ will be negative; hence T decreases with displacement. Since T cannot be negative, it will decrease till it becomes zero at some limiting configuration near the equilibrium configuration; the system will start coming back to an equilibrium configuration. Thus the system is in stable equilibrium. We conclude that for small displacements the condition for *stable equilibrium* is that the potential energy V_0 be minimum at the equilibrium configuration. Furthermore, at equilibrium dV/dq is zero, $V - V_0$ being positive means that d^2V/dq^2 is positive at stable equilibrium. At a position of unstable equilibrium, d^2V/dq^2 will be negative because $V - V_0$ is negative.

Applying the preceding discussion to a system with one degree of freedom, we may write

$$V = V(q) \tag{14.5}$$

and at an equilibrium configuration

$$F = -\frac{dV}{dq} = 0 \tag{14.6}$$

The stability condition may be written as

Stable equilibrium: V_0 is minimum $\dfrac{d^2V}{dq^2} > 0$ **(14.7)**

Unstable equilibrium: V_0 is maximum $\dfrac{d^2V}{dq^2} < 0$ **(14.8)**

For $d^2V/dq^2 = 0$, we must examine the higher-order derivatives. If the first nonvanishing derivative is odd, the system must be in unstable equilibrium. If, on the other hand, the nonvanishing derivative is of an even order, then the system may be in a stable or unstable equilibrium depending on the value of the derivative (whether it is greater than zero or less than zero).

If $\dfrac{d^nV}{dq^n} \neq 0$, $n > 2$ and odd   system is unstable   **(14.9)**

If $\dfrac{d^nV}{dq^n} > 0$, $n > 2$ and even   system is stable   **(14.10)**

If $\dfrac{d^nV}{dq^n} < 0$, $n > 2$ and even   system is unstable   **(14.11)**

A more general case of this situation will be discussed shortly.

Example 14.1

Show that a bat of length l suspended from point O with a center of mass at a distance d from O is in a stable equilibrium position as in (a) and an unstable equilibrium position as in (b).

Solution

The situation is as shown in Fig. Ex. 14.1. When the bat is displaced, the line OC makes an angle θ with the vertical as in Fig. Ex. 14.1(a). The center of mass is raised a distance h and the potential energy is given by

	(a)	(b)

Potential energy when the bat is displaced.

(a)

$$V = m \cdot g \cdot d \cdot (1 - \cos(\theta))$$

$$\frac{dV}{d\theta} = \frac{d}{d\theta}(m \cdot g \cdot d \cdot (1 - \cos(\theta)))$$

$$\frac{dV}{d\theta} = m \cdot g \cdot d \cdot \sin(\theta)$$

$$\theta = 0 \qquad \frac{dV}{d\theta} = 0$$

$$\left(\frac{dV^2}{d\theta^2}\right) = \frac{d}{d\theta} m \cdot g \cdot d \cdot \sin(\theta)$$

(b)

$$V = -m \cdot g \cdot d \cdot (1 - \cos(\theta))$$

$$\frac{dV}{d\theta} = \frac{d}{d\theta}(-m \cdot g \cdot d \cdot (1 - \cos(\theta)))$$

$$\frac{dV}{d\theta} = -m \cdot g \cdot d \cdot \sin(\theta)$$

$$\theta = 0 \qquad \frac{dV}{d\theta} = 0$$

$$\left(\frac{dV^2}{d\theta^2}\right) = \frac{d}{d\theta} -m \cdot g \cdot d \cdot \sin(\theta)$$

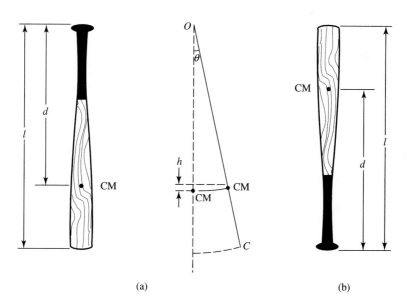

(a)   (b)

Figure Ex. 14.1

Taking the second derivative and evaluating at $\theta = 0$ reveals that **(a)** is in stable equilibrium while **(b)** is in unstable equilibrium

$$\frac{dV^2}{d\theta^2} \equiv m \cdot g \cdot d \cdot \cos(\theta)$$

$$\frac{dV^2}{d\theta^2} \equiv - m \cdot g \cdot d \cdot \cos(\theta)$$

$$\theta = 0 \quad \frac{dV^2}{d\theta^2} \equiv m \cdot g \cdot d > 0$$

$$\theta = 0 \quad \frac{dV^2}{d\theta^2} \equiv - m \cdot g \cdot d < 0$$

Stable   unstable

From our discussion, we may conclude that *if the center of mass lies below the point of suspension, the system will be in stable equilibrium; and if the center of mass lies above the center of suspension, the system will be in unstable equilibrium.*

EXERCISE 14.1 The spherical or cylindrical object shown in Fig. Exer. 14.1 is placed on a plane horizontal surface. The radius of curvature is a, and the center of mass is at a distance d, as shown. Show that the system is in stable equilibrium.

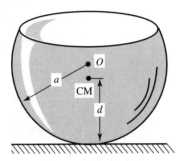

Figure Ex. 14.1

14.3 TWO COUPLED OSCILLATORS AND NORMAL COORDINATES

As a simple example of a coupled system, let us consider two harmonic oscillators coupled together by a spring, as shown in Fig. 14.2. Each harmonic oscillator has a particle of mass m, and the spring constant of one is k_1 and that of the other is k_2. The two are coupled together by another spring of spring constant k'. The motion of the two masses is restricted along the line joining the two masses, say along the X-axis. Thus the system has two degrees of freedom represented by the coordinates x_1 and x_2. The configuration of the sytem is represented by the displacements measured from the equilibrium positions O_1 and O_2, respectively. The displacements to the right are positive and those to the left are negative. If the two oscillators were not connected, each would vibrate independently of the other with frequencies

$$\omega_{10} = \sqrt{\frac{k_1}{m}} \quad \text{and} \quad \omega_{20} = \sqrt{\frac{k_2}{m}} \tag{14.12}$$

When these oscillators are connected by a spring of spring constant k', the system vibrates with different frequencies, which we wish to calculate now.

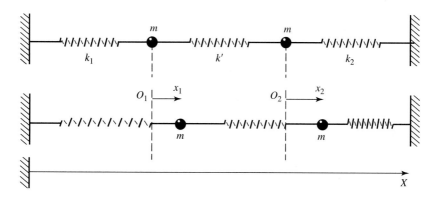

Figure 14.2   Two harmonic oscillators coupled together by a spring of spring constant k'.

The kinetic energy of the system is

$$T = \tfrac{1}{2}m\dot{x}_1^2 + \tfrac{1}{2}m\dot{x}_2^2 \tag{14.13}$$

and the potential energy of the system is

$$V = \tfrac{1}{2}k_1 x_1^2 + \tfrac{1}{2}k_2 x_2^2 + \tfrac{1}{2}k'(x_1 - x_2)^2 \tag{14.14}$$

Hence the Lagrangian function L of the system is

$$L = T - V = \tfrac{1}{2}m\dot{x}_1^2 + \tfrac{1}{2}m\dot{x}_2^2 - \tfrac{1}{2}k_1 x_1^2 - \tfrac{1}{2}k_2 x_2^2 - \tfrac{1}{2}k'(x_1 - x_2)^2 \tag{14.15}$$

The Lagrange equations of motion

$$\frac{d}{dt}\left(\frac{\partial L}{\partial \dot{x}_1}\right) - \frac{\partial L}{\partial x_1} = 0 \quad \text{and} \quad \frac{d}{dt}\left(\frac{\partial L}{\partial \dot{x}_2}\right) - \frac{\partial L}{\partial x_2} = 0 \tag{14.16}$$

take the form

$$m\ddot{x}_1 + k_1 x_1 + k'(x_1 - x_2) = 0 \tag{14.17a}$$

$$m\ddot{x}_2 + k_2 x_2 + k'(x_2 - x_1) = 0 \tag{14.17b}$$

The third term in each of these two equations is the result of coupling between the two oscillators. If there were no coupling, these oscillators would vibrate with frequencies given by Eq. (14.12). The preceding second-order linear differential equations may be written as

$$m\ddot{x}_1 + (k_1 + k')x_1 - k'x_2 = 0 \tag{14.18a}$$

$$m\ddot{x}_2 + (k_2 + k')x_2 - k'x_1 = 0 \tag{14.18b}$$

These equations will be independent of each other if the third term in each equation is not present. That is, if we hold the second mass at rest, $x_2 = 0$, and the frequency of vibrations of the first oscillator, from Eq. (14.18a), will be

$$\omega_1' = \sqrt{\frac{k_1 + k'}{m}} \qquad (14.19a)$$

On the other hand, if the first mass is at rest, that is, $x_1 = 0$, then the frequency of vibration of the second oscillator, from Eq. (14.18b), will be

$$\omega_1' = \sqrt{\frac{k_2 + k'}{m}} \qquad (14.19b)$$

The frequencies ω_1' and ω_2' given by Eqs. (14.19) are higher than ω_{10} and ω_{20} given by Eq. (14.12). The reason is that each mass is tied to two springs, not just one.

To obtain different possible modes of vibrations, we must solve simultaneously the second-order linear differential equations (14.18). The problem can be made somewhat simpler if we assume the two oscillators to be completely identical, that is,

$$k_1 = k_2 = k \qquad (14.20)$$

and Eqs. (14.18) take the form

$$m\ddot{x}_1 + (k + k')x_1 - k'x_2 = 0 \qquad (14.21)$$

$$m\ddot{x}_2 + (k + k')x_2 - k'x_1 = 0 \qquad (14.22)$$

The trial solution of these equations can take any one of the following three forms:

$$x = A \cos(\omega t + \phi) \qquad (14.23)$$

$$x = A_1 \cos \omega t + A_2 \sin \omega t \qquad (14.24)$$

$$x = A e^{i(\omega t + \delta)} \qquad (14.25)$$

where δ is the initial phase factor. Let us assume Eq. (14.25) to be a trial solution, so that

$$x_1 = A e^{i(\omega t + \delta_1)} \quad \text{and} \quad x_2 = B e^{i(\omega t + \delta_2)}$$

If we assume the initial phase factors to be zero, that is, $\delta_1 = \delta_2 = 0$, then these two solutions take the form

$$x_1 = A e^{i\omega t} \qquad (14.26)$$

and

$$x_2 = B e^{i\omega t} \qquad (14.27)$$

Substituting these in Eqs. (14.21) and (14.22), we obtain, after rearranging,

$$(k + k' - m\omega^2)A - k'B = 0 \qquad (14.28)$$

$$-k'A + (k + k' - m\omega^2)B = 0 \tag{14.29}$$

We have two algebraic equations with three unknowns A, B, and ω. These equations can be solved for the ratio A/B; that is,

$$\frac{A}{B} = \frac{k'}{k + k' - m\omega^2} = \frac{k + k' - m\omega^2}{k'} \tag{14.30}$$

We could solve for ω from the last equality in Eq. (14.30); or we could solve directly Eqs. (14.28) and (14.29) by assuming that the determinant of the coefficients of A and B is zero; that is,

$$\begin{vmatrix} k + k' - m\omega^2 & -k' \\ -k' & k + k' - m\omega^2 \end{vmatrix} = 0 \tag{14.31}$$

This is called the *secular equation*. This may be written as

$$(k + k' - m\omega^2)^2 - k'^2 = 0 \tag{14.32}$$

or
$$\left(\frac{k}{m} - \omega^2\right)\left(\frac{k + 2k'}{m} - \omega^2\right) = 0 \tag{14.33}$$

which yields the following two roots:

$$\omega = \pm\omega_1 = \pm\left(\frac{k}{m}\right)^{1/2} \tag{14.34a}$$

and
$$\omega = \pm\omega_2 = \pm\left(\frac{k + 2k'}{m}\right)^{1/2} \tag{14.34b}$$

In terms of the roots ω_1 and ω_2, the general solutions of Eqs. (14.21) and (14.22) may be written as

$$x_1 = A_1 e^{i\omega_1 t} + A_{-1} e^{-\omega_1 t} + A_2 e^{i\omega_2 t} + A_{-2} e^{-i\omega_2 t} \tag{14.35}$$

$$x_2 = B_1 e^{i\omega_1 t} + B_{-1} e^{-\omega_1 t} + B_2 e^{i\omega_2 t} + B_{-2} e^{-i\omega_2 t} \tag{14.36}$$

There are eight arbitrary constants for two differential equations, but these are not all independent. Substituting Eqs. (14.34a) and (14.34b) in Eqs. (14.28) and (14.29) or in Eq. (14.30), we can obtain the ratios of A/B for different values of ω to be

$$\text{If } \omega = \omega_1, \qquad A = +B \tag{14.37}$$

$$\text{If } \omega = \omega_2, \qquad A = -B \tag{14.38}$$

Combining Eqs. (14.37) and (14.38) with Eqs. (14.35) and (14.36), we obtain

$$x_1 = A_1 e^{i\omega_1 t} + A_{-1} e^{-i\omega_1 t} + A_2 e^{i\omega_2 t} - A_{-2} e^{i\omega_2 t} \tag{14.39}$$

$$x_2 = A_1 e^{i\omega_1 t} + A_{-1} e^{-i\omega_1 t} - A_2 e^{i\omega_2 t} - A_{-2} e^{i\omega_2 t} \tag{14.40}$$

Thus we have only four arbitrary constants, A_1, A_{-1}, A_2, A_{-2}, as expected from the general so-lution of two second-order differential equations. The actual values of the constants can be de-termined from initial conditions.

Normal Coordinates

After determining the constants in Eqs. (14.39) and (14.40), each coordinate (x_1 and x_2) may de-pend on two frequencies, ω_1 and ω_2. Hence it may not be so simple to interpret the type of mo-tion with which the system is oscillating. It is possible to find new coordinates X_1 and X_2, which are linear combinations of x_1 and x_2, such that each new coordinate oscillates with a single fre-quency. In the present situation, the sum and difference of x_1 and x_2 [using Eqs. (14.39) and (14.40) give us the new coordinates; that is,

$$X_1 = x_1 + x_2 = 2(A_1 e^{i\omega_1 t} + A_{-1} e^{-i\omega_1 t}) = C e^{i\omega_1 t} + D e^{-i\omega_1 t} \tag{14.41}$$

$$X_2 = x_1 - x_2 = 2(A_2 e^{i\omega_2 t} + A_{-2} e^{-i\omega_2 t}) = E e^{i\omega_2 t} + F e^{-i\omega_2 t} \tag{14.42}$$

where C, D, E, and F are the new constants. The new coordinates X_1 and X_2 correspond to new modes of oscillation, each mode oscillating with a single frequency. These are called the *nor-mal modes,* and the corresponding coordinates are called the *normal coordinates*. One out-standing characteristic of normal modes is that, for any given normal modes (X_1 or X_2), all the coordinates (x_1 and x_2 in this case) oscillate with the same frequency. Normally, all the normal coordinates are excited simultaneously, except under special circumstances. If, however, one mode is initially not excited, it will remain so throughout the motion.

The nature of any one of the normal modes can be investigated if all the other normal modes can be equated to zero. In the present situation, to study the appearance of the X_1 mode, we must have $X_2 = 0$; that is, if $X_1 \neq 0$,

$$X_2 = 0 = x_1 - x_2 \quad \text{or} \quad x_1 = x_2 \tag{14.43}$$

Thus X_1 is a *symmetric mode,* and, as shown in Fig. 14.3(a), both masses have equal displace-ments, have the same frequency $\omega_1 = (k/m)^{1/2}$, and are in phase. On the other hand, the appear-ance of the X_2 mode is made possible by letting $X_1 = 0$; that is, if $X_2 \neq 0$,

$$X_1 = 0 = x_1 + x_2 \quad \text{or} \quad x_1 = -x_2 \tag{14.44}$$

Thus X_2 is an *antisymmetric mode* and is as shown in Fig. 14.3(b). Both masses have equal and opposite displacement (out of phase), but vibrate with the same frequency $\omega_2 = [(k + k')/m]^{1/2}$. In short,

$$\text{Symmetric mode } X_1: \qquad \omega_1 = \sqrt{\frac{k}{m}}, \qquad\qquad X_2 = 0: x_1 = x_2 \tag{14.45}$$

$$\text{Antisymmetric mode } X_2: \qquad \omega_2 = \sqrt{\frac{k + 2k'}{m}}, \qquad X_1 = 0: x_1 = -x_2 \tag{14.46}$$

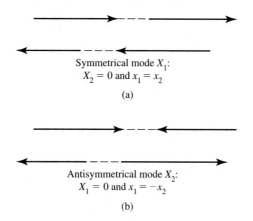

Symmetrical mode X_1:
$X_2 = 0$ and $x_1 = x_2$

(a)

Antisymmetrical mode X_2:
$X_1 = 0$ and $x_1 = -x_2$

(b)

Figure 14.3   Modes of vibration of the two coupled oscillators in Fig. 14.2: (a) symmetrical mode, and (b) antisymmetrical mode.

It is clear that in a symmetric mode the two oscillators vibrate as if there were no coupling between them, and their frequency is the same as the original frequency. In the antisymmetric mode, the result of the coupling is such that the oscillators oscillate out of phase, and their frequency is higher than their individual uncoupled frequency. In general, *the mode that has the highest symmetry will have the lowest frequency, while the antisymmetric mode has the highest frequency.* As the symmetry is destroyed, the springs must work harder, thereby increasing the frequency.

To excite a symmetric mode, the two masses should be pulled from their equilibrium positions by equal amounts in the same direction and released so that $x_1 = x_1(t)$ and $x_2 = x_2(t)$ take the form

$$x_1(0) = x_2(0) \quad \text{and} \quad \dot{x}_1(0) = \dot{x}_2(0) \tag{14.47}$$

For the excitation of an antisymmetric mode, the two masses are pulled apart equally in opposite directions and then released, so that

$$x_1(0) = -x_2(0) \quad \text{and} \quad \dot{x}_1(0) = -\dot{x}_2(0) \tag{14.48}$$

In general, the motion of the system will consist of a combination of these two modes.

Equations of Motion in Normal Coordinates

We obtain expressions for kinetic energy and potential energy in terms of normal coordinates. From Eqs. (14.41) and (14.42),

$$x_1 = \frac{X_1 + X_2}{2} \tag{14.49}$$

and

$$x_2 = \frac{X_1 - X_2}{2} \tag{14.50}$$

Substituting this in Eqs. (14.13) and (14.14),

$$T = \frac{m}{2}\left(\frac{\dot{X}_1 + \dot{X}_2}{2}\right)^2 + \frac{m}{2}\left(\frac{\dot{X}_1 - \dot{X}_2}{2}\right)^2 = m\left(\frac{\dot{X}_1}{2}\right)^2 + m\left(\frac{\dot{X}_2}{2}\right) \tag{14.51}$$

$$V = \frac{k}{2}\left(\frac{X_1 + X_2}{2}\right)^2 + \frac{k}{2}\left(\frac{X_1 - X_2}{2}\right)^2 + \frac{k'}{2}X_2^2$$

or

$$V = \frac{k}{2}\left(\frac{X_1^2}{2}\right) + \left(\frac{k + 2k'}{2}\right)\left(\frac{X_2^2}{2}\right) \tag{14.52}$$

whereas

$$L = T - V = \frac{m}{4}\dot{X}_1^2 + \frac{m}{4}\dot{X}_2^2 - \frac{k}{4}X_1^2 - \left(\frac{k + 2k'}{4}\right)X_2^2 \tag{14.53}$$

Note that the expressions for T, V, and L do not contain cross terms. Thus the Lagrange equations of motion in normal coordinates,

$$\frac{d}{dt}\left(\frac{\partial L}{\partial \dot{X}_1}\right) - \frac{\partial L}{\partial X_1} = 0 \quad \text{and} \quad \frac{d}{dt}\left(\frac{\partial L}{\partial \dot{X}_2}\right) - \frac{\partial L}{\partial X_2} = 0 \tag{14.54}$$

yield

$$\ddot{X}_1 + \omega_1^2 X_1 = 0, \quad \text{where } \omega_1 = \left(\frac{k}{m}\right)^{1/2} \tag{14.55}$$

and

$$\ddot{X}_2 + \omega_2^2 X_2 = 0, \quad \text{where } \omega_2 = \left(\frac{k + 2k'}{m}\right)^{1/2} \tag{14.56}$$

That is, an X_1 mode vibrates with frequency ω_1, and an X_2 mode vibrates with frequency ω_2 in agreement with the results derived previously. [Note that these equations can be obtained by directly substituting Eqs. (14.49) and (14.50) into Eq. (14.18).]

From our discussion, we can conclude the following about normal coordinates: *No cross terms are present when the kinetic and potential energies are expressed in terms of normal coordinates; that is, both T and V are homogeneous quadratic functions.* The differential equations are automatically separated; that is, there is one differential equation for each normal coordinate. The solution of each differential equation represents a separated mode of vibration. In the following, we shall establish the general procedure of transferring to normal coordinates and hence to normal modes of vibrations.

14.4 THEORY OF SMALL OSCILLATIONS

Consider a system of N interacting particles with $3n$ degrees of freedom and described by a set of generalized coordinates $(q_1, q_2, \ldots, q_{3n})$. Furthermore, let us assume that frictional forces are absent and that the forces between particles are conservative. We shall demonstrate that the method of Lagrange's equations can be used for the determination of the frequencies and amplitudes of small oscillations about positions of stable equilibrium in conservative systems.

For such a conservative system, let us express the potential energy by $V(q_1, q_2, \ldots, q_{3n})$. Small oscillations take place about an equilibrium point whose generalized coordiantes are $(q_{10}, q_{20}, \ldots, q_{3n0})$. Expanding the potential energy about an equilibrium point in a multidimensional Taylor series, we have

$$V(q_1, q_2, \ldots, q_{3n}) = V(q_{10}, q_{20}, \ldots, q_{3n0}) + \frac{1}{1!} \sum_{l=1}^{3n} \left(\frac{\partial V}{\partial q_l} \right) \bigg|_{q_l = q_{l0}} (q_l - q_{l0})$$

$$+ \frac{1}{2!} \sum_{l=1}^{3n} \sum_{m=1}^{3n} \left(\frac{\partial^2 V}{\partial q_l \, \partial q_m} \right) \bigg|_{\substack{q_l = q_{l0} \\ q_m = q_{m0}}} (q_l - q_{l0})(q_m - q_{m0}) + \cdots \tag{14.57}$$

Since the zero of the potential energy is arbitrary, the first term on the right is constant and may be equated to zero without affecting the equations of motion. Also, because the system is in equilibrium, the generalized forces Q_l must vanish; that is,

$$Q_l = - \frac{\partial V}{\partial q_l} = 0, \qquad l = 1, 2, \ldots, 3n \tag{14.58}$$

and the second term in the expansion vanishes. Thus, keeping the second-order term and dropping the higher-order terms, we may write the potential energy to be

$$V(q_1, q_2, \ldots, q_{3n}) = \frac{1}{2!} \sum_{l=1}^{3n} \sum_{m=1}^{3n} \left(\frac{\partial^2 V}{\partial q_l \, \partial q_m} \right) \bigg|_{\substack{q_l = q_{l0} \\ q_m = q_{m0}}} (q_l - q_{l0})(q_m - q_{m0}) \tag{14.59}$$

Introducing a new set of generalized coordinates η_l that represent the displacements from the equilibrium,

$$V = V(\eta_l) = \frac{1}{2!} \sum_{l=1}^{3n} \sum_{m=1}^{3n} V_{lm} \, \eta_l \, \eta_m \tag{14.60}$$

where $\eta_l = (q_l - q_{l0})$ and $\eta_m = (q_m - q_{m0})$

and

$$V_{lm} = \left(\frac{\partial^2 V}{\partial q_l \, \partial q_m} \right) \bigg|_{\substack{q_l = q_{l0} \\ q_m = q_{m0}}} = V_{ml} = \text{constant} \tag{14.61}$$

The constants V_{lm} form a symmetric matrix **V**. Since we are considering motioins about *stable* equilibrium, the potential energy must be minimum; that is, $V(\eta_l) > V(0)$; hence the homogeneous quadratic form of V given by Eq. (14.60) must be positive. [That is, for a one-dimensional case, $(\partial^2 V/\partial q^2)_{q=q_0} > 0$, the second derivative evaluated at equilibrium is greater than zero.] Thus for a multidimensional system the necessary and sufficient conditions that a homogeneous quadratic form be positive definite are (derivatives are evaluated about equilibrium)

$$\frac{\partial^2 V}{\partial q_l^2} > 0, \qquad\qquad\qquad\qquad l = 1, 2, \ldots, 3n \tag{14.62a}$$

$$
\begin{vmatrix}
\dfrac{\partial^2 V}{\partial q_1^2} & \dfrac{\partial^2 V}{\partial q_1\,\partial q_m} \\[2ex]
\dfrac{\partial^2 V}{\partial q_1\,\partial q_m} & \dfrac{\partial^2 V}{\partial q_m^2}
\end{vmatrix} > 0,
\qquad
\begin{aligned}
l &= 1, 2, \ldots, 3n \\
m &= 1, 2, \ldots, 3n \\
l &\neq m
\end{aligned}
\qquad \textbf{(14.62b)}
$$

$$
\begin{vmatrix}
\dfrac{\partial^2 V}{\partial q_1^2} & \dfrac{\partial^2 V}{\partial q_1 \partial q_2} & \cdots & \dfrac{\partial^2 V}{\partial q_1 \partial q_{3n}} \\[2ex]
\dfrac{\partial^2 V}{\partial q_2 \partial q_1} & \dfrac{\partial^2 V}{\partial q_2^2} & \cdots & \dfrac{\partial^2 V}{\partial q_2 \partial q_{3n}} \\[2ex]
\vdots & \vdots & & \vdots \\[1ex]
\dfrac{\partial^2 V}{\partial q_{3n} \partial q_1} & \dfrac{\partial^2 V}{\partial q_{3n} \partial q_2} & \cdots & \dfrac{\partial^2 V}{\partial q_{3n}^2}
\end{vmatrix} > 0
\qquad \textbf{(14.62c)}
$$

or, in terms of matrix notation, the coefficients $V_{lm} = V_{ml}$ must satisfy the conditions

$$
V_{11} > 0
$$

$$
\begin{vmatrix}
V_{11} & V_{12} \\
V_{21} & V_{22}
\end{vmatrix} > 0
$$

$$
\begin{vmatrix}
V_{11} & V_{12} & V_{13} \\
V_{21} & V_{22} & V_{23} \\
V_{31} & V_{32} & V_{33}
\end{vmatrix} > 0
$$

$$
\begin{vmatrix}
V_{11} & V_{12} & \cdots & V_{1m} \\
V_{21} & V_{22} & \cdots & V_{2m} \\
\vdots & \vdots & & \vdots \\
V_{1l} & V_{12} & \cdots & V_{1m}
\end{vmatrix} > 0
\qquad \textbf{(14.63)}
$$

where V_{lm} are given by Eq. (14.61) and each individual V_{lm} need not be positive.

If the derivative $V_{lm} = \partial^2 V/\partial q_l\, \partial q_m = 0$ for all values of l and m, stable equilibrium is still possible provided the first nonzero derivative of the potential is of an even order.

Let us now consider the kinetic energy of the system. In terms of Cartesian coordinates, the kinetic energy of the system is

$$
T = \frac{1}{2} \sum_{j=1}^{3n} m_j \dot{x}_j^2
\qquad \textbf{(14.64)}
$$

The transformation equations from Cartesian to generalized coordinates may be utilized to express T in terms of generalized coordinates; that is,

$$
x_j = x_j(q_1, q_2, \ldots, q_l, t)
$$

and
$$\dot{x}_j = \sum_{l=1}^{3n} \frac{\partial x_j}{\partial q_l} \dot{q}_l + \frac{\partial x_j}{\partial t}$$

Hence the kinetic energy given by Eq. (14.64) may be written as

$$T = \frac{1}{2} \sum_{j=1}^{3n} m_j \left(\sum_{l=1}^{3n} \frac{\partial x_j}{\partial q_l} \dot{q}_l + \frac{\partial x_j}{\partial t} \right) \left(\sum_{m=1}^{3n} \frac{\partial x_j}{\partial q_m} \dot{q}_m + \frac{\partial x_j}{\partial t} \right) \tag{14.65}$$

Upon expanding the right side, we find that T contains three types of terms: (1) terms that are quadratic in generalized velocities, (2) terms linear in generalized velocities, and (3) terms independent of generalized velocities. We are interested in transformation equations that do not contain time explicitly (terms such as $\partial x_j / \partial t$ contain time explicitly). This means that T from Eq. (14.65) should contain only those terms that are quadratic in generalized velocities. (The transformation equations involving other terms occur, for example, in rotating coordinate systems.) Hence Eq. (14.65) for kinetic energy takes the form

$$T = \frac{1}{2} \sum_{l=1}^{3n} \sum_{m=1}^{3n} \left(\sum_{j=1}^{3n} m_j \frac{\partial x_j}{\partial q_l} \frac{\partial x_j}{\partial q_m} \right) \dot{q}_l \dot{q}_m \tag{14.66}$$

For small oscillations about equilibrium, the term in parentheses may be expanded and written as

$$\sum_{j=1}^{3n} m_j \frac{\partial x_j}{\partial q_l} \frac{\partial x_j}{\partial q_m} = \sum_{j=1}^{3n} m_j \left(\frac{\partial x_j}{\partial q_l} \right)_{q_{l0}} \left(\frac{\partial x_j}{\partial q_m} \right)_{q_{m0}} + \sum_{j=1}^{3n} m_j \sum_{k=1}^{3n} \frac{\partial}{\partial q_k} \left(\frac{\partial x_j}{\partial q_l} \frac{\partial x_j}{\partial q_m} \right)_{\substack{q_{l0} \\ q_{m0}}} \eta_k + \cdots \tag{14.67}$$

where $\eta_k = (q_k - q_{k0})$. Since we are interested in small oscillations, we need keep only those $\dot{q}$ terms in T that are of the same order as q in V. Hence, from Eqs. (14.66) and (14.67), remembering that $\dot{q}_l = \dot{\eta}_l$ and $\dot{q}_m = \dot{\eta}_m$, we may write

$$T \simeq \frac{1}{2} \sum_{l=1}^{3n} \sum_{m=1}^{3n} T_{lm} \dot{\eta}_l \dot{\eta}_m \tag{14.68}$$

where
$$T_{lm} = \frac{1}{2} \sum_{j=1}^{3n} m_j \left(\frac{\partial x_j}{\partial q_l} \right)_{q_{l0}} \left(\frac{\partial x_j}{\partial q_m} \right)_{q_{m0}} = T_{ml} \tag{14.69}$$

and T_{lm} are the elements of a symmetric matrix **T**.

After obtaining the expressions for potential energy given by Eq. (14.60) and kinetic energy, Eq. (14.68), we are now in a position to write the Lagrangian:

$$L = T - V = \frac{1}{2} \sum_{l=1}^{3n} \sum_{m=1}^{3n} (T_{lm} \dot{\eta}_l \dot{\eta}_m - V_{lm} \eta_l \eta_m) \tag{14.70}$$

Hence the Lagrange equations

$$\frac{d}{dt} \left(\frac{\partial L}{\partial \dot{\eta}} \right) - \frac{\partial L}{\partial \eta} = 0 \tag{14.71}$$

take the form

$$\sum_{m=1}^{3n} (T_{lm} \ddot{\eta}_m + V_{lm} \eta_m) = 0, \qquad l = 1, 2, \ldots, 3n \qquad (14.72a)$$

or $\qquad T_{l1} \ddot{\eta}_1 + V_{l1} \eta_1 + T_{l2} \ddot{\eta}_2 + V_{l2} \eta_2 + \cdots + T_{l3n} \ddot{\eta}_{3n} + V_{l3n} \eta_{3n} = 0 \qquad (14.72b)$

Equations (14.72) represent $3n$ linear, coupled, second-order differential equations. From our experience with the solution of a one-dimensional case, we may write the solution of Eq. (14.72) to be

$$\eta_m = A_m \cos(\omega t + \phi_m) \qquad (14.73)$$

where the amplitude A_m and the phase angle ϕ_m are to be determined from initial conditions, while the natural frequency ω is determined from the system's constants. Substituting Eq. (14.73) into Eq. (14.72a), we obtain

$$\sum_{m=1}^{3n} [V_{lm} A_m \cos(\omega t + \phi_m) - \omega^2 T_{lm} A_m \cos(\omega t + \phi_m)] = 0, \qquad l = 1, 2, \ldots, 3n \qquad (14.74)$$

For a given ω, all ϕ_m must be the same, $\phi_m = \phi$; hence $\cos(\omega t + \phi)$ can be factored out; that is,

$$\cos(\omega t + \phi) \sum_{m=1}^{3n} [V_{lm} A_m - \omega^2 T_{lm} A_m] = 0, \qquad l = 1, 2, \ldots, 3n \qquad (14.75)$$

Since $\cos(\omega t + \phi)$ is *not*, in general, equal to zero, we must have

$$\sum_{m=1}^{3n} [V_{lm} A_m - \omega^2 T_{lm} A_m] = 0, \qquad l = 1, 2, \ldots, 3n \qquad (14.76)$$

Thus we have a total of $3n$ linear, homogeneous, algebraic equations in A_m and ω represented as

$$(V_{11} - \omega^2 T_{11})A_1 \ + \ (V_{12} - \omega^2 T_{12})A_2 \ + \cdots + \ (V_{1.3n} - \omega^2 T_{1.3n})A_{3n} = 0$$
$$\vdots \qquad\qquad\qquad\qquad \vdots \qquad\qquad\qquad \vdots$$
$$(V_{3n.1} - \omega^2 T_{3n.1})A_1 + (V_{3n.2} - \omega^2 T_{3n.2})A_2 + \cdots + (V_{3n.3n} - \omega^2 T_{3n.3n})A_{3n} = 0 \qquad (14.77)$$

For a nontrivial solution, the determinant of the coefficients of A_m in Eq. (14.77) must be zero; that is,

$$\begin{vmatrix} (V_{11} - \omega^2 T_{11}) & (V_{12} - \omega^2 T_{12}) & \cdots & (V_{1.3n} - \omega^2 T_{1.3n}) \\ \vdots & \vdots & & \vdots \\ (V_{3n.1} - \omega^2 T_{3n.1}) & (V_{3n.2} - \omega^2 T_{3n.2}) & \cdots & (V_{3n.3n} - \omega^2 T_{3n.3n}) \end{vmatrix} = 0 \qquad (14.78a)$$

$$|\mathbf{V} - \omega^2 \mathbf{T}| = 0 \qquad (14.78b)$$

This results in a secular equation of a $3n$-degree polynomial in ω^2. Each of the $3n$ roots of this equation represents a different frequency. Thus the general solution, for small amplitude of oscillations, is

$$\eta_l = \sum_{k=1}^{3n} A_{kl} \cos(\omega_k t + \phi_k) \qquad (14.79)$$

where the values of ω_k are known from the secular equation, Eq. (14.78), while A_{kl} and ϕ_k are determined from initial conditions.

If ω^2 is negative ($\omega^2 < 0$), ω will be complex and there will be no small oscillations. If $\omega^2 = 0$, the coordinate η remains constant, hence with no oscillations, only translation or rotation of the whole system. Only if $\omega^2 > 0$ will there be oscillation about the stable equilibrium. Thus

$$\text{If } \omega_k^2 > 0, \qquad \eta_k = A_k e^{i\omega_k t} + B_k e^{-i\omega_k t} \tag{14.80}$$

$$\text{If } \omega_k^2 = 0, \qquad \eta_k = C_k t + D_k \tag{14.81}$$

$$\text{If } \omega_k^2 < 0, \qquad \eta_k = E_k e^{\omega_k t} + F_k e^{-\omega_k t} \tag{14.82}$$

We have found the frequencies, while the task of calculating the amplitudes still remains. The amplitudes A_{kl} are related by the algebraic equations (14.77). Substituting each value of ω_k separately in Eq. (14.77), it is possible to determine all the coefficients A_{kl} except one, say A_{k1}. Thus it is possible to determine the coefficients A_{kl} in terms of A_{k1} in the form of ratios:

$$\frac{A_{k2}}{A_{k1}}, \frac{A_{k3}}{A_{k1}}, \dots, \frac{A_{k,3n}}{A_{k1}} \tag{14.83}$$

We must determine $6n$ constants ($3n$ are A_{k1} and $3n$ are ω_k) from initial conditions.

14.5 SMALL OSCILLATIONS IN NORMAL COORDINATES

Let us once again consider an arbitrary system with r degrees of freedom. The system has small oscillations about some stable equilibrium point. The potential energy is described in terms of generalized coordinates $(q_1', q_2', \dots, q_l')$, while the equilibrium configuration is described by the coordinates $(q_{10}', q_{20}', \dots, q_{l0}')$, where $l = 1, 2, \dots, r$. As explained in the previous section, for stable equilibrium the only nonzero coefficient in the expansion of the potential energy $V(q_1', q_2', \dots, q_l')$ is V_{lm} given by

$$V = \frac{1}{2} \sum_{l=1}^{r} \sum_{m=1}^{r} V_{lm} \, \eta_l' \, \eta_m' \tag{14.84}$$

where

$$\eta_l' = q_l' - q_{l0}'$$

$$\eta_m' = q_m' - q_{m0}'$$

and

$$V_{lm} = \left(\frac{\partial^2 V}{\partial q_l' \, \partial q_m'} \right) \Bigg|_{\substack{q_l' = q_{l0}' \\ q_m' = q_{m0}'}} = V_{ml} = \text{constant} \tag{14.85}$$

Thus the potential energy expression, as stated earlier, is not only a homogeneous quadratic but is also positive definite for stable equilibrium. It cannot be negative and is zero only if all the coordinates are zero. For such a system, the potential energy V may be written as

$$V = a_{11} \, \eta_1'^2 + a_{22} \, \eta_2'^2 + \dots + a_{rr} \, \eta_r'^2 + 2a_{12} \, \eta_1'\eta_2' + \dots \tag{14.86}$$

where every term is quadratic in the coordinates and the coefficients $a_{11}, a_{22}, \ldots, a_{rr}, a_{12}, \ldots,$ and so on, are all constant. Note the presence of square terms as well as cross terms.

Similarly, we have seen that, if the kinetic energy T does not contain time explicitly, it will be homogeneous in velocity coordinates and may be written as

$$T = b_{11}\,\dot{\eta}_1'^{\,2} + b_{22}\,\dot{\eta}_2'^{\,2} + \cdots + b_{rr}\,\dot{\eta}_r'^{\,2} + b_{12}\,\dot{\eta}_1'\dot{\eta}_2' + \cdots \qquad (14.87)$$

For small oscillations, the quantities $b_{11}, b_{22}, \ldots, b_{rr}, b_{12}, \ldots,$ and so on, are constant; hence T is positive definite. Once again, note the presence of cross terms.

It is possible to cause a linear transformation to new generalized coordinates $\eta_1, \eta_2, \ldots, \eta_r$, in which V and T will not contain cross terms. The original coordinates $\eta_1', \eta_2', \ldots, \eta_r'$ by their linear combination can result in new generalized coordinates $\eta_1, \eta_2, \ldots, \eta_r$.

$$\begin{aligned}
\eta_1 &= e_{11}\,\eta_1' + e_{12}\,\eta_2' + \cdots + e_{1r}\,\eta_r' \\
\eta_2 &= e_{21}\,\eta_1' + e_{22}\,\eta_2' + \cdots + e_{2r}\,\eta_r' \\
&\;\;\vdots \qquad\quad \vdots \qquad\qquad\quad \vdots \\
\eta_r &= e_{r1}\,\eta_1' + e_{r2}\,\eta_2' + \cdots + e_{rr}'\,\eta_r'
\end{aligned} \qquad (14.88)$$

so that V and T will take the following forms that do not contain cross terms:

$$V = \tfrac{1}{2}(\lambda_1^2\,\eta_1^2 + \lambda_2^2\,\eta_2^2 + \cdots + \lambda_r^2\,\eta_r^2) \qquad (14.89)$$

and

$$T = \tfrac{1}{2}(m_1\dot{\eta}_1^2 + m_2\,\dot{\eta}_2^2 + \cdots + m_r\,\dot{\eta}_r^2) \qquad (14.90)$$

where λ's and m's are constants. The new linear combination $\eta_1, \eta_2, \ldots, \eta_r$ is called the *normal coordinates* of the system.

Now the Lagrangian equations for the normal coordinates η_l are

$$\frac{d}{dt}\left(\frac{\partial L}{\partial \dot{\eta}_l}\right) - \frac{\partial L}{\partial \eta_l} = 0 \qquad (14.91)$$

where $L = T - V$. If V and T are given by Eqs. (14.89) and (14.90), the resulting equations of motion for η_l are

$$\ddot{\eta}_l + \omega_l^2 \eta_l = 0 \qquad (14.92)$$

where ω_l are the normal frequencies given by

$$\omega_l^2 = \frac{\lambda_l^2}{m_l} \qquad (14.93)$$

The quantities $\eta_1, \eta_2, \ldots, \eta_r$ are *normal coordinates,* and $\omega_1, \omega_2, \ldots, \omega_r$ are the corresponding *normal frequencies.* The solution of Eq. (14.92) is

$$\text{If } \omega_l^2 > 0, \qquad \eta_l = A_l e^{i\omega_l t} + B_l e^{-i\omega_l t} \qquad (14.94a)$$

$$\text{or } \quad \eta_l = A_l'\cos(\omega_l t + \phi_l) \qquad (14.94b)$$

$$\text{If } \omega_l^2 = 0, \qquad \eta_l = C_l t + D_l \tag{14.95}$$

$$\text{If } \omega_l^2 < 0, \qquad \eta_l = E_l e^{\omega_l t} + F_l e^{-\omega_l t} \tag{14.96}$$

where A_l, B_l, A_l', ϕ_l, C_l, D_l, E_l, and F_l are all constants.

As pointed out earlier and as is clear from Eq. (14.92), each normal coordinate varies with only one normal frequency ω_l; hence these are called *normal modes* of vibration and each normal coordinate η_l is given by Eq. (14.94). It is necessary to note that if a normal coordinate η_l for which the associated frequency ω_l^2 is not greater than zero, such a coordinate does not correspond to oscillatory motion about the equilibrium. Thus, if $\omega_l^2 = 0$, as is obvious from the solution in Eq. (14.95), the mode of motion is that of translation motion; that is, if the particle is slightly displaced, there will be no restoring force, and the particle will simply translate about the center of mass. On the other hand, if $\omega_l^2 < 0$, as is clear from Eq. (14.96), the motion is nonoscillatory; it consists of increasing and decreasing exponentials, with the result that the motion grows without bounds.

14.6 TENSOR FORMULATION FOR THE THEORY OF SMALL OSCILLATIONS

The problems of small oscillations discussed in the two previous sections can be presented and solved more elegantly by using the techniques of tensor analysis similar to the one used in describing rigid body motion in Chapter 13.

For a system with $3n$ degrees of freedom, the expression for small oscillation about a stable equilibrium, the Lagrangian equations according to Eq. (14.76), are

$$\sum_{m=1}^{3n} [V_{lm} A_m - \omega^2 T_{lm} A_m] = 0, \qquad l = 1, 2, \ldots, 3n \tag{14.97}$$

where

$$V_{lm} = \left(\frac{\partial^2 V}{\partial q_l \, \partial q_m} \right) \Bigg|_{\substack{q_l = q_{l0} \\ q_m = q_{m0}}} = V_{ml} \tag{14.98}$$

$$T_{lm} = \frac{1}{2} \sum m_j \left(\frac{\partial x_j}{\partial q_l} \right)_{q_{l0}} \left(\frac{\partial x_j}{\partial q_m} \right)_{q_{m0}} = T_{ml} \tag{14.99}$$

Equation (14.97) is equivalent to the $3n$ linear equations of the form

$$
\begin{aligned}
(V_{11} - \omega^2 T_{11})A_1 \;+\; & (V_{12} - \omega^2 T_{12})A_2 + \cdots + (V_{1.3n} - \omega^2 T_{1.3n})A_{3n} = 0 \\
\vdots \qquad\qquad & \qquad\qquad\qquad \vdots \qquad\qquad\qquad\qquad \vdots \\
(V_{3n.1} - \omega^2 T_{3n.1})A_1 \;+\; & (V_{3n.2} - \omega^2 T_{3n.2})A_2 + \cdots + (V_{3n.3n} - \omega^2 T_{3n.3n})A_{3n} = 0
\end{aligned}
\tag{14.100}
$$

The quantities V_{lm} are the elements of symmetric matrix **V** given by

$$
\mathbf{V} = \begin{bmatrix}
V_{11} & V_{12} & \cdots & V_{1,3n} \\
V_{21} & V_{22} & \cdots & V_{2,3n} \\
\vdots & \vdots & & \vdots \\
V_{3n,1} & V_{3n,2} & \cdots & V_{3n,3n}
\end{bmatrix}
\tag{14.101}
$$

and the quantities T_{lm} are the elements of a symmetric matrix **T** given by

$$\mathbf{T} = \begin{bmatrix} T_{11} & T_{12} & \cdots & T_{1,3n} \\ T_{21} & T_{22} & \cdots & T_{2,3n} \\ \vdots & \vdots & & \vdots \\ T_{3n,1} & T_{3n,2} & \cdots & T_{3n,3n} \end{bmatrix} \tag{14.102}$$

while the Lagrange equations, Eqs. (14.97) and (14.100), may be written in tensor form as

$$(\mathbf{V} - \omega^2 \mathbf{T})\mathbf{A} = 0 \tag{14.103}$$

where **A** is a column vector:

$$\mathbf{A} = \begin{bmatrix} A_1 \\ \vdots \\ A_{3n} \end{bmatrix} \tag{14.104}$$

For each frequency ω_k, there corresponds a vector A_k: hence, as before, the general solution will be the linear combinations of individual solutions.

The next task is to determine the normal coordinates corresponding to each normal frequency, that is, to determine the normal modes of vibrations. This involves transferring both **V** and **T** to a new set of generalized coordinates in which both **V** and **T** matrices will be diagonal (so that the off-diagonal elements will be zero). The existence of such a coordinate transformation that will cause simultaneous diagonalization of **V** and **T** is possible only if both the **V** and **T** matrices are symmetrical with real elements, and **V** as well as **T** is positive definite (determinant is greater than zero). Such a process of simultaneous diagonalization will change Eq. (14.103) into

$$(\mathbf{V}' - \omega^2 \mathbf{T}')\mathbf{A} = 0 \tag{14.105}$$

where

$$(\mathbf{V}' - \omega^2 \mathbf{T}') = \begin{bmatrix} V'_{11} - \omega^2 T'_{11} & 0 & 0 & \cdots & 0 \\ 0 & V'_{22} - \omega^2 T'_{22} & 0 & \cdots & 0 \\ 0 & 0 & V'_{33} - \omega^2 T'_{33} & \cdots & 0 \\ \vdots & \vdots & \vdots & \vdots & \vdots \\ 0 & 0 & 0 & \cdots & V'_{3n,3n} - \omega^2 T'_{3n,3n} \end{bmatrix} \tag{14.106}$$

The diagonalization can be achieved in a manner explained in Chapter 13.

For each normal frequency ω_m, there exists a solution of the form

$$\eta_l = C_m a_{lm} \cos(\omega_m t + \phi_m) \tag{14.107}$$

where C_m is the scale factor, a_{lm} is the coefficient, and ϕ_m is the phase angle. This solution is a linear combination of two independent functions $\cos \omega_m t$ and $\sin \omega_m t$. Thus the most general solution will be

$$\eta_l(t) = \sum_{m=1}^{n} a_{lm} C_m \cos(\omega_m t + \phi_m) \tag{14.108}$$

which is a linear combination of $2n$ functions. Equation (14.108) may be written as

$$\eta_l(t) = \sum_{m=1}^{n} [a_{lm}(C_m \cos \phi \cos \omega_m t - C_m \sin \phi \sin \omega_m t)] \tag{14.109}$$

Defining

$$A_m = C_m \cos \phi \quad \text{and} \quad B_m = -C_m \sin \phi$$

we may write Eq. (14.109) as

$$\eta_l(t) = \sum_{m=1}^{n} [a_{lm}(A_m \cos \omega_m t + B_m \sin \omega_m t)] \tag{14.110}$$

where the coefficients a_{lm} form a set associated with the frequency ω_m or the mth normal mode.

The constants in Eq. (14.110) may now be determined by the following procedure. First, calculate the normal frequencies ω_m from the characteristic equation

$$\det |V - \omega^2 T| = 0$$

Second, replace ω^2 in

$$\sum_{l=1}^{n} [V_{lm} - \omega^2 T_{lm}]a_{lm} = 0, \qquad m = 1, 2, \ldots, n \tag{14.111}$$

by ω_m and calculate the n sets of solutions (a_{lm}), one for each m. (One of the factors a_{lm} must be assigned a unit value; otherwise, only the ratios of the coefficients will be calculated.) Third, A_m and B_m may be calculated by using the initial conditions of the systems.

$$\eta_l(0) \equiv \eta_{l0} = \sum_{m=1}^{n} a_{lm} A_m \tag{14.112}$$

$$\dot{\eta}_l(0) \equiv \dot{\eta}_{l0}(= v_{l0}) = \sum_{m=1}^{n} a_{lm} \omega_m B_m \tag{14.113}$$

In a special case, if the number of degrees is very large and we impose the condition $a_{lm} = \delta_{lm}$; then Eqs. (14.110), (14.112), and (14.113) become

$$\eta_l(t) = A_l \cos \omega_l t + B_l \omega_l t$$

$$\eta_l(0) \equiv \eta_{l0} = A_l$$

and
$$\dot{\eta}_l(0) \equiv v_{l0} = \omega_l B_l$$

This holds for normal coordinates; that is, it is possible to find a coordinate transformation such that all $\eta_l(t)$ are normal coordinates as represented by this equation.

 Example 14.2 _____

Consider the situation of two coupled pendula, as shown in Fig. Ex. 14.2. Using matrix notation calculate **(a)** the components V_{lm} of **V**, **(b)** the components T_{lm} of **T**, **(c)** the normal frequencies, and **(d)** the normal modes. **(e)** Find the equations of motion and **(f)** the general solution.

Solution

As shown in Fig. Ex. 14.2, each pendulum is of length l and mass m, and equilibrium is where both are vertical in which position $x_1 = x_2 = 0$. The two masses are tied by a spring of spring constant k. The displacements x_1 and x_2 to the right are positive, while θ_1 and θ_2 are positive in a counterclockwise direction.

(a) The potential energy of the system is given by

$$V = mgl(1 - \cos\theta_1) + mgl(1 - \cos\theta_2) + \tfrac{1}{2}k(x_1 - x_2)^2$$

For a small angle,

$$mgl(1 - \cos\theta) = mgl\left[1 - \left(1 - \frac{\theta^2}{2} + \cdots\right)\right] \simeq mgl\,\frac{\theta^2}{2}$$

$$\simeq \frac{mgl}{2}\left(\frac{x}{l}\right)^2 \simeq \frac{mg}{2l}\,x^2$$

Therefore,

$$V = \frac{mg}{2l}\,x_1^2 + \frac{mg}{2l}\,x_2^2 + \frac{1}{2}k(x_1 - x_2)^2$$

$$= \frac{1}{2}\left(k + \frac{mg}{l}\right)x_1^2 + \frac{1}{2}\left(k + \frac{mg}{l}\right)x_2^2 - kx_1x_2 \tag{i}$$

$$\left.\frac{\partial V}{\partial x_1}\right|_{\substack{x_1 = 0 \\ x_2 = 0}} = \left(k + \frac{mg}{l}\right)x_1 - kx_2\bigg|_{\substack{x_1 = 0 \\ x_2 = 0}} = 0 \tag{ii}$$

$$\left.\frac{\partial V}{\partial x_2}\right|_{\substack{x_1 = 0 \\ x_2 = 0}} = \left(k + \frac{mg}{l}\right)x_2 - kx_1\bigg|_{\substack{x_1 = 0 \\ x_2 = 0}} = 0$$

$$\left.\frac{\partial^2 V}{\partial x_1^2}\right|_{\substack{x_1 = 0 \\ x_2 = 0}} = k + \frac{mg}{l} \quad\text{and}\quad \left.\frac{\partial^2 V}{\partial x_2^2}\right|_{\substack{x_1 = 0 \\ x_2 = 0}} = k + \frac{mg}{l} \tag{iii}$$

$$\left.\frac{\partial^2 V}{\partial x_2\,\partial x_1}\right|_{\substack{x_1 = 0 \\ x_2 = 0}} = -k \quad\text{and}\quad \left.\frac{\partial^2 V}{\partial x_1 \partial x_2}\right|_{\substack{x_1 = 0 \\ x_2 = 0}} = -k \tag{iv}$$

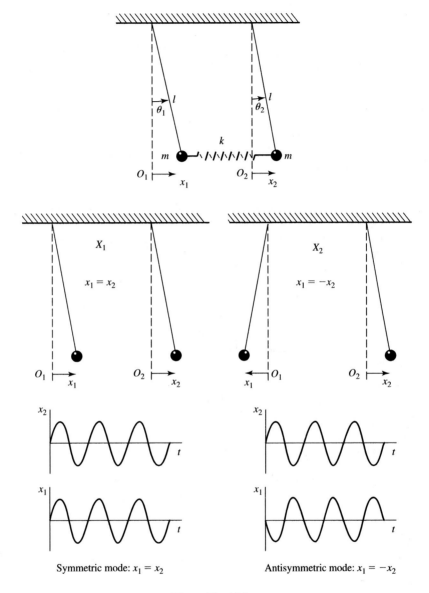

Figure Ex. 14.2

Thus the required matrix for the potential energy is

$$
\mathbf{V} =
\begin{bmatrix}
k + \dfrac{mg}{l} & -k \\[2ex]
-k & k + \dfrac{mg}{l}
\end{bmatrix}
\tag{v}
$$

Since this gives

$$\begin{vmatrix} V_{11} & V_{12} \\ V_{21} & V_{22} \end{vmatrix} > 0$$

the associated homogeneous quadratic form is positive definite.

(b) The expression for kinetic energy is

$$T = \tfrac{1}{2}m\dot{x}_1^2 + \tfrac{1}{2}m\dot{x}_2^2 \tag{vi}$$

The components T_{ll} and T_{lm} are coefficients of $\tfrac{1}{2}\dot{x}_l^2$ and $\dot{x}_l\dot{x}_m$. Hence

$$\mathbf{T} = \begin{bmatrix} m & 0 \\ 0 & m \end{bmatrix} \tag{vii}$$

Thus the Lagrangian for the system is

$$L = T - V = \sum_{l=1}^{2}\sum_{m=1}^{2}\frac{1}{2}\,(T_{lm}\dot{x}_l\dot{x}_m - V_{lm}x_lx_m \tag{viii}$$

while the Lagrangian equations are

$$\sum_{l=1}^{2}(T_{lm}\ddot{x}_l + V_{lm}x_l) = 0, \qquad m = 1, 2 \tag{ix}$$

That is,

$$T_{11}\ddot{x}_1 + V_{11}x_1 + T_{12}\ddot{x}_2 + V_{12}x_2 = 0$$

$$T_{21}\ddot{x}_1 + V_{21}x_1 + T_{22}\ddot{x}_2 + V_{22}x_2 = 0$$

Using Eqs. (v) and (vii) in the preceding equations, we get

$$m\ddot{x}_1 + \left(k + \frac{mg}{l}\right)x_1 - kx_2 = 0 \tag{x}$$

$$m\ddot{x}_2 + \left(k + \frac{mg}{l}\right)x_2 - kx_1 = 0 \tag{xi}$$

These are two coupled equations.

(c) To determine the normal or characteristic frequencies, we use Eq. (14.78b), that is,

$$|\mathbf{V} - \omega^2\mathbf{T}| = 0$$

Thus

$$\begin{vmatrix} k + \dfrac{mg}{l} - m\omega^2 & -k \\ -k & k + \dfrac{mg}{l} - m\omega^2 \end{vmatrix} = 0$$

or

$$\left(k + \frac{mg}{l} - m\omega^2\right)^2 - k^2 = 0 \tag{xii}$$

$$\left(k + \frac{mg}{l} - m\omega^2 - k\right)\left(k + \frac{mg}{l} - m\omega^2 + k\right) = 0$$

Either
$$k = \frac{mg}{l} - m\omega^2 - k = 0$$

which gives

$$\omega^2 = \omega_1^2 = \frac{g}{l} \quad \text{or} \quad \omega_1 = \pm\left(\frac{g}{l}\right)^{1/2} \tag{xiii}$$

or
$$k + \frac{mg}{l} - m\omega^2 + k = 0$$

which gives

$$\omega^2 = \omega_2^2 = \left(\frac{g}{l} + \frac{2k}{m}\right) \quad \text{or} \quad \omega_2 = \pm\left(\frac{g}{l} + \frac{2k}{m}\right)^{1/2} \tag{xiv}$$

As before, we try the solutions

$$x_1 = Ae^{i\omega t} \quad \text{and} \quad x_2 = Be^{i\omega t} \tag{xv}$$

Substituting these in Eqs. (x) and (xi), we get

$$\left(k + \frac{mg}{l} - m\omega^2\right)A - kB = 0$$

$$\left(k + \frac{mg}{l} - m\omega^2\right)B - kA = 0$$

$$\text{If } \omega^2 = \omega_1^2 = \frac{g}{l}, \quad \text{we get } A = B \tag{xvi}$$

$$\text{If } \omega^2 = \omega_2^2 = \frac{g}{l} + \frac{2k}{m}, \quad \text{we get } A = -B \tag{xvii}$$

Hence, using these, the general solution becomes

$$x_1 = A_1 e^{i\omega_1 t} + A_{-1} e^{i\omega_1 t} + A_2 e^{i\omega_2 t} + A_{-2} e^{-i\omega_2 t} \tag{xviii}$$

$$x_2 = A_1 e^{i\omega_1 t} + A_{-1} e^{i\omega_1 t} - A_2 e^{i\omega_2 t} - A_{-2} e^{-i\omega_2 t} \tag{xix}$$

These two equations contain four constants, as they should for two linear differential equations. These constants are determined from initial conditions.

(d) We now proceed with Eq. (14.103) or (14.76) to determine the normal coordinates

$$(\mathbf{V} - \omega^2 \mathbf{T})\mathbf{A} = 0$$

or
$$\sum_{m=1}^{2} |V_{lm} - \omega^2 T_{lm}| A_m = 0, \quad l = 1, 2$$

That is, for $\omega^2 = \omega_1^2 = g/l$,

$$\begin{pmatrix} k + \dfrac{mg}{l} - \dfrac{mg}{l} & -k \\[2mm] -k & k + \dfrac{mg}{l} - \dfrac{mg}{l} \end{pmatrix} \begin{pmatrix} a_{11} \\ a_{12} \end{pmatrix} = 0$$

or
$$\begin{pmatrix} k & -k \\ -k & k \end{pmatrix} \begin{pmatrix} a_{11} \\ a_{12} \end{pmatrix} = 0$$

gives

$$\text{If } a_{11} = 1, \qquad a_{12} = 1 \qquad \textbf{(xx)}$$

Similarly, for $\omega^2 = \omega_2^2 = (g/l) + (2k/m)$,

$$\begin{pmatrix} -k & -k \\ -k & -k \end{pmatrix} \begin{pmatrix} a_{21} \\ a_{22} \end{pmatrix} = 0$$

That is,

$$\text{If } a_{21} = 1, \qquad a_{22} = -1 \qquad \textbf{(xxi)}$$

Thus the normal modes are

$$\eta_1 = a_{11}x_1 + a_{12}x_2$$

$$\eta_2 = a_{21}x_1 + a_{22}x_2$$

Substituting these values of a_{11}, a_{12}, a_{21}, and a_{22} from Eqs. (xx) and (xxi) and x_1 and x_2 from Eqs. (xviii) and (xix), we get

$$\eta_1 = x_1 + x_2 = 2(A_1 e^{i\omega_1 t} + A_{-1} e^{-i\omega_1 t}) \qquad \textbf{(xxii)}$$

$$\eta_2 = x_1 - x_2 = 2(A_2 e^{i\omega_2 t} + A_{-2} e^{-i\omega_2 t}) \qquad \textbf{(xxiii)}$$

Thus each normal mode depends only on one frequency. Furthermore, we can see the physical meaning of these modes as before.

For the η_1 mode, we take $\eta_2 = 0$; therefore,

$$x_1 - x_2 = 0 \quad \text{or} \quad x_1 = x_2 \qquad \textbf{(xxiv)}$$

In order to really understand and illustrate the natural modes and normal modes of vibrations, we graph for arbitrary numerical values.

normal modes: X1 with frequency $\omega 1$ and X2 with frequency $\omega 2$
natural modes: x1 (= X1 + X2) and x2 (= X1 - X2)

We will first graph the normal modes and then the natural modes.

X1 and X2 (or $\eta 1$ and $\eta 2$) determine the normal coordinates with the characteristic frequencies $\omega 1$ and $\omega 2$.

$$X1 \equiv A22 \cdot \exp(-i \cdot \omega 1 \cdot t) + A11 \cdot \exp(i \cdot \omega 1 \cdot t) \qquad X2 \equiv A12 \cdot \exp(-i \cdot \omega 2 \cdot t) - A21 \cdot \exp(i \cdot \omega 2 \cdot t)$$

Let us now follow the reverse process, that is, find the values of natural displacements x1 and x2 from the relation X1 = x1 + x2 and X2 = x1 – x2 and then make the plots of x1 and x2. Note that we are going to use prime (') for the variables; otherwise we will get the numerical results because the values of the constants are already given.

Given

$$x1 + x2 \equiv 2 \cdot \left(A12' \cdot e^{i \cdot \omega 1' \cdot t'} + A12' \cdot e^{-i \cdot \omega 1' \cdot t'} \right) \qquad x1 - x2 \equiv 2 \cdot \left(A21' \cdot e^{i \cdot \omega 2' \cdot t'} + A22' \cdot e^{-i \cdot \omega 2' \cdot t'} \right)$$

$$\text{Find}(x1, x2) \rightarrow \begin{pmatrix} A22' \cdot \exp(-i \cdot \omega 2' \cdot t') + A12' \cdot \exp(i \cdot \omega 1' \cdot t') + A12' \cdot \exp(-i \cdot \omega 1' \cdot t') + A21' \cdot \exp(i \cdot \omega 2' \cdot t') \\ -A22' \cdot \exp(-i \cdot \omega 2' \cdot t') + A12' \cdot \exp(i \cdot \omega 1' \cdot t') + A12' \cdot \exp(-i \cdot \omega 1' \cdot t') - A21' \cdot \exp(i \cdot \omega 2' \cdot t') \end{pmatrix}$$

$$x1 \equiv A22' \cdot \exp(-i \cdot \omega 2' \cdot t') + A12' \cdot \exp(i \cdot \omega 1' \cdot t') + A12' \cdot \exp(-i \cdot \omega 1' \cdot t') + A21' \cdot \exp(i \cdot \omega 2' \cdot t')$$

$$x2 \equiv -A22' \cdot \exp(-i \cdot \omega 2' \cdot t') + A12' \cdot \exp(i \cdot \omega 1' \cdot t') + A12' \cdot \exp(-i \cdot \omega 1' \cdot t') - A21' \cdot \exp(i \cdot \omega 2' \cdot t')$$

This is the same result that we obtained earlier. We will use the original equations with the arbitrary numerical values and graph them.

Let us now make the graphs, as shown in Figure Ex.(14.2) (b) and (c), by using the arbitrary values given below.

$$A11 := 4 \quad A12 := 2 \quad A21 := 2 \quad A22 := 4 \qquad g := 9.8 \qquad 1 := 2 \qquad k := 1 \qquad m := 1 \qquad i := \sqrt{-1}$$

$$N := 200 \quad n := 0..N \quad t_n := \frac{n}{10} \qquad \omega 1 := \sqrt{\frac{g}{1}} \qquad \omega 1 = 2.214 \qquad \omega 2 := \sqrt{\frac{g}{1} + 2 \cdot \frac{k}{m}} \qquad \omega 2 = 2.627$$

$$X1_n := A11 \cdot \exp\left(i \cdot \omega 1 \cdot t_n\right) - A12 \cdot \exp\left(-i \cdot \omega 1 \cdot t_n\right) \qquad X2_n := \left(A21 \cdot \exp\left(-i \cdot \omega 2 \cdot t_n\right) + A22 \cdot \exp\left(i \cdot \omega 2 \cdot t_n\right)\right)$$

$$x1_n := A22 \cdot \exp\left(-i \cdot \omega 2 \cdot t_n\right) + A11 \cdot \exp\left(i \cdot \omega 1 \cdot t_n\right) + A12 \cdot \exp\left(-i \cdot \omega 1 \cdot t_n\right) + A21 \cdot \exp\left(i \cdot \omega 2 \cdot t_n\right)$$

$$x2_n := -A22 \cdot \exp\left(-i \cdot \omega 2 \cdot t_n\right) + A11 \cdot \exp\left(i \cdot \omega 1 \cdot t_n\right) + A12 \cdot \exp\left(-i \cdot \omega 1 \cdot t_n\right) - A21 \cdot \exp\left(i \cdot \omega 2 \cdot t_n\right)$$

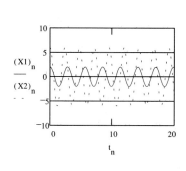
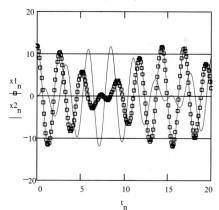

Figure Ex. 14.2(b)

Normal symmetric modes X1 and X2 and natural symmetric modes x1 and x2

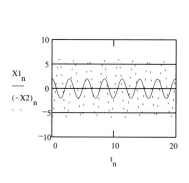
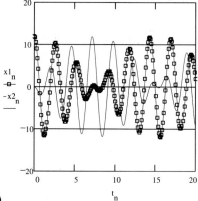

Figure Ex. 14.2(c)

Normal antisymmetric modes X1 and –X2 and natural antisymmetric modes x1 and –x2

Answer the following by looking at the two graphs.

Do the symmetric modes repeat themselves for each mass?

Do the antisymmetric modes repeat themselves for each mass?

What is the difference between the two types of modes with respect to their frequencies and the amplitudes?

EXERCISE 14.2 For the system shown in Fig. 14.2 and discussed in Section 14.3, find the normal frequencies and normal modes using the matrix method discussed above.

Example 14.3 _____

Find the frequencies of small oscillation for a double pendulum, as shown in Fig. Ex. 14.3(a). We may assume that

$$m_1 = m_2 = m \quad \text{and} \quad l_1 = l_2 = l$$

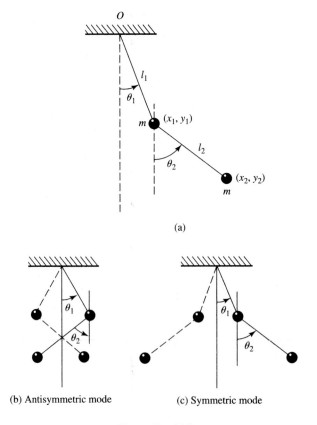

(a)

(b) Antisymmetric mode   (c) Symmetric mode

Figure Ex. 14.3

Solution

Let (x_1, y_1) and (x_2, y_2) be the coordinates of the two masses of the pendulums such that the lengths of the pendulums make angles θ_1 and θ_2 as shown. From Fig. Ex. 14.3(a),

$$x_1 = l_1 \sin \theta_1$$

$$x_2 = l_1 \sin \theta_1 + l_2 \sin \theta_2$$

$$y_1 = l_1 \cos \theta_1$$

$$y_2 = l_1 \cos \theta_1 + l_2 \cos \theta_2$$

Thus the potential energy of the system is

$$V = -mgy_1 - mgy_2 = -mgl \cos \theta_1 - mgl(\cos \theta_1 + \cos \theta_2) \tag{i}$$

$$\frac{\partial V}{\partial \theta_1}\bigg|_{\substack{\theta_1 = 0 \\ \theta_2 = 0}} = 0 \quad \text{and} \quad \frac{\partial V}{\partial \theta_2}\bigg|_{\substack{\theta_1 = 0 \\ \theta_2 = 0}} = 0$$

The components V_{lm} are

$$V_{11} = \frac{\partial^2 V}{\partial \theta_1^2}\bigg|_{\substack{\theta_1 = 0 \\ \theta_2 = 0}} = mgl + mgl = 2mgl$$

$$V_{22} = \frac{\partial^2 V}{\partial \theta_2^2}\bigg|_{\substack{\theta_1 = 0 \\ \theta_2 = 0}} = mgl$$

and

$$V_{12} = V_{21} = 0$$

Thus

$$V = \begin{pmatrix} 2mgl & 0 \\ 0 & mgl \end{pmatrix} \tag{ii}$$

Since

$$\begin{vmatrix} V_{11} & V_{12} \\ V_{21} & V_{22} \end{vmatrix} > 0 \tag{iii}$$

Therefore, the associated homogeneous quadratic form is positive.

The components T_{lm} of T are calculated as follows:

$$T = \tfrac{1}{2}m(\dot{x}_1^2 + \dot{y}_1^2) + \tfrac{1}{2}m(\dot{x}_2^2 + \dot{y}_2^2)$$

$$= \tfrac{1}{2}m[l \cos \theta_1\, \dot{\theta}_1]^2 + \tfrac{1}{2}m[l(-\sin \theta_1)\dot{\theta}_1]^2$$

$$+ \tfrac{1}{2}m[l \cos \theta_1\, \dot{\theta}_1 + l \cos \theta_2\, \dot{\theta}_2]^2 + \tfrac{1}{2}m[l(-\sin \theta_1)\dot{\theta}_1]^2 + [l(-\sin \theta_2)\dot{\theta}_2]^2$$

$$= \tfrac{1}{2}ml^2\dot{\theta}_1^2 + \tfrac{1}{2}m[l^2\dot{\theta}_1^2 + l^2\dot{\theta}_2^2 + 2l^2 \cos(\theta_1 - \theta_2)\dot{\theta}_1\, \dot{\theta}_2] \tag{iv}$$

At the equilibrium point, $\theta_1 = \theta_2 = 0$,

$$T = \tfrac{1}{2}(2ml^2)\dot{\theta}_1^2 + \tfrac{1}{2}ml^2\dot{\theta}_2^2 + ml^2\dot{\theta}_1\, \dot{\theta}_2 \tag{v}$$

The components T_{ll} and T_{lm} are the coefficients of $\frac{1}{2}\dot{\theta}_1^2$ and $\dot{\theta}_l\dot{\theta}_m$; that is,

$$T_{11} = 2ml^2, \qquad T_{22} = ml^2, \qquad T_{12} = T_{21} = ml^2$$

Therefore,

$$\mathbf{T} = \begin{pmatrix} 2ml^2 & ml^2 \\ ml^2 & ml^2 \end{pmatrix} \tag{vi}$$

The normal frequencies of the double pendulum are given by

$$|\mathbf{V} - \omega^2\mathbf{T}| = 0 \tag{vii}$$

$$\begin{vmatrix} 2mgl - \omega^2 2ml^2 & -\omega^2 ml^2 \\ -\omega^2 ml^2 & mgl - \omega^2 ml^2 \end{vmatrix} = 0$$

which gives

$$\omega_1^2 = (2 - \sqrt{2})\frac{g}{l} \quad \text{and} \quad \omega_2^2 = (2 + \sqrt{2})\frac{g}{l} \tag{viii}$$

The normal modes for a double pendulum for $\omega^2 = \omega_1^2$ are

$$\begin{pmatrix} 2mgl - (2 - \sqrt{2})\frac{g}{l}2ml^2 & -(2 - \sqrt{2})\frac{g}{l}ml^2 \\ -(2 - \sqrt{2})\frac{g}{l}ml^2 & mgl - (2 - \sqrt{2})\frac{g}{l}ml^2 \end{pmatrix}\begin{pmatrix} a_{11} \\ a_{21} \end{pmatrix} = 0 \tag{ix}$$

which reduces to

$$(2 - 2\sqrt{2})a_{11} + (2 - \sqrt{2})a_{21} = 0 \tag{x}$$

$$(2 - \sqrt{2})a_{11} + (1 - \sqrt{2})a_{21} = 0 \tag{xi}$$

and

$$\text{If } a_{11} = 1, \qquad a_{21} = \sqrt{2} \tag{xii}$$

Similarly, for $\omega^2 = \omega_2^2$, we get

$$\text{If } a_{12} = 1, \qquad a_{22} = -\sqrt{2} \tag{xiii}$$

a_{11} and a_{12} correspond to particle 1, and a_{21} and a_{22} correspond to particle 2. The two modes are

$$\eta_1 = a_{11}x_1 + a_{12}x_2 = x_1 + x_2 \tag{xiv}$$

$$\eta_2 = a_{21}x_1 + a_{22}x_2 = \sqrt{2}(x_1 - x_2) \tag{xv}$$

In mode η_1, the particles oscillate out of phase and it is an antisymmetric mode, as shown in Fig. Ex. 14.3(b). In mode η_2, they oscillate in phase and it is a symmetric mode, as shown in Fig. Ex. 14.3(c).

The above remarks are illustrated using numerical values. Below are graphed the natural modes x1, x2; normal modes X1, X2; and the sum of natural modes x1 + x2 and sum of the normal modes X1 + X2.

$$A11 := 2 \qquad A12 := 4 \qquad A21 := 3 \qquad A22 := 6 \qquad k1 := 5 \qquad k2 := 15 \qquad m := 2$$

$$N := 50 \quad t := 0..N \quad \omega 1 := \sqrt{\frac{k1}{m}} \qquad \omega 1 = 1.581 \qquad \omega 2 := \sqrt{\frac{k1 + k2}{m}} \qquad \omega 2 = 3.162 \qquad i := \sqrt{-1}$$

$$x1_t := A11 \cdot e^{i \cdot \omega 1 \cdot t} + A12 \cdot (e)^{-i \cdot \omega 1 \cdot t} + A21 \cdot e^{i \cdot \omega 2 \cdot t} - A22 \cdot (e)^{-i \cdot \omega 2 \cdot t} \qquad X1_t := 2 \cdot A11 \cdot e^{i \cdot \omega 1 \cdot t} + 2 \cdot A12 \cdot e^{-i \cdot \omega 1 \cdot t}$$

$$x2_t := \left(A11 \cdot e^{i \cdot \omega 1 \cdot t} + A12 \cdot e^{-i \cdot \omega 1 \cdot t}\right) - A21 \cdot e^{i \cdot \omega 2 \cdot t} - A22 \cdot e^{-i \cdot \omega 2 \cdot t} \qquad X2_t := 2 \cdot A21 \cdot e^{i \cdot \omega 2 \cdot t} + 2 \cdot A22 \cdot e^{-i \cdot \omega 2 \cdot t}$$

(a) In each of the graphs, explain the differences (in terms of frequencies, amplitudes, and phase differences) between:
x1 and x2
X1 and X2
x1 + x2 and X1 + X2

(b) What are the outstanding features of normal modes as shown by these graphs?

(c) What is the significance of the maximum and minimum values of the two graphs?

(d) What do you conclude from these graphs?

EXERCISE 14.3 Consider the situation shown in Fig. Exer. 14.3. Mass M is constrained to move on a smoother frictionless track AB. Another mass m is connected to M by a massless inextensible string of length l. Calculate the frequencies of small oscillations. Draw graphs similar to those in Exercise 14.3.

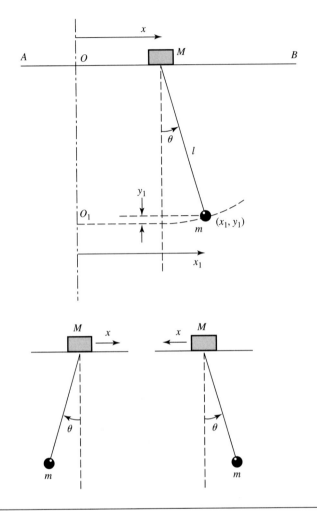

Figure Ex. 14.3

14.7 SYMPATHETIC VIBRATIONS AND BEATS

Let us consider two simple oscillators each of length l and mass m that are coupled by a spring constant k, as shown in Fig. Ex. 14.2. If the spring offers a small resistance to the relative motion of the two pendulums, we say that the system has *weak coupling*, whereas if the spring offers a greater resistance, the system is said to have *strong coupling*. If the pendulums are not exactly equal in length or in mass, we say that the two pendulums are *out of tune* or *detuned*.

For the present, let us assume that the two pendulums are exactly of equal length and mass, and they are weakly coupled by a spring. We assume that they oscillate in the same plane. Let us further assume that the one pendulum is excited by giving an initial displacement while the other pendulum is at rest. As time passes, the resulting oscillations of the two pendulums are as

Figure 14.4 _____

Below the resonance between two weakly coupled oscillators such as pendulums is shown.
We may use Eqs. (14.119) and (14.120) or (14.121).

$$n := 200 \qquad i := 0..n \qquad t_i := \frac{i}{20} \qquad A := 10 \qquad A11 := 10 \qquad A12 := 10$$

$$\omega1 := 88 \qquad \omega2 := 90 \qquad \omega0 := \frac{\omega2 - \omega1}{2} \qquad \omega0 = 1 \qquad T := \frac{2 \cdot \pi}{\omega0} \qquad T = 6.283$$

$$x1_i := A11 \cdot \cos(\omega1 \cdot t_i) - A12 \cdot \cos(\omega2 \cdot t_i) \qquad\qquad x2_i := \left(A11 \cdot \cos(\omega1 \cdot t_i) + A12 \cdot \cos(\omega2 \cdot t_i)\right)$$

First oscillator:

$t=0 \quad x1=0 \quad v1=0$

$x1_0 = 0$

$\max(x1) = 19.947$

$\min(x1) = -19.604$

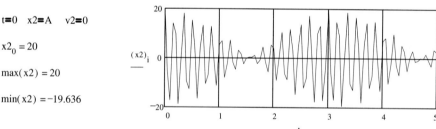

Second oscillator:

$t=0 \quad x2=A \quad v2=0$

$x2_0 = 20$

$\max(x2) = 20$

$\min(x2) = -19.636$

(a) What determines the amplitude of the oscillations in the two cases?
(b) In the two graphs draw the envelope of the oscillations.
(c) How will the increase or decrease in frequency affect the resonance?
(d) How will your increase or decrease in the amplitude affect the resonance?
(e) How do the above graphs illustrate the transfer of energy from one oscillator to the other and vice versa?

shown in Fig. 14.4. As is clear, the oscillations are modulated, and the energy is continuously being transferred from one pendulum to the other. When one pendulum is oscillating with maximum amplitude, the other pendulum is at rest, and vice versa. This is the phenomenon of *resonance* or sympathetic vibration between two systems. The alternation of energy between the two pendulums can be shown mathematically as explained next. This is the *theory of resonance,* as illustrated in Fig. 14.4. A slight detuning leads to the phenomenon of beats, as we shall see later.

Suppose, for the case in Fig. Ex. 14.2, at $t = 0$, we have $x_1 = 0$, $\dot{x}_1 = 0$, $x_2 = A$, and $\dot{x}_2 = 0$. Applying these conditions to Eqs. (xviii) and (xix) in Example 14.2, that is, we get [or for the system shown in Fig. 14.2, resulting in Eqs. (14.39) and (14.40)]

$$x_1(t) = A_1 e^{i\omega_1 t} + A_{-1} e^{-i\omega_1 t} + A_2 e^{i\omega_2 t} + A_{-2} e^{-i\omega_2 t} \qquad \textbf{(14.114)}$$

$$x_2(t) = A_1 e^{i\omega_1 t} + A_{-1} e^{-i\omega_1 t} - A_2 e^{i\omega_2 t} - A_{-2} e^{-i\omega_2 t} \qquad \textbf{(14.115)}$$

we obtain, at $t = 0$,

$$A_1 + A_{-1} + A_2 + A_{-2} = 0 \qquad \textbf{(14.116a)}$$

$$A_1 + A_{-1} - A_2 - A_{-2} = A \qquad \textbf{(14.116b)}$$

$$i\omega_1(A_1 - A_{-1}) + i\omega_2(A_2 - A_{-2}) = 0 \qquad \textbf{(14.117a)}$$

$$i\omega_1(A_1 - A_{-1}) - i\omega_2(A_2 - A_{-2}) = 0 \qquad \textbf{(14.117b)}$$

Solving these equations yields

$$A_1 = A_{-1} = \frac{A}{4} \quad \text{and} \quad A_2 = A_{-2} = -\frac{A}{4} \qquad \textbf{(14.118)}$$

Substituting these in Eqs. (14.114) and (14.115), we obtain

$$x_1(t) = \frac{A}{4} [(e^{i\omega_1 t} + e^{-i\omega_1 t}) - (e^{i\omega_2 t} + e^{-i\omega_2 t})]$$

$$x_2(t) = \frac{A}{4} [(e^{i\omega_1 t} + e^{-i\omega_1 t}) + (e^{i\omega_2 t} + e^{-i\omega_2 t})]$$

Since $2\cos\theta = e^{i\theta} + e^{-i\theta}$, we may write

$$x_1 = \frac{A}{2}(\cos \omega_1 t - \cos \omega_2 t) \qquad \textbf{(14.119)}$$

$$x_2 = \frac{A}{2}(\cos \omega_1 t + \cos \omega_2 t) \qquad \textbf{(14.120)}$$

Equations (14.119) and (14.120) may also be written as

$$x_1 = A \sin\left(\frac{\omega_2 - \omega_1}{2} t\right) \sin\left(\frac{\omega_1 + \omega_2}{2} t\right) \qquad \textbf{(14.121)}$$

$$x_2 = A \cos\left(\frac{\omega_2 - \omega_1}{2} t\right) \cos\left(\frac{\omega_1 + \omega_2}{2} t\right) \qquad \textbf{(14.122)}$$

Let $(\omega_1 + \omega_2)/2 = \omega_0$ and $\omega_2 \simeq \omega_1$; then we may write

$$x_1 = A \sin\left(\frac{\omega_2 - \omega_1}{2} t\right) \sin \omega_0 t \qquad (14.123)$$

$$x_2 = A \cos\left(\frac{\omega_2 - \omega_1}{2} t\right) \cos \omega_0 t \qquad (14.124)$$

Note that, at $t = 0$, $x_1 = 0$, and $x_2 = A$, as it should be. These equations state that x_1 and x_2 are executing oscillatory motions $\sin \omega_0 t$ and $\cos \omega_0 t$, with their slowly varying amplitudes given respectively by

$$A \sin\left(\frac{\omega_2 - \omega_1}{2} t\right) \qquad (14.125)$$

and

$$A \cos\left(\frac{\omega_2 - \omega_1}{2} t\right) \qquad (14.126)$$

This implies that, as the amplitude of x_1 becomes larger, that of x_2 becomes smaller and smaller, and vice versa. This is demonstrated in Fig. 14.4. This means that there is a transfer of energy back and forth. The period T of this energy transfer is

$$T = \frac{2\pi}{\omega} = \frac{4\pi}{\omega_2 - \omega_1} \qquad (14.127)$$

If the two pendulums are slightly detuned (have slightly different frequencies), the energy exchange will still take place, but this exchange is not complete. The initially excited second, pendulum reaches a certain minimum amplitude, but not zero amplitude. The first pendulum initially at rest, does reach zero amplitude during its oscillations. This results in the phenomenon of *beats,* as shown in Fig. 14.5. Thus sympathetic vibration or resonance is upset by slight detuning. We can apply these considerations to another example, that of the double pendulum, as discussed in Example 14.3. If the two masses and the two lengths are equal, we still can have sympathetic resonance vibrations. But suppose the upper mass (and hence weight) is much larger than the lower mass. This leads to slight detuning and to the formation of beats. Suppose we set the pendulum in motion by pulling the upper mass slightly away from the vertical and releasing it. In the subsequent motion, at regular intervals, the lower mass will come to rest, while the upper mass will have a maximum amplitude, or the upper mass will have a minimum amplitude (different from zero) when the lower mass has maximum amplitude. This is the phenomenon of beats, as illustrated in Fig. 14.5. Once again, due to slight detuning, there is an incomplete transfer of energy.

If instead of looking at the normal modes, we look at the motion of the two separately, the resulting natural modes of the two are as was shown in Fig. 14.4. It is clear that when one has maximum displacement, the other has minimum and vice versa.

If in the preceding examples, both pendulums were set in motion simultaneously either (1) in the same direction or (2) in opposite directions, we would find that there would be no energy exchange between the two pendulums. We get the normal modes of vibrations as discussed in Section 14.3 and in Example 14.2.

Figure 14.5 _____

Below the phenomenon of beats resulting from two slightly detuned, weakly coupled oscillators (pendulums in this case) are shown.

$$n := 200 \qquad i := 0..n \qquad \omega1 := 12 \qquad \omega2 := 13$$

$$t_i := \frac{i}{20} \qquad A := 10 \qquad \omega0 := \frac{\omega2 + \omega1}{2} \qquad \omega0 = 12.5$$

$$T := \frac{2 \cdot \pi}{\omega0} \qquad |T| = 0.503$$

$$x1_i := (A) \cdot \left[\sin\left[\frac{\omega2 - \omega1}{2} \cdot (t_i + 5) \right] \cdot \sin(\omega0 \cdot t_i + 5) \right] \qquad x2_i := A \cdot \cos\left(\frac{\omega2 - \omega1}{2} \cdot t_i \right) \cdot \cos(\omega0 \cdot t_i)$$

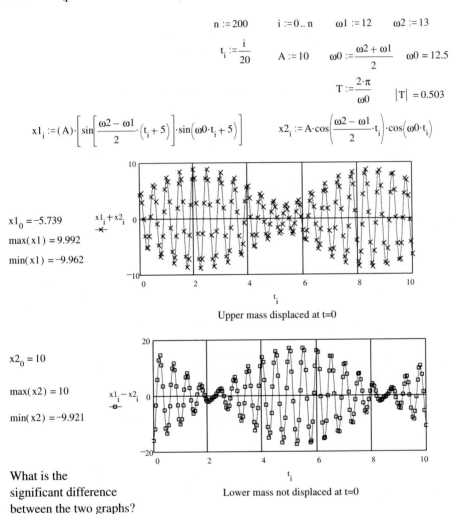

$x1_0 = -5.739$

$\max(x1) = 9.992$

$\min(x1) = -9.962$

Upper mass displaced at t=0

$x2_0 = 10$

$\max(x2) = 10$

$\min(x2) = -9.921$

Lower mass not displaced at t=0

What is the significant difference between the two graphs?

The preceding discussion for coupled *mechanical oscillating systems* can be extended to electrical systems. Sympathetic oscillations are of great importance in electrical circuits. In electrical systems, we have a primary and a secondary circuit that are usually inductively coupled with each other. Thus, if the primary circuit is excited, the secondary circuit will also oscillate

Figure 14.5 (continued) _____

The transfer of displacement is equivalent to transfer of energy, between two lightly coupled oscillators. Thus if we graph x1 and x2 separately, it illustrates the trasfer of energy between the two lightly coupled oscillators as shown below.

$$n := 100 \qquad i := 0..n \qquad t_i := i \qquad A := 10 \qquad \omega1 := 40 \qquad \omega2 := 42$$

$$\omega0 := \frac{\omega2 + \omega1}{2} \qquad \omega0 = 41$$

$$x1_i := A \cdot \sin\left(\frac{\omega2 - \omega1}{2} \cdot t_i\right) \cdot \sin\left(\omega0 \cdot t_i\right) \qquad x2_i := A \cdot \cos\left(\frac{\omega2 - \omega1}{2} \cdot t_i\right) \cdot \cos\left(\omega0 \cdot t_i\right)$$

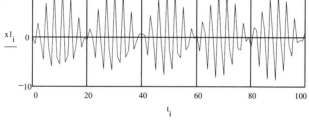

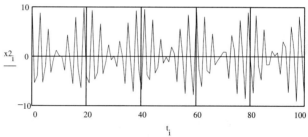

(a) How do you explain that when x1 is minimum x2 is maximum and vice versa?
(b) What is the phase relation between x1 and x2 and how do you explain it?

strongly if there is a resonance. Unlike the coupled pendulums considered previously, in electrical circuits damping must be included. As discussed in Chapter 4, damping is equivalent to ohmic resistance, mass corresponds to the self-inductance, and restoring force to the capacitance effects. Furthermore, in electrical oscillations, we deal not only with "position coupling" but also with "velocity and acceleration coupling."

14.8 VIBRATION OF MOLECULES

We shall consider possible modes of vibrations for diatomic and triatomic molecules. A typical diatomic molecule may be regarded as equivalent to two masses m_1 and m_2 connected by a massless spring of spring constant k and of unstretched length a, vibrating along the line joining the two masses, as shown in Fig. 14.6. Let x_1 and x_2 be the coordinates of m_1 and m_2 measured from a fixed point O. The potential energy and kinetic energy of the system are

$$V = \tfrac{1}{2}k(x_2 - x_1 - a)^2 \tag{14.128}$$

$$T = \tfrac{1}{2}m_1\dot{x}_1^2 + \tfrac{1}{2}m_2\dot{x}_2^2 \tag{14.129}$$

The expression for the potential energy is not a homogeneous quadratic function; hence a linear transformation to normal coordinates is not possible. But this difficulty can be overcome by making the substitution

$$u = x_2 - a \quad \text{and} \quad \dot{u} = \dot{x}_2 \tag{14.130}$$

Substituting these in Eqs. (14.128) and (14.129),

$$V = \tfrac{1}{2}k(u - x_1)^2 \tag{14.131}$$

$$T = \tfrac{1}{2}m_1\dot{x}_1^2 + \tfrac{1}{2}m_2\dot{u}^2 \tag{14.132}$$

By using x_1 and u as generalized coordinates, we can solve the Lagrangian equation for x_1 and u. By using proper linear combinations of x_1 and u, we can find the normal coordinates X_1 and X_2 corresponding to ω_1 and ω_2 respectively. Thus

$$X_1 = \frac{m_1}{m_2}x_1 + u \quad \text{and} \quad X_2 = x_1 - u \tag{14.133}$$

If mode X_1 is excited, then X_2 must be suppressed; that is,

$$\text{For mode } X_1: \quad X_2 = x_1 - u = 0$$

or

$$x_1 = u = x_2 - a \tag{14.134}$$

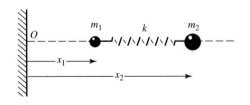

Figure 14.6   Schematic of a system equivalent to a diatomic molecule.

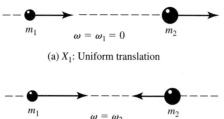

(a) X_1: Uniform translation

Figure 14.7 Two possible normal modes of vibration of the system of Fig. 14.6.

(b) X_2: Oscillations

which corresponds to uniform translation motion of the system, as shown in Fig. 14.7(a). Similarly, if mode X_2 is excited, then X_1 must be suppressed; that is,

$$\text{For mode } X_2: \qquad X_1 = \frac{m_1}{m_2} x_1 + u = 0$$

or

$$x_1 = -\frac{m_2}{m_1} u = -\frac{m_2}{m_1}(x_2 - a) \qquad \textbf{(14.135)}$$

which indicates that the two masses oscillate relative to the center of mass, as shown in Fig. 14.7(b).

The results obtained can be arrived at by an inspection of the situation and recognizing the basic physical problem. Let us demonstrate this in the case of a triatomic molecule such as CO_2, as shown in Fig. 14.8. CO_2 is a linear molecule, and if the motion is constrained along a line, it will have three degrees of freedom and hence three normal coordinates.

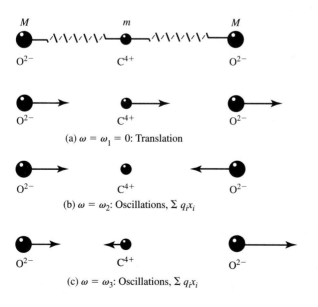

(a) $\omega = \omega_1 = 0$: Translation

(b) $\omega = \omega_2$: Oscillations, $\Sigma \, q_i x_i$

(c) $\omega = \omega_3$: Oscillations, $\Sigma \, q_i x_i$

Figure 14.8 A triatomic molecule and its three possible normal modes of vibration.

14.9 DISSIPATIVE SYSTEMS AND FORCED OSCILLATIONS

So far in the discussion of small oscillations, we neglected the effects of viscous or frictional forces. A common situation is one in which the viscous damping forces are proportional to the first power of the velocity. In such situations, the motion of the ith particle may be described by Newton's second law as

$$m_i \ddot{\mathbf{r}} = \mathbf{F}_i - c_i \dot{\mathbf{r}}_i \tag{14.136}$$

which in component form may by written as

$$m_i \ddot{x}_i = F_{ix} - c_i \dot{x}_i \tag{14.137a}$$

$$m_i \ddot{y}_i = F_{iy} - c_i \dot{y}_i \tag{14.137b}$$

$$m_i \ddot{z}_i = F_{iz} - c_i \dot{z}_i \tag{14.137c}$$

where c_i are constants and F_{ix}, F_{iy}, and F_{iz} are the components of a resultant force $\mathbf{F}_i$ that are derivable from a potential, and the potential is a homogeneous quadratic function of the coordinates.

Suppose the system has l degrees of freedom and is described by l independent coordinates:

$$q_1', q_2', \ldots, q_l' \tag{14.138}$$

The relations between these and the x, y, and z coordinates are given by the following $3n$ equations for n particles.

$$x_i = x_i(q_1', q_2', \ldots, q_l')$$

$$y_i = y_i(q_1', q_2', \ldots, q_l')$$

$$z_i = z_i(q_1', q_2', \ldots, q_l') \tag{14.139}$$

Note that there is no explicit dependence on time t because kinetic energy T is a homogeneous quadratic function of time. Multiply each of Eqs. (14.137), respectively, by the quantities $\partial x_i/\partial q_j'$, $\partial y_i/\partial q_j'$, and $\partial z_i/\partial q_j'$; adding all three and summing over all the n particles yields

$$\sum_{i=1}^{n} m_i \left(\ddot{x}_i \frac{\partial x_i}{\partial q_j'} + \ddot{y}_i \frac{\partial y_i}{\partial q_j'} + \ddot{z}_i \frac{\partial z_i}{\partial q_j'} \right)$$

$$= \sum_{i=1}^{n} \left(F_{ix} \frac{\partial x_i}{\partial q_j'} + F_{iy} \frac{\partial y_i}{\partial q_j'} + F_{iz} \frac{\partial z_i}{\partial q_j'} \right) - \sum_{i=1}^{n} c_i \left(\dot{x}_i \frac{\partial x_i}{\partial q_j'} + \dot{y}_i \frac{\partial y_i}{\partial q_j'} + \dot{z}_i \frac{\partial z_i}{\partial q_j'} \right) \tag{14.140}$$

where

$$\text{First term on the left} \equiv \frac{d}{dt} \left(\frac{\partial T}{\partial \dot{q}_j'} \right) - \frac{\partial T}{\partial q_j'}$$

$$\text{First term on the right} \equiv -\frac{\partial V}{\partial q_j'} \equiv Q_i, \quad \text{the generalized force, excluding the dissipative forces}$$

$$\text{Second term on the right} \equiv -\frac{\partial}{\partial \dot{q}_j'} \left[\frac{1}{2} \sum_{i=1}^{n} c_i (\dot{x}_i^2 + \dot{y}_i^2 + \dot{z}_i^2) \right] = -\frac{\partial F_r}{\partial \dot{q}_j'}$$

and $F_r = \frac{1}{2}\Sigma c_i(\dot{x}_i^2 + \dot{y}_i^2 + \dot{z}_i^2)$ is the dissipative function named by Rayleigh and represents one-half the rate at which the energy is being dissipated through the action of frictional forces. Thus Eq. (14.140) may be written as

$$\frac{d}{dt}\left(\frac{\partial T}{\partial \dot{q}'_j}\right) - \frac{\partial T}{\partial q'_j} = -\frac{\partial V}{\partial q'_j} = -\frac{\partial F_r}{\partial \dot{q}'_j} \tag{14.141}$$

Since $L = T - V$, Eq. (14.137) or (14.141) takes form

$$\frac{d}{dt}\left(\frac{\partial L}{\partial \dot{q}'_j}\right) = \frac{\partial L}{\partial q'_j} + Q_{rj} \tag{14.142}$$

where Q_{rj} is the generalized damping force

$$Q_{rj} = -\frac{\partial F_r}{\partial \dot{q}'_j} \tag{14.143}$$

For sufficiently small motions, the expressions for V, T, and F_r may be written as

$$V = a_{11}q_1'^2 + \cdots + a_{ll}q_1'^2 + 2a_{12}q_1'q_2' + \cdots \tag{14.144a}$$

$$T = b_{11}\dot{q}_1'^2 + \cdots + b_{ll}\dot{q}_1'^2 + 2b_{12}\dot{q}_1'\dot{q}_2' + \cdots \tag{14.144b}$$

$$F_r = c_{11}\dot{q}_1'^2 + \cdots + c_{ll}\dot{q}_1'^2 + 2c_{12}\dot{q}_1'\dot{q}_2' + \cdots \tag{14.144c}$$

where $a_{ll}, \ldots, b_{ll}, \ldots,$ and $c_{ll}, \ldots,$ are constants.

The resulting differential equations of motion obtained from Eq. (14.141) or (14.142) are similar to the undamped case, except that terms of the form $\dot{q}$ are present. To calculate normal modes, we must find new coordinates that are linear combinations of $q_1', q_2', \ldots, q_l'$ so that V, T, and F_r, when expressed in terms of coordinates $\eta_1, \eta_2, \ldots, \eta_j$, do not contain cross terms; that is, they contain the sum of the squares of the new coordinates and their time derivatives. Because of the presence of F_r, it is not always possible to find such new coordinates. In some situations it is possible to find a normal coordinate transformation, and the resulting differential equations are of the form

$$m_j\ddot{\eta}_j + c_j\dot{\eta}_j + k_j\eta_j = 0 \tag{14.145}$$

which have solutions of the form

$$\eta_j = A_j e^{-\lambda_j t}\cos(\omega_j t + \phi_j) \tag{14.146}$$

Thus, unlike the case of undamped motion in which one observes oscillations, in the present case the motion may be underdamped, critically damped, or overdamped, as the case may be; hence the motion may be nonoscillatory. The normal coordinates and their phases are the same as in the corresponding problem of undamped motion. The amplitude decreases exponentially with time, while the frequencies are different from the ones in the undamped case.

First, we must assume that the driving forces are small enough so that the squares of the displacements and velocities will be such that the equations of motion are still linear. If the forces are constant, such as a system under gravitational force, the only change is in the equilibrium position about which the oscillations take place. If the driving force is periodic, it is possible to discuss motion in terms of normal coordinates. For convenience, let us assume that a single harmonic force of the type $Q_{jext} \cos \omega t$ or $Q_{jext} e^{i\omega t}$ is applied. The resulting equation of motion in normal coordinates is of the form (in the presence of a linear restoring force, dissipative force, and driving force)

$$m_j \ddot{\eta}_j + c_j \dot{\eta}_j + k_j \eta_j = Q_{jext} e^{i\omega t} \tag{14.147}$$

If the driving frequency is equal to one of the normal frequencies of the system, the corresponding normal mode will assume the largest amplitude in the steady state. Furthermore, if the damping constants are small, not all normal modes are excited to any appreciable extent; only one normal mode that has the same frequency as the driving force will be excited.

 Example 14.4

Let us consider once again the situation of two coupled pendula, as discussed in Example 14.2. Let us assume that the driving force is $F \cos \omega t$, and the frictional force proportional to velocity is $c\dot{x}$, where c is a constant. Discuss the solution of this problem.

Solution

The equations describing the system are

$$m\ddot{x}_1 + \frac{mg}{l} x_1 + k(x_1 - x_2) = -c\dot{x}_1 + F \cos \omega t$$

$$m\ddot{x}_2 + \frac{mg}{l} x_2 - k(x_1 - x_2) = -c\dot{x}_2 + F \cos \omega t$$

Equations involving normal coordinates X_1 and X_2 are ($\eta_1 = X_1 = x_1 + x_2$ and $\eta_2 = X_2 = x_1 - x_2$)

$$\ddot{X}_1 + \frac{c}{m}\dot{X}_1 + \frac{g}{l}X_1 = \frac{2F}{m} \cos \omega t$$

$$\ddot{X}_2 + \frac{c}{m}\dot{X}_2 + \left(\frac{g}{l} + \frac{2k}{m}\right)X_2 = 0$$

We should be able to recognize these differential equations, which have the following solutions:

$$X_1 = e^{-(c/2m)t}(A_1 e^{i\omega_1' t} + A_{-1} e^{-i\omega_1' t}) + \frac{2F \cos(\omega t - \phi)}{[m^2(\omega_0^2 - \omega^2)^2 + \omega^2 c^2]^{1/2}}$$

and

$$X_2 = e^{-(c/2m)t}(A_2 e^{i\omega_2' t} + A_{-2} e^{-i\omega_2' t})$$

where

$$\omega_0 = \left(\frac{g}{l}\right)^{1/2}, \quad \omega_1' = \left(\frac{g}{l} - \frac{c^2}{4m^2}\right)^{1/2}, \quad \omega_2' = \left[\left(\frac{g}{l} + \frac{2k}{m}\right) - \frac{c^2}{4m^2}\right]^{1/2}$$

$$\tan \phi = \frac{\omega c}{[m(\omega_0^2 - \omega^2)]}, \quad \text{for } g/l > c^2/4$$

Both X_1 and X_2 contain transient terms. Only X_1 possesses a steady-state term, and only X_1 will remain excited (for any initial conditions) with the same frequency as the driving frequency, which is similar to a system having one degree of freedom, X_2 will decay in a short interval. These points are illustrated in the following graphs.

Assuming the following values and graphing with and without the driving force, X1 and X2, respectively, gives

$$g := 9.8 \qquad l := 1 \qquad c := 0.5 \qquad m := 1 \qquad k := 2 \qquad n := 0..60 \qquad t_n := n \qquad i := \sqrt{-1}$$

$$A1 := 4 \qquad A12 := 2 \qquad A2 := 15 \qquad A21 := 10 \qquad F := 5$$

$$\omega 0 := \sqrt{\frac{g}{l}} \qquad \omega 1 := \left(\frac{g}{l} - \frac{c^2}{4 \cdot m^2}\right)^{\frac{1}{2}} \qquad \omega 2 := \left[\left(\frac{g}{l} + \frac{2 \cdot k}{m}\right) - \frac{c^2}{4 \cdot m^2}\right]^{\frac{1}{2}}$$

$$\omega := 3 \qquad \phi := \frac{\pi}{2}$$
$$\omega 0 = 3.13$$
$$\omega 1 = 3.12$$
$$\omega 2 = 3.706$$

$$X1_n := e^{-\left(\frac{c}{2 \cdot m}\right) \cdot t_n} \cdot \left(A1 \cdot e^{i \cdot \omega 1 \cdot t_n} + A12 \cdot e^{-i \cdot \omega 1 \cdot t_n}\right) + \frac{2 \cdot F \cdot \cos\left(\omega \cdot t_n - \phi\right)}{\sqrt{m^2 \cdot \left(\omega 0^2 - \omega^2\right) + \omega^2 \cdot c^2}}$$

$$X2_n := e^{-\left(\frac{c}{2 \cdot m}\right) \cdot t_n} \cdot \left(A2 \cdot e^{i \cdot \omega 2 \cdot t_n} + A21 \cdot e^{-i \cdot \omega 2 \cdot t_n}\right)$$

Both X1 and X2 contain transient terms. Only X1 possesses a steady-state term and remains excited for any initial condition with the same frequency as the driving frequency. X2 will decay away in a short time.

EXERCISE 14.4 Repeat the above example with the applied force equal to $F \sin(\omega t)$. What are the similarities and differences between the two?

PROBLEMS

14.1. A cube of side $2a$ is balanced on top of a rough spherical surface of radius R. Show that the equilibrium is stable if $R > a$ and unstable if $R = a$. What happens if $R = a$? Find the frequency of small oscillations.

14.2. In Problem 14.1, if the cube is replaced by a homogeneous solid hemisphere of radius r, show that for stable equilibrium $r < \frac{3}{5}R$.

14.3. A homogeneous rectangular slab of thickness d is placed atop and at right angles to a fixed cylinder of radius R with its axis horizontal. Assuming no slipping, show that the condition for stable equilibrium is $R < d/2$. Draw a potential energy function versus the angular displacement θ, and show that there is a minimum at $\theta = 0$ for $R > d/2$ but not for $R < d/2$. Find the frequency of small oscillations about equilibrium.

14.4. A homogeneous disk of mass M and radius R rolls without slipping on a horizontal surface and is attracted toward a point that lies at a distance d below the surface. The attractive force is proportional to the distance between the center of mass and the force center. Is the disk in stable equilibrium? If so, find the frequency of small oscillations.

14.5. Two identical springs each of natural length l_0 and stiffness constant k have their upper ends tied at two points A and B, which are a distance $2a$ apart. The two lower ends are tied together at C, and a mass m hangs it, as shown in Fig. P14.5. Find the position of equilibrium. Is it a position of stable equilibrium? Find the frequency of small oscillations.

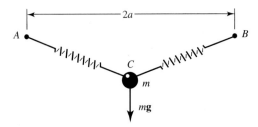

Figure P14.5

14.6. A mass m is subject to a force whose potential energy function is

$$V = V_0 \exp[(5x^2 + 5y^2 + 8z^2 - 8yz - 26ya - 8za)/a^2]$$

where V_0 and a are constants. Find, if any, positions of stable or unstable equilibrium. Find the normal frequencies of vibration about the minimum.

14.7. A particle of mass m moves along the X-axis under the influence of a potential energy given by $V(x) = -Axe^{-kx}$, where A and k are constants. Make a plot of $V(x)$ versus x. Find the position of equilibrium. Also calculate the frequency of small oscillations.

14.8. Consider a rod of length L and mass m supported by two springs, as shown in Fig. P14.8. Assuming that the rod remains in the vertical plane, calculate the normal frequencies of oscillation. Graph the normal modes.

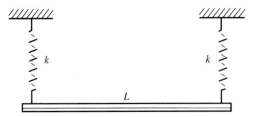

Figure P14.8

14.9. For the configuration of two masses and two springs as shown in Fig. P14.9, calculate the normal frequencies and normal coordinates, assuming that the motion is restricted to the vertical plane. Graph the natural as well as normal modes.

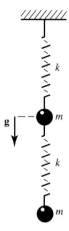

Figure P14.9

14.10. Three identical masses and four identical springs are connected as shown in Fig. P14.10. If the system is displaced from its equilibrium position along the line joining the masses, calculate the normal frequencies and normal coordinates for small oscillations. The unstretched length of each spring is a and k is its spring constant. Graph the natural as well as normal modes.

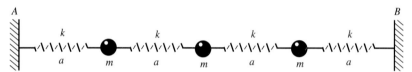

Figure P14.10

14.11. In Problem 14.10, there is a tension T in the spring at points A and B. Calculate the normal frequencies and normal coordinates for small transverse oscillations. Graph the tension versus displacement.

14.12. A light rod OA of length r is fixed at O, and a mass M is attached to the other end, as shown in Fig. P14.12. It is forced to move in the XY-plane. A pendulum of length l and mass m attached at A can oscillate in the YZ-plane. Find the normal frequencies and normal modes of vibration. Graph the normal modes of vibrations.

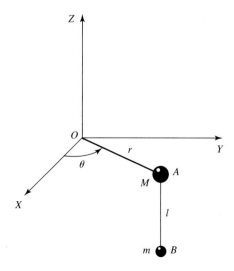

Figure P14.12

14.13. Three oscillators of mass m each are coupled in such a way that the force between them is given by the potential energy function

$$V = \tfrac{1}{2}[k_1(x_1^2 + x_3^2) + k_2 x_2^2 + k_3(x_1 x_2 + x_2 x_3)]$$

where $k_3 = (2k_1 k_2)^{1/2}$. Find the points of equilibrium and their stability. Find the normal frequencies of the system and normal modes of vibration. Graph the normal modes. Is there any physical significance of the null mode?

14.14. Three masses M, m, and m are connected by identical springs of stiffness constant k and placed on a fixed circular loop in space, as shown in Fig. P14.14. Calculate the normal frequencies and normal coordinates. What happens if $M = m$? Also describe the type of motion of these masses. Draw the polar graphs of the motion of the mass m and M.

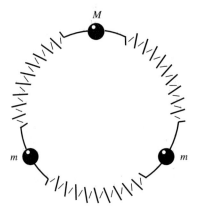

Figure P14.14

14.15. A particle of mass m is moving in a force field that is represented by the potential energy given by

$$V(x) = (1 - \alpha x)e^{-\alpha x}, \quad x \geqslant 0$$

where α is a positive constant. Find **(a)** the equilibrium points, **(b)** the nature of the equilibrium points, and **(c)** the frequency for small oscillations about equilibrium. Graph V and F versus x and displacement versus time.

14.16. A mass m is attached to a mass M by a light string of length l. The mass M slides without friction on a table, while the other mass hangs vertically through a hole in the table, as shown in Fig. P14.16. Find the steady-state motion, normal frequencies, and normal modes for small oscillations. Make appropriate polar graphs to describe the motion of masses m and M. What happens when M touches the whole in the table?

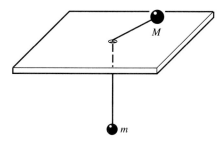

Figure P14.16

14.17. Suppose two identical harmonic oscillations are coupled via a force that is proportional to the relative velocity of the two masses (instead of a force proportional to distance). Find the normal frequencies and normal modes of vibrations.

14.18. A thin wire of mass M bent in the form of circle of radius R is suspended from a point on its circumference. A bead of mass m is attached to the wire and constrained to move on it (frictionless). Find the normal frequencies and normal modes of vibrations if the wire is free to swing in its own plane. If $M = m$, show that

$$\omega_1 = \sqrt{2}\sqrt{\frac{g}{R}} \quad \text{and} \quad \omega_2 = \frac{1}{\sqrt{2}}\sqrt{\frac{g}{R}}$$

Do the normal modes describe any physical situation?

14.19. Consider a double pendulum that consists of one pendulum of length l_1 and mass m_1, and the other of length l_2 and mass m_2. Calculate the normal frequencies. Also find the normal modes. For what initial conditions will the system oscillate in its normal modes? Draw the appropriate graphs to describe the motion.

14.20. Find the normal frequencies and normal modes for the system shown in Fig. P14.20, which consists of three springs and two masses forming a right-angled triangle. What type of motion is expected if one of the masses is displaced in the XY-plane?

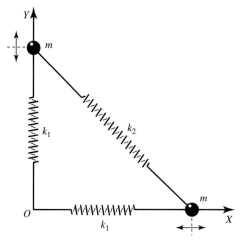

Figure P14.20

14.21. Consider a symmetrical rigid body mounted in weightless, frictionless gimbal rings. One ring exerts a torque $-k\phi$ (ϕ is the Euler angle) about the Z-axis. (This is done by attaching a hair spring.) Investigate the steady-state motion for small vibrations.

14.22. Derive expressions for the normal frequencies and the normal modes of vibration for the triatomic molecule discussed in Section 14.8, that is, CO_2. The mass of carbon is m_1 and that of oxygen is m_2, and assume that the force between adjacent atoms can be represented by a spring of spring constant k and that there is no interaction between the end atoms.

14.23. Consider a plane triatomic molecule consisting of equal masses at the vertices of an equilateral triangle, as shown in Fig. P14.23 (stretched and unstreched). The unstretched length of each spring is a, and the spring constant is k. Consider small oscillations in the plane of the triangle. How many normal modes do you expect and how many of these have zero normal frequencies? Find the frequency of small oscillations for a mode in which all three springs stretch symmetrically, as shown.

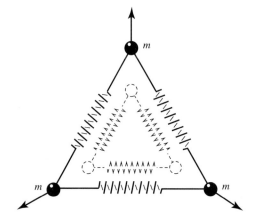

Figure P14.23

14.24. Two simple pendulums are coupled by a weak attractive force given by K/r^2, where r is the distance between the two particle masses. Show that, for a small displacement from equilibrium, the Lagrangian has the same form as that for two coupled oscillators. Furthermore, if one pendulum is set into oscillations while the other is at rest, then eventually the second pendulum will oscillate, and the first will come to rest. The process will repeat with time. Draw graphs that describe this motion.

14.25. As in Problem 14.24, let us once again consider the problem of two linearly coupled pendula, except that the lengths of the two are not equal. Find the normal frequencies and normal modes of vibrations. Show that, unlike the situation discussed in Problem 14.24, the energy of the system is never completely transferred to either of the pendula. Draw graphs to demonstrate this.

14.26. Two identical pendula coupled by a spring as shown in Fig. Ex. 14.2 are moving in a viscous medium that produces a retarding force proportional to its velocity. Find the normal frequencies and normal coordinates. Draw graphs to show the motion.

14.27. Three equal masses m are joined by two identical springs of spring constant k. The system is free to oscillate and move along the line joining the masses. The system is placed in a viscous medium that exerts a retarding force proportional to its velocity. Find the normal frequencies and the normal modes of oscillations. Draw graphs describing the motion.

14.28. Two equal masses and three identical springs are connected as shown in Fig. P14.28 and surrounded by a viscous medium which exerts a retarding force proportional to velocity. There is a tension of T in the springs at A and B. One mass is held and the second is displaced a distance d vertically, and then both are released. Find the normal frequencies and the normal modes of vibration. Draw graphs that describe the motion of the masses.

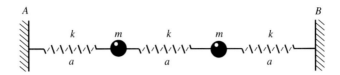

Figure P14.28

14.29. In Problem 14.24, the system is surrounded by a viscous medium that produces a retarding force proportional to its velocity. Find the normal frequencies and the normal modes of vibrations. Assume proper initial conditions.

14.30. Consider the system shown in Fig. P14.28. The unstretched length for each spring is a. **(a)** Find the normal frequencies and normal modes of vibration. **(b)** Suppose each mass m is subjected to a force $F = F_0 \sin \omega t$ at time $t = 0$ when the system is at rest. Discuss the motion using normal coordinates and draw graphs of this motion.

14.31. In Problem 14.30, each mass is subject to a frictional force $-bm\dot{x}_1$. Discuss the motion of the system and draw a graph of it.

SUGGESTIONS FOR FURTHER READING

BECKER, R. A., *Introduction to Theoretical Mechanics,* Chapter 14. New York: McGraw-Hill Book Co., 1954.

CORBEN, H. C., and STEHLE, P., *Classical Mechanics,* Chapter 8. New York: John Wiley & Sons, Inc., 1960.

FOWLES, G. R., *Analytical Mechanics,* Chapter 11. New York: Holt, Rinehart and Winston, Inc., 1962.

FRENCH, A. P., *Vibrations and Waves,* Chapter 5. New York: W. W. Norton and Co., Inc., 1971.

*GOLDSTEIN, H., *Classical Mechanics,* 2nd ed., Chapter 6. Reading, Mass.: Addison-Wesley Publishing Co., 1980.

HAUSER, W., *Introduction to the Principles of Mechanics,* Chapter 11. Reading, Mass.: Addison-Wesley Publishing Co., 1965.

*LANDU, L. D., and LIFSHITZ, E. M., *Mechanics,* Chapter 5. Reading, Mass.: Addison-Wesley Publishing Co., 1960.

MARION, J. B., *Classical Dynamics,* 2nd ed., Chapters 6 and 7. New York: Academic Press, Inc., 1970.

*MOORE, E. N., *Theoretical Mechanics,* Chapter 7. New York: John Wiley & Sons, Inc., 1983.

ROSSBERG, K., *Analytical Mechanics,* Chapter 10. New York: John Wiley & Sons. Inc., 1983.

SLATER, J. C., *Mechanics,* Chapter 7. New York: McGraw-Hill Book Co., 1947.

STEPHENSON, R. J., *Mechanics and Properties of Matter,* Chapter 5. New York: John Wiley & Sons, Inc., 1962.

SYMON, K. R., *Mechanics,* 3rd ed., Chapter 12. Reading, Mass.: Addison-Wesley Publishing Co., 1971.

*The asterisk indicates works of an advanced nature.

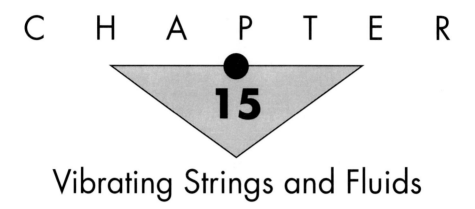

C H A P T E R

15

Vibrating Strings and Fluids

15.1 INTRODUCTION

This chapter is a continuation of the study of mechanics of continuous media such as strings and fluids (gases and liquids). Because a large number of particles are involved, it is cumbersome to apply the laws of mechanics and investigate the resultant motion. Some simplifying assumptions must be made and an overall picture of the motion obtained. We shall divide our discussion into three parts. First, we investigate transverse vibrations of strings in one dimension. To start, we shall consider a simple case but then generalize it by the methods of Lagrange formulation. Second, we study sound waves, that is, longitudinal waves in a gaseous medium. In both of these cases, the main problem involves setting up a wave equation describing the given situation, followed by solving these differential equations by applying appropriate boundary conditions.

Third, we investigate fluids at rest and in motion. We close the chapter by investigating the motion of fluids in the presence of frictional forces (viscous forces).

15.2 VIBRATING STRING

We investigate the propagation of waves along vibrating strings. Our discussion is divided in two parts: the equation of motion and the general solution (normal modes of vibration).

Equation of Motion

Consider a homogeneous string of length L that is fixed at both ends: $x = 0$ and $x = L$. The string has a linear density (mass per unit length) μ and is under tension T throughout the string. The string is in equilibrium along the X-axis, as shown in Fig. 15.1(a). We are interested in investi-

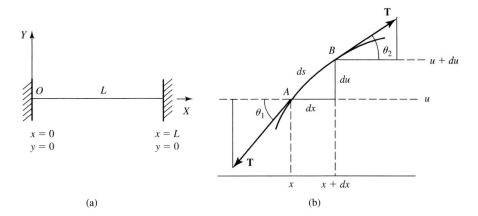

Figure 15.1   **(a)** A string of length L is horizontal when in equilibrium. **(b)** A small portion ds of a string under a small displacement results in transverse vibrations.

gating the motion of such a string following an initial lateral displacement from its equilibrium position. Also, the displacements of the string are not large enough to change the tension T appreciably. Furthermore, we assume that the force due to gravity $(=\mu Lg)$ is small compared to the tension T and may be neglected.

To obtain a differential equation that describes the motion of the string, consider a small portion AB of length ds and of horizontal length dx between x and $x + dx$, as shown in Fig. 15.1(b). Since for small displacements the tension remains the same, we may write the X and Y components of the tensions acting on this small element to be

$$\sum F_x = T \cos \theta_2 - T \cos \theta_1 \qquad \text{(15.1a)}$$

$$\sum F_y = T \sin \theta_2 - T \sin \theta_1 \qquad \text{(15.1b)}$$

If θ_1 and θ_2 are small, $\cos \theta_1 \simeq \cos \theta_2$; hence there is no net horizontal force. This means there is no longitudinal displacement of the string. That is, for small displacements of the string, we are concerned only with the lateral or transverse motion (motion perpendicular to the length of the string). That is, the string vibrates in the XY plane. Also, for small displacements or small angles, we may replace the sine by the tangent; that is,

$$\sin \theta_1 \simeq \tan \theta_1 \quad \text{and} \quad \sin \theta_2 \simeq \tan \theta_2 \qquad \text{(15.2)}$$

Thus the resultant force in the Y direction is

$$\sum F_y \simeq T \tan \theta_1 - T \tan \theta_2 \qquad \text{(15.3)}$$

The motion of the string is described by a displacement function $u(x, t)$ of each point x and at an instant of time t.

Let the vertical displacement of the string be u at x and $u + du$ at $x + dx$. According to Newton's second law,

$$\sum F_y = ma_y = m\frac{\partial^2 u}{\partial t^2} \tag{15.4}$$

where m is the mass of a string of length AB, $m = \mu\, dx$, and $u = u(x, t)$, is the lateral displacement of the string at position x and instant of time t. (Partial derivatives are used because u is a function of both x and t.) Thus combining the preceding equations and assuming $ds \simeq dx$,

$$\mu\, dx\, \frac{\partial^2 u}{\partial t^2} = T\tan\theta_2 - T\tan\theta_1 \tag{15.5}$$

Using
$$T\tan\theta = T\frac{\partial u}{\partial x} \tag{15.6}$$

we may write the net vertical force as

$$T\tan\theta_2 - T\tan\theta_1 = T\left(\frac{\partial u}{\partial x}\right)_B - T\left(\frac{\partial u}{\partial x}\right)_A \tag{15.7}$$

The slope of the string at B may be expanded by using a Taylor series; that is,

$$\left(\frac{\partial u}{\partial x}\right)_B = \left(\frac{\partial u}{\partial x}\right)_A + \left(\frac{\partial^2 u}{\partial x^2}\right)_A dx + \cdots \tag{15.8}$$

which, on substituting in Eq. (15.7) and combining with Eq. (15.5), yields

$$\mu\, dx\, \frac{\partial^2 u}{\partial t^2} = T\frac{\partial^2 u}{\partial x^2}\, dx \tag{15.9}$$

or
$$\mu\, \frac{\partial^2 u}{\partial t^2} = T\frac{\partial^2 u}{\partial x^2} \tag{15.10}$$

$$\frac{\partial^2 u}{\partial x^2} = \frac{\mu}{T}\frac{\partial^2 u}{\partial t^2} \tag{15.11}$$

Since the dimensions of μ are $[ML^{-1}]$ and the dimensions of T are of force $[MLT^{-2}]$, the dimension of μ/T are $[L^{-2}T^2]$, that is, the reciprocal of velocity squared. Hence the *wave equation* of the vibrating string is

$$\frac{\partial^2 u}{\partial x^2} - \frac{1}{v^2}\frac{\partial^2 u}{\partial t^2} = 0 \tag{15.12}$$

where
$$v = \sqrt{\frac{T}{\mu}} \tag{15.13}$$

v is not simply a velocity of propagation; it has a much deeper physical interpretation, which we shall seek later. Here v may be identified as the *wave velocity* with which the wave propagates along the string.

If there were an external vertical force F_e per unit length acting on the string, Eq. (15.9) would take the form

$$\mu \, dx \, \frac{\partial^2 u}{\partial t^2} = T \frac{\partial^2 u}{\partial x^2} \, dx + F_e \, dx$$

or

$$\mu \frac{\partial^2 u}{\partial t^2} = T \frac{\partial^2 u}{\partial x^2} + F_e \tag{15.14}$$

We shall not deal with these situations and shall return to Eq. (15.12) for further discussion.

General Solution: Normal Modes of Vibration

Equation (15.12) is a partial differential equation for the function $u(x, t)$ that describes the motion of a vibrating string. To evaluate the function $u(x, t)$, we make use of initial and boundary conditions. Suppose at $t = 0$ the function $u(x, t)$ satisfies the following initial conditions:

$$u(x, 0) = u_0(x) \tag{15.15}$$

$$\left[\frac{\partial u}{\partial t} \right]_{t=0} = \dot{u}_0(x) \tag{15.16}$$

where $u_0(x)$ is the displacement and $\dot{u}_0(x)$ is the velocity of the string at $t = 0$, and both are functions of position x. Since the string is tied at the ends, it must satisfy the following boundary conditions:

$$u(0, t) = u(L, t) = 0 \tag{15.17}$$

That is, the displacement at the ends is zero at all times.

We now proceed to find the solution $u(x, t)$ of the differential equation (15.12). We make use of the method of separation of variables. Let

$$u(x, t) = X(x)\Theta(t) \tag{15.18}$$

where X is a function of x alone and Θ is a function of t alone. From Eq. (15.18),

$$\frac{\partial^2 u}{\partial x^2} = \Theta \frac{d^2 X}{dx^2} \quad \text{and} \quad \frac{\partial^2 u}{\partial t^2} = X \frac{d^2 \Theta}{dt^2} \tag{15.19}$$

Substituting these in Eq. (15.12) and rearranging, we obtain

$$\frac{v^2}{X} \frac{d^2 X}{dx^2} = \frac{1}{\Theta} \frac{d^2 \Theta}{dt^2} \tag{15.20}$$

The left side of this equation is a function of x only, while the right side is a function of t only. This is possible for all values of x and t only if each side is equal to a constant. Let this constant be $-\omega^2$. The minus sign indicates that the acceleration of the element of the string is always directed toward the equilibrium position (position of the string when it is along the X-axis), that is, the acceleration is opposite to the displacement. Thus, from Eq. (15.20),

$$\frac{v^2}{X} \frac{d^2 X}{dx^2} = -\omega^2 \quad \text{or} \quad \frac{d^2 X}{dx^2} + \frac{\omega^2}{v^2} X = 0 \tag{15.21}$$

and

$$\frac{1}{\Theta}\frac{d^2\Theta}{dt^2} = -\omega^2 \quad \text{or} \quad \frac{d^2\Theta}{dt^2} + \omega^2\Theta = 0 \tag{15.22}$$

where ω may be interpreted as the angular frequency. The solution of Eq. (15.21) is

$$X(x) = C\cos\frac{\omega}{v}x + D\sin\frac{\omega}{v}x \tag{15.23}$$

and that of Eq. (15.22) is

$$\Theta(t) = E\cos\omega t + F\sin\omega t \tag{15.24}$$

where C, D, E, and F are the four constants of integration to be evaluated by using the initial and boundary conditions given by Eqs. (15.15) to (15.17).

Thus, by substituting for $X(x)$ and $\Theta(t)$ from Eqs. (15.23) and (15.24) into Eq. (15.18), we get the general solution:

$$u(x, t) = \left(C\cos\frac{\omega}{v}x + D\sin\frac{\omega}{v}x\right)(E\cos\omega t + F\sin\omega t) \tag{15.25}$$

We may now apply the boundary conditions to evaluate the constants C and D. At $x = 0$, $u(0, t) = 0$ for all values of t; that is $X(0) = 0$ in Eq. (15.23):

$$0 = C\cos\left(\frac{\omega}{v}0\right) + D\sin\left(\frac{\omega}{v}0\right)$$

which is possible only if $C = 0$; thus

$$X(x) = D\sin\frac{\omega}{v}x \tag{15.26}$$

At $x = L$, $u(L, t) = 0$ for all values of t; that is, $X(L) = 0$ in Eq.(15.26):

$$0 = D\sin\frac{\omega}{v}L \tag{15.27}$$

Since $C = 0$ and D cannot be zero because that would give a trivial solution, to satisfy Eq. (15.27), we must have

$$\sin\frac{\omega}{v}L = 0 \quad \text{or} \quad \frac{\omega}{v}L = n\pi \tag{15.28}$$

where $n = 1, 2, 3, \ldots$, or, replacing ω by ω_n, and $v = \sqrt{T/\mu}$,

$$\omega_n = \frac{n\pi v}{L} = \frac{n\pi}{L}\sqrt{\frac{T}{\mu}} \tag{15.29}$$

Thus, with $C = 0$ and letting $DE = A_n$ and $DF = B_n$, we may write Eq. (15.25) to be

$$u(x, t) = (A_n\cos\omega_n t + B_n\sin\omega_n t)\sin\frac{n\pi}{L}x \tag{15.30}$$

where $\omega_n = 2\pi\nu_n$, ν_n being the normal frequencies of vibrations. For a given value of n, we may write

$$u(x, t) = A_n \sin\frac{n\pi x}{L} \cos\frac{n\pi v}{L}t + B_n \sin\frac{n\pi x}{L} \sin\frac{n\pi v}{L}t \qquad (15.31)$$

Equation (15.30) or (15.31) represents *normal mode of vibration* of the string in particular, the *n*th mode. The velocity of the normal mode can be obtained by differentiating Eq. (15.31):

$$\dot{u}(x, t) = \frac{d}{dt}u(x, t)$$

$$= A_n \sin\left(\frac{n\pi x}{L}\right)\left(-\frac{n\pi v}{L}\right)\sin\frac{n\pi v}{L}t + B_n \sin\left(\frac{n\pi x}{L}\right)\left(\frac{n\pi v}{L}\right)\cos\frac{n\pi v}{L}t \qquad (15.32)$$

We can now evaluate the constants A_n and B_n of the *n*th mode of vibration by using the initial conditions that, at $t = 0$,

$$u(x, 0) = u_0(x) \quad \text{and} \quad \dot{u}(x,0) = \dot{u}_0(x) \qquad (15.33)$$

Using these conditions in Eqs. (15.31) and (15.32), respectively, we obtain

$$u_0(x) = A_n \sin\frac{n\pi x}{L} \qquad (15.34)$$

$$\dot{u}_0(x) = \frac{n\pi v}{L} B_n \sin\frac{n\pi x}{L} \qquad (15.35)$$

We know from the theory of differential equations that if $u_1(x, t)$ and $u_2(x, t)$ are any two solutions that satisfy the boundary conditions given by Eq. (15.17), then $u(x)$, which is a linear combination of $u_1(x, t)$ and $u_2(x, t)$, that is

$$u(x, t) = u_1(x, t) + u_2(x, t) \qquad (15.36)$$

is also a solution. A more general solution is obtained by adding together all the n particular solutions using different constants A_n and B_n corresponding to different frequencies ω_n. Thus the general solution of motion of a vibrating string is a linear combination of a large number of normal modes [from Eq. (15.30)] and is given by

$$u(x, t) = \sum_{n=1}^{\infty}\left(A_n \sin\frac{n\pi x}{L}\cos\omega_n t + B_n \sin\frac{n\pi x}{L}\sin\omega_n t\right) \qquad (15.37)$$

where

$$\omega_n = \frac{n\pi v}{L}, \quad n = 1, 2, 3, \dots$$

which is a solution containing an infinite number of arbitrary constants. In initial conditions corresponding to different modes are given, that is, at $t = 0$

$$u(x, 0) = u_0(x) \quad \text{and} \quad \dot{u}(x, 0) = \dot{u}_0(x) \qquad (15.38)$$

then from Eq. (15.37) we obtain

$$u_0(x) = \sum_{n=1}^{\infty} A_n \sin \frac{n \pi x}{L} \tag{15.39}$$

$$\dot{u}_0(x) = \sum_{n=1}^{\infty} \frac{n \pi v}{L} B_n \sin \frac{n \pi x}{L} \tag{15.40}$$

Before we get involved in evaluating the constants, we show in Fig. 15.2 plots of $u(x, t)$ versus x for $n = 1, 2, 3, 4$. The mode of vibration in which $n = 1$ is called the *fundamental* or *first harmonic*. The mode of vibration for $n = 2$ is called the *first overtone* or *second harmonic*; similarly, $n = 3$ corresponds to the *second overtone* or *third harmonic*. The frequency of the nth harmonic is n times that of the fundamental frequency. In general, a string vibrates with several modes simultaneously.

The general solution given by Eq. (15.37) consisting of sums of sines and/or cosines is called a *Fourier series*. The general solution is completely known if the coefficients A_n and B_n are known. These coefficients can be evaluated if the initial conditions, that is, the values of $u_0(x)$ and $\dot{u}_0(x)$, are known. We use the Fourier technique to evaluate these constants. Multiply both sides of Eq.(15.39) by $\sin(m \pi x/L)$, where m is an integer; and integrate from $x = 0$ to $x = L$.

$$\int_0^L u_0(x) \sin \frac{m \pi x}{L} \, dx = \int_0^L \sum_{n=1}^{\infty} A_n \left(\sin \frac{n \pi x}{L} \right) \left(\sin \frac{m \pi x}{L} \right) dx \tag{15.41}$$

But all the terms on the right will vanish unless $m = n$. Thus integration yields

$$\int_0^L u_0(x) \sin \frac{n \pi x}{L} \, dx = A_n \int_0^L \sin^2 \frac{n \pi x}{L} \, dx = A_n \frac{L}{2}$$

or

$$A_n = \frac{2}{L} \int_0^L u_0(x) \sin \frac{n \pi x}{L} \, dx \tag{15.42}$$

Similarly, multiplying both sides of Eq. (15.40) by $\sin(m \pi x/L)$ and integrating from $x = 0$ to $x = L$, that is,

$$\int_0^L \dot{u}_0(x) \sin \frac{m \pi x}{L} \, dx = \int_0^L \sum_{n=1}^{\infty} \frac{n \pi v}{L} B_n \sin \frac{n \pi x}{L} \sin \frac{m \pi x}{L} \, dx$$

yields, as before,

$$B_n = \frac{2}{n \pi v} \int_0^L \dot{u}(x) \sin \frac{n \pi x}{L} \, dx \tag{15.43}$$

Thus Eqs. (15.42) and (15.43) state that, if displacements $u_0(x)$ and velocity $\dot{u}_0(x)$ are given for all points of the string at one time, A_n and B_n can be evaluated. Once these constants are known, the motion of the string at all subsequent times is known.

Figure 15.2 _____

Below some possible modes of vibration of a string are shown. In general, a string vibrates in a combination of several modes simultaneously.

$$N := 9 \qquad n := 1..N \qquad I := 100 \qquad i := 0..I \qquad x_i := i \qquad t_i := i \qquad L := 100 \qquad v := 1$$

N = number of modes

$$A_n := 10 \qquad B_n := 10 \qquad\qquad \omega_n := n \cdot \frac{\pi}{2} \cdot \frac{v}{L}$$

$$u_{i,n} := \left(A_n \cdot \sin\left(\frac{n \cdot \pi \cdot x_i}{L \cdot 2} \right) \cdot \cos\left(\omega_n \cdot t_i \right) + B_n \cdot \sin\left(\frac{n \cdot \pi \cdot x_i}{L \cdot 2} \right) \cdot \sin\left(\omega_n \cdot t_i + \frac{\pi}{2} \right) \right) \qquad ut_{i,n} := u_{i,1} + u_{i,2} + u_{i,3} + u_{i,4}$$

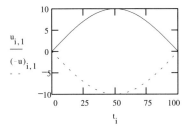

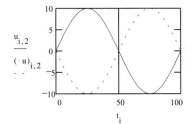

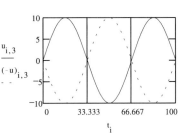

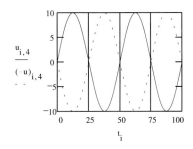

The first four graphs show four different modes, n = 1, 2, 3, and 4. The fifth graph shows the combination of all four modes excited simultaneously.

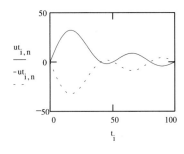

Example 15.1 _____

A string of length L of mass per unit length μ is fixed at two ends and is under tension T. The string is initially displaced a distance h ($h \ll L$) at the middle of the string and then released. Evaluate the Fourier coefficients for the subsequent motion of the string.

Solution

Figure Ex. 15.1(a) shows the initial configuration of the string. Hence the initial conditions are

$$\text{For } 0 < x < \frac{L}{2}, \quad \frac{u}{x} = \frac{h}{L/2} \quad \text{or} \quad u = \frac{2h}{L} x \tag{i}$$

$$\text{For } \frac{L}{2} < x < L, \quad \frac{u}{L - x} = \frac{h}{L/2} \quad \text{or} \quad u = \frac{2h}{L}(L - x) \tag{ii}$$

$$\text{At } t = 0, \quad \frac{du_0(x)}{dt} = \dot{u}_0(x) = 0 \tag{iii}$$

Substituting the value of $\dot{u}_0(x)$ from Eq. (iii) in Eq. (15.43) reveals that $B_n = 0$ for all n. The values of A_n can be determined by using the initial conditions given by Eqs. (i) and (ii). Substituting these in Eq.(15.42),

$$A_n = \frac{2}{L}\left[\frac{2h}{L}\int_0^{L/2} x \sin\frac{n\pi x}{L}\,dx + \frac{2h}{L}\int_{L/2}^{L}(L - x)\sin\frac{n\pi x}{L}\,dx\right] \tag{iv}$$

Evaluating integrals for different values of n, we obtain

$$A_n = 0, \qquad\qquad \text{if } n \text{ is even} \tag{v}$$

$$A_n = \frac{8h}{n^2\pi^2}\sin\frac{n\pi}{L}, \qquad \text{if } n \text{ is odd} \tag{vi}$$

Substituting these in Eq. (15.37), we obtain the general solution of the form

$$u(x, t) = \frac{8h}{\pi^2}\sin\frac{\pi x}{L}\cos\frac{n\upsilon t}{L} + \frac{1}{3^2}\sin\frac{3\pi x}{L}\cos\frac{3\pi\upsilon t}{L} + \frac{1}{5^2}\frac{8h}{\pi^2}\sin\frac{5\pi x}{L}\cos\frac{5\pi\upsilon t}{L} + \cdots \tag{vii}$$

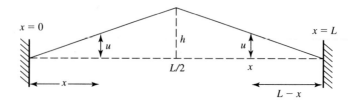

Figure Ex. 15.1(a)

Note that only the odd harmonics have been excited. The plots of the first three odd harmonics are shown in Fig. Ex. 15.1(b), and it is clear that none of the harmonics have been excited that would have a node at the midpoint. the modes shown and others are usually excited simultaneously.

The integration of expression of Eq. (iv) yields the value of A_n. Substituting different values of n, we can calculate the values of different A_n.

$$A_n = \frac{2}{L} \left[\frac{2 \cdot h}{L} \cdot \int_0^{\frac{L}{2}} x \cdot \sin\left(\frac{n \cdot \pi \cdot x}{L}\right) dx + \frac{2 \cdot h}{L} \cdot \int_{\frac{L}{2}}^{L} (L - x) \cdot \sin\left(\frac{n \cdot \pi \cdot x}{L}\right) dx \right]$$

$$A_n = 4 \cdot h \cdot \frac{\left(2 \cdot \sin\left(\frac{1}{2} \cdot n \cdot \pi\right) - \sin(n \cdot \pi)\right)}{\left(n^2 \cdot \pi^2\right)}$$

$$A_1 = 8 \cdot \frac{h}{\pi^2}$$

$n := 1..5 \quad i := 6..19 \quad t_i := i$

$L := 50 \quad h := 2 \quad Ae := 0$

$$A_n := 4 \cdot h \cdot \frac{\left(2 \cdot \sin\left(\frac{1}{2} \cdot n \cdot \pi\right) - \sin(n \cdot \pi)\right)}{\left(n^2 \cdot \pi^2\right)}$$

$$A_3 = \frac{8}{9} \cdot \frac{h}{\pi^2}$$

$v := 4 \cdot \pi \quad x_i := \frac{2 \cdot h}{L} \cdot (L - i)$

$$U1_i := \frac{8 \cdot h}{\pi^2} \cdot \sin\left(\frac{\pi \cdot x_i}{L}\right) \cdot \cos\left(\frac{v \cdot t_i}{L}\right)$$

$$A_5 = \frac{8}{25} \cdot \frac{h}{\pi^2}$$

n	A_n
1	1.621
2	0
3	-0.18
4	0
5	0.065

$$U3_i := \frac{8 \cdot h}{3^2 \cdot \pi^2} \cdot \sin\left(\frac{3 \cdot \pi \cdot x_i}{L}\right) \cdot \cos\left(\frac{3 \cdot v \cdot t_i}{L}\right)$$

$$U5_i := \frac{8 \cdot h}{(5 \cdot \pi)^2} \cdot \sin\left(\frac{5 \cdot \pi \cdot x_i}{L}\right) \cdot \cos\left(5 \cdot v \cdot \frac{t_i}{L}\right)$$

$\max(U1) = 0.305$

$\max(U3) = 0.091$

$\max(U5) = 0.046$

Explain why there are only odd numbers of segments.

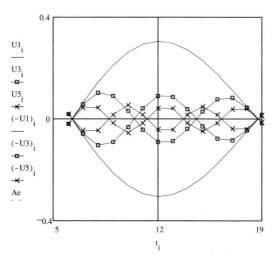

Fig. Ex 15.1(b)

15.3 WAVE PROPAGATION IN GENERAL

Wave motion is not limited to the vibration of strings. It is a phenomenon that occurs in many different branches of physics and involves such cases as sound waves, waves on a liquid surface, and electromagnetic waves. With the advent of quantum mechanics, in which we deal with such abstract ideas as probability waves, the study of wave motion has assumed a much more important and fundamental role. One may be tempted to say that wave motion deals with those phenomena that exhibit periodicity or oscillations. But this is not always true in general. For example, a pulse that travels on a rope or a tidal wave does not exhibit periodicity.

A better definition of wave motion is discussed in terms of energy transport; when a wave reaches a portion of a medium, it sets the particles of the medium into motion. After the wave has passed the particles come to rest, while neighboring particles are set in motion. From this we may conclude that one of the common characteristics of all wave motion is the following:

Wave motion provides a mechanism for transfer of energy from one point to another without physical transfer of any material between the points.

It may be pointed out that wave motions in solids, liquids, and gases do need a medium to transfer energy, while electromagnetic waves can transport energy without requiring a medium to carry them. Thus it is essential that we adopt a more basic viewpoint of wave motion (a kinematical viewpoint instead of the dynamical one stated previously).

Let us discuss the propagation of a single pulse in one dimension. Consider a stretched rope that has been shaken at one end, resulting in a pulse traveling along its length and taking the form shown in Fig. 15.3. This pulse, wave, or disturbance travels along the rope, say along the X-axis, without distortion in form; that is, it has the same shape at t_1 as at any other later time, t_2. We have assumed an ideal case in which the form of the pulse does not change. In actual practice, because of damping, there will be some change in form. The pulse form travels with a constant velocity. The same remarks can be made about any wave disturbance or wave

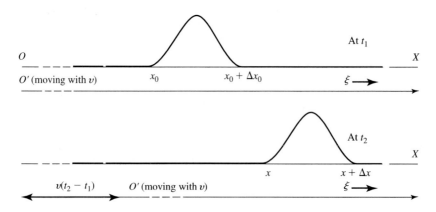

Figure 15.3   Pulse in a rope traveling to the right and viewed by an observer moving with velocity v along an axis parallel to the rope.

motion. Thus we may define it as follows:

> ***Wave motion*** *is a disturbance that propagates itself with constant velocity without chang-*
> *ing its form or pattern.*

Suppose a pulse or disturbance is traveling along the $+X$-axis with constant velocity v. Now we view this pulse from the ξ-axis, which is moving with velocity v along and parallel to the X-axis. Furthermore, if the origins of the X-axis and ξ-axis coincide at $t = 0$, we may write

$$\xi = x - vt \tag{15.44}$$

Thus, to an observer in the ξ system, the form and position of the disturbance remains unchanged; that is, the disturbance has such a time dependence that it is a function of ξ alone. Thus the wave propagating to the right is

$$u(x, t) = f(\xi) \equiv f(x - vt) \tag{15.45}$$

where $f(\xi)$ is a completely arbitrary function. Equation (15.45) guarantees that it is a wave traveling to the right. Thus, as t increases, x must increase so that ξ remains constant; hence $f(\xi)$ represents a wave traveling to the right. Similarly, we define

$$\eta = x + vt \tag{15.46}$$

and a wave propagating to the left is given by

$$u(x, t) = g(\eta) \equiv g(x + vt) \tag{15.47}$$

where $g(\eta)$ is another arbitrary function. Once again, as t increases, x must decrease so that η is constant, and then $g(\eta)$ represents a wave propagating to the left. f and g given by Eqs. (15.45) and (15.47) are referred to as *wave forms* and represent the most general type of one-dimensional motion.

By direct substitution of f and g from Eqs. (15.45) and (15.47) into Eq. (15.12), we can show that these satisfy the wave equation.

The general expression for u is a combination of two functions, one of which depends only on ξ and the other only on η; that is, the sum of the two linear functions of ξ and η [individual functions $f(\xi)$ and $g(\eta)$ are also solutions as long as they are linear] is

$$u(x, t) = f(\xi) + g(\eta) = f(x - vt) + g(x + vt) \tag{15.48}$$

Thus the most general solution of the wave equation, Eq. (15.12), is given by Eq. (15.48) or any other *linear* combination of $f(\xi)$ and $g(\eta)$. That Eq. (15.48) is a general solution is consistent with the fact *that the general solution of a second-order differential equation contains two arbitrary functions.*

Let us now proceed to evaluate these functions using initial conditions; that is, at $t = 0$,

$$u = u_0(x) \quad \text{and} \quad \dot{u} = \dot{u}_0(x) \tag{15.49}$$

gives
$$u(x, 0) = f(x) + g(x) = u_0(x) \tag{15.50}$$

and
$$\left[\frac{\partial u}{\partial t} \right]_{t=0} = \left[-v \frac{df}{d\xi} + v \frac{dg}{d\eta} \right]_{t=0} \dot{u}_0(x) \tag{15.51}$$

At $t = 0$, $\xi = \eta = x$, and Eq. (15.51) takes the form

$$v \frac{d}{dx}[-f(x) + g(x)] = \dot{u}_0(x) \tag{15.52}$$

which on integration gives

$$-f(x) + g(x) = \frac{1}{v} \int_0^x \dot{u}_0(x)\, dx + C \tag{15.53}$$

where C is a constant of integration. Adding and subtracting Eqs. (15.50) and (15.53), we obtain

$$f(x) = \frac{1}{2}\left[u_0(x) - \frac{1}{v} \int_0^x \dot{u}_0(x)\, dx - C \right] \tag{15.54}$$

$$g(x) = \frac{1}{2}\left[u_0(x) + \frac{1}{v} \int_0^x \dot{u}_0(x)\, dx + C \right] \tag{15.55}$$

Since these solutions hold for any value of x, we may replace x by ξ or η. Also, the constant C may be dropped because it may be eliminated in linear combinations of solutions. Thus

$$f(\xi) = \frac{1}{2}\left[u_0(\xi) - \frac{1}{v} \int_0^\xi \dot{u}_0(\xi)\, d\xi \right] \tag{15.56}$$

$$g(\eta) = \frac{1}{2}\left[u_0(\eta) + \frac{1}{v} \int_0^\eta \dot{u}_0(\eta)\, d\eta \right] \tag{15.57}$$

Our next step is to see the connection between the general solution obtained in this section and those in the previous section concerning vibrations of strings. The partial differential equation Eq. (15.12) was separated into two differential equations, Eqs. (15.21) and (15.22); that is,

$$\frac{d^2X}{dx^2} + \frac{\omega^2}{v^2} X = 0 \tag{15.21}$$

$$\frac{d^2\Theta}{dt^2} + \omega^2\Theta = 0 \tag{15.22}$$

Instead of writing the solutions in the form of sines and cosines as given by Eqs. (15.23) and (15.24), we may write the solutions of these equations in the following alternative form:

$$X(x) = Ce^{i(\omega/v)x} + De^{-i(\omega/v)x} \tag{15.58}$$

$$\Theta(t) = Ee^{i\omega t} + Fe^{-i\omega t} \tag{15.59}$$

where C, D, E, and F are the constants to be determined from the boundary conditions. Thus the general solution will be of the form

$$u(x, t) = X(x)\Theta(t) = Ae^{\pm i(\omega/v)x}\, e^{\pm i\omega t}$$

$$= Ae^{\pm i(\omega/v)(x \pm vt)} \tag{15.60}$$

where A is a constant. This states that the general solution $u(x, t)$ is a linear combination of the following terms:

$$\pm i(\omega/v)(x \pm vt) \tag{15.61}$$

Note that these solutions already contain the quantities that are functions of $x + vt$ and $x - vt$. By taking the real part or by adding the complex conjugate and dividing by 2, we obtain the solutions

$$u(x, t) = A \cos \frac{\omega}{v}(x - vt) \tag{15.62}$$

$$u(x, t) = A \cos \frac{\omega}{v}(x + vt) \tag{15.63}$$

and by adding the imaginary parts or subtracting the complex conjugate and dividing by $2i$, we obtain

$$u(x, t) = A \sin \frac{\omega}{v}(x - vt) \tag{15.64}$$

$$u(x, t) = A \sin \frac{\omega}{v}(x + vt) \tag{15.65}$$

The solutions containing $x - vt$ represent waves traveling to the right. While those containing $x + vt$ represent waves traveling to the left. These solutions do not satisfy the boundary conditions because they represent *traveling waves* down the string or medium.

Furthermore, these equations are not satisfied by only one particular value of $-\omega^2$; many more are possible. Thus the general solution is not only a linear combination of harmonic terms given by Eq. (15.60), but also must be summed over all possible frequencies. Thus the most general solution is

$$u(x, t) = \sum_n A_n e^{\pm i(\omega_n/v)(x \pm vt)} \tag{15.66}$$

Once the boundary conditions are known, the constants can be evaluated in a manner similar to the case of evaluation of coefficients in infinite Fourier series. For our discussion, we shall write the solution in the following form, it being understood that the complete solution is summed over all frequencies:

$$u(x, t) = A e^{i(\omega/v)(x - vt)} \tag{15.67}$$

The quantity k, called the *propagation constant* or *angular wave number* or simply *wave number* (number of waves per unit length) has dimensions of reciprocal length and is defined as

$$k^2 \equiv \frac{\omega^2}{v^2} \quad \text{or} \quad |k| = \frac{\omega}{v} \tag{15.68}$$

Thus the wave equation for X, Eq. (15.21) and its general solution Eq. (15.67) take the forms

$$\frac{d^2X}{dx^2} + k^2X = 0 \tag{15.69}$$

and
$$u(x, t) = Ae^{ik(x-vt)} = Ae^{i(kx-\omega t)} \tag{15.70}$$

If ν is the frequency of vibration so that $\omega = 2\pi\nu$, then the *wavelength* λ is defined as the distance for one complete vibration of the wave

$$\lambda = \frac{v}{\nu} = \frac{2\pi v}{2\pi\nu} = \frac{2\pi v}{\omega} \tag{15.71}$$

Combining this with Eq. (15.68), we get

$$k = \frac{2\pi}{\lambda} \tag{15.72}$$

Let us see what happens if we superimpose two waves, both of the same frequency and amplitude, but one traveling to the right and the other to the left. Thus

$$u = u_1 + u_2 = Ae^{i(kx-\omega t)} + Ae^{i(kx+\omega t)} \tag{15.73}$$

$$u = 2Ae^{-i\omega t} \cos kx \tag{15.74}$$

The real part of this equation yields

$$u(x, t) = 2A \cos kx \cos \omega t \tag{15.75}$$

This wave has the property that it does not propagate forward with time. This superposition of waves leads to the formation of *standing waves*. There are certain points where there is no motion at all because of the cancellation of one wave by the other. Such points are called *nodes*. Since at nodes no motion is possible, no energy is transmitted from one side to the other; hence the pattern is named standing waves. From Eq. (15.75), we can obtain the condition for the position of the nodes to be

$$x = (2n + 1)\frac{\lambda}{4} = (2n + 1)\frac{\pi}{2k} \tag{15.76}$$

Before concluding this section, let us talk about phase velocity and dispersion. To start, let us say that we have a wave motion of a single wave (or frequency) given by Eq. (15.70):

$$u(x, t) = Ae^{i(kx-\omega t)} \tag{15.70}$$

The quantity $kx - \omega t$ is defined as the *phase* ϕ of the wave represented by $u(x, t)$; that is,

$$\phi \equiv kx - \omega t \tag{15.77}$$

The wave pattern or form will remain unchanged in time if ϕ remains constant. For ϕ to remain constant, we must have

$$d\phi = 0 \quad \text{or} \quad k\,dx - \omega\,dt = 0 \tag{15.78}$$

That is, we define the *phase velocity* v_p to be the velocity with which the wave pattern travels; it is given by

$$v_p = \frac{dx}{dt} = \frac{\omega}{k} = v \tag{15.79}$$

That is, for a simple wave the phase velocity v_p is equal to the wave velocity v. This is not true in general. The phase velocity is usually a function of frequency; that is, in a given medium the phase velocity is frequency dependent, $v_p = v_p(k)$. Such a medium is called a *dispersive medium*. In a dispersive medium the phase velocity is not equal to the wave velocity. (As an example, for electromagnetic waves in a given refractive medium, the velocity of the waves is a function of the wavelength.) Thus, in such cases, the wave pattern is modified; it does not remain constant. But even such a pattern will appear unchanged to an observer who is moving with a velocity v_g given by (ω being a function of k)

$$v_g = \frac{d\omega(k)}{dk} \tag{15.80}$$

where v_g is called the *group velocity*.

15.4 LAGRANGE FORMULATION OF A VIBRATING STRING: ENERGY AND POWER

If we calculate the kinetic energy and potential energy of a vibrating string, we can set up the Lagrangian L and the Lagrange equations; hence can calculate the normal modes of a vibrating string. Furthermore, we know the total energy stored in the string and also the rate at which the energy is being transferred from one portion of the string to the other.

Let us reconsider the vibrating string shown in Fig. 15.1, which has length L and is fixed at both ends. As shown in Fig. 15.1(b), the element of length dx when in equilibrium is stretched to length ds when vibrating. The tension in the string is T when it is vibrating. Thus the amount of potential energy stored in this vibrating element of the string is, assuming the potential energy to be zero when the string is unstretched,

$$dV = T(ds - dx) = T\left(\frac{ds}{dx} - 1\right)dx \tag{15.81}$$

where
$$\frac{ds}{dx} = \left[1 + \left(\frac{\partial u}{\partial x}\right)^2\right]^{1/2} \tag{15.82}$$

Substituting this in Eq. (15.81), assuming $\partial u/\partial x \ll 1$, and using the binomial theorem for expansion, we obtain

$$dV \simeq \frac{T}{2}\left(\frac{\partial u}{\partial x}\right)^2 dx \qquad (15.83)$$

Thus the total potential energy stored in the string may be obtained by integrating Eq. (15.83); that is,

$$V = \frac{T}{2}\int_0^L \left(\frac{\partial u}{\partial x}\right)^2 dx \qquad (15.84)$$

The mass of an element of length dx is $\mu\, dx$; hence its kinetic energy is (in order to avoid confusion we will start using K for kinetic energy instead of T, which we are using for tension)

$$dK = \frac{1}{2}\,\mu\, dx\left(\frac{\partial u}{\partial t}\right)^2 \qquad (15.85)$$

while the total kinetic energy of the string is obtained by integrating Eq. (15.85):

$$K = \frac{\mu}{2}\int_0^L \left(\frac{\partial u}{\partial t}\right)^2 dx \qquad (15.86)$$

To evaluate V and K, we make use of the solution given by Eq. (15.30),

$$u(x, t) = \Theta_n(t)\sin\frac{\omega_n x}{v} \qquad (15.87)$$

where

$$\Theta_n(t) = A_n \cos \omega_n t + B_n \sin \omega_n t \qquad (15.88)$$

and we have used the relation given in Eq. (15.29),

$$\omega_n = \frac{n\pi v}{L} = \frac{n\pi}{L}\sqrt{\frac{T}{\mu}} \qquad (15.29)$$

Thus, from Eqs. (15.87) and (15.29), we obtain (for all solutions)

$$\frac{\partial u}{\partial x} = \frac{\pi}{L}\sum_{n=1}^{\infty} n\Theta_n \cos\frac{n\pi x}{L} \qquad (15.89)$$

$$\frac{\partial u}{\partial t} = \sum_{n=1}^{\infty} \dot{\Theta}_n \sin\frac{n\pi x}{L} \qquad (15.90)$$

Substituting Eq. (15.89) into Eq. (15.84), we get

$$V = \frac{\pi^2 T}{2L^2}\sum_{n=1}^{\infty}\sum_{m=1}^{\infty}\left(nm\dot{\Theta}_n\Theta_m\int_0^L \cos\frac{n\pi x}{L}\cos\frac{m\pi x}{L}\,dx\right) \qquad (15.91)$$

On integrating, we find that only those terms are nonzero for which $m = n$, and each of these terms on integration yields $L/2$. Thus

$$V = \frac{\pi^2 T}{4L} \sum_{n=1}^{\infty} n^2 \Theta_n^2, \quad n = 1, 2, 3, \ldots \tag{15.92}$$

Similarly, substituting Eq. (15.90) into Eq. (15.86), we obtain

$$K = \frac{\mu}{2} \sum_{n=1}^{\infty} \sum_{m=1}^{\infty} \left(\dot{\Theta}_n \dot{\Theta}_m \int_0^L \sin \frac{n\pi x}{L} \sin \frac{m\pi x}{L} dx \right) \tag{15.93}$$

Once again, on integrating we find that only those terms are nonzero for which $m = n$, and each of those terms on integration yields

$$K = \frac{\mu L}{4} \sum_{n=1}^{\infty} \dot{\theta}_n^2 \tag{15.94}$$

while the Lagrangian of the system may be written as

$$L = K - V = \tfrac{1}{4} \sum_{n=1}^{\infty} \left(\mu L \dot{\Theta}_n^2 - \frac{\pi^2 T}{L} n^2 \theta_n^2 \right) \tag{15.95}$$

Note that the potential energy is the sum of quantities of the form $A_n \theta_n^2$, and the kinetic energy has terms of the form $B_n \dot{\theta}_n^2$. The Lagrangian equations

$$\frac{d}{dt} \left(\frac{\partial L}{\partial \dot{\Theta}_n} \right) - \frac{\partial L}{\partial \Theta_n} = 0 \tag{15.96}$$

take the form

$$\ddot{\Theta}_n + \frac{\pi^2 T}{\mu L^2} n^2 \Theta_n = 0 \tag{15.97}$$

where Θ_n is the dependent variable and t is the independent variable. The solutions of these yield the normal coordinates Θ_n. Since n varies from 1 to ∞, the number of normal coordinates for a vibrating string is infinite.

It is now a simple matter to write the total energy E by using Eqs. (15.92) and (15.94):

$$E = K + V = \frac{1}{4} \sum_{n=1}^{\infty} \left(\mu L \dot{\Theta}_n^2 + \frac{\pi^2 T}{L} n^2 \theta_n^2 \right) \tag{15.98}$$

Since $\mu L = M$ (the mass of the string), and from Eq. (15.29),

$$T = \frac{\mu L^2 \omega_n^2}{n^2 \pi^2} = \frac{ML}{n^2 \pi^2} \omega_n^2 \tag{15.99}$$

and, using Eq. (15.88), we may write Eq. (15.98) in the form

$$E = \frac{M}{4} \sum_{n=1}^{\infty} [\omega_n^2 (A_n^2 + B_n^2)] \tag{15.100}$$

where A_n and B_n are constants (see Problem 15.16).

    Finally, let us calculate the rate of flow of energy, that is, power P, delivered from the left to right across any point x along the string. To calculate power, we make use of the definition that $P = F\dot{u}$, where F is the magnitude of the driving force **F**. F is equal in magnitude to tension T and must be applied in a direction tangent to the string. Thus the component of **F** in the direction of transverse displacement at point x is

$$F_y = -T \sin\theta \simeq -T \tan\theta = -T \frac{\partial u}{\partial x} \tag{15.101}$$

while the component of velocity $\dot{u}$ at the point x is $\partial u/\partial t$. Therefore,

$$P = F_y \dot{u} = \left(-T \frac{\partial u}{\partial x}\right)\left(\frac{\partial u}{\partial t}\right) \tag{15.102}$$

The value of P can be evaluated by using the values of $\partial u/\partial x$ and $\partial u/\partial t$ given by Eqs. (15.89) and (15.90), respectively.

    Let us calculate P for a particular case. Consider a wave traveling to the right and given by

$$u = f(x - vt) = f(\xi) \tag{15.103}$$

Suppose f is a sinusoidal function of the form

$$u = f(\xi) = A \cos(kx - \omega t) \tag{15.104}$$

Evaluating $\partial u/\partial x$ and $\partial u/\partial t$ and substituting in Eq. (15.102) yields

$$P = k\omega T A^2 \sin^2(kx - \omega t) \tag{15.105}$$

Since the average value of $\sin^2(kx - \omega t)$ is $\frac{1}{2}$, the average power P transmitted from left to right will be

$$\langle P \rangle = \tfrac{1}{2} k\omega T A^2 \tag{15.106}$$

15.5 SYSTEM OF PARTICLES: THE LOADED STRING

In previous discussions we considered an idealized string that is characterized by its linear mass density μ. Actually, a string is made up of a finite number of particles. We can view the situation as a number of identical particles each of mass m placed at regular intervals on an elastic string, as shown in Fig. 15.4(a). There are N particles where the equilibrium distance between adjacent particles is d, and the attractive force between adjacent particles is T. Thus the length

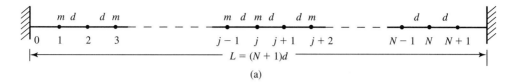

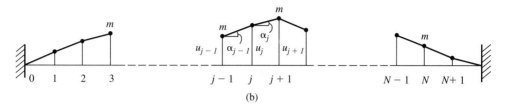

Figure 15.4   (a) A large number of identical particles each of mass m placed at regular intervals constitutes an elastic string. (b) Transverse displacements of point masses.

of the string, as shown, is $L = (N + 1)d$. Such a string is tied at both ends and is horizontal when in equilibrium. We are interested in investigating a small transverse displacement and hence the oscillations of the particles about equilibrium positions.

Consider small vertical displacements of particles $j - 1, j$, and $j + 1$, each of mass m, and vertical displacements u_{j-1}, u_j, and u_{j+1}. Assuming the displacements to be small means that the angles α_j are small, and the slopes are small; hence we may replace $\sin \alpha_j$ by $\tan \alpha_j$. For a small displacement, the resultant X component of the force on the jth particle is

$$-T \cos \alpha_{j-1} + T \cos \alpha_j \simeq \tfrac{1}{2} T(\alpha_{j-1}^2 - \alpha_j^2) \simeq 0$$

The resultant Y component of force on the jth particle for small displacements may be written as

$$F_y = -T \sin \alpha_{j-1} + T \sin \alpha_j \simeq -T \tan \alpha_{j-1} + T \tan \alpha_j$$

$$= -T \frac{u_j - u_{j-1}}{d} + T \frac{u_{j+1} - u_j}{d} \tag{15.107}$$

Since
$$F_j = m\ddot{u}_j = m \frac{d^2 u_j}{dt^2} \tag{15.108}$$

the equation of motion of the jth particle is $(F_j = F_y)$

$$m \frac{d^2 u_j}{dt^2} = T \left(\frac{u_{j+1} - u_j}{d} - \frac{u_j - u_{j-1}}{d} \right) \tag{15.109}$$

If the number of particles is taken to be very large, we may then assume the string to be smooth and write

$$u(jd, t) = u_j(t) = u(x, t)$$

while the linear mass density is $\mu = m/d$. Thus Eq. (15.109) takes the form

$$\frac{d^2 u_j}{dt^2} = \frac{T}{\mu d}\left(\frac{u_{j+1} - u_j}{d} - \frac{u_j - u_{j-1}}{d}\right) \tag{15.110}$$

Before proceeding to solve Eq. (15.110), we shall show that it represents a wave equation. For the right side of Eq. (15.110),

$$\frac{1}{d}\left(\frac{u_{j+1} - u_j}{d} - \frac{u_j - u_{j-1}}{d}\right) = \frac{1}{d}\left[\left(\frac{\partial u}{\partial x}\right)_{(j+1/2)d} - \left(\frac{\partial u}{\partial x}\right)_{(j-1/2)d}\right] = \left(\frac{\partial^2 u}{\partial x^2}\right)_{jd} \tag{15.111}$$

Substituting in Eq. (15.110), we obtain the familiar wave equation

$$\frac{\partial^2 u}{\partial x^2} - \frac{1}{v^2}\frac{\partial^2 u}{\partial t^2} = 0 \tag{15.12}$$

Let us look at Eq. (15.110), which describes the motion of the jth particle, and try to find a possible solution. Let

$$\frac{T}{md} = \omega_0^2 \tag{15.112}$$

and write Eq. (15.110) as the general wave equation:

$$\ddot{u}_j + 2\omega_0^2 u_j - \omega_0^2(u_{j+1} + u_{j-1}) = 0 \tag{15.113}$$

Since these are N particles, we can write a set of N differential equations, each being similar to Eq. (15.113). Note that we have assumed that $u_0 = 0$ and $u_{j+1} = 0$.

Before solving the general equation, Eq. (15.113), we shall first consider some simple cases. Suppose there is only one particle; that is, $N = 1$. Then Eq. (15.113) takes the form

$$\frac{d^2 u_1}{dt^2} + 2\omega_0^2 u_1 = 0 \tag{15.114}$$

which represents transverse harmonic motion of a single particle oscillating with an angular frequency of $\sqrt{2}\omega_0$. This situation is shown in Fig. 15.5(a). If we had two particles, that is, $N = 2$. Eq. (15.113) would yield

$$\frac{d^2 u_1}{dt^2} + 2\omega_0^2 u_1 - \omega_0^2 u_2 = 0 \tag{15.115}$$

$$\frac{d^2 u_2}{dt^2} + 2\omega_0^2 u_2 - \omega_0^2 u_1 = 0 \tag{15.116}$$

These are coupled equations, similar to those for two coupled oscillators or pendula, having the same natural frequency ω_0. Thus there are two normal modes for $N = 2$. The lower mode has an angular frequency $\omega = \omega_0$, and the higher mode has an angular frequency $\omega = \sqrt{3}\omega_0$, as shown in Fig. 15.5(b)(i) and (ii), respectively.

Let us go back to Eq. (15.113) and try to find the normal modes of oscillation for N particles. Basically, we apply the same technique as used for two particles. For each normal mode,

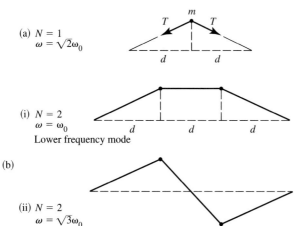

(a) $N = 1$
$\omega = \sqrt{2}\omega_0$

(i) $N = 2$
$\omega = \omega_0$
Lower frequency mode

(b)

(ii) $N = 2$
$\omega = \sqrt{3}\omega_0$
Higher frequency mode

Figure 15.5   Normal modes of vibration for (a) $N = 1$, and (b) $N = 2$ particles.

we seek a sinusoidal solution so that each particle oscillates with the same frequency. Let such a solution be

$$u_j(t) = A_j \cos \omega t, \qquad j = 1, 2, 3, \ldots, N \qquad \textbf{(15.117)}$$

where A_j and ω are the amplitude and frequency of the jth particle. We could have equally started with a solution of the form

$$u_j(t) = A_j e^{i\omega t} \qquad \textbf{(15.118)}$$

(see Problem 15.23). Thus, if we know A_j and ω, a set of differential equations for Eq. (15.113) can be solved. Furthermore, a solution of the type given by Eq. (15.117) assumes that each particle has zero velocity at $t = 0$. This is obvious if we differentiate Eq. (15.117), which gives

$$\frac{du_j}{dt} = -\omega A_j \sin \omega t, \qquad j = 1, 2, 3, \ldots, N \qquad \textbf{(15.119)}$$

Thus, if $t = 0$, $\dot{u}_j = 0$. Substituting the trial solution, Eq. (15.117), in Eq. (15.113), we obtain

$$(-\omega^2 + 2\omega_0^2)A_j - \omega_0^2(A_{j-1} + A_{j+1}) = 0, \qquad j = 1, 2, 3, \ldots, N \qquad \textbf{(15.120)}$$

which is equivalent to the following set of equations:

$$
\begin{aligned}
(-\omega^2 + 2\omega_0^2)A_1 - \omega_0^2(A_0 + A_2) &= 0 \\
(-\omega^2 + 2\omega_0^2)A_2 - \omega_0^2(A_1 + A_3) &= 0 \\
\vdots \qquad\qquad \vdots & \\
(-\omega^2 + 2\omega_0^2)A_j - \omega_0^2(A_{j-1} + A_{j+1}) &= 0 \\
\vdots \qquad\qquad \vdots & \\
(-\omega^2 + 2\omega_0^2)A_N - \omega_0^2(A_{N-1} + A_{N+1}) &= 0
\end{aligned}
\qquad \textbf{(15.121)}
$$

To have a nontrivial solution, the determinant of the coefficients in Eqs. (15.121) must be zero. That is,

$$
\begin{vmatrix}
(-\omega^2 + 2\omega_0^2) & -\omega_0^2 & 0 & 0 & 0 & \cdots \\
-\omega_0^2 & (-\omega^2 + 2\omega_0^2) & -\omega_0^2 & 0 & 0 & \cdots \\
0 & -\omega_0^2 & (-\omega^2 + 2\omega_0^2) & -\omega_0^2 & 0 & \cdots \\
0 & 0 & -\omega_0^2 & (-\omega^2 + 2\omega_0^2) & -\omega_0^2 & \cdots \\
\vdots & \vdots & \vdots & \vdots & \vdots & \cdots
\end{vmatrix} = 0
$$

(15.122)

For $N = 1$, we get

$$\left| -\omega^2 + 2\omega^2 0 \right| = 0 \;\rightarrow\; \omega = \sqrt{2}\omega_0$$

For $N = 2$, we get

$$
\begin{vmatrix}
(-\omega^2 + 2\omega_0^2) & -\omega_0^2 \\
-\omega_0^2 & (-\omega^2 + 2\omega_0^2)
\end{vmatrix} = 0
$$

which gives the frequencies of the two normal modes to be

$$\omega = \omega_0 \quad \text{or} \quad \omega = \sqrt{3}\omega_0$$

These are the results we predicted. This method is simple enough for calculating the frequencies of normal modes as long as we are dealing with a small number of particles. For a very large number of particles, this method is cumbersome. The following alternative approach is desirable.

Let us refer back to Eq. (15.120). The requirement that both ends are fixed leads to the boundary conditions

$$A_0 = 0 \quad \text{and} \quad A_{N+1} = 0 \tag{15.123}$$

The existence of normal modes, that for each mode all particles vibrate with the same frequency, imposes certain restrictions on the ratios of the amplitudes. Equation (15.120) may be written as

$$\frac{A_{j-1} + A_{j+1}}{A_j} = \frac{-\omega^2 + 2\omega_0^2}{\omega_0^2}, \qquad j = 1, 2, 3, \ldots, N \tag{15.124}$$

Since for a given mode ω^2 is constant, the right side is constant. Thus, if A_{j-1} and A_j are given, A_{j+1} can be evaluated. For example, if $A_0 = 0$ and A_1 is given, A_2 can be calculated.

Furthermore, Eq. (15.124) implies that since the right side is constant the left side must be constant. Let us not forget that we want to get the value of ω^2. A neat method of doing this is to assume the following form of a solution for A_j:

$$A_j = C \sin j\theta \tag{15.125}$$

where θ is some angle. With similar expressions for A_{j-1}, we may write

$$A_{j-1} + A_{j+1} = C \sin(j-1)\theta + C \sin(j+1)\theta = 2C \sin j\theta \cos \theta = 2A_j \cos \theta \tag{15.126}$$

which may be rewritten as

$$\frac{A_{j-1} + A_{j+1}}{A_j} = 2\cos\theta \qquad (15.127)$$

The right side is independent of j. Thus, if we could evaluate θ, we would have the value of the constant needed for $2\cos\theta$, which we can substitute in Eq. (15.124) and thus evaluate ω. To do this, we make use of the boundary conditions that $A_j = 0$ for $j = 0$ and $j = N + 1$. From Eq. (15.125), we see that if $j = 0$ and for $j = N + 1$, A_j will be zero only if $(N + 1)\theta$ is an integer multiple of π: that is,

$$(N + 1)\theta = n\pi, \qquad n = 1, 2, 3, \ldots \qquad (15.128\text{a})$$

or

$$\theta = \frac{n\pi}{N + 1} \qquad (15.128\text{b})$$

Substituting this in Eq. (15.125),

$$A_j = C\sin\left(\frac{nj\pi}{N + 1}\right) \qquad (15.129)$$

Using Eqs. (15.124), (15.127), and (15.128), we get the frequencies of the possible normal modes:

$$\frac{A_{j-1} + A_{j+1}}{A_j} = \frac{-\omega^2 + 2\omega_0^2}{\omega_0^2} = 2\cos\left(\frac{n\pi}{N + 1}\right) \qquad (15.130)$$

Therefore, the relation for the frequencies of the normal mode (independent of j) is obtained by solving for ω^2:

$$\omega^2 = 2\omega_0^2\left[1 - \cos\left(\frac{n\pi}{N + 1}\right)\right] = 4\omega_0^2\sin^2\left(\frac{n\pi}{2(N + 1)}\right) \qquad (15.131)$$

Taking the square root, we have the required frequencies:

$$\omega = 2\omega_0\sin\left(\frac{n\pi}{2(N + 1)}\right) \qquad (15.132)$$

From Eq. (15.112), substituting the value of ω_0, and since different values of n correspond to a different normal mode with the corresponding frequency, we may replace ω by ω_n and write the *normal mode frequencies* as

$$\omega_n = 2\sqrt{\frac{T}{md}}\sin\left(\frac{n\pi}{2(N + 1)}\right) \qquad (15.133)$$

The same type of procedure can be carried out for the longitudinal oscillations where T/d is replaced by k, the spring constant. After replacing T/d by k. Eq. (15.133) yields the frequencies of the two normal modes of two coupled oscillators after substituting $N = 2$ and $n = 1, 2$.

For all practical purposes, we have solved the problem of N coupled oscillators. We must look closely at the motion these equations describe so that we can obtain a physical interpretation of the situation. To describe the displacement of the jth particle when a collection of N particles is oscillating in the nth mode, Eq. (15.117) must be written in the following form:

$$u_{jn} = A_{jn}\cos\omega_n t \qquad (15.134)$$

where A_{jn} is obtained by substituting Eq. (15.128b) in Eq. (15.125) and replacing A_j by A_{jn} and C by C_n, we obtain

$$A_{jn} = C_n \sin\left(\frac{jn\pi}{N+1}\right) \tag{15.135}$$

A_{jn} represents the amplitude of the jth particle in the nth mode of the system, ω_n represents the frequency of the nth mode and is given by Eqs. (15.135) or (15.133); that is,

$$\omega_n = 2\omega_0 \sin\left(\frac{n\pi}{2(N+1)}\right) \tag{15.136}$$

The solution given in Eq. (15.134) assumes that at time $t = 0$ the particle is at rest. But this difficulty can be overcome and any arbitrary initial conditions can be satisfied by adding a phase factor ϕ_n; that is,

$$u_{jn} = A_{jn} \cos(\omega_n t - \phi_n) \tag{15.137}$$

First, we would like to know the number of possible normal modes. We shall now show that for N oscillators there are only N independent modes; that is, $n = N$ and the corresponding amplitudes and frequencies are A_{jn} and ω_n. For modes beyond $n = N$, that is, for $n = N + 1$, $N + 2, \ldots$, and so on, the preceding equations do not lead to new physical situations. We shall show that for values greater than $n = N$ the amplitudes and frequencies of the normal modes repeat themselves.

Figure 15.6 shows a plot of mode frequency ω_n (always taken to be positive) versus mode number n [for convenience written as $n\pi/2(N + 1)$ instead of n]. If we substitute $n = 0$ or $N + 1$ in Eq. (15.135), the amplitude factors A_{jn} turn out to be zero. These values of n are called *null modes*. For $n = 1$ to $n = N$, we have N different characteristic frequencies as discussed, reaching a maximum value of $\omega_{max} = 2\omega_0$, from Eq. (15.136), for $n = N + 1$, because $\sin(\pi/2)$ becomes 1. But for this maximum value of the characteristic frequency, the corresponding amplitude for $n = N + 1$ from Eq. (15.135) is zero. Let us calculate the characteristic frequency for the mode $n = N + 2$. From Eq. (15.136), we get

$$\omega_{N+2} = 2\omega_0 \sin\left[\frac{(N+2)\pi}{2(N+1)}\right] = 2\omega_0 \sin\left[\pi - \frac{N\pi}{2(N+1)}\right] = 2\omega_0 \sin\left[\frac{N\pi}{2(N+1)}\right] = \omega_N$$

That is,
$$\omega_{N+2} = \omega_N$$
Similarly,
$$\omega_{N+3} = \omega_{N-1}$$
$$\omega_{N+4} = \omega_{N-2}$$
$$\vdots \qquad \vdots \tag{15.138}$$

Thus there are only $n = N$ number of independent modes; for any further values of n, the modes repeat themselves.

The same is true for the amplitudes as well; that is, the relative amplitudes of the particles in a normal mode repeat themselves. That is, from Eq. (15.135)

$$A_{j(N+2)} = A_{jN}$$
$$A_{j(N+3)} = A_{j(N-1)}$$
$$A_{j(N+4)} = A_{j(N-2)}$$

Figure 15.6

Three frequency spectrums are shown below:
(a) the graph of the mode frequency ω versus the mode number n,
(b) the graph of the square of the mode frequency ω versus n, and
(c) the graph (b) for N = 5 (string loaded with 4 masses).

Equation (15.136) gives the frequency of the n modes

$$n := 0..100 \qquad N := 5 \qquad j := 1..N \qquad t_j := j \qquad \omega0 := 2$$

$$\omega_n := 2 \cdot \omega0 \cdot \sin\left[\frac{n \cdot \pi}{2 \cdot (N+1)}\right] \qquad \max(\omega) = 4$$

Explain the difference in graphs (a) and (b).

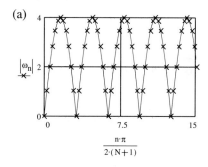

(a)

$$\omega_1 = 1.035 \qquad \left(\omega_1\right)^2 = 1.072$$

$$\omega_2 = 2 \qquad \left(\omega_2\right)^2 = 4$$

$$\omega_3 = 2.828 \qquad \left(\omega_3\right)^2 = 8$$

$$\omega_4 = 3.464 \qquad \left(\omega_4\right)^2 = 12$$

(b)   (c) Showing only a small region of (b)

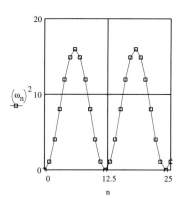

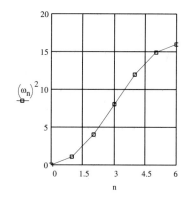

Note that $n = 2N + 2$ gives the next null mode. Thus there are only N distinct modes, and if n increases beyond N, it simply duplicates the normal mode for smaller N. The conclusion of this discussion is illustrated in Fig. 15.7, which shows the normal modes of a vibrating string for $N = 12$. Note that $n = 13$ is a null mode, while modes for $n = 14, 15, 16, 17, 18$ repeat the pattern of $n = 12, 11, 10, 9, 8$ with opposite sign. The sinusoidal curves represent the variation in the amplitude A_{jn} for various values of n. One must be careful to note that the frequencies of these sine curves have no relation to the frequencies of the vibrating particles.

Figure 15.7 _____

Below the normal modes of a vibrating string for N = 12 particles are shown.

$$N := 12 \qquad j := 1 .. N \qquad t_j := j \qquad n := 0 .. 25 \qquad C_n := n \qquad \omega 0 := 1$$

$$A_{j,n} := C_n \cdot \sin\left(\frac{j \cdot n \cdot \pi}{N+1}\right) \qquad \omega_n := 2 \cdot \omega 0 \cdot \sin\left[\frac{n \cdot \pi}{2 \cdot (N+1)}\right] \qquad u_{j,n} := A_{j,n} \cdot \cos\left[\left(\omega_n \cdot t_j\right) - \frac{\pi}{2}\right]$$

The null point is n = N + 1 = 12 + 1 = 13. Hence the modes for:

n = 14, 15, 16, 17, 18, . . . repeat the pattern for
n = 12, 11, 10, 9, 8, . . . but with opposite signs.

We have shifted the position of the graphs
by adding 20 to one of them.

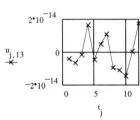

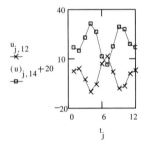

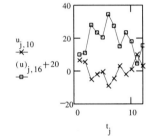

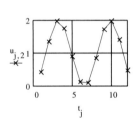

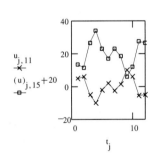

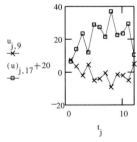

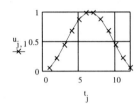

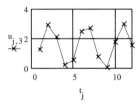

Let us now discuss specific modes of vibrations, assuming that there is a large number of particles N. The particle displacement corresponding to the mode $n = 1$ is, from Eqs. (15.134) and (15.135),

$$u_{j1} = C_1 \sin\left(\frac{j\pi}{N+1}\right) \cos \omega_1 t, \qquad j = 1, 2, 3, \dots , N \qquad \textbf{(15.139)}$$

This equation implies that at any given time the $C_1 \cos \omega_1 t$ factor is the same for all particles, while the displacements of different particles are given by the factor $\sin[j\pi/(N+1)]$. The bold-face curve in Fig. 15.8(a) is a plot of $\sin[j\pi/(N+1)]$ versus j for $j = 0$ to $N + 1$ and gives the amplitudes of different particles. As time passes, the particles have different displacements and oscillate with frequency ω_1, as shown in Fig. 15.8(b). The dotted curves give the positions of the particles at different times. For the $n = 2$ mode, the situation is as shown in Fig. 15.9, where the amplitudes are given by the boldface curve, while the dotted curves give the positions of the particles vibrating with frequency ω_2; that is,

$$u_{j2} = C_2 \sin\left(\frac{2j\pi}{N+1}\right) \cos \omega_2 t, \qquad j = 1, 2, 3, \dots , N \qquad \textbf{(15.140)}$$

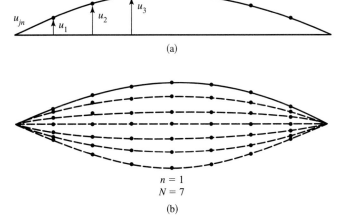

(a)

n = 1
N = 7

(b)

Figure 15.8 (a) Plot of $\sin[j\pi/(N + 1)]$ versus j for $j = 0$ to $N + 1$ shown by the boldface curve (for seven particles). (b) The dotted curves give the positions of the particles at different times vibrating with frequency ω_1.

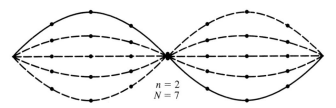

n = 2
N = 7

Figure 15.9 For $n = 2$ modes, the amplitudes are shown by a boldface curve, while the dotted curves give the positions of the particles vibrating with frequency ω_2.

15.6 BEHAVIOR OF A WAVE AT DISCONTINUITY: ENERGY FLOW

As an example of discontinuity, consider two semi-infinite strings of different linear mass densities tied together at $x = 0$, as shown in Fig. 15.10. The string that extends over $-\infty \leqslant x \leqslant 0$ has a linear mass density μ_1, and the wave traveling along this string has a velocity v_1, while the string that extends over $0 \leqslant x \leqslant \infty$ has a linear mass density μ_2, and the wave traveling along this string has a velocity v_2. Let the tension in the string be T. We want to investigate the effect of a sudden change in density at $x = 0$ on a continuous harmonic wave.

Let an incident wave traveling from the left for $x < 0$ be represented by

$$u_I = A_I \cos(k_1 x - \omega t) \qquad (15.141)$$

where A_I is the amplitude of the incident wave, $k_1 = \omega/v_1$, v_1 being the wave velocity. When this wave reaches $x = 0$, the point where the two strings join (the point of discontinuity), part of the wave is reflected back along the first string, while the remaining wave is transmitted. The reflected wave is represented by

$$u_R = A_R \cos(k_1 x + \omega t) \qquad (15.142)$$

where A_R is the amplitude of the reflected wave. The transmitted wave is given by

$$u_T = A_T \cos(k_2 x - \omega t) \qquad (15.143)$$

where A_T is the amplitude of the transmitted wave and $k_2 = \omega/v_2$, v_2 being the velocity of the wave on the second string to the right of $x = 0$. It may be noted that we could have used solutions of the following form:

$$u_I = \text{Re } A_I e^{i(k_1 x - \omega t)} \qquad (15.144)$$

where Re stands for the real part of the expression.

Our aim is to evaluate the reflected and transmitted amplitudes A_R and A_T in terms of the incident amplitude A_I. This can be done by imposing the boundary conditions that at the junction of the two string ($x = 0$) the displacement u and its derivative $\partial u/\partial x$ must be continous. These are the *continuity conditions* and are valid for any other types of wave motion, such as sound waves. The first condition satisfies the requirement that there is no break in the string,

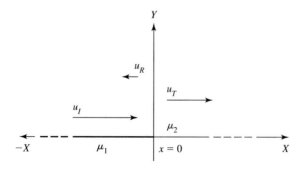

Figure 15.10   Two semi-infinite strings of different linear mass densities tied together at $x = 0$.

while the second condition implies that the restoring force resulting from a displacement y is the same on each side of the junction. If this were not true, then a finite force acting on a vanishing small mass element would produce an infinite acceleration. Thus the boundary conditions may be written as

$$(u_I + u_R)\big|_{x=0} = u_T\big|_{x=0} \tag{15.145}$$

and

$$\left(\frac{\partial u_1}{\partial x} + \frac{\partial u_R}{\partial x}\right)\bigg|_{x=0} = \left(\frac{\partial u_T}{\partial x}\right)\bigg|_{x=0} \tag{15.146}$$

Using Eqs. (15.141), (15.142), and (15.143), the continuity of u, Eq. (15.145), yields

$$A_I + A_R = A_T \tag{15.147}$$

while the continuity of $\partial u/\partial x$, given by Eq. (15.146), yields

$$k_1(A_I - A_R) = k_2 A_T \tag{15.148}$$

Solving these two equations for A_R/A_I and A_T/A_I,

$$\frac{A_R}{A_I} = \frac{k_1 - k_2}{k_1 + k_2} \tag{15.149}$$

$$\frac{A_T}{A_I} = \frac{2k_1}{k_1 + k_2} \tag{15.150}$$

Since $k = \omega/v$ and $v = \sqrt{T/\mu}$, we may write these results as

$$\frac{A_R}{A_I} = \frac{v_2 - v_1}{v_1 + v_2} = \frac{\sqrt{\mu_1} - \sqrt{\mu_2}}{\sqrt{\mu_1} + \sqrt{\mu_2}} \tag{15.151}$$

$$\frac{A_T}{A_I} = \frac{2v_2}{v_1 + v_2} = \frac{2\sqrt{\mu_1}}{\sqrt{\mu_1} + \sqrt{\mu_2}} \tag{15.152}$$

It is clear that the ratio A_T/A_I is always positive; hence the transmitted wave is always in phase with the incident wave. If the second medium is lighter, $v_2 > v_1$ or $\mu_2 < \mu_1$, the ratio A_R/A_I will be positive; hence the reflected wave will be in phase with the incident wave. On the other hand, if the second medium is denser than the first, $v_2 < v_1$ or , $\mu_2 > \mu_1$, A_R/A_I will be negative. This means that the reflected wave is out of phase by π with respect to the incident wave. This type of behavior is typical of many kinds of wave motion.

The intensity, the rate of energy flow, for any type of wave motion is proportional to the square of the amplitude. For this purpose, we define the *reflection coefficient, R,* to be the fraction of the incident energy that is reflected back; that is,

$$R \equiv \left(\frac{A_R}{A_I}\right)^2 \left(\frac{k_1 - k_2}{k_1 + k_2}\right)^2 = \left(\frac{v_2 - v_1}{v_1 + v_2}\right)^2 \tag{15.153}$$

While the *transmission coefficient, T* defined as the fraction of the incident energy that is transmitted, must satisfy the condition

$$R + T = 1 \tag{15.154}$$

or
$$T \equiv |1 - R| \equiv \frac{4v_1 v_2}{(v_1 + v_2)^2} \tag{15.155}$$

(*Note:* From Eq. (15.153), R becomes larger and larger as the difference between v_1 and v_2 becomes larger, while correspondingly T becomes smaller.]
    Finally, let us calculate the rate of energy flow dE/dt across the junction at $x = 0$. This is equal to the work done by the adjacent portion of the string on the particle at $x = 0$ and is equal to the product of the restoring force $-T(\partial u/\partial x)$ and the velocity of the particle $\partial u/\partial t$ both evaluated at $x = 0$. Then

$$\frac{dE}{dt} = \left(-T\frac{\partial u}{\partial x}\right)_{x=0}\left(\frac{\partial u}{\partial t}\right)_{x=0} \tag{15.156}$$

If we want to calculate the energy transmitted to the left of the string at $x = 0$, we let

$$u = u_I + u_R$$

$$= A_I \cos(k_1 x - \omega t) + A_R \cos(k_1 x + \omega t) \tag{15.157}$$

Substituting this in Eq. (15.156), we get

$$\left(\frac{dE}{dt}\right)_- = \omega k_1 T(A_I^2 - A_R^2) \sin^2 \omega t \tag{15.158}$$

Similarly, if we use

$$u = u_T = A_T \cos(k_2 x - \omega t)$$

in Eq. (15.156), we get energy transmitted to the right as

$$\left(\frac{dE}{dt}\right)_+ = \omega k_2 T A_T^2 \sin^2 \omega t \tag{15.159}$$

Since the average value of $\sin^2 \omega t$ over one complete cycle is $\frac{1}{2}$, we may write Eqs. (15.158) as

$$\left[\left(\frac{dE}{dt}\right)_-\right]_{ave} = \frac{1}{2}\omega k_2 T A_I^2 - \frac{1}{2}\omega k_1 T A_R^2 \tag{15.160}$$

where the first term on the right is the mean rate at which the energy is incident on the junction, while the second term is the mean rate at which the energy is reflected back. Similarly, the mean rate at which the energy is transmitted, from Eq. (15.159), is

$$\left[\left(\frac{dE}{dt}\right)_{+}\right]_{ave} = \frac{1}{2}\omega k_2 T A_T^2 \qquad (15.161)$$

This is the net rate at which the energy is supplied to the junction from left to right (see Problem 15.28).

15.7 SOUND WAVES: LONGITUDINAL WAVES

So far we have been dealing with transverse waves in solids. These waves consist of crests and troughs. We now start with the discussion of *sound waves,* which are basically longitudinal in nature and consist of compressions and rarefaction. Sound waves can travel in solids and fluids (liquids and gases) and propagate in general in three dimensions. For simplicity, we will deal with sound waves in fluids traveling only in one dimension, say along the X-axis. We use a simple procedure using the results derived already.

We showed that the net upward force acting on a small element of length of a string is, from Eq. (15.101) (replacing F_y by F),

$$F = -T\frac{\partial u}{\partial x} \qquad (15.162)$$

while the upward velocity at a point of such an element is

$$\dot{u} = \frac{\partial u}{\partial t} \qquad (15.163)$$

Note that $\dot{u}$ is the particle velocity and is not to be confused with the wave velocity v. Also, from Eq. (15.9),

$$\mu \, dx \, \frac{\partial^2 u}{\partial t^2} = \frac{\partial}{\partial x}\left(T\frac{\partial u}{\partial x}\right) dx \qquad (15.9)$$

Using the preceding three equations, we can show

$$\frac{\partial F}{\partial t} = -T\frac{\partial \dot{u}}{\partial x} \qquad (15.164)$$

and $$\frac{\partial \dot{u}}{\partial t} = -\frac{1}{\mu}\frac{\partial F}{\partial x} \qquad (15.165)$$

Equation (15.164) states that the time rate of change of F is proportional to $\partial \dot{u}/\partial x$ (= the difference in the velocities at the ends of the line segment divided by the length of the line segment). Equation (15.165) states that the acceleration of the string is proportional to $\partial F/\partial x$ (= the difference in the forces at the ends of a line segment divided by the length of the line segment).

We can conclude that, for a small amplitude, *for the quantities F and $\dot{u}$, the time rate of change of either is proportional to the space derivative of the other.*

Starting with Eqs. (15.164) and (15.165), and taking derivatives,

$$\frac{\partial F}{\partial t} = -T\frac{\partial \dot{u}}{\partial x}, \qquad \frac{\partial^2 F}{\partial t^2} = -T\frac{\partial^2 \dot{u}}{\partial t\,\partial x}$$

$$\frac{\partial \dot{u}}{\partial t} = -\frac{1}{\mu}\frac{\partial F}{\partial x}, \qquad \frac{\partial^2 \dot{u}}{\partial x\,\partial t} = -\frac{1}{\mu}\frac{\partial^2 F}{\partial x^2}$$

Combining these equations, we get

$$\frac{\partial^2 F}{\partial x^2} = \frac{1}{v^2}\frac{\partial^2 F}{\partial t^2}, \qquad v = \sqrt{\frac{T}{\mu}} \tag{15.166}$$

and, similarly

$$\frac{\partial^2 \dot{u}}{\partial x^2} = \frac{1}{v^2}\frac{\partial^2 \dot{u}}{\partial t^2} \tag{15.167}$$

Thus, instead of writing the usual wave equation, Eq. (15.12), where the displacement $u(x, t)$ is the variable, here we have two wave equations with F and $\dot{u}$ as the two independent variables.

As an application of Eqs. (15.166) and (15.167), let us consider plane sound waves traveling in air in the X direction. This will be equivalent to, as an example, sound waves traveling in an organ pipe. In Eq. (15.166), F is replaced by p, the pressure in excess of atmospheric pressure, while $\dot{u}$ represents the velocity of the volume element of air at any point, μ by ρ, the density of air, and T by B, the bulk modulus. Thus Eqs. (15.164) and (15.165) take the form

$$\frac{\partial p}{\partial t} = -B\frac{\partial \dot{u}}{\partial x} \tag{15.168}$$

and

$$\frac{\partial \dot{u}}{\partial t} = -\frac{1}{\rho}\frac{\partial p}{\partial x} \tag{15.169}$$

Both p and $\dot{u}$ satisfy the equations

$$\frac{\partial^2 p}{\partial x^2} = \frac{1}{v}\frac{\partial^2 p}{\partial t^2} \tag{15.170}$$

$$\frac{\partial \dot{u}}{\partial x^2} = \frac{1}{v^2}\frac{\partial^2 \dot{u}}{\partial t^2} \tag{15.171}$$

where

$$v = \sqrt{\frac{B}{\rho}} \tag{15.172}$$

The power transmitted in the X direction from left to right may be written as (making use of the definition $P = F\dot{u}$)

$$P = p\dot{u} \tag{15.173}$$

Note that p is pressure in excess of atmosphere pressure, while $\dot{u}$ is the particle velocity (and not the wave velocity).

Suppose a pipe of infinite length and cross-sectional area A is placed along the X-axis. When air (or any other fluid) in the pipe is undisturbed, the pressure at any point is p_0 and the mass density is ρ_0. If the pressure changes to $p = p_0 + \Delta p$ while the density changes to $\rho = \rho_0 + \Delta \rho$, then according to the definition of the bulk modulus, B, we may write

$$B = \frac{\Delta p}{\Delta \rho / \rho_0} \quad \text{or} \quad \frac{dp}{d\rho} = \frac{B}{\rho_0} = \frac{1}{K\rho_0} \tag{15.174}$$

where K is the compressibility. Thus the velocity of wave propagation

$$v = \sqrt{\frac{dp}{d\rho}} = \sqrt{\frac{B}{\rho_0}} = \sqrt{\frac{1}{K\rho_0}} \tag{15.175}$$

This relation is good only for propagation of waves in liquids. In gases, the situation is quite different. A small change in pressure will cause a considerable change in temperature. The compressions and rarefactions take place so rapidly that there is no time for heat to flow out or in; hence the process may be assumed to be adiabatic. For such situations, assuming an ideal gas, we have the relation

$$p = B\rho^\gamma \tag{15.176}$$

where γ is the ratio of the specific heat of the gas at constant pressure to the specific heat of the gas at constant volume; that is, $\gamma = C_p / C_v$. Combining the above equations and using the approximation $|\partial \eta / \partial x| \ll 1$, we get

$$\frac{dp}{d\rho} = \frac{\gamma p_0}{\rho_0} \tag{15.177}$$

Using the ideal gas equation,

$$p = \frac{RT}{M} \rho \tag{15.178}$$

where R is the gas constant, M the molecular weight, and T the absolute temperature, and using Eq. (15.177), we may write the wave propagation velocity to be, from Eq. (15.175),

$$v = \sqrt{\frac{\gamma p_0}{\rho_0}} = \sqrt{\frac{\gamma RT}{M}} \tag{15.179}$$

which clearly indicates that the temperature T alone determines the velocity of the propagation of sound waves in an ideal gas.

If we extend out discussion of sound waves to their propagation in three dimensions, we get the following equations (as compared to Eqs. (15.170) and (15.171) for one dimension):

$$\frac{\partial^2 \rho}{\partial t^2} = \frac{dp}{dt} \nabla^2 \rho \tag{15.180}$$

$$\frac{\partial^2 p}{\partial t^2} = \frac{dp}{dt}\nabla^2 p \qquad (15.181)$$

where, for example, $p = p(x, y, z, t) = \Psi(x, y, z)\Theta(t)$ and $\Psi(x, y, z) = X(x)\,Y(y)\,Z(z)$.

15.8 FLUID STATICS

A fluid is a substance that does not have a fixed shape. It consists of a continuum of matter and will undergo a finite displacement when an infinitesimal shear stress is applied. A small volume of a fluid can be treated as a continuum if it contains such a large number of molecules that the average distance traveled by molecules between collisions is much smaller than the size of the volume of the fluid. Fluids are characterized by physical and mechanical properties, such as density, pressure, temperature, and velocity. Both liquids and gases are fluids, but there are fundamental differences between the two. Liquids are not easily compressed and hence may be considered to have fixed volumes and densities. This is not so with gases, which can be easily compressed. Gases do not have any fixed shape; they simply fill up any container. Liquids do not have any definite shape, but they do have a distinct surface.

Hydrostatics or fluid statics deals with fluids at rest and *hydrodynamics* or *fluid dynamics* deals with fluids in motion. If the fluid flow is time independent, it is said to be *steady*. The fluid flow is *laminar* or *streamlined* if different layers of fluids move past each other with no mixing. If mixing between layers takes place, the flow is said to be *turbulent*.

In this section, we limit our discussion to fluid statics, while fluid dynamics will be investigated in the following sections. Newton's laws and conservation laws will be applied to fluids, since after all fluids are merely a collection of a large number of particles.

Let us consider a fluid in static equilibrium. Thus each elemental volume of fluid is at rest and the velocity at each point in the fluid is zero. We now discuss two characteristics of fluid statics: (1) A fluid exerts equal pressure in all directions, and (2) pressures at equal depths are the same.

Consider a very small triangular prism, as shown in Fig. 15.11. Let F, F_y, and F_z be the forces acting on the three surfaces of areas A, A_y, and A_z, respectively, as shown. These forces must act normal to the surfaces. If any forces acted tangent to the prism surfaces, the fluid would be set in motion, which is contrary to our assumption of a static fluid.

Thus the only forces acting are the normal forces and the weight W of the fluid. For equilibrium, the forces acting in the Y and Z directions are zero. (We have assumed no forces along the X-direction.)

$$\sum F_y = F\sin\theta - F_y = 0 \qquad (15.182)$$

$$\sum F_z = F_z - F\cos\theta - W = 0 \qquad (15.183)$$

If P, P_y, and P_z are the pressures (normal force per unit area) acting on the three surfaces, and $W = \rho g[(dx\,dy\,dz)/2]$, we may write Eqs. (15.182) and (15.183) as

$$P\frac{dx\,dz}{\sin\theta}\sin\theta - P_y\,dx\,dz = 0 \qquad (15.184)$$

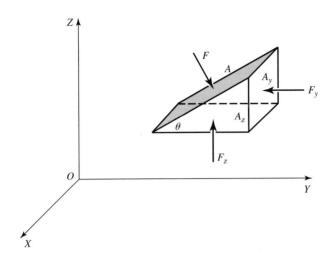

Figure 15.11 Small fluid element (in the shape of a small triangular prism) in static equilibrium.

and
$$P_z \, dx \, dy - P \, \frac{dx \, dz}{\sin \theta} \, \cos \theta - \rho g \, \frac{dx \, dy \, dz}{2} = 0 \qquad \textbf{(15.185)}$$

where ρ is the density of the fluid and $dx \, dy \, dz/2$ is the volume of the prism. As dx, dy, and dz go to zero, the last term in Eq. (15.185), the weight term, becomes negligible compared to the pressure term. Hence, from the preceding equations, we conclude

$$P = P_y = P_z \qquad \textbf{(15.186)}$$

$$P = \frac{F}{A} = \frac{F_y}{A_y} = \frac{F_z}{A_z} \qquad \textbf{(15.187)}$$

which states that the pressure is independent of the direction; that is, *pressure is the same in all directions* and is a scalar quantity. Equation (15.186) is the statement of *Pascal's law*.

Let us now derive an expression for the variation of pressure with vertical position in a static fluid. Consider an infinitesimal volume $dx \, dy \, dz$ of fluid, as shown in Fig. 15.12. Since the

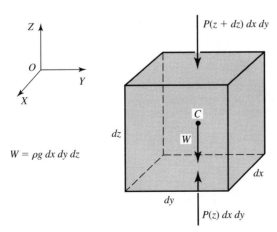

Figure 15.12 Infinitesimal volume $dx \, dy \, dz$ of a fluid in equilibrium.

fluid is in equilibrium, the sum of the forces in the Z direction must add up to zero; that is,

$$\sum F_z = P(z)\, dx\, dy - P(z + dz)dx\, dy - \rho g\, dx\, dy\, dz = 0 \qquad (15.188)$$

The second term, when expanded in a Taylor series about z to the first order, takes the form

$$P(z + dz) = P(z) + \frac{dP}{dz}\, dz + \cdots \qquad (15.189)$$

Substituting this in Eq. (15.188) yields

$$\frac{dP(z)}{dz} = -\rho g \qquad (15.190)$$

which, on integration, assuming the fluid to be incompressible and $P = P_0$ at $z = 0$, gives

$$P(z) = P_0 - \rho g z \qquad (15.191)$$

Since z is taken to be positive upward, this equation states that P increases as z decreases. Also, it states that the pressure at any depth of a column is equal to the sum of the pressure P_0 at the top of the column and the weight of the liquid column.

An alternative approach to this treatment is the following. Let $\mathbf{w} = \rho \mathbf{g}$ be the weight density, that is, the weight per unit volume acting in the direction of $\mathbf{g}$. Consider two points 1 and 2 in a fluid where the pressures are P_1 and P_2 and separated by an infinitesimal distance $d\mathbf{r}$. Let us construct a right circular cylinder of cross-sectional area dA and length dr, as shown in Fig. 15.13. The only forces acting on the cylinder are due to the liquid pressure and gravity. Since the liquid is in equilibrium, the sum of the components of the forces along $d\mathbf{r}$ must be zero; that is,

$$P_1\, dA - P_2\, dA + \mathbf{w} \cdot d\mathbf{r}\, dA = 0 \qquad (15.192)$$

where $dr\, dA = dV$ is the volume of fluid inside the cylinder. Thus the differences in pressure ΔP between the two points is

$$\Delta P = P_2 - P_1 = \mathbf{w} \cdot d\mathbf{r} \qquad (15.193)$$

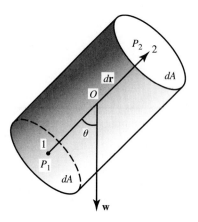

Figure 15.13 In a fluid, point 1 at pressure P_1 is at a distance $d\mathbf{r}$ from point 2 at pressure P_2.

If the two points 1 and 2 are located at distances $\mathbf{r}_1$ and $\mathbf{r}_2$, then the pressure between the two points is obtained by integrating Eq. (15.193); that is,

$$P_2 - P_1 = \int_{\mathbf{r}_1}^{\mathbf{r}_2} \mathbf{w} \cdot d\mathbf{r} \tag{15.194}$$

This line integral implies that the pressure difference between two points in a fluid depends on gravitational forces and the spatial orientation of the two points. Furthermore, this equation states that "any change in pressure at one point will be transmitted to every point in the fluid," which is the statement of *Pascal's law*.

Integration of Eq. (15.194), assuming $P_1 = P_0$ at $z = 0$ and $P_2 = P(z)$ is the pressure at a distance z above point 1, yields ($\mathbf{w} \cdot d\mathbf{r} = -\rho g \, dz$)

$$P(z) = P_0 - \rho g z \tag{15.191}$$

which is the same as Eq. (15.191). Thus, if at the surface of a lake or pond the atmospheric pressure is $P_a [= 1.103 \times 10^5$ Pa (1 Pa = 1 N/m^2)], the pressure at a depth h below will be ($z = -h$)

$$P(h) = P_a + \rho g h \tag{15.195}$$

In Eq. (15.191), we have assumed that the density of the fluid is constant. This is not true, especially in the case of gases. Thus, if there is a change in pressure, it will result in a change of volume. If B is the bulk modulus of the gas,

$$B = -\frac{dP}{dV/V} \quad \text{or} \quad \frac{dV}{V} = -\frac{dP}{B} \tag{15.196}$$

The minus sign is due to the fact that as P increases V decreases. If m is the mass of a gas of volume V, the density is $\rho = m/V$. Hence

$$d\rho = -\frac{m}{V^2} \, dV = \rho\left(-\frac{dV}{V}\right)$$

or
$$\frac{dV}{V} = -\frac{d\rho}{\rho} \tag{15.197}$$

Combining Eq. (15.197) with Eq. (15.196), we have

$$\frac{d\rho}{\rho} = \frac{dP}{B} \tag{15.198}$$

For an ideal gas, the equation of state is

$$PV = nRT \tag{15.199}$$

where n is the number of moles given by $n = m/M$, M being the molecular mass, $R = 8.134$ J/mol-K, and T is the absolute temperature of the gas. Thus

$$\rho = \frac{m}{V} = \frac{mP}{nRT} = \frac{MP}{RT} \tag{15.200}$$

Let us apply this expression for ρ in calculating the variation of pressure in our atmosphere as a function of altitude z. From Eq. (15.190), substituting the value of ρ from Eq. (15.200) yields

$$\frac{dP}{dz} = -\rho g = -\frac{Mg}{RT} P \tag{15.201}$$

Thus, if P_0 is the pressure at sea level, integrating Eq. (15.201) yields

$$P(z) = P_0 e^{-(Mg/RT)z} \tag{15.202}$$

We may define, providing the temperature remains constant, for an isothermal atmosphere, the *atmosphere scale height H* as

$$H = \frac{RT}{Mg} \tag{15.203}$$

Thus Eq. (15.202), the atmospheric pressure variation with z, takes the form

$$P(z) = P_0 e^{-z/H} \tag{15.204}$$

while the variation in the density takes the form

$$\rho(z) = \rho_0 e^{-z/H} \tag{15.205}$$

Thus H may be defined as the distance in which the density or the pressure decreases by $1/e$ of its initial value. Note that, for a constant density, Eq. (15.204) reduces to the familiar expression for $P(z)$. Assuming z to be small, so that ρ will not change, expanding Eq. (15.204) yields, [using Eq. (15.201)],

$$P(z) = P_0\left(1 - \frac{z}{H} + \cdots\right) \simeq P_0 - P_0\frac{z}{H} + \cdots = P_0 - \rho g H \frac{z}{H} \simeq P_0 - \rho g z$$

Archimedes' Principle

Let us consider the weight of a fluid of volume V, so that

$$\mathbf{W} = \int\int\int_V \mathbf{w}\, dV = \int\int\int_V \rho \mathbf{g}\, dV \tag{15.206}$$

Since the fluid is at rest, the weight (or force) is balanced by the forces of pressure exerted by the surrounding fluid on the surface of this volume; that is,

$$\mathbf{F}_b = \int\int_S \hat{\mathbf{n}}\, P\, dA \tag{15.207}$$

For a fluid at rest $\mathbf{F}_b$ must be equal and opposite to $\mathbf{W}$; that is,

$$\mathbf{F}_b = -\mathbf{W} = -\int\int\int_V \rho\mathbf{g}\,dV = -\rho\mathbf{g}V \qquad (15.208)$$

Thus the buoyant force $\mathbf{F}_b$ on a volume V in a fluid is equal to the weight of the fluid inside the volume V. This is *Archimedes' principle,* which states that *the buoyant force on a body immersed in a fluid is equal to the weight of the volume of the fluid displaced.*

15.9 FLUIDS IN MOTION

The study of fluids in motion can be divided into two parts: fluid kinematics and fluid dynamics. We shall first investigate kinematics. There are two approaches, both suggested by Euler, by which fluids in motion may be investigated. The first approach is the direct application of Newtonian mechanics to a system of particles. Time t is considered to be the only independent coordinate, and the coordinates (x, y, z) are expressed in terms of the initial coordinate (x_0, y_0, z_0) at time t_0 and time t. The resulting equations are called *Lagrangian equations* (this approach is also called Lagrange's method). The resulting equations are so numerous that this method of keeping track of each fluid particle becomes cumbersome. The second approach, also due to Euler, is equally cumbersome, but manageable.

According to the Eulerian system for fluids, we describe such properties of fluids as density $\rho(x, y, z, t)$, velocity $\mathbf{v}(x, y, z, t)$, and pressure P, at different positions (x, y, z) and time t along the path of the fluid. Thus we are focusing our attention on a point in space where the fluid is flowing, instead of the fluid particles themselves. This leads us to define two different time rates of change for any quantity such as ρ, $\mathbf{v}$, or P. The *partial time derivative* $(\partial/\partial t)$ is the time rate of change of a quantity measured at a point fixed in space. The *total time derivative* is the time rate of change of a quantity as measured with respect to a particle moving with the fluid.

As an example, for the velocity vector $\mathbf{v}$,

$$\mathbf{v} = \mathbf{v}(x, y, z) \qquad (15.209)$$

the change in velocity vector is given by

$$d\mathbf{v} = \mathbf{v}(x + dx, y + dy, z + dz, t + dt) - \mathbf{v}(x, y, z, t)$$

$$\simeq \frac{\partial\mathbf{v}}{\partial x}\,dx + \frac{\partial\mathbf{v}}{\partial y}\,dy + \frac{\partial\mathbf{v}}{\partial z}\,dz + \frac{\partial\mathbf{v}}{\partial t}\,dt \qquad (15.210)$$

In the limit as $dt \to 0$, we may write the total time derivative of $\mathbf{v}$ as

$$\frac{d\mathbf{v}}{dt} = v_x\frac{\partial\mathbf{v}}{\partial x} + v_y\frac{\partial\mathbf{v}}{\partial y} + v_z\frac{\partial\mathbf{v}}{\partial z} + \frac{\partial\mathbf{v}}{\partial t} \qquad (15.211)$$

Similarly, for the total time derivative of pressure P, we may write

$$\frac{dP}{dt} = v_x\frac{\partial P}{\partial x} + v_y\frac{\partial P}{\partial y} + v_z\frac{\partial P}{\partial z} + \frac{\partial P}{\partial t} \qquad (15.212)$$

Quantities v_x, v_y, and v_z ($\equiv dx/dt$, dy/dt, and dz/dt) are the components of the fluid velocity **v** at any point (x, y, z) and time t. Relations of the form of Eqs. (15.211) or (15.212) hold for any quantity describing the fluid. These equations in concise form may be written as

$$\frac{d\mathbf{v}}{dt} = (\mathbf{v} \cdot \nabla)\mathbf{v} + \frac{\partial \mathbf{v}}{\partial t} \qquad (15.213)$$

and

$$\frac{dP}{dt} = (\mathbf{v} \cdot \nabla)P + \frac{\partial P}{\partial t} \qquad (15.214)$$

From these two equations, we may reduce a common operator

$$\frac{d}{dt} = (\mathbf{v} \cdot \nabla) + \frac{\partial}{\partial t} \qquad (15.215)$$

called the *substantial derivative*. This operator is applicable to both vector and scalar quantities. We now apply these ideas by dividing our discussion into three parts:

1. Continuity equation
2. Equation of motion for an ideal fluid flow
3. Bernoulli's equation

Continuity Equation

We can arrive at the continuity equation by applying the law of conservation of mass to a Euler-ian system. Consider a small differential volume element $dx\, dy\, dz$ of fluid surrounding a point (x, y, z), as shown in Fig. 15.14. The velocities of the fluid at different faces are as shown. The mass flowing in across face I (shown shaded) in the time dt is

$$dm_I = \rho(x, y, z, t)v_x(x, y, z, t)\, dy\, dz\, dt \qquad (15.216a)$$

where ρ is the mass density, and v_x is the x component of the velocity, which is normal to the area $dy\, dz$. The mass flowing out from face II in time dt is

$$dm_{II} = \rho(x + dx, y, z, t)v_x(x + dx, y, z, t)\, dy\, dz\, dt \qquad (15.216b)$$

Thus the net mass of the fluid leaving the volume element in the X direction is

$$dm_{II} - dm_I = [\rho(x + dx, y, z, t)v_x(x + dx, y, z, t) - \rho(x, y, z, t)v_x(x, y, z, t)]\, dy\, dz\, dt$$
$$(15.217)$$

Using the following expansions,

$$\rho(x + dx, y, z, t) = \rho(x, y, z, t) + \frac{\partial \rho(x, y, z, t)}{\partial x}\, dx + \cdots$$

$$v_x(x + dx, y, z, t) = v_x(x, y, z, t) + \frac{\partial v_x(x, y, z, t)}{\partial x}\, dx + \cdots$$

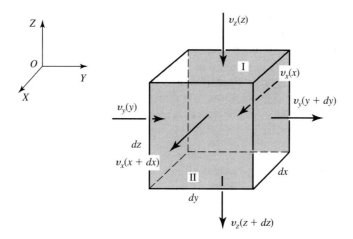

Figure 15.14   Motion of fluid across a small differential volume element $dx\,dy\,dz$ of fluid surrounding a point (x, y, z).

in Eq. (15.217) and neglecting the higher-order terms, we obtain

$$dm_{II} - dm_I = \frac{\partial \rho}{\partial x} v_x\, dx\, dy\, dz\, dt + \rho \frac{\partial v_x}{\partial x}\, dx\, dy\, dz\, dt \qquad \textbf{(15.218)}$$

Applying the same procedure to the remaining faces, the net total mass leaving the volume element $dx\,dy\,dz$ in time dt is

$$dm = \left(\frac{\partial \rho}{\partial x} v_x + \frac{\partial \rho}{\partial y} v_y + \frac{\partial \rho}{\partial z} v_z + \rho \frac{\partial v_x}{\partial x} + \rho \frac{\partial v_y}{\partial y} + \rho \frac{\partial v_z}{\partial z} \right) dx\, dy\, dz\, dt \qquad \textbf{(15.219)}$$

This net mass leaving the volume element must be equal to the decrease in mass within the element so as to conserve mass; that is,

$$dm = -\left(\frac{\partial \rho}{\partial t} \right) dx\, dy\, dz\, dt \qquad \textbf{(15.220)}$$

Equating Eqs. (15.219) and (15.220), we obtain

$$\frac{\partial \rho}{\partial x} v_x + \frac{\partial \rho}{\partial y} v_y + \frac{\partial \rho}{\partial z} v_z + \rho \frac{\partial v_x}{\partial x} + \rho \frac{\partial v_y}{\partial y} + \rho \frac{\partial v_z}{\partial z} = - \frac{\partial \rho}{\partial t} \qquad \textbf{(15.221)}$$

or, in vector notation, we may write this as

$$\boldsymbol{\nabla} \cdot \rho \mathbf{v} + \frac{\partial \rho}{\partial t} = 0 \qquad \textbf{(15.222)}$$

Equations (15.221) and (15.222) are the statements of the *continuity equation* and simply represent the law of conservation of mass. Matter is nowhere created or destroyed, and the mass density in any volume element dV ($=dx\ dy\ dz$) moving with the fluid remains constant. The quantity $\rho\mathbf{v}$ is the *mass flux* (also called the momentum density or mass current), defined as the mass of the fluid leaving the volume element in a unit time through a unit area. Thus Eq. (15.222) states that the divergence of the mass flux leaving a volume is equal to the rate at which the mass density decreases.

We now look at a further interpretation of Eq. (15.222). The mass flow can be determined by integrating over a fixed volume V bounded by a surface A with outward normal $\hat{\mathbf{n}}$; that is,

$$\int \int_V \int \mathbf{\nabla} \cdot (\rho\mathbf{v})\, dV + \int \int_V \int \frac{\partial \rho}{\partial t}\, dV = 0 \qquad (15.223)$$

We rewrite the first term by using Gauss's divergence theorem, Eq. (5.129), and we can take the time differentiation outside the second term because V is a fixed volume. Thus Eq. (15.223) takes the form

$$\int_A \int \hat{\mathbf{n}} \cdot (\rho\mathbf{v})\, dA = -\frac{d}{dt} \int_V \int \int \rho\, dV \qquad (15.224)$$

This equation states that the outward flow of mass across the surface is equal to the rate of decrease of mass inside the volume V.

For a steady flow, $\partial \rho / \partial t = 0$; hence the mass entering is exactly equal to the mass leaving. Thus Eq. (15.224) takes the form

$$\int_A \int \hat{\mathbf{n}} \cdot (\rho\mathbf{v})\, dA = 0 \qquad (15.225)$$

Furthermore, if in addition to a constant fluid density the velocity is constant at the flow areas and is perpendicular to such areas, Eq. (15.225) yields

$$\rho_1 v_1 A_1 = \rho_2 v_2 A_2 \qquad (15.226)$$

and if the fluid is incompressible so that $\rho_1 = \rho_2$,

$$v_1 A_1 = v_2 A_2 = \text{constant} \qquad (15.227)$$

That is, the volume flux vA is constant for incompressible fluid and steady flow.

Let us consider again the case for an incompressible fluid flow, that is, $\rho = \text{constant}$; Eq. (15.222) takes the form

$$\mathbf{\nabla} \cdot \mathbf{v} = 0 \qquad (15.228)$$

We know that the divergence of the **curl** of the vector is zero. Therefore, $\mathbf{v}$ is derivable from a vector potential $\mathbf{\Phi}$. That is, if

$$\mathbf{v} = \mathbf{\nabla} \times \mathbf{\Phi} \qquad (15.229)$$

then

$$\mathbf{\nabla} \cdot (\mathbf{\nabla} \times \mathbf{\Phi}) = 0 \qquad (15.230)$$

These equations are similar to the equations for vector potentials associated with magnetic fields.

In describing fluid flow, the **curl** of the velocity, $\nabla \times \mathbf{v}$, is useful, as we explain now. Consider the relation

$$\iint\limits_{A} \hat{\mathbf{n}} \cdot (\nabla \times \mathbf{v}) \, dA = \oint \mathbf{v} \cdot d\mathbf{r} \tag{15.231}$$

The expression on the left represents the integral over the surface area A of the normal component of **curl v**, while the right side is obtained by using Stokes' theorem [see Eq. (5.224)]. Figure 15.15 shows two examples of fluid flow, a vortex and a transverse velocity gradient. In both examples, the line integral $\oint \mathbf{v} \cdot d\mathbf{r}$ is nonzero. Hence the **curl** of the velocity ($=\nabla \times \mathbf{v}$) must be nonzero. The quantity $(\nabla \times \mathbf{v})$ may be considered to be a measure of the rate of rotation of the fluid per unit area. In Fig. 15.15(a), the **curl v** has a nonzero value around a vortex. In Fig. 15.15(b), even though there is no vortex and the fluid does not actually circle a point, but because of the transverse velocity gradient, the **curl v** is nonzero. The fluid motion is said to have rotational properties.

If the **curl v** is zero everywhere in the fluid, the motion is said to be irrotational. That is, if about a given point

$$\nabla \times \mathbf{v} = 0 \tag{15.232}$$

the particles of the fluid will have no angular velocity about that point. Furthermore, if the **curl v** is zero, then **v** is derivable from a scalar potential ϕ. Since the **curl** of a gradient of a scalar is zero, we must have

$$\mathbf{v} = -\nabla \Phi \tag{15.233}$$

Substituting in Eq. (15.232) gives

$$\nabla \times \nabla \Phi = 0 \tag{15.234}$$

The equation represents the irrotational flow.

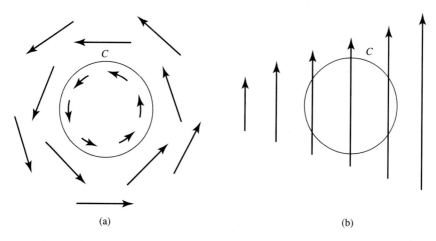

(a)   (b)

Figure 15.15   Two examples of fluid flow: (a) a vortex, and (b) a transverse velocity gradient. In both cases, $\nabla \times \mathbf{v}$ is nonzero; hence both have rotational flow.

Equation of Motion for an Ideal Fluid Flow

Once again we assume that we are dealing with an ideal fluid; that is, the fluid does not support any shear stress when in equilibrium. But any flowing fluid has viscosity, no matter how small, and hence will have some shearing stress. Thus we assume that an ideal fluid will have no viscosity.

A fluid in motion must not only satisfy the continuity equation, but must satisfy Newton's laws as well. Consider a fluid of volume $dx\,dy\,dz$, as shown in Fig. 15.16, for which the net force acting on the body is not zero. Let us assume that, in addition to pressure, the fluid is acted on by a general body force of $\mathbf{f}$ per unit volume. Thus the total body force acting on the volume element is $\mathbf{f}\,dx\,dy\,dz$. The force due to the pressure on face I is $p(x, y, z, t)\,dy\,dz$, and that due to face II is $p(x + dx, y, z, t)\,dy\,dz$. Thus, applying Newton's second law in the X direction,

$$dF_x = f_x\,dx\,dy\,dz + p(x, y, z, t)\,dy\,dz - p(x + dx, y, z, t)\,dy\,dz$$

Expanding $p(x + dx, y, z, t)$ to the first order in $dx\,dy\,dz$, we obtain

$$dF_x = \left(f_x - \frac{\partial p}{\partial x} \right) dx\,dy\,dz \tag{15.235}$$

Also, from Newton's second law,

$$dF_x = \frac{d(mv_x)}{dt} = \frac{d}{dt}\,(\rho v_x)\,dx\,dy\,dz \tag{15.236}$$

Equating these two equations, we get

$$f_x - \frac{\partial p}{\partial x} = \frac{d}{dt}\,(\rho v_x) \tag{15.237}$$

with similar expressions for the other two directions.

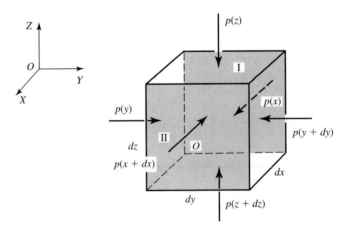

Figure 15.16 A fluid of volume $dx\,dy\,dz$ on which the net force acting is not zero. Besides pressure, the body force per unit is $\mathbf{f}$.

In vector notation, these three equations can be combined into one as

$$\mathbf{f} - \nabla p = \frac{d}{dt}\,(\rho \mathbf{v}) \tag{15.238}$$

Making use of the relation in Eq. (15.215), that is,

$$\frac{d}{dt} = (\mathbf{v} \cdot \nabla) + \frac{\partial}{\partial t} \tag{15.215}$$

Eq. (15.238) may take the form

$$\mathbf{f} - \nabla p = (\mathbf{v} \cdot \nabla)(\rho \mathbf{v}) + \frac{\partial}{\partial t}\,(\rho \mathbf{v}) \tag{15.239}$$

or

$$\frac{\partial \mathbf{v}}{\partial t} + \mathbf{v} \cdot \nabla \mathbf{v} + \frac{1}{\rho}\nabla p = \frac{\mathbf{f}}{\rho} \tag{15.240}$$

Equation (15.238), (15.239), or (15.240) is *Euler's equation of motion* for a fluid. The quantity $\mathbf{f}/\rho$ is the body force per unit mass. If the density ρ depends only on pressure p, the fluid is said to be homogeneous.

   If the body force $\mathbf{f}$ is given, we still have five unknowns: density, pressure, and the three components of velocity. The continuity equation and Euler equation provide us with four scalar equations only. If the density (or one other unknown quantity) is known, the problem can be solved.

Bernoulli's Equation

The law of conservation of energy when applied to the motion of the fluid as given by the Euler equation results in Bernoulli's equation. The scalar product of Euler's equation, Eq. (15.238), with velocity vector $\mathbf{v}$ gives

$$\mathbf{f} \cdot \mathbf{v} - \nabla p \cdot \mathbf{v} = \rho \frac{d\mathbf{v}}{dt} \cdot \mathbf{v} \tag{15.241}$$

The product $\mathbf{f} \cdot \mathbf{v}$ (force per unit volume times velocity) is the power per unit volume supplied by the body force $\mathbf{f}$. The second term may be written as

$$-\nabla p \cdot \mathbf{v} = -\frac{\partial p}{\partial x}\frac{dx}{dt} - \frac{\partial p}{\partial y}\frac{dy}{dt} - \frac{\partial p}{\partial z}\frac{dz}{dt} = -\frac{dp}{dt} + \frac{\partial p}{\partial t} \tag{15.242}$$

and the last term in Eq. (15.241) may be written as

$$\rho \frac{d\mathbf{v}}{dt} \cdot \mathbf{v} = \rho \frac{d}{dt}\left(\frac{1}{2}v^2\right) = \frac{d}{dt}\left(\frac{1}{2}\rho v^2\right) - \frac{1}{2}v^2\frac{d\rho}{dt} \tag{15.243}$$

Thus Eq. (15.241) takes the form

$$\mathbf{f} \cdot \mathbf{v} - \frac{dp}{dt} + \frac{\partial p}{\partial t} = \frac{d}{dt}\left(\frac{1}{2}\rho v^2\right) - \frac{1}{2}v^2\frac{d\rho}{dt} \qquad (15.244)$$

Since the fluid is incompressible ($d\rho/dt = 0$) and the flow is steady ($\partial p/\partial t = 0$), Eq. (15.244) takes the form

$$\mathbf{f} \cdot \mathbf{v} - \frac{dp}{dt} - \frac{d}{dt}\left(\frac{1}{2}\rho v^2\right) = 0 \qquad (15.245)$$

Now Euler's equation is in a form that can be integrated. Multiplying both sides by dt and integrating, we get

$$\int (\mathbf{f} \cdot \mathbf{v})\, dt - p - \frac{1}{2}\rho v^2 = \text{constant} \qquad (15.246)$$

The first term on the left is the work done by the body force per unit volume. If the body force $\mathbf{f}$ is derivable from a *scalar potential* Φ so that

$$\mathbf{f} = -\nabla\Phi \qquad (15.247)$$

where Φ is the *potential energy per unit volume,* we may write the first term in Eq. (15.246) as

$$W = \int (\mathbf{f} \cdot \mathbf{v})\, dt = \int \mathbf{f} \cdot d\mathbf{r} = \int (-\nabla\Phi) \cdot d\mathbf{r} = -\Phi \qquad (15.248)$$

Then Eq. (15.246) takes the form

$$p + \tfrac{1}{2}\rho v^2 + \Phi = \text{constant} \qquad (15.249)$$

which is the general form of Bernoulli's equation.

If the body force is the gravitational force, $\Phi = \rho gz$, Eq. (15.249) takes the form

$$p + \tfrac{1}{2}\rho v^2 + \rho gz = \text{constant} \qquad (15.250)$$

This equation, which is a statement of the conservation of energy, is known as *Bernoulli's equation* and is applicable to steady flow of incompressible fluid in a gravitational field. The first term, pressure p, represents the work done per unit volume by the fluid, the second term $\tfrac{1}{2}\rho v^2$ represents the kinetic energy per unit volume of the fluid, and the last term ρgz is the potential energy per unit volume of the fluid.

15.10 VISCOSITY AND VISCOUS FLOW

In previous discussions, we have assumed that the fluid was nonviscous; hence there was no friction between different layers of fluid when in motion. When adjacent layers of fluids are moving, the shearing force tends to reduce their relative motion. The existence of frictional force is illustrated as follows.

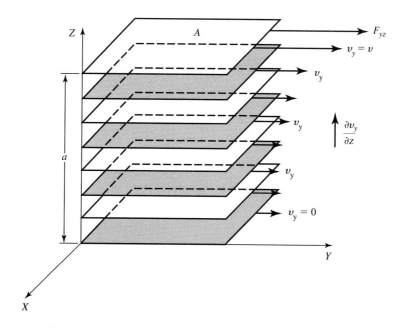

Figure 15.17   Velocity distribution in the case of viscous fluid flow.

Let us assume that the velocity of the fluid is in the Y direction. The fluid is flowing in layers that are parallel to the XY plane, as shown in Fig. 15.17, while the velocity v_y is a function of z only; that is, $v_y = f(z)$. Suppose plate A is in contact with the upper layer of a fluid and is moving with velocity v in the Y direction. A constant force $\mathbf{F}$ is needed to maintain a constant velocity, indicating the presence of a frictional force within the fluid. A layer that is in contact with the moving plate moves with the velocity of the plate so that there is no relative velocity between them. Similarly, a fluid layer next to a stationary layer will be at rest. That is, there is zero relative velocity between the solid-fluid interface, leading to zero slip at these surfaces.

As shown in Fig. 15.17, the velocity gradient is $\partial v_y/\partial z$ and is positive to the right. The viscous friction produces a positive shearing stress F_{yz} acting from left to right across an area A and parallel to the XY plane such that the normal to this plane is parallel to the Z-axis. The *coefficient of viscosity* η is defined as the ratio of the shearing stress to the velocity gradient; that is,

$$\eta = \frac{F_{yz}/A}{\partial v_y/\partial z} \qquad (15.251)$$

Actually, the presence of a velocity gradient implies the existence of a shearing force acting on different layers of the fluid. Equation (15.251) takes a simple form if $F_{yz} = F$ and $\partial v_y/\partial z = v/a$; thus

$$\eta = \frac{F/A}{v/a} \qquad (15.252)$$

This definition implies a simple type of distribution in which the shear stress is proportional to the first power of the velocity gradient. This is *Newtonian flow*. In most situations, flow is non-Newtonian and viscosity is a much more complicated function, resulting in a complicated shearing stress. We shall limit our discussion to Newtonian flow and illustrate the preceding definition by applying it to a laminar flow (fluid flows in layers) in circular pipes.

Consider a steady flow of fluid through a circular pipe of cross-sectional area $A = \pi r_0^2$, where r_0 is the radius of the pipe. The velocity everywhere is parallel to the axis of the pipe. As shown in Fig. 15.18, the axis of the pipe is taken along the Y-axis, and the velocity v_y is a function only of the distance r from the axis of the pipe; that is, the velocity gradient is dv_y/dr. Consider a fluid cylinder of radius r and length L so that $A = (2\pi r)L$. Thus the force exerted on this cylinder from the fluid outside this cylinder is

$$F = \eta(2\pi rL)\frac{dv_y}{dr} \tag{15.253}$$

The only forces acting on these fluids are the viscous force and the pressure difference ΔP between the two ends that are a distance L apart. In the absence of a body force and no acceleration, the sum of these two forces must be zero; that is,

$$\Delta P(\pi r^2) + F = 0 \tag{15.254}$$

Substituting for F from Eq. (15.253) and rearranging, we get

$$\frac{dv_y}{dr} = -\frac{\Delta P}{2\eta L} r \tag{15.255}$$

We integrate this outward from the cylinder axis, assuming $v = v_0$ at $r = 0$, and $v = v_y$ at $z = r$:

$$\int_{v_0}^{v_y} dv_y = -\frac{\Delta P}{2\eta L}\int_0^r r\, dr \tag{15.256}$$

$$v_y - v_0 = -\frac{\Delta P}{4\eta L} r^2 \tag{15.257}$$

If we assume that the fluid is at rest at the walls, that is, $v_y = 0$ at $r = r_0$, we get the maximum velocity:

$$v_0 = \frac{\Delta P}{4\eta L} r_0^2 \tag{15.258}$$

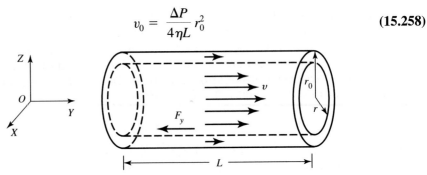

Figure 15.18   Laminar flow in a cylindrical pipe.

Substituting this in Eq. (15.257) yields

$$v_y = \frac{\Delta P}{4\eta L}(r_0^2 - r^2) \tag{15.259}$$

Since $A = \pi r^2$ and $dA = 2\pi r\, dr$, the total fluid current I or mass flow through the pipe is given by

$$I = \int\!\!\int_A \rho v_y\, dA = 2\pi\rho \int_0^{r_0} v_y r\, dr \tag{15.260}$$

Substituting for v_y from Eq. (15.259) and integrating, we obtain

$$\frac{I}{\rho} = \frac{\pi r_0^4}{8\eta L}\Delta P \tag{15.261}$$

which is the statement of *Poiseuille's law*. Equation (15.261) contains measurable quantities; hence η can be calculated from it.

We can find the average velocity $\bar{v}$ of the fluid by using the definition of mass flow. Consider the expression

$$\rho\bar{v}A = \rho \int v_y\, dA = \text{mass flow} \tag{15.262}$$

Substituting for v_y from Eq. (15.259), $dA = 2\pi r\, dr$, and integrating from $r = 0$ to $r = r_0$, we obtain

$$\bar{v} = \frac{\Delta P}{2\eta L r_0^2}\int_0^{r_0}(r_0^2 - r^2)r\, dr = \frac{r_0^2}{8\eta L}\Delta P \tag{15.263}$$

which gives the relation between the pressure drop and the average velocity.

Laminar (Streamline) and Turbulent Motions

Let us now investigate the motion of an object in a fluid and its relation to frictional forces. Suppose a sphere of radius r is moving with a small constant velocity v in a liquid of viscosity η. It is assumed that the velocity is small enough so that we can have a streamlined motion. Since the sphere is moving with uniform velocity, the applied forces must be equal to the frictional force F. We can evaluate F by means of dimensional analysis. Let us assume that the frictional force F is a function of r, v, and η. Thus we may write

$$F = Kr^a v^b \eta^c \tag{15.264}$$

where K is a dimensionless constant that cannot be evaluated from dimensional analysis. Substituting the dimensions of various quantities, we get

$$[MLT^{-2}] = [L]^a[LT^{-1}]^b[ML^{-1}T^{-1}]^c \tag{14.265}$$

which gives

$$a = b = c = 1 \qquad (15.266)$$

Hence

$$F = Kr\upsilon\eta \qquad (15.267)$$

The value of K can be determined experimentally. This is done by measuring the force required to pull a sphere of known radius through a liquid of known viscosity. K is found to be 6π. Thus Eq. (15.267) takes the form

$$F = 6\pi r\upsilon\eta \qquad (15.268)$$

which is known as *Stokes' law*.

We can now discuss the motion of a small sphere falling through a viscous fluid at constant velocity. According to Archimedes' principle, the net weight of the sphere is

$$F_{net} = \frac{4\pi}{3} r^3(\rho_s - \rho_l)g \qquad (12.269)$$

where ρ_s and ρ_l are the densities of the material of the sphere and that of the liquid, respectively. This force must be equal to the frictional force given by Eq. (15.268). That is,

$$6\pi r\upsilon\eta = \frac{4\pi}{3} r^3(\rho_s - \rho_l)g$$

That is,

$$\upsilon = \frac{2g}{9\eta}(\rho_s - \rho_l)r^2 \qquad (15.270)$$

Thus, by measuring υ, since all the other quantities are known, we can calculate η. It is important to remember that the preceding results are applicable only if the motion is laminar or streamlined. For example, a stone falling through glycerine may have streamlined motion, but not if falling through water.

Sir Osborne Reynolds found that, as the velocity of an object through any liquid increases, there is a critical velocity when a sudden change from laminar motion to turbulent motion occurs. This critical velocity υ_c depends on the density ρ of the liquid, its viscosity η, and diameter d of the cylindrical tube in which the liquid is flowing. Thus we may once again make use of dimensional analysis and write

$$\upsilon_c = R_e\rho^a\eta^b d^c \qquad (15.271)$$

where R_e is a dimensionless quantity called the *Reynolds number*. Substituting dimensions for different quantities, we obtain

$$[LT^{-1}] = [ML^{-3}]^a[ML^{-1}T^{-1}]^b[L]^c$$

which yields $a = -1, \quad b = 1, \quad \text{and} \quad c = -1$

Therefore,

$$v_c = R_e \frac{\eta}{\rho d} \qquad (15.272)$$

and

$$R_e = \frac{\rho d v_c}{\eta} \qquad (15.273)$$

Thus, using cylindrical tubes, knowing d, ρ, and η, we can measure v_c and hence calculate R_e. Since the velocity of a liquid in a tube varies from a maximum along the axis to zero at the edges, we must use average velocities over the whole cross section in order to calculate critical velocity. From his experimental work using the flow of liquids through glass tubes, Reynolds concluded that the flow of liquids is laminar if $R_e < 2000$, whereas the flow of liquids is turbulent if $R_e > 4000$. For a liquid where the predominantly viscous forces damp out any fluctuations, Reynolds numbers have low values. On the contrary, if the viscous forces are significant, Reynolds numbers will be large, indicating the existence of turbulent flow.

When an object is moving below the critical velocity v_c, the motion is laminar, and the frictional force is caused by viscosity. As soon as the velocity is greater than the critical velocity, the motion is turbulent; eddies are set up in front of the moving object. The frictional force now mainly depends on the pressure difference between the front and back of the object and very slightly on the viscosity. Since the pressure difference depends on the cross-sectional area of the object, we may write the frictional force as

$$F = K v^a \rho^b A^c \qquad (15.274)$$

Once again dimensional analysis yields

$$F = K \rho A v^2 \qquad (15.275)$$

where K depends on the shape of the body and may have a value varying from 0.9 to 0.01.

PROBLEMS

15.1. Derive Eqs. (15.34) and (15.35).

15.2. A string of length L and mass m is tied at both ends. The midpoint of the string is pulled a distance $h = L/10$ in the vertical direction and released. Find an expression that describes the motion of the string.

15.3. A uniform string of length L and linear mass density μ under tension T, is displaced initially ($h \ll L$), as shown in Fig. P15.3. Find the general solution of the equation that describes the motion of the vibrating string and evaluate the coefficients by using initial conditions.

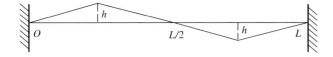

Figure P15.3

15.4. A uniform string of length L and linear mass density μ, under tension T, is initially in an equilibrium position but has a velocity given by

$$v = ax, \qquad 0 < x < \frac{L}{2}$$

$$v = a(x - L), \qquad \frac{L}{2} < x < L$$

where a is a constant. Find the general solution of the equation that describes the motion of the vibrating string and evaluate the coefficients by using initial conditions. Make graphs to describe the nature of the vibrating string.

15.5. A string of length L and mass m is tied at $x = 0$ and the end $x = L$ is tied to a ring that slides without friction on a vertical rod. Show that the boundary condition at end $x = L$ is $(\partial u/\partial x)_{x=L} = 0$, and find the normal frequencies and normal modes of vibrations. Make the graphs to describe the nature of the vibrating string.

15.6. Find the general solution for the equation of motion and normal modes of vibrations of a string with the following initial conditions:

$$u(x, 0) = A \sin \frac{3\pi x}{L} \quad \text{and} \quad \dot{u}(x, 0) = 0$$

15.7. A stretched string of length L and mass m is set into vibration by striking it over a length $2a$ at the center. The situation is described by the following initial conditions:

$$u(x, 0) = 0$$

$$\dot{u}(x, 0) = 0, \qquad \text{for } x < \frac{L}{2} - a \quad \text{and} \quad x > \frac{L}{2} + a$$

Describe the motion of the string.

15.8. Calculate the characteristic frequencies and its amplitudes for different modes for a vibrating string under the following initial conditions:

$$u(x, 0) = \frac{4x(L - x)}{L^2} \quad \text{and} \quad \dot{u}(x, 0) = 0$$

15.9. A string of length L and mass m, under tension T, is fixed at both ends. If the string is pulled in such a way that it has a parabolic shape given by $y = a(L - x)x$ and then released, investigate the motion of the string by graphing different vibrating modes.

15.10. Consider a string of length L and mass m. The end $x = L$ is tied, while at the end $x = 0$ a sinusoidal force of a $\sin \omega_0 t$ (a and ω_0 are constants) is applied. Find the solution in which all portions of the string vibrate with the same frequency ω_0; that is, find the solution of the equation for steady-state motion. Discuss the nature of vibrations by graphing.

15.11. Consider a stretched string of L and mass m tied at both ends. A force proportional to the position of the string, that is,

$$F(x, t) = F_0 \sin \frac{n\pi x}{L} \cos \omega t$$

where n is an integer, is applied along the length of the string. Investigate the steady-state motion of the string by assuming a similar dependence for $u(x, t)$.

15.12. Solve Problem 15.11 for a more general situation in which the applied force has the form $F(x, t) = F_0(x) \cos \omega t$, where $F_0(x)$ is zero at both ends of the string.

15.13. If the wave function $u(x, t) = Ae^{i(\omega t - kx)}$ is such that the quantities ω and v are complex, while k is real, then such a wave is damped in time. (Assume $\omega = \alpha + i\beta$ and $v = u + iw$, where α, β, u, and w are real.)

15.14. Solve Problem 15.2 (pulled $L/10$ at the midpoint) by superimposing waves $f(x - ct)$ and $g(x + ct)$. Draw wave forms at different time intervals.

15.15. Derive Eqs. (15.92) and (15.94).

15.16. Derive Eq. (15.100).

15.17. Evaluate P given in Eq. (15.102) by using Eqs. (15.89) and (15.90).

15.18. Derive Eq. (15.109) by first evaluating K, V, and L; then use the Lagrange equation.

15.19. In a vibrating string fixed at both ends, if p_n is the generalized momentum conjugate to θ_n, what is the Hamiltonian function H?

15.20. Consider a uniform string of length L and linear mass density μ, tied at both ends and under tension T. A force of $a \cos \omega t$ (a and ω being constants) is applied at $x = L/2$. Initially, the string is at rest with its middle point having a displacement of h ($=L/20$). The retarding frictional force is proportional to the velocity ($=-b\dot{x}$) and acts all along the string. Find the solution of the motion of the vibrating string for the underdamped case.

15.21. In Problem 15.20, suppose the string is vibrating in the nth mode. Calculate (**a**) the rate at which the driving force is doing work, and (**b**) the average rate of doing this work.

15.22. Consider a string fixed at both ends and vibrating in a viscous medium. The damping force on any portion of the string is directly proportional to the velocity ($=-b\dot{x}$). Show that the general solution of the motion of the vibrating string is satisfied by

$$u(x, t) = e^{-\alpha x} \sin \omega \left(\frac{x}{v} - t \right)$$

where α is a constant.

15.23. Obtain the solution of Eq. (15.113), starting with Eq. (15.128) instead of Eq. (15.127).

15.24. Show that the equations of motion for longitudinal vibrations of a loaded string are exactly of the same form as transverse vibrations, provided we replace T/d by k, the force constant of the string.

15.25. Discuss the wave propagation along a string loaded with two different types of particle masses that alternate in their positions:

$$m_j = \begin{cases} m_1, & \text{for } j \text{ even} \\ m_2, & \text{for } j \text{ odd} \end{cases}$$

Show that the $\omega - k$ curve has two branches in this case.

15.26. Consider the situation discussed in Problem 15.5, where the right end of the string is attached to a ring around a vertical rod. Once the string is set into vibration, discuss the reflection of waves from the right end for (**a**) a massless and frictionless ring, (**b**) a ring of small mass and little friction, (**c**) a ring of small mass and large friction, and (**d**) a very heavy ring.

15.27. Suppose, in Section 15.6, the incident wave in coming from the right and meets a junction between two strings of different densities. Calculate the relative amplitudes and intensities for the reflected and transmitted waves. Also calculate the energy being transmitted from one string to the other, and vice versa.

15.28. In connection with Eq. (15.161), show that this is the rate at which the energy is supplied to the junction from left to right.

15.29. Consider a infinitely long string, as shown in Fig. P15.29. For $x < 0$ and $x > L$, the linear mass density of the string is μ_1, and for $0 < x < L$, the linear mass density is $\mu_2 (> \mu_1)$. A wave of amplitude A_0 and frequency ω is incident from the left side. Find the reflected and transmitted intensities at A and B. How do these values change for a relative change in the values of μ_1 and μ_2?

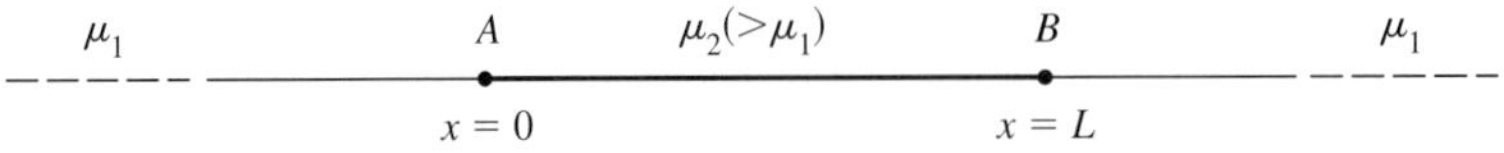

Figure P15.29

15.30. A stretched string of infinite length is under tension T. A wave of frequency ω and velocity ω/k is incident from the left. Calculate the reflected and transmitted amplitudes when mass M is attached (a) at $x = 0$. and (b) at $x = L$.

15.31. An electrical transmission line has a uniform inductance L per unit length and a uniform capacitance C per unit length. Show that the alternating current i in such a transmission line satisfies the wave equation

$$\frac{\partial^2 i}{\partial x^2} = \frac{1}{v^2} \frac{\partial^2 i}{\partial t^2}, \qquad \text{where } v = 1/\sqrt{LC}$$

15.32. Show that the spherical wave $p = f(r - vt)/r$ satisfies the longitudinal wave equation for sound waves.

15.33. Consider a right circular cone of half-angle ϕ, height h, and mass density ρ_l. The cone is floating in a liquid of density ρ_l. Show that the cone will be in stable equilibrium only if the vertex points vertically upward. Determine the frequency of small oscillations for this system.

15.34. We know that the density of air in the atmosphere varies with the altitude. Let us assume that the density is constant and equal to 1.3 kg/m^3, that is, the density at standard temperature and pressure at sea level. What would be the total thickness of the atmosphere?

15.35. For an incompressible fluid, if two components of velocity are given, how would you determine the third? If $v_x = 3x^2 y^2 z t^3$ and $v_y = x^2 y^2 z^2 t^3$, calculate v_z.

15.36. Rewrite Euler's equations (15.238) and (15.240) for the motion of the fluids in cylindrical polar coordinates.

15.37. Rewrite Euler's equations (15.238) and (15.240) for the motion of the fluids in spherical polar coordinates.

15.38. Show that the velocity $\mathbf{v}$ for an ideal incompressible fluid that experiences irrotational flow may be derived from a scalar potential satisfying Laplace's equation; that is, $\nabla^2 \phi = 0$, where $\mathbf{v} = -\nabla \phi$.

15.39. Using Gauss's divergence theorem where appropriate, write (a) the continuity equation in integral form, and (b) Euler's equation in integral form.

15.40. Consider a sealed cubical container half filled with water and half with air. A small hole is made in the base of the container. Determine the velocity of efflux of the water as a function of the water level. How does this compare with the results obtained for an open container? You may assume that the water is incompressible and that the entire process is isothermal.

15.41. A container of cross-sectional area A and height H is filled with an ideal incompressible fluid. The fluid is drained through a small hole of cross-sectional area a. Calculate the time required to drain half of the fluid.

15.42. The velocity distribution v for an incompressible fluid in a turbulent flow through a circular pipe of radius r_0 is given by

$$v = v_0\left(1 - \frac{r}{r_0}\right)^{1/7}$$

where v_0 is the velocity at the axis. Calculate the volumetric flow rate.

15.43. Using Euler's equation, Eq. (15.240), derive an expression for the conservation of angular momentum.

15.44. The function $\phi = c/r$, where c is a constant and r is a distance from a fixed point, satisfies Laplace's equation $\nabla^2 \phi = 0$ except at $r = 0$. Discuss the nature of the fluid flow if ϕ represents the velocity potential.

15.45. A circular pipe of length l and radius r_0, open at both ends, is held vertically. Air is blown across the top open end. What is the differential equation for the sound waves set up in the pipe? Using the appropriate boundary conditions, determine the normal modes of vibrations along the axis of the pipe.

15.46. Repeat Problem 15.45 if the lower end of the pipe is closed.

15.47. Show that the force F resulting from the fluid viscosity may be written as

$$\mathbf{F} = \int\int_A \eta \nabla \mathbf{v} \cdot d\mathbf{A}$$

Calculate the body force $\mathbf{f}$, that is, the force per unit volume.

15.48. Find the increase in the density of water 30 m below the surface of a lake. The bulk modulus of water is 2×10^4 atm and its density is 1000 kg/m^3. For each 10-m depth, the pressure increases by 1 atm.

15.49. Let P_0 be the pressure at sea level and P at the top of a column of air H meter in height. Assuming a uniform temperature of T K, show that $\log_{10} P_0 - \log_{10} P = C[H/T]$, where C is a constant. Assuming pressure at sea level to be 1.013×10^5 N/m^2 and the density of air at 0°C to be 1.29 kg/m^3.

15.50. Consider a streamlined flow of water. At some fixed point, the velocity of the water is 60 cm/s and the rate of change of velocity with distance is 12 cm/s/cm. Calculate the acceleration of the water at the fixed point.

15.51. A horizontal tube 20-cm long and 1.0 cm in diameter is connected at one end to a water tank whose constant-level height is 2 m. Calculate the coefficient of viscosity if 500 cm^3 flows through the tube in 5 min.

15.52. Consider a horizontal capillary tube of length L and radius R connected to an airtight vessel of volume V. Air escapes from the vessel through the capillary tube, and in time t the air pressure reduces from P_1 to P_2. If the atmospheric pressure is P_0, show that the coefficient of viscosity is given by

$$\eta = \frac{\pi R^4 P_0 t}{8LV \log_e \dfrac{(P_1 - P_0)(P_2 + P_0)}{(P_2 - P_0)(P_1 + P_0)}}$$

(*Hint:* Consider a small section of a capillary tube of length dx where pressure is P, and then integrate over the whole length and use Boyle's law.)

15.53. Using the expression derived in Problem 15.52, calculate the coefficient of viscosity for the following data: $L = 1$ m, $R = 0.05$ m, $V = 0.5$ m^3, the original pressure is 81 cm of mercury, the final pressure after 30 s is 79.5 cm of mercury, the atmospheric pressure is 76 cm of mercury, and the temperature is assumed to be constant throughout.

15.54. Assuming a Reynolds number of 1200 for a cylindrical pipe of 2.5-cm radius, calculate the critical velocity of (**a**) water and (**b**) glycerin, both at 20°C.

SUGGESTIONS FOR FURTHER READING

ARTHUR, W., and FENSTER, S. K., *Mechanics*, Chapter 13. New York: Holt, Rinehart and Winston, Inc., 1969.

BECKER, R. A., *Introduction to Theoretical Mechanics*, Chapter 15. New York: McGraw-Hill Book Co., 1954.

BRILLOUIN, L., *Wave Propagation and Group Velocity*. New York: Academic Press, Inc., 1960.

DAVIS, A. DOUGLAS, *Classical Mechanics*, Chapter 12. New York: Academic Press, Inc., 1986.

FRENCH, A. P., *Vibrations and Waves*, Chapters 5, 6, and 7. New York: W. W. Norton and Co., Inc., 1971.

HAUSER, W., *Introduction to the Principles of Mechanics*, Chapter 12. Reading, Mass.: Addison-Wesley Publishing Co., 1965.

MARION, J. B., *Classical Dynamics*, 2nd ed., Chapters 14 and 15. New York: Academic Press, Inc., 1970.

*MOORE, E. N., *Theoretical Mechanics,* Chapter 7. New York: John Wiley & Sons, Inc., 1983.

MORSE, P. M., *Vibration and Sound*, 2nd ed. New York: McGraw-Hill Book Co., 1948.

PEARSON, J. M., *A Theory of Waves*. Needham Heights, Mass.: Allyn and Bacon, Inc., 1966.

ROSSBERG, K., *Analytical Mechanics*, Chapter 10. New York: John Wiley & Sons, Inc., 1983.

SLATER, J. C., *Mechanics,* Chapters 8–11 and 13. New York: McGraw-Hill Book Co., 1947.

STEPHENSON, R. J., *Mechanics and Properties of Matter*, Chapters 8 and 9. New York: John Wiley & Sons, Inc., 1962.

SYMON, K. R., *Mechanics*, 3rd ed., Chapter 8. Reading, Mass.: Addison-Wesley Publishing Co., 1971.

SYNGE, J. L. and GRIFFITH, B. A., *Principles of Mechanics*, Chapter 17. New York: McGraw-Hill Book Co., 1959.

*The asterisk indicates works of an advanced nature.

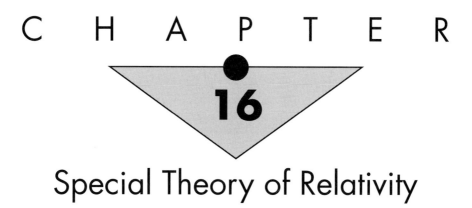

C H A P T E R

16

Special Theory of Relativity

16.1 INTRODUCTION

By the close of the nineteenth century, classical mechanics as formulated by Newton and electromagnetic theory as formulated by Maxwell had been well established. Newtonian mechanics assumes a complete separability of space and time, as well as the concept of the absoluteness of time. It was the breakdown of these concepts that led to the formulation of the special theory of relativity. Questions that were to be answered were:

1. How do we transform the description of a system from one inertial coordinate system to another?
2. What happens to the equations that describe the system when such a transformation is made?
3. Do Newton's laws of motion describe the motion of the system at all speeds?

In the next section, we shall discuss Galilean transformations that answer the first two questions. But two difficulties are encountered with Galilean transformations:

1. They do not correctly transform electromagnetic equations.
2. These transformations are not valid even for Newtonian mechanics when the systems are moving with very high speeds approaching that of light (3×10^8 m/s).

By the close of the nineteenth century, there were several experimental results that could not be explained by Newtonian mechanics. The final blow was struck by the experimental results of A. A. Michelson and E. W. Morley (1881–1887). They failed in their attempt to measure the absolute velocity of Earth with respect to "stationary ether." That is, they did not succeed in their search for an absolute reference frame or inertial frame. Because of these difficulties, fundamental changes in the structure of dynamics were carried out by H. Poincare, H. A. Lorentz, and A. Einstein. The resulting new mathematical structure, as formulated by Einstein in 1905, is known as the *special theory of relativity,* which deals with inertial systems moving with uniform velocities with respect to each other. In 1916, Einstein introduced another mathematical structure known as the *general theory of relativity,* which deals with accelerated, or noninertial systems. We shall limit our discussion to the special theory of relativity.

16.2 GALILEAN TRANSFORMATIONS AND GALILEAN INVARIANCE

Consider two observers located in two different inertial systems S and S', called the unprimed and primed systems, respectively. Let $X_1X_2X_3$ and $X_1'X_2'X_3'$ be the coordinate axes located in systems S and S', respectively. Inertial system S' is moving with velocity v with respect to system S along the X_1–X_1' axes (or S is moving with velocity $-v$ with respect to S'), as shown in Fig. 16.1. To describe an event (a physical phenomenon) in an inertial system, an observer must specify, in addition to three space coordinates, a fourth coordinate, time. Let the event at P be described by the two sets of coordinates (x_1, x_2, x_3, t) and (x_1', x_2', x_3', t'), corresponding to two observers in inertial systems S and S', respectively. According to Newton's concept, this fourth coordinate time has the same value in all inertial systems; that is, time is an absolute quantity independent of the reference frame. This implies that $t = t'$ and leads to the *invariance* of the laws of mechanics under the following Galilean transformations. Later we shall see that this is not true, and it is through the relaxation of this condition, that is, $t \neq t'$, that an altogether different theory results.

An event taking place at P when observed by two observers is related by the equations:

$$x_1' = x_1 - vt \qquad x_2' = x_2 \qquad x_3' = x_3 \qquad t' = t \qquad \textbf{(16.1)}$$

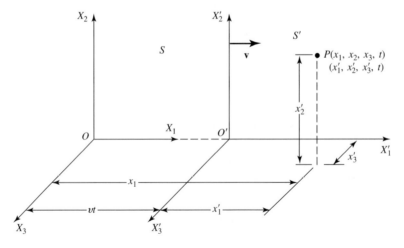

Figure 16.1   An inertial system S' is moving relative to inertial system S with velocity v along the X_1–X_1' axes.

These equations, which relate the space and time coordinates in these two systems moving with uniform relative velocity, are called *Galilean* (or *Newtonian*) *transformations.* According to Eqs. (16.1), an element of length in the two systems is the same and is given by

$$ds^2 = \sum_{i=1}^{3} dx_i^2 = \sum_{i=1}^{3} dx_i'^2 = ds'^2 \qquad \text{(16.2)}$$

As stated earlier, the fact that Newton's laws are invariant with respect to Galilean transformations is called the *principle of Newtonian relativity* or *Galilean invariance.* We demonstrate this now. Galilean velocity transformations may be obtained by differentiating Eqs. (16.1) and noting that $t = t'$, and the operators d/dt and d/dt' are identical.

$$\frac{dx_1'}{dt'} = \frac{dx_1}{dt} - v, \qquad \frac{dx_2'}{dt'} = \frac{dx_2}{dt}, \qquad \frac{dx_3'}{dt'} = \frac{dx_3}{dt} \qquad \text{(16.3)}$$

Similarly, by differentiating Eq. (16.3), we get the corresponding Galilean acceleration transformations.

$$\frac{d^2x_1'}{dt'^2} = \frac{d^2x_1}{dt^2}, \qquad \frac{d^2x_2'}{dt'^2} = \frac{d^2x_2}{dt^2}, \qquad \frac{d^2x_3'}{dt'^2} = \frac{d^2x_3}{dt^2} \qquad \text{(16.4a)}$$

That is,
$$\ddot{x}_j' = \ddot{x}_j \qquad \text{(16.4b)}$$

or
$$\mathbf{a}' = \mathbf{a} \qquad \text{(16.5)}$$

That is, accelerations are the same as viewed from either inertial system. Thus Newton's equations of motion for a force acting on a particle of mass m at P as viewed from two inertial systems are

$$F_j = m\ddot{x}_j = m\ddot{x}_j' = F_j' \qquad \text{(16.6)}$$

or
$$\mathbf{F} = m\mathbf{a} = m\mathbf{a}' = \mathbf{F}' \qquad \text{(16.7)}$$

When the equations have the same form in different inertial systems, we say that the law of motion is *invariant* under Galilean transformations. It may be pointed out that individual terms are not invariant, but each term transfers according to the same scheme. When different terms transform according to the same scheme, they are said to be *covariant.* Mathematically, Galilean invariance may be stated as follows. Suppose a mechanical situation is described in system S by an equation of the form

$$f(x_1, x_2, x_3, t) = 0 \qquad \text{(16.8)}$$

Then in the system S' it will have the form

$$f(x_1', x_2', x_3', t') = 0 \qquad \text{(16.9)}$$

A key question to ask at this stage is: Are these Galilean transformations valid for electromagnetism as well? Unfortunately, the laws of electromagnetism are not invariant under

Galilean transformations. For example, a spherical electromagnetic wave propagating with constant speed c in system S may be given by

$$x_1^2 + x_2^2 + x_3^2 - c^2t^2 = 0 \tag{16.10}$$

For this equation to be invariant, its form in the system S' should be

$$x_1'^2 + x_2'^2 + x_3'^2 - c^2t'^2 = 0 \tag{16.11}$$

Substituting for x_1, x_2, x_3, and t from Eq. (16.1) into Eq. (16.10) yields (taking c to be constant in all inertial systems, an experimental fact)

$$(x_1' + vt)^2 + x_2'^2 - x_3'^2 - c^2t'^2 = 0 \tag{16.12}$$

which implies

$$x_1^2 + x_2^2 + x_3^2 - c^2t^2 \neq x_1'^2 + x_2'^2 + x_3'^2 - c^2t'^2 = 0 \tag{16.13}$$

or

$$\sum_{i=1}^{3} x_i^2 - c^2t^2 \neq \sum_{i=1}^{3} x_i'^2 - c^2t'^2 \tag{16.14}$$

Thus Galilean invariance does not hold for electromagnetism.

16.3 EINSTEIN'S POSTULATES AND LORENTZ TRANSFORMATIONS

All experimental evidence accumulated by the close of the nineteenth century led to the conclusion that there is no special frame of reference, and that all frames of references moving with uniform relative velocities are equivalent. The solution to this and many other problems in physics was provided by Einstein in 1905 in his paper on the electrodynamics of moving bodies. The assumptions made by him can be stated in the form of two basic postulates called *Einstein's postulates of the special theory of relativity.*

> *Postulate I: Principle of Special Relativity.*   *The laws of physics, including those of mechanics and electromagnetism, are the same in all inertial systems, even though these inertial frames may be in relative uniform translational motion.*

> *Postulate II: Principle of the Constancy of the Speed of Light.*   *The speed of light in free space (vacuum) is always constant, equal to c ($= 3 \times 10^8$ m/s), independent of the inertial system and the relative motion of the source and observer.*

Postulate I is a generalization of the fact that all physical laws (including those of mechanics and electromagnetism) are invariant under all transformations. This brings about a degree of unity between the two. It also states that an absolute (or preferred) inertial system does not exist. Postulate II is merely a statement of an experimental fact.

Although these two postulates look very simple, they have far-reaching consequences. First, let us derive transformation equations for inertial systems moving with uniform relative velocity. Let us at the same time require that these transformations be applicable to both Newtonian mechanics as well as electromagnetism. Einstein derived these transformations in 1905, and they are called *Lorentz transformations,* because H. A. Lorentz, in 1890, originated them as part of his theory of electromagnetism. These Lorentz transformations are discussed next.

Once again, consider two uniformly moving inertial systems S and S' whose origins coincide at time $t = 0$, as shown in Fig. 16.1. Suppose, at $t = 0$, a source of light at the origin of the unprimed inertial system S emits a pulse of light. An observer in this system will see a spherical wavefront propagating with speed c. The equation of this wavefront according to an observer in S is

$$x_1^2 + x_2^2 + x_3^2 - c^2 t^2 = 0 \qquad (16.15)$$

Because of the invariance of the speed of light, to an observer in a moving system S' (with respect to the source), light propagates as a spherical wave from the observer's origin. The equation of the wavefront is

$$x_1'^2 + x_2'^2 + x_3'^2 - c^2 t'^2 = 0 \qquad (16.16)$$

Note that $t \neq t'$ implies the possibility that the time scale may also transform in going from one inertial system to another. Thus the invariance of a physical law requires that

$$x_1^2 + x_2^2 + x_3^2 - c^2 t^2 = x_1'^2 + x_2'^2 + x_3'^2 - c^2 t'^2 \qquad (16.17)$$

or
$$\sum_{i=1}^{3} x_i^2 - c^2 t^2 = \sum_{i=1}^{3} x_i'^2 - c^2 t'^2 \qquad (16.18)$$

Let us now introduce a fourth imaginary coordinate:

$$x_4 = ict \quad \text{and} \quad x_4' = ict' \qquad (16.19)$$

Thus Eq. (16.18) takes the form

$$\sum_{\mu=1}^{4} x_\mu^2 = \sum_{\mu=1}^{4} x_\mu'^2 \qquad (16.20)$$

(Note that Roman letters, $i, j, k,$ and so on, are used when the range of indexes is 1 to 3, and Greek letters, $\mu, \nu, \lambda,$ and so on, are used when the range of indexes is 1 to 4.) Comparison of Eq. (16.20) with Eq. (16.2) shows that it is analogous to the three-dimensional, distance-preserving *orthogonal transformation* for rotation from one system to another. Thus Eq. (16.20) implies that the transformation we are seeking corresponds to a rotation in a four-dimensional space consisting of three dimensions of ordinary space and a fourth imaginary dimension proportional to time (ict). This four-dimensional space is known as *Minkowski space* or *world space.* Therefore, *Lorentz transformations are simply orthogonal transformations in Minkowski space.*

The inertial system S' is moving with uniform velocity **v** with respect to inertial system S. Thus we are seeking a pure Lorentz transformation, one that is not concerned with spatial rotation. A pure Lorentz transformation is concerned only with uniformly moving inertial systems. There is no loss of generality if we assume that the direction of velocity **v** of S' with respect to S is along one of the axes, say along the X_1–X_1' axes. Thus, if the transformations are orthogonal, we are interested in obtaining the matrix elements $\lambda_{\mu\nu}$ of the transformation matrix λ between X and X'; that is, in matrix notation

$$\mathbf{X'} = \lambda\mathbf{X} \tag{16.21}$$

or

$$x_\mu' = \sum_{\nu=1}^{4} \lambda_{\mu\nu}x_\nu \tag{16.22}$$

where $\lambda_{\mu\nu}$ are the elements of the Lorentz transformation matrix. Since the velocity is along the X_1–X_1' axes, we have

$$x_2 = x_2' \quad \text{and} \quad x_3 = x_3'$$

so that the transformation matrix λ is of the form

$$\lambda = \begin{pmatrix} \lambda_{11} & 0 & 0 & \lambda_{14} \\ 0 & 1 & 0 & 0 \\ 0 & 0 & 1 & 0 \\ \lambda_{41} & 0 & 0 & \lambda_{44} \end{pmatrix} \tag{16.23}$$

The matrix elements must satisfy the orthogonality conditions

$$\sum_{\nu=1}^{4} \lambda_{\mu\nu}\lambda_{\eta\nu} = \delta_{\mu\eta} \tag{16.24}$$

which result in the following three independent equations:

$$\lambda_{11}^2 + \lambda_{14}^2 = 1 \tag{16.25}$$

$$\lambda_{41}^2 + \lambda_{44}^2 = 1 \tag{16.26}$$

$$\lambda_{11}\lambda_{41} + \lambda_{14}\lambda_{44} = 0 \tag{16.27}$$

Unlike spatial orthogonal transformations, the elements in the present case may be imaginary. Since the x_1', x_2', and x_3' coordinates are real, the elements λ_{14}, λ_{24}, and λ_{34} must be imaginary. On the other hand, x_4' is imaginary, which requires λ_{41}, λ_{42}, and λ_{43} to be imaginary, while λ_{44} must be real.

To solve for the four independent matrix elements, we make use of the three relations given by Eqs. (16.25), (16.26), and (16.27), but we still need another one. A fourth condition is supplied from the fact that the origin of the prime system ($x_1' = 0$) is moving with uniform velocity along the X_1–X_1' axes so that at time t

$$x_1 = vt = -i\frac{v}{c}\,ict = -i\beta x_4, \qquad \text{for } x_1' = 0 \tag{16.28}$$

where

$$\beta \equiv \frac{v}{c} \tag{16.29}$$

From the relations of Eqs. (16.21) and (16.22), we may write

$$x_1' = \lambda_{11}x_1 + \lambda_{14}x_4 \qquad\qquad (16.30)$$

which must satisfy the condition that, when $x_1' = 0$, x_1 must be equal to vt. Then we can substitute, from Eq. (16.28), the value of $x_1 = -i\beta x_4$ in Eq. (16.30); that is,

$$0 = \lambda_{11}(-i\beta x_4) + \lambda_{14}x_4$$

or
$$\lambda_{14} = i\beta\lambda_{11} \qquad\qquad (16.31)$$

Substituting this in Eq. (16.25) yields

$$\lambda_{11}^2 + (i\beta\lambda_{11})^2 = 1$$

or
$$\lambda_{11} = \frac{1}{\sqrt{1 - \beta^2}} \qquad\qquad (16.32)$$

and, using this in Eq. (16.31), we obtain

$$\lambda_{14} = \frac{i\beta}{\sqrt{1 - \beta^2}} \qquad\qquad (16.33)$$

Using Eqs. (16.32) and (16.33) in Eqs. (16.26) and (16.27), we obtain

$$\lambda_{44} = \frac{1}{\sqrt{1 - \beta^2}} \qquad\qquad (16.34)$$

and
$$\lambda_{41} = \frac{-i\beta}{\sqrt{1 - \beta^2}} \qquad\qquad (16.35)$$

Thus, with the results obtained in the preceding four equations together with the notation

$$\gamma \equiv \frac{1}{\sqrt{1 - \beta^2}} \qquad\qquad (16.36)$$

the Lorentz transformation matrix $\boldsymbol{\lambda}$, Eq. (16.23), takes the form

$$\boldsymbol{\lambda} = \begin{pmatrix} \gamma & 0 & 0 & i\beta\gamma \\ 0 & 1 & 0 & 0 \\ 0 & 0 & 1 & 0 \\ -i\beta\gamma & 0 & 0 & \gamma \end{pmatrix} \qquad\qquad (16.37)$$

while, by using the relation $\mathbf{X'} = \lambda\mathbf{X}$, Eq. (16.21), the *Lorentz transformation equations* are

$$x_1' = \gamma(x_1 - vt) = \frac{x_1 - vt}{\sqrt{1 - \beta^2}}$$

$$x_2' = x_2$$

$$x_3' = x_3$$

$$t' = \gamma\left(t - \frac{\beta}{c}x_1\right) = \frac{t - (v/c^2)x_1}{\sqrt{1 - \beta^2}} \qquad (16.38)$$

The inverse transformation can be obtained by simply transposing the matrix (or by replacing v by $-v$ and interchanging the primed and unprimed coordinates); that is,

$$x_1 = \gamma(x_1' + vt')$$

$$x_2 = x_2'$$

$$x_3 = x_3'$$

$$t = \gamma\left(t' + \frac{\beta}{c}x_1'\right) \qquad (16.39)$$

These Lorentz transformations are covariant to the laws of mechanics as well as electromagnetism. Furthermore, these equations reduce to Galilean transformations as $v \to 0$.

16.4 SOME CONSEQUENCES OF LORENTZ TRANSFORMATIONS

Applying Lorentz transformations to a variety of situations leads to many interesting results and apparent paradoxes. We do not intend to get involved in these, but simply discuss some of the famous consequences of Lorentz transformations:

>   length contraction: Fitzgerald–Lorentz contraction
>   time dilation: slowing down of clocks
>   simultaneity
>   velocity addition

Length Contraction: Fitzgerald–Lorentz Contraction

According to classical physics, the length of an object, say a meter stick, is the same for all observers even though these observers may be in relative motion with respect to each other. But this is not so in relativity, and by making use of the Lorentz transformation equations, we can show that the length of an object is different for different observers.

Consider a rigid rod at rest in the unprimed system S lying along the X-axis so that its length is $L_0 = x_2 - x_1$. A moving observer in system S' (moving with velocity v along the X_1–X_1' axes) measures the length of the rod by locating the positions of both end points in his system at time t' to be x_1' and x_2'. From the inverse Lorentz transformations, Eq. (16.39), we obtain

$$x_1 = \frac{x_1' + vt'}{\sqrt{1 - \beta^2}} \tag{16.40}$$

$$x_2 = \frac{x_2' + vt'}{\sqrt{1 - \beta^2}} \tag{16.41}$$

Note that we have used the inverse transformations so that both ends of the rod are measured simultaneously by a moving observer, implying that $t_1' = t_2' = t'$; thus

$$x_2 - x_1 = \frac{x_2' - x_1'}{\sqrt{1 - \beta^2}} \tag{16.42}$$

Since $x_2 - x_1 = L_0$, and letting $x_2' - x_1' = L$, we may write

$$L = L_0\sqrt{1 - \beta^2} \tag{16.43}$$

Since $\sqrt{1 - \beta^2}$ is always less than 1, a rod of length L_0 at rest seems to be contracted to length L as measured by a moving observer. Thus the length is not absolute, but depends on the observer. Also the effect is reciprocal. If the rod were at rest in system S', It would look contracted as measured by an observer in system S, which is moving with velocity $-v$ with respect to the system S'. Equation (16.43) represents the *Fitzgerald–Lorentz contraction*.

We may caution the reader to recognize that the contraction of length is not a physical process. In relativity, it is the definition of measuring the length of an object in relative motion with respect to the observer that results in the contraction of length. One may think that, because of contraction, the objects viewed from systems in relative motion will look distorted—a sphere may look like an ellipse. But this is not so, the objects are only rotated, and we may see some otherwise nonvisible portions of the moving objects. This was first pointed out by James Terrell in 1959.

Time Dilation: Slowing Down of Clocks

Like length, a time interval also is not absolute. The time interval elapsed between two events depends on the relative motion of the observers.

Consider a clock located in an unprimed system S at point x_1. Suppose, according to an observer in S, the clock reads t_1, but an observer in a moving system S' notes the time t_1' to be

$$t_1' = \frac{t_1 - (v/c^2)x_1}{\sqrt{1 - \beta^2}} \tag{16.44}$$

Similarly, when an observer in system S notes time t_2, an observer in S' notes

$$t_2' = \frac{t_2 - (v/c^2)x_1}{\sqrt{1 - \beta^2}} \tag{16.45}$$

(Note that the clock is fixed at x_1 in system S for both observations.) An event lasting for a time interval $\Delta t = t_2 - t_1$ in system S will last for $\Delta t' = t_2' - t_1'$ in system S'. By subtracting Eq. (16.44) from Eq. (16.45), we obtain

$$t_2' - t_1' = \frac{t_2 - t_1}{\sqrt{1 - \beta^2}} \tag{16.46}$$

Thus the apparent time interval $\Delta t' = T$ in terms of the stationary time interval $\Delta t = T_0$ is

$$T = \frac{T_0}{\sqrt{1 - \beta^2}} \tag{16.47}$$

Because $\sqrt{1 - \beta^2}$ is always less than unity, the time interval T_0 on a stationary clock in S is recorded in a moving system S' as a *long time interval T* on a clock. Equation (16.47) is the relation showing the phenomenon of time dilation or the slowing down of clocks. For example, if $v = 0.98c$, from Eq. (16.47), $T = 5T_0$. This means, according to an observer in the S' system, that a clock in the S system is moving much slower and takes a time interval of five times longer for the same two events. The effect is reciprocal; when an observer in system S looks at a clock in system S', the observer thinks the clock in S' is running slow. Thus we may conclude:

If an observer is at rest in a reference frame, this frame is called the *proper frame;* the time interval measured between events taking place in the same frame is called the *proper time;* the length of an object located and measured in this system is called the *proper length.* If the observer is moving with respect to the inertial frame in which the object is located and in which the events are taking place, the measured length and time intervals are called the *improper length* and *improper time,* respectively. There is no such thing as an absolute proper frame, length, or time. Figure 16.2 illustrates length contraction [Eq. (16.43)] and time dilation [Eq. (16.47)].

Simultaneity

One fundamental difference between Newtonian mechanics and relativistic mechanics is that in relativistic mechanics space and time coordinates are so much interrelated that it is not possible to talk of one without the other; thus the definition of absolute simultaneity in the classical sense is meaningless. For example:

1. Two events that are happening at x_1 and x_2 in system S and are simultaneous in S ($t_1 = t_2$) will not be simultaneous in S' ($t_1 \neq t_2$).
2. Two events that are happening in system S at the same position ($x_1 = x_2$) but at different times will appear, as observed from system S', to be happening not only at different times but also happening at different locations ($x_1' \neq x_2'$ and $t_1' \neq t_2'$). These results are mathematically obvious from the Lorentz transformation equations.

The Lorentz transformation equations also indicate that one cannot have velocity greater than c. If a system could have $v > c$, the transformations from this system to a system at rest would not be possible because that will involve Lorentz transformations with imaginary coordinates, which is not possible.

Figure 16.2 _____

The Lorentz transformation for length contraction and time dilation. For given values of proper length L0 and proper time interval T0, we calculate the improper values by using Eqs. (16.43) and (16.47), respectively.

$$i := 0..15 \qquad c := 3 \cdot 10^8 \qquad L0 := 10^7 \qquad T0 := 2 \cdot 10^7$$

$$v0 := \frac{c}{15} \qquad vc_i := i \cdot v0 \qquad \beta_i := \frac{vc_i}{c}$$

$$L_i := L0 \cdot \sqrt{1 - \left(\beta_i\right)^2} \qquad T_i := \frac{T0}{\sqrt{1 - \left(\beta_i\right)^2} + 0.001}$$

vc_i	L_i	T_i
0	$1 \cdot 10^7$	$1.998 \cdot 10^7$
$2 \cdot 10^7$	$9.978 \cdot 10^6$	$2.002 \cdot 10^7$
$4 \cdot 10^7$	$9.911 \cdot 10^6$	$2.016 \cdot 10^7$
$6 \cdot 10^7$	$9.798 \cdot 10^6$	$2.039 \cdot 10^7$
$8 \cdot 10^7$	$9.638 \cdot 10^6$	$2.073 \cdot 10^7$
$1 \cdot 10^8$	$9.428 \cdot 10^6$	$2.119 \cdot 10^7$
$1.2 \cdot 10^8$	$9.165 \cdot 10^6$	$2.18 \cdot 10^7$
$1.4 \cdot 10^8$	$8.844 \cdot 10^6$	$2.259 \cdot 10^7$
$1.6 \cdot 10^8$	$8.459 \cdot 10^6$	$2.362 \cdot 10^7$
$1.8 \cdot 10^8$	$8 \cdot 10^6$	$2.497 \cdot 10^7$
$2 \cdot 10^8$	$7.454 \cdot 10^6$	$2.68 \cdot 10^7$
$2.2 \cdot 10^8$	$6.799 \cdot 10^6$	$2.937 \cdot 10^7$
$2.4 \cdot 10^8$	$6 \cdot 10^6$	$3.328 \cdot 10^7$
$2.6 \cdot 10^8$	$4.989 \cdot 10^6$	$4.001 \cdot 10^7$
$2.8 \cdot 10^8$	$3.59 \cdot 10^6$	$5.555 \cdot 10^7$
$3 \cdot 10^8$	0	$2 \cdot 10^{10}$

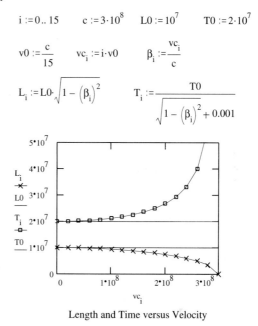

Length and Time versus Velocity

Looking at these graphs and the tables explain the changes in length and time with increasing velocity. How do these values differ from the classical theory values?

Velocity Addition

As before, let us consider two inertial systems S and S' moving with relative velocity **v** along the X_1–X_1' axes. Consider a particle P, which is moving in space and has velocity $\mathbf{u}(u_1, u_2, u_3)$ as measured by an observer in system S, and velocity $\mathbf{u}'(u_1', u_2', u_3')$ as measured by an observer in system S'. The aim is to find the relation between

$$u_1 = \frac{dx_1}{dt}, \qquad u_2 = \frac{dx_2}{dt}, \qquad u_3 = \frac{dx_3}{dt}$$

$$u_1' = \frac{dx_1'}{dt}, \qquad u_2' = \frac{dx_2'}{dt}, \qquad u_3' = \frac{dx_3'}{dt}$$

Differentiating Eqs. (16.38), we obtain

$$dx_1' = \gamma(dx_1 - v\,dt) \qquad dx_2' = dx_2 \qquad dx_3' = dx_3 \qquad dt' = \gamma\left(dt - \frac{v}{c^2}\,dx_1\right) \qquad \textbf{(16.48)}$$

Thus the components of $\mathbf{u'}$ are given by

$$u_1' = \frac{dx_1'}{dt'} = \frac{\gamma(dx_1 - v\,dt)}{\gamma[dt - (v\,dx_1/c^2)]} = \frac{(dx/dt) - v}{1 - (v/c^2)(dx_1/dt)}$$

or
$$u_1' = \frac{u_1 - v}{1 - (vu_1/c^2)} \qquad \textbf{(16.49a)}$$

Similarly,

$$u_2' = \frac{u_2}{\gamma[1 - (vu_1/c^2)]} \quad \text{and} \quad u_3' = \frac{u_3}{\gamma[1 - (vu_1/c^2)]} \qquad \textbf{(16.49b)}$$

These are called *Lorentz velocity transformations*. The inverse transformations $\mathbf{u}$ in terms of $\mathbf{u'}$ can be obtained by replacing v by $-v$ and interchanging the prime and unprimed coordinates.

Using these velocity transformations, it becomes quite clear that (1) nothing moves in a vacuum with velocity greater than the velocity of light, and (2) the velocity of light is independent of the relative motion of a source or an observer. Suppose in the inverse transformation

$$u_1 = \frac{u_1' + v}{1 + (vu_1'/c^2)} \qquad \textbf{(16.50)}$$

u_1' and v are both greater than $c/2$. According to classical theory, relative velocity u_1 will be greater than c; but according to Eq. (16.50), u_1 will be less than c. Similarly, if $u_1' = c$ and $v = c$, then $u_1 = 2c$ according to classical theory, but according to Eq. (16.50), u_1 is still equal to c. It must be noted that even though signal velocities cannot exceed c, there are certain other situations, such as phase velocity, in which v exceeds c.

Let us generalize our discussion in the following fashion. Consider three inertial systems S, S', and S'' in collinear motion along their respective X_1–X_1'–X_1'' axes. Let the velocity of S'' relative to S' be u_2 and that of S' relative to S be u_1. We want to find the relative velocity u of S'' with respect to S. This can be achieved by using the Lorentz transformation matrix matrix $\boldsymbol{\lambda}$. Thus

$$\boldsymbol{\lambda}_{s''\to s} = \boldsymbol{\lambda}_{s''\to s'}\boldsymbol{\lambda}_{s'\to s}$$

$$= \begin{pmatrix} \gamma_1 & 0 & 0 & i\beta_1\gamma_1 \\ 0 & 1 & 0 & 0 \\ 0 & 0 & 1 & 0 \\ -i\beta_1\gamma_1 & 0 & 0 & \gamma_1 \end{pmatrix} \begin{pmatrix} \gamma_2 & 0 & 0 & i\beta_2\gamma_2 \\ 0 & 1 & 0 & 0 \\ 0 & 0 & 1 & 0 \\ -i\beta_2\gamma_2 & 0 & 0 & \gamma_2 \end{pmatrix}$$

$$= \begin{pmatrix} \gamma_1\gamma_2(1 + \beta_1\beta_2) & 0 & 0 & i\gamma_1\gamma_2(\beta_1 + \beta_2) \\ 0 & 1 & 0 & 0 \\ 0 & 0 & 1 & 0 \\ -i\gamma_1\gamma_2(\beta_1 + \beta_2) & 0 & 0 & \gamma_1\gamma_2(1 + \beta_1\beta_2) \end{pmatrix} \qquad \textbf{(16.51)}$$

For this to represent a direct transformation matrix $\boldsymbol{\lambda}$ from S'' to S we must have

$$\boldsymbol{\lambda}_{s'' \to s} = \begin{pmatrix} \gamma & 0 & 0 & i\beta\gamma \\ 0 & 1 & 0 & 0 \\ 0 & 0 & 1 & 0 \\ -i\beta\gamma & 0 & 0 & \gamma \end{pmatrix} \tag{16.52}$$

Comparing Eqs. (16.51) and (16.52) gives

$$\gamma = \gamma_1\gamma_2(1 + \beta_1\beta_2)$$

$$\beta\gamma = \gamma_1\gamma_2(\beta_1 + \beta_2) \tag{16.53}$$

which when solved for β gives

$$\beta = \frac{\beta_1 + \beta_2}{1 + \beta_1\beta_2} \tag{16.54}$$

Since $\beta = u/c$, $\beta_1 = u_1/c$, $\beta_2 = u_2/c$, multiplying Eq. (16.54) by c results in

$$u = \frac{u_1 + u_2}{1 + (u_1 u_2/c^2)} \tag{16.55}$$

which is similar to the result obtained earlier. Note that, as before, if $u_1 < c$, $u_2 < c$, then $u < c$; and if $u_1 = u_2 = c$, then $u = c$.

16.5 COVARIANT FORMULATIONS AND FOUR VECTORS

From our previous discussions we can conclude that the laws of physics are invariant in form under Lorentz transformations. So our first task is to find the modifications required for the laws of mechanics so that they will have the same form in all inertial systems moving with uniform relative velocities. Modification of such equations is convenient in terms of four-dimensional space. Furthermore, such modifications should not change the physical contents of the law under any particular orientation of the spatial axes. Thus the laws of physics must be invariant in form under rigid body rotations, that is, under spatial orthogonal transformations. We pursue this discussion of invariance under spatial rotation a bit further.

In constructing any equation representing a physical law, it is required that the terms of the equation must be *all scalars,* or *all vectors,* or, in general, *all tensors of the same rank.* Thus, if two vectors **A** and **B** have the relation

$$\mathbf{A} = \mathbf{B} \tag{16.56}$$

it implies that their components are related by

$$A_i = B_i \tag{16.57}$$

When spatial rotation is performed, the values of these components are *not invariant,* but they do transfer to new components A_i' and B_i' in such a way that

$$A_i' = B_i' \tag{16.58}$$

and
$$\mathbf{A}' = \mathbf{B}' \tag{16.59}$$

We say that the terms of the equation are *covariant,* while the invariance in Eq. (16.59) is due to the fact that both sides of Eq. (16.56) transfer as vectors. In general, if we consider two tensors of second rank that are equal,

$$\mathbf{C} = \mathbf{D} \tag{16.60}$$

then the tensors transform covariantly under a spatial rotation and we may write

$$\mathbf{C}' = \mathbf{D}' \tag{16.61}$$

Thus we may conclude:

> *For a physical law to be invariant under rotation of the spatial coordinates requires that the terms of the equation must be covariant under a three-dimensional orthogonal transformation.*

The Lorentz transformations are orthogonal transformations in world space. The different quantities involved may be world scalars, world vectors, or world tensors of any rank. If we express a given law in a *covariant four-dimensional form,* its invariance under a Lorentz transformation can be concluded from inspection.

Before we can proceed to modify the laws of classical mechanics we must familiarize ourselves with a four-dimensional concept, since four-vectors are used in relativity.

We are familiar with quantities that are vectors, each vector having three components and so called a *three-vector.* In the present chapter we started by dealing with quantities with four components, such as (x_1, x_2, x_3, ict). Quantity **A** is called a *four-vector* if it consists of four components and each component transforms according to the relation

$$A_\mu' = \sum_\nu \lambda_{\mu\nu} A_\nu \tag{16.62}$$

where $\lambda_{\mu\nu}$ are the elements of a Lorentz transformation matrix. Thus the quantity (x_1, x_2, x_3, ict) is a four-vector in *Minkowski space* represented as

$$x_\mu \equiv \mathbf{x} = (x_1, x_2, x_3, ict) = (\mathbf{x}, ict) \tag{16.63}$$

where the first three components of **x** are the components of vector **x** in ordinary three-dimensional space. Similarly, the differential element of $d\mathbf{x}$ is also a four-vector given by

$$dx_\mu \equiv d\mathbf{x} = (d\mathbf{x}, ic\, dt) \tag{16.64}$$

An element of length ds in four-dimensional Minkowski space is invariant under Lorentz transformations and may be written as

$$ds = \sqrt{\sum_{\mu} dx_{\mu}^2} = \sqrt{\sum_{i} dx_i^2 - c^2 dt^2} = \text{invariant} \tag{16.65}$$

Furthermore, an element of *proper time* or *world time, $d\tau$*, in Minkowski space, defined next, is also invariant:

$$d\tau = \sqrt{dt^2 - \frac{\Sigma_i dx_i^2}{c^2}} = \sqrt{dt^2 + \frac{i^2}{c^2} \sum_i dx_i^2} \tag{16.66}$$

$$= \frac{i}{c} \sqrt{\sum_i dx_i^2 - c^2 \, dt^2} = \frac{i}{c} \sqrt{\sum_i dx_i^2 + (ic \, dt)^2}$$

That is,
$$d\tau = \frac{i}{c} \sqrt{\sum_{\mu} dx_{\mu}^2} = \frac{i}{c} ds = \text{constant} \tag{16.67}$$

Since ds is constant or invariant, as shown in Eq. (16.65), and $d\tau$ is i/c times ds, $d\tau$ must be constant or invariant. From Eq. (16.66),

$$d\tau = dt \sqrt{1 - \frac{1}{c^2} \sum_i \left(\frac{dx_i}{dt}\right)^2} \tag{16.68}$$

but
$$\sum_i \left(\frac{dx_i}{dt}\right)^2 = \left(\frac{dx_1}{dt}\right)^2 + \left(\frac{dx_2}{dt}\right)^2 + \left(\frac{dx_3}{dt}\right)^2 = v^2 \tag{16.69}$$

Hence
$$d\tau = dt \sqrt{1 - \frac{v^2}{c^2}} = dt \sqrt{1 - \beta^2} \tag{16.70}$$

or
$$dt = \frac{d\tau}{\sqrt{1 - \beta^2}} \tag{16.71}$$

Note that this is the result we obtained for time dilation, where the proper time $d\tau$ was the time interval on a clock attached to a particle or object itself, while dt is the time interval measured by an observer stationed in the laboratory.

Since one of the components of a four-vector is imaginary, it is no longer necessary that the square of the four-vector always be positive definite. If the magnitude of the square of the four-vector is greater than or equal to zero, it is called *spacelike;* and if the square of the magnitude is negative, it is called *timelike*. These magnitudes are world scalars; hence they are not affected by Lorentz transformations. The difference vector between two world points can be either spacelike or timelike; that is,

$$dx_{\mu} = x_{1\mu} - x_{2\mu} \tag{16.72}$$

$$\sum_{\mu} dx_{\mu}^2 = |\mathbf{r}_1 - \mathbf{r}_2|^2 - c^2(t_1 - t_2)^2 \tag{16.73}$$

Thus
$$|\mathbf{r}_1 - \mathbf{r}_2|^2 \geq c^2(t_1 - t_2)^2 , \qquad \text{spacelike} \tag{16.74}$$

and
$$|\mathbf{r}_1 - \mathbf{r}_2|^2 < c^2(t_1 - t_2)^2 , \qquad \text{timelike} \tag{16.75}$$

Two world points separated by a spacelike difference vector *cannot be connected* by any wave traveling with speed c, while timelike world points can be connected by a light signal traveling with speed v.

Suppose the preceding two points lie on the X_1-axis. This can be achieved by a Lorentz transformation. For a spacelike dx_μ,

$$c(t_1 - t_2) < x_1 - x_2$$

and a point in world space corresponds to something happening at a given point $\mathbf{x}$ and time t; that is, it describes an *event*. It is possible to cause transformations such that dx_4' is zero; that is, two events will be simultaneous in a spacelike situation.

Knowing the basic definition of differential elements of distance and time, we can cite many more examples of a four-vector. Thus a four-vector velocity $\mathbf{V} \equiv u_\mu$ is defined as the rate of change of a position vector of a particle with respect to its proper time; that is, dividing the four-vector element dx_u by the proper time $d\tau$, we obtain

$$u_\mu \equiv \mathbf{V} = \frac{dx_u}{d\tau} = \left(\frac{dx_i}{d\tau} , ic \frac{dt}{d\tau} \right) \tag{16.76}$$

Substituting for $d\tau$ from Eq. (16.70), we obtain

$$u_\mu \equiv \mathbf{V} = \frac{dx_u}{d\tau} = \left(\frac{dx_i}{dt\sqrt{1 - \beta^2}} , \frac{ic}{\sqrt{1 - \beta^2}} \right) \tag{16.77}$$

Thus the space and time components are

$$u_i = \frac{v_i}{\sqrt{1 - \beta^2}} \quad \text{and} \quad u_4 = \frac{ic}{\sqrt{1 - \beta^2}} \tag{16.78}$$

The world velocity has a constant magnitude as follows:

$$\sum_{\mu=1}^{4} u_\mu^2 = \sum_{i=1}^{3} u_i^2 + u_4^2 = \frac{v^2}{1 - \beta^2} - \frac{c^2}{1 - \beta^2} = -c^2 \tag{16.79}$$

Since it is negative, it is *timelike*.

16.6 RELATIVISTIC DYNAMICS

We have seen that Newton's equations of motion are invariant under Galilean transformations. Newton's second law

$$\frac{d}{dt}(m_0 v_i) = F_i \tag{16.80}$$

itself is not Lorentz invariant, but its relativistic generalization will be. It will be a four-vector equation whose three spatial components reduce to Eq. (16.80) in the limit $\beta \to 0$. The desired generalization of Eq. (16.80) using a relativistic treatment may be written as

$$\frac{d}{d\tau} m_0 u_\mu = F_\mu \equiv \mathbf{F} \tag{16.81}$$

where $\mathbf{F}$ is a four-vector force known as a *Minkowski force*. The form of $\mathbf{F}$ must be such that in the limit $\beta \to 0$, $F_\mu \to F_i$ [as given in Eq. (16.80)]. A suitable form of F_μ is

$$F_\mu = \frac{F_i}{\sqrt{1 - \beta^2}} \tag{16.82}$$

Using the value of $d\tau$, F_i, and the definition of world velocity u_μ, we may, using Eq. (16.80), define the rate of change of momentum as

$$\frac{d}{dt} \left(\frac{m_0 v_i}{\sqrt{1 - \beta^2}} \right) = F_\mu \sqrt{1 - \beta^2} \tag{16.83}$$

which on comparison with

$$\frac{dp_i}{dt} = F_i \tag{16.84}$$

yields momentum p_i:

$$p_i = \frac{m_0 v_i}{\sqrt{1 - \beta^2}} \tag{16.85}$$

So far we have defined only p_i, the three spatial components of a four-vector. From the definition of four-vector velocity $\mathbf{V}$ we may define the four-vector $\mathbf{P} \equiv p_\mu$ as

$$p_\mu = \mathbf{P} = m_0 \mathbf{V} = m_0 \left(\frac{\mathbf{v}}{\sqrt{1 - \beta^2}}, \frac{ic}{\sqrt{1 - \beta^2}} \right) = \left(\frac{m_0 \mathbf{v}}{\sqrt{1 - \beta^2}}, ip_4 \right) \tag{16.86}$$

where the fourth component is

$$p_4 = \frac{m_0 c}{\sqrt{1 - \beta^2}} \tag{16.87}$$

and the first three spatial components are [the same as defined in Eq. (16.85)]

$$p_i = m v_i = \frac{m_0 v_i}{\sqrt{1 - \beta^2}}, \qquad i = 1, 2, 3 \tag{16.85}$$

and

$$m = \frac{m_0}{\sqrt{1 - \beta^2}} \tag{16.88}$$

Equation (16.88) reveals that if we want to interpret momentum in the strict classical sense, then the mass is no longer invariant. The mass is a function of the velocity of the reference frame. That is, for momentum to have a covariant formulation the inertial mass must be given by Eq. (16.88). Taking the time derivative of p_i in Eq. (16.85) yields the three spatial components of the four-vector force

$$F_i = \frac{d}{dt}\left(\frac{m_0 v_i}{\sqrt{1-\beta^2}}\right)$$

or in vector form

$$\mathbf{F} = \frac{d}{dt}\left(\frac{m_0 \mathbf{v}}{\sqrt{1-\beta^2}}\right) \tag{16.89}$$

We may now write the generalized equation of motion of a particle as

$$\frac{dp_\mu}{d\tau} = F_\mu \tag{16.90}$$

Using the value of $\mathbf{F}$ given by Eq. (16.89), other relations in relativistic kinematic may be obtained. We can obtain an expression for the relativistic kinetic energy T. The work done, $\mathbf{F} \cdot \mathbf{v}$, on a particle is equal to the rate of change of its kinetic energy, dT/dt; that is,

$$\frac{dT}{dt} = \mathbf{F} \cdot \mathbf{v} = \mathbf{v} \cdot \mathbf{F} = \mathbf{v} \cdot \frac{d}{dt}\left(\frac{m_0 \mathbf{v}}{\sqrt{1-\beta^2}}\right) \tag{16.91}$$

which may also be written as

$$\frac{dT}{dt} = v\frac{d}{dt}(m_0 \gamma v) \tag{16.92}$$

Integrating this expression yields

$$\int_0^T dT = \int vd(m_0 \gamma v) = m_0 \int_0^v \frac{v\,dv}{[1-(v^2/c^2)]^{3/2}}$$

That is,
$$T = \frac{m_0 c^2}{\sqrt{1-\beta^2}} - m_0 c^2 \tag{16.93}$$

or
$$T = mc^2 - m_0 c^2 \tag{16.94}$$

Each term is an energy term. Since the last term contains m_0, we may name $m_0 c^2$ to be the *rest mass energy,* while it is appropriate to call mc^2 the total energy E so that

$$E = T + m_0 c^2 \tag{16.95a}$$

$$\text{Total energy} = \text{kinetic energy} + \text{rest mass energy} \tag{16.95b}$$

where
$$E = mc^2 \qquad (16.96)$$

which may also be written as

$$\Delta E = \Delta mc^2 \qquad (16.97)$$

where ΔE is the change in energy due to the change in mass: $\Delta m = m - m_0$.

   If Eq. (16.93) is truly an expression for kinetic energy, it must reduce to $\frac{1}{2}mv^2$ as $\beta \to 0$. Thus

$$\lim_{\beta \to 0} T = \lim_{\beta \to 0} m_0 c^2 \left[\frac{1}{\sqrt{1 - (v^2/c^2)}} - 1 \right]$$

$$= \lim_{\beta \to 0} m_0 c^2 \left[1 + \frac{1}{2}\frac{v^2}{c^2} + \cdots - 1 \right] \simeq \frac{1}{2} m_0 v^2 \qquad (16.98)$$

Before proceeding further, let us draw plots (Figures 16.3, 16.4, and 16.5) showing variations in mass, momentum, force, kinetic energy, and total energy versus velocity using Equations (16.82), (16.85), (16.88), and (16.95).

Figure 16.3 _____

The Lorentz transformations for mass, momentum, and force versus velocity are graphed below by using Eqs. (16.88), (16.85), and (16.82).

$$i := 0..25 \qquad c := 3 \cdot 10^8 \qquad M0 := 0.1 \qquad v0 := \frac{c}{25} \qquad vc_i := i \cdot v0$$

$$p0_i := M0 \cdot vc_i \qquad \beta_i := \frac{vc_i}{c} \qquad F0_i := 10$$

$$M_i := \frac{M0}{\sqrt{1 - (\beta_i)^2} + 0.0001} \qquad p_i := \frac{M0 \cdot vc_i}{\sqrt{1 - (\beta_i)^2} + 0.0001} \qquad F_i := \frac{F0_i}{\sqrt{1 - (\beta_i)^2} + 0.0001}$$

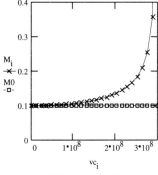

Mass versus velocity

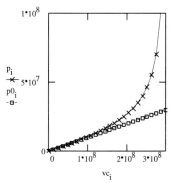

Momentum versus velocity

Figure 16.3 (continued)

$$\frac{M_1}{M0} = 1.001 \qquad \frac{M_{10}}{M0} = 1.091 \qquad \frac{M_{20}}{M0} = 1.666 \qquad \frac{M_{25}}{M0} = 1 \cdot 10^4$$

$$\frac{P_1}{p0_1} = 1.001 \qquad \frac{P_{10}}{p0_{10}} = 1.091 \qquad \frac{P_{20}}{p0_{20}} = 1.666 \qquad \frac{P_{25}}{p0_{25}} = 1 \cdot 10^4$$

$$\frac{F_1}{F0_1} = 1.001 \qquad \frac{F_{10}}{F0_{10}} = 1.091$$

$$\frac{F_{15}}{F0_{15}} = 1.25 \qquad \frac{F_{20}}{F0_{20}} = 1.666$$

$$\frac{F_1}{F0_1} = 1.001 \qquad \frac{F_{25}}{F0_{25}} = 1 \cdot 10^4$$

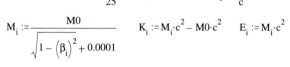

Force versus velocity

 Figure 16.4

The Lorentz transformations for kinetic energy, rest mass energy, and total energy versus velocity are graphed below using Eqs. (16.94) and (16.96).

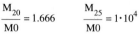

$$i := 0 .. 25 \qquad c := 3 \cdot 10^8 \qquad M0 := 10^{-10} \qquad v0 := \frac{c}{25} \qquad vc_i := i \cdot v0 \qquad \beta_i := \frac{vc_i}{c}$$

A constant value of 1,000,000 has been added to E so that the two graphs, M and E, will not overlap.

$$M_i := \frac{M0}{\sqrt{1 - \left(\beta_i\right)^2} + 0.0001} \qquad K_i := M_i \cdot c^2 - M0 \cdot c^2 \qquad E_i := M_i \cdot c^2$$

$$(M0 \cdot c)^2 + \max(K) = 8.999 \cdot 10^{10}$$

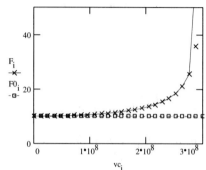

Energy versus velocity

Explain how the above value and the figure satisfy the conservation of energy relation.

Figure 16.5 _____

The relative variations in length, time, mass, and momentum versus velocity vc = v/c
(where vc is the velocity of the object and c is the velocity of light) are graphed below.

$$LO := 3 \qquad TO := 2 \qquad MO := 1 \qquad i := 0..20 \qquad vc_i := \frac{i}{20} \qquad p0_i := MO \cdot vc_i$$

As the velocity of the
object changes, which
quantity changes the
fastest, the slowest, or
doesn't change at all?

$$L_i := LO \cdot \sqrt{1 - \left(vc_i\right)^2} \qquad T_i := \frac{TO}{\sqrt{1 - \left(vc_i\right)^2} + 0.0001}$$

$$M_i := \frac{MO}{\sqrt{1 - \left(vc_i\right)^2} + 0.0001} \qquad p_i := \frac{MO \cdot vc_i}{\sqrt{1 - \left(vc_i\right)^2} + 0.0001}$$

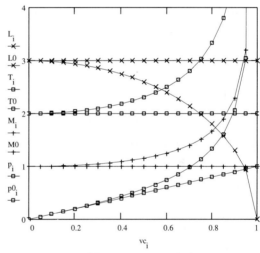

Momentum versus velocity

$L_0 = 3$	$L_5 = 2.905$	$L_{10} = 2.598$	$L_{15} = 1.984$	$L_{20} = 0$
$T_0 = 2$	$T_5 = 2.065$	$T_{10} = 2.309$	$T_{15} = 3.023$	$T_{20} = 2 \cdot 10^4$
$M_0 = 1$	$M_5 = 1.033$	$M_{10} = 1.155$	$M_{15} = 1.512$	$M_{20} = 1 \cdot 10^4$
$p_0 = 0$	$p_5 = 0.258$	$p_{10} = 0.577$	$p_{15} = 1.134$	$p_{20} = 1 \cdot 10^4$

Let us now continue our discussion to derive a four-vector momentum expression and the
Lorentz transformation matrix using momentum and energy (instead of space and time). Equation (16.87) for the fourth component of momentum may be written, using Eq. (16.96), as

$$p_4 \equiv \frac{m_0 c}{\sqrt{1 - \beta^2}} = mc = \frac{E}{c} \tag{16.99}$$

Thus a four-vector momentum may be written as

$$p_\mu = \mathbf{p} = m_0 \mathbf{v} = \left(\mathbf{p}, i\frac{E}{c}\right) \tag{16.100}$$

The theory of relativity combines the concepts of momentum and energy into one, just as it does for space and time. We can treat p_μ as consisting of three momentum components $\mathbf{p}_i(p_1, p_2, p_3)$ and the fourth energy component iE/c as one four-vector, and use the Lorentz transformations as we did for space and time. Thus

$$\mathbf{p}' = \lambda \mathbf{p} \tag{16.101}$$

or, in component form,

$$\left(\mathbf{p}', i\frac{E'}{c}\right) = \lambda\left(\mathbf{p}, i\frac{E}{c}\right) \tag{16.102}$$

where λ is the Lorentz transformation matrix given by Eq. (16.37); that is,

$$\lambda = \begin{pmatrix} \gamma & 0 & 0 & i\beta\gamma \\ 0 & 1 & 0 & 0 \\ 0 & 0 & 1 & 0 \\ -i\beta\gamma & 0 & 0 & \gamma \end{pmatrix} \tag{16.37}$$

Thus we obtain the following *Lorentz momentum energy transformations*:

$$p_1' = \frac{p_1 - (v/c^2)E}{\sqrt{1-\beta^2}}$$

$$p_2' = p_2$$

$$p_3' = p_3 \tag{16.103}$$

$$E' = \frac{E - vp_1}{\sqrt{1-\beta^2}}$$

The inverse transformations can be obtained by replacing v by $-v$ and interchanging the primed and unprimed quantities.

As we showed in Eq. (16.79), the square of the four-vector velocity is invariant and is equal to $-c^2$; we can also show that the square of the four-vector momentum is invariant; that is,

$$\mathbf{p}^2 \equiv \sum_{\mu=1}^{4} p_\mu^2 = m_0^2 \mathbf{v}^2 = -m_0^2 c^2 \tag{16.104}$$

But

$$\mathbf{p}^2 = p_\mu p_\mu = \left(\mathbf{p}, i\frac{E}{c}\right) \cdot \left(\mathbf{p}, i\frac{E}{c}\right) = p^2 - \frac{E^2}{c^2} \tag{16.105}$$

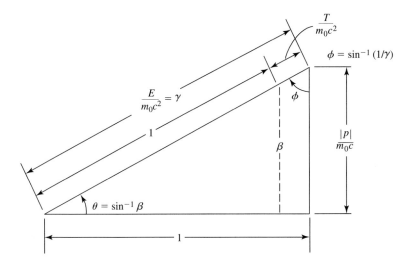

Figure 16.6 Relativistic triangle representing a geometrical relation between m_0c^2, p, T, and E normalized in energy units of m_0c^2 and momentum units of m_0c.

where $p^2 = p_1^2 + p_2^2 + p_3^2$. From these two equations we get the most important relation between mass, momentum, and energy to be

$$E^2 = p^2c^2 + m_0^2c^4 \tag{16.106}$$

A convenient geometrical representation of the relation among m_0c^2, p, $T(= mc^2 - m_0c^2)$, and E may be had by first writing Eq. (16.106) in the normalized form; that is,

$$\left(\frac{E}{m_0c^2}\right)^2 = \left(\frac{|p|}{m_0c}\right)^2 + 1 \tag{16.107}$$

The geometrical relation represented in Fig. 16.6 is called a *relativistic triangle* with angles θ and ϕ defined as (since $E = mc^2 = m_0\gamma c^2$)

$$\sin\theta = \beta \tag{16.108}$$

and

$$\sin\phi = \sqrt{1 - \beta^2} = \frac{1}{\gamma} \tag{16.109}$$

Note that Eq. (16.107) represents energy in units of m_0c^2 and momentum in units of m_0c.

16.7 LAGRANGIAN AND HAMILTONIAN FORMULATION OF RELATIVISTIC MECHANICS

In Chapter 13, while discussing Lagrangian and Hamiltonian dynamics, it was assumed that the mass remains constant; hence such formulations are correct only in the nonrelativistic limit. We wish to extend such a formulation to objects moving with high speed, approaching the speed of light. We can arrive at the Lagrangian function of a single particle moving in a velocity-independent potential as follows.

According to Eq. (16.85), the relativistic expression for the space momentum components of a particle of rest mass m_0 and velocity v_i are ($\beta = v/c$)

$$p_i = \frac{m_0 v_i}{\sqrt{1 - \beta^2}} \tag{16.110}$$

If L is the Lagrangian of a particle and for a velocity-independent potential, we still may define the canonical momentum to be

$$p_i = \frac{\partial L}{\partial \dot{q}_i} = \frac{\partial L}{\partial v_i} \tag{16.111}$$

From these two equations, we get

$$\frac{\partial L}{\partial v_i} = \frac{m_0 v_i}{\sqrt{1 - \beta^2}} \tag{16.112}$$

Thus, in the expression for the Lagrangian,

$$L = T_R(v_i) - V(x_i) \tag{16.113}$$

where T_R must take such a form so that Eq. (16.112) is valid. A suitable expression for T_R is

$$T_R = -m_0 c^2 \sqrt{1 - \beta^2} \tag{16.114}$$

Hence the *relativistic Lagrangian* is

$$L = -m_0 c^2 \sqrt{1 - \beta^2} - V \tag{16.115}$$

(Note that the velocity-independent part T_R is no longer kinetic energy $T_R \neq T$.)

We can show that the expression for L given by Eq. (16.115) is correct by using it in Eq. (12.38) for the Lagrange equation of a single particle: that is,

$$\frac{d}{dt}\left(\frac{\partial L}{\partial \dot{x}}\right) - \frac{\partial L}{\partial x} = 0 \tag{12.38}$$

takes the form (where $\dot{x} = v_i$)

$$\frac{d}{dt}\left(\frac{m_0 v_i}{\sqrt{1 - \beta^2}}\right) + \frac{\partial V}{\partial x} = 0 \tag{16.116}$$

or, since $F_x = -\partial V/\partial x$, and using Eq. (16.110),

$$\frac{dp_i}{dt} = F_x \tag{16.117}$$

as required for space components.

We also showed in Chapter 12 that if L does not contain time explicitly there exists a constant of motion, the Hamiltonian,

$$H = \sum \dot{q}_i p_i - L \tag{16.118}$$

This is still true as we show now. We may notice that $L \neq T - V$ and $\sum \dot{q}_i p_i \neq 2T$. If we use L given by Eq. (16.115), we still get H to be a constant of motion.

$$H \equiv \sum_i \dot{q}_i p_i - L = \sum_i \frac{m_0 v_i^2}{\sqrt{1 - \beta^2}} + m_0 c^2 \sqrt{1 - \beta^2} + V \tag{16.119}$$

reduces to

$$H = \frac{m_0(\dot{x}_1^2 + \dot{x}_2^2 + \dot{x}_3^2)}{\sqrt{1 - \beta^2}} + m_0 c^2 \sqrt{1 - \beta^2} + V = \frac{m_0 c^2}{\sqrt{1 - \beta^2}} + V$$

$$= mc^2 + V = m_0 c^2 + T + V = E = \text{constant} \tag{16.120}$$

(We have used the fact that $\dot{x}_1^2 + \dot{x}_2^2 + \dot{x}_3^2 = v^2$ and $\beta = v/c$.) Thus the relativistic Hamiltonian is still equal to the total energy, but it must include the rest mass energy as well.

 Example 16.1 _____

Show that the amount of energy needed to produce an antiproton by the following reaction is about 6 BeV.

$$p + p \rightarrow p^+ + p^+ + p^+ + \bar{p}^-$$

Solution

Suppose the initial momentum of the system is p_i. After the reaction, this momentum is equally distributed among the four particles:

$$p_{1f} = \frac{p_i}{4} \tag{i}$$

and

$$T_i = 4T_{1f} \tag{ii}$$

which is related to the minimum initial bombarding energy T_i of the incident proton by the relation

$$T_i = 4T_{1f} + 2M_0 C^2 \tag{iii}$$

where M_0 is the rest mass of the proton.

But the final momentum p_{1f}, kinetic energy T_f, and total energy E_T of the product particle are given by the relativistic relation

$$E_T^2 = (T_{1f} + M_0 c^2)^2 = p_{1f}^2 c^2 + M_0^2 c^4 \tag{iv}$$

or

$$p_{1f}^2 c^2 = (T_{1f} + M_0 c^2)^2 - M_0^2 c^4 \tag{v}$$

Similarly,

$$p_{1i}^2 c^2 = (T_i + M_0 c^2)^2 - M_0^2 c^4 \tag{vi}$$

Substituting Eqs. (v) and (vi) in Eq. (i), after rearranging, we get

$$(T_i + M_0 c^2)^2 - M_0^2 c^4 = 16[(T_{1f} + M_0 c^2)^2 - M_0^2 c^4] \tag{vii}$$

Substituting for T_i from Eq. (iii) into Eq. (vii) and solving for T_{1f}, we get

$$T_{1f} = M_0 c^2 \tag{viii}$$

Thus, from Eq. (iii), using Eq. (viii), we obtain

$$T_i = 4M_0 c^2 + 2M_0 c^2 = 6M_0 c^2 = 6(938 \text{ MeV}) = 5628 \text{ MeV} \tag{ix}$$

That is, an incident proton needs about 6 BeV of energy to produce an antiproton.

EXERCISE 16.1 Show that the minimum kinetic energy needed to produce a positron (e^+) by the following collision of two electrons is about 3 MeV (the rest mass energy of an electron or positron is 0.51 MeV).

$$e^- + e^- \rightarrow e^- + e^- + e^- + e^+$$

PROBLEMS

16.1. Consider an elastic collision between two masses. Making use of the Galilean transformation equations, show that if momentum and kinetic energy are conserved in one inertial system, they are conserved in all inertial systems.

16.2. If Ψ represents a scalar function of position and time, show that the wave equation (Ψ could be a component of an E or B field)

$$\frac{\partial^2 \Psi}{\partial x^2} + \frac{\partial^2 \Psi}{\partial y^2} + \frac{\partial^2 \Psi}{\partial z^2} - \frac{1}{c^2} \frac{\partial^2 \Psi}{\partial t^2} = 0$$

is not invariant under Galilean transformation; that is,

$$\frac{\partial^2 \Psi}{\partial x^2} + \frac{\partial^2 \Psi}{\partial y^2} + \frac{\partial^2 \Psi}{\partial z^2} - \frac{1}{c^2} \frac{\partial^2 \Psi}{\partial t^2} \neq \frac{\partial^2 \Psi}{\partial x'^2} + \frac{\partial^2 \Psi}{\partial y'^2} + \frac{\partial^2 \Psi}{\partial z'^2} - \frac{1}{c^2} \frac{\partial^2 \Psi}{\partial t'^2}$$

16.3. If $\Psi(x, y, z, t)$ represents a scalar function of position and time, show that the wave equation

$$\frac{\partial^2 \Psi}{\partial x^2} + \frac{\partial^2 \Psi}{\partial y^2} + \frac{\partial^2 \Psi}{\partial z^2} - \frac{1}{c^2} \frac{\partial^2 \Psi}{\partial t^2} = 0$$

is invariant under a Lorentz transformation.

16.4. Show that the Lorentz transformation equations for the S and S' systems in uniform relative motion may be expressed as

$$x' = x \cosh \alpha - ct \sinh \alpha$$

$$y' = y$$

$$z' = z$$

$$t' = t \cosh \alpha - \frac{x}{c} \sinh \alpha$$

where $\tanh \alpha = v/c$.

16.5. Two events taking place in the S' system have coordinates P_1 (200 m, 0, 0, 5×10^{-7} s) and P_2 (200 m, 0, 0, 2×10^{-7} s). The velocity of the S' system with respect to the S system is $\frac{2}{3}c$ along the $X-X'$ axes. What are the coordinates of these events in the S system? Do this using Eqs. (16.21), (16.36), and (16.37).

16.6. An S' system is moving with respect to system S with a velocity v along the $X-X'$ axes, while system S'' is moving with velocity v' with respect to system S' along the $X'-X''$ axes. Express the coordinates (x'', y'', z'', t'') in terms of (x, y, z, t).

16.7. A meter stick makes an angle θ with the X-axis in the S system. What is the length of this stick as observed from an S' system moving with velocity $\frac{3}{4}c$ along the $X-X'$ axes? What is the angle θ' the stick makes with the X'-axis? Calculate θ' if θ is $60°$.

16.8. A small box containing a clock is moving with velocity v. The volume of the box is dV_0 when measured from within the box and dV when measured from outside in LCS coordinates. The time interval between two events happening inside the box and measured by the clock inside the box is dt_0, while measured from outside it is dt. Show that $dV_0 \, dt_0 = dV \, dt$.

16.9. "We conclude that a balanced clock at the Equator must go more slowly, by a very small amount, than a precisely similar clock situated at one of the poles under otherwise identical conditions," according to Einstein's paper of 1905. Show that after a century of the two clocks will differ by approximately 0.003 second. Neglect the fact that the equator clock does not undergo uniform motion.

16.10. In system S, two events take place at two points separated by a distance Δx and at the same time t. Show that, as observed from the S' system, the two events are not simultaneous, but are separated by a time interval $\Delta t' = -\gamma \, \Delta x (v/c^2)$.

16.11. Two events $P_1(0, 0, 0, 0)$ and $P_2(x, 0, 0, t)$ take place in system S. Find the velocity of the S' system for which these two events will be simultaneous. What does the square of the distance between the two events in the S' system represent?

16.12. As observed from Earth, two distant galaxies A and B are receding in opposite directions each with a speed of $\frac{3}{4}c$. What will be the speed of recession of galaxy B as observed from galaxy A?

16.13. Prove the law of addition of the two parallel velocities [Eq. (16.54)] by considering two Lorentz transformations as successive rotations in the X_3X_4 plane.

16.14. A light beam is emitted at an angle θ' with respect to the X'-axis in system S'. Show that the angle θ as measured in system S is

$$\cos \theta = \frac{\cos \theta' + (v/c)}{1 + (v/c)\cos \theta'}$$

where v is the speed of system S' along the $X-X'$ axes. If $\theta' = 90°$, what should be the value of v so that θ will be only $5°$?

16.15. With a speed almost equal to that of light, a source is emitting electrons equally in all directions in the S' system, which is moving with velocity v along $X-X'$ axes relative to the S system. What is the value of v so that half the electrons are emitted in a cone of 4° as viewed from the S system?

16.16. Suppose a frame of reference S' is moving with velocity v_1 along the $X-X'$ axes and S'' is moving with a velocity v_2 relative to S along the $X'-X''$ axes. Calculate the velocity of the S'' system relative to the S' system.

16.17. Show that it is not possible to find an inertial system by consecutive Lorentz transformations in which the speed of a particle will be greater than the speed of light if the speed of the particle is less than the speed of light in any inertial system.

16.18. Using Eqs. (16.49), make graphs of **(a)** u_1' versus u_1 for different values of $u_1 = c/n$ where $n = 1$, $2, \ldots 9$ and $v = 0.9c$, and **(b)** u_1' versus v for different values of v and for a given value of u.

16.19. A mu meson (μ meson) carries a charge of one electron and has a mass about 207 times the rest mass of an electron. Fast-moving mu mesons form a part of cosmic radiation and can be produced in the laboratory as well. The mean life of mu mesons at rest in the laboratory is found to be 2.2×10^{-6} s, while the average life of fast-moving mu mesons observed in cosmic rays is 1.1×10^{-5} s. Calculate the following: **(a)** the speed of cosmic ray μ mesons; **(b)** the distance traveled by the cosmic ray μ mesons through the atmosphere in its average lifetime according to an observer **(i)** in the laboratory frame of reference, and **(ii)** in the μ meson frame of reference.

16.20. For the motion of an electron with different velocities, using Lorentz transformations, make plots of mass, momentum, kinetic energy, rest mass energy, and total energy versus velocity.

16.21. For the motion of a proton with different velocities, using Lorentz transformations, make plots of mass, momentum, kinetic energy, rest mass energy, and total energy versus velocity.

16.22. Consider a cube of side L_0 placed at the origin of an inertial system S. The rest mass of the cube is m_0 and its density is $\rho_0 = m_0/L_0^3$. Calculate the mass, volume, and density as viewed by an observer in an inertial system S' that has a velocity v along the $X-X'$ axes. Graph the mass and volume versus velocity v.

16.23. In Newtonian mechanics

$$dT = \mathbf{p} \cdot \frac{d\mathbf{p}}{m} = \mathbf{v} \cdot d\mathbf{p}$$

Show that this result also holds in relativistic mechanics.

16.24. A particle of rest mass m_0 and speed v has a perfect inelastic collision with a particle of mass M at rest. Show that the speed of the composite particle is $v = \gamma m_0 v/(\gamma m_0 + M)$.

16.25. A particle of mass m_0 moving with velocity v collides with another particle of mass m_0 moving with velocity v in the opposite direction. If the two masses stick together after collision, what is the change in the final mass? If $m_0 = 1$ kg and $v = 10$ km/s, calculate Δm.

16.26. A π meson of mass m_π comes to rest and disintegrates into a μ meson of rest mass m_μ and a neutrino of rest mass zero. Show that the kinetic energy T_μ of the μ meson is

$$T_\mu = \frac{(\mu_\pi - m_\mu)^2}{2m_\pi} c^2$$

16.27. If we assume that the mass changes with velocity, the expression for force is given by

$$\mathbf{F} = \frac{d}{dt} \frac{m_0 \mathbf{v}}{\sqrt{1 - \beta^2}}$$

If $\mathbf{v}$ is taken to be along the X_1-axis, show that the components of the force are $F_1 = m_l \dot{v}_1$, $F_2 = m_t \dot{v}_2$, and $F_3 = m_t \dot{v}_3$, where

$$m_l = \frac{m_0}{(1 - \beta^2)^{3/2}} \quad \text{and} \quad m_t = \frac{m_0}{(1 - \beta^2)^{1/2}}$$

where m_l and m_t are called the longitudinal mass and the transverse mass, respectively.

16.28. The average solar energy reaching the surface of Earth is 1.4×10^3 W/m². How much solar mass is being converted into energy per second?

16.29. (a) Show that if the quantity $p_0 (= \gamma m_0 c)$ is positive in any inertial system, it will be positive in all inertial systems.
(b) Show that the rest mass m_0 of a particle remains constant in magnitude and sign under a Lorentz transformation of momentum.

16.30. In an experiment with colliding beams, two separate proton beams, each with protons whose energy is 10 BeV, are directed against each other in a head-on collision.
(a) What are the velocities of the protons as seen by an observer in the LCS coordinates?
(b) What is the relative velocity of one proton with respect to the other?
(c) In the extreme relativistic case, the velocity v of the protons is very nearly equal to c. Let us define $\delta \simeq c - v$. Show that the energy E of the proton is approximately equal to

$$E = \frac{E_0}{\sqrt{2\delta/c}}$$

where $E_0 = m_0 c^2$ is the rest mass energy (or self-energy).

16.31. What is the minimum speed for a particle so that its total energy E can be identified as pc without causing an error of more than 1%. Calculate the kinetic energy of an electron and a proton at this speed.
(a) Show that the momentum of a particle may be written as

$$p = \frac{1}{c} (T^2 + 2E_0 T)^{1/2}$$

(b) When a particle of charge q moves at right angles to a uniform magnetic field of flux density B, it travels in a circle of radius

$$r = \frac{(T^2 + 2E_0 T)^{1/2}}{qcB}$$

where T is the kinetic energy and E_0 is the rest mass energy. Prove the relationship.

16.32. The total momentum of a system of particles in an inertial frame is Σp, while the total relativistic energy is ΣE. Show that the velocity v_c of the center of mass is $v_c = c^2 (\Sigma p / \Sigma E)$.

16.33. A pion of 1600-MeV kinetic energy is incident on a proton at rest and produces a number of pions by the reaction $\pi + p \rightarrow p + n\pi$. Using the center-of-mass coordinates system, calculate the maximum number n of pions produced in this reaction. The rest mass energy of a proton is 938 MeV and that of a pion is 140 MeV.

16.34. Show that the equation of motion of a relativistic rocket is given by

$$m_0 \frac{dv}{dt} + V \frac{dm_0}{dt} (1 - \beta^2) = 0$$

where $\beta = v/c$, v is the velocity of the rocket with respect to an inertial system, V is the velocity of the exhaust gases with respect to the rocket, and m_0 is the rest mass of the rocket.

16.35. Calculate the minimum energy of a π^- meson needed to initiate the following reaction:

$$\pi^- + p \rightarrow \pi^- + \pi^- + \pi^+ + p$$

where the rest mass of a π^- or π^+ is 139.56 MeV.

16.36. Show that the relativistic motion of a particle in an inverse square law of force is a precessing ellipse. Graph this precessing ellipse.

16.37. Consider a relativistic "photon rocket," a rocket that will eject photons at speed c to provide thrust. This rocket accelerates from rest, decelerates to a stop at its destination, and then returns to the initial position by the same procedure. If the initial rest mass of the rocket is m_0, show that the final payload is $m_0/16\gamma^4$, where $\gamma \gg 1$ when the velocity of the rocket is maximum.

SUGGESTIONS FOR FURTHER READING

American Association of Physics Teachers, *Special Relativity Theory, Selected Reprints.* New York: American Institute of Physics, 1963.

ARTHUR, W., and FENSTER, S. K., *Mechanics,* Chapter 10. New York: Holt, Rinehart and Winston, Inc., 1969.

BERGMANN, P. G., *Introduction to the Theory of Relativity.* Englewood Cliffs, N.J.: Prentice-Hall, Inc., 1946.

DAVIS, A. DOUGLAS, *Classical Mechanics,* Chapter 13. New York: Academic Press, Inc., 1986.

FOWLES, G. R., *Analytical Mechanics,* Chapter 12. New York: Holt, Rinehart and Winston, Inc., 1962.

*GOLDSTEIN, H., *Classical Mechanics,* 2nd ed., Chapter 7. Reading, Mass.: Addison-Wesley Publishing Co., 1980.

HAUSER, W., *Introduction to the Principles of Mechanics,* Chapter 15. Reading, Mass.: Addison-Wesley Publishing Co., 1965.

KITTEL, C., KNIGHT, W. D., and RUDERMAN. M. A., *Mechanics,* Berkeley Physics Course, Volume 1. Chapters 11–14. New York: McGraw-Hill Book Co., 1965.

KLEPPNER, D., and KOLENKOW, R. J., *An Introduction to Mechanics,* Chapters 11–14. New York: McGraw-Hill Book Co., 1973.

MARION, J. B., *Classical Dynamics,* 2nd ed., Chapter 10. New York: Academic Press, Inc., 1970.

*MOORE, E. N., *Theoretical Mechanics,* Chapter 9. New York: John Wiley & Sons, Inc., 1983.

NEY, E. P., *Electromagnetism and Relativity.* New York: Harper & Row, 1962.

RESNICK, R., and HALLIDAY, D., *Basic Concepts in Relativity and Early Quantum Theory,* 2nd ed. New York: John Wiley & Sons, Inc., 1985.

ROSSER, W. G. V., *An Introduction to the Theory of Relativity.* London: Butterworth, 1964.

SYMON, K. R., *Mechanics,* 3rd ed., Chapters 13 and 14. Reading, Mass.: Addison-Wesley Publishing Co., 1971.

TAYLOR, E. F., *Introductory Mechanics,* Chapters 11–13. New York: John Wiley & Sons, Inc., 1963.

——, and WHEELER, J. A., *Spacetime Physics.* San Francisco: W. H. Freeman, 1966

*The asterisk indicates works of an advanced nature.

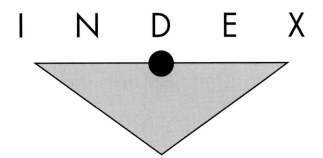

I N D E X